Springer Series in Optical Sciences

Volume 250

Springer Series in Optical Sciences is led by Editor-in-Chief William T. Rhodes, Florida Atlantic University, USA, and provides an expanding selection of research monographs in all major areas of optics:

- lasers and quantum optics
- ultrafast phenomena
- optical spectroscopy techniques
- optoelectronics
- information optics
- applied laser technology
- industrial applications and
- synchrotron radiation and X-ray optics
- other topics of contemporary interest.

With this broad coverage of topics the series is useful to research scientists and engineers who need up-to-date reference books.

Mário F. S. Ferreira • Mukul Chandra Paul
Editors

Optical Fiber Lasers

Recent Advances and Applications

Springer

Editors
Mário F. S. Ferreira
Department of Physics
University of Aveiro
Aveiro, Portugal

Mukul Chandra Paul
Fiber Optics and Photonics Division
Central Glass and Ceramic Research Institute
Kolkata, West Bengal, India

ISSN 0342-4111 ISSN 1556-1534 (electronic)
Springer Series in Optical Sciences
ISBN 978-3-032-04109-8 ISBN 978-3-032-04110-4 (eBook)
https://doi.org/10.1007/978-3-032-04110-4

This Springer imprint is published by the registered company Springer Nature Switzerland AG
The registered company address is: Gewerbestrasse 11, 6330 Cham, Switzerland

I dedicate this book to my early professors and colleagues at the Faculty of Sciences of Porto University, with whom I was introduced to the field of fiber optic technology. This book is also dedicated to the memory of my parents.

(Mário F. S. Ferreira)

This book is dedicated to my guide, Dr. Tarun Kumar Bandyopadhyay; Senior Colleagues (Dr. Kamal Dasgupta, Dr. Shyamal Bhadra, Dr. Ranjan Sen and others) of the Fiber Optics and Photonics Division (FOPD) at CGCRI, Kolkata, India, whose unwavering support and guidance—from the inception of my Ph.D. research journey to its concluding phases—have been a constant source of inspiration. Their mentorship, guidance, and deep expertise in fiber optics have been invaluable throughout my research career. This book is also dedicated to all those passionate about the science and technology of fiber lasers, and to the ground breaking innovations the future of this field holds. This book is also dedicated to the memory of my parents.

(Mukul Chandra Paul)

Acknowledgments

We are extremely satisfied with the results achieved through the collective efforts of the outstanding team of authors who have contributed their expertise to shape this book. The contributors are leading scientists in the field of fiber lasers and its diverse applications. First and foremost, we express our sincere thanks to all the chapter authors for their enthusiasm and dedication. We are also deeply grateful to the many reviewers around the world who, despite their demanding schedules, provided high-quality feedback that greatly enhanced the quality of the chapters.

We strongly believe that this book provides a comprehensive overview of the latest advancements in optical fiber laser technology, encompassing fundamental principles, cutting-edge developments, and a wide range of applications. Furthermore, we are firmly aware that this book properly addresses the growing importance of optical fiber lasers, which have become vital tools across scientific, military, medical, industrial, and commercial sectors—driving rapid innovation and expanding their impact in numerous fields. In our opinion, each chapter offers valuable insights into the most advanced and intriguing developments in the field.

We would like to extend our appreciation to Elsevier, OSA, and IEEE for granting permission to reuse diagrams and images from their publications. Our special thanks go to Nivetha Moorthi (Ms.), Production Editor, Springer Nature; Dr. Sam Harrison, Senior Editor, Springer Nature; and Mr. Jeffrey Taub, Project Coordinator at Springer Nature. Their prompt support, unwavering dedication, and exceptional organizational skills were crucial in helping us meet our deadlines. Their consistent encouragement also made our experience as editors truly fulfilling and rewarding.

Elias Snitzer invented the world's first optical fiber laser in 1961, which took nearly two decades to mature for commercial use. In 1988, he contributed to the invention of the double-clad fiber laser, enabling higher power output. Today, double-clad fiber lasers are widely used in various configurations and are primarily made of rare-earth elements like erbium, neodymium, and ytterbium. Fiber lasers are used in various industries, including manufacturing, medicine, telecommunications, and consumer electronics. They are used in diagnostic and surgical instruments, consumer electronics, additive manufacturing, and 3D printing due to their precision and minimal invasiveness. The integration of Artificial Intelligence (AI) in fiber laser systems enhances performance by optimizing parameters, enabling predictive maintenance, and delivering highly accurate cutting and welding with minimal error. The global fiber laser market reached a valuation of USD 3.0 billion in 2024 and is expected to grow at a compound annual growth rate of 7.8% between 2025 and 2033.

We are fully satisfied with the outcome achieved through the collective efforts of our dedicated team of authors, whose shared knowledge and collaboration have shaped this book into its final form. The contributors include researchers, academics, Ph.D. students, and professionals from both academia and industry, all of whom are engaged in the field of fiber lasers.

This book aims to present various aspects of different types of fiber lasers, along with some of the latest advancements and innovations in this rapidly evolving field. The book's structure consists of:

In **Chap. 1**, this book, *Optical Fiber Lasers: Recent Advances and Applications*, discusses the growing importance of fiber lasers in modern society and their diverse applications. **Chapter 2** covers the use of passive saturable absorbers (SAs) in erbium-doped fiber lasers, specifically Alq_3 and SMS fiber structures, which generate stable, tunable Q-switched pulses for industrial and scientific applications. **Chapter 3** also explores the dynamics of dissipative solitons in mode-locked fiber lasers using the Ginzburg-Landau equation, providing insights into their potential for ultrafast photonics technologies**. Chapter 4** discusses the generation of high-energy, ultrashort optical pulses in passively mode-locked fiber laser.

Chapter 5 discusses the development of multiplexed dissipative soliton fiber lasers, which offer increased complexity, flexibility, and functionality in optical communications, signal processing, and ultrafast photonics. **Chapter 6** describes the development of all-fiber mode-locked lasers, which achieved stable solitonic emission at 2070 nm, extending the emission wavelength beyond 1837 nm. **Chapter 7** analyzes ultrafast Raman fiber lasers, highlighting their technological advancements and increasing importance in generating short optical pulses across a broad wavelength range. **Chapter 8** explores gas-filled hollow-core fiber lasers (HCFLs), a hybrid laser system that combines conventional gas and fiber lasers, particularly useful in the mid-infrared spectral region. **Chapter 9** discusses recent advancements in fiber lasers based on distributed feedback (DFB), focusing on regular and random refractive index modulation structures. **Chapter 10** discusses future research and applications of Random fiber lasers (RFLs), highlighting their potential in telecommunications, medical diagnostics, and materials processing. Key areas for future research include AI/ML integration, nonlinear optical phenomena, quantum communication, bio-photonics, medical imaging, and portable systems.

Chapter 11 delves into Self-Induced Laser-Line Sweeping (SLLS), a unique phenomenon in fiber lasers, and its historical evolution, technological advancements, and versatility in tunable laser sources and spectroscopic systems. It calls for further research into SLLS integration with artificial intelligence and adaptive control systems for real-time optimization. **Chapter 12** discusses advanced technologies for wavelength tuning in mode-locked polarization-maintaining fiber lasers, focusing on materials, fiber designs, and modulation techniques. **Chapter 13** discusses nonlinear frequency conversion technologies in fiber laser systems, extending output to visible and mid-infrared wavelengths. **Chapter 14** discusses the use of two-dimensional materials (2DMs) in ultrafast laser technology, emphasizing their nonlinear optical behavior. **Chapter 15** discusses the evolution and technological advancements of large mode area (LMA) fibers, emphasizing the importance of maintaining single-mode beam quality to reduce power density and suppress nonlinear and thermal effects.

Chapter 16 discusses the growing demand for high-power narrow linewidth (HPNL) linearly polarized all-fiberized lasers in various industries. The authors emphasize ongoing research to improve laser brightness, overcome limitations, and enhance operational efficiency. In **Chap. 17** authors also discuss the development of Er/Yb co-doped optical fibers for high-power laser and optical amplifier applications, achieving over 41% lasing efficiency and stable output power. **Chapter 18** analyzes Spectral Beam Combining (SBC) technology's role in high-power laser systems and its potential for power scaling. **Chapter 19** discusses the transformative impact of AI on HPFL development, enhancing optical fiber design, laser property measurement, and coherent beam combining. **Chapter 20** explores random lasers, a new class of laser systems using disordered media for optical feedback, offering unique properties for specific applications.

In summary, this book comprises 20 chapters, including an introductory chapter, each authored by leading international scientists who have dedicated years to advancing the field of optical fiber laser technology. Every chapter is meticulously crafted, providing comprehensive insights into this dynamic and rapidly evolving research domain. Together, these chapters provide a rich and authoritative resource for researchers, engineers, and students engaged in the field of optical fiber lasers, offering both foundational knowledge and insights into future directions.

The content reflects significant progress across a diverse range of optical fiber laser systems, covering both fundamental principles and state-of-the-art fabrication technologies. Key areas of focus include various types of optical fiber lasers and their extensive applications in the photonics industry. Notable innovations discussed throughout the chapters include:

- Introduction About Optical Fiber Laser
- Q-Switched Erbium-Doped Fiber Lasers
- Dissipative Solitons in Passively Mode-Locked Lasers
- Ultra-Short, High-Amplitude Dissipative Solitons in Fiber Lasers
- Multiplexed Dissipative Soliton Fiber Lasers
- Thulium-Doped, Passively Mode-Locked, All Polarization-Maintaining Fiber Lasers
- Ultrafast Raman Fiber Lasers
- Gas-Filled Hollow-Core Fiber Lasers
- Distributed Feedback Fiber Lasers Based on Regular and Random Refractive-Index Structures
- Random Fiber Lasers: From Demonstration to Application
- Self-Sweeping Fiber Lasers
- Wavelength-Tunable Mode-Locked Polarization-Maintaining Fiber Lasers

- Nonlinear Frequency Conversion of Fiber Lasers
- Development of Pulse Fiber Lasers Based on Two-Dimensional Materials
- Large Mode Area (LMA) Fibers for High-Power Lasers
- High-Power, Narrow Linewidth, Linearly Polarized Fiber Lasers
- Multielement-Doped Silica Glass-Based Specialty Optical Fibers for High-Power Lasers and Amplifiers
- Advances in High-Power Fiber Laser Spectral Beam Combining Technology
- The Role of Artificial Intelligence in High-Power Fiber Laser Development
- Innovations in Plasmonic Fiber Random Lasers

Contents

Editors and Contributors

About the Editors

Mário F. S. Ferreira is a Professor at the Department of Physics at the University of Aveiro, Portugal. His research interests are concerned with the modeling and characterization of multi-section semiconductor lasers for coherent systems, quantum well lasers, optical fiber amplifiers and lasers, optical sensors, soliton propagation, and polarization and nonlinear effects in optical fibers. He has been the leader of the Optics and Optoelectronics Group of the *I3N—Institute of Nanostructures, Nanomodelling and Nanofabrication*. He has written about 400 scientific journal and conference publications and several books, namely: "*Optics and Photonics*" (Lidel, 2003, in Portuguese), "*Topics of Mathematical Physics*" (Editora Ciência Moderna, 2018, Brazil, in Portuguese), "*Optical Fibers: Technology, Communications and recent Advances*" (NOVA Science Publishers, 2017), "*Advances in Optoelectronic Technology and Industry Development*" (CRC Press, 2019), "*Nonlinear Effects in Optical Fibers*" (Wiley, 2011), "*Optical Signal Processing in Highly Nonlinear Fibers*" (CRC Press, 2020), "*Optical Fiber Technology and Applications: Recent Advances*" (IOP Publishing, 2021), "*Solitons in Optical Fiber Systems*" (Wiley, 2022), "*Dissipative Optical Solitons*" (Springer, 2022), "*Photonic Crystal Fibers: Recent Advances*" (NOVA Science Publishers, 2023), "*Specialty Optical Fibers: Materials, Fabrication Technology and Applications*" (Elsevier, 2024),

"*Optical Fiber Lasers*" (Springer, 2026), and "*Lasers Optics and Photonics*" (Oxford University Press, 2026). He was also the Guest Editor of several special issues.

He is a Fellow and a Traveling Lecturer of both OPTICA (Optical Society of America) and SPIE, The International Society for Optics and Photonics, and a Member of IEEE. He has served on various committees of OSA/OPTICA and SPIE, as well as on the organizing and scientific committees of various international conferences. Currently, he also serves as an Associate Editor or as an Advisor Board Member of several international journals in the area of optics and photonics.

Mukul Chandra Paul received his Ph.D. from Jadavpur University, Kolkata, in 2003. He served as Chief Scientist, Divisional Chair of the Fiber Optics and Photonics Division, and HOD of the Instrumentation and Information Technology Division at CSIR–Central Glass and Ceramic Research Institute, Kolkata. He has authored over 250 peer-reviewed papers and 15 book chapters, holds eight U.S. patents on rare-earth-doped specialty optical fibers, and has edited three books on fiber lasers and specialty fibers. His research has advanced the field through collaborations with institutions in Malaysia, China, Taiwan, the UK, France, Russia, Vietnam, and Portugal. His honors include the BOYSCAST Fellowship (2006), MOSTI Award (2010), IEEE Photonics Distinguished Lecture Award (2011), CSIR Technology Awards (2012, 2015), DST-UKIERI Award (2013), Indo-Russian (RFBR) Grant (2016), Indo-Portugal (FCT) Grant (2017), Senior Visiting Scientist Award, NTUT (2020), CK Somany Award (2020), Adjunct Professorship, Airlangga University (2021), Platinum Jubilee Distinguished Scientist (2025), and Senior IEEE Membership (2025). He is a Senior Member of IEEE and OSA, a Life Member of MRSI and the Indian Ceramic Society, and serves on Technical Committees CHD C and LITD 11 under the Bureau of Indian Standards. Dr. Paul has supervised 6 Ph.D. scholars and 10 M.Sc. projects and has served as PI for 17 sponsored projects totaling ₹416 lakhs. Overall, as PI/Co-PI/member, he has completed 34 projects with ₹46.0 crores in funding.

He has delivered over 50 invited talks and served as Guest Editor or Editor for special issues of leading journals, including *IEEE JSTQE* (2014), *Current Nanoscience* (2015), *Optical Materials Express* (2019), and *IEEE Photonics Journal* (2019). He edited *Fiber Laser* (InTech, 2016), *Fiber Laser: Advances in Research and Applications* (Nova, 2017), and most recently *Specialty Optical Fibers* (Elsevier, 2024). Post-retirement, his research interest will be developing next-generation specialty optical fibers using advanced optical materials for photonic devices in collaboration with industry and academia.

Contributors

S. R. Abdullina Institute of Automation and Electrometry, Siberian Branch, Russian Academy of Sciences, Novosibirsk, Russia

Novosibirsk State University, Novosibirsk, Russia

Yi An Test Center, National University of Defense Technology, Xi'an, Shaanxi, China

Yu An Key Laboratory of Optoelectronic Technology and Systems (Ministry of Education), Chongqing University, Chongqing, China

M. V. Andres Departamento de Física Aplicada y Electromagnetismo, ICMUV, Universidad de Valencia, Burjassot, Valencia, Spain

S. A. Babin Institute of Automation and Electrometry, Siberian Branch, Russian Academy of Sciences, Novosibirsk, Russia

Novosibirsk State University, Novosibirsk, Russia

S. K. Bhadra Raman Center for Atomic, Molecular and Optical Sciences, IACS, Kolkata, India

A. Carrascosa Departamento de Física Aplicada y Electromagnetismo, ICMUV, Universidad de Valencia, Burjassot, Valencia, Spain

Chen Chao State Key Laboratory of Transient Optics and Photonics, Xi'an Institute of Optics and Precision Mechanics, Chinese Academy of Sciences (CAS), Xi'an, Shaanxi, China

Guo Chen State Key Laboratory of Transient Optics and Photonics, Xi'an Institute of Optics and Precision Mechanics, Chinese Academy of Sciences (CAS), Xi'an, Shaanxi, China

Center of Materials Science and Optoelectronics Engineering, University of Chinese Academy of Sciences (UCAS), Beijing, China

Yisha Chen College of Advanced Interdisciplinary Studies, National University of Defense Technology, Changsha, Hunan, China

Yu Chen Photonics Engineering Laboratory, Department of Electrical Engineering, University of Malaya, Kuala Lumpur, Malaysia

J. L. Cruz Departamento de Física Aplicada y Electromagnetismo, ICMUV, Universidad de Valencia, Burjassot, Valencia, Spain

C. Cuadrado-Laborde Instituto de Física Rosario (CONICET-UNR), Rosario, Argentina

Maolin Dai Department of Electrical Engineering and Information Systems, The University of Tokyo, Bunkyo-ku, Tokyo, Japan

Research Center for Advanced Science and Technology, The University of Tokyo, Meguro-ku, Tokyo, Japan

Shyamal Das CSIR-Central Glass and Ceramic Research Institute, Kolkata, India

Anirban Dhar CSIR-Central Glass and Ceramic Research Institute, Kolkata, India

A. Diez Departamento de Física Aplicada y Electromagnetismo, ICMUV, Universidad de Valencia, Burjassot, Valencia, Spain

Chenchen Fan College of Advanced Interdisciplinary Studies, National University of Defense Technology, Changsha, Hunan, China

Yan Feng Hangzhou Institute for Advanced Study, University of Chinese Academy of Sciences, Hangzhou, China

Mário F. S. Ferreira i3N—Institute of Nanostructures, Nanomodelling and Nanofabrication & Department of Physics, University of Aveiro, Aveiro, Portugal

Aziz Gohar School of Physics and Optoelectronic Engineering, Beijing University of Technology, Beijing, China

Bing Han College of Information Science and Engineering, Northeastern University, Shenyang, China

Hebei Key Laboratory of Micro-Nano Precision Optical Sensing and Measurement Technology, Qinhuangdao, China

Sulaiman Wadi Harun Photonics Engineering Laboratory, Department of Electrical Engineering, University of Malaya, Kuala Lumpur, Malaysia

Liangjin Huang College of Interdisciplinary Studies, National University of Defense Technology, Changsha, Hunan, China

Naeem Iqbal School of Physics and Optoelectronic Engineering, Beijing University of Technology, Beijing, China

Min Jiang Test Center, National University of Defense Technology, Xi'an, Shaanxi, China

Pan Jiapeng State Key Laboratory of Transient Optics and Photonics, Xi'an Institute of Optics and Precision Mechanics, Chinese Academy of Sciences (CAS), Xi'an, Shaanxi, China

Center of Materials Science and Optoelectronics Engineering, University of Chinese Academy of Sciences (UCAS), Beijing, China

Yanzhao Ke College of Advanced Interdisciplinary Studies, National University of Defense Technology, Changsha, Hunan, China

Alexander V. Kir'yanov Centro de Investigaciones en Óptica, León, Guanajuato, Mexico

Sofia C. V. Latas i3N-Institute of Nanostructures, Nanomodelling and Nanofabrication, Department of Physics, University of Aveiro, Aveiro, Portugal

Jun Li College of Interdisciplinary Studies, National University of Defense Technology, Changsha, Hunan, China

Shuai Li Center for Advanced Laser Technology, Hebei University of Technology, Tianjin, China

Hebei Key Laboratory Advanced Laser Technology and Equipment, Tianjin, China

Junrui Liang College of Interdisciplinary Studies, National University of Defense Technology, Changsha, Hunan, China

Zijian Liang Changchun Institute of Optics, Fine Mechanics and Physics, Chinese Academy of Sciences, Changchun, China

Guanyu Liu School of Physics & Photoelectric Engineering, Guangdong University of Technology, Guangzhou, China

Haoguang Liu Key Laboratory of Optoelectronic Technology and Systems (Ministry of Education), Chongqing University, Chongqing, China

School of Optical and Electronic Information, Huazhong University of Science and Technology, Wuhan, China

Wei Liu College of Advanced Interdisciplinary Studies, National University of Defense Technology, Changsha, Hunan, China

Xuanyi Liu State Key Laboratory of Optoelectronic Materials and Technologies, School of Physics, Sun Yat-sen University, Guangzhou, China

Yiyang Luo Key Laboratory of Optoelectronic Technology and Systems (Ministry of Education), Chongqing University, Chongqing, China

Pengfei Ma College of Advanced Interdisciplinary Studies, National University of Defense Technology, Changsha, Hunan, China

Xiaoya Ma College of Advanced Interdisciplinary Studies, National University of Defense Technology, Changsha, Hunan, China

H. Munoz-Marco FYLA LASER SL, Paterna, Valencia, Spain

Mukul Chandra Paul Fiber Optics & Photonics Division, CSIR-Central Glass & Ceramic Research Institute, Kolkata, West Bengal, India

CSIR-Central Glass and Ceramic Research Institute, Kolkata, India

Wang Pengfei State Key Laboratory of Transient Optics and Photonics, Xi'an Institute of Optics and Precision Mechanics, Chinese Academy of Sciences (CAS), Xi'an, Shaanxi, China

Center of Materials Science and Optoelectronics Engineering, University of Chinese Academy of Sciences (UCAS), Beijing, China

P. Perez-Millan FYLA LASER SL, Paterna, Valencia, Spain

Ramadas Pillai Vinvish Technologies Pvt. Ltd., Trivandrum, Kerala, India

E. V. Podivilov Institute of Automation and Electrometry, Siberian Branch, Russian Academy of Sciences, Novosibirsk, Russia

Novosibirsk State University, Novosibirsk, Russia

Andrey Pryamikov Prokhorov General Physics Institute of the Russian Academy of Sciences, Moscow, Russia

Shuai Ren School of Information and Communications, National University of Defense Technology, Wuhan, Hubei, China

Ahmad Haziq Aiman Rosol Photonics Engineering Laboratory, Department of Electrical Engineering, University of Malaya, Kuala Lumpur, Malaysia

Wan Rui State Key Laboratory of Transient Optics and Photonics, Xi'an Institute of Optics and Precision Mechanics, Chinese Academy of Sciences (CAS), Xi'an, Shaanxi, China

Center of Materials Science and Optoelectronics Engineering, University of Chinese Academy of Sciences (UCAS), Beijing, China

Sameer Salam Department of Communications Technology and Engineering, Technical College, Imam Ja'afar Al-Sadiq University, Baghdad, Iraq

L. A. Sanchez Departamento de Física Aplicada y Electromagnetismo, ICMUV, Universidad de Valencia, Burjassot, Valencia, Spain

D. A. Shapiro Institute of Automation and Electrometry, Siberian Branch, Russian Academy of Sciences, Novosibirsk, Russia

Novosibirsk State University, Novosibirsk, Russia

Xiaoyu Shi School of Physics and Optoelectronic Engineering, Beijing University of Technology, Beijing, China

E. Silvestre Departamento de Óptica, ICMUV, Universidad de Valencia, Burjassot, Spain

M. I. Skvortsov Institute of Automation and Electrometry, Siberian Branch, Russian Academy of Sciences, Novosibirsk, Russia

Novosibirsk State University, Novosibirsk, Russia

Yinglin Song School of Physics, Harbin Institute of Technology, Harbin, China

Junhua Tong School of Physics and Optoelectronic Engineering, Beijing University of Technology, Beijing, China

Pius Varughese Vinvish Technologies Pvt. Ltd., Trivandrum, Kerala, India

Cong Wang School of Microelectronics, Shenzhen Institute of Information Technology, Shenzhen, P. R. China

Tsinghua Shenzhen International Graduate School, & Center of Double Helix, Tsinghua University, Shenzhen, P. R. China

Hanbin Wang Weihai Institute of Beijing Jiaotong University, Weihai, China

School of Physical Science and Engineering, Beijing Jiaotong University, Beijing, China

Han Wu College of Electronics and Information Engineering, Sichuan University, Chengdu, China

Li Xianda State Key Laboratory of Transient Optics and Photonics, Xi'an Institute of Optics and Precision Mechanics, Chinese Academy of Sciences (CAS), Xi'an, Shaanxi, China

Center of Materials Science and Optoelectronics Engineering, University of Chinese Academy of Sciences (UCAS), Beijing, China

Jiangming Xu College of Advanced Interdisciplinary Studies, National University of Defense Technology, Changsha, Hunan, China

Tianfu Yao College of Advanced Interdisciplinary Studies, National University of Defense Technology, Changsha, Hunan, China

Jun Ye College of Advanced Interdisciplinary Studies, National University of Defense Technology, Changsha, Hunan, China

Tianrui Zhai School of Physics and Optoelectronic Engineering, Beijing University of Technology, Beijing, China

Hanwei Zhang College of Advanced Interdisciplinary Studies, National University of Defense Technology, Changsha, Hunan, China

Jiaqi Zhou Shanghai Institute of Optics and Fine Mechanics, Chinese Academy of Sciences, Shanghai, China

Pu Zhou College of Advanced Interdisciplinary Studies, National University of Defense Technology, Changsha, Hunan, China

College of Interdisciplinary Studies, National University of Defense Technology, Changsha, Hunan, China

Chapter 1
Optical Fiber Lasers: Recent Advances and Applications

Mário F. S. Ferreira and Mukul Chandra Paul

Abstract The ongoing research in optical fiber laser technology represents an ever-evolving field, driven by the pursuit of enhanced performance and the creation of new generations of specialty optical fibers, including the exploration of novel materials and designs. This book comprises 20 chapters, dedicated to the recent advances in the field of optical fiber laser technology, ensuring a comprehensive coverage of this dynamic research area. The content reflects significant progress in various types of optical fiber lasers, addressing topics that range from fundamental principles to cutting-edge fabrication technologies, along with their wide-ranging applications. The findings and insights presented in this book serve to underscore the significance and potential of various kinds of optical fiber laser technology in modern science, paving the way for future advancements.

Keywords Fiber lasers · Specialty optical fibers · Q-switching · Passive mode-locking · Dissipative solitons · Distributed feedback lasers · Wavelength-tunable lasers · Frequency conversion · High-power lasers · Random lasers

Optical fiber lasers represent one of the most significant advances in laser technology in recent history, serving a critical role across a diverse range of applications that include scientific research, military operations, medical procedures, and various industrial and commercial functions. These applications span from spectroscopy to high-precision material processing such as cutting, welding, and marking. The global fiber laser market has experienced remarkable growth, with a valuation

M. F. S. Ferreira (✉)
i3N—Institute of Nanostructures, Nanomodelling and Nanofabrication & Department of Physics, University of Aveiro, Aveiro, Portugal
e-mail: mfernando@ua.pt

M. C. Paul (✉)
Fiber Optics and Photonics Division, Central Glass and Ceramic Research Institute, Kolkata, West Bengal, India
e-mail: mcpal@cgcri.res.in

M. F. S. Ferreira, M. C. Paul (eds.), *Optical Fiber Lasers*, Springer Series in Optical Sciences 250, https://doi.org/10.1007/978-3-032-04110-4_1

of $3.3 billion in 2022 and an expected increase to $9.2 billion by 2032. This impressive growth trajectory is forecasted at a Compound Annual Growth Rate (CAGR) of 11% from 2023 to 2032. The expansion of this market is fueled by continuous research and development efforts from academics, researchers, PhD candidates, and scientists, who are dedicated to innovating in several key areas. Over the past two decades, significant advancements have been made in the design of waveguide-based fiber structures, as well as in the exploration of new fiberglass materials and active dopants. These innovations have led to the creation of a variety of fiber lasers that operate across visible to mid-infrared wavelengths. Key enhancements have included improvements in heat dissipation, compact structures, conversion efficiency, output power levels, and beam quality.

In the realm of high-power fiber lasers, ongoing efforts to explore new doping host compositions aim to enhance lasing efficiency while minimizing photodarkening effects. Furthermore, ultrafast fiber lasers have entered a new phase of development, driven by advancements in materials, photonic devices, specialty optical fibers, and measurement technologies. Researchers are now focused on the generation of shorter laser pulses, which are becoming increasingly vital for a wide spectrum of scientific, technological, medical, and industrial applications. High-power narrow-linewidth fiber lasers with excellent beam quality are highly sought after for their ability to enable spectral and coherent beam-combining techniques. Linearly polarized fiber lasers are particularly advantageous in practical applications such as parametric conversion, coherent detection, and spectrally beam-combining laser subsystems. These technologies are essential for enhancing output power, notably in high-power laser sources utilized in Directed Energy Weapons (HPL-DEW). The global strategic sector has already established the technology for deploying laser-based DEW on the ground, providing effective air defense against threats like rockets and missiles.

In recent years, artificial intelligence (AI) has been increasingly integrated into fiber laser systems, particularly in industrial manufacturing settings. AI-enabled fiber laser systems offer significant benefits, enhancing efficiency for end-users and decreasing their reliance on operator skill. This is especially pertinent as industries face a shortage of experienced workers. By minimizing operator input requirements while maximizing quality and productivity, these cutting-edge systems facilitate usability for operators of all skill levels.

The evolution of optical fiber laser technology demonstrates its pivotal role in modern society, from defense applications to everyday manufacturing processes. As research and innovations continue to unfold, the future of fiber lasers appears bright, paving the way for advancements that will further expand their capabilities and impact across various sectors. With ongoing developments in materials science, photonics, and AI integration, the potential of fiber lasers remains boundless, ensuring their relevance in our technologically driven world.

1.1 Target Audience

This book is tailored for graduate students, engineers, and researchers who are either entering the field of fiber laser technology or seeking to deepen their existing knowledge. Its structured approach, covering foundational concepts alongside cutting-edge advancements, makes it suitable for users at various expertise levels. We hope this book serves as a valuable resource, guiding readers through the fascinating world of fiber lasers and empowering them to contribute to its future. This proposed structure provides an organized and in-depth pathway to explore the multifaceted aspects of fiber laser technology, fulfilling the purpose of being an important learning material in this exciting and rapidly evolving field.

1.2 The Core Content of the Book

This book aims to provide a thorough overview of the latest developments in fiber laser technology, considering various parameters such as lasing wavelengths, output power, beam characteristics, pulse structure, cavity design, fiber structure, and the transformative impact of artificial intelligence (AI) in this domain. The potential of fiber laser technology is vast, driven by a combination of fundamental principles and ongoing innovations. Understanding these elements can not only unlock current applications but also pave the way for future developments, pushing the boundaries of what fiber lasers can achieve in various fields. As research continues, it's likely that new applications will emerge, further solidifying fiber lasers' role as a transformative technology. The chapters written in this book cover a broad range of advanced fiber laser topics. Such topics concern the advancements and innovations in fiber laser technology, with a focus on various types of lasers, mode-locking techniques, and novel materials. Here's an overview of key concepts related to each of these categories.

Chapter 1 is an introduction to the main theme of this book, entitled 'Optical Fiber Lasers – Recent Advances and Applications," reporting on the importance of various kinds of optical fiber lasers in our modern society, market demand, categorization of optical fiber laser lasing wavelengths, output power, beam characteristics, pulse structure, cavity design, fiber structure, and the transformative impact of artificial intelligence (AI) in this domain. Along with their application, before going to the discussion about the whole content of this book.

In **Chap. 2**, S.W. Harun et al. explore the development of Q-switched pulse generation using two innovative types of passive saturable absorbers (SAs), focusing on their application within erbium-doped fiber lasers (EDFLs). Both the organic SA based on Alq3 and the SMS structure proved effective for generating Q-switched pulses in EDFLs, with each method offering unique advantages. They have demonstrated stability, tunability, and straightforward integration of these SAs into fiber laser systems, highlighting their potential for a wide range of applications in both industrial and scientific settings.

In **Chap. 3**, Mário F. S. Ferreira et al. delve into the intriguing realm of dissipative solitons within mode-locked fiber lasers, utilizing the cubic-quintic complex Ginzburg-Landau equation (CGLE) as a key mathematical framework. Dissipative solitons are unique localized structures that manifest from a delicate interplay between competing processes: the balance of gain and loss in the system, alongside the nonlinear effects against dispersion or diffraction. They have discussed the main characteristics of both stationary and pulsating CGLE dissipative solitons, as well as the formation of dissipative soliton molecules. They have elaborated several types of pulsating dissipative solitons, namely plain pulsating, creeping, chaotic, and erupting solitons. An overview of recent experimental results concerning the observation of pulsating solitons and dissipative soliton molecules in mode-locking fiber lasers is also provided.

Sofia C. V. Latas et al. in **Chap. 4** delve into the generation of high-energy ultrashort optical pulses through passively mode-locked fiber lasers, with a particular focus on a specific type of pulse known as very high amplitude (VHA) soliton solutions of the cubic-quintic complex Ginzburg-Landau equation (CGLE). These VHA solutions are intriguing phenomena that arise from a singularity initially described by Akhmediev and his collaborators. This chapter begins by providing an overview of the characteristics of VHA solitons, leveraging both numerical simulations and analytical techniques such as the method of moments. The authors highlight the unique properties that distinguish these solitons, particularly in terms of their amplitude and stability.

In **Chap. 5**, Haoguang Liu et al. explored the advancements and applications of multiplexed dissipative soliton fiber lasers, contrasting them with conventional mode-locked lasers that typically feature a single output. Multiplexed soliton lasers open up a broader spectrum of potential applications due to their capacity for greater complexity and functionality.

C. Cuadrado-Laborde et al. in **Chap. 6** highlight their research findings on all-fiber mode-locked lasers that emit radiation around the 2 μm wavelength range. Although these lasers were engineered for diverse applications, they exhibit several common features: the implementation of an all-fiber polarization-maintaining cavity, the utilization of thulium-doped fiber as the gain medium, the incorporation of a semiconductor saturable absorber mirror for mode-locking, and resonant pumping in the C-telecommunication band. They have developed a low-repetition rate all-polarization-maintaining thulium-doped passively mode-locked fiber laser, which operated at a frequency of 2.3 MHz, and then explored the potential for generating emissions at a wavelength significantly exceeding the traditional maximum for the 3F4 → 3H6 transition in thulium-doped fibers, which is approximately 1837 nm. This exploration aimed to assess applications in spectroscopy. Consequently, we succeeded in designing a passively mode-locked thulium-doped fiber laser that emits at 2070 nm, marking it as one of the longest wavelengths documented for solitonic emission in thulium-doped fibers. This chapter concludes with a comprehensive performance review of these lasers, including an analysis of their stability.

Jiaqi Zhou et al. in **Chap. 7** provide a detailed overview of ultrafast Raman fiber lasers, highlighting their advancements and significance in generating short optical pulses at various wavelengths. This chapter begins with an introduction to the evolution of these lasers, acknowledging the superior pulse energy and duration of conventional rare-earth-doped laser systems while noting the unique advancements in Raman technologies. This chapter summarizes the technological advancements in ultrafast Raman fiber lasers, encouraging further exploration and application of these innovative tools in various fields. This overview serves as a valuable resource for understanding the current trends and future potential of ultrafast Raman fiber laser technologies.

Prof. Andrey Pryamikov in **Chap. 8** described gas-filled hollow-core fiber lasers (HCFLs) as a significant advancement in laser technology by harnessing the benefits of both fiber lasers and gas lasers. These systems are particularly compelling for generating high-energy outputs in the mid-infrared (mid-IR) spectrum, which is essential for a variety of applications, including remote sensing, environmental monitoring, and medical therapies. This chapter aims to provide a comprehensive overview of these key advancements and explore the current landscape of gas-filled HCFLs, highlighting both the challenges faced and the exciting possibilities that lie ahead. By examining the latest developments, this review will serve as a valuable resource for researchers and practitioners looking to deepen their understanding of this cutting-edge technology and its potential applications in the mid-IR spectral range.

S.A. Babin et al. in **Chap. 9** provide an overview of the advancements in fiber lasers utilizing distributed feedback (DFB), emphasizing both regular and random refractive-index structures. This chapter reviews various experimental setups featuring passive and active fibers, where different types of reflective structures are created. These include π-shifted gratings inscribed using femtosecond (fs) pulse techniques, gratings exhibiting random amplitude and phase variations, and numerous random arrays of gratings. The theoretical foundations concerning Bragg reflectors and Bragg gratings, particularly those with π-shifts, are discussed to provide insight into the underlying principles of DFB lasers. Overall, the chapter encapsulates a significant leap in fiber laser technology, demonstrating both theoretical insights and practical applications.

Jiangming Xu et al. in **Chap. 10** discuss future research directions and unexplored applications of random fiber lasers (RFLs), suggesting that advancements in RFL technology could lead to breakthroughs in fields like telecommunications, medical diagnostics, and materials processing. Key areas for exploration include integrating RFLs with artificial intelligence and machine learning to optimize performance and automate tuning. Additionally, studying nonlinear optical phenomena in RFLs may uncover novel effects and lead to new laser sources. RFLs offer scalability and easy integration, making them promising for quantum communication and secure networks. Their potential applications in biophotonics and medical imaging could enhance imaging quality, while portable RFL systems may revolutionize point-of-care diagnostics. Overall, ongoing research in RFLs is expected to lead to new innovations and broaden their impact in science and engineering.

Alexander V. Kir'yanov in **Chap. 11** provides a comprehensive overview of Self-Induced Laser-Line Sweeping (SLLS), a significant phenomenon in fiber lasers that has developed since its discovery in the early 2000s. SLLS is characterized by a sweeping action of the emitted laser line and has gained attention due to its unique properties and wide-ranging applications, such as in telecommunications and material processing. The key historical milestones include the initial studies focusing on the mechanisms behind SLLS, the importance of optical feedback and laser cavity design, and advancements in fiber fabrication that have improved SLLS efficiency. This review emphasizes the ongoing significance of SLLS in fiber laser technology, highlighting its potential for future breakthroughs and applications. Future investigations are encouraged, particularly into integrating SLLS with emerging technologies like artificial intelligence. The narrative serves as an informative resource for both established researchers and newcomers, emphasizing the importance of SLLS in advancing fiber laser research.

Prof. Dr. Xuanyi Liu in **Chap. 12** not only reviews the state-of-the-art technologies for wavelength tuning in mode-locked PM fiber lasers but also highlights the future directions for research and development in this field. The ongoing advancements in materials science, fiber design, and modulation techniques are expected to further enhance the performance and versatility of these lasers, paving the way for their widespread adoption in both industrial and scientific applications.

Bing Han et al. in **Chap. 13** delve into the advancements and technologies involved in the nonlinear frequency conversion of fiber lasers, focusing on their capabilities to produce visible and mid-infrared (MIR) wavelengths. They will explore several key processes, including second harmonic generation (SHG), sum-frequency mixing (SFM), optical parametric oscillation (OPO), and different frequency generation (DFG). Such processes can be effectively implemented in various types of fiber lasers, such as rare-earth-doped fiber lasers, Raman fiber lasers, and random fiber lasers.

Cong Wang et al. in **Chap. 14** discuss the role of two-dimensional materials (2DM) in ultrafast laser technology, focusing on their unique properties and applications as saturable absorbers (SAs). 2DM, which includes graphene, topological insulators, transition metal dichalcogenides (TMDs), black phosphorus, bismuthene, antimonene, and MXenes, exhibits exceptional optoelectronic and nonlinear optical characteristics, making them valuable in photonics. This chapter outlines various fabrication methods for 2DM, including mechanical exfoliation and chemical vapor deposition, and discusses the advantages and challenges associated with each technique. It emphasizes the nonlinear properties of 2DM particularly their saturable absorption mechanisms, which improve ultrafast laser performance. In conclusion, the chapter highlights the transformative potential of 2DM in ultrafast lasers, showcasing their versatility and effectiveness as saturable absorbers and setting the foundation for ongoing innovations in laser applications.

Wang Pengfei et al. in **Chap. 15** examine the evolution and advancements in large mode area (LMA) fibers, which are essential for the continued scalability of high-power fiber lasers. The need for increased laser power while maintaining single-mode (SM) output is crucial for reducing power density, thereby minimizing nonlinear effects and thermal effects that can lead to transverse mode instability (TMI). It begins with a historical overview of LMA fibers, tracing their development from traditional step-index fibers to the current state of double-clad and microstructured optical fibers (MOFs). The chapter highlights key advancements in the structural design principles that have been instrumental in optimizing the performance of solid-core LMA fibers, primarily composed of silica-based glasses. Overall, this chapter serves as a comprehensive resource for understanding the intricate link between fiber design and laser performance, emphasizing future directions for research and development in the fiber lasing domain.

Shuai Ren et al. in **Chap. 16** highlighted the demand for high-power narrow linewidth (HPNL) linearly polarized all-fiberized lasers due to their applications across various fields like industrial processing, medical therapies, scientific research, and national defence. Enhancing the brightness of these fiber lasers, defined as the ratio of output power to beam area, is a critical area of research that directly impacts the effectiveness and applicability of these systems. Although significant progress has been made in the development of HPNL linearly polarized all-fiberized lasers, ongoing research is essential to overcome existing limitations and explore new opportunities in various applications, contributing to the enhancement of laser brightness and overall performance.

M.C. Paul et al. in **Chap. 17** present a new class of Er/Yb specialty optical fibers designed with a focus on high-power laser and optical amplifier applications. Utilizing an octagonal cladding design and a composition of multielement-doped silica glass (Al, Ge, P, Ce, B, F), they achieved a remarkable lasing efficiency of over 41.0%, alongside stable output power and commendable radiation resistance. The optimization involved adjusting key parameters, including glass composition and the deposition temperature of the porous fluorinated phospho-silica core layer, through a modified chemical vapor deposition (MCVD) method combined with a solution doping technique to ensure uniform integration of the dopants, effectively mitigating clustering issues. The optimized doping ratios (7.5–11.1) of various elements, including fluorine, GdO2, CeO2, Al2O3, B2O3, P2O5, Er2O3, and Yb2O3, significantly enhanced fiber performance. Performance characterization revealed a gain of approximately 28.0 dB with little variation (less than 0.45 dB flatness) over a wavelength range of 1540–1570 nm for a 10 dBm input signal. Additionally, under prolonged operation exceeding 150 h, the fiber maintained an average output power of around 16.0 W, with parasitic lasing below 1.0 μm and less than 10% degradation in output after exposure to 6.0 Krad of cumulative 60Co gamma radiation. These results underscore the potential of this multielement-doped silica glass fiber in applications requiring robust performance and stability in challenging environments, such as free-space communication systems.

Hanbin Wang et al. in **Chap. 18** provide a comprehensive overview of spectral beam-combining (SBC) technology, highlighting its significance in high-power laser systems. SBC technology is crucial for advancing high-power lasers, requiring continuous innovation to meet challenges in power scaling and promising transformative applications across various sectors. The chapter aims to enhance understanding and inspire further developments in laser engineering.

Jun Li et al. in **Chap. 19** explore the transformative impact of artificial intelligence (AI) on the development of high-power fiber lasers (HPFL) through three pivotal lenses: optical fiber design, measurement of laser properties, and coherent beam combining. In summary, AI is driving a significant evolution in HPFL technology by enhancing the design process, refining measurement techniques, and optimizing complex systems like coherent beam combining. This synergy between AI and HPFL represents a promising trajectory that stands to further improve the capabilities and applications of fiber laser technology in the future.

Naeem Iqbal et al. in **Chap. 20** represent random lasers as a novel class of laser devices that diverge from traditional laser technologies by utilizing disordered media to achieve optical feedback. This unique approach leads to several distinctive characteristics, including random emission wavelengths and directions, which set them apart from conventional lasers. This chapter delves into the fundamental concepts, key characteristics, and control mechanisms of random lasers. It also examines design strategies, fabrication techniques, emission properties, and the applications of random lasers with fiber-embedded cavities, providing examples in imaging and sensing. Furthermore, it briefly discusses WGM-coupled random hybrid cavities and their applications in information security.

These topics illustrate the dynamic growth and diverse applications of fiber laser technology, showcasing innovations that have significant implications across industries such as telecommunications, manufacturing, medical devices, and environmental monitoring.

1.3 Interpretation of the Book

The introduction outlines a detailed chapter structure for exploring fiber laser technology, emphasizing both fundamental principles and innovative advancements. The proposed chapters cover various topics, starting with an overview of laser technology and its applications, followed by the specifics of fiber lasers, including their mechanisms, design, and advantages. Subsequent chapters address manufacturing processes, applications in industries and medicine, and emerging technologies impacting efficiency and performance. The text also discusses challenges and limitations, future growth opportunities, and showcases case studies of successful implementations. This book will encourage further research and engagement in the field. Overall, this structure is designed to provide a thorough understanding of fiber laser technology and facilitate exploration and innovation.

1.4 Conclusions

In conclusion, this book comprises 20 chapters, including the introduction chapter, authored by leading international scientists who have dedicated years to advancing the field of optical fiber laser technology. Each chapter has been meticulously crafted, ensuring comprehensive coverage of this dynamic research area. The content reflects significant progress in various types of optical fiber lasers, addressing topics that range from fundamental principles to cutting-edge fabrication technologies.

Key areas of focus include various kinds of optical fiber lasers, along with their wide-ranging applications within the photonic industry. Among the innovations discussed are: Q-switched erbium-doped fiber lasers; dissipative solitons in passively mode-locked lasers; ultra-short high amplitude dissipative solitons in fiber lasers; multiplexed dissipative soliton fiber lasers; thulium-doped passively mode-locked all polarization-maintaining fiber lasers; ultrafast Raman fiber lasers; gas-filled hollow core fiber lasers; distributed feedback fiber lasers based on regular and random refractive-index structures; random fiber laser: from demonstration to applications; self-sweeping fiber laser; wavelength-tunable mode-locked polarization-maintaining fiber laser; nonlinear frequency conversion of fiber laser; development of pulse fiber laser based on two-dimensional materials; large mode area (LMA) fibers for high-power lasers' high-power narrow-linewidth linearly polarized fiber laser; multielements doped silica glass-based specialty optical fibers for high-power optical lasers and optical amplifiers; the journey, current status, and challenges of high-power fiber laser spectral beam combining technology; high-power fiber laser in the AI era; and advancement in plasmonic fiber random lasers.

The ongoing research in optical fiber laser technology represents an ever-evolving field, driven by the pursuit of enhanced performance and the creation of new generations of specialty optical fibers (SOFs). This pursuit entails an exploration of novel materials and designs, which are crucial for developing innovative photonic devices and components across a diverse array of applications. The findings and insights presented in this book serve to underscore the significance and potential of various kinds of optical fiber laser technology in modern science, paving the way for future advancements.

Chapter 2
Q-Switched Erbium-Doped Fiber Lasers

Sulaiman Wadi Harun, Ahmad Haziq Aiman Rosol, Sameer Salam, Yu Chen, and Mukul Chandra Paul

Abstract Q-switched laser sources have obtained substantial attention over recent decades due to their extensive applications across various industries and scientific research fields. Advancements in finding new saturable absorber (SA) materials to achieve efficient Q-switching remain a key focus in fiber laser research. This study demonstrates the generation of Q-switched pulses utilizing two distinct passive SA types: an organic film absorber incorporating Alq_3 and an artificial SA based on a single-mode fiber (SMF), step-index multimode fiber (SIMF), single-mode fiber (SMS) structure. Initially, we employed an organic SA by embedding Alq_3 into a polyvinyl alcohol (PVA) film to create a Q-switched pulse train with a broad tuning range. An integrated tunable bandpass filter (TBF) within the Erbium-doped fiber laser (EDFL) cavity allowed for continuous wavelength tuning from 1520 nm to 1563.3 nm, covering nearly the entire C-band region. The fiber laser exhibited excellent stability, with a signal-to-noise ratio (SNR) of approximately 53 dB across various wavelengths. These findings highlight the broad absorption capabilities of Alq_3, confirming its suitability for compact, stable, and widely tunable fiber lasers. In a second approach, we fabricated an SMS structure by fusion-splicing a 158 mm length of 105 μm core diameter SIMF into the EDFL cavity. This structure effectively served as a passive SA, utilizing multimode interference and the Kerr effect to modulate the cavity loss for Q-switching. The repetition rate ranged from 87 to 158 kHz, achieved with a pump power between 115 and 343 mW. The minimum pulse width of 2.5 μs was obtained at a pump power of 343 mW. The SNR of the

S. W. Harun (✉) · A. H. A. Rosol · Y. Chen
Photonics Engineering Laboratory, Department of Electrical Engineering, University of Malaya, Kuala Lumpur, Malaysia
e-mail: swharun@um.edu.my

S. Salam
Department of Communications Technology and Engineering, Technical College, Imam Ja'afar Al-Sadiq University, Baghdad, Iraq

M. C. Paul
Fiber Optics & Photonics Division, CSIR-Central Glass & Ceramic Research Institute, Kolkata, West Bengal, India

M. F. S. Ferreira, M. C. Paul (eds.), *Optical Fiber Lasers*, Springer Series in Optical Sciences 250, https://doi.org/10.1007/978-3-032-04110-4_2

fundamental mode reached 72 dB at a central wavelength of 1530 nm. Our results demonstrate that the SMS structure is an effective SA for generating stable Q-switched EDFLs with a straightforward cavity configuration.

Keywords Fiber laser · Q-switching · Passive saturable absorber · Multimode interference

2.1 Introduction

Over the years, scientists worldwide have made numerous efforts to develop reliable and powerful lasers for various industrial applications [1, 2]. While bulk solid-state lasers have been prevalent in recent decades [3], their limitations have spurred research into alternative laser types, such as fiber lasers, which use rare-earth-doped fibers as the gain medium. Among these, erbium-doped fiber lasers (EDFLs) have garnered significant interest due to their remarkable attributes [4]. They offer exceptional flexibility, substantial single-pass gain, high beam quality, and powerful output concentrated within mode field diameters of just a few micrometers. Additionally, all-fiber lasers are alignment-free and exhibit low cavity losses, which are crucial for efficient Q-switching performance.

Q-switched fiber lasers have garnered substantial interest recently due to their potential in various applications, especially where low repetition rates are needed, such as in optical communication, medical surgery, remote sensing, defense, and material processing [5, 6]. There are two primary methods for generating Q-switched fiber lasers: active methods, which use a pulse modulator [7], and passive methods, which employ saturable absorbers (SAs) [8]. Passive methods are often preferred for their simplicity, cost-effectiveness, and compactness. Commonly used SAs include semiconductor saturable absorber mirrors (SESAMs) [9], carbon nanotubes [10], graphene [5], black phosphorus [11], doped fibers [12], and transition metal dichalcogenides [13]. Currently, SESAMs are the most widely used SAs in the industry due to the mature technology in semiconductor electronics. However, SESAMs are limited by their narrow bandwidth. Graphene, while offering an ultra-broad operational wavelength range, has relatively high losses. Each of these SAs has its own limitations, including complex fabrication processes, lower damage thresholds, and higher manufacturing costs.

Conversely, concerns have been growing about the potential health impacts of nanomaterials, particularly with long-term exposure [14]. Despite these concerns, researchers remain motivated to identify high-performance saturable absorbers (SAs) for use in passive pulsed fiber lasers. Organic materials, in this context, offer intriguing optical properties. They provide broad spectral tunability and ultrafast nonlinear responses [15, 16]. Additionally, these materials are lightweight, mechanically flexible, and cost-effective. Their biocompatibility and environmental friendliness make them promising candidates for medical applications [17]. Organic materials have been successfully utilized in various applications due to their tailored properties for optoelectronics, including bistable memory devices [18], thin-film transistors [19], solar cells [20], and light-emitting diodes [21]. Among these, Tris-(8-Hydroxyquinoline) aluminum (Alq_3) stands out as one of the most extensively studied organic semiconductors [22]. Alq_3 offers a simple and cost-effective fabrication process, coupled with a high damage threshold and excellent optical properties, including a non-zero optical bandgap [23]. Moreover, Alq_3 is distinguished by its long operational lifetime, exceptional thermal stability, and ease of fabrication and purification, thanks to its stable thin-film formation.

Similarly, the single-mode multimode single-mode (SMS) fiber structure has garnered increased interest as an artificial SA due to its inherent wavelength independence, high power damage threshold, and cost-effectiveness. The SMS structure operates as an SA through nonlinear effects within the optical fiber-based laser cavity [24]. Specifically, self-phase modulation (SPM) and cross-phase modulation (XPM) alter the self-imaging beat length at high-intensity power compared to low-intensity power. This variation in beat length modulates the cavity loss, leading to the generation of Q-switched pulses [25]. While numerous reports have demonstrated the successful application of SMS structures in mode-locked lasers [26, 27], there are only a few documented instances of Q-switched lasers utilizing SMS structures as SAs [25].

In this chapter, we explore the generation of Q-switched pulses in an EDFL cavity using two different approaches: a passive real SA based on an Alq_3 thin film and an artificial SA based on an SMS structure. The first approach employs Alq_3, an organic SA consisting of one aluminum ion (Al^{3+}) and three 8-hydroxyquinoline molecules. The aluminum ion, with its electron structure similar to that of inert gas atoms, provides excellent stability in dry environments. Alq_3 has a bandgap energy of approximately 2.7 eV and offers remarkable flexibility and high photoconductivity. The second approach involves using an SMS structure as an artificial SA. It is created by fusion-splicing a segment of 105/125 μm diameter step-index multimode fiber (SIMF) between two single-mode fibers (SMFs) before integrating it into the EDFL cavity. The SMS structure is advantageous compared to other types of SAs for generating Q-switched EDFLs because it can withstand higher power levels, potentially allowing for pulse widths to reach the nanosecond range. The C-band region, which is extensively studied in the literature, is particularly noted for its applications in spectroscopy, sensing, and telecommunications.

2.2 Erbium-Doped Fiber Laser

Since its development in the late 1980s, erbium-doped fiber (EDF) has demonstrated remarkable versatility across various applications, including wide-band optical amplifiers, broadband light sources, and tunable lasers. Amplified spontaneous emission (ASE) in EDF has been instrumental in creating broadband light sources used in fields such as optical coherence tomography, gyroscopes, and optical device characterization. The erbium-doped fiber amplifier (EDFA) was a groundbreaking innovation that revolutionized the optical communication industry starting in the early 1990s. Today, EDFAs are integral to a variety of fiber communication systems, particularly in dense wavelength division multiplexing (DWDM) networks. Moreover, erbium-doped fiber lasers (EDFLs) have seen substantial advancements recently, offering benefits like wide tunable wavelengths, excellent beam quality, low cost, and compact size.

An EDF features a core infused with erbium ions (Er^{3+}). The energy level diagram of Er^{3+} is depicted in Fig. 2.1. The energy levels are broadened by the DC-Stark effect, resulting in a relatively wide emission bandwidth. When a 975 nm pump laser diode is introduced into the fiber, it excites Er^{3+} ions from the ground state (E_1) to a higher energy level (E_3). The excited Er^{3+} ions then rapidly decay to the intermediate level (E_2) through nonradiative processes. The ions eventually return to the ground state (E_1) through spontaneous emission, generating photons in the 1520–1570 nm wavelength range.

When a laser signal with a wavelength between 1520 and 1570 nm is combined with a 975 nm pump laser in an EDF, there are three potential outcomes for the signal photon: (i) stimulated absorption, where the signal photon excites an Er^{3+} ion from E_1 to E_2, resulting in the photon's annihilation; (ii) stimulated emission, where the signal photon stimulates an Er^{3+} ion at E_2 to decay to E_1, creating an additional photon and thereby amplifying the signal; and (iii) the signal photon can travel through the fiber without being affected. Simultaneously, spontaneous emission occurs between levels E_2 and E_1. When the pump laser power is sufficiently high to achieve population inversion between these levels, the input laser signal is amplified. This process utilizes the EDF and pump laser to construct an optical amplifier known as the EDFA. ASE, which is always present in EDFA and acts as a primary noise source, can also be amplified by the pump laser. The efficiency of an EDFA tends to decrease as the dopant concentration increases, primarily due to ion-ion interactions among nearby Er^{3+} ions, leading to a loss in the excited-state

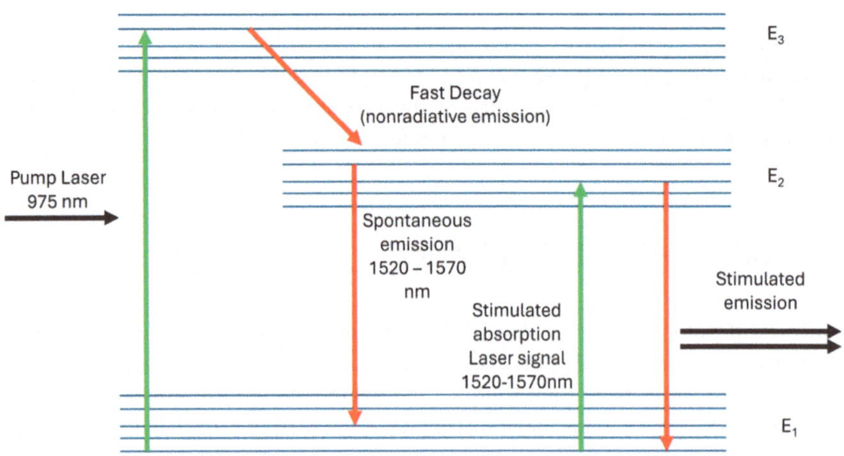

Fig. 2.1 The energy levels of Er^{3+} ions in an EDF

population. To address this, Quimby et al. proposed a method to reduce ion-ion interactions without lowering dopant concentration. By incorporating aluminum (Al) into the core glass, this approach minimizes clustering effects between Er^{3+} ions, thereby improving the amplifier's efficiency.

A laser can be viewed as an optical amplifier with positive feedback. When the output of an EDFA is fed back into its input to create a fiber loop, and pump laser power is introduced into this loop, the EDFA becomes an EDFL. The operating wavelength of the laser is determined by the cavity loss, meaning that by adjusting the cavity loss, the EDFL's wavelength can be tuned.

2.3 Q-Switching Principle

Pulse generation in the nanosecond to microsecond time domain is achieved by manipulating intracavity losses within a laser resonator, a technique known as Q-switching or Q-factor modulation, which can produce pulse energies in the range of a few hundred nanojoules. The Q-factor, in general, measures the damping strength of the resonator modes. In the context of energy storage, it represents the ratio of energy stored to the energy dissipated during one complete oscillation. However, in laser applications, the term takes on a slightly different meaning, focusing on two key characteristics: round trip losses and laser gain, which compensate for these losses during continuous-wave operation.

Q-switching operates by initially pumping the gain medium to maintain high resonator losses (low Q-factor), thus preventing energy extraction from the resonator. Subsequently, the Q-factor is increased due to a higher gain relative to resonator losses. This leads to an exponential rise in intracavity power until it reaches a peak, at which point the power decreases as gain saturation occurs. The light energy generated in this process comes from the energy stored in the gain medium, making the choice of gain medium crucial for producing a Q-switched laser with high pulse energy. An ideal gain medium should feature a high population of laser-active ions, a long upper-state lifetime, and moderate gain efficiency. If the gain medium is too efficient, it may lead to excessive ASE, which can limit the initial loss and energy storage of the Q-switched laser.

To better understand Q-switched laser operation, consider the mechanism illustrated in Fig. 2.2, which shows the single pulse energy of a Q-switched laser. Imagine a saturable absorber with an initial bleach state of "0," unbleached conditions "q_0," a saturable absorber loss coefficient "$q(t)$," and total cavity loss per round trip "l." During phase 1, depicted in Fig. 2.2, when the total cavity loss (low Q) matches the gain, a pulse emerges at the absorber's unbleached condition. As the laser intensity increases, the gain starts to bleach the absorber, causing a rapid rise in intracavity power $P(t)$ once the gain $g(t)$ approximates "l." This saturation of the absorber leads to a significant increase in intracavity power.

In phase 2, as the pump power increases, the absorber becomes fully bleached while the gain remains high (high Q). At this point, the gain is reduced to the level

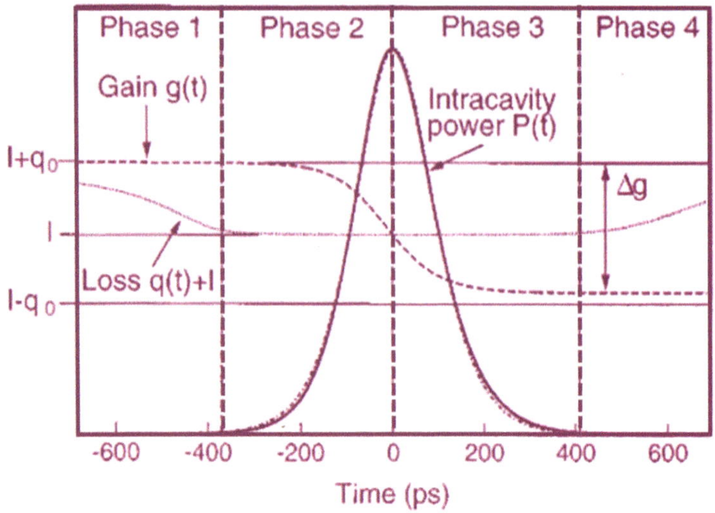

Fig. 2.2 Evolution of loss, gain, and power on the time scale of pulse width plotted from the numerical integration of the rate equation [28]

of total cavity loss, $q(t) + l$. When the gain g(t) matches the total cavity loss l under the bleached condition, the pulse reaches its maximum intensity. In phase 3, the gain continues to decline and eventually turns negative, causing the intracavity power P(t) to decrease. Finally, in phase 4, the absorber recovers to its unbleached state due to its shorter recovery time compared to the gain. Consequently, the pumped gain medium continues to provide gain to the laser, initiating the next pulse cycle.

The gain medium is responsible for storing the energy required for the entire lasing process, which can be expressed by the following equation:

$$E_{\text{stored}} = AL_G N_2 h\nu_1 \tag{2.1}$$

where $h\nu_1$ represents the photon energy at the pumping wavelength, N_2 denotes the excitation energy in the gain medium, and AL_G is the volume of the pumped gain medium.

2.4 Q-Switched Pulse Generation with Organic Material-Based Thin Film

In this section, Alq_3 was utilized as the SA in a compact all-fiber EDFL cavity to generate a Q-switched pulse train with a tunable wavelength. The wavelength could be adjusted continuously from 1520 to 1563.3 nm, covering nearly the entire C-band region and achieving a very high signal-to-noise ratio (SNR). Notably, the self-Q-switched operation was accomplished without the need for a polarizer controller, and the optical fiber's coiling did not impact the Q-switching performance. Alq_3 demonstrated several advantages over other materials such as CNT, graphene, TMD, WS_2, and TiO_2 [29], including a broad operational wavelength range, non-zero optical bandgap, excellent resistance to photobleaching, substantial modulation depth, and a high damage threshold.

2.4.1 Preparation of SA and Laser Configuration

The SA thin film was created by embedding Alq_3 into polyvinyl alcohol (PVA). The Alq_3 thin film was fabricated using a drop-casting technique at room temperature, as detailed in Fig. 2.3. First, 1 mL of distilled water was mixed with 10 mg of Alq_3. powder. This solution was stirred at 50 °C for 1 h, after which two drops of acetone were added to ensure uniform distribution of Alq_3. Concurrently, 1 g of PVA was dissolved in 100 mL of distilled water and subjected to ultrasonic treatment for about 1 h at room temperature. Next, 5 mL of the PVA solution was combined with the Alq_3 solution and stirred for approximately 3 hours at room temperature. The resulting mixture was then poured into a 60 mm diameter borosilicate glass petri dish to form the thin film, which was allowed to dry for 3 days at room temperature. The film, with a thickness of approximately 50 µm as measured by a Mitutoyo micrometer screw gauge, was then cut into a small piece and attached to a fiber ferrule. Increasing the thickness of the SA film by adding more Alq_3 enhances the modulation depth due to the higher number of absorbing particles. However, this also increases the non-saturable loss and insertion loss.

The fabricated film was characterized to identify the chemical groups present in the materials using a Perkin Elmer Spectrum 400 Fourier Transform Infrared (FTIR) Spectrometer. Both pure PVA and Alq_3 thin films were analyzed in transmittance mode with a resolution of 2 cm^{-1}. Figure 2.4 presents the FTIR spectra in the range of 450–4000 cm^{-1}. Both spectra reveal absorption peaks at 3280 cm^{-1}, 1725 cm^{-1}, and 1651 cm^{-1}. These peaks correspond to O-H stretching from hydrogen bonds,

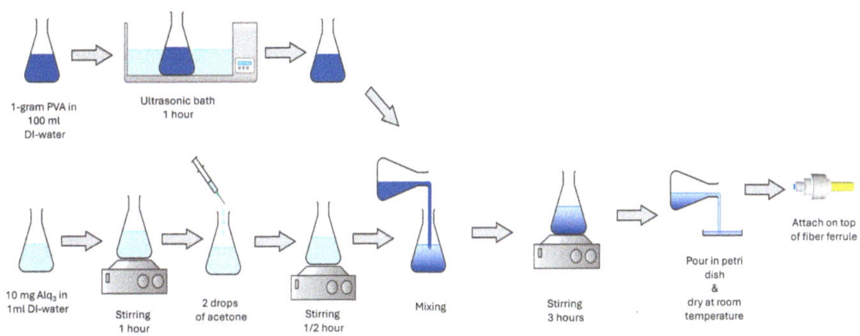

Fig. 2.3 Step-by-step procedure to prepare Alq_3 PVA thin film

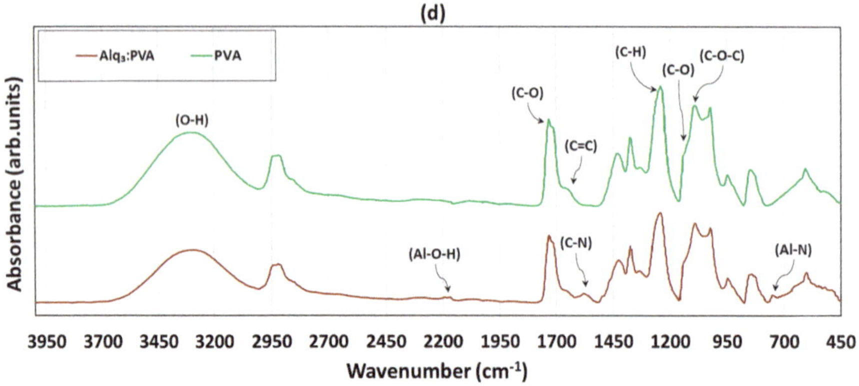

Fig. 2.4 FTIR absorbances of Alq_3:PVA and PVA thin films

C=O stretching from acetate, and C=C stretching, respectively [30, 31] [32]. The C-H aliphatic scissoring and C-H stretching from the alkyl group are observed between 1465 cm^{-1} and 1350 cm^{-1} and between 3000 cm^{-1} and 2810 cm^{-1}, respectively [33]. Peaks at 1245 cm^{-1}, 1141 cm^{-1}, and 1086 cm^{-1} are attributed to C-H [34], C-O [31], and C-O-C vibrations [30], respectively. Additionally, vibrations in the range of 900 cm^{-1} to 600 cm^{-1} are associated with C-H wagging [33]. While the FTIR spectra for Alq_3 and pure PVA thin films are largely similar, the Alq_3 film displays additional peaks at 750 cm^{-1}, 1580 cm^{-1}, and 2176 cm^{-1}, which correspond to Al-N, C=N, and Al-O-H bonding modes of Alq_3, respectively [35–37].

Figure 2.5(a) displays the nonlinear absorption characteristics of the Alq_3:PVA film. The measurement was conducted utilizing a home-built mode-locked laser operating at a wavelength of 1572 nm, with a pulse duration of 2.8 ps and a repetition rate of 1 MHz as the pulse source. The SA demonstrated a saturation intensity of 3 MW/cm^2 and a modulation depth of approximately 8.1%. Figure 2.5(b) shows the linear absorption of the SA, indicating a flat transmission of around 1.5 dB in the 1550 nm region. This linear transmission measurement was performed by directing light from a white light source (WLS) through the SA device and into an optical spectrum analyzer (OSA). The results were compared with measurements taken when the WLS was connected directly to the OSA. Figure 2.5(c) presents the optical absorbance of the SA across a range of 100–1950 nm. Linear absorbance was measured using a Perkin Elmer Lambda 750 spectrophotometer, which covers the wavelength range from 250 to 2000 nm. The unsaturated bonds of PVA and Alq_3 were observed at 300 nm and 385 nm, respectively [38–40]. The SEM image of the Alq_3 thin film, shown in Fig. 2.5(d), reveals a smooth surface with a uniform distribution of Alq_3 within the PVA polymer. The observed dots on the surface are likely contaminants, possibly resulting from aggregated Alq_3 particles that are challenging to disperse evenly within the PVA matrix. Finally, Fig. 2.5(e) depicts the SA device, which was assembled by attaching a 1 mm^2 piece of Alq_3 thin film between two fiber ferrules.

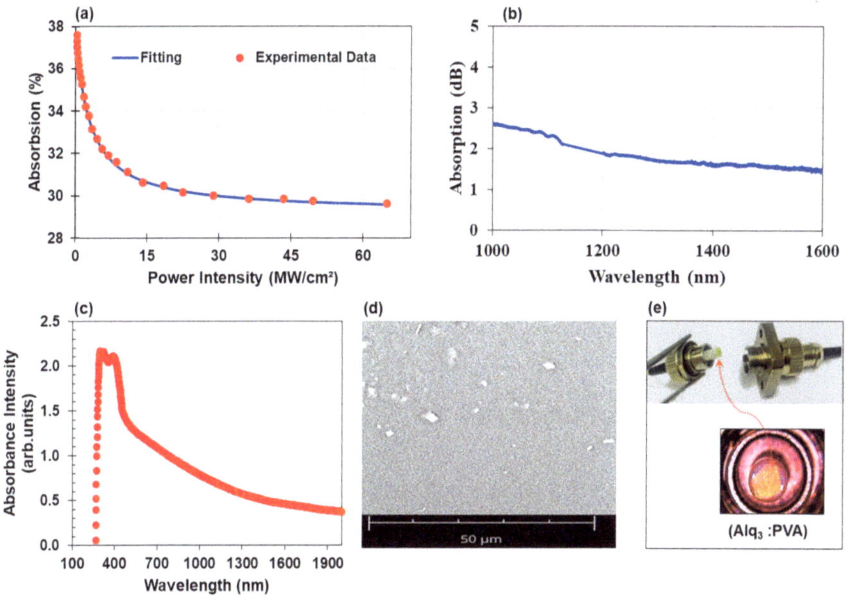

Fig. 2.5 Alq_3: PVA SA characteristics: (**a**) nonlinear absorption, (**b**) linear absorption within 1.0 to 1.6 µm, (**c**) the absorption spectrum measured by a spectrophotometer, (**d**) SEM image, and (**e**) the SA device features a small Alq_3: PVA film sandwiched between two ferrules

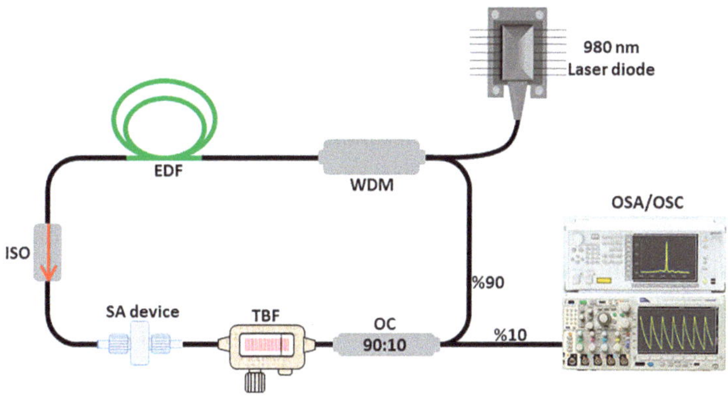

Fig. 2.6 Configuration of the Q-switched EDFL with tunable operational wavelength

The configuration of the tunable Q-switched EDFL is illustrated in Fig. 2.6. A 980 nm pump light was used to energize the fiber laser through a 980/1550 WDM coupler. The system incorporated a 2-meter-long single-mode EDF as the gain medium. The EDF features a core diameter of 4 µm, a cladding diameter of 125 µm, a numerical aperture of 0.16, a group velocity dispersion of 27.6 ps^2/km, and an erbium ion absorption of 23 dB/m at 980 nm. To prevent backward reflection, the EDF was connected to an optical isolator. The output end of the isolator was linked to the SA device. For wavelength tuning, a tunable bandpass filter (TBF) with a 1 nm bandwidth was employed. The TBF is chosen for its practicality, high stability, wide tuning range, and low loss, as well as its insensitivity to temperature changes. It enhances the fiber laser's mode operation by allowing only a single wavelength to pass through, thus mitigating timing jitter and mode-beating issues. The TBF offers a resolution of 0.1 nm and a tuning range of 43.3 nm. A 90:10 optical coupler was used to extract 10% of the output while retaining 90% within the cavity. The output optical spectrum, optical pulse train, radio frequency (RF) spectrum, and average output power were monitored with an optical spectrum analyzer (OSA), digital oscilloscope (OSC), RF spectrum analyzer, and optical power meter, respectively.

2.4.2 *Laser Performance*

Initially, the TBF was set to a wavelength of 1563.3 nm, and the 980 nm laser diode (LD) power was gradually increased. At 25 mW LD power, continuous-wave (CW) operation was observed, as confirmed by the OSC, which showed no pulsing. When the LD power was raised to 38 mW, a self-starting and stable pulsing operation was achieved. This pulsing was maintained up to the maximum input LD power of 206 mW. It is suggested that the range for generating pulsing could be extended by using a thicker thin-film SA, which would require higher power to achieve saturation. However, a thicker film would also introduce higher insertion loss, thereby increasing both the pulsing threshold power and the maximum input LD power. By tuning the TBF, the central wavelength was continuously adjusted from 1520 nm to 1563.3 nm, spanning a range of 43.3 nm and covering nearly the entire C-band. The output optical spectrum had a 3 dB bandwidth of 0.14 nm. The TBF allows a minimum increment of 0.1 nm, the smallest adjustment possible with the micrometer head of the TBF. Figure 2.7(a) displays spectra that are approximately 1 nm apart (to ensure clarity) at an LD power of 206 mW. The samples exhibit consistent spectral power of −3.2 dBm ± 1.2 dB and identical patterns and shapes, indicating that the pulsing operation was highly stable.

Figure 2.7(b) illustrates how the tunable wavelength affects pulse duration and pulse repetition rate. As the TBF was adjusted from 1520 nm to 1529.65 nm, the pulse duration decreased from 3 μs to 2.12 μs before increasing back to 3 μs as the TBF was tuned to 1563.3 nm. The pulse repetition rate, which is inversely related to pulse width duration, peaked at 111.6 kHz at 1529.65 nm. This peak is attributed to the higher EDF gain at 1529.65 nm compared to other wavelengths, which results in quicker saturation of the SA, thereby shortening pulse width duration and increasing the pulse repetition rate. Figure 2.7(c) examines the effects of wavelength tuning on pulse energy and average output power. The average output power decreased at shorter wavelengths, even with the LD power maintained at 206 mW, due to higher loss in the TBF at shorter wavelengths. Conversely, pulse energy, which is inversely related to the repetition rate, increased from 15 nJ at 1530.4 nm to 37.3 nJ at 1563.3 nm.

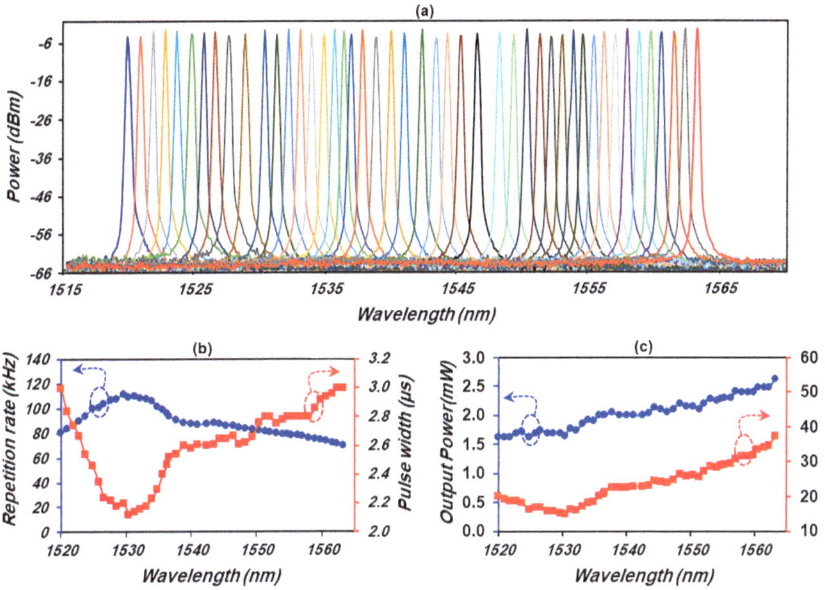

Fig. 2.7 (**a**) Output spectra at different operating wavelengths. (**b**) Pulse frequency and duration vs. wavelength. (**c**) Output power and pulse energy vs. wavelength

The output pulse train at the maximum LD power of 206 mW was measured using a 350 MHz oscilloscope with a 2 GHz photodetector, as shown in Fig. 2.8(a–c). The pulse shapes were uniform and free from amplitude fluctuations at wavelengths of 1529.6 nm (Fig. 2.8(a)), 1545.4 nm (Fig. 2.8(b)), and 1563.3 nm (Fig. 2.8(c)). The figures confirm that the pulsing operation was stable, without pulse jitter or self-mode-locking effects. The insets show pulse spacing for each pulse train, with the shortest spacing being 8.96 μs at 1529.65 nm, increasing to 11.56 μs at 1545.4 nm, and reaching 14.18 μs at 1563.3 nm. Figures 2.8(d) and (e) present the output RF spectrum at spans of 1810 kHz and 75 kHz, respectively, with the TBF tuned to 1551.4 nm and the LD power at maximum. The data, recorded using an Anritsu-MS2683A with a resolution bandwidth (RBW) of 300 Hz, shows a high signal-to-noise ratio (SNR) of 53.2 dB. Similar high SNR values were observed at various wavelengths (1563.3, 1551.4, 1542, 1530.5, and 1522.8 nm), further confirming the stability of the fiber laser. It is noted that recording over a long duration for a single wavelength does not necessarily reflect performance at other wavelengths. The TBF also filters out any distortion or fluctuations in the output spectrum. The experiment, which lasted approximately 4 h, demonstrated consistent performance with stable output pulse trains and SNR throughout. The laser maintained the same tuning range (1520–1563.3 nm) across different input pump power values.

Figure 2.9(a) shows the relationship between pulse repetition rate and pulse width duration with LD power. As expected, pulse duration decreased from 10 μs to

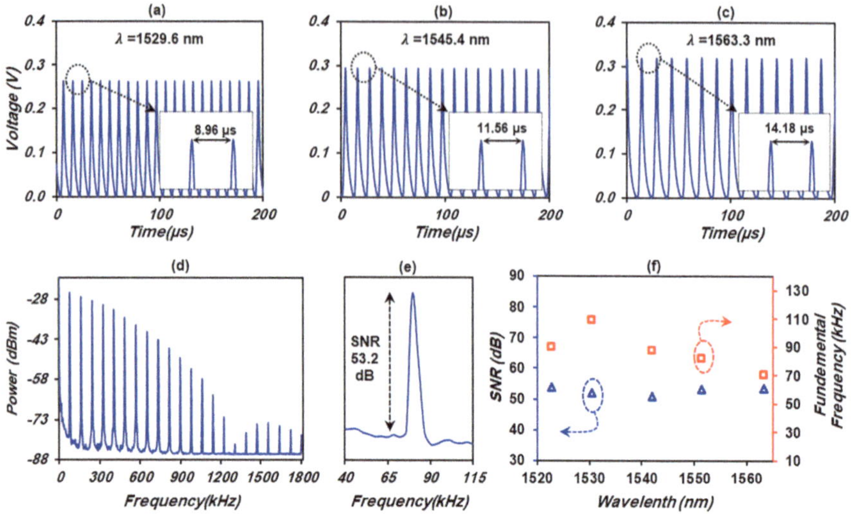

Fig. 2.8 Typical oscilloscope traces at wavelengths of 1529.6 nm, 1545.4 nm, and 1563.3 nm are shown in (**a**), (**b**), and (**c**), respectively. The insets display the corresponding dual-pulse envelopes. (**d**) and (**e**) present the RF spectrum with spans of 1810 kHz and 75 kHz, respectively. (**f**) illustrates the SNR and fundamental frequency as functions of the operating wavelength

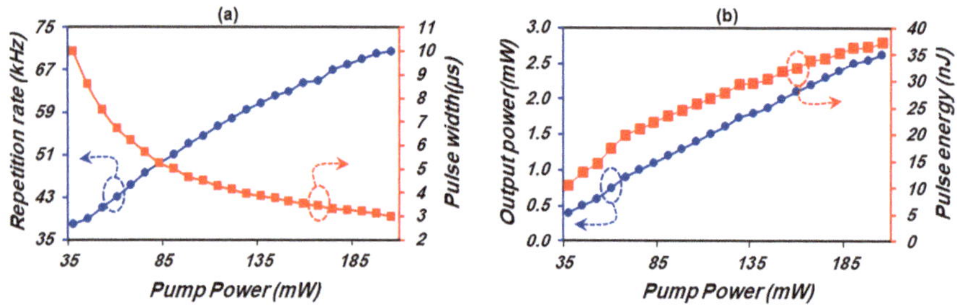

Fig. 2.9 (**a**) Pulse-rate and pulse width as functions of LD power. (**b**) Average output power and single pulse energy as functions of LD power

3 μs, while pulse repetition rate increased from 38 kHz to 71 kHz. A shorter pulse width of 1.2 μs was achieved in previous work [41], but in this case, the pulse width was slightly wider due to the cavity losses introduced by the TBF and SA. It is believed that pulse width can be reduced to the nanosecond range by shortening the cavity length to around 1 m and further optimizing the TBF and SA to minimize losses. Increasing the LD power to 350 mW caused some fluctuation in the output pulse trains, which eventually disappeared, indicating that the SA had become fully saturated. Further investigation is needed to assess the energy density and power in the cavity and compare these with the nonlinear properties of the SA to confirm saturation. When LD power was reduced below 206 mW, pulsing operation was

Table 2.1 Tunable-wavelength Q-switched fiber lasers with different SAs

SA	Wavelength tuning range (nm)	SNR (dB)	Maximum pulse energy (nJ)	Minimum pulse width (μs)	Ref.
Graphene	1519.3–1569.9	–	82.6	4.6	[44]
Bi_2Se_3	1493.6–1508.9	58	6.2	27.3	[45]
SnO_2	1521–1565	35.76	100.9	2	[46]
TiO_2	1534–1570	45	3.9	1.1	[47]
MoS_2	1519.6–1567.7	50	160	3.3	[48]
MWCNT	1519–1569	60	52.13	2.83	[49]
Alq_3	1520–1563.3	53	37.3	2.12	This work

restored. The optical damage threshold is thus concluded to be above 350 mW, as no Q-switching operation was observed beyond this point, and no deformation of the SA was noted during or after the experiment.

Figure 2.9(b) illustrates that both the average output power and pulse energy increase linearly with the input LD power. Specifically, the average output power rises from 0.4 mW to 2.63 mW, while pulse energy increases from 10.53 nJ to 37 nJ. This linear relationship in output pulse energy indicates the excellent performance of the Q-switching laser [42]. To confirm that Alq_3 is the sole contributor to the pulsing operation, we replaced the Alq_3 thin film with a pure PVA thin film. Under these conditions, no Q-switching was observed on the oscilloscope, even after extensive adjustments such as bending and twisting the optical fibers and varying the input power across a wide range. This result underscores that Alq_3 is indeed responsible for the Q-switching effect. The experiment demonstrated that Alq_3 functions effectively as a broadband passive Q-switcher and that the Q-switching pulsing can be tuned across different wavelengths. The broad absorption spectrum of the SA is crucial for achieving a widely tunable pulsed fiber laser [43]. Alq_3 has proven to be a promising broadband SA, showing performance comparable to other SAs such as graphene, Bi_2Se_3, SnO_2, TiO_2, MoS_2, and MWCNT in terms of wavelength tuning range, SNR, pulse energy, and pulse width, as summarized in Table 2.1.

This experiment demonstrates significant potential for practical applications requiring a broad tunable-wavelength range with highly stable performance, such as environmental sensing, biomedical diagnostics, and metrology. The observed tuning range was limited by the TBF rather than the SA, suggesting that a broader tuning range could be achieved with a TBF featuring a wider tuning capability. Furthermore, improvements in pulse energy and average output power could be realized by utilizing a high-gain fiber [50] and optimizing the cavity parameters [51]. It is also noteworthy that removing the TBF and substituting it with 52 m of single-mode fiber (SMF) resulted in a mode-locking operation, achieving a pulse duration of 1.6 ps and a repetition rate of 1.8 MHz [52]. Additionally, the SA was employed to develop a high-energy Q-switched laser at 1067 nm [53].

2.5 Q-Switched Pulse Generation with SMS Structure-Based Artificial SA

In this section, we introduce a new Q-switched EDFL utilizing an SMS (Single-Mode Fiber-Step-Index Multimode Fiber-Single-Mode Fiber) structure as an artificial SA. This approach is garnering increased attention due to its high power damage threshold, wavelength independence, and cost-effectiveness. The SMS structure functions as an SA through nonlinear effects within the optical fiber-based laser cavity. Specifically, self-phase modulation (SPM) and cross-phase modulation (XPM) cause variations in the self-imaging beat length depending on the intensity of the light. These intensity-dependent changes in the beat length modulate the cavity loss, leading to the generation of Q-switched pulses [25]. In this work, a segment of step-index multimode fiber (SIMF) with a 105/125 μm diameter was fusion spliced into the cavity, with single-mode fibers (SMF) on either side, to serve as the SA.

2.5.1 *Multimode Interference in SMS Structure*

When light propagates from an SMF into a SIMF that is spliced to it, multiple modes are excited within the MMF, leading to periodic self-imaging along the fiber where there is no phase difference between the excited modes. The light field distribution $E_{mm}(r, \varnothing, z)$ along the SIMF can be expressed as:

$$E_{MM}\left(r,\varphi,z\right)=\sum_{n=1}^{N}C_{n}e_{n}\left(r,\varphi,0\right)e^{-i\beta_{n}z}=e^{-i\beta_{1}z}\sum_{n=1}^{N}C_{n}e_{n}\left(r,\varphi,0\right)e^{-i(\beta_{n}-\beta_{1})z} \tag{2.2}$$

where C_n is the mode expansion coefficient, $e_n(r, \varnothing, 0)$ represents the n-th guided mode of the MMF, and β_1 and β_n are the propagation constants of the fundamental mode and the n-th excited mode in the SIMF, respectively. Self-imaging occurs when the following condition is met for all modes:

$$\left(\beta_n-\beta_1\right)\mathrm{z}=\mathrm{m_n}2\pi \tag{2.3}$$

where m_n is an integer. This can be rewritten in terms of the effective refractive index $\Delta n_{eff,n}$ of the optical fiber and the length of the MMF segment L as:

$$\Delta n_{eff,n}L=m_n\lambda_0 \tag{2.4}$$

where λ_0 is the center wavelength and $\Delta n_{eff,n}=\frac{\left(\beta_n-\beta_1\right)\lambda_0}{2\pi}$. Equation (2.4) indicates that the performance of the SMS structure as an SA is influenced by the effective refractive index, the length of the MMF, and the center wavelength of the light.

To minimize the required length of the SIMF, bending or applying stress to the fiber can be used to induce birefringence. This introduces additional nonlinear phase shifts as the light passes through the fiber. The magnitude of these phase shifts depends on the degree of birefringence, which can be adjusted by altering the polarization controller (PC).

2.5.2 Laser Configuration and SMS Characterization

Figure 2.10 presents the schematic diagram of the SMS-based laser configuration. The continuous wave LD, with a maximum output power of 343 mW at 975 nm (II-VI, Model LC96Z400-74), was coupled into a 2-m-long EDF (Fibercore I25) via a 980/1550 nm wavelength division multiplexer (WDM). The EDF served as the gain medium in the cavity. To ensure unidirectional operation of the laser light, the output from the gain medium was connected to a polarization-insensitive isolator. Following this, the SMS structure, acting as an artificial SA, was fusion spliced into the cavity. The SIMF component of the SA was positioned within an in-line polarization controller (PC), allowing for the application of stress and strain to the SIMF and control of the polarization of the oscillating laser. An 80:20 optical coupler was placed after the SMS structure, serving both as a feedback mechanism to the cavity and an output port. The total length of the cavity, including the EDF, SIMF, and surrounding single-mode fibers, was approximately 9.6 m.

The SMS structure is shown schematically in Fig. 2.11(a). In this study, when light propagates from a single-mode fiber (SMF) into a step-index multimode fiber (SIMF), it exhibits a periodic interference pattern due to multimode interference. This interference arises from the excitation of multiple modes as the laser light travels through the SIMF. The core diameter of the SIMF is significantly larger than that of the SMF, leading to multimode interference at the splicing point where the

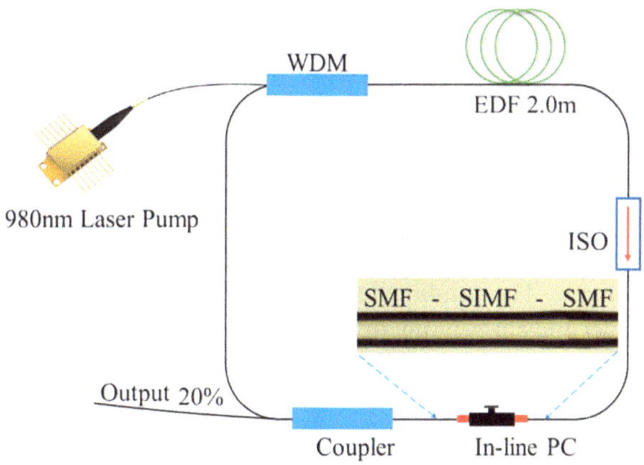

Fig. 2.10 Schematic diagram of the experimental setup of the ring cavity

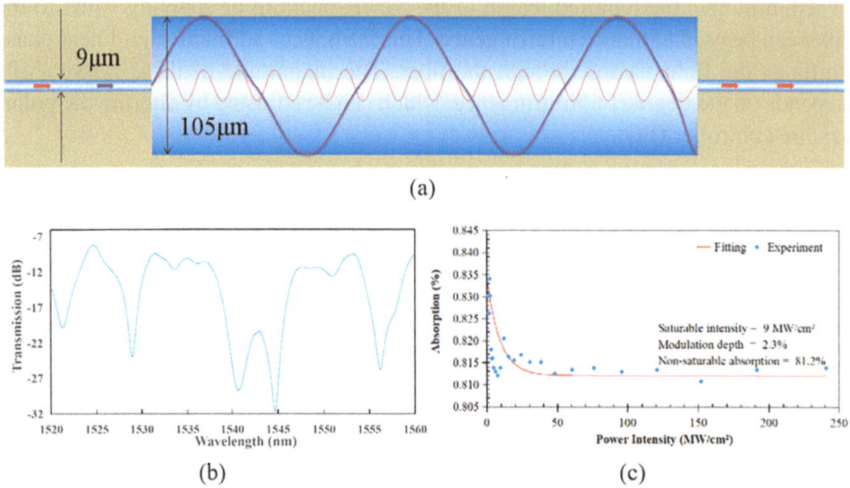

Fig. 2.11 (**a**) The schematic diagram of SMS structure with 158 mm length of SIMF (**b**) The transmission spectrum characteristic of the structure. (**c**) Nonlinear absorption curve of the structure

cross-sections of the fibers are mismatched. This mismatch results in mode excitation and coupling at the splice, causing wavelength-dependent transmission as the light oscillates unidirectionally within the ring cavity. The multimode interference observed is strongly influenced by the effective refractive index within the fiber's core [25, 54]. The Kerr effect plays a crucial role here; it causes the refractive index of the optical fiber to vary with light intensity. As a result, the transmission through the SMS structure is high at high power intensities and low at low intensities [55]. This behavior mimics that of an SA, allowing the SMS structure to function as an artificial SA that modulates cavity loss for Q-switching applications [56].

Figure 2.11(b) illustrates the spectral transmission characteristics of the SMS structure, revealing a periodic interference pattern caused by multimode interference. This pattern is the result of multiple modes being excited as the laser light propagates through the SIMF. For this work, a 158 mm long SIMF was employed to construct the SMS structure. The SMF, originating from the isolator shown in Fig. 2.10, was fused to one end of the 158 mm SIMF using a commercial fusion splicer (Fujikura 90S+ model). The other end of the SIMF was connected to an SMF at the input end of an 80/20 coupler. The splice losses were measured at 0.01 dB and 0.02 dB, respectively. The SIMF has core and cladding diameters of 105 µm and 125 µm, respectively. The combined segments of SMF and SIMF thus formed the SMS structure. To prevent bending of the SIMF, which could alter the self-imaging length, the entire SMS structure was positioned in an in-line polarization controller (PC), as depicted in Fig. 2.10.

A balanced twin-detector system was employed to measure the modulation depth and saturation intensity of the SMS structure [57]. A custom laser, operating at a center wavelength of 1567 nm with a pulse width of 700 fs and a repetition rate of

21.73 MHz, served as the light source. The nonlinear transmission curve of the SMS structure was analyzed and fitted using the following equation:

$$T(I) = 1 - q \bullet e^{-I/I_{sat}} - a_{ns} \tag{2.5}$$

where q represents the modulation depth, I_{sat} is the saturation intensity, and a_{ns} denotes the nonsaturable loss [58]. The experimental results, depicted in Fig. 2.11(c), indicate that $q = 2.3\%$, $I_{sat} = 9\ mW/cm^2$, and $a_{ns} = 81.2\%$. These results illustrate that the absorption characteristics of the SMS structure vary with the intensity of the incident light. At lower intensities, the device exhibits high transmission, whereas at higher intensities, the absorption saturates, leading to a decrease in transmission. This characteristic is vital for Q-switched lasers, where the ability to generate short pulses depends on such dynamic absorption properties.

2.5.3 Q-Switching Performance with the Artificial SA

The performance of the Q-switched laser, utilizing the SMS structure as an artificial SA, was thoroughly examined. The Q-switched laser pulses were successfully observed at a threshold LD power of 115 mW. This relatively high threshold can be attributed to the significant loss inherent in the SMS structure. The Q-switched operation was stable up to the maximum LD power of 343 mW. Notably, the absence of pulses when the SMS structure was removed from the cavity underscores its critical role in the Q-switching process. The Kerr effect plays a significant role in this setup. As the input light intensity increases, the refractive index of the SIMF's core undergoes changes. At low intensities, the refractive index remains relatively stable. However, at higher intensities, the nonlinear change in the refractive index alters the behavior of the SIMF's core. This change can shift the conditions for interference, affecting the coupling of light into specific modes. Consequently, this shift results in increased transmission at high-intensity levels, allowing the SMS structure to function effectively as an SA.

The characteristics of the Q-switched pulse laser are summarized in Fig. 2.12. The laser operated at a consistent wavelength of 1530 nm across varying input pump powers, as depicted in Fig. 2.12(a). This wavelength was chosen due to its optimal gain, which compensates for the losses incurred by the SMS structure. A typical pulse waveform at an input power of 343 mW is shown in Fig. 2.12(b). The waveform illustrates a stable output pulse train with minimal intensity fluctuations. Figure 2.12(c) provides an enlarged view of the pulse train, revealing a pulse period of 6.3 μs and a pulse duration of 2.5 μs. This pulse period corresponds to a repetition rate of 157.7 kHz. The RF spectrum of the output laser at 343 mW pump power is shown in Fig. 2.12(d). It demonstrates an SNR of 72 dB at the fundamental frequency of 157.7 kHz, indicating the laser's high stability. The elevated SNR signifies a significant reduction in noise within the laser cavity, a testament to the effective modulation of intracavity loss by the SMS structure. Additionally, at least 13

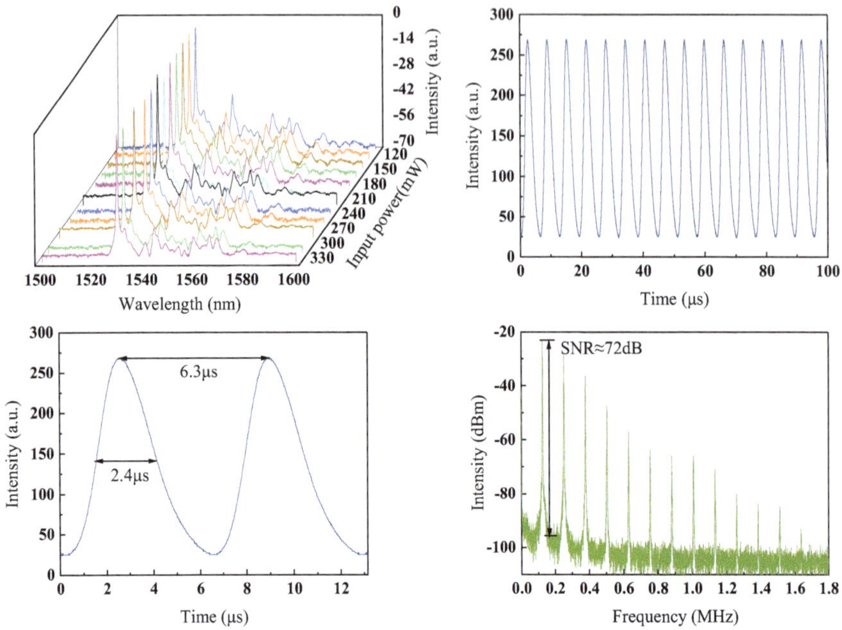

Fig. 2.12 Spectral, temporal, and frequency performance of the Q-switched laser (**a**) evolvement of output spectrum with the pump power increment, (**b**) typical pulse train when pumped at 343 mW, (**c**) the corresponding two-pulse envelop, and (**d**) typical RF spectrum when pumped at 343 mW

harmonics were detected within a 1.8 MHz frequency band, further confirming the stability of the Q-switching operation. The laser maintained stable performance in laboratory conditions for over 24 h with no noticeable degradation.

To further investigate the extended stability of the Q-switched pulsed laser, we operated the laser at its maximum power output of 343 mW in a controlled laboratory environment for a duration of 5 h. During this period, spectral data were collected at half-hour intervals, as shown in Fig. 2.13. The collected spectra revealed minimal fluctuations in optical intensity, central wavelength, and 3 dB bandwidth throughout the measurement period. This consistent behavior, evidenced by the stable spectral readings, highlights the reliability and stability of the laser pulse operation over extended periods. The minimal variation in key performance metrics underscores the effectiveness of the SMS structure in maintaining stable Q-switched pulse generation under continuous operation conditions. This long-term stability is a crucial factor for practical applications requiring reliable and consistent laser performance.

Figure 2.14(a) illustrates how the repetition rate and pulse width of the Q-switched fiber laser vary with pump power. Specifically, as the pump power increased from 115 mW to 343 mW, the repetition rate rose from 87.4 kHz to 157.7 kHz, while the pulse width decreased from 4.2 μs to 2.5 μs. This trend is characteristic of Q-switched fiber lasers, where increased pump power leads to gain compression in the cavity,

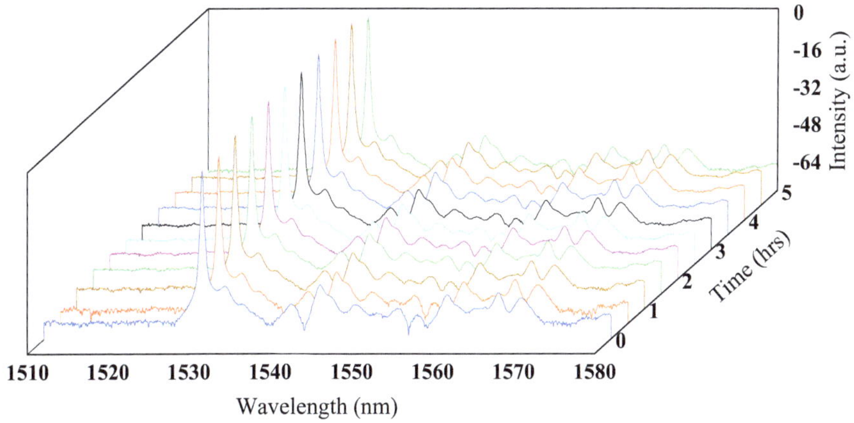

Fig. 2.13 The output spectrum changes over time

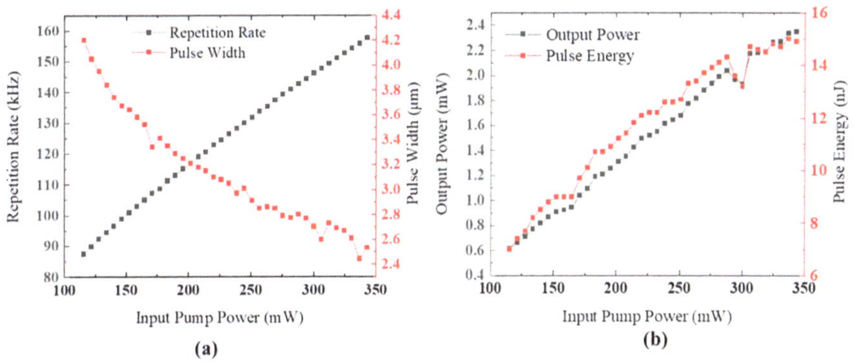

Fig. 2.14 (**a**) Repetition rate and pulse width as functions of pump power, and (**b**) output power and pulse energy as functions of pump power

thereby shortening the pulse width. Figure 2.14(b) depicts the relationship between the average output power and pulse energy as a function of pump power. The output power increased linearly from 0.61 mW to 2.35 mW, and the pulse energy rose from 7.0 nJ to 15.0 nJ within the same range of pump power. Despite tuning the pump power up to its maximum of 343 mW, no mode-locked pulses were observed. This lack of mode-locking is attributed to the use of a single in-line polarization controller (PC), which limits the ability to adjust polarization and thereby restricts the number of oscillating longitudinal modes. To achieve mode-locking, it is recommended to modify the laser cavity by adding a three-pedal PC, which would enhance polarization control and facilitate mode-locked operation. Additionally, optimizing the cavity length can help balance nonlinearity and dispersion, creating favorable conditions for mode-locked pulse generation.

Overall, we successfully demonstrated a Q-switched fiber laser utilizing an SMF-SIMF-SMF structure within the EDFL cavity. This method is straightforward, cost-effective, and yields stable laser performance. However, the SMS structures proved to be more sensitive to environmental fluctuations compared to thin-film SAs, which could potentially impact long-term stability.

2.6 Conclusion

Extensive research has been conducted on Q-switched fiber lasers utilizing SAs to achieve short-pulse laser emission. In our study, we successfully demonstrated Q-switched EDFLs employing two distinct types of passive SAs: an organic film absorber based on Alq_3 and an artificial SA based on an SMS structure. Initially, we used an organic SA, made by embedding the Alq_3 compound into a PVA film, to generate a Q-switched pulse train with a broad wavelength tunability. The fiber laser exhibited a tuning range of 43.3 nm, covering wavelengths from 1520 nm to 1563.5 nm. Q-switching was achieved with input LD power ranging from 35 mW to 206 mW. The fiber laser demonstrated an exceptionally high SNR of approximately 53 dB across various tuning wavelengths. These results confirm that Alq_3 provides broad absorption characteristics, making it a viable option for compact, stable, and widely tunable fiber lasers. In our second approach, we utilized a 158 mm length of a 105 μm core diameter SIMF to construct an SMS structure. This SMS structure served as an artificial SA, effectively modulating cavity loss through nonlinear multimode interference. At the maximum pump power of 343 mW, the laser achieved a repetition rate of 157.7 kHz, a minimum pulse width of 2.5 μs, a maximum output power of 2.35 mW, and a pulse energy of 15.0 nJ. Overall, our results demonstrate that both the organic film and SMS structure are practical SAs for generating Q-switched pulses. Their simplicity and cost-effectiveness make them suitable for applications in metrology, environmental sensing, and biomedical fields.

References

1. Gafner, M., et al.: Ultrafast pulsed laser high precision micromachining of rotational symmetric parts. J. Laser Appl. **33**(1), 012053 (2021)
2. Shi, W., et al.: Fiber lasers and their applications: introduction. JOSA B. **34**(3), FLA1 (2017)
3. Sennaroglu, A.: Solid-State Lasers and Applications. CRC Press (2017)
4. Li, X., et al.: Hydrazone Organics Modulation for H-Shaped Nanosecond Pulse Generation in 1.3-Km Ultra-Long Mode-Locked Erbium-Doped Fiber Laser, vol. 169, p. 110060. Optic. Laser Technol. (2024)
5. Martinez, A., Sun, Z.: Nanotube and graphene saturable absorbers for fibre lasers. Nat. Photonics. **7**(11), 842–845 (2013)
6. Piao, Z., et al.: Q-switched erbium-doped fiber laser at 1600 nm for photoacoustic imaging application. Appl. Phys. Lett. **108**(14), 143701 (2016)
7. Zhang, A., et al.: Multi-pulse operation in an actively Q-switched Er-doped fiber laser based on electro-optic modulator. Optoelectron. Lett. **17**(12), 729–733 (2021)
8. Najm, M.M., et al.: Effect of MAX phase chromium aluminum carbide thin film thickness on Q-switched erbium-doped fiber lasers. Opt. Fiber Technol. **70**, 102853 (2022)

9. Zhang, Y., et al.: Study of the influence of SESAM parameters on the evolution of mode-locked pulses at different repetition rates. Appl. Phys. B. **127**(8), 1–10 (2021)
10. Ahmed, M., et al.: Q-switched erbium doped fiber laser based on single and multiple walled carbon nanotubes embedded in polyethylene oxide film as saturable absorber. Opt. Laser Technol. **65**, 25–28 (2015)
11. Markom, A.M., et al.: Dark pulse mode-locked fibre laser with zirconia-based erbium-doped fibre (Zr-EDF) and Black phosphorus saturable absorber. Optik. **223**, 165635 (2020)
12. Zhang, P., et al.: Switchable single/dual-wavelength Q-switched all-fiber laser utilizing thulium-doped fiber saturable absorber. J. Lumin. **258**, 119809 (2023)
13. Wei, B.-N., Jiao, Z.-H., Liu, W.-J.: Ternary transition metal dichalcogenides for passively Q-switched Er-doped fiber laser applications. Optik. **248**, 168096 (2021)
14. Malakar, A., et al.: Nanomaterials in the environment, human exposure pathway, and health effects: a review. Sci. Total Environ. **759**, 143470 (2021)
15. Salam, S., et al.: Hybrid organic small molecules as a saturable absorber for passive Q-switching in erbium-doped fiber laser. OSA Contin. **3**(2), 177–185 (2020)
16. Park, J.-M., et al.: Ultrafast nonlinear transparency driven at a telecom wavelength in an organic semiconductor system. AIP Adv. **9**(2) (2019)
17. Zhao, N., et al.: Versatile types of organic/inorganic nanohybrids: from strategic design to biomedical applications. Chem. Rev. **119**(3), 1666–1762 (2018)
18. Ramana, C.V., et al.: Fabrication of stable low voltage organic bistable memory device. Sensors Actuators B Chem. **161**(1), 684–688 (2012)
19. Jia, X., et al.: Stable organic thin-film transistors. Sci. Adv. **4**(1), eaao1705 (2018)
20. Zheng, Z., et al.: Tandem organic solar cell with 20.2% efficiency. Joule. **6**(1), 171–184 (2022)
21. Abbasi, F., et al.: Investigating the effect of copper phthalocyanine-emitting dopant on the production of near-infrared organic light-emitting diodes. J. Mater. Sci. Mater. Electron. **33**(13), 10494–10503 (2022)
22. Cölle, M., Brütting, W.: Thermal, structural and photophysical properties of the organic semiconductor Alq_3. Phys. Stat. Solid. **201**(6), 1095–1115 (2004)
23. Ahmad, A.A., et al.: Photoluminescence, optical, and electrical properties of Bis (8-Hydroxyquinoline) zinc and Tris-(8-Hydroxyquinoline) aluminum organometallics and their films. J. Electron. Mater. **53**(1), 338–346 (2024)
24. Nazemosadat, E., Mafi, A.: Nonlinear multimodal interference and saturable absorption using a short graded-index multimode optical fiber. J. Opt. Soc. Am. B. **30**(5), 1357 (2013)
25. Fu, S., et al.: Passive Q-switching of an all-fiber laser induced by the Kerr effect of multimode interference. Opt. Express. **23**(13), 17255–17262 (2015)
26. Cheng, P., et al.: Generation of different mode-locked states in a Yb-doped fiber laser based on nonlinear multimode interference. Opt. Express. **30**(20), 35911–35922 (2022)
27. Wu, Q.-C., et al.: Femtosecond soliton erbium-doped fiber laser with a symmetrical GIMF–SIMF–GIMF saturable absorber. IEEE Photon. J. **11**(6), 1–9 (2019)
28. Spühler, G.J., et al.: Experimentally confirmed design guidelines for passively Q-switched microchip lasers using semiconductor saturable absorbers. J. Opt. Soc. Am. B. **16**(3), 376–388 (1999)
29. Wang, X., et al.: Titanium dioxide Langmuir–Blodgett film saturable absorber for passively Q-switched Nd: GdVO 4 laser. IEEE Photon. J. **11**(2), 1–10 (2019)
30. Mansur, H.S., et al.: FTIR spectroscopy characterization of poly (vinyl alcohol) hydrogel with different hydrolysis degree and chemically crosslinked with glutaraldehyde. Mater. Sci. Eng. C. **28**(4), 539–548 (2008)
31. Andrade, G., et al.: Design of novel hybrid organic–inorganic nanostructured biomaterials for immunoassay applications. Biomed. Mater. **1**(4), 221 (2006)
32. Awada, H., Daneault, C.: Chemical modification of poly (vinyl alcohol) in water. Appl. Sci. **5**(4), 840–850 (2015)
33. Choi, J., et al.: Organic/inorganic hybrid composites from cubic silsesquioxanes. J. Am. Chem. Soc. **123**(46), 11420–11430 (2001)

34. Wang, T., Turhan, M., Gunasekaran, S.: Selected properties of pH-sensitive, biodegradable chitosan–poly (vinyl alcohol) hydrogel. Polym. Int. **53**(7), 911–918 (2004)
35. Laukamp, C., Salama, W., González-Álvarez, I.: Proximal and remote spectroscopic characterisation of regolith in the Albany–Fraser Orogen (Western Australia). Ore Geol. Rev. **73**, 540–554 (2016)
36. Quintero, O.M.S., et al.: Influence of the microstructure on the electrochemical properties of Al-Cr-N coatings deposited by co-sputtering method from a Cr-Al binary target. Mater. Res. **16**(1), 204–214 (2013)
37. Correia, F.C., et al.: Synthesis and characterization of a new semiconductor oligomer having quinoline and fluorene units. J. Braz. Chem. Soc. **26**(1), 84–91 (2015)
38. Elashmawi, I., Hakeem, N., Selim, M.S.: Optimization and spectroscopic studies of CdS/poly (vinyl alcohol) nanocomposites. Mater. Chem. Phys. **115**(1), 132–135 (2009)
39. Aziz, S.B., et al.: Synthesis and characterization of CdS nanoparticles grown in a polymer solution using in-situ chemical reduction technique. Int. J. Electrochem. Sci. **12**, 3263–3274 (2017)
40. Cuba, M., Muralidharan, G.: Improved luminescence intensity and stability of thermal annealed ZnO incorporated Alq 3 composite films. J. Fluoresc. **25**(6), 1629–1635 (2015)
41. Salam, S., et al.: Tris-(8-hydroxyquinoline) aluminium thin film as saturable absorber for passively Q-switched erbium-doped fibre laser. IET Optoelectron. **13**(5), 247–253 (2019)
42. Svelto, O., Hanna, D.C.: Principles of Lasers. Springer US (2013)
43. Huang, Y., et al.: Widely-tunable, passively Q-switched erbium-doped fiber laser with few-layer MoS 2 saturable absorber. Opt Exp. **22**(21), 25258–25266 (2014)
44. Cao, W.J., et al.: Graphene-based, 50 nm wide-band tunable passively Q-switched fiber laser. Laser Phys. Lett. **9**(1), 54–58 (2011)
45. Ahmad, H., et al.: Tunable S-band Q-switched fiber laser using Bi 2 Se 3 as the saturable absorber. IEEE Photon. J. **7**(3), 1–8 (2015)
46. Siddiq, N., et al.: C-band tunable performance of passively Q-switched erbium-doped fiber laser using Tin (IV) oxide as a saturable absorber. Opt. Commun. **442**, 1–7 (2019)
47. Ahmad, H., et al.: Tunable Q-switched erbium-doped fiber laser in the C-band region using nanoparticles (TiO2). Opt. Commun. **435**, 283–288 (2019)
48. Huang, Y., et al.: Widely-tunable, passively Q-switched erbium-doped fiber laser with few-layer MoS2 saturable absorber. Opt. Express. **22**(21), 25258–25266 (2014)
49. Ahmad, H., et al.: Broadband tuning in a passively Q-switched erbium doped fiber laser (EDFL) via multiwall carbon nanotubes/polyvinyl alcohol (MWCNT/PVA) saturable absorber. Opt. Commun. **365**, 54–60 (2016)
50. Luo, Z., et al.: Topological-insulator passively Q-switched double-clad fiber laser at 2 μm wavelength. IEEE J. Select. Topic. Quant. Electron. **20**(5), 1–8 (2014)
51. Spühler, G., et al.: Experimentally confirmed design guidelines for passively Q-switched microchip lasers using semiconductor saturable absorbers. JOSA B. **16**(3), 376–388 (1999)
52. Salam, S., et al.: Soliton mode-locked Er-doped fiber laser by using Alq_3 saturable absorber. Opt. Laser Technol. **123**, 105893 (2020)
53. Salam, S., et al.: High-energy Q-switched ytterbium-doped all-fiber laser with tris-(8-hydroxyquinoline) aluminum as saturable absorber. Opt Mater Express. **9**(8), 3215–3225 (2019)
54. Yang, F., et al.: Saturable absorber based on a single mode fiber—graded index fiber—single mode fiber structure with inner micro-cavity. Opt. Express. **26**(2), 927–934 (2018)
55. Zhao, W., et al.: All-fiber saturable absorbers for ultrafast fiber lasers. IEEE Photon. J. **11**(5), 1–19 (2019)
56. Kim, B., et al.: All-fiber integrated saturable absorber-tunable wavelength filter for Q-switching laser in both C- and L-bands. Opt. Express. **29**(9), 13183–13192 (2021)
57. Qi, Y., et al.: Tunable all fiber multi-wavelength mode-locked laser with a large dynamic range using polarization controller coiled SMF-GIMF-SMF structure as both saturable absorber and comb filter. Opt. Fiber Technol. **74**, 103055 (2022)
58. Zhang, K., Rochette, M.: All-fiber saturable absorber using nonlinear multimode interference in a chalcogenide fiber. J. Lightwave Technol. **38**(22), 6321–6326 (2020)

Chapter 3
Dissipative Solitons in Passively Mode-Locked Lasers

Mário F. S. Ferreira and Sofia C. V. Latas

Abstract We introduce the concept of dissipative solitons, which are the result of a balance among dispersion, nonlinearity, gain, and loss. We focus our discussion on dissipative solitons in passively mode-locked fiber systems, which can be described by the cubic-quintic complex Ginzburg-Landau equation (CGLE). The conditions to have stable solutions of the CGLE are discussed using the perturbation theory. We use also the variational approach to find approximate pulse solutions for the CGLE. The existence of both stationary and pulsating soliton solutions is indicated by such approach and confirmed through some numerical examples. In particular, we describe the dynamics of plain pulsating, creeping, and erupting soliton solution. Moreover, we discuss the impact that some higher-order effects, namely the third-order dispersion, intrapulse Raman scattering, and self-steepening, can have on these pulses. We show that, for some range of the parameter values, these higher-order effects can transform the pulsating solitons into fixed-shape solitons.

Keywords Dissipative solitons · Pulsating solitons · Complex Ginzburg-Landau equation · Variational approach · Higher-order effects · Soliton molecules

3.1 Introduction

The term "soliton" was coined in 1965, to reflect the remarkable particle-like nature of solitary waves that remain intact even after mutual collisions [1]. These waves correspond to localized solutions of integrable equations such as the Korteveg de Vries and nonlinear Schrödinger equations (NLSE). Given these circumstances, solitons were initially attributed only to integrable systems. Such solutions have been discovered and investigated in various branches of physics, such as

M. F. S. Ferreira (✉) · S. C. V. Latas
i3N-Institute of Nanostructures, Nanomodelling and Nanofabrication, Department of Physics, University of Aveiro, Aveiro, Portugal
e-mail: mfernando@ua.pt; sofia.latas@ua.pt

M. F. S. Ferreira, M. C. Paul (eds.), *Optical Fiber Lasers*, Springer Series in Optical Sciences 250, https://doi.org/10.1007/978-3-032-04110-4_3

Bose-Einstein condensates [2], hydrodynamics [3], astronomy [4], and optics [5]. The existence of solitons was first predicted and experimentally observed in optical fibers, which could be interpreted as a result of an interplay between dispersive and nonlinear Kerr effects [6, 7].

A dramatic turning point occurred when it was found that solitary wave solutions did exist also in a wide range of nonintegrable dissipative systems. Self-organized solutions in nonlinear systems far from equilibrium were termed dissipative solitons or auto-solitons [8–10]. In contrast to conventional conservative solitons, which result from a balance between nonlinearity and dispersion, dissipative solitons require a continuous exchange of energy with an external source, in order to satisfy an additional balance between gain and loss. The concept of temporal and spatial dissipative solitons can be applied for solid-state lasers and fiber lasers as well as for optical pulses propagating in fiber-optic communication systems in which real losses are compensated with fiber-optic amplifiers [5, 8, 11–22]. Actually, Dissipative optical solitons have many unique properties which differ from those of their conservative counterparts. For example, except for very few cases [17], they form zero-parameter families and their properties are completely determined by the external parameters of the optical system. They can exist indefinitely in time, as long as these parameters stay constant. These features of dissipative solitons are highly desirable for several applications, such as in-line regeneration of optical data streams and generation of stable trains of laser pulses by mode-locked fiber lasers.

The basic mathematical model for the description of all nonlinear phenomena occurring in mode-locked fiber lasers and long-haul soliton transmission lines is the cubic-quintic complex Ginzburg-Landau equation (CGLE) [8–10, 17, 18]. However, the CGLE has been also used to describe a number of phenomena including second-order phase transitions, superconductivity, superfluidity, Bose-Einstein condensation, liquid crystals, and the string theory [23]. Actually, the areas of application and the number of solutions to this equation are so large that the sphere of this knowledge by itself has been called "the world of the Ginzburg–Landau equation" [24].

In spite of various analytical and semianalytical techniques that have been applied to investigate its soliton solutions, massive numerical simulations appear as the only way to fully explore the CGLE. Different types of soliton solutions were obtained numerically, which can be divided in two main classes: localized fixed-shape solutions and localized pulsating solutions [25–28]. Examples of localized fixed-shape solutions are the plain stationary pulses, the flat-top pulses, the composite pulses (CPs), and the moving pulses [17]. Among the localized pulsating solutions, we may refer the plain pulsating and the creeping solions, as well as the erupting solitons, which belong to the class of chaotic solutions [29, 30].

The stability and other properties of a dissipative soliton are linked to the stability and characteristics of its attractor, in the terminology of nonlinear dynamics. For example, a stationary soliton is associated to a fixed point. By tuning the system parameters, we can transform such fixed point into a limit cycle, which is an oscillatory attracting state. Thus, a previously stationary soliton becomes a pulsating soliton. Further changing the system parameters may result in irregular and chaotic dynamics.

In this Chapter, we will discuss different types of dissipative soliton solutions in mode-locked fiber lasers. We begin in Sect. 3.2 by presenting the cubic-quintic CGLE, after which we describe, in Sect. 3.3, the variational approach to find approximate solutions to this equation. In Sect. 3.4, we discuss some types of stationary soliton solutions; whereas in Sect. 3.5, several types of pulsating soliton solutions are described. Section 3.6 discusses the formation of soliton molecules in mode-locked fiber lasers. Finally, we summarize the main conclusions in Section 3.7.

3.2 The Cubic-Quintic Complex Ginzburg-Landau Equation

The cubic-quintic complex Ginzburg Landau equation can be written in the form [8–10, 17]:

$$i\frac{\partial q}{\partial Z}+\frac{1}{2}\frac{\partial^2 q}{\partial T^2}+|q|^2 q=i\delta q+i\beta\frac{\partial^2 q}{\partial T^2}+i\varepsilon|q|^2 q+i\mu|q|^4 q-\nu|q|^4 q, \qquad (3.1)$$

where Z is the normalized propagation distance, T is the retarded time, q is the normalized envelope of the electric field respect β stands for spectral filtering ($\beta > 0$), δ is the linear gain or loss coefficient, ε accounts for nonlinear gain-absorption processes (e.g., two-photon absorption), μ represents a higher-order correction to the nonlinear gain-absorption, and ν is a higher-order correction term to the nonlinear refractive index.

Equation (3.1) has been used to model the propagation of optical pulses in soliton transmission systems [8–10, 17]. It has been also used as a master equation to describe the behavior of solid-state lasers [12] and mode-locked fiber lasers [8, 9, 31–33]. The relations between the physical parameters describing a ring fiber laser mode locking through nonlinear polarization rotation (NPR) and the coefficients of CGLE have been derived in [34].

Equation (3.1) becomes the standard nonlinear Schrödinger equation (NLSE) when the right-hand side is set to zero. When this does not happen, Eq. (3.1) is non-integrable, and only particular solutions can be obtained. The soliton perturbation theory can be used for small values of the parameters [25]. In the case of the cubic CGLE$\mu = \nu = 0$ (), exact solutions can be obtained using a special ansatz [35], Hirota bilinear method [36], or reduction to systems of linear PDEs [37]. Concerning the quintic CGLE, the existence of soliton-like solutions in the case $\varepsilon > 0$ as well as sources, sinks, and fronts with fixed velocity have been demonstrated both analytically and numerically [17, 38, 39]. Analytical solutions can be presented explicitly only for certain relations between the parameters of the equation. Furthermore, so far, only stationary solutions of the CGLE are known in analytical form. Actually, several types of localized pulsating solutions of the CGLE have been also found numerically [29, 30, 40–43]. Approximate expressions for some of these localized solutions can be derived for arbitrary values of the CGLE parameters by reducing this equation to finite-dimensional dynamical models. The reduced models can be obtained by applying the method of moments [44], or Lagrangian techniques [27, 45–47].

In spite of having no known conserved quantities, there are two rate equations which can be written for the CGLE: one for the energy,

$$Q = \int_{-\infty}^{\infty} |q|^2 \, dT \tag{3.2}$$

and the other for the momentum,

$$M = Im\left(\int_{-\infty}^{\infty} q \frac{dq^*}{dT} dT \right). \tag{3.3}$$

For a dissipative system, Q is a function of Z rather than a conserved quantity. The rate of change of the energy is [8].

$$\frac{d}{dZ} Q = F[q] \tag{3.4}$$

where the real functional $F[q]$ is given by

$$F[q] = 2\int_{-\infty}^{\infty} \left[\delta |q|^2 + \varepsilon |q|^4 + \mu |q|^6 - \beta \left| \frac{\partial q}{\partial T} \right|^2 \right] dT \tag{3.5}$$

For localized solutions, Q is finite and it takes a constant value when the solution is stationary. The rate of change of the momentum is given by [8]:

$$\frac{d}{dZ} M = J[q], \tag{3.6}$$

where the real functional $J[q]$ is given by

$$J[q] = 2Im\int_{-\infty}^{\infty} \left[\left(\delta + \varepsilon |q|^2 + \mu |q|^4 \right) q + \beta \frac{\partial^2 q}{\partial T^2} \right] \frac{\partial q^*}{\partial T} dT \tag{3.7}$$

The rate equations (3.4) and (3.6) can be used for studying different problems related to CGLE solitons. For example, assuming that the coefficients on the right-hand side of Eq. (3.1) are all small, they can be used to derive the evolution equations for the amplitude and the frequency of the perturbed soliton solution of the NLSE:

$$q(T,Z) = \eta sech\left[\eta (T + \kappa Z) \right] \exp\left[-i\kappa T + i\left(\eta^2 - \kappa^2 \right) Z / 2 \right] \tag{3.8}$$

The equation for the evolution of $\eta(Z)$ can be derived from the equation for the energy (3.4), whereas the equation for $\kappa(Z)$ can be obtained from the equation for the momentum (3.6). It turns out that both results are given by [25]:

$$\frac{\partial \eta}{\partial Z} = 2\delta\eta - 2\beta\eta\left(\frac{1}{3}\eta^2 + \kappa^2\right) + \frac{4}{3}\varepsilon\eta^3 + \frac{16}{15}\mu\eta^5 \tag{3.9}$$

$$\frac{d\kappa}{dZ} = -\frac{4}{3}\beta\eta^2\kappa \tag{3.10}$$

As can be seen from Eq. (3.10), the soliton frequency approaches asymptotically to $\kappa = 0$ (stable fixed point) if $\eta \neq 0$. It has been found that, in order to have a stable pulse propagating in a stable background, the following conditions must be satisfied [25]:

$$\delta \left\langle 0, \mu \left\langle 0, \varepsilon \right\rangle \beta / 2, 15\delta \right\rangle 8\mu\eta_s^4 \tag{3.11}$$

where

$$\eta_s^2 = \frac{-5(\varepsilon - \varepsilon_S) - 5\sqrt{(\varepsilon - \varepsilon_S)^2 - 24\delta\mu / 5}}{8\mu} \tag{3.12}$$

is the stationary value for the soliton amplitude and $\varepsilon_S = \beta/2$ for small values of β.

To maintain a stationary pulse profile, dissipative solitons require an internal energy redistribution among the various pulse parts, as due to nonlinear dissipation, some parts will be amplified, whereas others will be attenuated. Remembering that energy flows through phase gradients, such energy redistribution therefore maps with a nonuniform phase profile, which will generally combine linear and nonlinear frequency chirping.

The continuity relation for the CGLE equation, which is given by [8]:

$$\frac{\partial \rho}{\partial Z} + \frac{\partial j}{\partial T} = P \tag{3.13}$$

where $\rho = |q|^2$ is the energy density,

$$j = \frac{i}{2}\left(q q_T^* - q_T q^*\right) \tag{3.14}$$

is the energy flux, and

$$P = 2\delta|q|^2 + 2\varepsilon|q|^4 + 2\mu|q|^6 - 2\beta\left|\frac{\partial q}{\partial T}\right|^2 + \beta\frac{\partial^2\left(|q|^2\right)}{\partial T^2} \tag{3.15}$$

is the density of energy generation.

For a stationary dissipative soliton, the first term in Eq. (3.13) is zero, as the energy density remains constant along the propagation. Such stationary solution results from the continuous exchange of energy with the environment along with the concomitant redistribution of energy between different parts of the pulse. Let us consider that the stationary solution corresponds to a higher-order Gaussian-type pulse, given by:

$$q(Z,T) = A(m)\sqrt{\frac{Q}{w}}\exp\left(-\frac{T^2}{w^2}-\frac{T^4}{mw^4}+iCT^2-i\kappa Z\right) \tag{3.16}$$

where Q, w, and C represent the pulse energy, width, and chirp factor, respectively, m is an adjustment parameter, and A is a constant, such that the total pulse energy is equal to Q. Using Eq. (3.16), the flux and the density of energy generation become:

$$j(T) = P_0 T\exp\left(-2\frac{T^2}{w^2}-2\frac{T^4}{mw^4}\right) \tag{3.17}$$

$$P(T) = P_0\left(1-4\frac{T^2}{w^2}-\frac{8}{m}\frac{T^4}{w^4}\right)\exp\left(-2\frac{T^2}{w^2}-2\frac{T^4}{mw^4}\right) \tag{3.18}$$

where

$$P_0 = \frac{2A^2(m)CQ}{w} \tag{3.19}$$

Figure 3.1 illustrates the normalized density of energy generation, $\hat{P} = P/P_0$, and the normalized energy flux, $\hat{j} = j/(wP_0)$, for $m = 1$ and $C > 0$. In this case, $\hat{P} < 0$ in the wings and $\hat{P} > 0$ around the pulse center. This means that energy is generated in the middle and flows toward the wings, where it is dissipated. However, if $C < 0$, energy is generated in the wings and flows toward the middle where it is lost. This flux of energy leads to a dynamic equilibrium which allows the existence of the dissipative soliton.

3.3 The Variational Approach to CGLE Solitons

The variational technique is particularly useful for describing the propagation of chirped pulses in nonlinear media. Its use to describe pulse propagation in nonlinear dispersive media, such as optical fibers, was first realized by Anderson [48]. This approach can be quite accurate even when the effect of nonlinearity is large [49]. The accuracy depends essentially on the ansatz chosen to describe the evolution of the field experiencing nonlinearity.

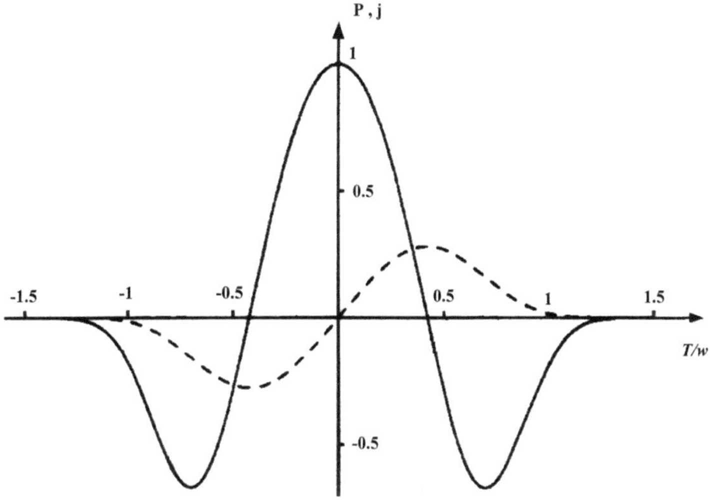

Fig. 3.1 Normalized density of energy generation $\hat{P} = P / P_0$ (full line) and normalized energy flux $\hat{j} = j / (wP_0)$ (dashed line) for the case of a higher-order Gaussian pulse ($m = 1$)

The standard variational approach can be modified to allow for dissipative terms in the CGLE. Using such approach, Eq. (3.1) can be obtained from the Euler-Lagrange equation [27]:

$$\frac{\partial}{\partial \tau}\left(\frac{\partial L_d}{\partial q_T^*}\right)+\frac{\partial}{\partial z}\left(\frac{\partial L_d}{\partial q_Z^*}\right)-\frac{\partial L_d}{\partial q^*}=S \tag{3.20}$$

where L_d is the Lagrangian density, given by

$$L_d=\frac{i}{2}\left(q\frac{\partial q^*}{\partial Z}-q^*\frac{\partial q}{\partial Z}\right)+\frac{1}{2}\left|\frac{\partial q}{\partial T}\right|^2-\frac{1}{2}|q|^4-\frac{\nu}{3}|q|^6 \tag{3.21}$$

and S represents the dissipative terms:

$$S=i\delta q+i\beta\frac{\partial^2 q}{\partial T^2}+i\varepsilon|q|^2 q+i\mu|q|^4 q \tag{3.22}$$

Considering the Lagrangian

$$L=\int_{-\infty}^{\infty} L_d dT \tag{3.23}$$

and a trial solution containing parameters ν_j, $j = 1,2,\ldots$, the modified Euler-Lagrange equations are given by [27]:

$$\frac{d}{dz}\left(\frac{\partial L}{\partial v_z}\right)-\frac{\partial L}{\partial v}=2\,\mathrm{Re}\left(\int_{-\infty}^{\infty}S\frac{\partial q^*}{\partial v}\right)dT \tag{3.24}$$

for each parameter ν.

Let us consider a soliton solution given by:

$$q(Z,T)=B(Z,T)\exp(-i\theta(Z)) \tag{3.25}$$

where $B(Z, T)$ is the complex amplitude and $\theta(Z)$ gives the phase evolution on propagation. Then, using Eq. (3.24), we have for a general parameter ν:

$$\frac{\partial L}{\partial v}-\frac{d}{dz}\left(\frac{\partial L}{\partial v_z}\right)=2\,\mathrm{Re}\int_{-\infty}^{\infty}\frac{1}{iq^*}\frac{\partial q^*}{\partial v}\left(\delta|B|^2+\varepsilon|B|^4+\mu|B|^6+\beta B^*\frac{\partial^2 B}{\partial T^2}\right)dT \tag{3.26}$$

If $\nu = \theta(Z)$ in Eq. (3.26), we have.

$$\frac{\partial L}{\partial \theta}=0,\quad \frac{\partial L}{\partial \theta'}=-Q(Z),\quad \frac{1}{iq^*}\frac{\partial q^*}{\partial \theta}=1 \tag{3.27}$$

and

$$\frac{dQ}{dZ}=2\,\mathrm{Re}\int_{-\infty}^{\infty}\left[\delta|B|^2+\varepsilon|B|^4+\mu|B|^6+\beta B^*\frac{\partial^2 B}{\partial T^2}\right]dT \tag{3.28}$$

Equation (3.28) corresponds to the balance equation for the energy Q, given by Eqs. (3.4) and (3.5). Obviously, the right-hand side is zero in the case of a stationary soliton solution.

3.3.1 Sech-Type Ansatz

Let us assume that

$$B(Z,T)=A\,\mathrm{sech}\left(\frac{T}{w(Z)}\right)\exp\left(iC(Z)T^2\right) \tag{3.29}$$

where $A(Z)$, $w(Z)$, and $C(Z)$ are the amplitude, width, and chirp parameter, respectively. In this case, we have

$$Q=2A^2w \tag{3.30}$$

and Eq. (3.28) becomes

$$\frac{dQ}{dZ} = 2Q\left[\delta - \beta\left(\frac{1}{3w^2} + \frac{\pi^2}{3}C^2w^2\right) + \frac{\varepsilon}{3}\frac{Q}{w} + \frac{2\mu}{15}\frac{Q^2}{w^2}\right] \tag{3.31}$$

The Lagrangian corresponding to Eqs. (3.25) and (3.29) is given by:

$$L = Q\left[\frac{\pi^2C^2w^2}{6} + \frac{1}{6w^2} - \frac{Q}{6w} - \frac{2vQ^2}{45w^2} + \frac{\pi^2w^2C'(Z)}{12} - \theta'(Z)\right] \tag{3.32}$$

If $\nu = C(Z)$ in Eq. (3.26), we obtain the following equation for the pulse width

$$\frac{dw}{dZ} = 2cw + \beta\left(\frac{8}{\pi^2 w} - \frac{16\pi^2}{15}C^2w^3\right) - \frac{2\varepsilon}{\pi^2}Q - \frac{\mu}{\pi^2}\frac{Q^2}{w} \tag{3.33}$$

The equation for the chirp parameter $C(Z)$ is obtained using the parameter $\nu = w(Z)$ in Eq. (3.26). We have

$$\frac{dC}{dZ} = 2\left(\frac{1}{\pi^2w^4} - C^2\right) - \frac{1}{\pi^2}\frac{Q}{w^3} - 4\left(\frac{1}{3} + \frac{1}{\pi^2}\right)\beta\frac{C}{w^2} - \frac{8v}{15\pi^2}\frac{Q^2}{w^4} \tag{3.34}$$

Finally, using $\nu = Q(Z)$ in Eq. (3.26), we obtain the following equation for the phase evolution

$$\frac{d\theta}{dZ} = \frac{1}{3w^2} - \beta C\left(\frac{1}{3} + \frac{\pi^2}{9}\right) - \frac{5}{12}\frac{Q}{w} - \frac{8v}{45}\frac{Q^2}{w^2} \tag{3.35}$$

3.3.2 Gaussian-Type Ansatz

As a second example, let us consider a simple Gaussian-type ansatz:

$$B(Z,T) = A(Z)\exp\left(-\frac{T^2}{w^2(Z)} + iC(Z)T^2\right) \tag{3.36}$$

where $A(Z)$, $w(Z)$, and $C(Z)$ are the amplitude, width, and chirp parameter, respectively. In this case, the pulse energy Q is given by

$$Q(Z) = \sqrt{\frac{\pi}{2}}A^2(Z)w(Z) \tag{3.37}$$

whereas the Lagrangian can be written as:

$$L = Q\left[\frac{C^2 w^2}{2} + \frac{1}{2w^2} - \frac{Q}{2\sqrt{\pi}w} - \frac{2\nu Q^2}{3\sqrt{3}\pi w^2} + \frac{1}{4}w^2 C'(Z) - \theta'(Z)\right] \tag{3.38}$$

Considering $\nu = \theta$, C, w, Q, in Eq. (3.36), we obtain the following evolution equations for the pulse parameters:

$$\frac{dQ}{dZ} = 2Q\left(\delta - \frac{\beta}{w^2} - \beta C^2 w^2 + \frac{1}{\sqrt{\pi}}\frac{\varepsilon Q}{w} + \frac{2}{\pi\sqrt{3}}\frac{\mu Q^2}{w^2}\right) \tag{3.39}$$

$$\frac{dw}{dZ} = 2\frac{\beta}{w} + 2Cw - 2\beta C^2 w^3 - \frac{1}{2\sqrt{\pi}}\varepsilon Q - \frac{4}{3\pi\sqrt{3}}\frac{\mu Q^2}{w} \tag{3.40}$$

$$\frac{dC}{dZ} = \frac{2}{w^4} - 2C^2 - 8\frac{\beta C}{w^2} - \frac{1}{\sqrt{\pi}}\frac{Q}{w^3} - \frac{8}{3\pi\sqrt{3}}\frac{\nu Q^2}{w^4} \tag{3.41}$$

$$\frac{d\theta}{dZ} = \frac{1}{w^2} - 2\beta C - \frac{5}{4\sqrt{\pi}}\frac{Q}{w} - \frac{8}{3\pi\sqrt{3}}\frac{\nu Q^2}{w^2} \tag{3.42}$$

We observe that Eqs. (3.39)–(3.42) and Eqs. (3.31)–(3.35) are similar. The only difference lies in the numerical values of the coefficients of the terms.

A system of evolution equations for the soliton parameters similar to those presented above, considering either the sech or the Gaussian ansatz, can be obtained using the method of moments [44]. Actually, we can conclude that both the method of moments and the variational approach provide the possibility of reducing the cubic-quintic CGLE to the same finite-dimensional dynamical system.

An analysis of the dynamical system given by Eqs. (3.31)–(3.35) has demonstrated a clear correspondence between attractors and various types of localized waves governed by the cubic-quintic CGLE [44]. Both fixed points and limit cycles can be found for such dynamical system. A fixed point corresponds to a stable focus attractor, which guaranties that all initial pulse conditions that belong to the attractor's basin of attraction will converge toward a llfixed pulse profile. Thus, a fixed point corresponds to a stationary dissipative soliton. On the other hand, a limit cycle corresponds to a pulsating soliton. A transformation from a stationary soliton to a pulsating soliton appears as the result of a Hopf bifurcation in the reduced dynamical system [44]. Some examples of stationary soliton solutions are given in Sect. 3.4. Different types of pulsating soliton solutions will be discussed in Sect. 3.5.

3.4 Stationary Dissipative Solitons

In general, if the result of the numerical calculation converges to a stationary solution, it can be considered as a stable one, and the chosen set of parameters can be deemed to belong to the class of those which permit the existence of solitons. The

CGLE model shows that dissipative soliton attractors of focus type, corresponding to stationary dissipative solitons, exist for a wide range of system parameters. The amplitude pulse profile of the dissipative soliton can vary widely according to the system parameters. For small values of the parameters in the right-hand side of Eq. (3.1), the stationary soliton solutions of the CGLE have a sech profile, similar to the soliton solutions of the NLSE, and correspond to the so-called plain pulses (PPs). However, rather different pulse profiles can be obtained for nonsmall values of those parameters. For example, Fig. 3.2 illustrates the formation and stable propagation of a flat-top soliton, starting from an initial pulse with a sech profile and assuming the following parameter values: $\delta = -0.1$, $\beta = 0.5$, $\varepsilon = 0.66$, and $\mu = \nu = -0.01$.

Figure 3.3 shows the amplitude and the spectrum of a plain pulse, considering the parameter values: $\delta = -0.01$, $\beta = 0.5$, $\mu = -0.03$, $\nu = 0$, and $\varepsilon = 1.5$. The plain pulse resembles a conventional soliton, exhibiting a bell-shaped amplitude profile and a single peak spectrum.

Another example is given in Fig. 3.4, which shows (a) the amplitude profiles and (b) the spectra of a plain pulse, as well as of two composite pulses (CPs) [35]. The following parameter values were considered in this case: $\delta = -0.01$, $\beta = 0.5$, $\mu = -0.03$, $\nu = 0$, $\varepsilon = 1.5$ (plain pulse), $\varepsilon = 2.0$ (narrow composite pulse), and $\varepsilon = 2.15$ (wide composite pulse). Figure 3.4c illustrates the formation and propagation of the wide composite pulse starting from the plain pulse solution represented in (a) and (b). A composite pulse exhibits a dual-peak but symmetric spectrum (Fig. 3.4b) and can be considered as a bound state of a plain pulse and two fronts attached to it from both sides [17]. The "hill" between the two fronts should be counted as a source, because it follows from the phase profile that energy flows from the center to the CP wings.

It has been found numerically that the energy of a dissipative soliton solution of the CGLE increases indefinitely when the equation parameters converge to a given region of the parameter space [50]. Such set of parameters was called a dissipative soliton resonance (DSR) [51]. Found in the normal dispersion regime, it initially required a positive quintic reactive nonlinearity to appear [50, 51], but then DSR was also revealed in the chromatic dispersion-free regime, along with negative

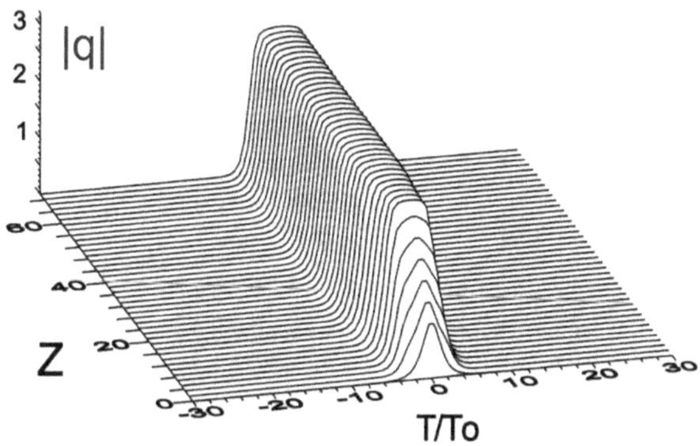

Fig. 3.2 Formation and stable propagation of a flat-top soliton, starting from an initial pulse with a sech profile for: $\delta = -0.1$, $\beta = 0.5$, $\varepsilon = 0.66$, and $\mu = \nu = -0.01$

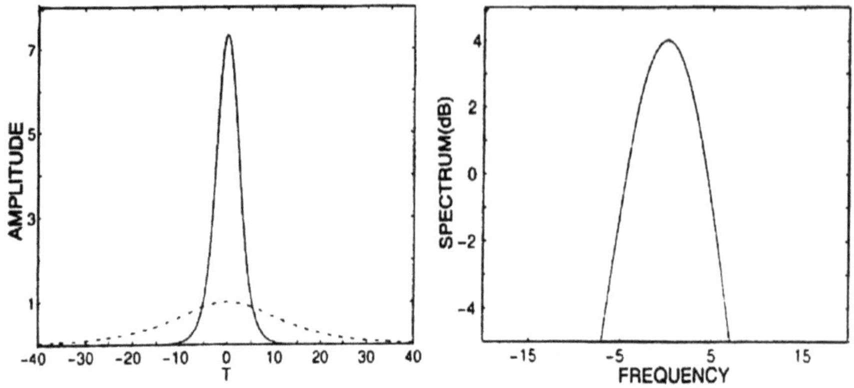

Fig. 3.3 Amplitude profile and spectrum of a plain pulse for $\delta = -0.01, \beta = 0.5, \mu = -0.03, \nu = 0,$ and $\varepsilon = 1.5$

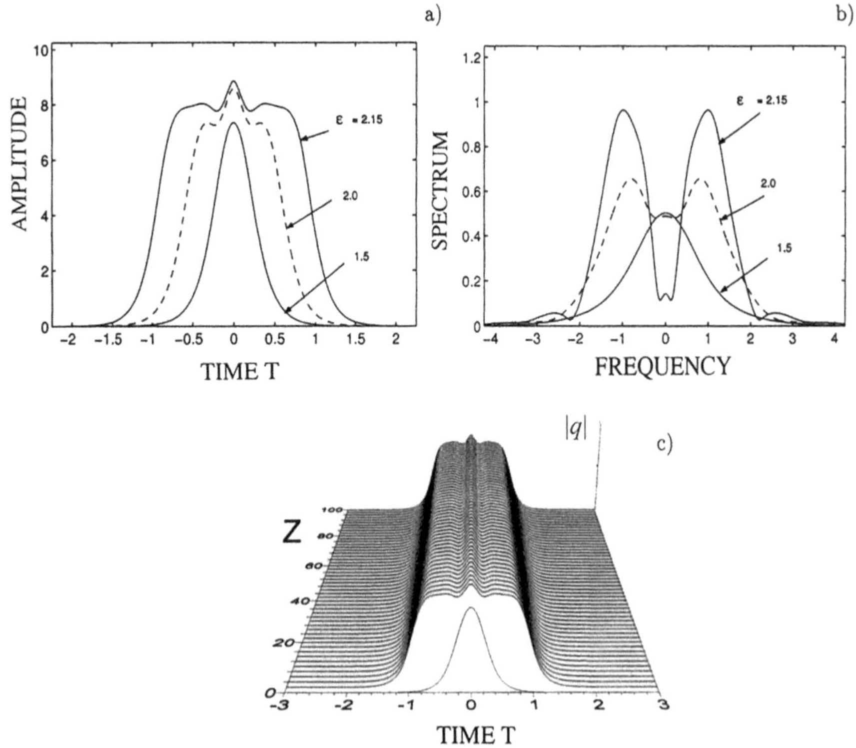

Fig. 3.4 Amplitude profiles (**a**) and spectra of a plain pulse, as well as of two composite pulses (**b**) for $\delta = -0.01, \beta = 0.5, \mu = -0.03, \nu = 0, \varepsilon = 1.5$ (plain pulse), $\varepsilon = 2.0$ (narrow composite pulse), and $\varepsilon = 2.15$ (wide composite pulse). Figure 3.4c illustrates the formation and propagation of the wide composite pulse starting from the plain pulse solution represented in (**a**) and (**b**)

quintic reactive nonlinearity [52]. The concept of dissipative soliton resonance has soon been proposed to achieve high-energy wave-breaking-free pulses [50–55].

The energy of a DSR pulse increases mainly due to the increase of the pulse width, while keeping the amplitude at a constant level [50–53]. Nevertheless, as the DSR pulse is highly chirped, this can be used to compress the pulse [56] and hence obtain high-energy ultrashort pulses. Actually, the DSR phenomenon has been frequently observed and reported in the area of fiber lasers [57–76].

A different kind of stationary high-energy ultrashort pulses, corresponding to the very high amplitude (VHA) soliton solutions of the CGLE, have been reported recently [77, 78]. Such VHA solutions occur due to a singularity first predicted in Refs. [79, 80], namely, as the nonlinear gain saturation effect tends to vanish. The increase in energy of these pulses is mainly due to the increase of the pulse amplitude, whereas the pulse width becomes narrower. VHA pulses are discussed in Chap. 4.

3.5 Pulsating Dissipative Solitons

Besides the stationary solutions, the cubic-quintic complex-Ginzburg-Landau equation (CGLE) presents also different types of localized solutions with periodically or quasi-periodically varying amplitude, width, and energy. Among such pulsating solitons, we may refer the plain pulsating, creeping, erupting, and chaotic solutions [28–30, 40–43]. Actually, in order to observe these pulsating solitons, the parameters of the CGLE must be far enough from the nonlinear Schrödinger equation. The regions of existence of pulsating solitons have been presented by Akhmediev et al. [30].

3.5.1 Plain Pulsating Solitons

A plain pulsating soliton is a localized solution of the CGLE that changes its width, amplitude, and other parameters periodically during propagation. It corresponds to a limit cycle of the infinite-dimensional dynamical system and can be described by a closed loop in the phase space of the system.

Plain pulsating soliton solutions of the cubic-quintic CGLE were first found by Deissler and Brand [81] in the normal dispersion regime. The profile of such solitons shows a periodic variation only in its tails, whereas the energy remains almost constant. In contrast, the plain pulsating solitons found by Akhmediev et al. [29, 30] were observed in the anomalous dispersion regime and both their shape and energy change quite significantly.

Figure 3.5(a) shows the amplitude evolution of a plain pulsating soliton assuming the following parameter values: $\delta = -0.1$, $\beta = 0.08$, $\varepsilon = 0.66$, $\mu = -0.1$, and $\nu = -0.1$. The pulse shows a periodic behavior, with a period $Z \approx 14$. It has different

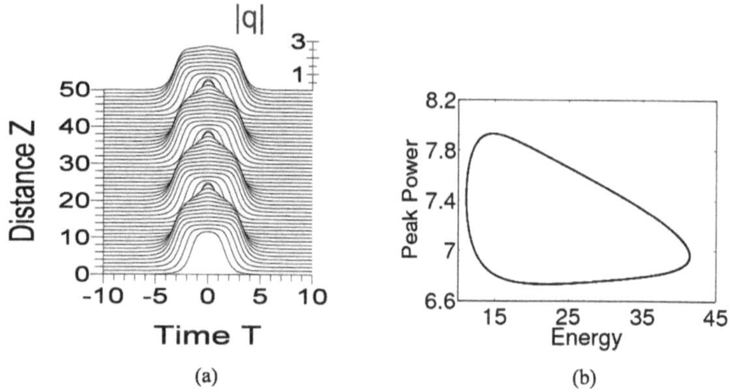

Fig. 3.5 (**a**) Amplitude evolution and (**b**) peak power versus energy, for a plain pulsating soliton, assuming the following parameters values: $\delta = -0.1$, $\beta = 0.08$, $\varepsilon = 0.66$, $\mu = -0.1$, and $\nu = -0.1$

shapes at each position Z, but it recovers exactly its initial shape after a period. Figure 3.5(b) shows the soliton peak power, $|q|^2$, against the soliton energy, given by Eq. (3.2). The trajectory corresponds to a cycle repeating itself indefinitely, which means that the pulse energy and peak power also have a periodic behavior. The pulse energy changes from 10 to 40, while the pulse peak power changes between 6.8 and 8, approximately.

The first experimental observation of soliton pulsations was reported by Soto-Crespo et al. in a mode-locked fiber ring laser [9]. Nevertheless, the lack of a high-resolution real-time diagnostic method has precluded detailed characterization of pulsating behaviors. This limitation has been broken through by the development of a novel powerful real-time spectra measurement technique called dispersive Fourier transform (DFT) [10]. Using this technique, the spectrum of the soliton could be mapped into a temporal waveform by using dispersive element with enough group-velocity dispersion. Thus, the ultrafast spectral signals can be captured by a real-time oscilloscope with a high-speed photodetector. The DFT technique made it possible to observe the real-time spectral dynamics of pulsating solitons [82–87].

Soliton self-organization and pulsation in a passively mode-locked fiber laser was observed experimentally in 2018 [82]. The soliton pulsation process showed a period corresponding to tens of the cavity round trip time. In another 2018 experiment, pulsating dissipative solitons in a mode-locked fiber laser at normal dispersion were observed for the first time using the DFT technique [83]. In that experiment, the artificial saturable absorbers, as well as the birefringent filter formed by the nonlinear polarization rotation, made the polarization controller an effective component to adjust the laser state from stationary to pulsating. The pulsating dissipative solitons were accompanied with the spectrum breathing and oscillating structures due to the nonlinear pulse propagation.

In a 2019 experiment, three types of soliton pulsations were observed in an L-band normal-dispersion mode-locked fiber laser via the DFT technique [85]. They were classified as single-periodic pulsating soliton, double-periodic pulsating soliton, and soliton explosion. These pulsations exhibited common features such as energy oscillation, bandwidth breathing, and temporal shift. However, the pulse was repeated every two oscillations for double-periodic pulsating soliton.

Most phenomena related to soliton pulsations have been observed in mode-locked fiber lasers based on the nonlinear polarization rotation (NPR). As a polarizer is required for the implementation of the NPR technique, the polarization degree of freedom is frozen, and pulsating solitons formed in these lasers are scalar ones. In a 2020 experiment, the generation of pulsating group-velocity-locked vector solitons (GVLVSs) in a net-normal dispersion fiber laser mode locked by nonlinear multimode interference has been reported [88]. In another 2020 experiment, the vector nature of various pulsating solitons in an ultrafast fiber laser with single-wall carbon nanotubes has been investigated [89]. By virtue of the DFT technique, the polarization-resolved spectral evolution of pulsating vector solitons was measured in real time. Double-periodic pulsation in the cavity was also observed. Also in 2020, the vector features of pulsating solitons were studied in an erbium-doped fiber (EDF) laser [90].

3.5.2 *Creeping Solitons*

A creeping soliton is a pulsating localized solution that changes its shape periodically and shifts a finite distance in the transverse direction after each pulsation. It has a rectangular pulse with two fronts and one sink on the top [30], presenting a shape that resembles that of the composite soliton [17, 91, 92]. Creeping solitons were first observed numerically in [29, 30].

Figure 3.6 shows (a) the amplitude evolution and (b) the soliton peak power versus the energy of a creeping pulsating pulse, assuming the following set of parameters: $\delta = -0.1$, $\beta = 0.101$, $\varepsilon = 1.3$, $\mu = -0.3$, and $\nu = -0.101$. We observe that the

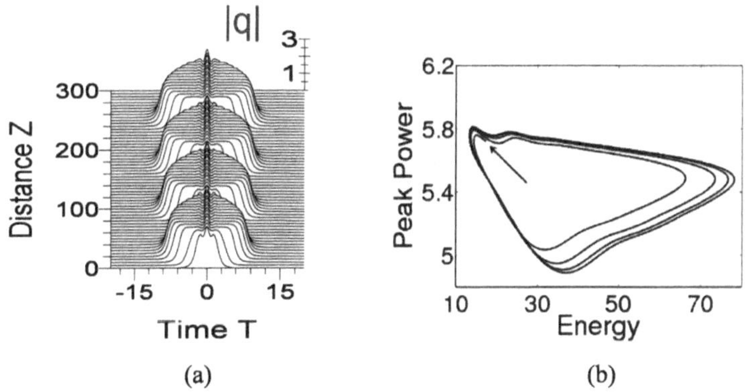

Fig. 3.6 (**a**) Amplitude and (**b**) soliton peak power versus energy for a creeping soliton, assuming the following set of parameters: $\delta = -0.1$, $\beta = 0.101$, $\varepsilon = 1.3$, $\mu = -0.3$, and $\nu = -0.1$

two fronts pulsate back and forth relatively to the sink asymmetrically at the two sides of the pulse. Due to this asymmetry, the position of the center of mass of the whole soliton shifts a finite distance in the transverse direction after each period of oscillation, which generally determines a nonzero constant velocity. The trajectories represented in Fig. 3.6(b) seem to evolve to a limiting cycle. In this process, the pulse energy maximum grows slightly, while the minimum pulse energy remains almost constant. On the other hand, the minimum value of the pulse peak power tends to decrease, whereas its maximum value remains nearly constant.

In a 2020 experiment, the motion dynamics of the creeping soliton have been demonstrated in a passively mode-locked fiber laser in the negative-dispersion regime by utilizing a Raman-assisted temporal magnifier system together with the DFT technique [93]. The periodical variation of pulse width, peak power, and motion range could be observed in real time, while the corresponding spectral evolution exhibited breathing dynamics.

3.5.3 *Chaotic Solitons*

Among the pulsating CGLE solutions, chaotic but localized pulses can also be found [8, 30]. Their profiles evolve during propagation, but never repeat. These pulses can arise as a result of the general dynamics of a passively mode-locked fiber laser, namely if there are several periods involved in these dynamics.

An example of a chaotic soliton is illustrated in Fig. 3.7(a), from which we observe that the pulse shape remains localized and is symmetric. However, in other cases, an asymmetric soliton profile has been observed. In spite of being chaotic, the pulse shape is constrained within certain limits, which means that its parameters remain confined to a given region in a phase space. This is illustrated by Fig. 3.7(b), which shows the soliton peak power against the soliton energy. The trajectory was obtained from a simulation over 200 distances, for an initial condition given by a

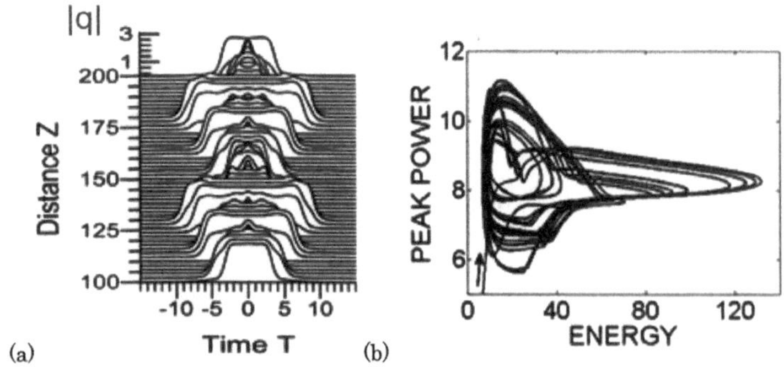

Fig. 3.7 (**a**) Amplitude and (**b**) peak power against pulse energy of a chaotic soliton, assuming the following parameter values: $\delta = -0.1$, $\beta = 0.04$, $\varepsilon = 0.75$, $\mu = -0.1$, and $\nu = -0.08$

unitary sech profile. It can be seen that such trajectory is not a cycle that repeats itself, as should be expect for a periodic solution. Instead, it occupies a certain region in the plot, which tends to be filled if the propagation distance is increased [94].

The first experimental evidence of the pulsating soliton with chaotic behavior in an ultrafast fiber laser was reported in a 2018 experiment [95]. Using the DFT technique, the chaotic behavior of soliton pulsation was visualized by the fact that the mode-locked spectrum collapsed abruptly in an unpredictable way during the pulsating process.

3.5.4 Erupting Solitons

One of the most striking forms of pulsating solitons is the erupting soliton which has been found numerically for the first time in [29], where the soliton eruption manifests itself as a chaotic and quasi-periodic process when the dissipative system is in a metastable state. The soliton in the metastable dissipative system abruptly erupts into pieces in temporal domain and gradually recover its original state after the eruption, which is similar to exploding behavior and thus regarded as the so-called soliton explosion [29]. Several numerical investigations have been carried out, trying to better understand the intrinsic mechanisms involved in soliton explosions and subsequent recovering of the soliton [40–43, 96–99].

Figure 3.8 shows (a) the amplitude and (b) the peak power versus energy, of a CGLE soliton for the following set of parameters: $\delta = -0.1$, $\beta = 0.125$, $\varepsilon = 1.0$, $\mu = -0.1$, and $\nu = -0.6$. Erupting soliton starts from a stationary localized solution. After a while, it becomes covered with small ripples which seem to move downward along the two slopes of the soliton, such that after a short distance, the pulse becomes covered with this seemingly chaotic structure. When the ripples increase in size, the soliton cracks into pieces, after which it evolves in order to restore its shape. The erupting process repeats forever, but never exactly in equal periods, as can be seen

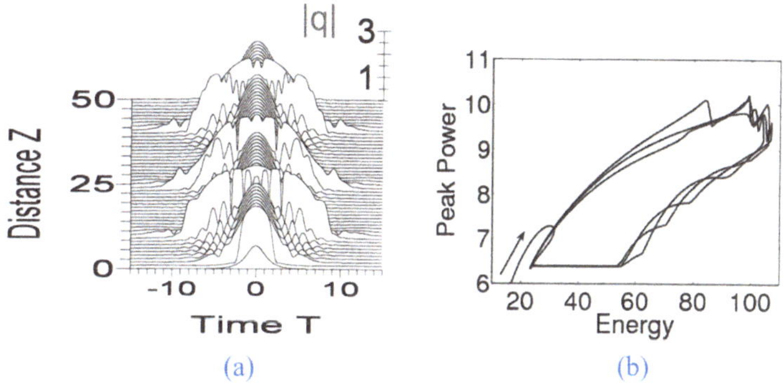

(a) (b)

Fig. 3.8 (**a**) Amplitude, and (**b**) peak power against pulse energy of an erupting soliton, for the following parameter values: $\delta = -0.1$, $\beta = 0.125$, $\varepsilon = 1.0$, $\mu = -0.1$, and $\nu = -0.6$

from Fig. 3.8(b). In fact, the pulse energy and the pulse peak power have a periodic behavior, but each trajectory is different from one explosion to another, having in common only the stationary regime of evolution. In this stationary regime, the pulse energy seems to grow as the ripples develop, while the pulse peak power remains almost constant.

The existence of the exploding solitons has been experimentally confirmed in a passively mode-locked Ti:sapphire, by virtue of a fast spectra measurement technique realized by spectrally dispersing the output of the laser across an array of six detectors and measuring the corresponding temporally resolved spectrum [100].

The use of the DFT technique has boosted the experimental investigations of soliton explosions in mode-locked fiber lasers during the recent years [101–111]. In a 2016 experiment [101], it was shown that the probability with which explosions occur depends on the pump power and that the explosion characteristics depend critically on the position in the cavity where the output coupler is located. The first investigation of soliton explosions in an ultrafast fiber laser in the multisoliton regime was reported in 2018 [104]. It was demonstrated that explosion of one soliton could be induced by another one through the soliton interactions mediated by the transient gain response of an erbium-doped fiber.

Distinct dynamical diversity of pulsating solitons in a fiber laser was revealed in a 2019 experiment [106]. Weak to strong explosive behaviors of pulsating solitons as well as the rogue wave (RW) generation during explosions were observed. Also, in 2019, up to seven nonlinear regimes were observed successively in a mode-locked fiber laser by increasing the pump power, including single-pulse mode locking, standard soliton explosions, noise-like mode locking, stable double pulsing, soliton collision induced explosions, soliton molecules, and double-pulse noise-like mode locking [108].

The first observation of the breathing dissipative soliton explosion in a carbon nanotube mode-locked bidirectional ultrafast fiber was reported in a 2020 experiment [112]. Rogue waves (RWs) were also generated owing to the breathing soliton explosion, and the amplitude of RW was significantly enhanced by the collision of the bidirectional breathing solitons. The observation of dissipative RWs generation induced by single and multisoliton explosions in a mode-locked fiber laser was also reported in 2022 [113]. The dynamics of soliton explosions in a transient chaotic state between a single and double pulsing state as well as periodic explosions induced by soliton collisions were observed in another 2022 experiment [114], in which a modified DFT technique relying on second-harmonic generation was used.

It was reported in 2023 that the exploding dynamics of dissipative solitons in a mode-locked fiber laser can be manipulated by dynamically engineering the cavity dispersion [115]. Moreover, pure-quartic solitons have been realized experimentally and numerically by introducing negative fourth-order dispersion (FOD) to balance with positive Kerr nonlinearity [116, 117]. Exploding solitons were found in pure-quartic soliton laser [118]. The influence of third-order dispersion (TOD) on the soliton dynamics was also experimentally investigated in 2024 [119]. In particular, it was found that, by increasing the TOD values, an initially stable dissipative soliton can split into two pulses, enter in the soliton explosion regime, and become finally a chaotic state [119].

3.6 Dissipative Soliton Molecules

Ultrafast lasers are prone to generate multiple identical dissipative solitons, when driven at high pump power. This tendency is exacerbated in fiber lasers operated in the anomalous dispersion regime, owing to the lower pulse energy limitation. These soliton pulses will always interact, as they have virtually an infinite time to do so within the bounded cavity propagation path.

Two laser pulses, interacting through their pulse tails while they propagate round the laser cavity, can reach a stable equilibrium consisting of a stationary pattern, corresponding to a multisoliton bound state. Stable multiple-light-pulse bound states have been called soliton molecules, due to their interesting analogies with matter molecules [120–125]. Actually, they are robust multisoliton structures that can be perfectly stationary and whose energy is proportional to the number of single-soliton constituents. Such structures differ significantly from higher-order soliton solutions of the nonlinear Schrödinger equation. In fact, these are nonlinear superpositions of fundamental solitons, which oscillate periodically in both temporal and spectral domains, and are easily destroyed by external perturbations or additional effects, such as third-order dispersion or Raman scattering.

The central case for study is that of two interacting pulses, which can form a stable dissipative soliton pair, characterized by a stationary relative separation and a self-locked phase between the two dissipative solitons. The potential to form a soliton pair is related to the existence of a point attractor in the dynamical system. An example of stable propagation of a soliton pair is given in Fig. 3.9. Figure 3.10 shows (a) the initial (dashed curve) and final (full curve) profiles, (b) phase, and (c) spectrum of the bound state. Phase-locked two-soliton molecules were experimentally observed in 2002 using a stretched-pulse fiber laser [126]. After that, they have been found numerically and experimentally considering different laser configurations and dispersion regimes [125, 127–130].

The two-soliton molecule can be assumed as the building block to construct various multisoliton molecules [130]. An example is given in Fig. 3.11 of the four-soliton solution with a phase difference of $\pi/2$ between adjacent plain pulses. As observed in the case of the two-soliton solution, multisoliton solutions formed by plain pulses also move with the same constant velocity along the T axis. Soliton molecules like that illustrated in Fig. 3.11 may increase the capacity of telecommunication and can be also attractive in all-optical information storage [131, 132].

Soliton fiber lasers constitute an ideal testbed for the study on soliton molecules. Actually, there have been during the recent years many reports on these structures in fiber lasers mode locked with the carbon nanotubes, graphene, black phosphorus, and so on [133–138]. As these saturable absorbers may be polarization independent, soliton molecules formed by vector solitons were also reported. The experimental

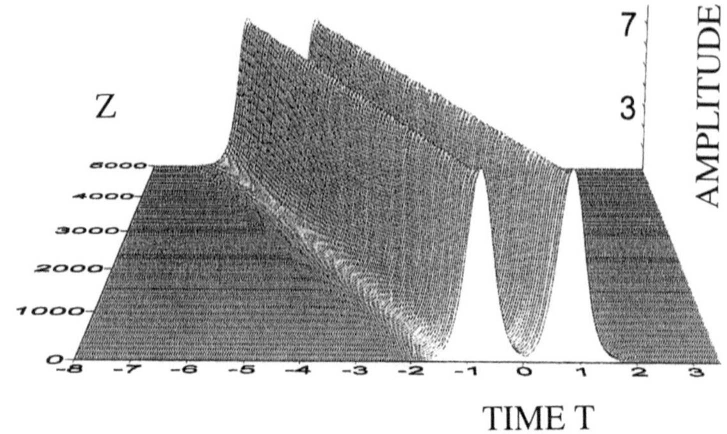

Fig. 3.9 Propagation of a soliton pair, considering the following parameter values: $\delta = -0.01$, $\beta = 0.5$, $\varepsilon = 1.5$, $\nu = 0$, and $\mu = -0.03$

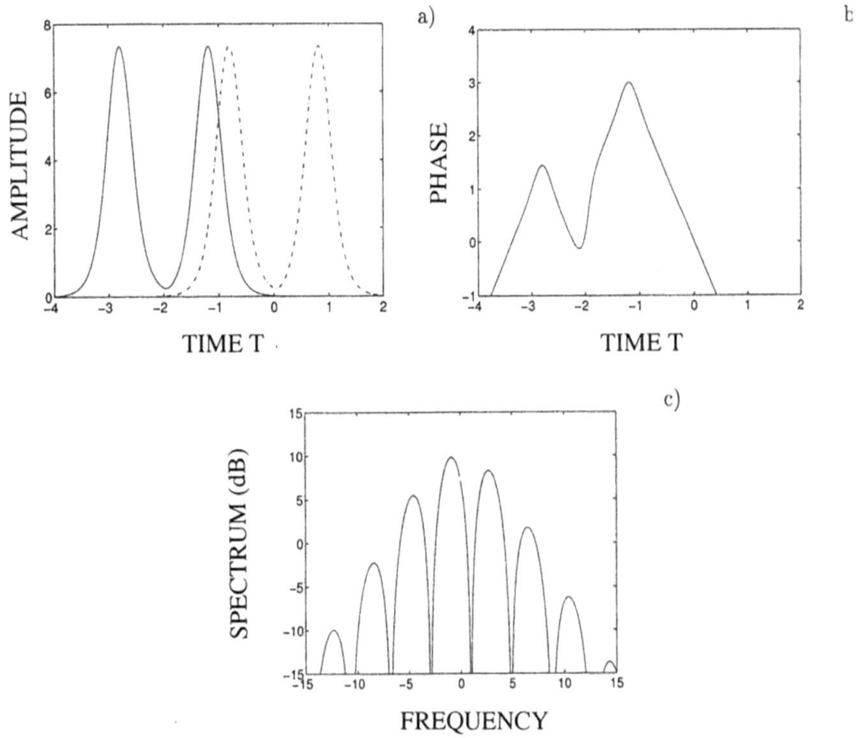

Fig. 3.10 (**a**) Initial (dashed curve) and final (full curve) profiles, (**b**) phase, and (**c**) spectrum of a bound state of two plain pulses with a phase difference of $\pi/2$ between them

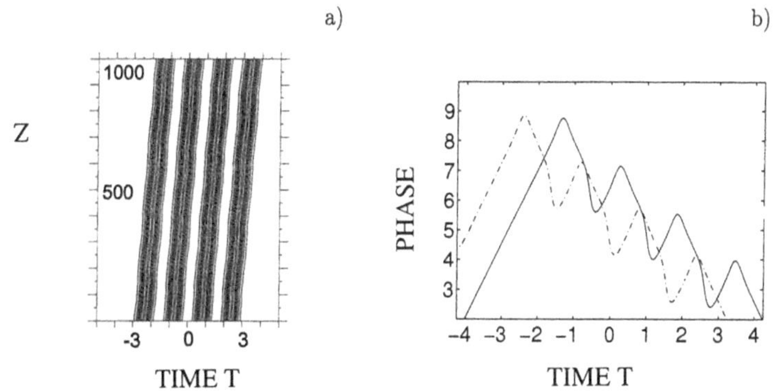

Fig. 3.11 Four-plain pulse solution with a phase difference of $\pi/2$ between adjacent pulses. The dash-dotted (full) lines in (**b**) correspond to the initial (final) phase profiles

observation of tightly and loosely bound states of vector solitons in a carbon nanotube mode-locked fiber laser in the anomalous dispersion regime was reported in 2013 [139]. In a latter experiment, localized interactions among vector solitons, vector soliton with bound vector solitons, and vector soliton with a bunch of vector solitons were observed in a fiber laser passively mode locked by graphene [140]. Owing to the bound soliton interactions, bound-bound solitons were also found experimentally [141]. The first experimental observation of group-velocity-locked vector soliton (GVLVS) molecules has been reported in 2017 in a semiconductor saturable absorber mirror (SESAM) mode-locked fiber laser using a birefringence-enhanced fiber [142].

3.7 Conclusions

We introduced the concept of dissipative solitons, focusing in passively mode-locked fiber lasers which can be described by the cubic-quintic complex Ginzburg-Landau equation. The perturbation theory was used to discuss the conditions to have stable soliton solutions and the variational approach to find some properties of the dissipative solitons. We found that the evolution equations for the soliton parameters derived using such approach are similar to those obtained by the method of moments. The existence of both stationary and pulsating soliton solutions is indicated by these approaches and confirmed numerically. We described the main properties of three types of pulsating pulses: plain pulsating, erupting, and creeping solitons. We demonstrated the dramatic impact that some higher-order effects, namely the intrapulse Raman, third-order dispersion and self-steepening effects, can have on the such pulsating solitons. In particular, we have shown the possibility of converting them in fixed-shape pulses by using a proper combination of the above effects. The formation of soliton molecules in passively mode-locked fiber lasers was also discussed.FundingFundação para a Ciência e a Tecnologia (FCT) (Project UID/CTM/50025/2022).

References

1. Zabusky, N., Kruskal, M.D.: Phys. Rev. Lett. **15**, 240 (1965)
2. Kartashov, Y.V., Konotop, V.V.: Phys. Rev. Lett. **118**, 190401 (2017)
3. Wu, J., Keolian, R., Rudnick, I.: Phys. Rev. Lett. **52**, 1421 (1984)
4. Antipov, S.V., Nezlin, M.V., Snezhkin, E.N., et al.: Nature. **323**, 238 (1986)
5. Grelu, P., Akhmediev, N.: Nat. Photonics. **6**, 84 (2012)
6. Hasegawa, A., Tappert, F.: Appl. Phys. Lett. **23**, 142 (1973)
7. Mollenauer, L.F., Stolen, R.H., Gordon, J.P.: Phys. Rev. Lett. **45**, 1095 (1980)
8. Akhmediev, N.N., Ankiewicz, A.A. (eds.): Dissipative Solitons. Springer-Verlag, Berlin Heidelberg (2005)
9. Ferreira, M.F. (ed.): Dissipative Optical Solitons. Springer, Cham (2022)
10. Ferreira, M.F.: Solitons in Optical Fiber Systems. John Wiley & Sons, Hoboken, NJ (2022)
11. Haus, H.A.: J. Appl. Phys. **46**, 3049 (1975)
12. Haus, H.A., Fujimoto, J.G., Ippen, E.P.: J. Opt. Soc. Am. B. **8**, 2068 (1991)
13. Moores, J.D.: Opt. Commun. **96**, 65 (1993)
14. Kärtner, F.X., Ausder Au, J., Keller, U.: IEEE J. Sel. Top. Quantum Electron. **4**, 159 (1998)
15. Matsumoto, M., Ikeda, H., Uda, T., Hasegawa, A.: J. Lightwave Technol. **13**, 658 (1995)
16. Kodama, Y., Romagnoli, M., Wabnitz, S.: Electron. Lett. **28**, 1981 (1992)
17. Akhmediev, N., Ankiewicz, A.: Solitons: Nonlinear Pulses and Beams. Chapman and Hall, London (1997)
18. Kalashnikov, V.L.: In: Al-Kursan, A. (ed.) Solid state lasers, pp. 145–184. IntechOpen (2012)
19. Rozanov, N.: Optical Bistability and Hysteresis in Distributed Nonlinear Systems. Physical and Mathematical Literature Publishing Company, Moscow, Nauka (1997)
20. Taranenko, V.B., Staliunas, K., Weiss, C.O.: Phys. Rev. A. **56**, 1582 (1997)
21. Slekys, G., Staliunas, K., Weis, C.O.: Opt. Commun. **149**, 113 (1998)
22. Staliunas, U., Sanchez-Morcillo, V.J.: Transvers Patterns in Nonlniear Optical Resonators Springer tracts in modern physics, vol. 183, pp. 7–37. Springer-Verlag, Berlin, Heidelberg (2003)
23. Cross, M.C., Hohenberg, P.C.: Rev. Mod. Phys. **65**, 851 (1993)
24. Aranson, I.S., Kramer, L.: Rev. Mod. Phys. **74**, 99 (2002)
25. Ferreira, M.F.: Nonlinear Effects in Optical Fibers. John Wiley & Sons, Hoboken, NJ (2011)
26. Mancas, S.C., Choudhury, S.R.: Chaos. Solitons & Fractals. **40**, 91 (2009)
27. Ferreira, M.F.: IET Optoelectron. **12**, 122 (2018)
28. Chang, W., Ankiewicz, A., Akhmediev, N., Soto-Crespo, J.: Phys. Rev. E. **76**, 016607 (2007)
29. Soto-Crespo, J., Akhmediev, N., Ankiewicz, A.: Phys. Rev. Lett. **85**, 2937 (2000)
30. Akhmediev, N., Soto-Crespo, J., Town, G.: Phys. Rev. E. **63**, 056602 (2001)
31. Nelson, L.E., Jones, D.J., Tamura, K., Haus, H.A., Ippen, E.P.: Appl. Phys. B Lasers Opt. **65**, 277 (1997)
32. Wise, F.W., Chong, A., Renninger, W.H.: Laser Phot. Rev. **2**, 58 (2008)
33. Renninger, W.H., Chong, A., Wise, F.W.: Opt. Express. **19**, 22496 (2011)
34. Komarov, A., Leblond, H., Sanchez, F.: Phys. Rev. E. **72**, 025604(R) (2005)
35. Pereira, N., Stenflo, L.: Phys. Fluids. **20**, 1733 (1977)
36. Nozaki, K., Bekki, N.: Phys. Soc. Japan. **53**, 1581 (1984)
37. Conte, R., Musette, M.: Physica D. **69**, 1 (1993)
38. Akhmediev, N., Afanasjev, V., Soto-Crespo, J.: Phys. Rev. E. **53**, 1190 (1996)
39. Thual, O., Fauvre, S.: J. Phys. **49**, 1829 (1988)

40. Akhmediev, N., Soto-Crespo, J.: Phys. Lett. A. **317**, 287 (2003)
41. Akhmediev, N., Soto-Crespo, J.: Phys. Rev. E. **70**, 036613 (2004)
42. Soto-Crespo, J., Akhmediev, N.: Phys. Rev. Lett. **95**, 024101 (2005)
43. Soto-Crespo, J., Akhmediev, N.: Math. Comp. Simulation. **69**, 526 (2005)
44. Tsoy, E., Ankiewicz, A., Akhmediev, N.: Phys. Rev. E. **73,036621-1-10** (2006)
45. Ankiewicz, A., Akhmediev, N., Devine, N.: Opt. Fiber Technol. **13**, 91 (2007)
46. Manousakis, M., Papagiannis, P., Moshonas, N., Hizanidis, K.: Opt. Commun. **198**, 351 (2001)
47. Mancas, S., Choudhury, S.: Theor. Math. Phys. **152**, 1160 (2007)
48. Anderson, D.: Phys. Rev. A. **27**, 3135 (1983)
49. Anderson, D., Lisak, M., Reichel, T.: Phys. Rev. A. **38**, 1618 (1988)
50. Akhmediev, N., Soto-Crespo, J.-M., Grelu, P.: Phys.Lett. A. **372**, 3124 (2008)
51. Chang, W., Ankiewicz, A., Soto-Crespo, J.M., Akhmediev, N.: Phys. Rev. A. **78**, 023830 (2008)
52. Chang, W., Soto-Crespo, J.M., Ankiewicz, A., Akhmediev, N.: Phys. Rev. A. **79**, 033840 (2009)
53. Grelu, P., Chang, W., Ankiewicz, A., Soto-Crespo, J., Akhmediev, N.: J. Opt. Soc. Am. B. **27**, 2336 (2010)
54. Komarov, A., Amrani, F., Dmitriev, A., Komarov, K., Sanchez, F.: Phys. Rev. A. **87**, 023838 (2013)
55. Du, W., Li, H., Li, J., Wang, Z., Wang, P., Zhang, Z., Liu, Y.: Opt. Express. **27**, 8059 (2019)
56. Grischkowsky, D., Balant, A.C.: Appl. Phys. Lett. **41**, 1 (1982)
57. Wu, X., Tang, D.Y., Zhang, H., Zhao, L.M.: Opt. Express. **17**, 5580 (2009)
58. Liu, X.: Phys. Rev. A. **81**, 053819 (2010)
59. Ding, E., Grelu, P., Kutz, J.N.: Opt. Lett. **36**, 1146 (2011)
60. Wang, S.-K., Ning, Q.-Y., Luo, A.-P., Lin, Z.-B., Luo, Z.-C., Xu, W.-C.: Opt. Express. **21**, 2402 (2013)
61. Li, D., Tang, D., Zhao, L., She, D.: J. Lightwave Technol. **33**, 3781 (2015)
62. Lin, W., Wang, S., Xu, S., Luo, Z.-C., Yang, Z.: Opt. Express. **23**, 14860 (2015)
63. Xu, Y., Song, Y., Du, G., Yan, P., Guo, C., Zheng, G., Ruan, S.: IEEE Photonics J. **7**, 1502007 (2015)
64. Krzempek, K., Sotor, J., Abramski, K.: Opt. Lett. **41**, 4995 (2016)
65. Semaan, G., Braham, F., Fourmont, J., Salhi, M., Bahloul, F., Sanchez, F.: Opt. Lett. **41**, 4767 (2016)
66. Krzempek, K.: Opt. Express. **23**, 30651 (2015)
67. Krzempek, K., Abramski, K.: Opt. Express. **24**, 22379 (2016)
68. Lee, J.S., Koo, J.H., Lee, J.H.: Opt. Eng. **55**, 081309 (2016)
69. Li, D., Li, L., Zhou, J., Zhao, L., Tang, D., Shen, D.: Sci. Rep. **6**, 23631 (2016)
70. Rivera, A., Laborde, C., Carrascosa, A., Kuzin, E., Beltrán-Perez, G., Díez, A., Andrés, M.: Opt. Express. **24**, 9966 (2016)
71. Krzempek, K., Tomaszewska, D., Abramski, K.: Opt. Express. **25**, 24853 (2017)
72. Lyu, Y., Zou, X., Shi, H., Liu, C., Wei, C., Li, J., Li, H., Liu, Y.: Opt. Express. **25**, 13286 (2017)
73. Wang, P., Zhao, K., Xiao, X., Yang, C.: Opt. Express. **25**, 30708 (2017)
74. Zhao, G.-K., Lin, W., Chen, H.-J., Lv, Y.-K., Tan, X.-M., Yang, Z.-M., Mashinsky, V.M., Krylov, A., Luo, A.-P., Cui, H., Luo, Z.-C., Xu, W.-C., Dianov, E.M.: Opt. Express. **25**, 20923 (2017)
75. Du, T., Li, W., Ruan, Q., Wang, K., Chen, N., Luo, Z.: Appl. Phys. Express. **11**, 052701 (2018)
76. Wang, N., Cai, J.-H., Qi, X., Chen, S.-P., Yang, L.-J., Hou, J.: Opt. Express. **26**, 1689 (2018)
77. Latas, S.C., Ferreira, M.F.S., Facão, M.: J. Opt. Soc. Am. B. **34**, 1033 (2017)
78. Latas, S.C., Ferreira, M.F.: J. Opt. Soc. of Am. B. **36**, 3016 (2019)
79. Soto-Crespo, J.M., Akhmediev, N.N., Afanasjev, V.V., Wabnitz, S.: Phys. Rev. E. **55**, 4783 (1997)
80. Afanasjev, V.V.: Opt. Lett. **20**, 704 (1995)
81. Deissler, R., Brandt, H.: Phys. Rev. Lett. **72**, 478 (1994)
82. Wang, Z., Wang, Z., Liu, Y., He, R., Zhao, J., Wang, G., Yang, G.: Opt. Lett. **43**, 478 (2018)

83. Du, Y., Xu, Z., Shu, X.: Opt. Lett. **43**, 3602 (2018)
84. Wei, Z.-W., Liu, M., Ming, S.-X., Luo, A.-P., Xu, W.-C., Luo, Z.-C.: Opt. Lett. **43**, 5965 (2018)
85. Wang, X., Liu, Y.-G., Wang, Z., Yue, Y., He, J., Mao, B., He, R., Hu, J.: Opt. Express. **27**, 17729 (2019)
86. Chen, H.-J., Tan, Y.-J., Long, J.-G., Chen, W.-C., Hong, W.-Y., Cui, H., Luo, A.-P., Luo, Z.-C., Xu, W.-C.: Opt. Express. **27**, 28507 (2019)
87. Du, W., Li, H., Li, J., Wang, Z., Zhang, Z., Zhang, S., Liu, Y.: Opt. Lett. **45**, 5024
88. Luo, Y., Xiang, Y., Shum, P.P., Liu, Y., Xia, R., Ni, W., Lam, H.Q., Sun, Q., Tang, X.: Opt. Express. **28**, 4216 (2020)
89. Du, W., Li, H., Li, J., Wang, Z., Zhang, Z., Zhang, S., Liu, Y.: Opt. Lett. **45**, 5024 (2020)
90. Li, T., Liu, M., Luo, A., Luo, Z., Xu, W.: Opt. Express. **28**, 32010 (2020)
91. Latas, S., Ferreira, M.: Opt. Fiber Technol. **11**, 292 (2005)
92. Latas, S., Ferreira, M.: Appl. Phys. B Lasers Opt. **105**, 863 (2011)
93. Zhang, Y., Cui, Y., Huang, L., Tong, L., Liu, X.: Op. Lett. **45**, 6246 (2020)
94. Latas, S.C., Ferreira, M.F.: Opt. Lett. **37**, 3897 (2012)
95. Wei, Z., Liu, M., Ming, S., Luo, A., Xu, W., Luo, Z.: Opt. Lett. **43**, 5965 (2018)
96. Descalzi, O., Brand, H.R.: Phys. Rev. E. **82**, 026203 (2010)
97. Descalzi, O., Cartes, C., Cisternas, J., Brand, H.R.: Phys. Rev. E. **83**, 056214 (2011)
98. Cartes, C., Descalzi, O., Brand, H.: Phys. Rev. E. **85**, 015205 (2012)
99. Cisternas, J., Descalzi, O.: Phys. Rev. E. **88**, 022903 (2013)
100. Cundiff, S., Soto-Crespo, J., Akhmediev, N.: Phys. Rev. Lett. **88**, 073903 (2002)
101. Runge, A., Broderick, N., Erkinatalo, M.: J. Opt. Soc. Am. B. **33**, 46 (2016)
102. Liu, M., Luo, A., Yan, Y., Hu, S., Liu, Y., Cui, H., Luo, Z., Xu, W.: Opt. Lett. **41**, 1181 (2016)
103. Wang, P., Xiao, X., Zhao, H., Yang, C.: IEEE Photonics J. **9**, 1 (2017)
104. Yu, Y., Luo, Z., Kang, J., Wong, K.: Opt. Lett. **43**, 4132 (2018)
105. Suzuki, M., Boyraz, O., Asghari, H., Trinh, P., Kuroda, H., Jalali, B.: Opt. Lett. **43**, 1862 (2018)
106. Chen, H., Tan, Y., Long, J., Chen, W., Hong, W., Cui, H., Luo, A., Luo, Z., Xu, W.: Opt. Express. **27**, 28507 (2019)
107. Wei, Z., Liu, M., Ming, S., Cui, H., Luo, A., Xu, W., Luo, Z.: Opt. Lett. **45**, 531 (2019)
108. Peng, J., Zeng, H.: Commun. Phys. **2**, 34 (2019)
109. Liu, M., Li, T., Luo, A., Xu, W., Luo, Z.: Photonics Res. **8**, 246 (2020)
110. Zhao, K., Gao, C., Xiao, X., Yang, C.: Photonics Res. **9**, 289 (2021)
111. Wang, Y., Wang, C., Zhang, F., Guo, J., Ma, C., Huang, W., Song, Y., Ge, Y., Liu, J., Zhang, H.: Rep. Prog. Phys. **83**, 116401 (2020)
112. Zhou, Y., Ren, Y., Shi, J., Wong, K.: Photonics Res. **8**, 1566 (2020)
113. Luo, M., Zhang, Z., Liu, M., Luo, A., Xu, W., Luo, Z.: Opt. Express. **30**, 22143 (2022)
114. Zeng, J., Sander, M.: Opt. Express. **30**, 7894 (2022)
115. Luo, M., Zhang, Z.X., Chen, N.M., et al.: Adv. Phys. Res. **33**, 2200103 (2023)
116. Redondo, A., Sterke, C., Sipe, J., et al.: Nat. Commun. **7**, 10427 (2016)
117. Runge, A., Hudson, D., Tam, K., et al.: Nat. Photonics. **14**, 492 (2020)
118. Han, Y., Gao, B., Hu, G., Huo, J., Wen, H., Li, Y., Liu, L., Ma, C.: Opt. Express. **31**, 1787 (2023)
119. Luo, M., et al.: Opt. Express. **32**, 10059 (2024)
120. Malomed, B.A.: Phys. Rev. E. **47**, 2874 (1993)
121. Afanasjev, V., Akhmediev, N.: Phys. Rev. E. **53**, 6471 (1996)
122. Afanasjev, V.V., Malomed, B.A., Chu, P.L.: Phys. Rev. E. **56**, 6020 (1997)
123. Akhmediev, N., Ankiewicz, A., SotoCrespo, J.M.: Phys. Rev. Lett. **79**, 4047 (1997)
124. Akhmediev, N., Ankiewicz, A., Soto-Crespo, J.: J. Opt. Soc. Am. B. **15**, 515 (1998)
125. Latas, S.V., Ferreira, M.F.: Opt. Fiber Tech. **11**, 292 (2005)
126. Grelu, P., Belhache, F., Gutty, F., Soto-Crespo, J.: Opt. Lett. **27**, 966 (2002)
127. Tang, D., Man, W., Tam, H., Drummond, P.: Phys. Rev. A. **64**, 033814 (2001)
128. Seong, N., Kim, D.: Opt. Lett. **27**, 1321 (2002)
129. Grelu, P., Beal, J., Soto-Crespo, J.: Opt. Express. **11**, 2238 (2003)
130. Martel, G., Chédot, C., Réglier, V., Hideur, A., Ortaç, B., Grelu, P.: Opt. Lett. **32**, 343 (2007)

131. Stratmann, M., Pagel, T., Mitschke, F.: Phys. Rev. Lett. **95**, 143902 (2005)
132. Pang, M., He, W., Jiang, X., Russell, P.: Nat. Photonics. **10**, 454 (2016)
133. Luo, A., Zheng, X., Liu, M., Liu, H., Zhao, N., Ning, Q., Luo, Z., Xu, W.: Appl. Phys. Express. **8**, 042702 (2015)
134. Wang, P., Bao, C., Fu, B., Xiao, X., Grelu, P., Yang, C.: Opt. Lett. **41**, 2254 (2016)
135. Ma, C., Tian, X., Gao, B., Wu, G.: Laser Phys. **27**, 065102 (2017)
136. Liu, B., Xiang, Y., Luo, Y., Zhu, S., Yan, Z., Sun, Q., Liu, D.: Appl. Phys. B Lasers Opt. **124**, 151 (2018)
137. Niknafs, A., Rooholamininejad, H., Bahrampour, A.: Laser Phys. **28**, 045406 (2018)
138. Song, Y., Zhang, H., Zhao, L., Shen, D., Tang, D.: Opt. Express. **24**, 1814 (2016)
139. Qin, H.Q., Xiao, X., Wang, P., Yang, C.: Opt. Lett. **43**, 1982 (2018)
140. Mou, C., Sergeyev, S., Rozhin, A., Turitsyn, S.: Opt. Express. **21**, 26868 (2013)
141. Gui, L., Zhang, W., Li, X., Xiao, X., Zhu, H., Wang, K., Wu, D., Yang, C.: IEEE Photon. Technol. Lett. **23**, 1790 (2011)
142. Luo, Y., Cheng, J., Liu, B., Sun, Q., Li, L., Fu, S., Tang, D., Zhao, L., Liu, D.: Sci. Rep. **7**, 2369 (2017)

Chapter 4
Ultrashort High-Amplitude Dissipative Solitons on Optical Fiber Lasers

Sofia C. V. Latas and Mário F. S. Ferreira

Abstract The cubic-quintic complex Ginzburg-Landau equation (CQCGLE) is a master equation that has been used to describe passively mode-locked fiber lasers. The main characteristics of very high amplitude (VHA) soliton solutions of the CQCGLE are described numerically and using the method of moments, both in the normal and anomalous dispersion regimes of the fiber laser. The region of existence of such pulses is found numerically in the plane defined by the dispersion and the nonlinear gain saturation parameters. High-energy ultrashort pulses are found mainly in the normal dispersion region, which agrees with the experimental observations reported by other authors. The impact of the higher-order reactive nonlinearity on VHA solitons is also described, focusing mainly in the normal dispersion regime. In the presence of this effect, the amplitude and the energy of the VHA pulses decrease, whereas pulse formation is observed for much higher absolute values of dispersion. The region of existence of the VHA pulses under this effect is found in the semiplane defined by the dispersion and the nonlinear gain.

Keywords Optical solitons · Dissipative solitons · High-energy solitons · Complex Ginzburg-Landau equation · Method of moments · Reactive Nonlinearity

4.1 Introduction

High-energy ultrashort optical pulses can be generated by passively mode-locked fiber lasers, and they are of great importance for a variety of applications [1–5]. In particular, they can be used to develop broadband supercontinuum light sources [2] that play a very important role in several biomedical and medical applications [4–8]. They can also be used, for example, in the acceleration of charged particles [9], in

S. C. V. Latas · M. F. S. Ferreira (✉)
i3N-Institute of Nanostructures, Nanomodelling and Nanofabrication, Department of Physics, University of Aveiro, Aveiro, Portugal
e-mail: sofia.latas@ua.pt; mfernando@ua.pt

M. F. S. Ferreira, M. C. Paul (eds.), *Optical Fiber Lasers*, Springer Series in Optical Sciences 250, https://doi.org/10.1007/978-3-032-04110-4_4

molecular finger printing [10], fabrication of optical couplers [11], optical communications [12], and optical microscopy [13].

A common approach to describe passively mode-locked lasers is an average model in which the effects of discrete laser components in the cavity are averaged over one round trip [14–17]. The resulting master equation is the cubic-quintic complex Ginzburg-Landau equation (CQCGLE) or a generalized version of it.

The study of pulsed dynamics in the framework of the CQCGLE has provided an adequate explanation of several unusual single-pulse and multiple-pulse phenomena that have been observed in mode-locked lasers, such as self-pulsations [18], explosions [19], and the formation of multisoliton complexes [20]. Such studies have been also of fundamental importance in the development of the concept of dissipative solitons (DSs) [21, 22]. Generally, a DS implies a composite balance among dispersion, nonlinearity, gain, and loss [23, 24].

It has been found numerically that the energy of a dissipative soliton solution of the CQCGLE increases indefinitely when the equation parameters converge to a given region of the parameter space [25]. Such set of parameters was called a dissipative soliton resonance (DSR) [26]. Found in the normal dispersion regime, it initially required a positive quintic reactive nonlinearity to appear [25, 26], but then DSR was also revealed in the chromatic dispersion-free ($D = 0$) regime, along with negative quintic reactive nonlinearity [27]. With the increase of pump power, the energy of a DSR pulse increases mainly due to the increase of the pulse width, while keeping the amplitude at a constant level [26, 27]. As a consequence, in order to obtain high-energy ultrashort pulses, a linear pulse compression technique has to be used outside the laser cavity.

The authors have investigated a different kind of high-energy ultrashort pulses, which correspond to the very high amplitude (VHA) soliton solutions of the CQCGLE [28, 29]. Such VHA solutions occur due to a singularity first predicted both numerically and using the soliton perturbation theory in Refs [30, 31], namely, as the nonlinear gain saturation effect tends to vanish. The increase in energy of these pulses is mainly due to the increase of the pulse amplitude, whereas the pulse width becomes narrower. The region of existence of VHA pulses was found in Ref. [29], considering both the normal and the anomalous dispersion regimes, but neglecting the higher-order reactive nonlinearity effect. In the same work, VHA pulses with high energy were found mainly in the normal dispersion region. This is in agreement with the large majority of experimental results, reporting the observation of high-energy pulses in passively mode-locked lasers [21–24]. Further investigations of the VHA pulses in the normal dispersion regime were carried out by the authors both numerically and using the method of moments (MMs). A special attention was paid to the impact of the higher-order reactive nonlinearity effect [32].

This chapter provides an overview of the main properties of very high amplitude (VHA) soliton solutions of the CQCGLE. In Sect. 4.2, we present this equation, which is the master equation describing passively mode-locked fiber lasers. In Section 4.3, we present the method of moments to find approximate solutions of the CQCGLE. In Sect. 4.4, we describe the VHA soliton solutions of the CQCGLE, whereas in Sect. 4.5, we analyze the impact of the higher-order reactive nonlinearity on these pulses. A region of existence of the VHA pulses is obtained in the plane defined by nonlinear gain and the dispersion. Finally, Section 4.6 summarizes the main results.

4.2 The Cubic-Quintic Complex Ginzburg-Landau Equation

The cubic-quintic complex Ginzburg Landau equation (CQCGLE) can be written in the form:

$$i\frac{\partial q}{\partial Z}+\frac{D}{2}\frac{\partial^2 q}{\partial T^2}+|q|^2 q = i\delta q + i\beta\frac{\partial^2 q}{\partial T^2}+i\varepsilon|q|^2 q + i\mu|q|^4 q - \nu|q|^4 q, \quad (4.1)$$

where Z is the normalized propagation distance, T is the retarded time, q is the normalized envelope of the electric field respect, β stands for spectral filtering ($\beta > 0$), δ is the linear gain or loss coefficient, ε accounts for nonlinear gain-absorption processes (e.g., two-photon absorption), μ represents a higher-order correction to the nonlinear gain-absorption, and ν is a higher-order correction term to the nonlinear refractive index. On the left-hand side, D represents the cavity group velocity dispersion, with $D > 0$ in the anomalous regime and $D < 0$ in the normal regime.

Equation (4.1) becomes the standard nonlinear Schrödinger equation (NLSE) when the right-hand side is set to zero. When this does not happen, Eq. (4.1) is non-integrable, and only particular solutions can be obtained. Various techniques have been applied to investigate the soliton solutions of the complex Ginzburg-Landau equation (CGLE). Exact solutions were found [33–38], but they can be presented explicitly only for certain relations between the parameters of the equation. Furthermore, so far, only stationary solutions of the CGLE are known in analytical form. The soliton perturbation theory can be used for small values of the parameters [38, 39]. Moreover, approximate expressions for some localized solutions can be derived for arbitrary values of the CGLE parameters by reducing this equation to finite-dimensional dynamical models. The reduced models can be obtained by applying the method of moments [40], or Lagrangian techniques [41–44].

4.3 Method of Moments

Approximate stationary solutions of Eq. (4.1) for arbitrary values of the parameter can also be obtained using the method of moments (MM) [23, 26, 45]. This method provides a reduction of the complete evolution problem with an infinite number of degrees of freedom to the evolution of a finite set of pulse characteristics [23]. For a localized solution with a single maximum, these characteristics include the peak-amplitude, pulse width, center-of-mass position, and phase parameters. A full description of this method can be found in Refs [23, 26] and references therein.

The method of moments makes use of the integrals for the energy and the momentum M, as given by

$$Q = \int_{-\infty}^{\infty} |q|^2 \, dT \tag{4.2}$$

$$M = \frac{1}{2} \int_{-\infty}^{\infty} \left(q q_T^* - q^* q_T \right) dT. \tag{4.3}$$

as well as of the higher-order generalized moments:

$$I_1 = \int_{-\infty}^{+\infty} T |q|^2 \, dT \tag{4.4}$$

$$I_2 = \int_{-\infty}^{+\infty} (T - T_0)^2 |q|^2 \, dT \tag{4.5}$$

$$I_3 = \int_{-\infty}^{+\infty} (T - T_0)\left(q^* q_T - q q_T^* \right) dT \tag{4.6}$$

where $T_0 = I_1/Q$.

The energy and the momentum, as well as the above three higher-order moments, are conserved quantities of the nonlinear Schrödinger equation. For Eq. (4.1), they are not conserved but satisfy the following truncated set of first-order ordinary differential equations [23, 26, 45]:

$$\frac{dQ}{dZ} = i \int_{-\infty}^{+\infty} \left(q R^* - q^* R \right) dT \tag{4.7}$$

$$\frac{dM}{dZ} = -i \int_{-\infty}^{+\infty} \left(q_T R^* + q_T^* R \right) dT \tag{4.8}$$

$$\frac{dI_1}{dZ} = iDM + i \int_{-\infty}^{+\infty} T \left(q R^* - q^* R \right) dT \tag{4.9}$$

$$\frac{dI_2}{dZ} = -iDI_3 + i \int_{-\infty}^{+\infty} (T - T_0)^2 \left(q R^* - q^* R \right) dT \tag{4.10}$$

$$\frac{dI_3}{dZ} = 2M \frac{dT_0}{dz} + i \int_{-\infty}^{+\infty} \left(2D |q_T|^2 - |q|^4 \right) dT + 2i \int_{-\infty}^{+\infty} (T - T_0)\left(q_T R^* + q_T^* R \right) dT + i \int_{-\infty}^{+\infty} \left(q R^* + q^* R \right) dT \tag{4.11}$$

where q_T represents the derivative of q relative to T, q^* is the complex conjugate of q, and

$$R = i\delta q + i\beta \frac{\partial^2 q}{\partial T^2} + i\varepsilon |q|^2 q + (-\nu + i\mu)|q|^4 q \tag{4.12}$$

Let us consider the following commonly used higher-order Gaussian trial function [23, 26]:

$$q(T,Z) = A\exp\left(-\frac{T^2}{w^2} - \frac{T^4}{w^4}\right)\exp\left(iCT^2\right) \tag{4.13}$$

where A is the soliton amplitude, w is the soliton width, and C is the soliton chirp. As the solutions we are looking for are symmetric, with zero (transverse) velocity, some of the moments are identically zero, and the nonzero moments obtained for the above trial function are:

$$Q = 1.051A^2 w, \quad I_2 = 0.145Qw^2, \quad I_3 = 4iCI_2 \tag{4.14}$$

A second trial function is related to the previous one but it involves only a quartic term, which is considered more appropriate for pulses with extreme values of high energy [23, 26]:

$$q(T,Z) = A\exp\left(-\frac{T^4}{w^4}\right)\exp\left(iCT^2\right) \tag{4.15}$$

The nonzero moments obtained for the above trial function are:

$$Q = 1.524A^2 w, \quad I_2 = 0.239Qw^2, \quad I_3 = 4iCI_2 \tag{4.16}$$

Inserting (4.13) into the equations of the method of moments [23, 26], we obtain the following evolution equations for the soliton parameters [29]:

$$\frac{dQ}{dZ} = Q\left(2\delta - \frac{2.738\beta}{w^2} - 1.158C^2 w^2 \beta + \frac{1.433Q\varepsilon}{w} + \frac{1.146Q^2\mu}{w^2}\right) \tag{4.17a}$$

$$\frac{dw}{dZ} = w\left(2CD + \frac{2.142\beta}{w^2} - 0.874C^2 w^2 \beta - \frac{0.290Q\varepsilon}{w} - \frac{0.325Q^2\mu}{w^2}\right) \tag{4.17b}$$

$$\frac{dC}{dZ} = \frac{1}{w^2}\left(-2C^2 w^2 D + \frac{6.453D}{w^2} - \frac{1.237Q}{w} - \frac{1.319Q^2\nu}{w^2} - 19.624C\beta\right) \tag{4.17c}$$

Considering the trial function given by Eq. (4.15), we obtain the following evolution equations [29]:

$$\frac{dQ}{dZ}=Q\left(2\delta-\frac{2.868\beta}{w^2}-1.912C^2w^2\beta+\frac{1.103Q\varepsilon}{w}+\frac{0.654Q^2\mu}{w^2}\right) \quad (4.18a)$$

$$\frac{dw}{dZ}=w\left(2CD+\frac{0.388\beta}{w^2}-1.136C^2w^2\beta-\frac{0.162Q\varepsilon}{w}-\frac{0.138Q^2\mu}{w^2}\right) \quad (4.18b)$$

$$\frac{dC}{dZ}=\frac{1}{w^2}\left(-2C^2w^2D+\frac{3D}{w^2}-\frac{0.577Q}{w}-\frac{0.456Q^2\nu}{w^2}-16.74C\beta\right) \quad (4.18c)$$

It can be seen that both systems given by Eqs. (4.17) and (4.18) are structurally similar, differing only in the values of the coefficients of each term.

4.4 Very-High Amplitude CGLE Solitons

The following results are based on the method of moments (MMs), and on numerical simulations carried out with Eq. (4.1). Concerning these numerical simulations, we used a split-step Fourier symmetric method, in Fortran, for time steps ΔT between 0.0002 and 0.01 and distance steps ΔZ between 0.000005 and 0.0001, as described in [46]. Absorbing boundary conditions were used, as presented in [47]. An initial condition of the form 20 *sech (5 T),* has been considered. Some results were verified using a different computer and a different code with a Runge-Kutta 4th order pseudo-spectral method, in Matlab.

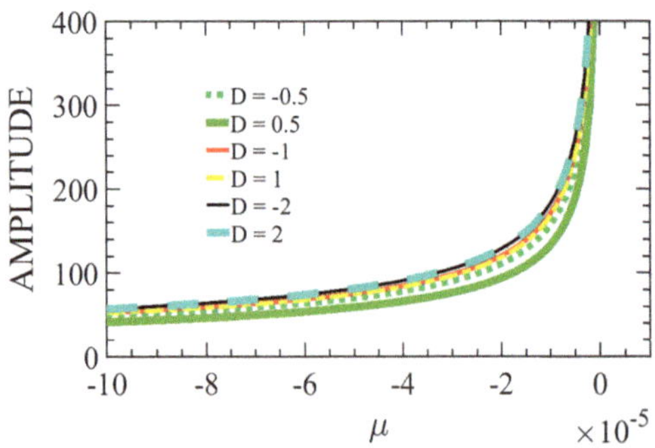

Fig. 4.1 Stationary pulse amplitude against the nonlinear gain saturation parameter, μ, for six different values of dispersion parameter: $D = 0.5$ (thick solid curve), $D = -0.5$ (dotted curve), $D = 1$ (dashed-dotted curve), $D = -1$ (medium solid curve), $D = -2$ (thin solid curve), and $D = 2$ (dashed curve). The other parameter values are: $\delta = -0.1$, $\beta = 0.2$, $\varepsilon = 0.35$, and $\nu = 0$ (Reprinted with permission from [29] © The Optical Society)

Figure 4.1 shows the stationary amplitudes against the nonlinear gain saturation parameter, μ, considering six different values of dispersion parameter D, in the absence of higher-order reactive nonlinearity ($\nu = 0$). These results were obtained by solving numerically the system of Eqs. (4.18), but similar results are provided by the system of Eqs. (4.17). The remaining parameter values are the following: $\delta = -0.1$, $\beta = 0.2$, and $\varepsilon = 0.35$. We observe from Fig. 4.1 that, as $\mu \to 0-$, the pulse amplitude tends to infinity, that is, a singularity is found, for all the values of D considered. However, pulse decay is observed for $D = 0$. Such very-high amplitude (VHA) solutions were first found by Afanasjev, in Ref. [31].

Figure 4.2 shows (a) the stationary pulse (SP) amplitude and (b) the energy against the dispersion parameter, D, considering four different values of nonlinear gain saturation parameter, μ. The results were obtained using the dynamical system given by Eqs. (4.17).

From Fig. 4.2(a), it can be seen that the pulses equilibrium amplitude evolves with D in a similar fashion for all the different values of μ considered. In general, as the magnitude of D decreases, the amplitude also decreases and a minimum is found nearby D ≈ +0.1. Each curve looks almost symmetric relatively to this minimum, such that for symmetric values of D the amplitude looks nearly the same.

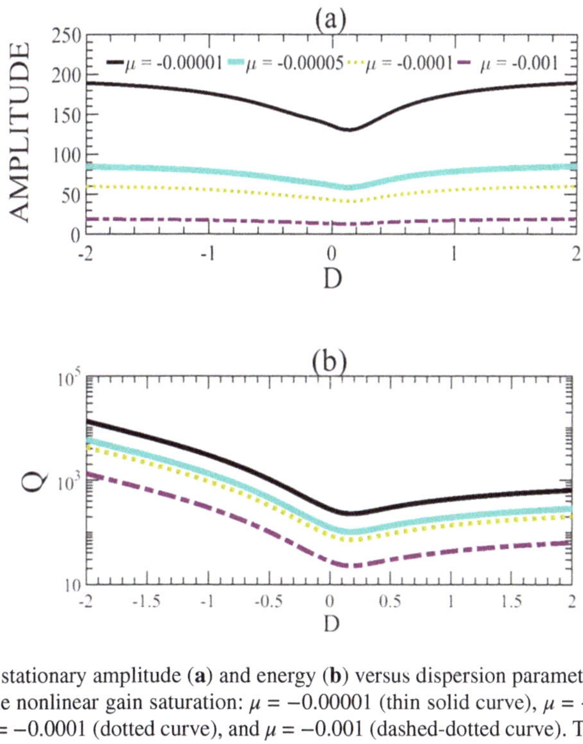

Fig. 4.2 Pulse stationary amplitude (**a**) and energy (**b**) versus dispersion parameter, D, for different values of the nonlinear gain saturation: $\mu = -0.00001$ (thin solid curve), $\mu = -0.00005$ (thick solid curve), $\mu = -0.0001$ (dotted curve), and $\mu = -0.001$ (dashed-dotted curve). The other parameter values are: $\delta = -0.1$, $\beta = 0.2$, $\varepsilon = 0.35$, and $\nu = 0$. These results were obtained using the dynamical system (4.17) (Reprinted with permission from [29] © The Optical Society)

Concerning the pulse energy, Fig. 4.2(b) shows that it decreases with the magnitude of D, until a minimum is reached nearby D ≈ +0.1. However, the energy is not symmetric around this minimum, as it reaches much higher values in the normal than in the anomalous dispersion regimes. For example, the pulse energy for $D = -2$ is higher by one order of magnitude than for $D = +2$.

Figure 4.3 shows (a) the stationary pulse amplitude and (b) the energy against the dispersion parameter, D, for the same cases considered in Fig. 4.2, but using the system of Eqs. (4.18). We observe that, in general, the amplitude and the energy depends on D in a similar fashion. However, a main difference occurs in the range – 0.1 < D < 0.4, where no stationary solutions are found in Fig. 4.3.

In order to clarify which of dynamical systems (4.17) or (4.18) are more accurate, Eq. (4.1) was numerically solved. The region of existence of stationary pulses (SPs) of the CGLE in the plane (μ, D) is illustrated in Fig. 4.4.

The region of existence of SPs shown in Fig. 4.4 is not as large as that predicted by the method of moments in Fig. 4.3. In the normal dispersion regime, front's expansion is observed for D < −1.65. On the other hand, on the upper right corner (anomalous dispersion regime), instability is found. Furthermore, for the assumed values of μ, stationary pulses are not found for D between −0.5 and 0.1. This result is in good qualitative agreement with the predictions of the method of moments, as given by the system of Eqs. (4.18).

Figures 4.5 and 4.6 illustrate some of the VHA stationary solutions found numerically in the plane (μ, D). In Fig. 4.5, several values of D are considered, and the

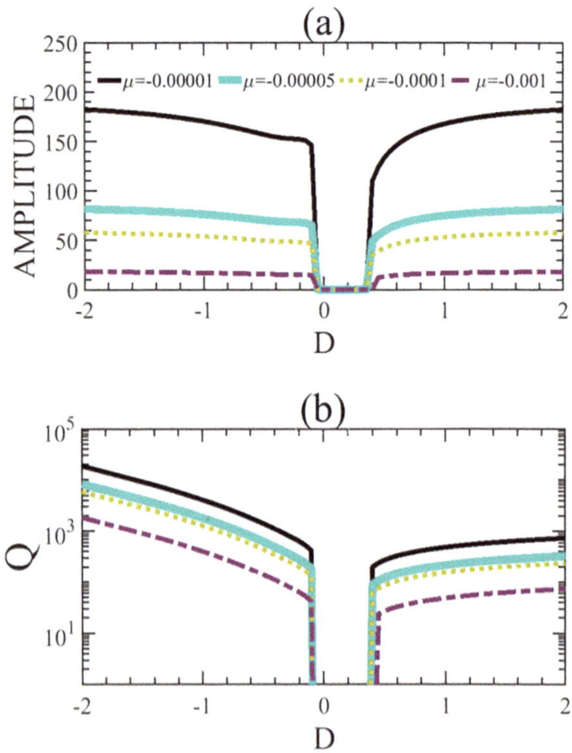

Fig. 4.3 Pulse stationary amplitude (**a**) and energy (**b**) versus dispersion parameter, D, for different values of the nonlinear gain saturation: $\mu = -0.00001$ (thin solid curve), $\mu = -0.00005$ (thick solid curve), $\mu = -0.0001$ (dotted curve), and $\mu = -0.001$ (dashed-dotted curve). The other parameter values are: $\delta = -0.1$, $\beta = 0.2$, $\varepsilon = 0.35$, and $\nu = 0$. These results were obtained using the dynamical system (4.18) (Reprinted with permission from [29] © The Optical Society)

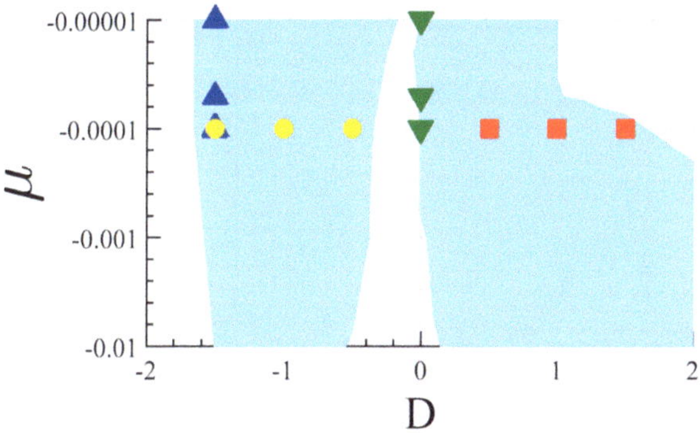

Fig. 4.4 Region of existence of stationary pulses (SPs) of the CGLE in the plane (μ, D). The other parameters are $\delta = -0.1$, $\beta = 0.2$, $\varepsilon = 0.35$, and $\nu = 0$. The circles and squares correspond to SPs profiles represented in Fig. 4.5(a) and (c), respectively. The triangles (down and up) correspond to SPs profiles represented in Figs. 4.6(a) and (c), respectively (Reprinted with permission from [29] © The Optical Society)

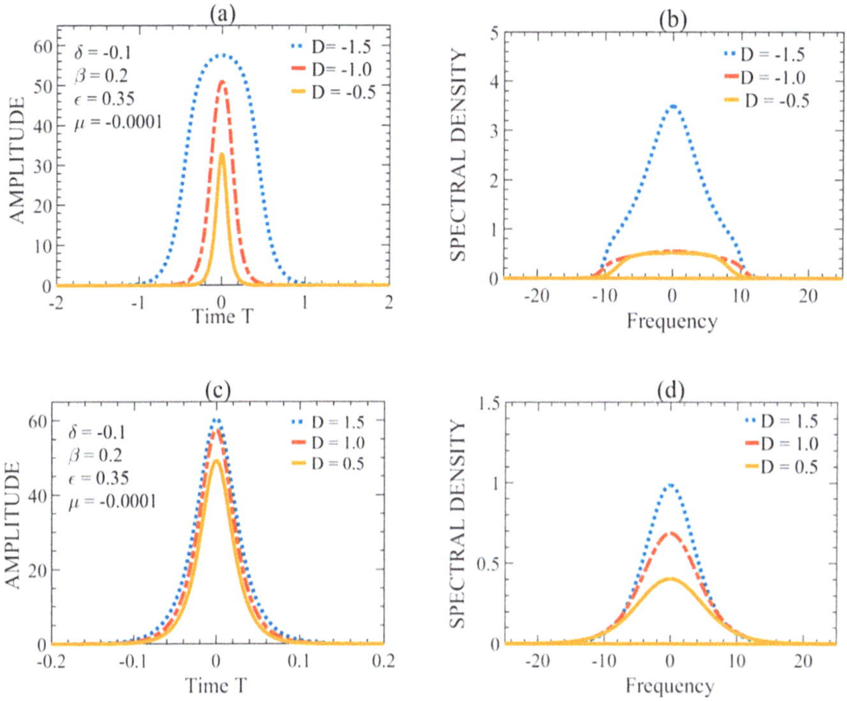

Fig. 4.5 Amplitude profiles, (**a**) and (**c**), and power spectral density, (**b**) and (**d**), in the normal (D < 0) and anomalous (D > 0) dispersion regimes, respectively, for D = ±1.5 (dashed curves), D = ±1 (dashed-dot curves), and D = ±0.5 (solid curves). The other parameter values are: $\delta = -0.1$, $\beta = 0.2$, $\varepsilon = 0.35$, and $\nu = 0$ (Reprinted with permission from [29] © The Optical Society)

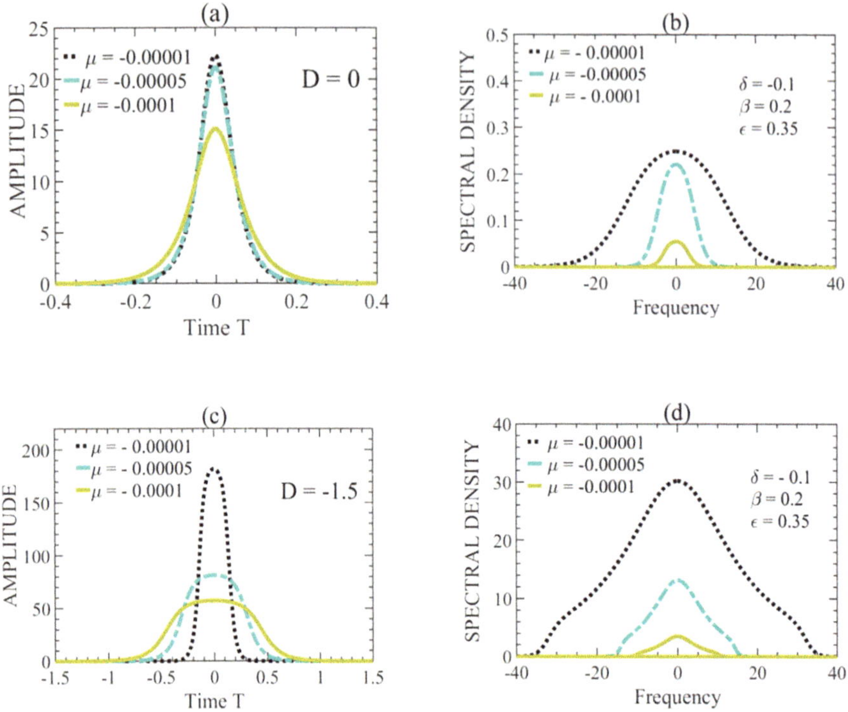

Fig. 4.6 Amplitude profiles, (**a**) and (**c**), and power spectral density, (**b**) and (**d**), for D = 0 and D = −1.5, respectively. The nonlinear gain saturation parameter assumes the values: $\mu = -0.00001$ (dashed curves), $\mu = -0.00005$ (dashed-dot curves), and $\mu = -0.0001$ (solid curves). The other parameter values are: $\delta = -0.1$, $\beta = 0.2$, and $\varepsilon = 0.35$ (Reprinted with permission from [29] © The Optical Society)

nonlinear gain saturation parameter is kept constant, with a value $\mu = -0.0001$. The other parameter values are $\delta = -0.1$, $\beta = 0.2$, and $\varepsilon = 0.35$.

Figures 4.5(a) and (c) illustrate the pulse profiles in the normal (D < 0) and anomalous (D > 0) dispersion regimes, respectively, for D = ±1.5, ±1, and ± 0.5. The pulse width changes significantly with D in the normal dispersion regime, whereas it remains almost unchanged in the anomalous dispersion regime. In general, pulses are wider in the first case when compared with second one. The pulse power spectral density is represented in Figs. 4.5(b) and (d), for D < 0 and D > 0, respectively. One main feature is that the spectral range is almost the same for all cases in both regimes. On the other hand, the spectra peak power achieves higher values when increasing the magnitude of D, which corresponds to higher values of the pulse energy.

Figure 4.6 presents several pulse amplitude profiles and the respective power spectral density, for different values of the nonlinear gain saturation parameter and for two values of the dispersion parameter: D = 0 and D = −1.5. The other parameter values are similar to those in Fig. 4.5.

Figures 4.6(a) and (c) show that the peak amplitude increases and the pulse becomes narrower as $\mu \to 0-$, which is more evident in the case D = -1.5 (normal dispersion regime). The pulses power spectra are represented in Figs. 4.9(b) and (d) for D = 0 and D = -1.5, respectively. We observe that both the peak values and the spectral range of such pulses increase as $\mu \to 0-$. Once more, the spectral range seems to be almost independent of D and strongly dependent of the nonlinear gain saturation parameter, μ. Decreasing the magnitude of this parameter allows the pulses in the normal dispersion regime to reach very high energies.

It is important to recognize a significant difference between the VHA pulses discussed in this chapter and the high-energy pulses found numerically in Ref. [25, 26]. Actually, it has been found numerically that the energy of a dissipative soliton solution of the CGLE increases indefinitely when the equation parameters converge to a given region of the parameter space. Such a set of parameters was called a dissipative soliton resonance (DSR) [25]. Found in the normal dispersion regime, it initially required a positive quintic reactive nonlinearity to appear [25, 26], but then DSR was also revealed in the chromatic dispersion-free (D = 0) regime, along with negative quintic reactive nonlinearity [27].

The energy of a DSR pulse increases mainly due to the increase of the pulse width, while keeping the amplitude at a constant level [25, 26]. Nevertheless, as the DSR pulse is highly chirped, this can be used to compress the pulse and hence obtain high-energy ultrashort pulses.

In contrast with the DSR pulse, the increase in energy of the pulses discussed in this chapter, which occurs when the absolute value of μ decreases, is mainly due to the increase of the pulse amplitude, whereas the pulse width becomes narrower. Clearly, high-energy ultrashort pulses can be obtained in this case without using any additional linear pulse compression technique. It might be noted, from Fig. 4.1, that there is no limit for the pulse amplitude. However, a limitation could be imposed by some higher-order effects, as seen in the next section.

4.5 Impact of the Higher-Order Reactive Nonlinearity

In this section, we describe the impact of the higher-order reactive nonlinearity on VHA pulses, focusing mainly in the normal dispersion regime, where high-energy pulses have been found both theoretically and experimentally [21–24].

Figure 4.7 shows (a, c) the pulse amplitude and (b, d) the pulse energy against the dispersion parameter, D, obtained by solving the dynamical system given by Eqs. (4.18). for values of D ≤ 0. Three different values of the nonlinear gain saturation parameter are considered, namely, $\mu = -0.0001$, $\mu = -0.00005$, and $\mu = -0.00001$, both in the absence of higher-order reactive nonlinearity (a, b) and in the presence of such effect (c, d), considering $\nu = 0.00035$. The other parameter values are $\delta = -0.1$, $\beta = 0.2$, and $\varepsilon = 0.25$.

From Figs. 4.7(a) and (b), it can be seen that, as D changes from -5 to 0, both the pulse amplitude and the pulse energy decrease in a nonlinear fashion, until

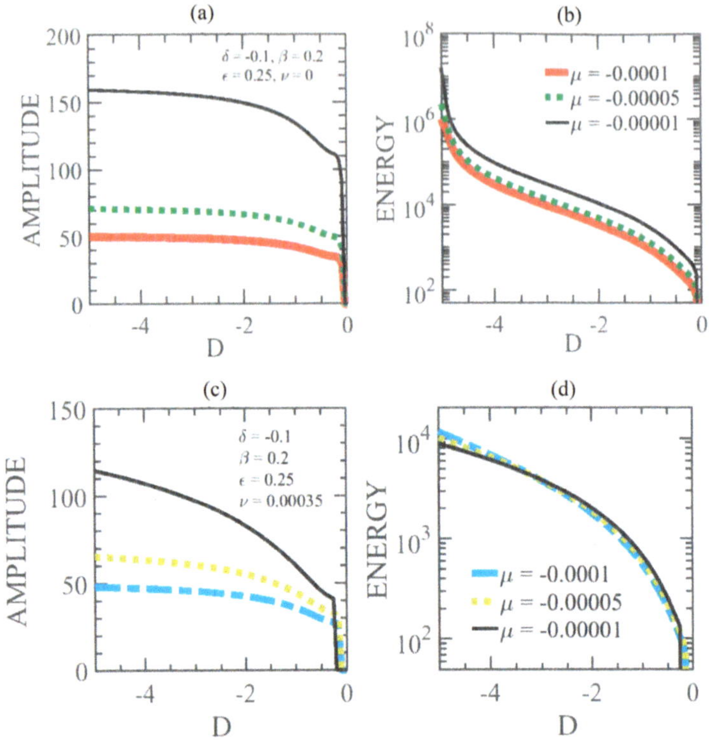

Fig. 4.7 Pulse equilibrium amplitude (**a**, **c**) and energy (**b**, **d**), versus the dispersion parameter, D, in the normal dispersion regime. Three different values of the nonlinear gain saturation are considered, namely, $\mu = -0.00001$ (thin solid curve), $\mu = -0.00005$ (dotted curve), and $\mu = -0.0001$ (dashed-dotted curve). The other parameter values are $\delta = -0.1$, $\beta = 0.2$, $\varepsilon = 0.25$, and $\nu = 0$ (**a**, **b**) or $\nu = 0.00035$ (**c**, **d**). These results were obtained using Eqs. (4.18), based on the method of moments

decay is reached for D ≈ 0. Both the amplitude and the energy reach higher values for lower absolute values of the nonlinear gain saturation parameter.

Comparing the Figs. 4.7(a) and (c), it is apparent that the rate of decrease of the amplitude with D is higher in the presence of higher-order reactive nonlinearity, for the three values of μ considered. Decay is observed in Fig. 4.7(c) for D ≈ −0.2, whereas it occurs for D ≈ 0 in Fig. 4.7(a). Comparing Figs. 4.7(d) and (b), we verify that the energy experiences also a significant reduction in the presence of higher-order reactive nonlinearity, especially for higher absolute values of D.

Figure 4.8 shows the VHA pulse amplitude profiles and the corresponding power spectral density found by solving numerically Eq. (4.1) in the presence of higher-order reactive nonlinearity ($\nu = 0.00035$), for D = −2.3. This is the same value of D, for which the highest pulse energy was found in the absence of the higher-order reactive nonlinearity effect.

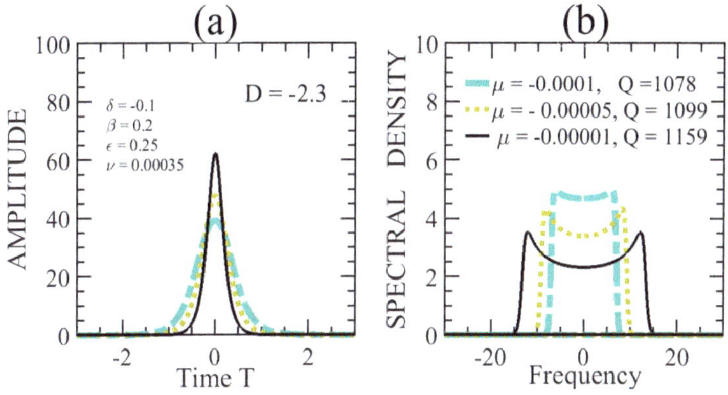

Fig. 4.8 (**a**) Amplitude profiles and (**b**) power spectral density for D = −2.3, in the presence of higher-order reactive nonlinearity term (ν = 0.00035). Three different values of nonlinear gain saturation parameter μ, have been considered, namely, −0.00001 (solid curves), −0.00005 (dashed curves), and −0.0001 (dashed-dotted curves). The other parameter values are $\delta = -0.1$, $\beta = 0.2$, and $\varepsilon = 0.25$

The higher-order reactive nonlinearity reduces significantly the pulse amplitude and the pulse energies, which become similar for the three values of μ: Q = 1078, 1099, and 1159. Actually, the pulse energies are not significantly enhanced when the nonlinear gain saturation parameter vanishes, as occurs in the absence of the higher-order reactive nonlinearity effect. Moreover, we observe from Fig. 4.8(b) that the power spectral density presents a dual peak in the presence of this effect. The pulses spectral range shrinks, but becomes larger as the nonlinear gain saturation parameter is reduced, as expected.

Assuming a value ν = 0.00035 for the higher-order reactive nonlinearity coefficient, pulse formation was observed until D = −5, D = −4.6, and D = -3.5, for μ = −0.0001, μ = −0.00005, and μ = −0.00001, respectively. For higher absolute values of D, expansion with instability was observed. The amplitude profiles and the corresponding power spectral densities in such limited cases are presented in Fig. 4.9.

Two qualitatively different kinds of pulses can be observed in Figs. 4.9(a) and (b). On one hand, we have a wide pulse, with a relatively high energy (Q = 10,367) and a narrow band (−10 < f < 10) single-peak spectrum for μ = −0.0001. On the other hand, there are two narrow pulses, with small energies (Q = 4615 and Q = 2356) and dual-peak spectra, with large spectral bands, for μ = −0.00005 and −0.00001, respectively.

Figure 4.10 shows the region of existence of VHA pulses in the plane defined by dispersion, D, and nonlinear gain, ε. Both stationary pulses (SPs) and multi-pulses (MPs) have been found in this plane. Above the upper ε limit of the regions shown in Fig. 4.10, the solutions tend to expand. Below the lower limit, the solutions tend to decay. When D = 0, no decay is observed for ε > 0.4. Moreover, when ε > 0.7, multi-pulses formation is observed.

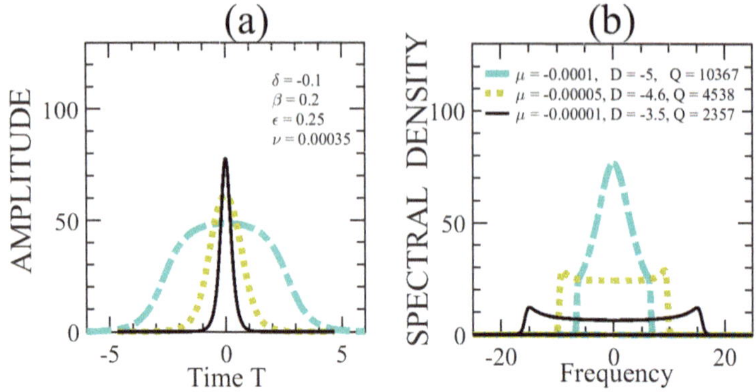

Fig. 4.9 (**a**) Amplitude profiles and (**b**) power spectral density for D = − 5 (μ = −0.0001), D = −4.6 (μ = −0.00005), and D = − 3.5 (μ = −0.00001), in the presence of higher-order reactive nonlinearity effect. The three different profiles correspond to three different values of the nonlinear gain saturation parameter μ, namely, −0.00001 (solid curves), −0.00005 (dashed curves), and −0.0001 (dashed-dotted curves), respectively. The other parameter values are δ = −0.1, β = 0.2, ε = 0.25, and ν = 0.00035

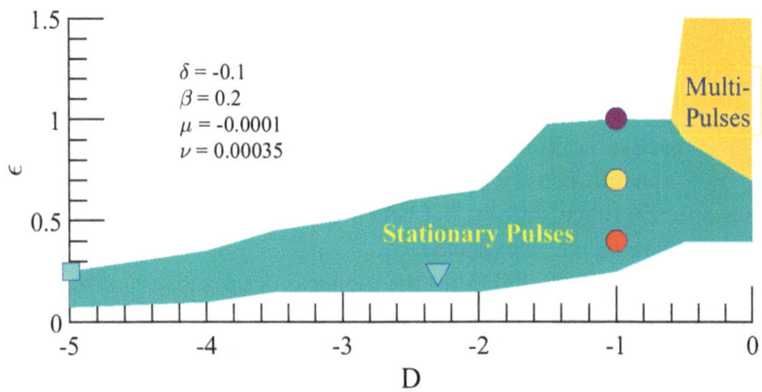

Fig. 4.10 Region of existence in the plane (ε, D) of VHA pulses in the presence of higher-order reactive nonlinearity effect (ν = 0.00035). The other parameter values are: δ = −0.1, β = 0.2, and μ = −0.0001. The triangle and the square correspond to the solitons represented in Fig. 4.8 (D = −2.3) and Fig. 4.9 (D = −5) for ε = 0.25, in that order. The circles correspond to the pulses represented in Fig. 4.11 (D = −1), for ε = 0.4, 0.7, and 1, respectively

Figure 4.11 shows (a) the amplitude profiles and (b) the power spectral density for three stationary pulses (SPs), found in the parameter plane (ε, D) at D = −1, corresponding to the three circles, for ε = 0.4 (dashed-dotted curve), 0.7 (dashed curve), and 1.0 (solid curve). We observe that, as ε increases, the pulses become narrower, whereas their energy and spectral range increase.

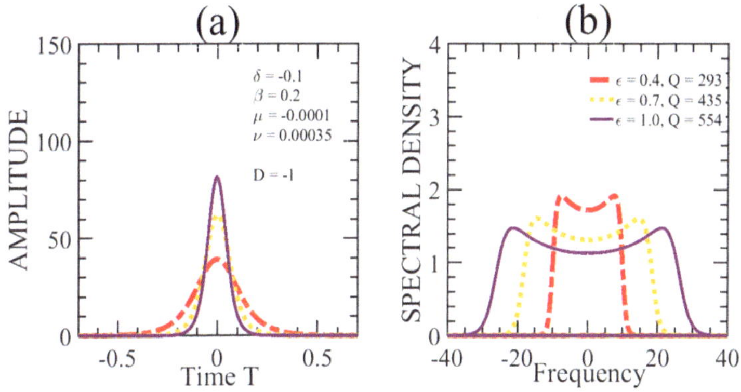

Fig. 4.11 (**a**) Amplitude profiles and (**b**) power spectral density for three stationary pulses (SPs), found in the parameter plane (ε, D) at D = −1, corresponding to the three circles, for $\varepsilon = 0.4$ (dashed-dotted curve), 0.7 (dashed curve), and 1.0 (solid curve). The other parameter values are $\delta = -0.1$, $\beta = 0.2$, $\mu = -0.0001$, and $\nu = 0.00035$

4.6 Conclusions

In this chapter, very high amplitude (VHA) pulses of the CGLE are presented, due to a singularity predicted by the soliton perturbation theory when the nonlinear gain saturation parameter tends to zero. Their existence is also predicted by the method of moments and confirmed numerically.

A region of existence of these VHA pulses was found numerically in the plane defined by the nonlinear gain saturation parameter and the dispersion parameter, D. A good agreement was found between the numerical simulations and the method of moments predictions, considering a trial function with a quartic term only. The pulses have higher energies in the normal dispersion regime than in the anomalous dispersion regime.

The increase in energy of the pulses discussed in this work, found in the normal dispersion regime when the magnitude of μ decreases, is mainly due to the increase of amplitude, whereas the pulse width becomes narrower. This is a different mechanism of the one observed in DSR, were the growth of energy is mainly due to the increase of the pulse width, while pulse amplitude and spectral range remain constant.

The impact of the higher-order reactive nonlinearity on the VHA pulses was also described, using the method of moments (MMs) and solving numerically solving the cubic-quintic CGLE. In the presence of this effect, the amplitude and the energy of the VHA pulses decrease, while their spectral range shrinks, for a given value of dispersion. Moreover, for higher absolute values of the normal dispersion, the pulse energy no longer grows significantly when the nonlinear gain saturation vanishes. A region of existence of VHA pulses was found in the plane defined by the normal dispersion and the nonlinear gain, in the presence of the higher-order reactive nonlinearity.FundingFundação para a Ciência e a Tecnologia (FCT) (Project UID/CTM/50025/2020).

References

1. Christov, I.P., Murnane, M.M., Kapteyn, H.C.: High-harmonic generation of attosecond pulses in the 'single-cycle' regime. Phys. Rev. Lett. **78**, 1251 (1997)
2. Genty, G., Coen, S., Dudley, J.M.: Fiber supercontinuum sources. J. Opt. Soc. Am. B. **24**, 1771–1785 (2007)
3. Maxwell, I., Chung, S., Mazur, E.: Nanoprocessing of subcellular targets using femtosecond laser pulses. Med. Laser Appl. **20**, 193–200 (2005)
4. Horikis, T.P., Ablowitz, M.J.: Passive mode-locking under higher order effects. J. Opt. Soc. Am. B. **31**(11), 2748–2753 (2014)
5. Fu, W., Wright, L.G., Sidorenko, P., Backus, S., Wise, F.W.: Several new directions for ultrafast fiber lasers. Opt. Exp. **26**(8), 9432–9463 (2018)
6. Kawagoe, H., Ishida, S., Aramaki, M., Sakakibara, Y., Omoda, E., Kataura, H., Nishizawa, N.: Development of a high power supercontinuum source in the 1.7 μm wavelength region for highly penetrative ultrahigh-resolution optical coherence tomography. Biomed. Opt. Express. **5**(3), 932–943 (2014)
7. Lindfors, K., Kalkbrenner, T., Stoller, P., Sandoghdar, V.: Detection and spectroscopy of gold nanoparticles using supercontinuum white light confocal microscopy. Phys. Rev. Lett. **93**(3), 037401 (2004)
8. Labruyère, A., Tonello, A., Couderc, V., Huss, G., Leproux, P.: Compact supercontinuum sources and their biomedical applications. Opt. Fiber Technol. **18**(5), 375–378 (2012)
9. Kando, M., et al.: Experimental results of laser wakefield acceleration using a femtosecond terawatt laser pulse. Jpn. J. Appl. Phys. **Part 2**(38), L967–L969 (1999)
10. Diddams, S.A., Hollberg, L., Mbele, V.: Molecular fingerprinting with the resolved modes of femtosecond laser frequency comb. Nature. **445**(7128), 627–630 (2007)
11. Suzuki, K., Sharma, V., Fujimoto, J.G., Ippen, E.P., Nasu, Y.: Characterization of symmetric (3*3) directional couplers fabricated by direct writing with a femtosecond laser oscillator. Opt. Express. **14**, 2335–2343 (2006)
12. Kang, J., Kong, C., Feng, P., Wei, X., Luo, Z.C., Lam, E.Y., Wong, K.K.Y.: Broadband high-energy all-fiber laser at 1.6 μm. IEEE Photon. Technol. Lett. **30**(4), 311–314 (2018)
13. Hu, S., Yao, J., Liu, M., Luo, A.P., Luo, Z.C., Xu, W.C.: Gain-guided soliton fiber laser with high-quality rectangle spectrum for ultrafast time-stretch microscopy. Opt. Express. **24**(10), 10786–10796 (2016)
14. Wang, S., Docherty, A., Marks, B.S., Menyuk, C.R.: Comparison of numerical methods for modeling laser mode locking with saturable gain. J. Opt. Soc. Am. B. **30**(11), 3064–3074 (2013)
15. Wang, S., Marks, B.S., Menyuk, C.R.: Comparison of models of fast saturable absorption in passively modelocked lasers. Opt. Express. **24**(18), 20228–20244 (2016)
16. Wang, S., Droste, S., Sinclair, L.C., Coddington, I., Newbury, N.R., Carruthers, T.F., Menyuk, C.R.: Wake mode sidebands and instability in mode-locked lasers with slow saturable absorbers. Opt. Lett. **42**(12), 2362–2365 (2017)
17. Liao, R., Song, Y., Chai, L., Hu, M.: Pulse dynamics manipulation by the phase bias in a nonlinear fiber amplifying loop mirror. Opt. Express. **27**(10), 14705–14715 (2019)
18. Soto-Crespo, J.M., Grapinet, M., Grelu, P., Akhmediev, N.: Bifurcations and multiple-period soliton pulsations in a passively mode-locked fiber laser. Phys. Rev. E. **70**, 066612 (2004)
19. Cundiff, S., Soto-Crespo, J.M., Akhmediev, N.: Experimental evidence for soliton explosions. Phys. Rev. Lett. **88**, 073903 (2002)

20. Akhmediev, N., Soto-Crespo, J.M., Grapinet, M., Grelu, P.: Dissipative soliton interactions inside a fiber laser cavity. Opt. Fiber Technol. **11**, 209–228 (2005)
21. Chong, A., Renninger, W.H., Wise, F.W.: All-normal-dispersion femtosecond fiber laser with pulse energy above 20 nJ. Opt. Lett. **32**(16), 2408–2410 (2007)
22. Zhao, L.M., Tang, D.Y., Wu, J.: Gain-guided soliton in a positive group-dispersion fiber laser. Opt. Lett. **31**(12), 1788–1790 (2006)
23. Akhmediev, N., Ankiewicz, A.: Dissipative Solitons: 'From Optics to Biology and Medicine' Lecture notes in Physics. Springer (2008)
24. Grelu, P., Akhmediev, N.: Dissipative solitons for mode-locked lasers. Nat. Photonics. **6**(2), 84–92 (2012)
25. Akhmediev, N., Soto-Crespo, J.-M., Grelu, P.: Roadmap to ultra-short record high-energy pulses out of laser oscillators. Phys. Lett. A. **372**, 3124–3128 (2008)
26. Chang, W., Ankiewicz, A., Soto-Crespo, J.M., Akhmediev, N.: Dissipative soliton resonances. Phys. Rev. A. **78**, 023830 (2008)
27. Chang, W., Soto-Crespo, J.M., Ankiewicz, A., Akhmediev, N.: Dissipative soliton resonances in the anomalous dispersion regime. Phys. Rev. A. **79**, 033840 (2009)
28. Latas, S.C., Ferreira, M.F.S., Facão, M.: Ultrashort high-amplitude dissipative solitons in the presence of higher-order effects. J. Opt. Soc. Am. B. **34**(5), 1033–1040 (2017)
29. Latas, S.C., Ferreira, M.F.: High-amplitude dissipative solitons in the normal and anomalous dispersion regimes. J. Opt. Soc. Am. B. **36**(11), 3016–3023 (2019)
30. Soto-Crespo, J.M., Akhmediev, N.N., Afanasjev, V.V., Wabnitz, S.: Pulse solutions of the cubic-quintic complex Ginzburg-Landau equation in the case of normal dispersion. Phys. Rev. E. **55**, 4783–4796 (1997)
31. Afanasjev, V.V.: Soliton singularity in the system with nonlinear gain. Opt. Lett. **20**, 704–705 (1995)
32. Latas, S.V., Ferreira, M.F.: Impact of the higher-order reactive nonlinearity on high-amplitude dissipative solitons. J. Nonlinear Math. Phys. **31**, 16 (2024)
33. Akhmediev, N.N., Afanasjev, V.V., Soto-Crespo, J.M.: Singularities and special soliton solutions of the cubic-quintic complex Ginzburg-Landau equation. Phys. Rev. E. **53**, 1190–1201 (1996)
34. Soto-Crespo, J.M., Akhmediev, N.N., Afanasjev, V.V.: Stability of the pulselike solutions of the quintic complex Ginzburg-Landau equation. J. Opt. Soc. Am. B. **13**, 1439–1448 (1996)
35. Pereira, N.R., Stenflo, L.: Nonlinear Schrödinger equation including growth and damping. Phys. Fluids. **20**, 1733–1734 (1977)
36. Nozaki, K., Bekki, N.: Exact solutions of the generalized Ginzburg-Landau equation. Phys. Soc. Japan. **53**, 1581–1582 (1984)
37. Akhmediev, N., Ankiewicz, A.: Solitons, Nonlinear Pulses and Beams. Chapman & Hall, London (1997)
38. Ferreira, M.: Nonlinear Effects in Optical Fibers. John Wiley & Sons, Hoboken, NJ (2011)
39. Ferreira, M.F., Facão, M.V., Latas, S.V.: Stable soliton propagation in a system with spectral filtering and nonlinear gain. Fiber Integrated Opt. **19**, 31–41 (2000)
40. Tsoy, E.N., Ankiewicz, A., Akhmediev, N.: Dynamical models for dissipative localized waves of the complex Ginzburg-Landau equation. Phys. Rev. E. **73**, 036621 (2006)
41. Manousakis, M., Papagiannis, P., Moshonas, N., Hizanidis, K.: Propagation of Pereira-Stenflo type pulses in a dispersion varying link: variational approach. Opt. Commun. **198**, 351–367 (2001)
42. Ankiewicz, A., Akhmediev, N., Devine, N.: Dissipative solitons with Lagrangian approach. Optical Fib. Technol. **13**, 91–97 (2007)
43. Mancas, S.C., Choudhury, S.R.: A novel variational approach to pulsating solitons in the cubic-quintic Ginzburg-Lanadu equation. Theor. Math. Phys. **152**, 1160–1172 (2007)
44. Mancas, S.C., Choudhury, S.R.: Spatiotemporal structure of pulsating solitons in the cubic-quintic Ginzburg-Landau equation: a novel variational formulation. Chaos, Solitons Fractals. **40**, 91–105 (2009)

45. Uzunov, I.M., Arabadzhiev, T.N., Georgiev, Z.D.: Influence of higher-order effects on pulsating solutions, stationary solutions and moving fronts in the presence of linear and nonlinear gain/loss and spectral filtering. Opt. Fib. Tech. **24**, 15–23 (2015)
46. Agrawal, G.P.: Nonlinear Fiber Optics, 4th edn. Academic & Elsevier (2007)
47. If, F., Berg, P., Christiansen, P.L., Skovgaard, O.: Split-step spectral method for nonlinear Schrodinger equation with absorbing boundaries. J. Comput. Phys. **72**, 501–503 (1987)

Chapter 5
Multiplexed Dissipative Soliton Fiber Lasers

Yiyang Luo, Haoguang Liu, and Yu An

Abstract In contrast to conventional mode-locked lasers with a single output, multiplexed soliton lasers exhibit a significantly broader range of potential applications. In this chapter, we summarize several commonly used multiplexed dissipative soliton fiber lasers, including wavelength multiplexing, bidirectional multiplexing, and multistate multiplexing. Multiwavelength mode locking can be utilized as multichannel light sources and microwave generation. Bidirectional and multistate multiplexing would be effective methods for the generation of dual-frequency combs. Finally, some novel methods to realize the multiplexed dissipative soliton fiber lasers are discussed.

Keywords Multiplexed dissipative soliton fiber lasers · Wavelength multiplexing · Bidirectional multiplexing · Multistate multiplexing · 2D material · Intracavity filtering effect · Artificial manipulation

5.1 Introduction

The concept of "soliton" was first introduced in 1965 [1] to describe localized wave solutions in a range of nonlinear systems, spreading fields such as fluid dynamics, plasma physics, Bose-Einstein condensates, and optics. In addition to serving as a high-quality pulse source, the mode-locked fiber laser can also act as a platform for

---

Y. Luo (✉) · Y. An
Key Laboratory of Optoelectronic Technology and Systems (Ministry of Education), Chongqing University, Chongqing, China
e-mail: yyluo@cqu.edu.cn

H. Liu
Key Laboratory of Optoelectronic Technology and Systems (Ministry of Education), Chongqing University, Chongqing, China

School of Optical and Electronic Information, Huazhong University of Science and Technology, Wuhan, China

M. F. S. Ferreira, M. C. Paul (eds.), *Optical Fiber Lasers*, Springer Series in Optical Sciences 250, https://doi.org/10.1007/978-3-032-04110-4_5

nonlinear optical systems to study various soliton dynamics. In dissipative soliton (DS) fiber lasers, solitons can be generated under the dual balance of gain-loss and dispersion nonlinearity [2]. Optical solitons have steadily established themselves as a key component in advanced applications, including laser processing [3], LiDAR [4], and optical communications [5].

In recent years, the growing multifunctionality of lasers has spurred the flourishing development of multiplexed laser systems. Similarly, multiplexed soliton fiber lasers have emerged in response to this trend. Various approaches have been proposed to enable the multiplexing of soliton lasers, including wavelength multiplexing, bidirectional multiplexing, multistate multiplexing, polarization multiplexing, among others. For wavelength multiplexing, as a prominent one of the ultrafast lasers, the multiwavelength mode-locked laser boasts several key advantages, including a wide range of output wavelengths, ultra-short pulse duration, high peak power, and an extensive operational bandwidth. To realize multiwavelength mode locking, classical approaches including specialized structural designs [6], intracavity filtering [7], and the utilization of specific 2D materials [8] have been proposed. For bidirectional multiplexing, in the absence of unidirectional components like isolators within the laser cavity, bidirectional oscillation occurs, opening up a new dimension for multiplexing. It is noteworthy that bidirectional mode locking has been successfully realized across a diverse range of gain media and dispersion regimes. Furthermore, the counterpropagation of solitons in both directions makes it an ideal platform for investigating the intricate dynamics of soliton collisions. From the perspective of light sources, bidirectional multiplexed lasers have also emerged as potential candidates for dual-comb sources.

Moreover, in recent decades, significant efforts have been dedicated to the multistate multiplexing of soliton fiber lasers. Through specialized design of the laser cavity or dispersion management, the laser can simultaneously output solitons in multiple states, thereby meeting the growing demands of various applications. Moreover, emerging technologies such as polarization multiplexing and dynamic artificial control are progressively offering novel strategies to further advancement in multiplexed soliton fiber lasers. In the following sections, we will introduce the above methods of multiplexed dissipative soliton fiber lasers.

5.2 Wavelength-Multiplexed Dissipative Soliton Fiber Lasers

Due to gain competition effects, passively mode-locked fiber lasers can generally only output a single wavelength. Dual- and multiwavelength ultrafast fiber lasers also play an important role in fields such as fiber sensing, fiber communication, and biological imaging. Under typical conditions, multiwavelength soliton output can be achieved through several approaches, including the dual-cavity configuration, two-dimensional materials, and intracavity filtering effects.

5.2.1 Dual-Cavity Configuration–Based Multiwavelength Soliton Fiber Laser

By designing a dual-resonator structure, stable dual-wavelength mode-locked fiber laser output can be achieved, in which two wavelengths of light propagate along distinct pathways within the system. A dual-wavelength ytterbium-doped fiber laser (YDFL) mode-locked using a semiconductor saturable absorber mirror (SESAM) has been reported in Ref. [9], in which the tunable dual-wavelength output can be achieved by independently adjusting the volume Bragg gratings (VBGs) in the two resonator cavities. By adjusting the angle of the VBGs, wavelength tuning can be achieved. Similarly, dual-wavelength output can be effectively realized by employing two fiber Bragg gratings (FBGs) with distinct central wavelengths [10]. A dual-wavelength synchronously mode-locked fiber laser oscillator based on the dual-branch nonlinear polarization rotation (NPR) technique has been reported in Ref. [11]. The central wavelengths can be tuned within the range of 1537.2 nm–1542.8 nm and 1556.6–1562.2 nm, respectively.

As the wavelength range of laser output is largely determined by the gain medium, the simultaneous utilization of different types of gain fibers within the cavity can directly facilitate the generation of multiwavelength output. In Ref. [12], a passively synchronized dual-color fiber laser system can simultaneously generate the solitons with central wavelengths of 1082.8 nm and 1569.7 nm (see Fig. 5.1(a)). Similarly, a passively synchronized thulium (Tm)- and erbium (Er)-doped fiber (EDF) laser mode-locked by a common graphene saturable absorber (GSA) has been also implemented as shown in Fig. 5.1(b) [13]. Compared to those dual-wavelength mode-locked fiber lasers based on a dual-cavity structure typically sharing a common gain medium, this scheme can expand the output wavelength separation to several hundred nanometers. The dual-cavity structure-based dual-wavelength mode-locked fiber laser typically demonstrates stable output characteristics. While the generated laser beams propagate along distinct branches, it will lead to a reduced coherence between the two output wavelengths.

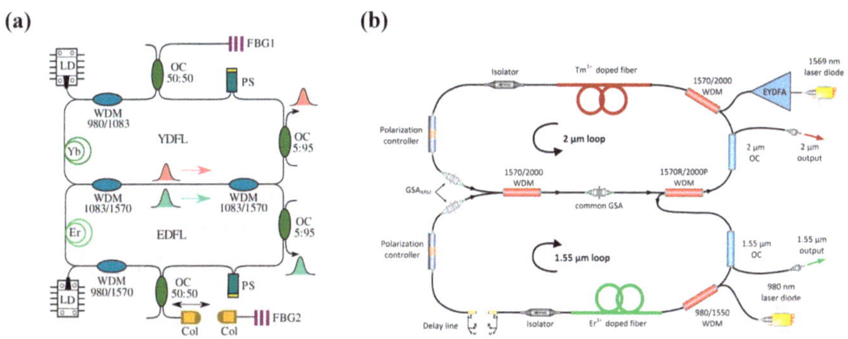

Fig. 5.1 (**a**) Dual-resonator structure with ytterbium-doped fiber and Er-doped fiber [12]. (**b**) Dual-resonator structure with Er-doped fiber and Tm-doped fiber [13]

5.2.2 *Intracavity Filter Effect-Based Multiwavelength Soliton Fiber Laser*

Utilizing cavity filtering effect is one of the commonly employed approaches for achieving multiwavelength multiplexing, in which adjusting the birefringence would be an effective method. The generation of multiwavelength dissipative soliton (DS) in an all-normal-dispersion fiber laser passively mode-locked with a semiconductor saturable absorber mirror (SESAM) was reported in Ref. [14]. The laser structure is depicted in Fig. 5.2(a). Despite the fact that no polarizing components were used in the cavity, the slight residual polarization asymmetry of the components used, such as the SESAM and the circulator, could still cause the formation of a linear artificial birefringent filter in the cavity. Via tuning the strength of the cavity birefringence, stable single-, dual-, and triple-wavelength DSs can be formed in the laser (see Fig. 5.4(b)). Similarly, in an abnormal-dispersion fiber laser, attributing to the nonlinear polarization rotation (NPR)-induced birefringence filtering effect inside the cavity, the fiber laser approaches dual-wavelength working regime, which simultaneously delivers modulation instability (MI) solitons and conventional solitons in different wavebands (see Fig. 5.2(c), (d), and (e)). Particularly, the coexistence of soliton singlets and molecules with dual wavelength is also obtained, as depicted in Fig. 5.2(f) and (g) in which the MI soliton singlets and conventional soliton molecules can be respectively emitted from these two lasing wavebands with asynchronous repetition rates [15].

In addition to the residual intracavity birefringence effects, spectral filtering can also be introduced by birefringence-induced comb filters. By introducing phase delays through a birefringent medium, comb filters are typically utilized within the resonator, typically with Lyot birefringent filters being employed to achieve multiwavelength laser output [16, 17]. A typical dual-wavelength mode-locked fiber laser employing a Lyot filter is demonstrated in Ref. [18]. Using the low birefringence induced by bending the single-mode fiber to form a Lyot filter, a dual-wavelength mode-locked fiber laser was generated, in which the dual-wavelength output centers at 1532 and 1556 nm. Another widely used approach for the formation of Lyot filter involves incorporating polarization-maintaining fiber (PMF) within the cavity. Liu S et al. [19] proposed a fiber laser based on a Lyot filter, as illustrated in the structural schematic in Fig. 5.3(a). By adjusting two PCs, the fiber laser was able to generate a stable and flat multiwavelength spectrum, with up to 17 wavelengths within the 1982–1998 nm spectral range. The tunability of the laser was achieved by varying the effective length of the intracavity polarization-maintaining fiber (PMF), enabling the tuning of the wavelength spacing of the laser output (see Fig. 5.3(b) and (c)). Xin Lingyi et al. [20] proposed a tunable and switchable multiwavelength fiber laser incorporating a Lyot filter and a nonlinear optical loop mirror (NOLM), in which the NOLM was employed to suppress mode competition in the erbium-doped fiber (EDF). The structural schematic is shown in Fig. 5.3(d). By adjusting the PCs, the number of wavelengths can be tuned as illustrated in Fig. 5.3(e) and (f).

Moreover, multiwavelength soliton output can be effectively realized through the cascading of Lyot filters. For example, Li Yuan et al. [21] proposed a tunable and

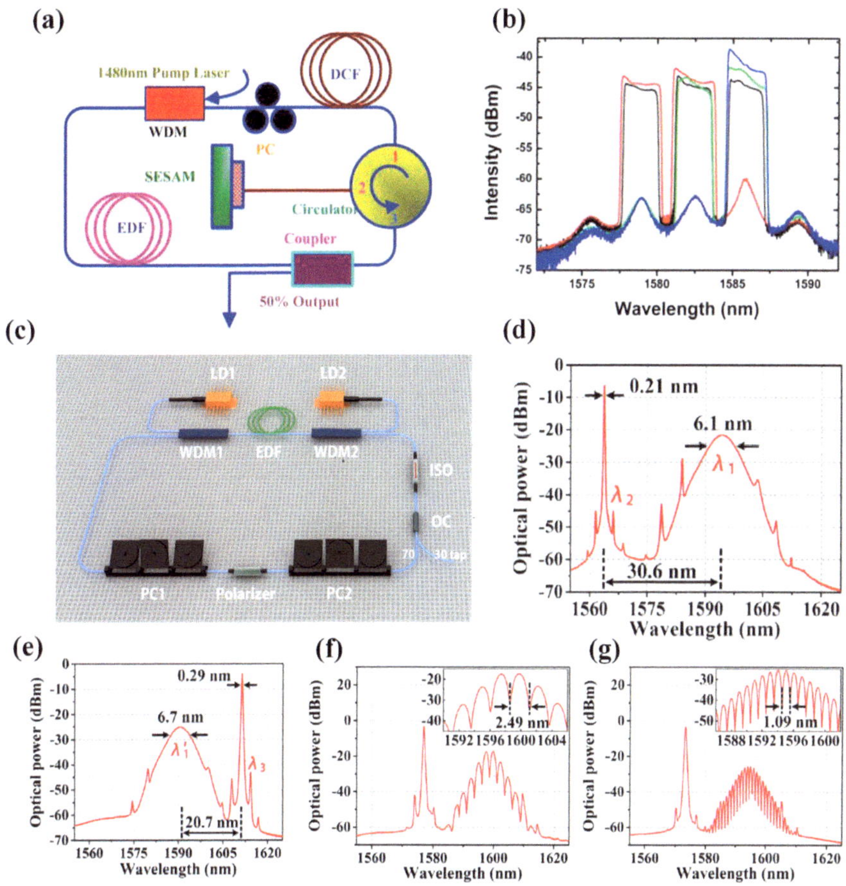

Fig. 5.2 (**a**) Schematic of the experimental setup of the SESAM-based passively mode-locked fiber laser. (**b**) Single/dual/triple-wavelength spectra obtained through rotating PCs but kept the pump strength fixed [14]. (**c**) Schematic diagram of the NPR-based dual-wavelength mode-locked fiber laser. (**d–e**) Optical spectra of the coexistence of modulation-instability (MI) solitons and conventional solitons. (**f–g**) Coexistence of soliton singlets and molecules with dual wavelength [15]

switchable fiber laser based on a two-stage Lyot filter and a NOLM that induces the IDL effect. The structural schematic is shown in Fig. 5.3(g). This laser enables independent tuning of multiple parameters, such as adjusting the wavelength spacing of the output, the number of central wavelengths, and the spectral output range. As in parallel depicted in Fig. 5.3(h), by adjusting the angle of the PCs in the Lyot filter, the wavelength spacing of the output can be switched between 0.53 nm, 0.78 nm, and 1.56 nm. Meanwhile, under the influence of the birefringence effect in the NOLM loop, the IDL modulation can control the number of central wavelengths and the spectral output range of the multiwavelength laser by adjusting the pump power. The mode-locked fiber laser based on birefringence-induced Lyot filtering

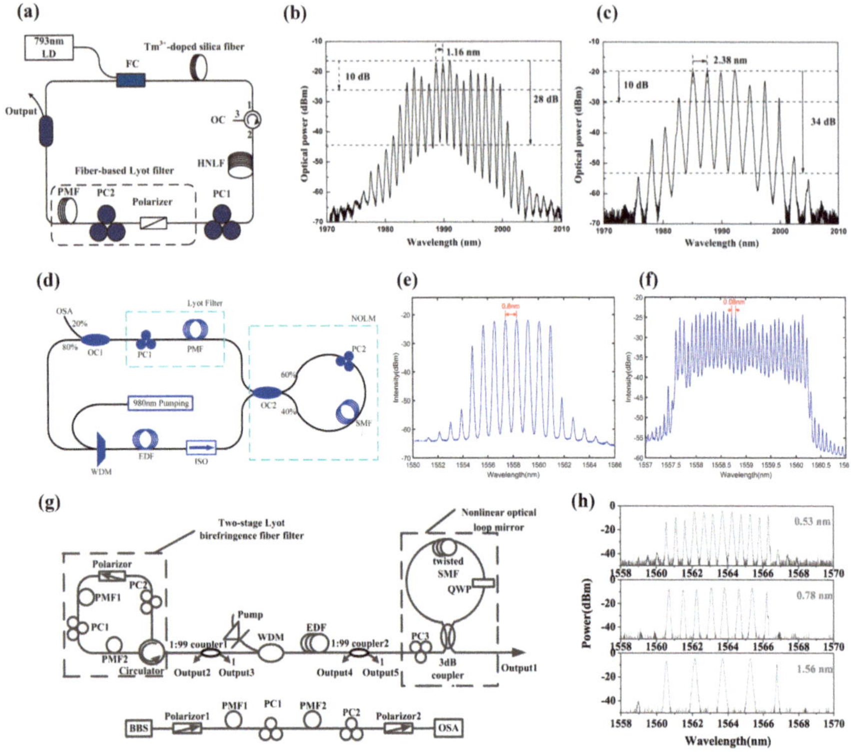

Fig. 5.3 (**a**) Schematic diagram of the dual-wavelength mode-locked fiber laser employing a Lyot filter by incorporating PMF within the cavity. (**b**) Output spectrum of (a) with a spectral spacing of 1.16 nm. (**c**) Output spectrum of (a) with a spectral spacing of 2.38 nm [19]. (**d**) Tunable and switchable multiwavelength fiber laser incorporating a Lyot filter and a NOLM. (**e**) Output spectrum of (d) with a spectral spacing of 0.8 nm. (**f**) Output spectrum of (d) with a spectral spacing of 0.08 nm [20]. (**g**) Tunable and switchable fiber laser based on a two-stage Lyot filter and a NOLM. (**h**) Output spectrum of (g) [21]

effects presents advantages such as low cost and low loss, making it a highly efficient solution for multiwavelength generation.

5.2.3 2D Material-Based Multiwavelength Soliton Fiber Laser

To further simplify the experimental setup, multiwavelength mode-locked lasers based on nonlinear effects have been developed. Based on such nonlinear effects, power is redistributed among different wavelengths, effectively suppressing mode competition caused by the uniform broadening of the gain fiber, thereby enabling multiwavelength pulse output. Since some 2D materials exhibit saturated absorption characteristics and high nonlinearity over a wide wavelength range, they can be used to achieve multiwavelength mode locking. Researchers have further utilized the high nonlinearity of graphene and graphene oxide to achieve multiwavelength Q-switching [22] and mode-locked pulses [23, 24] in fiber lasers. By simultaneously exploiting the graphene-induced nonlinear FWM and its super-broadband saturable absorption, the multiwavelength Q-switched Er 3+ -doped fiber laser is realized as demonstrated in Fig. 5.4(a–c) [25]. Besides, Zhao et al. employed graphene as a saturable absorber for mode locking and generated dual-wavelength pulses in an ytterbium-doped fiber laser, with central wavelengths of 1061.8 nm and 1068.8 nm [26]. The structural schematic is shown in Fig. 5.4(d), and the CW laser operation and mode-locked spectrum are respectively demonstrated in Fig. 5.4(e) and (f). It should be noted that graphene, as a mode-locking device, exhibits many advantages. However, there are two shortcomings: zero bandgap and weak absorption (the absorption coefficient for a single layer is only 2.3%), which severely limits its optical modulation capability and its application in fields involving strong light-matter interactions [27].

In addition to graphene, various other types of 2D materials, including topological insulators [28, 29], transition metal dichalcogenides [30], and black phosphorus [31], have also been explored for the development of multiwavelength ultrafast lasers. Bo Guo et al. experimentally demonstrated a multiwavelength ultrafast erbium-doped fiber laser incorporating a μm-scale topological insulator: Bi2Se3/polyvinyl alcohol film as both an excellent saturable absorber for mode locking and a high-nonlinear medium to induce a giant third-order optical nonlinear effect for mitigating the mode competition of erbium-doped fiber laser and stabilizing the multiwavelength oscillation [32]. The SEM image of the topological insulator: Bi2Se3/polyvinyl alcohol film is demonstrated in Fig. 5.5 (a). Through properly adjusting the PCs and the pump power, four-wavelength mode-locking performance can be achieved, of which the spectrum and temporal waveform are depicted in Fig. 5.5 (b) and (c). As for transition metal dichalcogenides, the WS_2-deposited microfiber can operate simultaneously as a mode locker and a high-nonlinear device, therefore facilitating the operation of dual-wavelength solitons within the resonator as shown in Fig. 5.5(d–f) [33]. Besides, black phosphorus possesses a unique structure and thickness-dependent direct bandgap, making it highly promising for the field of multiwavelength ultrafast lasers. Zhao et al. realized a tri-wavelength synchronously mode-locked erbium-doped fiber laser with black phosphorus as demonstrated in Fig. 5.5(g–i) [34]. The aforementioned studies suggest that 2D materials have widespread applications in multiwavelength mode-locked lasers, which offer advantages such as simple and compact structures.

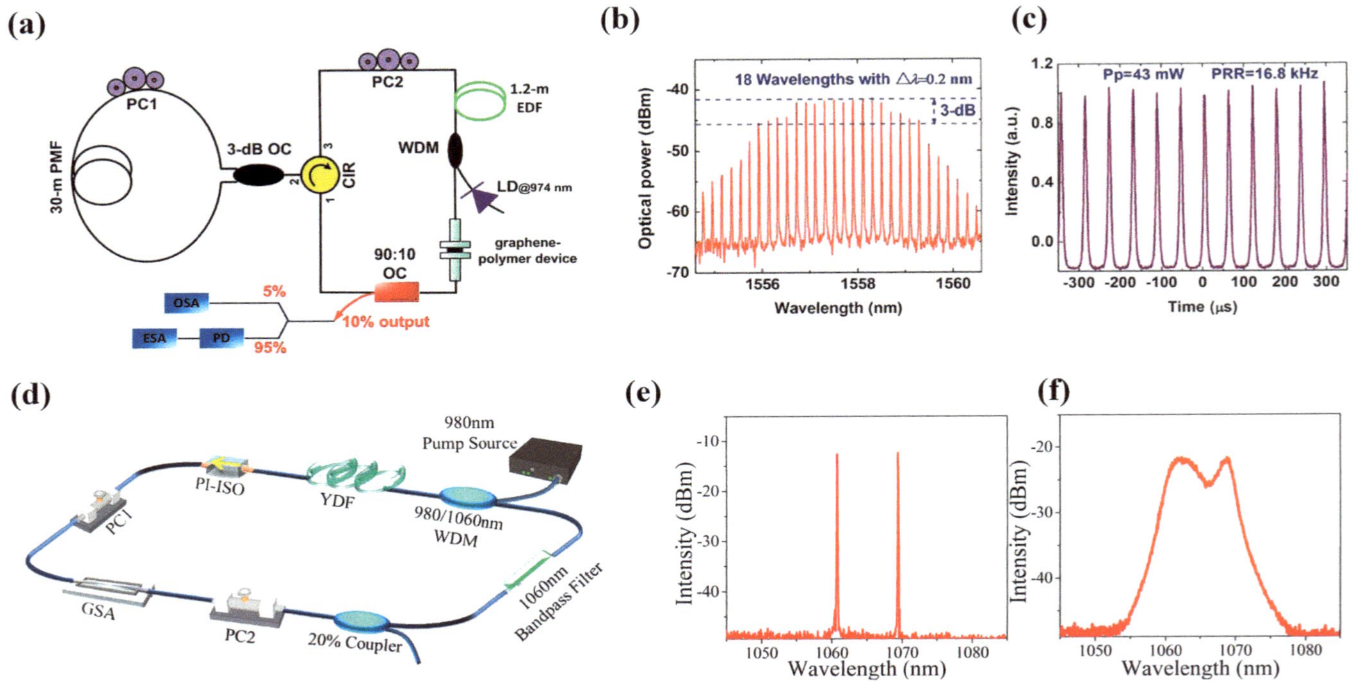

Fig. 5.4 (**a**) Structural schematic of the multiwavelength Q-switched Er 3 + -doped fiber laser based on the graphene. (**b**) and (**c**) are the spectrum and temporal waveform of the laser in (a) [25]. (**d**) Graphene-based dual-wavelength mode-locked fiber laser. (**e**) and (**f**) are CW laser operation and mode-locked spectrum of the graphene-based dual-wavelength mode-locked fiber laser [26]

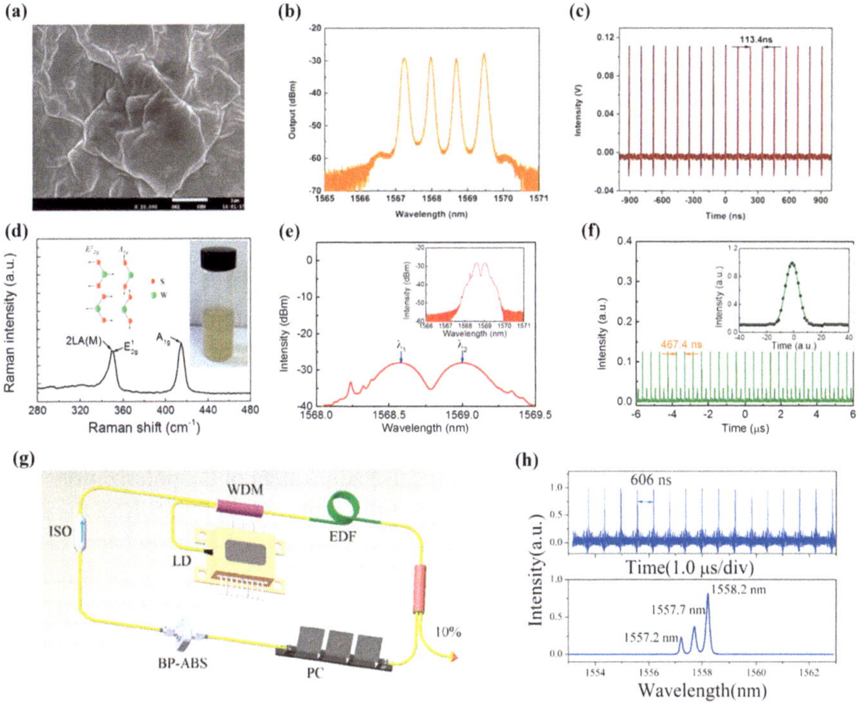

Fig. 5.5 (**a**) SEM image of TI: Bi2Se3/PVA film. (**b**) and (**c**) are the spectrum and pulse train of four-wavelength mode locking [32]. (**d**) Raman spectrum of the WS_2 nanosheets. (**e**) and (**f**) are the spectrum and pulse train of dual-wavelength mode locking based on the WS_2 [33]. (**g**) Schematic of mode-locked fiber laser based on black phosphorus-SA. (**h**) Three-wavelength mode-locking output of (g) [34]

5.3 Bidirectional Multiplexed Dissipative Soliton Fiber Lasers

The bidirectional mode-locked fiber laser is a laser configuration in which no optical isolator is incorporated into the resonant cavity, enabling the light to propagate simultaneously in both clockwise (CW) and counterclockwise (CCW) directions within the cavity. The bidirectional mode-locked fiber laser has attracted widespread attention in recent years due to its ability for multiplexing and the complex output characteristics resulting from soliton collisions. In recent years, bidirectional mode-locked fiber lasers have been successfully realized with a variety of gain media. In the following, we will provide a categorization and detailed introduction based on the different types of gain fibers.

5.3.1 Er-Doped Bidirectional Mode-Locked Fiber Laser

The erbium-doped bidirectional mode-locked fiber laser is the most thoroughly researched and widely investigated configuration. Bidirectional mode locking can be achieved across different dispersion regimes. In the anomalous dispersion regime, by utilizing a four-port circulator, Chunmei Ouyang et al. integrate two distinct semiconductor saturable absorber mirrors (SESAMs) into the laser cavity as demonstrated in Fig. 5.6(a), thereby facilitating bidirectional mode locking. Two countercirculating pulse trains with repetition rates of 21.3 and 15.2 MHz are obtained simultaneously. Temporal and spectral profiles coming out of the bidirectional mode-locked laser are respectively plotted in Fig. 5.6(b), in which the prominent Kelly sidebands indicate that solitons in both directions reside within the anomalous dispersion regime [35]. Besides, in Ref. [36], Y. Li et al. demonstrated a scheme for achieving bidirectional mode locking using CNT as the mode-locking element, shown in Fig. 5.7(a). Through precise adjustment of the polarization state, stable bidirectional mode locking is successfully achieved. In Fig. 5.7(b), the spectra of both the CW and CCW solitons are depicted, with the Kelly sidebands clearly indicating that conventional solitons are formed. Single-shot spectral evolutions of the CW and CCW operations are shown in Fig. 5.7(c), which as well demonstrates the synchronization between them. In Ref. [37], a bidirectional passively mode-locked soliton laser with graphene-based saturable absorbers is presented, in which a four-port circulator and two graphene-based saturable absorbers are inserted. Moreover, bidirectional soliton output can also be generated in mode-locked lasers based on multimode interference. J. Lin et al. reported an all-fiber bidirectional mode-locked laser based on nonlinear multimode interference [38], of which the configuration is shown in Fig. 5.7(d). The laser can generate asynchronous pulse trains in different directions with a repetition rate difference of 7 Hz; the pulse train and spectrum of CW and CCW solitons are respectively plotted in Fig. 5.7(e) and (f).

In normal-dispersion regime, bidirectional mode locking can be also achieved. For instance, K. Yang et al. realized the bidirectional operation of dissipative

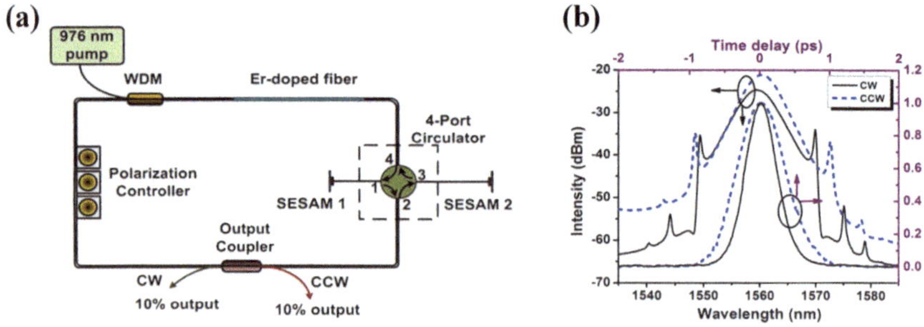

Fig. 5.6 (**a**) Schematic diagram of the SESAM-based erbium-doped bidirectional mode-locked fiber laser. (**b**) Spectrum of CW and CCW soliton [35]

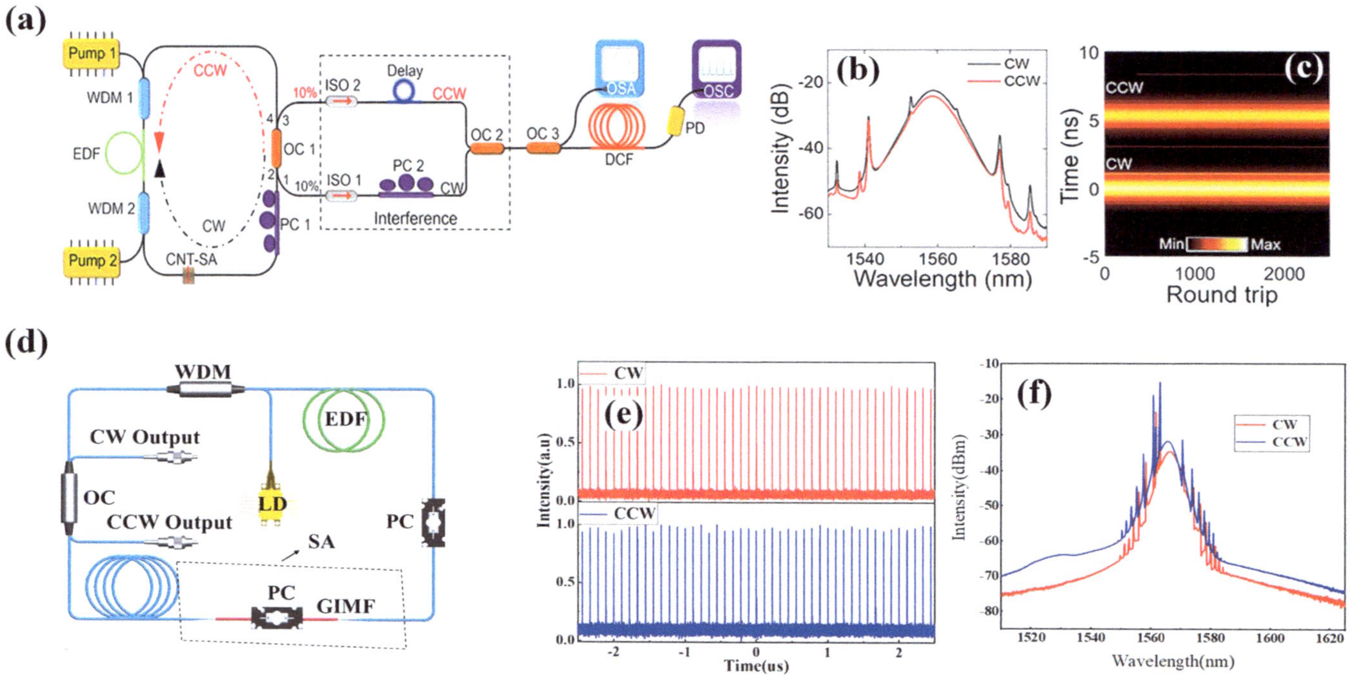

Fig. 5.7 (**a**) Schematic diagram of the CNT-based erbium-doped bidirectional mode-locked fiber laser. (**b**) Spectrum of CW and CCW soliton. (**c**) Single-shot spectral evolutions of the CW and CCW operations. (**d**) Configuration of the bidirectional mode-locked fiber laser based on the nonlinear multimode interference. (**e**) and (**f**) are temporal waveforms and spectrum of CW and CCW solitons

solitons within a cavity with the normal dispersion [39, 40]. A carbon nanotube (CNT) saturable absorber SA is utilized in the cavity as shown in Fig. 5.8(a), and the mode-locked spectra of the CW and CCW solitons are presented in Fig. 5.8(b) [31]. It is noteworthy that by carefully controlling the linear birefringence within the cavity, periodic polarization transitions between the two counterpropagating solitons can be distinctly observed. Similarly, Y. Zhou et al. shed insight into dynamics of dissipative soliton in erbium-doped bidirectional mode-locked fiber laser, and a breathing soliton pair with energy transfer is observed [41]. Besides, they reported on the first observation of a breathing dissipative soliton explosion in a net-normal-dispersion bidirectional ultrafast fiber laser [42].

In bidirectional ultrafast fiber lasers, apart from the soliton singlets, soliton molecules can be also generated. In Ref. [43], to avoid the adverse effect of the gain competition on the bidirectional mode-locking operation, a multiplexed fiber ring laser consisting of a bidirectional main cavity and two counterpropagating branches inserted with EDFs is constructed, as shown in Fig. 5.9(a). By properly adjusting the pump power and PCs, the counterpropagating soliton molecules can be obtained, of which the optical spectrum of CW and CCW soliton molecules is respectively shown in Fig. 5.9(b) and (c). Besides, L. Li et al. proposed and demonstrated an ultracompact, bidirectional passively mode-locked soliton Er-fiber laser which outputs soliton molecules, which had a near-zero net cavity dispersion with a simple linear cavity, and was mode-locked by a CNT-based saturable absorber (SA) [44]. Besides, soliton singlets and soliton molecules can be simultaneously generated in the bidirectional soliton laser [45]. In this work, soliton singlets and soliton molecules can be simultaneously generated in CW and CCW direction, as demonstrated in Fig. 5.9(d–f). The central wavelengths of the CW and CCW pulses can be synchronously tuned over a range of approximately 24 nm, while maintaining a relative tuning, facilitated by the bidirectional Lyot filter within the cavity.

It is also worth noting that, as a special type of chaotic pulse, noise-like pulses (NLPs) can also be generated in bidirectional mode-locked lasers. In Ref. [46], Y. Luo et al. develop a bidirectional NLP fiber laser mode-locked by a transmitted semiconductor saturable absorber (SESA) as depicted in Fig. 5.10(a). With appropriate settings, conventional solitons and NLPs can be both generated in this cavity, the optical spectrum of which is in parallel plotted in Fig. 5.10(b) and (c). Particularly, the evolution from CSs toward NLPs is experimentally observed in both the CW and the CCW directions, which enriches the dynamic exploration of bidirectional mode-locked soliton lasers.

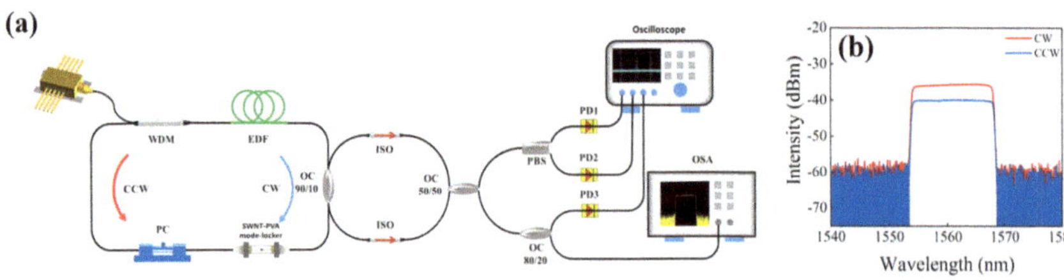

Fig. 5.8 (**a**) Schematic diagram of the CNT-based erbium-doped net-normal-dispersion bidirectional ultrafast fiber laser. (**b**) Spectrum of CW and CCW soliton [39]

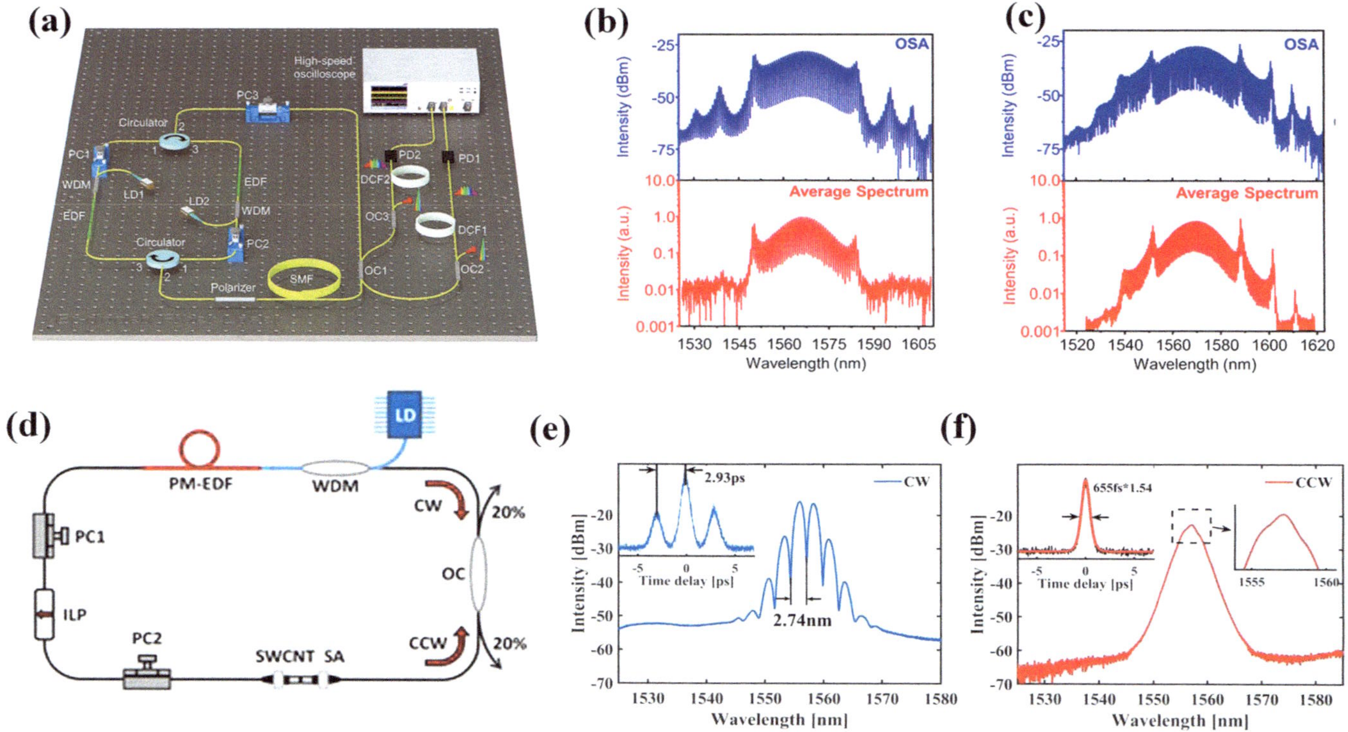

Fig. 5.9 (**a**) Schematic diagram of a multiplexed fiber ring laser consisting of a bidirectional main cavity and two counterpropagating branches. (**b**) Optical spectrum of a CW soliton molecule. (**c**) Optical spectrum of a CCW soliton molecule [43]. (**d**) Schematic diagram of the bidirectional mode-locked fiber laser with a Lyot filter. (**e**) and (**f**) are the optical spectrum emitted from the bidirectional mode-locked fiber laser in (d) [44]

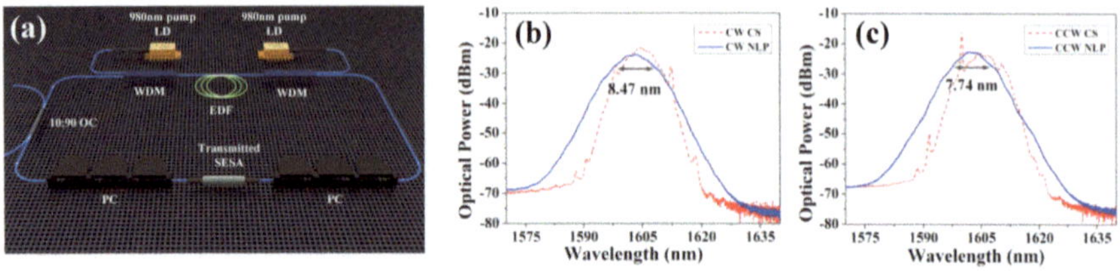

Fig. 5.10 (**a**) Schematic diagram of the SESA-based erbium-doped bidirectional ultrafast fiber laser. (**b**) and (**c**) are the optical spectrum of conventional solitons and NLPs emitted from the bidirectional ultrafast fiber laser in two directions [46]

5.3.2 Yb-Doped Bidirectional Mode-Locked Fiber Laser

The ytterbium-doped fiber laser, which is capable of emitting near-infrared laser, has also demonstrated the realization of bidirectional soliton output in recent years. In Ref. [47], the bidirectional all-normal-dispersion laser based on NPR was experimentally demonstrated, the schematic diagram of which is depicted in Fig. 5.11(a). A 2-m double-cladding ytterbium-doped fiber, pumped by a multimode 980-nm laser, was used as the gain medium. The pump was launched into the DC-YDF through a pump combiner that had 0.3-m double-cladding fiber pigtails on both ends. All other optical fiber components within the cavity are interconnected by HI 1060 fibers. With appropriate settings, bidirectional dissipative soliton output can be achieved, of which the spectrum and temporal waveform are respectively shown in Fig. 5.11(b) and (c). This all-normal-dispersion bidirectional soliton possesses higher pulse energy, therefore facilitating practical dual-comb applications. Also, in this laser, N. Prakash introduced a reference-free relative timing jitter characterization technique that enables studies to gain deeper insights into the relative timing jitter characteristics of bidirectional all-normal-dispersion soliton lasers [48]. In addition, In Ref. [49], utilizing a SMF-GIMF-SMF structure as the saturable absorber, a single-cavity bidirectional Raman-assisted mode-locked fiber laser was proposed. The variations in the repetition rate difference of the bidirectional pulses are induced by changes in dispersion and nonlinear effects. NLPs have also been observed in the bidirectional ytterbium-doped mode-locked laser [50, 51]. In Ref. [50], Y. Lu et al. proposed a bidirectional mode-locked fiber laser to generate cylindrical vector beam (CVB) asynchronous pulses based on a graded index multimode fiber, in which the NLPs can be generated in CW and CCW directions.

5.3.3 Thulium-Doped Bidirectional Mode-Locked Fiber Laser

Thulium-doped lasers possess a wide tunable spectral bandwidth, covering the range from 1800 to 2100 nm. It is considered an ideal light source for applications in medicine, eye safety, ultrafast optics, short-range remote sensing, and

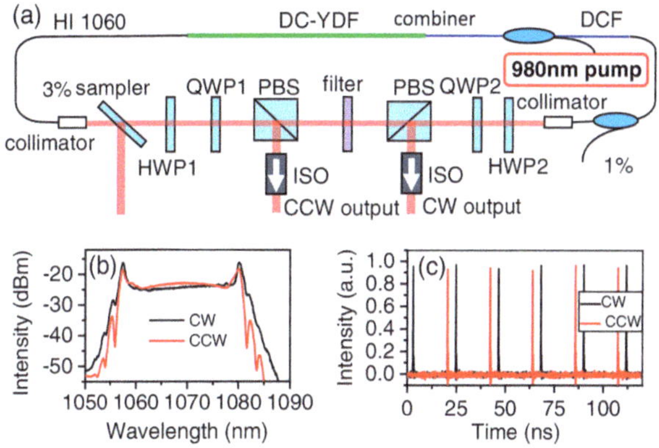

Fig. 5.11 (**a**) Experimental setup of the bidirectional mode-locked all-normal-dispersion laser. (**b**) Optical spectra and (**c**) oscilloscope trace of mode-locked pulses in CW and CCW direction [47]

biology. Recently, research on thulium-doped bidirectional mode-locked fiber lasers has also been gradually developing [52, 53]. In Ref. [54], N. Abdukerim et al. reported a bidirectional mode-locked thulium-doped fiber laser, of which the mode locking is enabled by the combination of SESAM saturable absorption and NPR (see Fig. 5.12(a)). Stable bidirectional mode locking with mutual coherence is achieved, of which the spectrum of CW and CCW soliton is plotted in Fig. 5.12(b). The output wavelength is tunable within a 35-nm range around a central wavelength of 1.917 μm. Furthermore, the repetition rates for each direction can be independently manipulated by varying the fiber lengths in the cavity's subpaths. Another work demonstrated an all-fiber bidirectional ultrafast thulium-doped fiber laser at 2 μm with a single-wall carbon nanotube presented as saturable absorber as shown in Fig. 5.12(c) [55]. Figure 5.12(d) and (e) shows the measured spectral characteristics of the bidirectional mode-locking operation, in which the Kelly sidebands are distinct. It is noteworthy that by adjusting the pump power, the bidirectional repetition frequency difference between CW and CCW solitons can be easily changed. In Ref. [56], based on the nonlinear polarization evolution (NPE), N. Li et al. experimentally demonstrated a stable unidirectional Tm^{3+}-doped mode-locked fiber laser in a bidirectional oscillation setup as depicted in Fig. 5.12(f). Mode-locking operation is achieved by adjusting the waveplate orientation to get appropriate cavity loss. Optical spectrum of the CW and CCW directions is respectively demonstrated in Fig. 5.12(g) and (i). For the CW solitons, the central wavelength is 1979.703 nm and the measured 3-dB spectral bandwidth is 6.9 nm, while for the CCW solitons, the central wavelength is 1979.258 nm and the measured 3-dB spectral bandwidth is 6.1 nm.

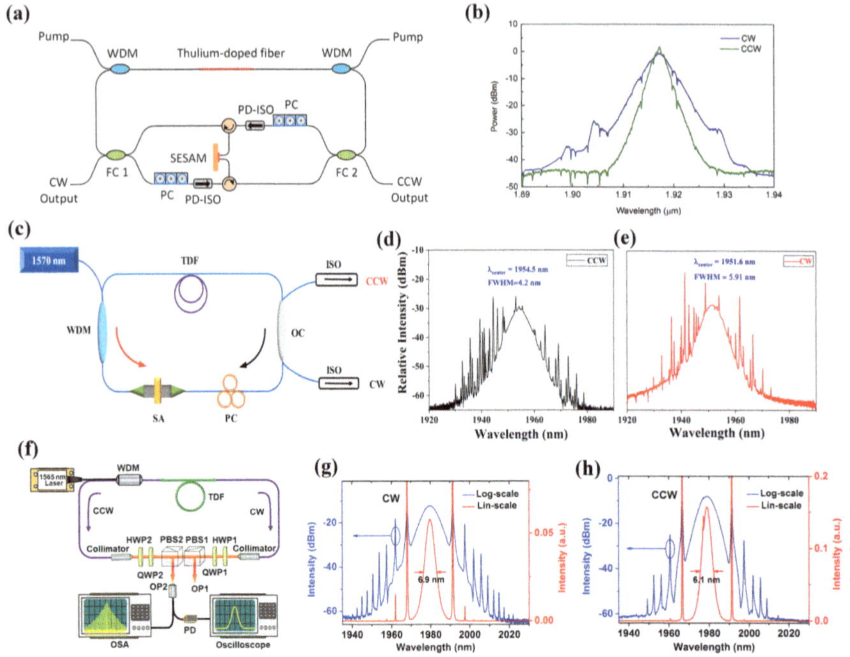

Fig. 5.12 (**a**) Schematic of the bidirectional mode-locked thulium-doped fiber laser based on the combination of SESAM saturable absorption and NPR. (**b**) Optical spectrum of the bidirectional mode-locked thulium-doped fiber laser in (a) [54]. (**c**) Single-wall carbon nanotube-based bidirectional mode-locked thulium-doped fiber laser. (**d**) and (**e**) are the CW and CCW spectrum [55]. (**f**) NPE-based bidirectional mode-locked thulium-doped fiber laser. (d) and (e) are the CW and CCW spectrum of the bidirectional mode-locked thulium-doped fiber laser in (f) [56]

5.4 Multistate Dissipative Soliton Fiber Lasers

In 1994, R. J. Deissler conducted theoretical research on the periodicity, quasi-periodicity, and local chaotic solutions of the Ginzburg-Landau equation, confirming that soliton pulses can exist in different states during their oscillation process in the laser cavity [57]. This opened the door to multistate soliton fiber lasers. In recent years, lasers capable of emitting multiple states have attracted increasing attention as a focal point of research owing to their multiplexing capability.

Through the innovative design of the resonator cavity, it becomes feasible for a single cavity to generate solitons in a variety of distinct forms. For instance, in Ref. [58], Y. Cui et al. ingeniously employed two three-port circulators to achieve bidirectional operation of a passively mode-locked fiber laser as shown in Fig. 5.13(a). By implementing different dispersion management in the two opposing transmission directions, a structure with net positive dispersion in the CW direction and net negative dispersion in the CCW direction is formed. This allows for the simultaneous output of dissipative soliton pulses and conventional soliton pulses within a single laser (see Fig. 5.13(b) and (c)). Similarly, B. Liu et al. proposed a NPR-based

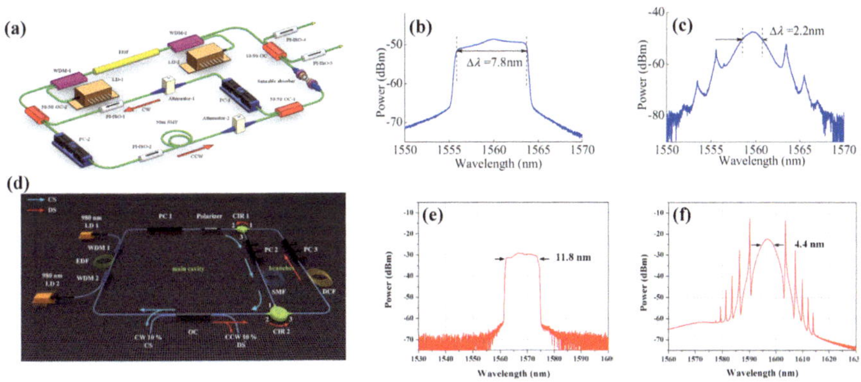

Fig. 5.13 (**a**) Schematic of the mode-locked fiber laser emitting DS and conventional soliton. (**b**) and (**c**) are the optical spectrum generated in (a) [58]. (**d–f**) NPR-based mode-locked fiber laser which can emit DS and conventional soliton [60]

mode-locked fiber laser consists of one main cavity and two counterpropagating branches with different dispersion distributions as demonstrated in Fig. 5.13(d) [59]. Consequently, stable conventional solitons and dissipative solitons are respectively formed in the CW and CCW directions of the same lasing oscillator (see Fig. 5.13(e) and (f)). Furthermore, B. Liu et al. proposed the multistate soliton generation in a chirped fiber Bragg grating (CFBG)-based multistate mode-locked fiber laser [60]. In particular, this cavity is designed to introduce reverse dispersion distributions in counterpropagation directions through incorporating the bidirectional reflection of the CFBG into the lasing oscillator. Similarly, it also enables the simultaneous output of traditional solitons and dissipative solitons. Besides, another advanced method was proposed by introducing convex-concave phase and spectral filtering with a programmable pulse shaper in the resonator, and the multicolor soliton compounds composed of chirp-free conventional solitons and chirped dissipative solitons can be produced [61].

Moreover, stretched pulse and dissipative soliton can be also simultaneously delivered in a single cavity [62, 63]. As the Er-doped fiber owns two gain peaks, broadband dual-wavelength pulses can be achieved. In Ref. [62], the weak third-order dispersion (TOD) promotes the generation of two pulses with distinct dispersion patterns—one being a stretched pulse and the other a dissipative soliton within the near-zero dispersion regime. In Ref. [63], a hybrid mode-locked fiber laser co-doped with thulium and holmium is experimentally demonstrated, capable of operating in two distinct modes. One mode involves switching between a stretched pulse and a dissipative soliton, while the other features dual-wavelength mode locking, where both the stretched pulse and dissipative soliton coexist simultaneously.

In Ref. [64], Y. Luo et al. proposed a scheme for generating multistate soliton outputs utilizing the breathing dynamics within the resonator cavity. A multioutput fiber laser mode-locked by NPR in net normal-dispersion regime is reported, of which the structure is demonstrated in Fig. 5.14(a). Four output ports are inserted

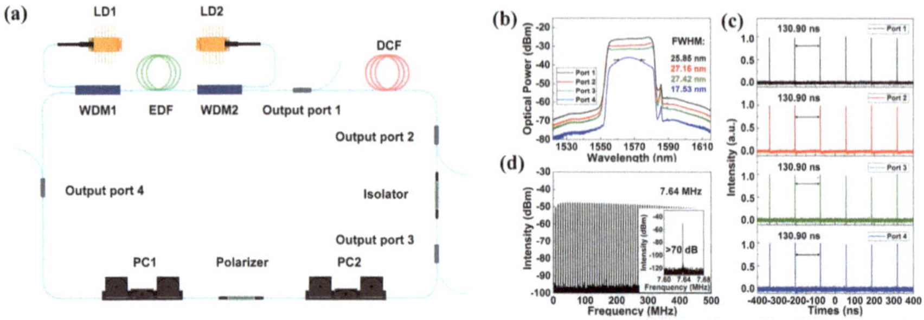

Fig. 5.14 (**a**) Schematic of the multioutput fiber laser mode-locked by NPR in net-normal-dispersion regime. (**b**) Optical spectrum output from different ports. (**c**) Oscilloscope traces from these output ports. (**d**) RF of the fundamental mode-locking state at port 1 [56]

into the cavity. As the asymmetry settings of the laser parameters can introduce the spectral reshaping and pulse breathing process, different spectral shapes from four output ports are observed as shown in Fig. 5.14(b). Both the transformation from dissipative solitons to similaritons and periodic stretching and compression of pulse widths are observed. Oscilloscope traces from these output ports and the RF of the fundamental mode-locking state at port 1 are respectively plotted in Fig. 5.14(c) and (d).

Once the constraints of scalar solitons are broken, vector solitons, with their rich dynamical phenomena, have gradually become a focus of research. In Ref. [65], Y. Xiang et al. innovatively proposed a scheme for the multiplexing of vector solitons with scalar solitons. In a bidirectional mode-locked fiber laser based on NPR, the simultaneous generation of scalar and vector solitons is realized (see Fig. 5.15(a)). Under bidirectional pumping, as a segment of PMF is placed into one of the branches, the strong local birefringence induced by PMF facilitates the fast polarization evolution, thus leading to the formation of vector solitons. In another branch, a segment of SMF is inserted, and the polarization discrimination effect based on NPR dominates the evolution of the linearly polarized pulse and limits the formation of vector solitons, the optical spectrum of which is demonstrated in Fig. 5.15(b). To validate the vector feature, the vector solitons are injected into a polarization beam splitter (PBS) together with a PC. By adjusting the PC3, the orientation of the polarization projection basis can be suitably operated. And the output laser field is projected onto two orthogonal polarization axes. Figure 5.15(c) depicts the total spectrum and the spectra in horizontal and vertical axis under two different states of the PC3. Although at the same central wavelength, the spectral intensity of two orthogonal polarization axes is different. This work realized the multiplexing of scalar and vector solitons, paving potential application scenarios of pulsed fiber lasers.

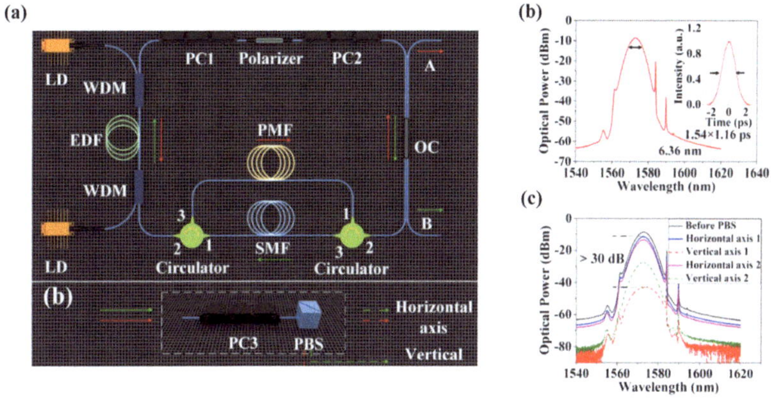

Fig. 5.15 (**a**) Schematic of the mode-locked fiber laser multiplexing the scalar and vector solitons. (**b**) Optical spectrum of scalar solitons. (**c**) Optical spectrum of vector solitons and the separate spectrum in two orthogonal polarization axes [65]

5.5 Conclusion and Outlook

In this chapter, we systematically summarize several technical approaches for generating multiplexed soliton fiber lasers. A variety of approaches to realize wavelength multiplexing, bidirectional multiplexing, and multistate multiplexing are respectively discussed. In addition to the aforementioned approaches, polarization-multiplexed dissipative soliton fiber laser is also a widely adopted multiplexing technique. As solitons propagate through the birefringent fiber, they split into two solitons with orthogonal polarization states, allowing each polarization state to be utilized independently. In Ref. [66], X. Zhao et al. investigated a polarization-multiplexed passively mode-locked fiber laser with intracavity birefringence. Dual asynchronous pulses with approximately orthogonal polarization states were obtained, exhibiting a pulse repetition frequency difference of several hundred hertz. This polarization-multiplexed single-cavity dual-comb laser can be applied in scenarios requiring a dual-comb system solution.

Moreover, the versatile intracavity dynamics provide a novel framework for exploring multiplexing in soliton fiber lasers. Coincidentally, emerging time-stretch dispersion Fourier transform (TS-DFT) technique offers powerful support for exploring the internal dynamics [67, 68]. Recently, the artificial manipulation of the optical degrees of freedom of solitons allows the laser to output different types of solitons on demand. For instance, In Ref. [69], Y. Liu et al. realized the phase tailoring of soliton molecule through programmable gain control. In Ref. [70], S. Liu et al. implemented the on-demand harnessing of the temporal spacing between soliton molecules via manipulating the second-order group-velocity dispersion and dispersion losses. In Ref. [71], via gain control, spectral period doubling of dissipative optical solitons can be realized. In these works, soliton fiber lasers can switch between multiple states and achieve customized encoding, thereby expanding their potential applications in optical communications and optical storage.

In conclusion, it is evident that a wide array of emerging multiplexed lasers is rapidly proliferating, providing numerous possibilities for their widespread applications. Particularly, high stability, multichannel functionality, and tunability are poised to be the defining directions for the future evolution of multiplexed soliton fiber laser.

References

1. Zabusky, N.J., Kruskal, M.D.: Interaction of 'solitons' in a collisionless plasma and the recurrence of initial states. Phys. Rev. Lett. **15**, 240–243 (1965)
2. Grelu, P., Akhmediev, N.: Dissipative solitons for mode-locked lasers. Nat. Photonics. **6**, 84–92 (2012)
3. Ilday, F.Ö.: Ultrafast burst-mode fiber lasers: source development and material processing. In: European Conference on Lasers and Electro-Optics 2015, OSA Technical Digest (Online), p. paper STh3L.5. Optica Publishing Group (2015)
4. Li, Z., Han, Y.Q., Wu, L.C., Zang, Z.H., Dai, M.L., Set, S., Yamashita, S., Li, Q., Fu, H.Y.: Towards an ultrafast 3D imaging scanning LiDAR system: a review. Photon. Res. **12**, 1709–1729 (2024)
5. Pang, M., He, W., Jiang, X., St, P., Russell, J.: All-optical bit storage in a fibre laser by optomechanically bound states of solitons. Nat. Photonics. **10**, 454–458 (2016)
6. Jiang, Y.F., Wu, J.M., Hao, Q., Huang, K., Zeng, H.P.: Experimental study on all-polarization-maintaining passive synchronization for dual-color mode-locked fiber lasers. Acta Opt. Sin. **40**, 0936001 (2020)
7. Zhao, X.L., Dong, M.L., Zhang, Y.M., Luo, F., Zhu, L.Q.: Switchable multi-wavelength and tunable wavelength spacing erbium-doped fiber laser based on a phase-shifted fiber Bragg grating combined with a Mach-Zehnder interferometer. Opt. Laser Technol. **112**, 500–507 (2019)
8. Liu, M., Zhao, N., Liu, H., Zheng, X.W., Luo, A.P., Luo, Z.C., Xu, W.C., Zhao, C.J., Zhang, H., Wen, S.C.: Dual-wavelength harmonically mode-locked fiber laser with topological insulator saturable absorber. IEEE Photon. Technol. Lett. **26**, 983–986 (2014)
9. Bai, Y.B., Xiang, W.H., Zu, P., Zhang, G.Z.: Tunable dual-wavelength passively mode-locked Yb-doped fiber laser using SESAM. Chin. Opt. Lett. **10**, 111405 (2012)
10. Dong, L.L., Xie, F., Ma, S., Wang, Y.Z., Chen, L.: Simple tunable dual-wavelength fiber laser and multiple self-mixing interferometry to large step height measurement. Opt. Express. **24**, 21880–21885 (2016)
11. Wang, S., Zhao, Z.G., Kobayashi, Y.: Wavelength-spacing controllable, dual-wavelength synchronously mode locked Er:Fiber laser oscillator based on dual-branch nonlinear polarization rotation technique. Opt. Express. **24**, 28228–28238 (2016)
12. Zeng, J., Li, B.W., Hao, Q., Yan, M., Huang, K., Zeng, H.P.: Passively synchronized dual-color mode-locked fiber lasers based on nonlinear amplifying loop mirrors. Opt. Lett. **44**, 5061–5064 (2019)
13. Sotor, J., Sobon, G., Tarka, J., Pasternak, I., Krajewska, A., Strupinski, W., Abramski, K.M.: Passive synchronization of erbium and thulium doped fiber mode-locked lasers enhanced by common graphene saturable absorber. Opt. Express. **22**, 5536–5543 (2014)

14. Zhang, H., Tang, D.Y., Wu, X., Zhao, L.M.: Multi-wavelength dissipative soliton operation of an erbium-doped fiber laser. Opt. Express. **17**, 12692–12697 (2009)
15. Liu, B.W., Liu, Y.S., Luo, Y.Y., Xiang, Y., Shum, P.P., Tang, X.H., Liu, D.M., Sun, Q.Z.: Coexistence of soliton singlets and molecules in a dual-wavelength mode-locked fiber laser. Opt. Commun. **457**, 124700 (2020)
16. Wang, B.S., Cao, Z.G., Ai, F., Chen, B., Zhen, S.L., Shi, X.Y., Yu, B.L.: Switchable quadruple-wavelength erbium-doped fiber ring laser based on two-segment Lyot-Sagnac filter. Opt. Laser Technol. **42**, 357–361 (2010)
17. Zhang, Z.X., Liang, P.S., San, M.H., Ye, Z.Q.: Wavelength-spacing switchable multiwavelength fiber lasers based on nonlinear polarization rotation with cascaded birefringence fibers. J. Mod. Opt. **58**, 82–86 (2011)
18. Zhu, Y., Xiang, F., Jin, L., Set, S.Y., Yamashita, S.: All-fiber dual-wavelength mode-locked laser using a bend-induced-birefringence Lyot-filter as gain-tilt equalizer. IEEE Photon. J. **11**, 1–7 (2019)
19. Liu, S., Yan, F.P., Ting, F., Zhang, L., Bai, Z., Han, W., Zhou, H.: Multi-wavelength thulium-doped fiber laser using a fiber-based Lyot filter. IEEE Photon. Technol. Lett. **28**, 864–867 (2016)
20. Xin, L.Y., Zhou, X.F., Chen, J.L.: Low threshold multi-wavelength erbium-doped fiber ring laser with NOLM and Lyot filter. Opt. Fiber Technol. **68**, 102800 (2022)
21. Li, Y., Tian, J.J., Quan, M.R., Yao, Y.: Tunable multiwavelength er-doped fiber laser with a two-stage Lyot filter. IEEE Photon. Technol. Lett. **29**, 287–290 (2017)
22. Huang, S.S., Wang, Y.G., Yan, P.G., Zhao, J.Q., Li, H.Q., Lin, R.Y.: Tunable and switchable multi-wavelength dissipative soliton generation in a graphene oxide mode-locked Yb-doped fiber laser. Opt. Express. **22**, 11417–11426 (2014)
23. Fu, L., Kane, C.L., Mele, E.J.: Topological insulators in three dimensions. Phys. Rev. Lett. **98**, 106803 (2007)
24. Zhou, J.J., Luo, A.P., Luo, Z.C., Wang, X.D., Feng, X.H., Guan, B.: Dual-wavelength single-longitudinal-mode fiber laser with switchable wavelength spacing based on a graphene saturable absorber. Photon. Res. **3**, A21–A24 (2015)
25. Luo, Z.Q., Zhou, M., Wu, D.D., Ye, C.C., Weng, J., Dong, J., Xu, H.Y., Cai, Z.P., Chen, L.J.: Graphene-induced nonlinear four-wave-mixing and its application to multiwavelength Q-switched rare-earth-doped fiber lasers. J. Lightwave Technol. **29**, 2732–2739 (2011)
26. Zhao, N., Liu, M., Liu, H., Zheng, X.W., Ning, Q.Y., Luo, A.P., Luo, Z.C., Xu, W.C.: Dual-wavelength rectangular pulse Yb-doped fiber laser using a microfiber-based graphene saturable absorber. Opt. Express. **22**, 10906–10913 (2014)
27. Guo, B.: Recent advances in multi-wavelength ultrafast lasers based on nonlinear effects of 2D materials (invited). Infr. Laser Eng. **48**, 103002 (2019)
28. Guo, B., Yao, Y., Yang, Y.F.: Tunable multi-wavelength mode-locked fiber laser with topological insulator: Bi2Se3/PVA solution. In: International Photonics and OptoElectronics, OSA Technical Digest (Online), p. paper OW2C.4. Optica Publishing Group (2015)
29. Guo, B., Yao, Y.: Tunable triple-wavelength mode-locked fiber laser with topological insulator Bi_2Se_3 solution. Optical Engineering. **55**, 081315 (2016)
30. Guo, B., Yao, Y., Yan, P.G., Xu, K., Liu, J.J., Wang, S.G., Li, Y.: Dual-wavelength soliton mode-locked fiber laser with a WS_2-based fiber taper. IEEE Photon. Technol. Lett. **28**, 323–326 (2016)
31. Yun, L.: Black phosphorus saturable absorber for dual-wavelength polarization-locked vector soliton generation. Opt. Express. **25**, 32380–32385 (2017)
32. Guo, B., Yao, Y., Yang, Y.F., Yuan, Y.J., Wang, R.L., Wang, S.G., Ren, Z.H., Yan, B.: Topological insulator: Bi_2Se_3/polyvinyl alcohol film-assisted multi-wavelength ultrafast erbium-doped fiber laser. J. Appl. Phys. **117**, 063108 (2015)
33. Guo, B., Li, S., Fan, Y., Wang, P.F.: Versatile soliton emission from a WS_2 mode-locked fiber laser. Opt. Commun. **406**, 66–71 (2018)

34. Zhao, R.W., Li, J., Zhang, B.T., Li, X.W., Su, X.C., Wang, Y.R., Lou, F., Zhang, H.K., He, J.L.: Triwavelength synchronously mode-locked fiber laser based on few-layered black phosphorus. Appl. Phys. Express. **9**, 092701 (2016)
35. Ouyang, C.M., Shum, P., Wu, K., Wong, J.H., Lam, H.Q., Aditya, S.: Bidirectional passively mode-locked soliton fiber laser with a four-port circulator. Opt. Lett. **36**, 2089–2091 (2011)
36. Li, Y.J., Wang, C., Huang, D.M., Chen, H.J., Li, F.: Mutual dynamics between synchronous solitons in a bidirectional mode-locked fiber laser. Opt. Lett. **47**, 2170–2173 (2022)
37. Mamidala, V., Woodward, R.I., Yang, Y., Liu, H.H., Chow, K.K.: Graphene-based passively mode-locked bidirectional fiber ring laser. Opt. Express. **22**, 4539–4546 (2014)
38. Lin, J.Q., Dong, Z.P., Dong, T.H., Ma, X.X., Dai, C.S., Yao, P.J., Gu, C., Xu, L.X.: Bidirectional mode-locked fiber laser based on nonlinear multimode interference. Opt. Laser Technol. **154**, 108269 (2022)
39. Yang, K., Li, T.J., Li, X.D., Chen, J.X., Liu, M., Cui, H., Luo, A.P., Xu, W.C., Luo, Z.C.: Mutually induced soliton polarization instability in a bidirectional ultrafast fiber laser. Opt. Lett. **46**, 4848–4851 (2021)
40. Yang, K., Luo, Z.R., Zhang, Z.X., Zhan, Z.Y., Wu, D.X., Liu, M., Luo, A.P., Xu, W.C., Luo, Z.C.: Anti-phase pulsation of counter-propagating dissipative solitons in a bidirectional fiber laser. Opt. Lett. **48**, 6464–6467 (2023)
41. Zhou, Y., Ren, Y.X., Shi, J.W., Wong, K.K.Y.: Dynamics of breathing dissipative soliton pairs in a bidirectional ultrafast fiber laser. Opt. Lett. **47**, 1968–1971 (2022)
42. Zhou, Y., Ren, Y.X., Shi, J.W., Wong, K.K.Y.: Breathing dissipative soliton explosions in a bidirectional ultrafast fiber laser. Photon. Res. **8**, 1566–1572 (2020)
43. Wang, Q.B., Wang, P., Xu, Q.Q., Shi, Z.F., He, J.Y., Wang, Z., Liu, Y.G.: Observation of the 'invisible' pulsation of soliton molecules in a bidirectional ultrafast fiber laser. Opt. Express. **31**, 19036–19047 (2023)
44. Li, L., Ruan, Q.J., Yang, R.H., Zhao, L.M., Luo, Z.Q.: Bidirectional operation of 100 fs bound solitons in an ultra-compact mode-locked fiber laser. Opt. Express. **24**, 21020–21026 (2016)
45. Tao, J.N., Song, P.Y., Hou, L., Lu, B.L., Bai, J.T.: Synchronously-tunable bidirectional mode-locked fiber laser based on Lyot filtering effect. Opt. Laser Technol. **163**, 109391 (2023)
46. Luo, Y.Y., Liu, B., Xiang, Y., Yan, Z.J., Qin, Y.X., Sun, Q.Z., Tang, X.H.: L-band noise-like pulse generation in a bidirectional mode-locked fiber laser. IEEE Photon. Technol. Lett. **30**, 1333–1336 (2018)
47. Li, B.W., Xing, J., Kwon, D., Xie, Y.J., Prakash, N., Kim, J., Huang, S.W.: Bidirectional mode-locked all-normal dispersion fiber laser. Optica. **7**, 961–964 (2020)
48. Prakash, N., Huang, S.W., Li, B.W.: Relative timing jitter in a counterpropagating all-normal dispersion dual-comb fiber laser. Optica. **9**, 717–723 (2022)
49. Luo, J., Lv, J.L., Liu, X.Z., Xu, L.X., Yao, P.J.: Single-cavity bidirectional Raman-assisted mode-locked fiber laser. Infrared Phys. Technol. **136**, 105048 (2024)
50. Lu, Y., Lin, J.Q., Lv, J.L., Luo, J., Liu, X.Z., Xu, L.X., Gu, C., Yao, P.J.: All-fiber bidirectional mode-locked laser to generate cylindrical vector beam asynchronous noise-like pulses. Opt. Express. **30**, 34012–34020 (2022)
51. Li, D., Shen, D.Y., Li, L., Chen, H., Tang, D.Y., Zhao, L.M.: Unidirectional dissipative soliton operation in an all-normal-dispersion Yb-doped fiber laser without an isolator. Appl. Opt. **54**, 7912–7916 (2015)
52. Feng, N., Hu, H., Zhou, R.L., Ji, E.C., Liu, X.X., Gu, H.C., Nakkeeran, K.: 2 μm soliton lasers in a bidirectional nonlinear polarization evolution Tm^{3+}-doped fiber oscillator. Results Phys. **41**, 105930 (2022)
53. Olson, J., Ou, Y.H., Azarm, A., Kieu, K.: Bi-directional mode-locked thulium fiber laser as a single-cavity dual-comb source. IEEE Photon. Technol. Lett. **30**, 1772–1775 (2018)
54. Abdukerim, N., Kayes, M.I., Rekik, A., Rochette, M.: Bidirectional mode-locked thulium-doped fiber laser. Appl. Opt. **57**, 7198–7202 (2018)
55. Li, Y.M., Yin, K., Zhang, X., Zheng, X., Cheng, X.G., Jiang, T.: All-fiber bidirectional mode-locked ultrafast fiber laser at 2 μm. IEEE Photon. J. **11**, 1–8 (2019)

56. Feng, N., Zhou, R.L., Liu, X.X., Hu, H., Li, L.: Switchable unidirectional Tm3+-doped soliton fiber laser. In: 2022 Conference on Lasers and Electro-Optics (CLEO), pp. 1–2. IEEE (2022)
57. Deissler, R.J., Brand, H.R.: Periodic, quasiperiodic, and chaotic localized solutions of the quintic complex Ginzburg-Landau equation. Phys. Rev. Lett. **72**, 478 (1994)
58. Cui, Y.D., Liu, X.M.: Graphene and nanotube mode-locked fiber laser emitting dissipative and conventional solitons. Opt. Express. **21**, 18969–18974 (2013)
59. Liu, B.W., Luo, Y.Y., Xiang, Y., Xiao, X.P., Sun, Q.Z., Liu, D.M., Shum, P.P.: Multiplexed ultrafast fiber laser emitting multi-state solitons. Opt. Express. **26**, 27461–27471 (2018)
60. Liu, B.W., Luo, Y.Y., Xiang, Y., Zhou, W., Shum, P.P., Yan, Z.J., Tang, X.H., Liu, D.M., Sun, Q.Z.: CFBG-based bidirectional mode-locked fiber laser emitting conventional and dissipative solitons. IEEE Photon. Technol. Lett. **31**, 1737–1740 (2019)
61. Zhang, H.Z., Mao, D., Du, Y.Q., Zeng, C., Sun, Z.P., Zhao, J.L.: Heteronuclear multicolor soliton compounds induced by convex-concave phase in fiber lasers. Commun. Phys. **6**, 191 (2023)
62. Jin, X.X., Zhang, M., Hu, G.H., Wu, Q., Zheng, Z., Hasan, T.: Broad bandwidth dual-wavelength fiber laser simultaneously delivering stretched pulse and dissipative soliton. Opt. Express. **28**, 6937–6944 (2020)
63. Xiao, L., Wang, T.S., Ma, W.Z., Zhu, H.F., Zhao, R.Y.: Switching and coexistence of stretched pulse and dissipative soliton in a hybrid mode-locked fiber laser. Opt. Commun. **522**, 128649 (2022)
64. Luo, Y.Y., Xiang, Y., Xia, R., Ni, W.J., Liu, B.W., Shum, P.P.: Breathing dynamics in a gain-guided dissipative soliton-similariton fiber laser. IEEE Photon. Technol. Lett. **32**, 481–484 (2020)
65. Xiang, Y., Luo, Y.Y., Liu, B.W., Xia, R., Shum, P.P., Tang, X.H.: Scalar and vector solitons in a bidirectional mode-locked fibre laser. J. Lightwave Technol. **37**, 5108–5114 (2019)
66. Zhao, X., Li, T., Liu, Y., Li, Q., Zheng, Z.: Polarization-multiplexed, dual-comb all-fiber mode-locked laser. Photon. Res. **6**, 853–857 (2018)
67. Goda, K., Jalali, B.: Dispersive Fourier transformation for fast continuous single-shot measurements. Nat. Photonics. **7**, 102–112 (2019)
68. Herink, G., Kurtz, F., Jalali, B., Solli, D.R., Ropers, C.: Real-time spectral interferometry probes the internal dynamics of femtosecond soliton molecules. Science. **356**, 50–54 (2017)
69. Liu, Y.S., Huang, S.Y., Liu, H.G., Sun, Y.X., Xia, R., Ni, W.J., Luo, Y.Y., Yan, L.S., Sun, Q.Z., Shum, P.P., Tang, X.H.: Unveiling manipulation mechanism on internal dynamics of soliton triplets via polarization and gain control. Opt. Lasers Eng. **168**, 107645 (2023)
70. Liu, S.L., Cui, Y.D., Karimi, E., Malomed, B.A.: On-demand harnessing of photonic soliton molecules. Optica. **9**, 240–250 (2022)
71. Yang, K.W., Zhou, Y., Ling, Y.Q., Tsia, K.K., Zeng, H., Wong, K.K.Y.: Spectral period doubling and encoding of dissipative optical solitons via gain control. PhotoniX. **5**, 26 (2024)

Chapter 6
Thulium-Doped Passively Mode-Locked All Polarization-Maintaining Fiber Lasers

C. Cuadrado-Laborde, H. Munoz-Marco, L. A. Sanchez, A. Carrascosa, P. Perez-Millan, A. Diez, J. L. Cruz, E. Silvestre, and M. V. Andres

Abstract This chapter discusses our findings in all-fiber mode-locked lasers emitting around 2 μm. Although these lasers were designed to serve different purposes, they share several characteristics, such as the use of an all-fiber polarization-maintaining cavity, the selection of a thulium-doped fiber as a medium gain, a semiconductor saturable absorber mirror as a mode locker, and, finally, resonant pumping in the C-telecommunication band. As a result, we were able to create first a low-repetition rate all-polarization-maintaining thulium-doped passively mode-locked fiber laser running at 2.3 MHz. Next, we have examined the possibility of emission at an optical wavelength well beyond the maximum for the $^3F_4 \rightarrow {}^3H_6$ transition for thulium-doped fibers (1837 nm) for potential applications in spectroscopy. As a result, we have designed a passively mode-locked thulium-doped fiber laser emitting at 2070 nm, one of the longest wavelengths reported for solitonic emission using thulium-doped fibers. Finally, we present the first thulium-doped mode-locked fiber laser that maintains polarization and operates at the ultra-high fundamental repetition rate of 1 GHz in a laser cavity built entirely from commercially available components. The entire performance of these lasers is described, including a stability analysis.

Keywords Mode-locking · Thulium-doped · Polarization-maintaining · All-fiber

C. Cuadrado-Laborde (✉)
Instituto de Física Rosario (CONICET-UNR), Rosario, Argentina
e-mail: cuadradolaborde@ifir-conicet.gov.ar

H. Munoz-Marco · P. Perez-Millan
FYLA LASER SL, Paterna, Valencia, Spain

L. A. Sanchez · A. Carrascosa · A. Diez · J. L. Cruz · M. V. Andres
Departamento de Física Aplicada y Electromagnetismo, ICMUV, Universidad de Valencia, Burjassot, Valencia, Spain

E. Silvestre
Departamento de Óptica, ICMUV, Universidad de Valencia, Burjassot, Spain

M. F. S. Ferreira, M. C. Paul (eds.), *Optical Fiber Lasers*, Springer Series in Optical Sciences 250, https://doi.org/10.1007/978-3-032-04110-4_6

6.1 Introduction

Passively mode-locked fiber lasers have gained popularity over the years due to their potential for research and industrial applications [1]. All-fiber passively mode-locked lasers have also been shown to be a cost-effective alternative to their solid-state counterpart due to their compact cavity structure, heat dissipation properties, free alignment, beam quality, and overall stability. Active mode-locked fiber lasers, on the contrary, require additional switching electronics and complex driven modulators. The ability of passively mode-locked fiber lasers to generate ultrafast solitons is crucial for practical applications like optical fiber sensing, biomedical diagnostics, material processing, nonlinear optics, terahertz generation, and ultrahigh-speed communications, among other applications [2].

In particular, there is a compelling reason to create trustworthy mode-locked oscillators with emission wavelengths around two microns, since they may find application in the following areas: laser scalpels [3]; eye-safe light detection and ranging (LIDAR); free-space optical communication [4]; low-latency hollow-core fiber interconnection [5]; cutting, engraving, and welding of plastics [6]; remote sensing; real-time spectroscopy; optical biopsy; and so on. Given that the emission cross section has a maximum centered at 1.837 μm, thulium-doped fiber (TDF) is the obvious choice below two microns [7]. On the other hand, holmium-doped fiber (HDF) is also a prime candidate as a gain medium since its emission cross section exhibits two distinct peaks centered symmetrically around two microns [8]. Nonetheless, there are additional practical justifications for the usage of TDFs. One of them is that TDFs, which can be pumped at 0.79 or 1.56 μm, have easier access to commercial pumping sources, while HDFs have fewer commercially available pumping sources at 1.1 or 1.9 μm. The second issue is reabsorption, which arises from the substantial overlap between the emission and absorption bands in HDFs but not in TDFs [9].

Although there are different mechanisms to mode-lock passively, the use of semiconductor saturable absorber mirrors (SESAMs) has been traditionally the preferred choice [10]. They are helpful optical components that enable mode-locking or Q-switching of light pulses. Its easy-to-use design stems from its straightforward assembly, i.e., a mirror with an integrated saturable absorber. SESAMs can be used with a variety of gain media operating over a wide range of wavelengths because their semiconductor layer can be adjusted to provide saturable absorption at different wavelengths. They can be tailored to a wide range of laser parameters, and if the device and operation parameters are selected correctly, they usually allow for reliable self-starting mode-locking. They can also be used at very high output power levels, as long as the resonator mode size on the absorber has been adequately selected.

In this context, although the use of an all-fiber cavity in the design of the resonator is an improvement, there is still the possibility of instabilities, especially in out-of-laboratory applications. In this sense, the possibility of using polarization-maintaining (PM) fibers and fiber components in the design of a mode-locked laser ensures maximum robustness against harsh and hostile environments [11].

This chapter focuses on our most recent findings in the area of all-fiber PM passively mode-locked lasers operating at two microns. Consequently, we first demonstrate a self-starting, low-repetition rate all-fiber laser that operates at 1951 nm and achieves mode-locking through the use of a SESAM [12]. Additionally, a fixed operating wavelength and a narrow spectral bandwidth of the output light pulses are ensured by the presence of a high-reflective, narrow-band fiber Bragg grating within the cavity. We have also contrasted the performance of this laser with that of a shorter, higher repetition rate version of this cavity. Second, we concentrate on TDF-based laser systems operating above 2 μm. We thus demonstrated a self-starting, fully PM, passively mode locked, TDF laser that emits at 2070 nm and is fully pumped in-band at 1561 nm [13]. These results are then compared with the same cavity emitting at 1951 nm. Finally, we demonstrate the first all-fiber laser with a fundamental repetition rate of 1 GHz that is passively mode-locked and fully polarization-retaining [14]. The entire cavity is made up of a very short, highly doped PM TDF that has a semiconductor saturable absorber mirror on one end and a dichroic mirror on the other.

6.2 Low-Repetition Rate PM TDF Laser Passively Mode Locked

These days, there is a larger demand for lasers with an emission wavelength of two microns [15]. One of their primary advantages, which is usually cited, is that they are eye-safe, as water has significant absorption lines at this wavelength [16]. Furthermore, this has opened doors for other applications [17, 18]. For these applications, low-repetition rate mode-locking methods might be useful. However, all-fiber mode-locked lasers often operate at repetition rates higher than ten megahertz, where chromatic dispersion, attenuation, and the build-up of nonlinear effects may be managed due to the comparatively short fiber cavity. This section presents a self-starting all-fiber laser with a low-repetition rate that operates at two microns, where mode-locking was achieved by a SESAM.

Figure 6.1 depicts the experimental setup, where the pump unit was custom-made in our laboratories and produces a continuous wave emission in the infrared C-band. This pump unit was able to deliver continuously variable power up to 820 mW at 1561 nm, with a measured slope efficiency of 10.57% and laser threshold of 47 mW, being in turn pumped by a laser diode at 980 nm. On the other hand, the setup of the oscillator was composed of a typical Fabry-Perot (FP) cavity starting with a PM FBG with high reflectivity (99%, Bragg wavelength of 1951 nm and -3 dB bandwidth of 0.14 nm), which in turn was fusion-spliced to a 2 m PM TDF (PM-TSF 9/125 from Nufern, numerical aperture 0.15, cut-off wavelength 1750 nm, and 9 ± 2 dB/m at 1180 nm of core absorption). The free end of the PM

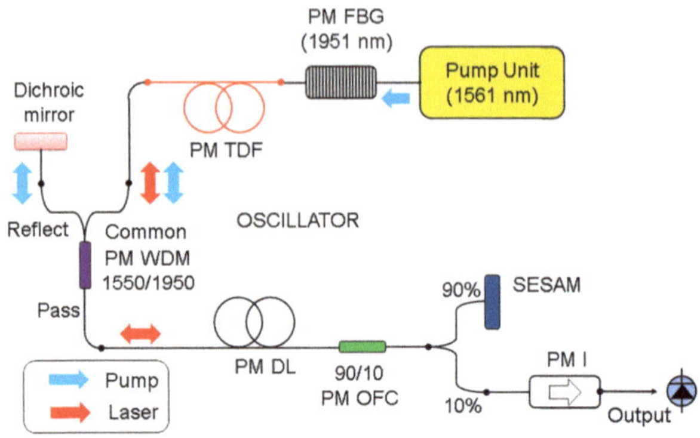

Fig. 6.1 Experimental setup

TDF was fusion-spliced to the common port of a 1550/1950 nm PM wavelength division multiplexer (PM WDM, insertion loss: pass to common <0.6 dB at 1950 nm and reflect to common <0.5 dB at 1550 nm), whose purpose is the protection of the SESAM from any residual pump. The residual pump power was returned to the cavity by connecting the reflect port of the PM WDM to a dichroic mirror. The passport of the PM WDM was in turn fusion-spliced to a 38.42 m long PM delay line (PM1550-XP from Nufern, numerical aperture 0.125, and cut-off wavelength 1380 ± 60 nm). The other end of the PM delay line (PM DL) was fusion-spliced to the input port of a PM optical fiber coupler (PM OFC, 90/10 with blocked fast axis). The presence of a blocked fast axis element (in this case, the PM OFC) guarantees a single polarization state in the resonator, even when CW pumping provides unpolarized light. Finally, the cavity FP was closed by connecting the 90% port of the PM OFC to a semiconductor saturable absorber mirror (SESAM, from BATOP Optoelectronics with high reflection bandwidth in the 1900–2080 nm range and relaxation time constant of 10 ps. The output signal from this oscillator was obtained through the 10% port of the PM OFC, which also incorporated two PM optical isolators in tandem (PM I, with a center wavelength of 1950 ± 20 nm and combined 56 dB peak isolation with a 1.2 dB insertion loss) to avoid unwanted reflections at the oscillator. The output light pulses were monitored with a 63 GHz sampling oscilloscope, a 13 GHz real-time oscilloscope, fast InGaAs photodetectors (rise time < 28 ps, fall time < 28 ps, and 0.95 A/W at 2 μm), an optical spectrum analyzer (optical range from 1200 nm to 2400 nm, resolution bandwidth 50 pm), a radio frequency (RF) spectrum analyzer (resolution bandwidth 1 Hz, DC to 10 GHz), and a pyroelectric detector for the measurement of the average output power.

The resulting cavity length was 44.67 m, with an estimated average anomalous dispersion parameter of 42.1 ps/nm × km and a group velocity dispersion of -85 ps^2/km at 1950 nm; i.e., a total chromatic dispersion per round-trip of -7.6 ps^2. We also measured the output of this laser when the PM DL was removed, i.e., for a shorter cavity configuration. In the last case, the total chromatic dispersion per round-trip was -1 ps^2. On the other hand, the attenuation at each round-trip of the light pulses is expected to be significant at the operating wavelength of this laser since the cavity is mainly composed of PM1550-XP, which has an attenuation of 20 dB/km at 1950 nm. In this case, the combined discrete-distributed power losses within a full round-trip are 7.4 dB (see Fig. 6.1) and 5.8 dB without the PM DL.

The laser exhibited self-starting stable mode-locking for a pump power of 170 mW. Fig. 6.2(a) shows the light pulse train with a temporal period of 0.435 μs; i.e., 2.3 MHz, obtained at the output, which corresponds to the expected round-trip time for an FP cavity of 44.67 m. At this high anomalous intracavity dispersion, the pump power setting is crucial because changing the pump power by less than 2 mW disables the mode-locking regime. Given that the repetition rate is low and that any change in pump power has a significant effect on pulse energy, this low-repetition rate could be the reason for the limited pumping range. The RF spectrum was also measured in order to assess the quality of this mode-locking regime. Using a resolution bandwidth of 10 Hz, the fundamental frequency was determined to be 2.3 MHz, and a high extinction ratio of up to 65 dB was observed, as shown in Fig. 6.2(b). It is clear from the RF spectrum that the laser was operating in a stable mode-locking mode. As long as the pump unit was running when mode-locking was initiated, the stable mode-locking regime could continue. This stability is typical of a fully polarization-retaining all-fiber design. By contrast, for the short-cavity setup without the PM DL, the laser was able to initiate stable mode-locking when the pump power was set to 250 mW. In this instance, the output light train operates at a consequently higher repetition rate of 15.6 MHz. In this case, any increase in pump power only results in the usual amplitude modulation of the output light train and does not destroy the mode-locking regime, at least until the maximum available pump power in our experiments.

Figure 6.3(a) provides a detailed representation of the temporal waveform of a single light pulse measured by the digital sampling oscilloscope. The best fit of the pulse is a secant hyperbolic function, with a full width at half maximum (FWHM) of 81 ps. We also measured the output of this laser without the PM delay line using the shortened version of this resonator; Fig. 6.3(b) also shows this waveform. Once

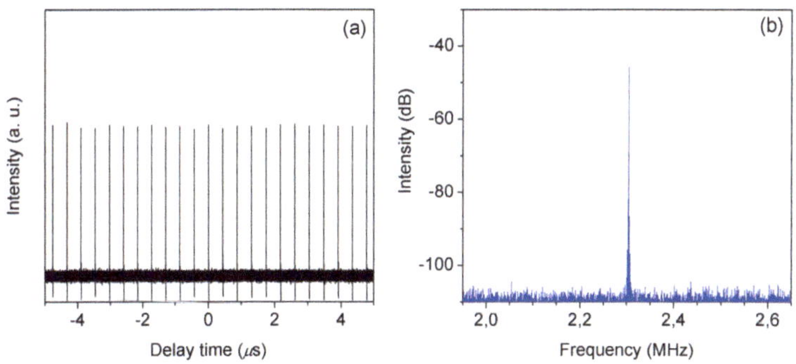

Fig. 6.2 (**a**) Oscilloscope traces showing the output train. (**b**) RF spectrum showing the fundamental frequency

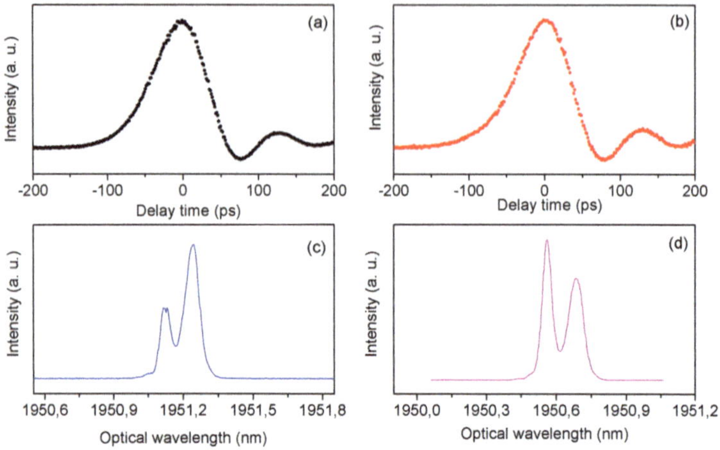

Fig. 6.3 Oscilloscope traces showing the output light pulses, with (**a**) and without (**b**) the PM DL. Measured optical spectra, with (**c**) and without (**d**) the PM DL

normalized in intensity, the two temporal waveforms superpose; naturally, the pulse also has an FWHM of 81 ps.

When comparing the shorter cavity resonator to the longer cavity resonator, there is no appreciable difference in the temporal waveform. We can use Martinez et al.'s work [19] as an example, where it is theoretically shown that the pulse width increases monotonically with the (anomalous) dispersion within the resonator for soliton mode-locking. It should be noted that the curves in the cited paper were produced assuming a modest and fixed self-phase modulation (SPM). Nevertheless, our results show that when the cavity length is greatly increased, this assumption is violated since the nonlinear SPM parameter rises linearly with length. Additionally, we measured the optical spectrum for the standard configuration with the PM DL, which is displayed in Fig. 6.3(c). There are two peaks in the spectrum, each having a 3-dB bandwidth of 50 pm in coincidence with the resolution of the optical spectrum analyzer. For the shorter cavity without the PM DL, we also carried out optical spectral measurements; the outcome is also displayed in Fig. 6.3(d). At the longer wavelength peak, the spectra overlap with our previous measurement of the long cavity and are nearly identical. Additionally, there is a slight wavelength shift of 0.5 nm between the two spectra, which could be explained by the combination of our wavelength accuracy with the thermal coefficient for an FBG at this wavelength (~15 pm/°C). On the other hand, the average output powers were 354 μW and 230 μW, with and without the PM DL. It is important to note that both CW and mode-locked pulse emission are included in these average output powers; the CW component accounts for 30% of the measured average power, and the pulsed emission for the remaining 70% for the long cavity. These ratios for the short cavity are 47% for the continuous wave and 53% for the pulsed emission. As a result, for the long and short cavities, the peak powers for the pulsed emission are approximately 1300 mW and 100 mW, respectively.

Finally, we used a specifically constructed PM all-fiber pre-amplifier, which can generate up to 1.1 W average power at 2 μm, to explore the nonlinear propagation of the amplified emission along a fiber coil. Here, our objective was to distinguish between the two peaks in Figs. 6.3(c–d) that, in fact, correspond to the contribution of a CW emission and a pulsed emission, respectively. Strong nonlinear effects would be anticipated when this preamplified output light train is propagated in a 656 m SMF-28 optical fiber for peak power pulses $P_0 = 500$ W, assuming a nonlinear coefficient $\gamma = 0.498$ W^{-1} km^{-1} (at 1950 nm). Strong nonlinear effects will be expected when this preamplified output light train is propagated in a 656 m SMF-28 optical fiber, since the nonlinear length of the SMF-28 fiber $L_{NL} = 1/(\gamma P_0) \geq 4$ m. In these experiments, the longer wavelength peak displayed the spectral changes typically associated with nonlinear processes, while the shorter wavelength peak did not show any significant change [12]. As a result, we deduce that the longer wavelength peak actually represents the spectrum linked to the pulse train, whereas the shorter wavelength peak represents a trace of continuous wave emission. We also employed a straightforward method that we had previously developed [20] and had been successfully implemented [21, 22] to measure the instantaneous angular frequency profile of the light pulses that were emitted. For this method, we employed a 656 m long SMF-28 optical fiber, whose GVD at 1950 nm is −83 ps^2/km, as a dispersive line [23]. These measurements led us to the conclusion that these light pulses are free of chirp right at the output of the cavity. The short-cavity pulses also exhibited this behavior, with no discernible variations. In conclusion, with the exception of repetition rate and average power, light pulses produced with and without the PM DL were nearly identical [12].

6.3 Passively Mode-Locked PM TDF Laser Emitting at 2.07 μm

Several studies have documented mode-locked all-fiber lasers utilizing TDF as the gain medium and emitting at wavelengths greater than 2 μm [24–26]. In contrast to these lasers, the mode-locked laser that is the subject of this section has one of the longest emission wavelengths (2070 nm), well beyond the $^3F_4 \rightarrow {}^3H_6$ transition at 1837 nm [13]. The emission wavelength may be further extended because of the emission cross section of TDFs, which spans the wavelength range of 1650–2200 nm [8]. Regarding the maximum value, the reported emission cross-sections for different TDFs at 2070 nm reach a value of only 28% of the maximum cross section at 1837 nm [27].

We display the experimental setup in Fig. 6.4. The same CW laser that was previously discussed in Sect. 6.2 pumped the resonator. The Fabry-Pérot resonator was essentially the same as described before, but with all its components adequately selected to operate at 2070 nm, see Fig. 6.1 as compared to Fig. 6.4. The polarization-maintaining fiber Bragg grating (PM FBG) was recorded at $\lambda_B = 2069$ nm with

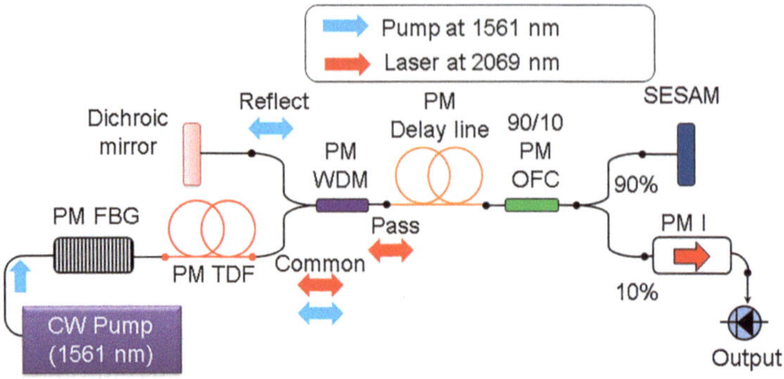

Fig. 6.4 Experimental setup

140 pm of bandwidth at −3 dB. A 3.2 m long PM TDF (PM-TSF 9/125 from Nufern®, numerical aperture 0.15, cut-off wavelength 1750 nm, and 9 ± 2 dB/m absorption at 1180 nm) was used, which in turn was fusion-spliced to the PM FBG. As compared with our previous work, Section 6.2, the PM TDF is longer in this case in order to compensate for the low gain of the TDF at this wavelength. A PM wavelength division multiplexer (WDM, 1550/2070 nm, insertion loss pass to common <0.6 dB at 2070) was also connected to the PM TDF via fusion splicing. The passport was connected to the input port of a PM fiber optic coupler (PM OFC, 90/10, central wavelength 1950 nm, and fast axis blocked), while the reflection port was connected to a dichroic mirror to return any leftover pump power to the cavity. Lastly, on the FC/PC connector, a SESAM (BATOP, high reflection bandwidth in the 1900–2080 nm range and relaxation time constant of 10 ps) was faced. In order to prevent undesired reflections in the cavity, a PM optical isolator (PM I) was inserted into the remaining 10% port of the PM OFC. This allowed for the extraction of the output. The characteristics of this laser were assessed through the use of a 63 GHz sampling oscilloscope, a 13 GHz real-time oscilloscope, fast InGaAs photodetectors (rise/fall times <28 ps), an optical spectrum analyzer (resolution <50 pm), a 10 GHz electrical spectrum analyzer (resolution <1 Hz), and a pyroelectric sensor to determine the average output power. Measured from the PM FBG to the SESAM, the resultant cavity length was 7.89 m, and the estimated group velocity dispersion (GVD) at 2069 nm was −100 ps^2/km, indicating a strongly anomalous cavity whose total round-trip dispersion for the resonator resulted in −1.58 ps^2. Both discrete and distributed losses were computed for this laser, resulting in a total loss of 5.3 dB per round-trip.

Mode-locked operation, albeit with some instabilities, self-starts when the pump power surpasses 200 mW, but at 317 mW of pump power, the instabilities totally disappeared. With a repetition rate of 13.08 MHz, the laser produces a single-pulse train at this pump power; this accurately matches the reciprocal round-trip time for an FP cavity with a 15.79 m total round-trip length; see Fig. 6.5(a). Afterward, an RF spectrum analyzer was also used to characterize the laser. The RF traces recorded for the first harmonic, centered at 13.084 MHz, with a resolution bandwidth of

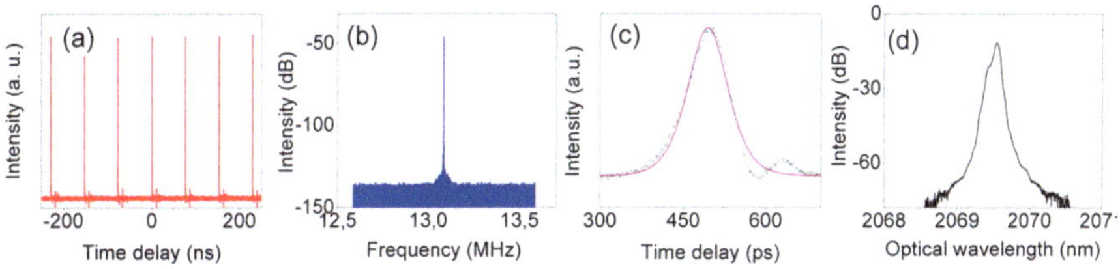

Fig. 6.5 (**a**) Pulse train. (**b**) Electrical spectrum of the first harmonic. (**c**) Measured output light pulse with its corresponding sech2 profile fitting (open dots and solid purple curve, respectively). (**d**) Optical spectrum

1 Hz, are displayed in Fig. 6.5(b); the data indicates a high extinction sideband suppression ratio of −80 dB. We infer from the RF spectra that the emission of this laser is very stable since there are no indications of low-frequency instabilities. The digital sampling oscilloscope trace for a single light pulse from the output train is displayed in Fig. 6.5(c). The trace has a full width at half maximum (FWHM) of 94 ps and can be fitted fairly well with a $sech^2$ profile (also shown). As compared with our previous work shown in Sect. 6.2, the pulses here are wider (94 ps as compared with our previous result of 81 ps), but for a correspondingly higher chromatic dispersion (−1.58 ps^2 as compared to the previous value of −1 ps^2). They probably have quite similar nonlinear parameters because the fundamental repetition rate was almost the same; in this instance, though, the dispersion has increased, leading to a larger pulse width. This outcome is in line with earlier theoretical research on soliton lasers, which predicted that for a fixed nonlinear parameter, an increase in pulse width would occur with an increase in chromatic dispersion [28].

Finally, Fig. 6.5(d) displays the optical spectrum. The −3 dB bandwidth of the spectrum is 70 pm at 2070 nm. The bandwidth of the output signal is relatively close to 140 pm (PM FBG bandwidth), which corresponds to the more restrictive filter element within the resonator. Consequently, the time-bandwidth product (TBP) that was obtained was 0.46, which is close to a transform-limited Gaussian shape (whose TBP $\geq$ 0.44) and slightly above a transform-limited $sech^2$ shape (whose TBP $\geq$ 0.315). Further, the laser entered a multi-pulse region when the pump power was increased. For instance, two pulses are produced every round-trip when the pump power is increased above 426 mW. Three pulses per round-trip occur if the pump power is increased further, and so on, until the maximum pump power is reached. One could say that soliton lasers typically exhibit this behavior. It should be mentioned that because the pulses are not distributed equally throughout the round-trip time, this regime is not precisely harmonic mode-locking.

6.4 One GHz Fundamental Repetition Rate PM TDF Laser Passively Mode-Locked

Ultrafast fiber lasers operating at two microns have recently been demonstrated to be an effective platform for accessing the mid/far infrared spectral region [29]. The demand for two-micron high-repetition rate fiber lasers is expected to be high. Harmonic mode-locking is a well-known method for achieving a high-repetition rate [30]; however, it has the disadvantage of producing more pulse noises, such as amplitude fluctuations and pulse-to-pulse timing jitter. Even worse, it is susceptible to environmental disruptions, which may alter the repetition rate. Fundamentally mode-locked fiber lasers operating at a high-repetition rate, on the other hand, are far superior because the emitted pulses have a period that is strictly fixed by the cavity length, resulting in significantly lower noise levels. Nevertheless, it can be difficult to mode-lock a fiber laser at a repetition rate higher than 200 MHz. Since in our case the cavity fiber is only ≤100 mm long, the active fiber needs to be heavily doped in order to provide sufficient gain, especially for repetition rates ≥1 GHz. Recently, a number of two-micron mode-locked fiber lasers with high fundamental repetition rates have been introduced [31, 32]. In this section, we present for the first time a passively mode-locked TDF laser that maintains polarization and operates at a fundamental repetition rate of 1 GHz, where the laser cavity uses only components that are available in the market.

The experimental setup is depicted in Fig. 6.6. The PM oscillator was a short, fully polarization-preserving Fabry-Perot (FP) resonator. The same continuous wave laser described before was used as a pump source, with emission at 1561 nm. A dichroic filter deposited on a fiber ferrule (transmissivity >90% at 1561 nm, reflectivity = 99% for $\lambda > 1780$ nm) allowed the pump power to enter the FP cavity. In turn, this fiber ferrule is one tip of a highly doped (~7%) PM TDF (PM-TSF-5/125, Panda-Type Tm-Doped optical fiber by Nufern; operating wavelength:

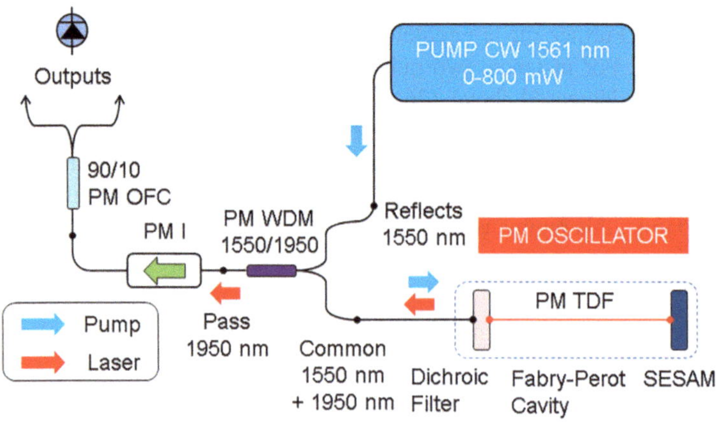

Fig. 6.6 Experimental setup

1900–2100 nm; numerical aperture of 0.24; absorption at 1560 nm of 362 dB/m). By facing a SESAM (BATOP, relaxation time constant of 10 ps) against a second fiber ferrule connected to the second end of the PM TDF, the FP cavity was eventually conformed. A PM wavelength division multiplexer (WDM, 1550/1950 nm, insertion loss: pass to common <0.6 dB at 1950 nm and reflection to common <0.5 dB at 1550 nm) outside the FP cavity receives the output of the laser. It is then connected to a PM fiber optic coupler (PM OFC, 90/10, central wavelength 1950 nm and peak isolation of 28 dB, with insertion loss of 0.6 dB). Consequently, the resonator round-trip length was 0.212 m since the PM TDF length was 0.106 m. The group velocity dispersion (GVD) at 1960 nm of the PM TDF and the group delay dispersion of the SESAM are -20 ps^2/km [33] and $-$ 600 fs^2 [34], respectively. As a result, the total group delay dispersion of the cavity per round-trip is -4840 fs^2; i.e., anomalous. A 63 GHz sampling oscilloscope, a 13 GHz real-time oscilloscope, fast InGaAs photodetectors (rise/fall times <28 ps), an optical spectrum analyzer (resolution <50 pm), and a 10 GHz electrical spectrum analyzer (resolution <1 Hz) were used to characterize the laser. Lastly, a pyroelectric sensor was used to measure the average output power.

Although there were significant amplitude instabilities, the mode-locked operation self-started at about 300 mW of pump power, as measured at the input of the dichroic mirror. As the pump power was raised to 339 mW, the instabilities reduced. The laser generates a steady single-pulse train at this pump power with a repetition rate of 947 MHz; this is completely compatible with the expected round-trip duration for an FP cavity with a round-trip length of 0.212 m; see Fig. 6.7(a). Figure 6.7(b), on the other hand, displays the optical spectrum; the spectrum is centered at 1960 nm and has a 3-dB bandwidth of 7 nm. The fast photodetectors

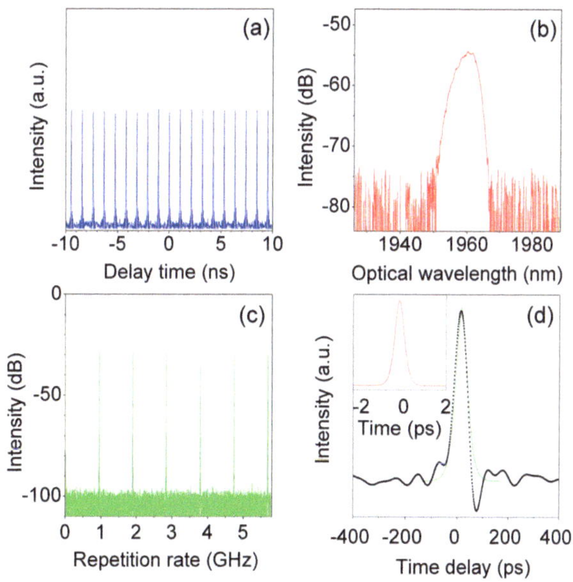

Fig. 6.7 (**a**) Oscilloscope trace of the pulse train. (**b**) Optical spectrum. (**c**) RF spectrum showing up to the 5th overtone. (**d**) Oscilloscope trace measured at the output of the dispersive line (solid scatter points), together with the simulation of a pulse profile dispersed through the same dispersive line (solid green curve); the inset shows the pulse profile reconstructed at the output of the oscillator (red curve)

were then connected to an RF spectrum analyzer; the trace recorded for a broad span up to the sixth harmonic is shown in Fig. 6.7(c) using a resolution bandwidth of 1 kHz. At 947 MHz, they are equally spaced and exhibit a high sideband suppression ratio (extinction >50 dB). We infer from the RF spectra that the emission of this laser is sufficiently stable because there are no low-frequency instabilities. Conversely, 15.2 μW was the average output power measured, which yielded a slope efficiency of 0.04‰. By measuring the output average power during continuous operation, the stability of this laser was also examined, resulting in a standard deviation of only 2% during a free-running operation of 4 h without room temperature control. This insensitivity to changes in temperature and mechanical vibrations is ensured by the use of a full polarization-maintaining cavity in conjunction with a SESAM as a mode locker. Lastly, we finished the experimental characterization of this laser by estimating its pulse width using an autocorrelator. Regretfully, the energy of this pulse was insufficient to break through the noise floor of the autocorrelator. Furthermore, we refrained from using optical amplification in order to maintain the original waveform of the pulse profile. Because of this, we employed an alternative technique in which we first measured the pulse broadened under a known dispersion, and then we recalculated the necessary pulse width at the input of the dispersive line in order to align the computed dispersed profile with our measurements. We did this by using a 210-meter-long optical fiber (Corning LEAF, numerical aperture 0.14) and 34.15 ps/(nm km) of chromatic dispersion at 1961 nm [35]. The temporal pulse profile, with FWHM of 54 ps, acquired at the output of the 210-meter-long dispersive line is shown in Fig. 6.7(d). On the other hand, for the same dispersive line, we simulated the propagation of unchirped sech^2 light pulses with different temporal widths. We could fit the measured dispersed light pulse to the simulated dispersed pulse only when the temporal width of the pulse profile at the input of the dispersive line was 0.53 ps; see the inset in Fig. 6.7(d). Since this laser is expected to emit a soliton-like pulse with a minimum time-bandwidth product (i.e., 0.315 for sech^2 pulses) in this regime, it was implicitly assumed that a transform-limited pulse [36]. In contrast, the spectrum at the output of the oscillator had a spectral bandwidth of 7 nm, which produced a time-bandwidth product of 0.29; that is in line with our original hypothesis.

6.5 Conclusions

A number of fiber lasers operating at the 2 μm optical wavelength have been described in this chapter. They are all distinguished by the use of a thulium-doped fiber as a medium gain, the presence of a SESAM acting as a passive mode locker, and the preservation of polarization.

In the first work, we experimentally characterized a long-cavity mode-locked PM TDF laser, resonantly pumped in the telecommunications *C*-band by a CW erbium-ytterbium doped all-fiber laser. A low-repetition rate train of transform-limited light pulses was generated at 2.3 MHz, with each light pulse having a temporal width of 81 ps and a spectral width of 50 pm. We also compare the performance of this laser with a shortened version of this cavity, 6.25 m long, emitting with a repetition rate of 15.6 MHz, with no appreciable differences in the output light pulses, except for the average power. This laser differs from previously presented lasers in its unique combination of transform-limited clean pulses with a low-repetition rate and narrow spectral linewidth.

Next, we presented a passively mode-locked thulium-doped polarization-maintaining laser resonantly pumped. The objective was to demonstrate the possibilities of emission at the longest wavelength possible using TDF. This laser was fully characterized, and as a result, a sequence of transform-limited light pulses was obtained at 13.084 MHz, with each light pulse having a temporal width of 94 ps and a spectral width of 70 pm, with a peak power of 7.94 W. The spectral emission was at 2070 nm, one of the longest wavelengths ever reported for solitonic emission based on the use of TDFs as medium gain. These characteristics make this laser a viable option for spectroscopy, LIDAR, and low-latency communications, among other uses.

Lastly, we demonstrated an all-fiber laser with a fundamental repetition rate of 1 GHz that is passively mode-locked and fully polarization-retaining. A 106-mm-long highly Tm-doped polarization-maintaining fiber with a semiconductor saturable absorber mirror on one end and a dichroic mirror on the other makes up the entire cavity. This laser produces a train of transform-limited light pulses with a temporal width of 520 fs at the central optical wavelength of 1.96 μm, which we experimentally characterized. As far as we are aware, this is the first study to report on a thulium-doped mode-locked fiber laser that maintains polarization and operates at a fundamental repetition rate of 1 GHz.

Acknowledgments This work was supported in part by the European Union, project IPN-Bio (Ref.: H2020-MSCA-RISE-2019-872049), and by the *Generalitat Valenciana* of Spain (Ref.: CIPROM/2022/30). C. Cuadrado-Laborde acknowledges financial assistance from CONICET (Ref.: PIP 11220220100052CO, Argentina).

References

1. Haus, H.A.: Mode-locking of lasers. IEEE J. Select. Topic. Quant. Electron. **6**, 1173–1185 (2000)
2. Shi, W., Fang, Q., Zhu, X., Norwood, R.A., Peyghambarian, N.: Fiber lasers and their applications [invited]. Appl. Opt. **53**, 6554–6568 (2014)
3. Xie, X., Xu, Q., Hu, W., Zhang, W., Dai, Q., Chen, J., Deng, J., Song, H., Wang, S.-X.: A brief review of 2 μm laser scalpel. In: 2020 IEEE 5th Optoelectronics Global Conference (OGC), pp. 63–67 (2020)
4. Lin, P., Wang, T., Ma, W., Yang, Q., Liu, Z.: Transmission characteristics of 1.55 and 2.04 μm laser carriers in a simulated smoke channel based on an actively mode-locked fiber laser. Opt. Express. **28**, 39216–39226 (2020)
5. Shen, W., Du, J., Sun, L., Wang, C., Zhu, Y., Xu, K., et al.: Low-latency and high-speed hollow-core fiber optical interconnection at 2-micron waveband. J. Lightwave Technol. **38**, 3874–3882 (2020)

6. Mingareev, I., Weirauch, F., Olowinsky, A., Shah, L., Kadwani, P., Richardson, M.: Welding of polymers using a 2 μm thulium fiber laser. Opt. Laser Technol. **44**, 2095–2099 (2012)
7. Khamis, M.A., Ennser, K.: Broadband amplified spontaneous emission thulium-doped fiber sources near 2 μm. In: Reimer, A. (ed.) Horizons in World Physics 299, Chapter 3. Nova Science Publishers (2019)
8. Hemming, A., Simakov, N., Haub, J., Carter, A.: A review of recent progress in holmium-doped silica fibre sources. Opt. Fiber Technol. **20**, 621–630 (2014)
9. Huang, C., Tang, Y., Wang, S., Zhang, R., Zheng, J., Xu, J.: Theoretical modeling of ho-doped fiber lasers pumped by laser-diodes around 1.125 μm. J. Light. Technol. **30**, 3235–3240 (2012)
10. Keller, U., Weingarten, K.J., Kärtner, F.X., Kopf, D., Braun, B., Jung, I.D., Fluck, R., Hönninger, C., Matuschek, N., Aus Der Au, J.: Semiconductor saturable absorber mirrors (SESAM'S) for femtosecond to nanosecond pulse generation in solid-state lasers. IEEE J. Sel. Top. Quantum Electron. **2**, 435–451 (1996)
11. Noda, J., Okamoto, K., Sasaki, Y.: Polarization-maintaining fibers and their applications. J. Lightwave Technol. **4**, 1071–1089 (1986)
12. Sánchez, L.A., Cuadrado-Laborde, C., Carrascosa, A., Díez, A., Cruz, J.L., Andrés, M.V.: Low-repetition-rate all-polarization maintaining thulium doped passively modelocked fiber laser. Opt. Laser Technol. **149**, 107856/1-6 (2022)
13. Cuadrado-Laborde, C., Díez, A., Cruz, J.L., Andrés, M.V.: All polarization-maintaining passively modelocked thulium doped fiber laser at 2.07 μm. IEEE Photon. J. **14**, 1–5 (2022)
14. Cuadrado-Laborde, C., Muñoz-Marco, H., Pérez-Millán, P., Díez, A., Cruz, J.L., Silvestre, E., Andrés, M.V.: 1 GHz fundamental repetition rate thulium-doped all-polarization maintaining modelocked fiber laser. J. Phys. Photonics 7, 015001 (2025)
15. Rudy, C.W., Digonnet, M.J.F., Byer, R.L.: Advances in 2-μm tm-doped mode-locked fiber lasers. Opt. Fiber Technol. **20**, 642–649 (2014)
16. Bertie, J.E., Lan, Z.: Infrared intensities of liquids XX: the intensity of the OH stretching band of liquid water revisited, and the best current values of the optical constants of H2O(l) at 25°C between 15,000 and 1 cm−1. Appl. Spectrosc. **50**, 1047–1057 (1996)
17. Polder, K.D., Bruce, S.: Treatment of melasma using a novel 1,927-nm fractional thulium fiber laser: a pilot study. Dermatologic Surg. **38**, 199–206 (2012)
18. Theisen, I.D., Ott, V., Bernd, H.W., Danicke, V., Keller, R., Brinkmann, R.: CW high power IR-laser at 2μm for minimally invasive surgery. In: Steiner, R. (ed.) Therapeutic Laser Applications and Laser-Tissue Interactions, vol. 5142. Optical Society of America (2003)
19. Martinez, O.E., Fork, R.L., Gordon, J.P.: Theory of passively mode-locked lasers including self-phase modulation and group-velocity dispersion. Opt. Lett. **9**, 156–158 (1984)
20. Cuadrado-Laborde, C., Carrascosa, A., Pérez-Millán, P., Díez, A., Cruz, J.L., Andrés, M.V.: Phase recovery by using optical fiber dispersion. Opt. Lett. **39**, 598–601 (2014)
21. Cuadrado-Laborde, C., Brotons-Gisbert, M., Serafino, G., Bogoni, A., Pérez-Millán, P., Andrés, M.V.: Phase recovery by using optical fiber dispersion and pulse pre-stretching. Appl. Phys. B Lasers Opt. **117**, 1173–1181 (2014)
22. Cuadrado-Laborde, C., Armas-Rivera, I., Carrascosa, A., Kuzin, E.A., Díez, A., Andrés, M.V.: Instantaneous frequency measurement of dissipative soliton resonance light pulses. Opt. Lett. **41**, 5704–5707 (2016)
23. Ciąćka, P., Rampur, A., Heidt, A., Feurer, T., Klimczak, M.: Dispersion measurement of ultra-high numerical aperture fibers covering thulium, holmium, and erbium emission wavelengths. J. Opt. Soc. Am. B. **35**, 1301–1307 (2018)
24. Kurkov, A.S., Sholokhov, E.M., Medvedkov, O.I., Dvoyrin, V.V., Pyrkov, Y.N., et al.: Holmium fiber laser based on the heavily doped active fiber. Laser Phys. Lett. **6**, 661–664 (2009)
25. Lee, J., Lee, J.H.: A passively Q-switched holmium-doped fiber laser with graphene oxide at 2058 nm. Appl. Sci. **11**, 407 (2021)
26. Wan, P., Yang, L.-M., Liu, J.: High power 2 μm femtosecond fiber laser. Opt. Express. **21**, 21374–21379 (2013)

27. Jackson, S.: The spectroscopic and energy transfer characteristics of the rare earth ions used for silicate glass fibre lasers operating in the shortwave infrared. Laser & Photon. Rev. **3**, 466–482 (2009)
28. Haus, H.A.: Mode-locking of lasers. IEEE J. Sel. Topic. Quantum Electron. **6**, 1173–1185 (2000)
29. Smolski, V.O., Yang, H., Gorelov, S.D., Schunemann, P.G., Vodopyanov, K.L.: Coherence properties of a 2.6–7.5 μm frequency comb produced as a subharmonic of a Tm-fiber laser. Opt. Lett. **41**, 1388–1391 (2016)
30. Luo, Z.-C., Liu, M., Liu, H., Zheng, X.-W., Luo, A.-P., Zhao, C.-J., Zhang, H., Wen, S.-C., Xu, W.-C.: 2 GHz passively harmonic mode-locked fiber laser by a microfiber-based topological insulator saturable absorber. Opt. Lett. **38**, 5212–5215 (2013)
31. Akosman, A.E., Sander, M.Y.: Low noise, mode-locked 253 mhz tm/ho fiber laser with core pumping at 790 nm. IEEE Photon. Technol. Lett. **28**, 1878–1881 (2016)
32. Kuan, P.-W., et al.: 0.5-Ghz repetition rate fundamentally tm-doped mode-locked fiber laser. IEEE Photon. Technol. Lett. **28**, 1525–1528 (2016)
33. Sotor, J., Bogusławski, J., Martynkien, T., Mergo, P., Krajewska, A., Przewłoka, A., StrupiŃski, W., SoboŃ, G.: All-polarization-maintaining, stretched-pulse tm-doped fiber laser, mode-locked by a graphene saturable absorber. Opt. Lett. **42**, 1592–1595 (2017)
34. https://www.batop.de/products/saturable-absorber/saturable-absorber-mirror/saturable-absorber-mirror-2000nm.html
35. Hlubina, P.: White-light spectral interferometry with the uncompensated Michelson interferometer and the group refractive index dispersion in fused silica. Opt. Commun. **193**, 1–7 (2001)
36. Siegman, A.E.: Lasers. University Science Books (1986)

Chapter 7
Ultrafast Raman Fiber Laser

Jiaqi Zhou and Yan Feng

Abstract Ultrafast Raman fiber lasers have emerged as one of the most effective technologies for generating short optical pulses at specific wavelengths, experiencing significant advancements over the past decades. However, achieving both high pulse energy and short pulse duration in Raman lasers remains challenging, a capability more readily realized in conventional rare-earth-doped ultrafast fiber lasers. This chapter provides an overview of three primary ultrafast Raman fiber laser technologies: mode-locking, synchronous pumping, and nonlinear optical gain modulation (NOGM). Each method is examined in detail, with a focus on its unique technical features and associated challenges. Notably, recent progress in NOGM technology is highlighted, as it offers several advantages, including structural simplicity, long-term stability, high pulse energy, and short pulse duration. This chapter aims to illustrate the developmental trajectory of ultrafast Raman fiber lasers and offer insights into the latest technologies, potentially fostering broader applications for these lasers.

Keywords Ultrafast fiber lasers · Raman fiber lasers · Mode-locking · Synchronously pumping · Nonlinear optical gain modulation

---

J. Zhou
Shanghai Institute of Optics and Fine Mechanics, Chinese Academy of Sciences, Shanghai, China

Y. Feng (✉)
Hangzhou Institute for Advanced Study, University of Chinese Academy of Sciences, Hangzhou, China
e-mail: yfeng@ucas.ac.cn

M. F. S. Ferreira, M. C. Paul (eds.), *Optical Fiber Lasers*, Springer Series in Optical Sciences 250, https://doi.org/10.1007/978-3-032-04110-4_7

7.1 Introduction

Fiber lasers offer a simple and robust solution for generating ultrafast optical pulses. Compared with solid-state counterparts like Ti-sapphire lasers, ultrafast fiber lasers provide advantages such as a compact structure, cost-effectiveness, ease of operation, and diffraction-limited beam quality [1, 2]. Rare-earth-doped fibers are commonly used as gain media in these lasers; however, the emission spectra of rare-earth ions restrict lasing to discrete spectral ranges. This limitation affects the utility of ultrafast fiber lasers in applications like bioimaging, fundamental scientific research, and optical metrology, where pulses at specific wavelengths are essential [3–5]. Consequently, driven by these advanced applications, research on ultrafast fiber lasers with high pulse energy and short pulse duration is progressing toward an expanded spectral range.

In most ultrafast fiber lasers, laser pulses are tightly confined within single-mode fibers over long interaction lengths, leading to significant nonlinear effects. Nonlinearity, in fact, represents the primary barrier to scaling up pulse energy in ultrafast fiber systems [6]. However, if we see this issue on the other side, nonlinearity can be manipulated to generate new spectral components. Several existing techniques exploit nonlinear optical effects in fibers to expand the output wavelength range of ultrafast fiber lasers, including four-wave mixing [7], self-phase modulation [8], and supercontinuum generation [9] et al. Nonetheless, each of these methods has specific limitations, such as restricted spectral tuning ranges, low spectral power density, and stringent requirements for fiber dispersion engineering.

In contrast, stimulated Raman scattering (SRS) in optical fibers can effectively address the limitations mentioned above. By using a pump laser with an appropriate wavelength, laser output at any target wavelength within the transparent window of conventional optical fibers can be achieved with high conversion efficiency in the near-infrared regime [10–12]. As a result, ultrafast Raman fiber lasers are recognized as an excellent solution for generating ultrafast pulses with wavelength agility. These lasers hold promise for producing high-power laser pulses at specialized wavelengths, which could expand applications across industry, biomedical fields, and scientific research. Beyond wavelength agility, Raman fiber lasers offer additional benefits over rare-earth-doped systems, such as lower quantum defect, higher damage thresholds, and resistance to photodarkening.

Various approaches have been developed for ultrafast Raman fiber lasers, including mode-locking, synchronous pumping, and nonlinear optical gain modulation (NOGM). With rapid advancements in both numerical and experimental demonstrations, ultrafast Raman fiber lasers have reached a level of maturity that supports their transition from the laboratory to real-world applications. This chapter reviews recent progress in ultrafast Raman fiber lasers, with a focus on three primary technologies for generating ultrafast Raman pulses in optical fibers.

In Sect. 2, we review mode-locked Raman fiber lasers utilizing different saturable absorbers, highlighting major technological challenges. Section 3 provides an overview of research on synchronously pumped Raman fiber lasers, which produce high-quality ultrafast Raman pulses with some trade-offs in technical simplicity. Section 4 discusses recent advancements in NOGM-based Raman fiber lasers, which offer a straightforward and effective approach for generating femtosecond Raman pulses with high pulse energy. Finally, Section 5 presents a summary and outlook on the future development directions for ultrafast Raman fiber laser technologies.

7.2 Mode-Locked Raman Fiber Laser

Passive mode-locking with saturable absorbers (SAs) is a highly effective method for generating high-quality ultrafast laser pulses in fiber oscillators [13, 14]. The SA provides passive modulation at a frequency that matches the cavity length, enabling periodic modulation. As illustrated in Fig. 7.1, the transmission of an SA is intensity dependent: continuous-wave (CW) lasers and low-intensity pulses experience higher losses, while high-intensity pulses achieve greater transmission. This mechanism allows CW components to be suppressed, establishing pulses in the temporal domain and locking phases between different longitudinal modes in the spectral domain. Although passive mode-locking is a mature technology for conventional rare-earth-doped ultrafast fiber lasers, new challenges arise when substituting rare-earth gain media with SRS.

A variety of SAs have been utilized in passive mode-locked Raman fiber lasers, generally categorized into two types: material-based SAs and artificial SAs. Table 7.1 summarizes the key performance characteristics of mode-locked Raman fiber lasers achieved in recent years.

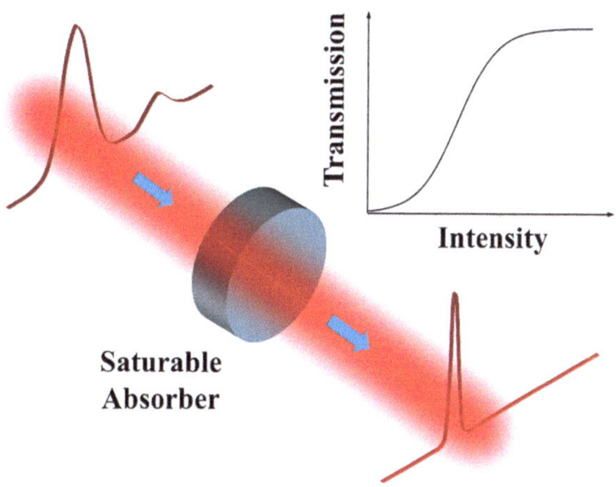

Fig. 7.1 Principle of SA in mode-locked lasers. The transmission of an SA is intensity dependent: CW lasers and pulses with low intensity would experience higher losses, while pulses with high intensity would achieve greater transmission

Table 7.1 Representative performances of mode-locked Raman fiber lasers

Laser	Center wavelength	Pulse energy	Pulse width	Repetition rate	SNR of the RF spectrum
Material-based SA: SESAM [15]	1590 nm	N/A	2.7 ps	220 kHz	N/A
Nanotubes [16]	1670 nm	3 nJ	2 ps	1.72 MHz	30 dB
Nanotubes [17]	1120 nm	20 nJ	236 ps	2.88 MHz	30 dB
Nanotubes [18]	1312 nm	0.1 nJ	2 ns	870 kHz	N/A
Graphene [19]	1180 nm	150 nJ	200 ns	400 kHz	56 dB
Artificial SA: NPR [20]	1380 nm	N/A	2 ps	300 kHz	N/A
NPR [21]	1651 nm	290 nJ	890 ps	378 kHz	41 dB
NPR [22]	1600 nm	193 nJ	180 ps	275 kHz	65 dB
NPR [23]	1120 nm	0.7 nJ	1 ps	2.47 MHz	85 dB
NOLM [24]	1410 nm	15 nJ	N/A	660 kHz	N/A
NOLM [25]	1534 nm	22 nJ	6 ps	64 kHz	N/A
NOLM [26]	1120 nm	64.1 nJ	25 ns	733 kHz	66 dB
NOLM [27]	1115 nm	1.23 nJ	63 ps	1.23 MHz	85 dB

7.2.1 Material-Based Saturable Absorber

Material-based saturable absorbers (SAs) are widely used to initiate and stabilize ultrafast pulses in fiber oscillators. The performance of different material-based SAs has been explored in Raman fiber lasers, including semiconductor saturable absorber mirrors (SESAM) [15], carbon nanotubes [16–18], and graphene [19]. For instance, Chamorovskiy et al. demonstrated a 1590 nm mode-locked Raman fiber laser using a SESAM as the SA and a 450 m long single-mode fiber to achieve sufficient Raman gain under continuous-wave (CW) pumping conditions [15]. Although pulsed operation was achieved, the resulting pulse train exhibited significant intensity fluctuations, indicating low inter-pulse stability. Similarly, Castellani et al. reported a 1670 nm ultrafast Raman fiber laser mode-locked by carbon nanotubes [16], requiring a long fiber cavity of over 100 m. The output pulses could be compressed to 2 ps with a peak power of 1.4 kW; however, the radio frequency (RF) spectrum showed a signal-to-noise ratio (SNR) of only 30 dB, again reflecting limited pulse stability.

In the previously mentioned studies, Raman fiber lasers mode-locked with material-based SAs exhibited unsatisfactory performance compared to their rare-earth-doped counterparts. The primary performance shortcomings, summarized in Table 7.1, include large pulse widths and poor pulse stability, as indicated by the low SNR in their radio frequency (RF) spectra. These deficiencies arise from the fact that Raman gain is a nonlinear process with an extremely fast response time. Typically, the response time of Raman scattering in optical fibers is shorter than the recovery time of material-based SAs. Consequently, under CW pumping conditions, real-time Raman effects occur when the SA becomes bleached. This results in stretched laser pulses and the generation and amplification of spontaneous Raman components between adjacent laser pulses, significantly degrading the mode-locking performance. Thus, while material-based SAs are relatively easy to implement in Raman fiber lasers, their slow recovery times present a major obstacle to achieving high-quality Raman mode-locking. This issue can be addressed by utilizing artificial SAs.

7.2.2 *Artificial Saturable Absorber*

In contrast to material-based saturable absorbers (SAs), artificial SAs exhibit nearly instantaneous response times on the order of femtoseconds, which is even shorter than the response time of the Raman process in optical fibers. This characteristic allows artificial SAs to be effectively functionalized throughout the entire Raman gain process, enabling the generation of mode-locked Raman pulses with high stability. The most commonly used artificial SAs in mode-locked fiber lasers are nonlinear polarization rotation (NPR) and nonlinear optical loop mirrors (NOLM).

7.2.2.1 Nonlinear Polarization Rotation

NPR is a technique that exploits Kerr nonlinearity in non-polarization-maintaining (PM) fibers [28]. The intensity-dependent nonlinear phase shift leads to variations in the polarization state as a function of light intensity. When a polarizer is aligned with the polarization state of high-intensity light, the laser pulses experience lower loss compared to CW light. As a result, NPR can function as an effective fast saturable absorber.

NPR mode-locked Raman fiber lasers have been extensively studied and reported by various research groups [20–23]. For instance, Kuang et al. demonstrated a passively mode-locked Raman fiber laser using an NPR-based configuration with a high-nonlinearity fiber (HNLF) cavity [21]. In their setup, NPR was implemented using two polarization controllers (PCs) and a PM isolator with the fast axis blocked. The pump source was a 1539 nm CW laser, coupled into a 500 m long HNLF as the Raman gain medium to generate 1650 nm Raman pulses. This Raman laser produced pulses of 890 ps with energies reaching up to 290.7 nJ. However, the large pulse width and low SNR of the RF spectrum (approximately 40 dB) remain unsatisfactory, limiting its applicability in many potential applications.

To reduce pulse width and enhance pulse stability, our group developed an improved NPR-based mode-locked Raman fiber laser, as illustrated in Fig. 7.2(a) [23]. First, we replaced traditional fiber oscillator pumps with an amplified spontaneous emission (ASE) pump source. Given that SRS is a nonlinear effect characterized by a fast response time, the temporal characteristics of the pump laser directly influence its Stokes component. Therefore, using a pump laser with poor temporal stability can degrade the performance of a mode-locked Raman fiber laser. In contrast to fiber oscillators, ASE sources exhibit significantly lower intensity

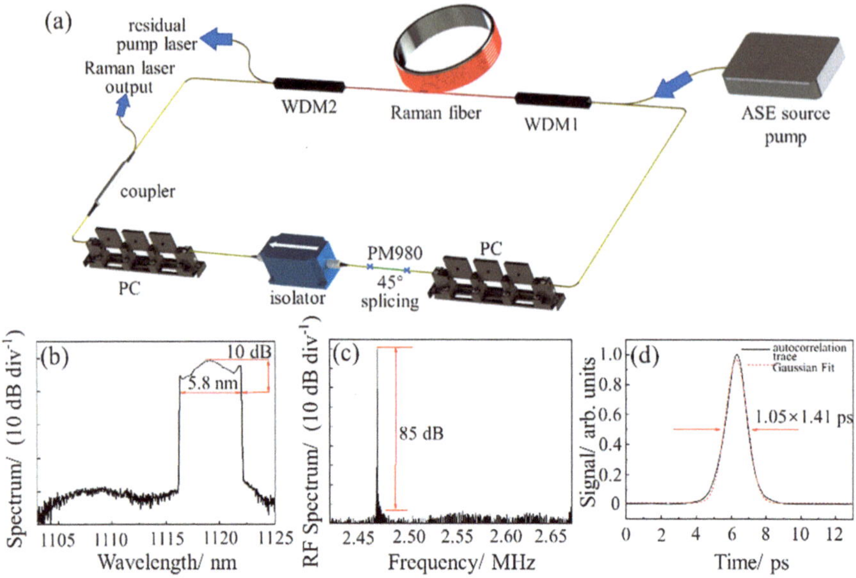

Fig. 7.2 (**a**) Scheme of the NPR-based mode-locked Raman dissipative soliton fiber laser with an ASE pump source. (**b**) Spectrum, (**c**) RF spectrum, and (**d**) autocorrelation trace of the Raman pulses [23]

fluctuations, making them more suitable as pumps for mode-locked Raman fiber lasers. Secondly, we incorporated an all-fiber Lyot filter into the cavity to initiate dissipative soliton (DS) formation, which is a more stable attractor capable of achieving higher pulse energy than conventional solitons. Lastly, we shortened the cavity length by utilizing a 70 m long fiber with a high Raman gain coefficient as the gain medium. With these enhancements, our laser successfully generates stable 1120 nm Raman pulses with a spectral bandwidth of 5.8 nm [Fig. 7.2(b)] and a pulse duration of 1.05 ps [Fig. 7.2(d)] at a repetition rate of 2.47 MHz. Additionally, we measured an SNR of 85 dB [Fig. 7.2(c)] in the RF spectrum, indicating excellent temporal stability compared to other reported systems.

However, due to the operational mechanism of NPR, non-PM fiber is typically required in an NPR-based oscillator. This dependency makes the performance of NPR-based systems sensitive to environmental factors such as temperature and humidity. As a result, NPR is seldom regarded as a viable option for achieving long-term stable mode-locking. This lack of long-term stability restricts their applicability in various fields and has driven research efforts toward alternative methods for generating long-term stable mode-locked Raman pulses.

7.2.2.2 Nonlinear Optical Loop Mirror

As another type of artificial SA, the NOLM can achieve mode-locking using an all-PM fiber structure. Typically, NOLM is based on a 2 × 2 fiber coupler, where the two arms of the coupler are configured to form a fiber loop mirror. When a laser beam is injected into the fiber loop from the input port, it is split into two beams that propagate in opposite directions. As these beams traverse the loop mirror, they accumulate different nonlinear phase shifts. The difference in nonlinear phase shifts is intensity dependent, resulting in variations in transmission and reflection based on light intensity. If the coupling ratio of the 2 × 2 fiber coupler and the length of the fiber loop are appropriately designed, laser pulses will exhibit higher transmission than CW light, encouraging the oscillator to operate in a pulsed regime [29].

NOLM-based mode-locked Raman fiber lasers have also been investigated for several years [24–27]. For instance, a NOLM-based mode-locked Raman fiber laser capable of generating parabolic pulses was studied both numerically and experimentally [25]. This NOLM consisted of a 45/55 coupler, with its two ends looped together using 727 m of HNLF to enhance the nonlinear phase shift difference. The pump source was a 1435 nm CW laser, coupled into a 2.4 km long dispersion compensating fiber, which served as the Raman gain medium, generating 1534 nm Raman pulses. This setup produced parabolic Raman pulses with a spectral bandwidth of 2.4 nm, a pulse duration of 450 ps, and pulse energy of up to 22 nJ. However, this NOLM-based Raman fiber laser utilized non-PM fibers, which limits the advantage of NOLM over NPR in terms of long-term stability.

The first all-PM NOLM-based mode-locked Raman fiber laser was subsequently developed by our group, as illustrated in Fig. 7.3 [27]. A PM fiber coupler with a splitting ratio of 20:80 was employed to connect the NOLM ring to the unidirectional ring. In the unidirectional ring, an 80 m long PM Raman fiber served as the Raman gain medium, and an all-fiber Lyot filter was incorporated into the cavity to initiate the formation of DS. To enhance inter-pulse stability, we utilized a 1064 nm ASE source with low-intensity fluctuations as the pump. The NOLM ring was constructed using 77 m of PM980 fiber. The large splitting ratio of the PM fiber coupler allowed for self-started mode-locking even with a significantly shorter NOLM fiber ring compared to previously reported works. This laser successfully generated stable 1120 nm Raman pulses with a spectral bandwidth of 6.6 nm [Fig. 7.3(b)] and a pulse duration of 69 ps [Fig. 7.3(d)] at a repetition rate of 1.23 MHz. We measured an SNR of 85 dB [Fig. 7.3(c)] in the RF spectrum, indicating excellent temporal stability. Thanks to the all-PM configuration, the mode-locked operation was maintained for weeks without any performance degradation.

Despite significant advancements, the performance of mode-locked Raman fiber lasers remains inferior to that of their rare-earth doped counterparts. One of the primary challenges is reducing the cavity length. As indicated in Table 7.1, the repetition rates of reported mode-locked Raman fiber lasers are all below 3 MHz, suggesting that their cavity lengths are either close to or exceed 100 m. Ultrafast pulses propagating through such lengthy fiber cavities experience considerable dispersion

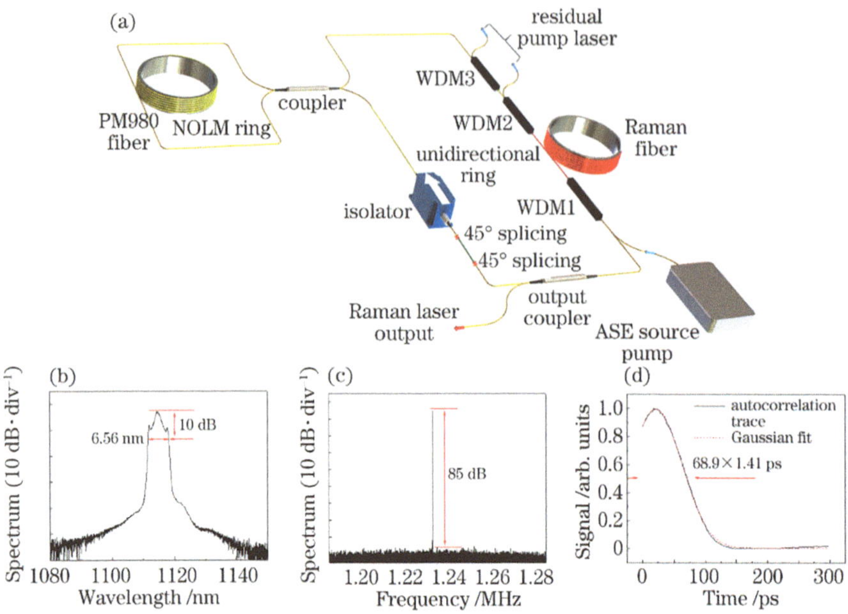

Fig. 7.3 (**a**) Scheme of the PM NOLM-based mode-locked Raman DS fiber laser. (**b**) Spectrum, (**c**) RF spectrum, and (**d**) autocorrelation trace of the Raman pulses [27]

and nonlinearity, which limits pulse energy, broadens pulse duration, and constrains the operational regime for single-pulse generation.

However, to achieve sufficient Raman gain, fiber cavities typically need to be tens or even hundreds of meters long under CW pumping conditions. Thus, replacing CW pumping with pulsed pumping can be an effective strategy for producing Raman pulses with improved performance while simultaneously shortening the cavity length and maintaining adequate Raman gain.

7.3 Synchronously Pumped Raman Fiber Laser

Since the Raman gain coefficient is intensity dependent [30], replacing CW pumping with pulsed pumping can significantly enhance Raman conversion efficiency, addressing the challenge of long fiber cavity lengths in mode-locked Raman fiber lasers. However, when using a pulsed pump, a critical matching condition must be fulfilled to achieve resonance enhancement: the repetition rate of the pulsed pump must correspond to the length of the Raman fiber cavity. As illustrated in Fig. 7.4, synchronization is essential; only when this condition is met can the generated Raman pulses align with the pump, allowing for continuous nonlinear optical gain within the cavity. This technique is referred to as synchronously pumping. Table 7.2

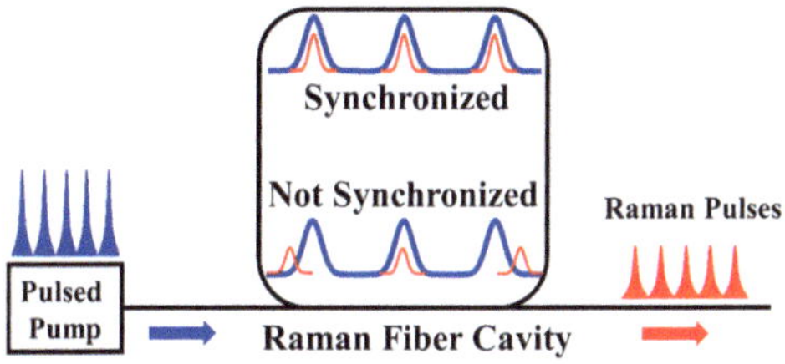

Fig. 7.4 Principle of synchronously pumped Raman fiber laser. When the pump is pulsed, a critical matching condition needs to be met to achieve resonance enhancement; that is, the repetition rate of the pulsed pump should match the length of the Raman fiber cavity

Table 7.2 Performances of some representative synchronously pumped Raman fiber lasers

Laser	Center wavelength	Pulse energy	Pulse width	Repetition rate	SNR of the RF spectrum
Fiber delay line [36]	1060 nm	12 nJ	400 fs	5 MHz	60 dB
Fiber delay line [37]	1080 nm	18 nJ	147 fs	43 MHz	55 dB
Fiber delay line [38]	1240 nm	1.6 nJ	570 fs	15 MHz	70 dB
Fiber delay line [39]	1270 nm	16 nJ	270 ps	19 MHz	N/A
Gain-switched diode [40]	1120 nm	73 nJ	18 ps	12 MHz	60 dB
Self-synchronization [41]	1120 nm	6 nJ	36 ps	28 MHz	60 dB

summarizes the performances of several representative synchronously pumped Raman fiber lasers developed in recent years.

In earlier studies, researchers often used nanosecond pulses as the pump to simplify synchronization challenges [31–35]. However, this approach makes it difficult to achieve Raman pulses in the few-picosecond or even femtosecond range due to the long pump pulse duration. A breakthrough was reported in 2014, where a Raman dissipative soliton (DS) was simultaneously generated alongside a traditional DS using a Yb-doped mode-locked fiber laser [36].

A key distinction of this laser compared to conventional designs is the inclusion of a fiber delay line within the cavity, which compensates for the differences in group velocities between the DS and the Raman pulses. Without this delay line, the generated Raman component could not be persistently amplified due to the walk-off effect between the DS and the Raman pulses, leading to noise-like Raman pulses with a triangular spectral shape. By incorporating a delay line as a Raman feedback loop, the Raman component can be synchronously amplified by the DS circulating in the cavity. Under these synchronized conditions, noise is minimized, and energy becomes concentrated within well-defined temporal and spectral ranges. The generated Raman pulse exhibits a spectral bandwidth of approximately 10 nm and a pulse duration of around 400 fs at a repetition rate of 5 MHz.

In the above-mentioned work, the DS and Raman DS share the same fiber cavity. Inspired by this work, researchers demonstrated a high-power synchronously pumped femtosecond Raman fiber laser by separating the Raman fiber cavity from the Yb-doped fiber oscillator [37]. In their setup [see Fig. 7.5(a)], a 4.7 m long all-PM fiber Raman cavity was pumped by a high-power picosecond fiber laser, mode-locked using a SESAM. A PM fiber-coupled variable delay line was integrated within the mode-locked oscillator to ensure synchronization with the Raman laser cavity. The Raman laser produced highly chirped coherent pulses with energies reaching up to 18 nJ, an average power of 0.76 W, and an efficiency of 88%. The pulses were then compressed to 147 fs using a pair of transmission gratings.

The performance of the Raman pulses was compared with and without synchronization, and the experimental results—depicting the spectrum and autocorrelation trace before and after compression—are illustrated in Figs. 7.5(b), (c), and (d), respectively. The findings confirm that achieving synchronization is crucial for generating highly coherent Raman pulses. Without the proper temporal alignment of the pump and Raman pulses, only noise-like Raman pulses can be obtained. The overall performance of the reported femtosecond Raman fiber laser is comparable to that of most rare-earth-doped lasers, effectively addressing the needs of various potential applications.

In addition to this work, delay line-assisted synchronously pumped Raman fiber lasers have also been reported using phosphosilicate fibers to enhance Raman frequency shifts [38, 39]. Although these results are promising, the incorporation of a delay line adds to the cost and complexity of the system. Furthermore, the length of the Raman fiber cavity is sensitive to environmental temperature variations, necessitating feedback control of the delay line to maintain long-term synchronization.

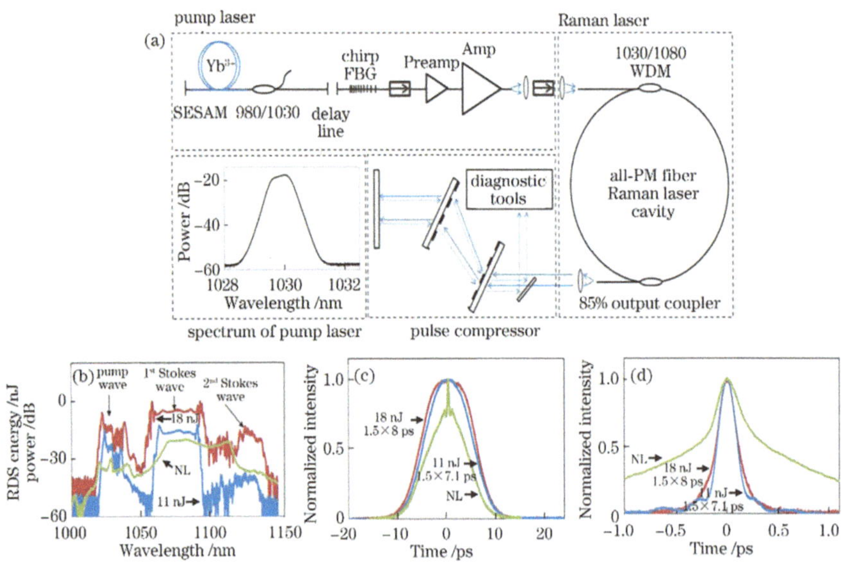

Fig. 7.5 (**a**) Scheme of the synchronously pumped Raman fiber laser with a delay line. (**b**) Output spectrum. (**c**), (**d**) Autocorrelation traces before and after compression [37]

Apart from using a delay line, another approach to achieve synchronous pumping involves gain-switched diodes (GSDs) [40]. In this method, the repetition rate of the diode pump can be continuously tuned by adjusting the electrical signal that controls the diode driver. The configuration of a synchronously pumped Raman fiber laser based on a GSD is illustrated in Fig. 7.6. A 1066 nm distributed feedback (DFB) laser diode is driven by a gain-switched electrical broadband, enabling a wide range of tunability for the repetition rate of the output pulses. To enhance the average power of the pump to 1.2 W, two stages of amplifiers are employed. The amplified pulses are subsequently injected into a Raman fiber cavity approximately 10 m in length. When the repetition rate of the pump diode is fixed at 12.41 MHz, the synchronization condition is achieved, allowing the laser to produce stable Raman pulses with a pulse duration of 78 ps, as shown in Figs. 7.6(b) and (c).

However, the generated Raman pulses exhibit low coherence, rendering them unsuitable for effective dechirping. This limitation arises from the poor coherence of the pump pulses produced by the gain-switching method. Therefore, to obtain highly coherent Raman pulses, further advancements in GSD technology are necessary. Nevertheless, compared to mode-locked Raman fiber lasers, the configuration

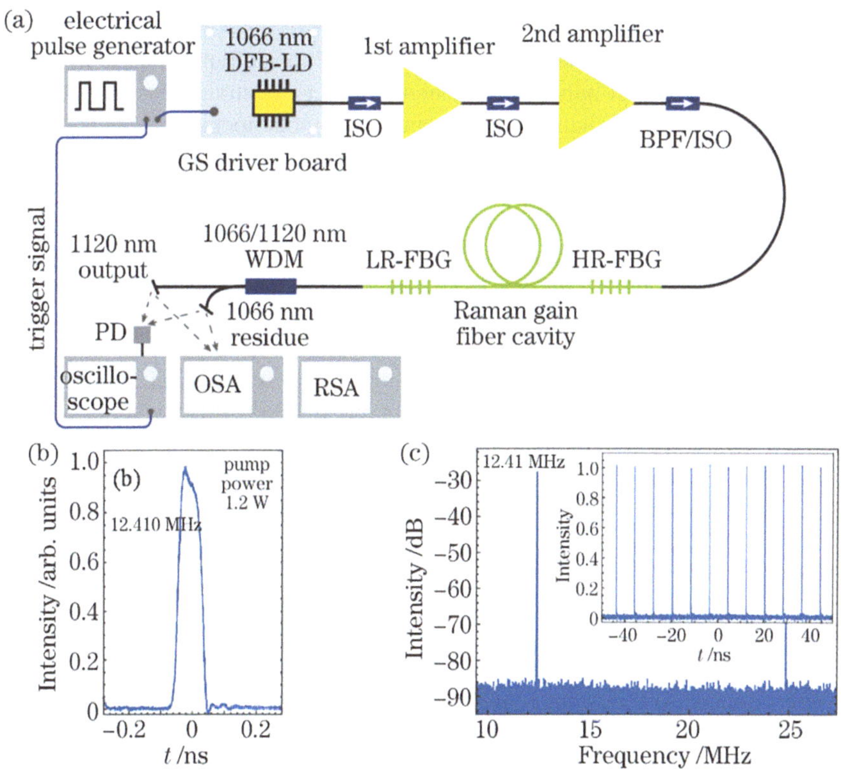

Fig. 7.6 (**a**) Scheme of the synchronously pumped Raman fiber laser with a gain-switched diode. (**b**) Temporal characteristic and (**c**) RF spectrum of the Raman pulses [40]

reported in this work can achieve an optical conversion efficiency of up to 80% with a short gain fiber, thereby highlighting the advantages of synchronous pumping.

From Table 7.2, it can be concluded that the output pulse energies of the reported synchronously pumped Raman fiber lasers are all below 100 nJ, given the current state of the art. One direct approach to increase the output pulse energy of these lasers is to boost the pump pulse energy. However, when the pump power surpasses a certain threshold, the excess energy begins to generate high-order Stokes pulses.

To address this, it is necessary to reduce the cavity length so that the nonlinear effects remain sufficiently controlled, allowing the energy to remain within the first-order Stokes pulses. Unfortunately, this adjustment may compromise the synchronization between the repetition rate of the pulsed pump and the length of the Raman fiber cavity. To achieve synchronously pumped Raman pulses with energies exceeding 100 nJ, it is crucial to employ a pulsed pump that offers both high pulse energy and high repetition rate. However, these stringent requirements on the pump will likely increase both the complexity and cost of the overall system.

7.4 Nonlinear Optical Gain Modulation

In terms of current technical maturity, generating ultrafast Raman pulses with high pulse energy and short pulse duration in a compact and straightforward structure remains a significant challenge, whether employing mode-locking or synchronously pumping techniques. As a result, the practical applications of ultrafast Raman fiber lasers are still limited. Therefore, the development of a new technology that surpasses existing methods holds great scientific significance.

Traditional methods for generating ultrafast Raman pulses primarily rely on fiber oscillators. This approach introduces challenges, such as the need for long Raman gain fibers under CW pumping or synchronization difficulties when using pulsed pumps. An alternative idea is to replace oscillators with single-pass amplifiers, which would eliminate the need for synchronization with pulsed pumps. Some exploratory works in this direction have been reported [42, 43]. A notable example involves the study of amplified spontaneous Raman emission under picosecond pulse pumping, investigated both experimentally and numerically [43]. However, due to the inherent poor coherence of spontaneous emission light, the resulting Raman pulses exhibit noise-like characteristics with partial coherence, making it impossible to properly dechirp them to achieve femtosecond pulse durations.

To enhance the coherence of the generated Raman pulses, a promising approach is to reduce the phase noise of the seed signal by replacing spontaneous Raman emission with a single-frequency laser. Motivated by this idea, our group proposed a novel method to transform a single-frequency CW laser into femtosecond-scale Raman pulses using NOGM [44, 45]. The concept of ultrafast Raman pulse generation via NOGM is illustrated in Fig. 7.7(a). In this method, the input signal is a single-frequency CW laser, which is amplified and temporally and spectrally reshaped within a piece of optical fiber. This optical fiber, which provides nonlinear

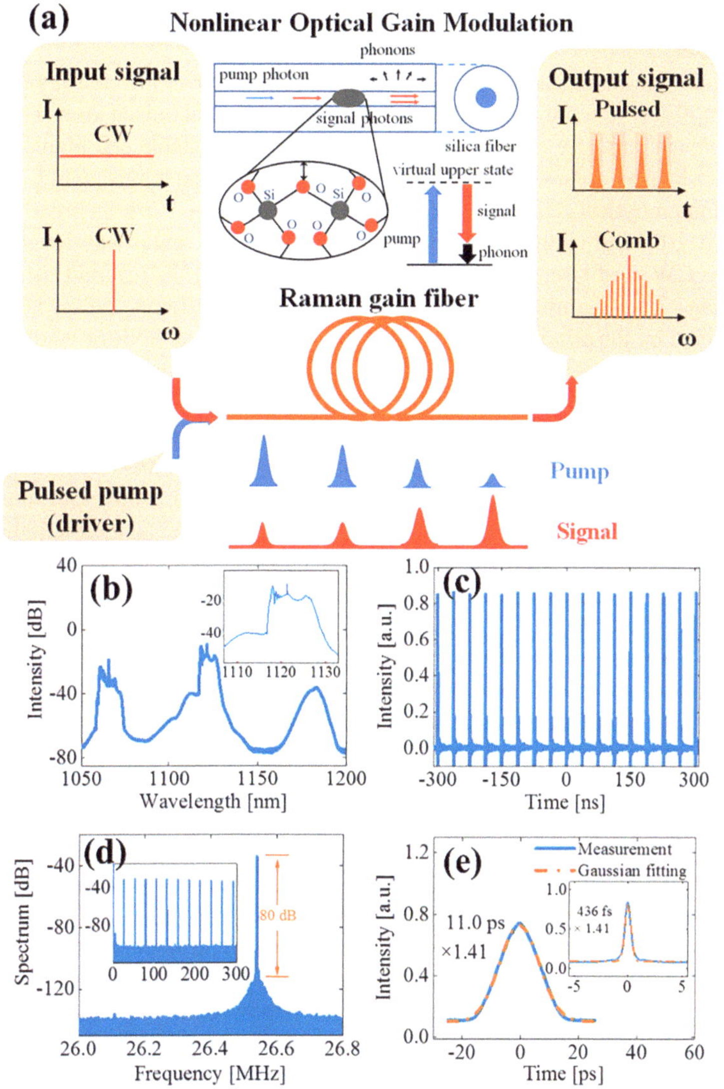

Fig. 7.7 Principle of NOGM (**a**). Characteristics of the experimentally generated NOGM Raman pulses: (**b**) spectrum, (**c**) pulse train, (**d**) RF spectrum, and (**e**) autocorrelation traces [44]

optical gain through SRS, is pumped by ultrafast pulses. The ultrafast pump laser acts as a modulation driver, offering rapid nonlinear optical gain modulation to reshape the CW signal. In the temporal domain, the CW laser is transformed into a pulsed laser, while in the spectral domain, the single longitudinal mode of the CW laser is modulated, generating new spectral components through nonlinear optical effects, resulting in a frequency comb shape. The repetition rate of the generated

Raman pulses, or the comb line spacing, is precisely determined by the pulsed pump laser.

In the experiment, a 1121 nm single-frequency laser is injected into a Raman fiber amplifier, which is pumped by a 1064 nm picosecond pulsed laser. This simple and compact structure enables the amplifier to generate stable and highly coherent 1121 nm Raman pulses with a pulse energy of 25.7 nJ and a pulse width of 436 fs, achieving an optical efficiency of 69.4%. As shown in Fig. 7.7(b), a broad spectrum with a 10-dB bandwidth of 9.5 nm is generated, with the center wavelength aligned with the CW seed laser. Both the oscilloscope trace (Fig. 7.7(c)) and the RF spectrum (Fig. 7.7(d)) confirm that the repetition rate of the Raman pulses is precisely determined by the pump laser. The RF spectrum exhibits an SNR of 80 dB, indicating high pulse-to-pulse stability. According to the autocorrelation measurement (Fig. 7.7(e)), the Raman pulse displays a Gaussian shape with a measured pulse width of 11.0 ps, which can be externally dechirped down to 436 fs, confirming the high coherence of the output Raman pulses. It has been demonstrated that this high pulse coherence arises from the single-frequency seed laser. Numerical simulations suggest that further scaling of the pulse energy to the μJ level is feasible if the power of the pulsed pump is increased.

To fill the spectral gaps left by the emission bands of rare-earth-doped fibers, cascaded Raman processes are necessary. Cascaded NOGM for coherent femtosecond pulse generation with wavelength versatility has also been proposed [46]. In the demonstration, both 1121 nm and 1178 nm single-frequency lasers are injected into a Raman fiber amplifier, which is pumped by a 1064 nm picosecond pulsed laser. The NOGM process transfers energy from 1064 nm to 1121 nm and then to 1178 nm, accomplished within this Raman fiber amplifier. The two-stage NOGM setup generates 1178 nm pulses with a pulse energy of 76 nJ and a pulse duration of 621 fs, achieving an optical conversion efficiency of 65%. Numerical simulations indicate that the walk-off effect between the pump and different-order Stokes pulses is the main factor limiting the performance of high-order Stokes pulses. This suggests that by designing the NOGM amplifier with optimal Raman fiber length, pump pulse energy, and pump pulse duration to minimize the walk-off effect, it is possible to obtain high pulse energy and wavelength-agile femtosecond pulses with higher conversion efficiency and shorter pulse durations. A similar two-order cascaded NOGM Raman fiber amplifier was also demonstrated using phosphosilicate fiber to increase the Raman frequency shift [47]. second Stokes pulses at 1319 nm with a pulse energy of 34 nJ and a pulse duration of 170 fs are generated, providing an efficient method to obtain high-energy sub-picosecond laser sources for applications such as multiphoton microscopy.

The previously mentioned NOGM Raman fiber amplifiers are consistently pumped by mode-locked fiber lasers with fixed repetition rates. This lack of flexibility in repetition rates can be limiting for applications such as nonlinear microscopy and micromachining, where pulse synchronization with scanning optics is essential. Recently, a GSD-pumped NOGM system has been demonstrated, effectively

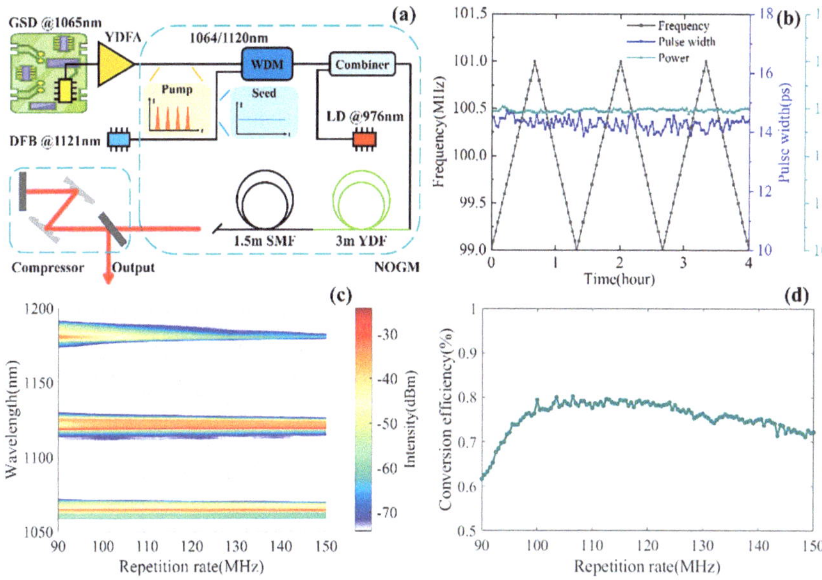

Fig. 7.8 (**a**) Schematic of the GSD-pumped NOGM Raman fiber amplifier. Experiment results of the output pulses at different repetition rate under the same pump power: (**b**) Variation of pulse width and output power versus the repetition rate within the tuning range from 99 to 101 MHz; (**c**) Spectral evolution, and (**d**) Raman conversion efficiency versus the repetition rate in the tuning range from 90 to 150 MHz [48]

generating femtosecond laser pulses with both repetition rate and wavelength agility [48]. The experimental setup for the GSD-pumped NOGM Raman fiber amplifier is illustrated in Fig. 7.8(a). In this setup, the conventional NOGM pump source is replaced with a commercial 1065 nm GSD capable of producing tens of picosecond laser pulses with a tunable repetition rate. In the experimental demonstration, the nonlinear fiber amplifier generates highly coherent 1121 nm femtosecond Raman pulses with up to 80.2% conversion efficiency. The results, shown in Fig. 7.8(b)–(d), confirm that the Raman pulses maintain high performance across a wide range of repetition-rate tunings.

The NOGM method presents several performance advantages over previous approaches. Unlike mode-locking, NOGM is not limited by long cavity lengths, enabling the generation of pulses with high energy and short duration. In comparison to synchronously pumping, NOGM eliminates the need for strict synchronization between the pump and Raman pulses, simplifying the process of generating high-performance ultrafast pulses. Furthermore, NOGM allows for scalability in pulse energy and expandability in wavelength, making it an effective solution for producing high-energy ultrafast pulses with wavelength agility. This capability could significantly enhance the technological adoption and application scope of ultrafast Raman fiber lasers in the coming years.

7.5 Conclusion and Prospect

Ultrafast Raman fiber lasers significantly expand the wavelength range of conventional ultrafast fiber lasers, offering promising applications in industry, biomedical fields, and scientific research. In these areas, ultrafast laser pulses with specific wavelengths play crucial roles.

Figure 7.9 summarizes the performances of the most representative ultrafast Raman fiber lasers reported to date in terms of pulse energy and duration. Currently, mode-locked Raman fiber lasers do not match the performance of their rare-earth-doped counterparts in terms of pulse energy and duration. This limitation primarily arises from the need for long optical fibers to provide adequate Raman gain under CW pumping conditions. Ultrafast pulses generated in such long fiber cavities encounter significant dispersion and nonlinearity, which restricts pulse energy, broadens pulse duration, and narrows the operational regime for single-pulse generation. This issue is particularly critical for mode-locked Raman fiber lasers, as highlighted in the blue elliptical area in the middle section of Fig. 7.9. Recently, spatiotemporal mode-locking and multimode laser-pumped Raman fiber lasers have been demonstrated [49, 50]. Both technologies apply graded-index fibers in their design. Therefore, the combination of these two technologies may address open scientific questions and pave the way toward a spatiotemporal mode-locked Raman fiber laser with a relatively short cavity.

As illustrated in Fig. 7.9, the performance of synchronously pumped Raman fiber lasers is concentrated in the middle-top area of the figure (orange circle). Pulsed pumping effectively addresses the issue of long cavity length, enabling femtosecond pulse operation. Currently, delay line-based synchronously pumping systems can produce Raman pulses with performance comparable to that of conventional

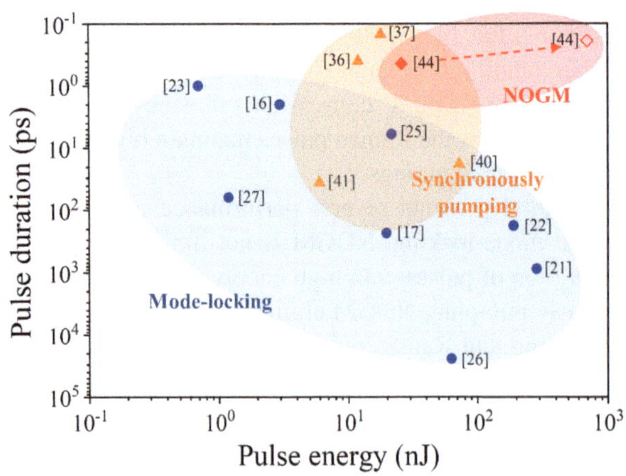

Fig. 7.9 Summary of the pulse energy and duration of the reported Raman fiber lasers, including mode-locking (blue circles), synchronously pumping (orange triangle), and NOGM (red diamond). The corresponding reference numbers of each work are marked nearby

rare-earth-doped ultrafast fiber lasers. However, the inclusion of a delay line adds complexity and cost to the system. In contrast, systems utilizing GSD present a simpler solution for achieving synchronization. Nevertheless, due to the poor coherence of gain-switched pulses, only noise-like Raman pulses have been reported thus far. One promising optimization approach involves using a Mamyshev regenerator to enhance the coherence of pulses generated by GSD [51]. If the coherence of gain-switched pulses can be improved to match that of pulses generated by mode-locked oscillators, it could enable the generation of high-performance Raman pulses similar to those produced by delay line systems.

It is important to emphasize that both mode-locked and synchronously pumped Raman fiber lasers are based on oscillator designs. Consequently, to balance dispersion, nonlinearity, gain, and loss within these fiber oscillators, their conventional output pulse energies are typically limited to the nanojoule level. In contrast, the NOGM method avoids the oscillator structure, offering several advantages, including a simpler design, long-term stability, higher pulse energy, and shorter pulse duration. Simulations have demonstrated that femtosecond Raman pulses with peak powers reaching several megawatts could be generated using NOGM, indicating the potential for achieving pulse energies on the order of several microjoules in an all-fiber NOGM configuration. This capability would position NOGM pulses in the upper right corner of Fig. 7.9, enhancing the performance of ultrafast Raman fiber lasers. High-quality ultrafast laser sources generated through NOGM are likely to expand the scope of applications for these technologies. Future research into NOGM-based frequency combs may further amplify the impact of this innovation. One primary technical challenge associated with NOGM is mitigating the walk-off effect between pump and Raman pulses caused by dispersion in the cascaded setup. Pulse walk-off can not only reduce conversion efficiency but also result in irregular pulse shapes and nonlinear pulse chirp, ultimately hindering the generation of Raman pulses with short durations. This walk-off issue can be addressed by either designing the pump and Raman wavelengths to operate around the zero-dispersion wavelength of specially designed fibers or by employing a dispersion-managed configuration, which consists of two fiber segments with normal and anomalous dispersion, respectively, in the Raman fiber amplifier.

References

1. Fermann, M.E., Hartl, I.: Ultrafast fibre lasers. Nat. Photonics. **7**, 868–874 (2013)
2. Zhou, J., Pan, W., Qi, W., Cao, X., Cheng, Z., Feng, Y.: Ultrafast Raman fiber laser: a review and prospect. PhotoniX. **3**, 18 (2022)
3. Hoover, E.E., Squier, J.A.: Advances in multiphoton microscopy technology. Nat. Photonics. **7**, 93–101 (2013)
4. Zipfel, W.R., Williams, R.M., Webb, W.W.: Nonlinear magic: multiphoton microscopy in the biosciences. Nat. Biotechnol. **21**, 1369–1377 (2003)
5. Chen, F., V'azquez de Aldana, J.R.: Optical waveguides in crystalline dielectric materials produced by femtosecond-laser micromachining. Laser Photonics Rev. **8**, 251–275 (2014)

6. Fu, W., Wright, L.G., Sidorenko, P., Backus, S., Wise, F.W.: Several new directions for ultrafast fiber lasers. Opt. Express. **26**, 9432–9463 (2018)
7. Murray, R.T., Kelleher, E.J.R., Popov, S.V., Mussot, A., Kudlinski, A., Taylor, J.R.: Widely tunable polarization maintaining photonic crystal fiber based parametric wavelength conversion. Opt. Express. **21**, 15826–15833 (2013)
8. Chung, H., Liu, W., Cao, Q., Kartner, F.X., Chang, G.: Er-fiber laser enabled, energy scalable femtosecond source tunable from 1.3 to 1.7 μm. Opt. Express. **25**, 15760–15771 (2017)
9. Yao, C., Jia, Z., Li, Z., Jia, S., Zhao, Z., Zhang, L., Feng, Y., Qin, G., Ohishi, Y., Qin, W.: High-power mid-infrared supercontinuum laser source using fluorotellurite fiber. Optica. **5**, 1264–1270 (2018)
10. Feng, Y.: Raman fiber lasers. Springer Nature (2017)
11. Supradeepa, V.R., Feng, Y., Nicholson, W.J.: Raman fiber lasers. J. Opt. **19**, 023001 (2017)
12. Zhang, L., Jiang, H., Cui, S., Hu, J., Feng, Y.: Versatile Raman fiber laser for sodium laser guide star. Laser Photonics Rev. **8**, 889–895 (2014)
13. Bao, Q., Zhang, H., Wang, Y., Ni, Z., Yan, Y., Shen, Z., Loh, K.P., Tang, D.Y.: Atomic-layer graphene as a Saturable absorber forUltrafast pulsed lasers. Adv. Funct. Mater. **19**, 3077–3083 (2009)
14. Okhotnikov, O., Grudinin, A., Pessa, M.: Ultrafast fibre laser systems based on SESAM technology: new horizons and applications. New J. Phys. **6**, 177 (2004)
15. Chamorovskiy, A., Rautiainen, J., Lyytikäinen, J., Ranta, S., Tavast, M., Sirbu, A., Kapon, E., Okhotnikov, O.G.: Raman fiber laser pumped by a semiconductor disk laser and mode locked by a semiconductor saturable absorber mirror. Opt. Lett. **35**, 3529–3531 (2010)
16. Castellani, C.E.S., Kelleher, E.J.R., Travers, J.C., Popa, D., Hasan, T., Sun, Z., Flahaut, E., Ferrari, A.C., Popov, S.V., Taylor, J.R.: Ultrafast Raman laser mode-locked by nanotubes. Opt. Lett. **36**, 3996–3998 (2011)
17. Castellani, C.E.S., Kelleher, E.J.R., Popa, D., Hasan, T., Sun, Z., Ferrari, A.C., Popov, S.V., Taylor, J.R.: CW-pumped short pulsed 1.12 μm Raman laser using carbon nanotubes. Laser Phys. Lett. **10**, 015101 (2013)
18. Steinberg, D., Saito, L.A.M., Rosa, H.G., Thoroh de Souza, E.A.: Single-walled carbon nanotube passively mode-locked O-band Raman fiber laser. Opt. Laser Technol. **79**, 55–61 (2016)
19. Zhang, L., Wang, G., Hu, J., Wang, J., Fan, J., Wang, J., Feng, Y.: Linearly polarized 1180-nm Raman fiber laser mode locked by graphene. IEEE Photon. J. **4**, 1809–1825 (2012)
20. Chamorovskiy, A., Rantamäki, A., Sirbu, A., Mereuta, A., Kapon, E., Okhotnikov, O.G.: 1.38-μm mode-locked Raman fiber laser pumped by semiconductor disk laser. Opt. Express. **18**, 23872–23877 (2010)
21. Kuang, Q., Zhan, L., Gu, Z., Wang, Z.: High-energy passively mode-locked Raman fiber laser pumped by a CW multimode laser. J. Lightwave Technol. **33**, 391–395 (2015)
22. Liu, J., Chen, Y., Tang, P., Miao, L., Zhao, C., Wen, S., Fan, D.: Duration switchable high-energy passively mode-locked Raman fiber laser based on nonlinear polarization evolution. IEEE Photon. J. **7**, 1503207 (2015)
23. Pan, W., Zhang, L., Zhou, J., Yang, X., Feng, Y.: Raman dissipative soliton fiber laser pumped by an ASE source. Opt. Lett. **42**, 5162–5165 (2017)
24. Chestnut, D.A., Taylor, J.R.: Wavelength-versatile sub-picosecond pulsed lasers using Raman gain in figure-of-eight fiber geometries. Opt. Lett. **30**, 2982–2984 (2005)
25. Aguergaray, C., Méchin, D., Kruglov, V., Harvey, J.D.: Experimental realization of a mode-locked parabolic Raman fiber oscillator. Opt. Express. **18**, 8680–8687 (2010)
26. Pan, W., Zhou, J., Zhang, L., Feng, Y.: Rectangular pulse generation from a mode locked Raman fiber laser. J. Lightwave Technol. **37**, 1333–1337 (2019)
27. Pan, W., Zhou, J., Zhang, L., Feng, Y.: Raman dissipative soliton fiber laser mode locked by a nonlinear optical loop mirror. Opt. Express. **27**, 17905–17911 (2019)
28. Hofer, M., Fermann, M.E., Haberl, F., Ober, M.H., Schmidt, A.J.: Mode locking with cross-phase and self-phase modulation. Opt. Lett. **16**, 502–504 (1991)
29. Doran, N.J., Wood, D.: Nonlinear-optical loop mirror. Opt. Lett. **13**, 56–58 (1988)

30. Agrawal, G.: Nonlinear Fiber Optics, 5th edn. Academic Press (2013)
31. Stolen, R.H., Lin, C., Jain, R.K.: A time-dispersion-tuned fiber Raman oscillator. Appl. Phys. Lett. **30**, 340–342 (1977)
32. Lin, C., French, W.G.: A near-infrared fiber Raman oscillator tunable from 1.07 to 1.32 μm. Appl. Phys. Lett. **34**, 666–668 (1979)
33. Nakazawa, M., Kuznetsov, M., Ippen, E.P.: Theory of the synchronously pumped fiber Raman laser. IEEE J. Quantum Electron. **22**, 1953–1966 (1986)
34. Smith, K., Kean, P.N., Crust, D.W., Sibbett, W.: An experimental study of a synchronously pumped fibre Raman oscillator. J. Mod. Opt. **34**, 1227–1233 (1987)
35. Lin, D., Alam, S., Teh, P.S., Chen, K.K., Richardson, D.J.: Tunable synchronously-pumped fiber Raman laser in the visible and near-infrared exploiting MOPA-generated rectangular pump pulses. Opt. Lett. **36**, 2050–2052 (2011)
36. Babin, S.A., Podivilov, E.V., Kharenko, D.S., Bednyakova, A.E., Fedoruk, M.P., Kalashnikov, V.L., Apolonski, A.: Multicolour nonlinearly bound chirped dissipative solitons. Nat. Commun. **5**, 4653 (2014)
37. Churin, D., Olson, J., Norwood, R.A., Peyghambarian, N., Kieu, K.: High-power synchronously pumped femtosecond Raman fiber laser. Opt. Lett. **40**, 2529–2532 (2015)
38. Kharenko, D.S., Efremov, V.D., Evmenova, E.A., Babin, S.A.: Generation of Raman dissipative solitons near 1.3 microns in a phosphosilicate-fiber cavity. Opt. Express. **26**, 15084–15089 (2018)
39. Kobtsev, S., Kukarin, S., Kokhanovskiy, A.: Synchronously pumped picosecond all-fibre Raman laser based on phosphorus-doped silica fibre. Opt. Express. **23**, 18548–18553 (2015)
40. Chen, H., Chen, S., Jiang, Z., Yin, K., Hou, J.: All-fiberized synchronously pumped 1120 nm picosecond Raman laser with flexible output dynamics. Opt. Express. **23**, 24088–24096 (2015)
41. Pan, W., Zhang, L., Jiang, H., Yang, X., Cui, S., Feng, Y.: Ultrafast Raman fiber laser with random distributed feedback. Laser Photonics Rev. **12**, 1700326 (2018)
42. Stolen, R.H., Johnson, A.M.: The effect of pulse walkoff on stimulated Raman scattering in fibers. IEEE J. Quantum Electron. **22**, 2154–2160 (1986)
43. Zhao, Q., Pan, W., Zeng, X., Feng, Y.: Partially coherent noise-like pulse generation in amplified spontaneous Raman emission. Appl. Opt. **57**, 2282–2286 (2018)
44. Qi, W., Zhou, J., Cui, S., Cheng, X., Zeng, X., Feng, Y.: Femtosecond pulse generation by nonlinear optical gain modulation. Adv. Photon. Res. **3**, 2100255 (2021)
45. Qi, W., Zhou, J., Cao, X., Cheng, Z., Jiang, H., Cui, S., Feng, Y.: Numerical simulation of nonlinear optical gain modulation in a Raman fiber amplifier. Opt. Express. **30**, 34848–34861 (2022)
46. Qi, W., Zhou, J., Cao, X., Cheng, Z., Jiang, H., Cui, S., Feng, Y.: Cascaded nonlinear optical gain modulation towards coherent femtosecond pulse generation with wavelength versatility. Opt. Express. **30**, 8889–8897 (2022)
47. Qi, W., Zhou, J., Cao, X., Cheng, Z., Li, S., Jiang, H., Cui, S., Feng, Y.: Generation of 1.3 μm femtosecond pulses by cascaded nonlinear optical gain modulation in phosphosilicate fiber. Opt. Lett. **48**, 1698–1701 (2023)
48. Cheng, Z., Zhou, J., Cao, X., Cui, S., Jiang, H., Feng, Y.: Repetition-rate and wavelength flexible femtosecond laser pulse generation. Laser Photonics Rev. **2400788** (2024)
49. Wright, L.G., Sidorenko, P., Pourbeyram, H., Ziegler, Z.M., Isichenko, A., Malomed, B.A., Menyuk, C.R., Christodoulides, D.N., Wise, F.W.: Mechanisms of spatiotemporal modelocking. Nat. Phys. **16**, 565–570 (2020)
50. Chen, Y., Fan, C., Yao, T., Xiao, H., Leng, J., Zhou, P., Nemov, I.N., Kuznetsov, A.G., Babin, S.A.: Brightness enhancement in random Raman fiber laser based on a graded-index fiber with high-power multimode pumping. Opt. Lett. **46**, 1185–1188 (2021)
51. Fu, W., Wright, L.G., Wise, F.W.: High-power femtosecond pulses without a modelocked laser. Optica. **4**, 831–834 (2017)

Chapter 8
Gas-Filled Hollow-Core Fiber Lasers

Andrey Pryamikov

Abstract Gas-filled hollow-core fiber lasers (HCFLs) not only combine the advantages of fiber lasers and gas lasers but also provide an efficient way to develop high-energy fiber lasers in the mid-infrared (IR) spectral range. The development of new designs and manufacturing technology for silica glass hollow-core fibers was key to success in obtaining laser generation at new wavelengths in the mid-IR spectral range. To date, gas-filled HCFLs based on stimulated Raman scattering and population inversion between rotational and vibrational energy levels of the gases filling the air-core have been demonstrated. This chapter reviews major aspects of these topical issues and outlines new promising approaches emerging in the field of gas-filled HCFLs.

Keywords Gas-filled fiber laser · Hollow-core photonic crystal fibers · Negative curvature hollow-core fiber · Stimulated Raman scattering · Population inversion

8.1 Introduction

More than 60 years have passed since Theodore Maiman created the world's first ruby laser, and it was demonstrated that masers could be made to operate in optical and infrared (IR) regions. The history of continuous-wave (CW) gas lasers dates back to the invention of helium-neon lasers in 1960 at Bell Lab. At that time, researchers studied two approaches to produce population inversion in gases, namely, optical pumping and discharge excitation of the gas. Despite the simplicity of the optical pumping approach, the main difficulty lays in its low efficiency for low-density gases. Therefore, the earlier studies of gas lasers were associated precisely with discharge pumping. The first continuous-wave helium–neon laser oscillated on a line at 1153 nm. After that, laser generation with gas discharge was obtained at other wavelengths and for mixtures of other gases such as neon–oxygen,

---

A. Pryamikov (✉)
Prokhorov General Physics Institute of the Russian Academy of Sciences, Moscow, Russia

M. F. S. Ferreira, M. C. Paul (eds.), *Optical Fiber Lasers*, Springer Series in Optical Sciences 250, https://doi.org/10.1007/978-3-032-04110-4_8

argon–oxygen, pure argon, krypton, and xenon [1]. As is known, the most powerful continuous-wave gas lasers were later developed based on molecular gases such as carbon dioxide that could convert more input energy into light. For example, gas dynamic lasers achieved a power level of a few hundred kilowatts in the 1970s. It is also worth noting the works that appeared in the 1970s of the last century and were dedicated to the development of pulsed rare-gas-halide lasers. Those were mainly electron-beam-excited systems.

On the other hand, in 1961, Elias Snitzer proposed a theoretical description of single-mode fibers whose core could carry light with only one waveguide mode. Then, Charles Cao and George Hockham published a paper demonstrating theoretically that light losses could be decreased dramatically in existing glass fibers by removing impurities. The first single-mode silica glass fiber with attenuation less than 20 dB/km was created in Corning Glass Works in 1970. Since then, solid-core silica glass fibers have found wide applications in a variety of industries. The big challenge lays in fabricating low-loss solid-core optical fibers for transmitting infrared (IR) to terahertz frequency radiation. This is precisely the spectral region in which gas lasers mainly operated, and thus, this encouraged the development of hollow-core fibers (HCFs). HCFs have also been used to transmit ultraviolet and soft X-ray radiation [2], for example, excimer laser radiation [3]. HCFs have several advantages over solid-core optical fibers, namely, low insertion loss, low nonlinear effects, high laser damage threshold, and low material absorption as most of the radiation is confined to the air core. Today, there are two types of HCFs: those HCFs that localize light based on total internal reflection such as silica solid-core fibers described above and, secondly, leaky-type HCFs. HCFs that guide light on the principle of total internal reflection have a major drawback, namely, the spectral range where the refractive index of the fiber cladding $n < 1$ is an anomalous dispersion region. In this case, the extinction coefficient of the fiber cladding is also high which leads to strong light attenuation in the fiber cladding. An example is the sapphire HCF that transmitted CO_2 laser radiation with a loss of 0.5 dB/m [4]. Further in this chapter, we will only consider leaky-type HCFs. Initially, leaky-type HCFs were either hollow-core metal tubes with a smooth metal core surface or glass–plastic tubes coated on the core surface with a metal coating [5, 6]. For example, the Ag/AgI metal coated hollow-core waveguide was used in conjunction with a water-cooling jacket to deliver 2.6-kW CO_2 laser radiation at a wavelength of 10.6 μm [7]. Bragg HCFs with multilayer cladding structure (alternating high/low refractive index dielectric layers) were demonstrated in [8]. They are successfully used in CO_2 laser delivery for medical applications.

It is obvious that these two concepts of lasers and hollow-core fibers would eventually lead to a new concept of a fiber laser without a solid core, but with gain provided by gases inside the fiber air-core. This solved the difficult problem of detrimental nonlinear optical processes and material damage in the solid-core gain medium. The available gas media could now cover a broad spectral range of emissions being spectrally narrow due to the nature of transitions in atomic and molecular gases.

The first hollow-core waveguide gas laser was reported by Smith [9]. The laser operated with He–Ne mixtures at 0.633 μm. The laser light propagated through a hollow-core dielectric tube with 430-μm inner diameter and a length of 20 cm. The tube also served to confine the discharge. The first He–Ne laser operating at a wavelength of 3.39 μm with 510 μm-inner diameter was demonstrated in [10]. It is also worth mentioning the development of CO_2 waveguide gas laser which had a discharge tube 23.7 cm length and 3.3 mm diameter [11]. The obtained output power was 2.5 W. The waveguide was located between the mirrors that form the optical resonator. The main reason why these waveguide gas lasers have not been widely used is due to the impossibility of reducing the inner air-core diameter without a significant increase in losses.

A new stage in the development of gas fiber lasers began with the advent of hollow-core photonic crystal fibers (HC PCF) [12]. They could host gases for very long interaction lengths with microscale modal areas. The optical properties of microstructured hollow-core fibers and their application to create new types of gas fiber lasers are discussed in the next Section.

8.2 Photonic Crystal Hollow-Core Fibers as a New Paradigm for Developing Gas-Filled Fiber Lasers

The first type of microstructured hollow-core fibers was HC PCFs whose guiding properties rely on existence of photonic bandgaps. Optical signals are guided in an air core surrounded by 2D photonic crystal cladding [13]. Although more than 99.5% of optical power is propagating in air core in this fiber, the loss was reduced to 1.2 dB/km at $\lambda = 1.62$ μm due to the intrinsic roughness of air–glass interfaces of the core-cladding boundary [14]. This result was obtained for a silica glass HC PCF with a 20-μm diameter core formed by omitting 19-unit cells from a cladding of 3.9 μm pitch, its attenuation being 1.7 dB/km at 1565 nm wavelength.

Increasing the diameter of the air core did not markedly reduce losses as the microbending and the longitudinal variations of the air core became significant for larger core sizes which led to additional losses [15]. In addition, it is necessary to take into account that the wavelength range in which HC PCFs, based on the photonic bandgap guiding, work is quite limited. In particular, this leads to suppressing the propagation of Raman-shifted light, which curbed developing gas fiber lasers based on Raman effect. However, in work [16], it was demonstrated that HC PCF enabled pure rotational stimulated Raman scattering (SRS) in hydrogen at power levels 6 orders of magnitude lower than previously reported in a single-pass configuration with a quantum efficiency close to 100%. The HC PCF transmitted over 1000–1150 nm spectral range with a loss <100 dB/km. The pump wavelength was 1064 nm, and the first Stokes wavelength was 1130 nm. Also, hydrogen and acetylene cells were produced based on HC PCF. The acetylene cells were made from HC PCF with a guidance band center at 1550 nm, and the hydrogen cells were made from HC PCF with a guidance band center at 1064 nm [17]. Generation of rotational SRS was obtained in a 5-m-long in-line hydrogen gas cell with quasi-continuous-wave pulses.

In order to increase a bandwidth of light guidance, a new class of HCFs, namely, Kagome–lattice HC PCFs was developed. The air-core guidance in those fibers did not rely on the formation of the photonic bandgaps. It is assumed that the cladding modes did not interact strongly with the air-core mode in question [13]. Thus, such a mechanism of light localization in hollow-core optical fibers arose due to an inhibited coupling between the air-core modes and the cladding, and the hollow-core fibers themselves came to be called inhibited coupling hollow-core photonic crystal fibers (IC HC PCFs). Kagome–lattice HC PCFs showed significantly greater losses compared to bandgap HC PCFs but guided light over much broader spectral ranges. One of the first achievements of Kagome–lattice IC HC PCFs was the generation of stimulated Raman scattering in 1-meter-long hollow-core fiber filled with hydrogen gas under pressure [18]. The pump-to-Stokes conversion efficiency was 30% at a pump wavelength of 532 nm and pulse duration of 6 ns. The first-order Stokes wavelength was 683 nm and the first-order anti-Stokes was 435.2 nm. A pulse energy was 4.5 microjoules. Thus, the Kagome–lattice cladding structure made it possible to obtain a wide spectral transmission band with a loss level of several dB/m. Which in turn gave the corresponding values of the Stokes shifts. Then, the first CW pumped Raman fiber–gas laser based on H_2-filled Kagome–lattice IC HC PCF was developed [19]. The pump wavelength was 1064 nm, and up to 99.99% of the output power was at the Stokes wavelength (1135 nm) even at atmospheric pressure. Another important achievement was the development of an optically pumped gas laser based on population inversion using Kagome–lattice IC HC PCF [20]. This time, the Kagome–lattice IC HC PCF with an air-core diameter of about 40 μm was filled with a few torr of acetylene ($^{12}C_2H_2$) and was pumped with 5-ns pulses at a wavelength of 1.52 μm. Lasing was observed at 3.12 μm and 3.16 μm in the mid-IR spectral region. The losses at the pump wavelength were 0.75 dB/m, while the losses at the laser output wavelengths were approximately 20 dB/m. Thus, despite confinement of the pump and laser over long interaction lengths in a compact configuration, silica glass Kagome–lattice IC PC HCFs had too much loss in the mid-IR spectral range. In addition to leakage losses, large material losses of silica glass in the mid-IR were added. Thus, a hollow-core fiber design was needed that would allow radiation to be transmitted with significantly lower losses in the mid-IR spectral range.

Losses in the mid-IR range of the spectrum were dramatically reduced after the invention of hollow-core fibers with a negative curvature of the core-cladding boundary. Those negative curvature hollow-core fibers (NCHCFs) have been developed by various scientific groups and have different designs. The most widely used NCHCFs to date have a simple design, consisting of silica glass capillaries located on a thick-walled tube made of the same material [21]. Others were based on Kagome–lattice IC HCFs or had more original designs [22, 23] (Fig. 8.1).

It is currently assumed that in NCHCFs with a cladding consisting of capillaries, the mechanism of light localization is associated with antiresonance scattering of

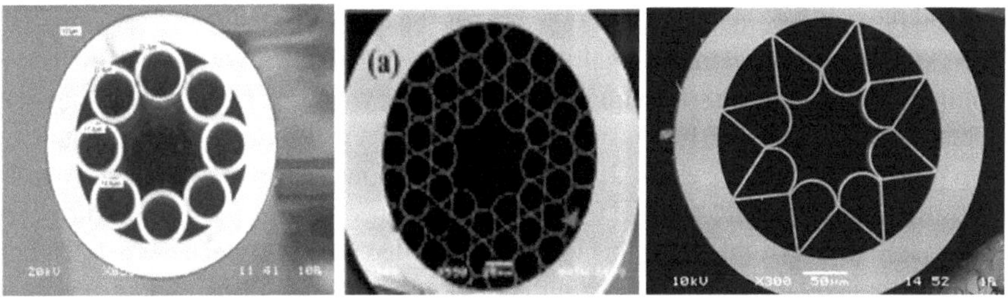

Fig. 8.1 (left) Silica glass NCHCF [21]; (middle) Kagome–lattice IC HCF with negative curvature of the core-cladding boundary [22]; (right) silica glass NCHCF [23]. Figures were taken from [21–23]

radiation from the walls of the cladding capillaries [24]. However, the authors of work [25] believe that for Kagome–lattice NCHCFs the inhibited coupling mechanism plays a decisive role. In addition, it is worth noting that complex vortex phenomena arise in NCHCFs both for the field components of the air-core modes and for their energy flows. As the authors [26] believe, this should undoubtedly affect the loss level in these fibers and possibly plays a decisive role in transmitting radiation with low losses in the mid-IR spectral range. As experimental studies have shown, losses in silica glass NCHCFs in the mid-IR can reach a level of several tens of dB/km [27]. For example, in silica glass NCHCF with a cladding consisting of 7 capillaries and a core diameter of 105 μm, the measured losses at a wavelength of 4 μm were 40 dB/km and the predicted losses at a wavelength of 5 μm were 0.69 dB/m. This is quite enough to develop gas-filled fiber lasers of various types operating in a wide range of the spectrum including the mid-IR. Advances in the development of these lasers based on silica glass NCHCFs will be discussed in the following sections.

8.3 Continuous-Wave Gas-Filled Fiber Lasers Based on NCHCFs

A continuous lasing mode means that key beam parameters such as power output, intensity, etc., remain constant throughout beam duration. Continuous-wave (CW) mid-IR fiber emission has been demonstrated for a variety of rare-earth-doped soft glasses. Fluoride, tellurite, and chalcogenide glasses have been used to extend the emission wavelengths further into the mid-IR. Due to an increase in the quantum defect between pump and laser photon energies, the output power greatly diminishes with an increasing emission wavelength. Nevertheless, although the traditional CW mid-IR fiber lasers based on rare-earth-doped fibers operate below a wavelength of 4 μm [28], CW Ce^{3+}-doped chalcogenide fiber was demonstrated which allowed to obtain up to 7 mW output at wavelength of 5.06 μm [29] (Fig. 8.2). The pump source was a CW Fe^{2+}:ZnSe laser emitting at a wavelength of 4.14 μm.

One of the first CW lasers developed on the basis of NCHCFs was Kagome–lattice NCHCF (Fig. 8.1(middle)) filled with the molecular iodine I_2 [30]. It was pumped at 532 nm with a frequency-doubled Nd:Vanadate laser. CW lasing was achieved in the 1280–1340 nm spectral region, and a maximum multiline output of 8 mW was obtained for about 250 mW of pump power. In work [31], a CW laser operating in the mid-IR spectral region was developed based on a hollow-core fiber shown in Fig. 8.1 (right). The fiber air-core was filled with acetylene gas at 0.6 mbar, and its diameter was 50 μm. The authors demonstrated 1.12 W output power at wavelength of 3 μm with 33% slope efficiency in 15-m-long NCHCF (Fig. 8.2). In 2019, the authors of [32] demonstrated the CW fiber laser operating well beyond 4 μm with optical power approximately 82 mW at wavelength of 4.3 μm based on a CO_2-filled silica glass NCHCF (Fig. 8.1 (left)). The pump was a homemade 2-μm laser diode and maximum laser efficiency was 19.3%. The emission wavelength of 4496 nm from a CW gas-filled fiber laser was demonstrated in [33], where narrow linewidth mid-IR emission from 3810 to 4496 nm was achieved with a maximum laser power of 500 mW and a slope efficiency of 18% (Fig. 8.2). The NCHCF was filled with HBr isotope gas and pumped with a self-developed thulium-doped fiber amplifier seeded by several diode lasers over the range of 1940–1983 nm. Continuing with this line of research, the authors of [34] published another work on a gas-filled fiber laser based on silica NCHCF filled with HBr. In this case, the laser source delivered output power of 3.1 W at a wavelength of 4.16 μm (Fig. 8.2).

A few promising studies on mid-IR CW gas-filled fiber lasers have been carried out in recent years. In [35], the authors reported a CW gas-filled fiber laser in CO_2-filled silica glass NCHCF with six cladding capillaries. Under 10 W pump power, at a wavelength of 2 μm, a densely step-tunable 46 lines from 4276 to 4428 nm were demonstrated. A maximum output power of 557 mW with a slope efficiency of 25.7% was obtained. In [36], it was shown that the use of CO_2-filled nested silica glass NCHCFs [24] in the mid-IR can lead to a CW output power of 6.6 W at wavelength of 4.3 μm with slope efficiency of 17%. The fiber air-core diameter was

Fig. 8.2 Summary of the state-of-art continuous-wave gas-filled lasers in terms of output and lasing wavelengths

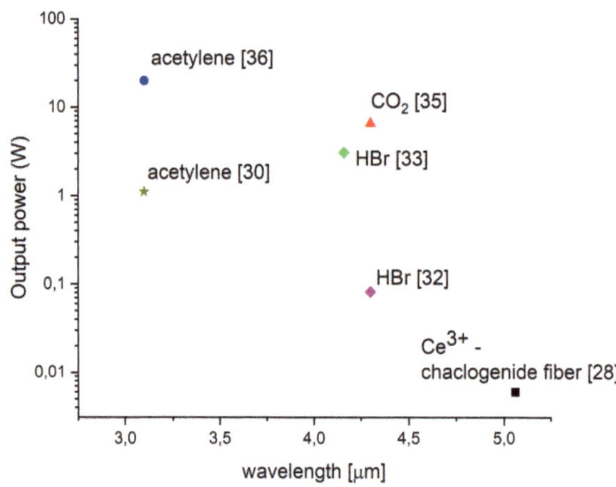

110 μm, and the transmission loss was 0.1 dB/m at a wavelength of 2 μm (the pump) and 0.24 dB/m at a wavelength of 4.3 μm. The pump source was a 100-W-level wavelength tunable single-frequency fiber laser (Fig. 8.2). In another work [37], 20-W-level acetylene-filled silica-glass-nested NCHCF amplified spontaneous emission source at a wavelength of 3.1 μm was demonstrated. The fiber air-core diameter was 120 μm, and the pump was 1535-nm 100-W single-frequency fiber laser. It was a record output power for amplified spontaneous emission sources (Fig. 8.2).

8.4 Pulsed Gas-Filled Fiber Lasers Based on NCHCFs

Regarding pulsed gas-filled fiber lasers, let us first review works on population inversion gas-filled fiber lasers. Two works are worth highlighting here. In the first one [38], the authors demonstrated a subwatt-level tunable acetylene gas-filled laser based on NCHCF (Fig. 8.1(left)) in both pulsed and CW regimes. An amplified tunable 1.5-μm diode laser was precisely tuned to match different acetylene lines, the laser output was step tunable in the range of 3.09–3.21 with a maximum pulse average power of 0.3 W (0.6 μJ pulse energy) (Fig. 8.3). In work [39], emission at a wavelength of 4.6 μm was demonstrated in Kagome–lattice silica glass NCHCF (Fig. 8.1(middle)) filled with N_2O. An 8-ns pump pulses at a wavelength of 1.57 μm excited vibrational overtone, which led to lasing on R and P branch fundamental transition from the upper pump state. Despite a high loss at a wavelength of 4.6 μm (30 dB/m), the slope efficiency was 3%. The fiber air-core diameter was 85 μm, which made it the longest wavelength ever generated by gas-filled fiber lasers.

An alternative way to generate pulses in the mid-IR is to use stimulated Raman scattering (SRS) in molecular gases. One of the advantages of using SRS is the fact that there is no need to precisely fit a gas narrow absorption line at some specific wavelength. In particular, near-IR sources are used to pump gas-filled fiber lasers in which a part of the pump photons energy excites molecular vibrations and the rest of their energy goes into a Stokes wave shifts at longer wavelengths. Taking into account the large vibrational Stokes shifts of H_2, D_2, and CH_4 gases (Ω_R = 4155, 2987, and 2917 cm^{-1}, respectively), mid-IR radiation can be generated in very few SRS cascades. Moreover, molecular rotations provide additional degrees of freedom in the process of SRS, which leads to enrichment of the output spectrum of the gas-filled hollow-core fibers.

The first works on SRS gas-filled fiber lasers were associated with lasing at a wavelength of 1.9 μm in NCHCFs filled with hydrogen [40, 41]. In the first one, efficient 1.9 μm emission by pure vibrational scattering was demonstrated in NCHCF (Fig. 8.1(right)) filled with hydrogen. It was pumped by 1064-nm microchip laser. In work [41], efficient Raman generation in NCHCF (Fig. 8.1(left)) filled with molecular hydrogen was demonstrated. In this case, the pump source was a Q-switched Nd:YAG laser with a pulse duration of 125 ns, and the maximum

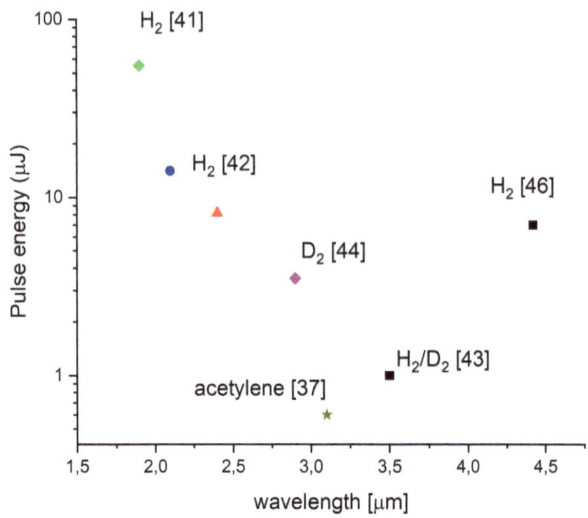

Fig. 8.3 Summary of the state-of-art pulsed gas-filled lasers in terms of pulse energy and lasing wavelengths

average output power at a wavelength of 1.9 μm was 300 mW. The authors of [42] managed to obtain a 100-kW power, GHz-linewidth, subpicosecond laser source at 1.9 μm by SRS in a silica glass NCHCF (Fig. 8.1(right)) filled with hydrogen (Fig. 8.3). A maximum peak power of 150 kW (pulse energy 55 μJ) was achieved with only 3 bar H_2 pressure. Subsequently, a high-pulse-energy multiwavelength Raman laser spanning from 1.53 μm to 2.4 μm was obtained by using the cascaded rotational SRS in a nested silica glass NCHCF [24] filled with hydrogen [43]. The Raman laser had lines at 1683 nm, 1868 nm, 2100 nm, and 2400 nm with pulse energies of 18. 25 μJ, 14.4 μJ, 14.1 μJ, and 8.2 μJ (Fig. 8.3).

The first mid-IR Raman gas-filled fiber laser was demonstrated using 11-m-long silica glass NCHCF (Fig. 8.1(left)) filled with H_2/D_2 mixture at a total pressure of 30 bar [44] (Fig. 8.3). The pump source (EDFL) at a wavelength of 1.56 μm emitted 3.4-ns pulses at an average power of 2.4 W. The maximum 20 mW of average power was measured at 2.9 μm when 400 mW of pump average power was coupled into the NCHCF. Single-pass generation was achieved due to SRS on vibrations of D_2 molecules. It was also possible to obtain spectral lines at 3.3 and 3.5 μm by increasing the pump power. In this case, the Stokes wave at 2.9 μm could serve as a pump for rotational SRS in D_2 and H_2. In recent work [45], the authors demonstrated a high-power Raman gas-filled fiber laser with a tunable wavelength range of 2851 to 2921 nm in D_2-filled NCHCF which was pumped by high-power pulsed erbium-doped fiber amplifiers. A maximum average power of 6 W at a pulse energy of 3.5 μJ was obtained in NCHCF with 0.9 MPa deuterium (Fig. 8.3). In [46], the first gas-filled fiber laser operating at a wavelength above 4 μm was realized. A silica glass NCHCF (Fig. 8.1(left)) had optical losses about 1 dB/m at 4.4 μm, and it was filled with pure H_2 at a pressure of 30 bar. In subsequent work [47], 1.4-W average power, ns-pulsed gas-filled Raman fiber laser based on a silica glass NCHCF at 4.42 μm was demonstrated (Fig. 8.3). It was shown that gas-filled Raman fiber lasers can generate watt-level average power despite the large quantum defect.

Important works were devoted to methane-filled Raman fiber lasers based on NCHCFs. Methane provides a single-step Raman conversion from 1.55 μm to 2.83 μm if it is pumped by erbium-doped fiber lasers. In work [48], 4-ns pulsed EDFL with an average power of 2.6 W as a pump source was used in the spectral range 1.54–1.56 μm. The emission at Stokes wavelength in the range of 2.796–2.863 μm was obtained. In order to use a high-peak-power laser as pump sources for fiber lasers filled with methane, for example, near 1 μm, it was necessary to use a cascade Raman conversion 1.06 → 1.54 → 2.81 μm. In work [49], 25 W of pump power at wavelength of 1.064 μm (15-ps pulses) was coupled into 1.2-m-long NCHCF filled with methane at a pressure of 10 bar which led to generation of an average power of 4 W at the second Stokes wavelength of 2.81 μm through the process of 1.06 → 1.54 → 2.81 μm conversion. Pulse duration at the first Stokes wavelength of 1.54 μm was 5 ps. The first 1.5 μm methane-filled fiber Raman laser pumped by a fiber laser was reported in [50]. The maximum power of 2.06 W at Stokes wavelength of 1543 nm was obtained in 15-m NCHCF filled with 2.5 bar methane. The pump source was a narrow-linewidth pulsed Yb-doped laser.

In conclusion, it is worth mentioning a very interesting result recently obtained in the field of developing new types of gas-filled fiber lasers. It was shown that the laser radiation can be generated directly inside NCHCF without a need for a pump laser [51]. The solution lay in using gas discharge to pump the gas-filled fiber laser, as it was in the case of the first gas lasers. Thus, the gas-discharge fiber lasers could combine the advantages of both gas lasers and fiber lasers in a single device. In [51], a helium–xenon gas-discharge fiber laser based on NCHCF (Fig. 8.1(left)) was demonstrated for the very first time. The laser operated at a wavelength of 2.03 μm in a quasi—continuous-wave mode with an output power of 2 mW. The pump was a pulsed microwave radiation at 2.45 GHz in a slot antenna. This finding has opened up new prospects for generating laser radiation from ultraviolet to the mid-IR spectral ranges, which is hardly accessible by any other methods.

8.5 Challenges and Future Prospects

It should be noted that the development of gas-filled fiber lasers is a dynamic and constantly changing process, also based on the advances in fiber optics, laser physics, and material science. Silica glass NCHCFs have a limit for transmitting radiation with losses that allow effective laser radiation in the mid-IR spectral range. This challenge may be overcome in two main ways, namely, either by creating hollow-core optical fibers from other glasses and materials with low material losses in the mid-IR spectral range or by proposing new geometric structures of the NCHCFs cross-section that allow for significantly greater localization of radiation even in silica glass hollow-core fibers. The first solution may involve using tellurite or chalcogenide glasses. Regarding the second solution, in our view, it should first be necessary to obtain a deeper and more accurate understanding of the light localization mechanism in NCHCFs, which will allow choosing the correct parameters of the fiber cladding (such as the shape of the core boundary, the number of the cladding elements, their thickness, etc.) necessary to optimize its design and reduce losses. In the future, this optimization of the cladding design might be possible due to machine learning methods, and, of course, the technology for manufacturing hollow-core fibers must be developed in order to make it possible to produce the fibers with more complex structure of the cross-section. And finally, the development gas-discharge fiber lasers based on NCHCFs will also be one of the main directions for the development of gas-filled fiber lasers in the near future.

References

1. Gordon, E.L., Labuda, E.F., Bridges, W.B.: Continuous visible laser action in singly ionized argon, krypton, and xenon. Appl. Phys. Lett. **4**, 178–180 (1964)
2. Matsuura, Y., Oyma, T., Miyagi, M.: Soft—X-ray hollow fiber optics with inner metal coating. Appl. Opt. **44**, 6193–6196 (2005)
3. Matsuura, Y., Akiyama, D., Miyagi, M.: Beam homogenizer for hollow—fiber delivery system of excimer laser light. Appl. Opt. **42**, 3505–3508 (2003)
4. Harrington, J.A.: Infrared Fibers and Their Applications. SPIE (2003)
5. Garmire, E., McMahon, T., Bass, M.: Flexible infrared waveguides for high—power transmission. IEEE J. Quant. Electron. **16**, 23–32 (1980)
6. Harrington, J.A., George, P., Pedersen, P., Mueller, E.: Hollow polycarbonate waveguides with inner cu coatings for delivery of terahertz radiaition. Opt. Express. **12**, 5263–5268 (2004)
7. Hongo, A., Morosawa, K., Matsumoto, K., Shiota, T., Hashimoto, T.: Transmission of kilowatt—class CO_2 laser light through dielectric—coated metallic hollow waveguides for material processing. Appl. Opt. **31**, 5114–5120 (1992)
8. Temelkuran, B., Hart, S.D., Benoit, G., Joannopoulos, J.D., Fink, Y.: Wavelength—scalable hollow optical fibres with large photonic band gaps for CO_2 laser transmission. Nature. **420**, 650–653 (2002)
9. Smith, P.W.: A waveguide gas laser. Appl. Phys. Lett. **19**, 132–134 (1971)
10. Gonchukov, S.A., Kornilov, S.T., Petrovskii, V.N., Protsenko, E.D.: Helium—neon waveguide laser. Sov. J. Quant. Electron. **5**, 232–233 (1975)
11. Jensen, R.E., Tobin, M.S.: CO_2 waveguide gas laser. Appl. Phys. Lett. **20**, 508–510 (1972)
12. Cregan, R.F., Mangan, B.J., Knight, J.C., Birks, T.A., Russell, P.S., J., Roberts, P. J., and Allan, D. C.: Single-mode photonic band gap guidance of light in air. Science. **285**(5433), 1537–1539 (1999)
13. Benabid, F., Roberts, P.J.: Linear and nonlinear optical properties of hollow core photonic crystal fiber. J. Mod. Opt. **58**, 87–124 (2011)
14. Roberts, P.J., Couny, F., Sabert, H., Mangan, B.J., Williams, D.P., Farr, L., Mason, M.W., Tomlinson, A., Birks, T.A., Knight, J.C., Russell, P.S., J.: Ultimate low loss of hollow—core photonic crystal fibres. Opt. Express. **13**, 236–244 (2005)
15. Fokoua, E.N., Mousavi, S.A., Jasion, G.T., Richardson and Poletti, F.: Loss in hollow—core optical fibers: mechanisms, scaling rules, and limits. Adv. Opt. Photon. **15**, 1–85 (2023)
16. Benabid, F., Bouwmans, G., Knight, J.C., Russell, P.S., J.: Ultrahigh efficiency laser wavelength conversion in a gas—filled hollow Core photonic crystal fiber by pure stimulated rotational Raman scattering in molecular hydrogen. PRL. **93**, 123903 (2004)
17. Benabid, F., Couny, F., Knight, J.C., Birks, T.A., Russell, P.S., J.: Compact, stable and efficient all—fibre gas cells using hollow—core photonic crystal fibres. Nature. **434**, 488–491 (2005)
18. Benabid, F., Knight, J.C., Antonopoulos, G., Russell, P.S., J.: Stimulated Raman scattering in hydrogen—filled hollow—Core photonic crystal fiber. Science. **298**, 399–402 (2002)

19. Couny, F., Benabid, F., Light, P.S.: Subwatt threshold cw Raman fiber—gas laser based on H_2—filled hollow—Core photonic crystal fiber. PRL. **99**, 143903 (2007)
20. Jones, A.M., Vasudevan Nampoothiri, A.V., Ratanavis, A., Fiedler, T., Wheeler, N.V., Couny, F., Kadel, R., Benabid, F., Washburn, B. R., Corwin, K. L., Rudolph, W.: Mid—infrared gas filled photonic crystal fiber laser based on population inversion. Opt. Express. **19**, 2309–2316 (2011)
21. Andrey D., Pryamikov Alexander S., Biriukov Alexey F., Kosolapov Victor G., Plotnichenko Sergei L., Semjonov Evgeny M., Dianov (2011) (2011) Demonstration of a waveguide regime for a silica hollow—core microstructured optical fiber with a negative curvature of the core boundary in the spectral region > 35 μm Optics Express **19**(2), 1441. https://doi.org/10.1364/OE.19.001441
22. Wang, Y.Y., Wheeler, N.V., Couny, F., Roberts, P.J., Benabid, F.: Low loss broadband transmission in hypocycloid—core Kagome hollow—core photonic crystal fiber. Opt. Lett. **36**, 669–671 (2011)
23. Yu, F., Wadsworth, W.J., Knight, J.C.: Low loss silica hollow core fibers for 3–4 μm spectral region. Opt. Express. **20**, 11153–11158 (2012)
24. Poletti, F.: Nested antiresonant nodeeless hollow core fiber. Opt. Express. **22**, 23807–23828 (2014)
25. Benabid, F., Gerome, F., Vincetti, L., Debord, B., Alharbi, M., Bradley, T.: Inhibited coupling hollow—core photonic crystal fiber. In: Proceedings of SPIE: Photonic and Phononic Properties of Engineered Nanostructures IV, vol. 89940. IEEE (2014)
26. Pryamikov, A.D., Gladyshev, A.V., Kosolapov, A.F., Bufetov, I.A.: Hollow—core optical fibers: current state and developments prospects. Physics—Uspekhi. **67**, 129–156 (2024)
27. Yu, F., Song, P., Wu, D., Birks, T., Bird, D., Knight, J.: Attenuation limit of silica—based hollow—core fiber at mid—IR wavelengths. APL Photon. **4**, 080803 (2019)
28. Maes, F., Fortin, V., Poulain, S., Poulain, M., Carree, J.-Y., Bernier, M., Vallee, R.: Room—temperature fiber laser at 3.92 μm. Optica. **5**, 761–764 (2018)
29. Koltashev, V.V., Frolov, M.P., Leonov, S.O., Sverchkov, S.E., Galagan, B.I., Korostelin, Y.V., Skasyrsky, Y.K., Snopatin, G.E., Sukhanov, M.V., Velmuzhov, A.P., Kozlovsky, V.I., Denker, B.I., Plotnichenko, V.G.: 5 μm CW Ce^{3+} – doped chalcogenide glass fiber laser with 17% slope efficiency. Opt. Lett. **49**, 2737–2740 (2024)
30. Nampoothiri, A.V.V., Debord, B., Alharbi, M., Gerome, F., Benabid, F., Rudolph, W.: CW hollow—core optically pumped I_2 fiber gas laser. Opt. Lett. **40**, 605–608 (2015)
31. Xu, M., Yu, F., Knight, J.: Mid—infrared 1 W hollow—core fiber gas laser source. Opt. Lett. **42**, 4055–4058 (2017)
32. Cui, Y., Huang, W., Wang, Z., Wang, M., Zhou, Z., Li, Z., Gao, S., Wang, Y., Wang, P.: 4.3 μm fiber laser in CO_2—filled hollow—core silica fibers. Optica. **6**, 951–954 (2019)
33. Zhou, Z., Wang, Z., Huang, W., Cui, Y., Li, H., Wang, M., Xi, X., Gao, S., Wang, Y.: Towards high—power mid—IR light source tunable from 3.8 to 4.5 μm by HBr—filled hollow—core silica fibres. Light Sci Appl. **11**, 15 (2022)
34. Zhou, Z., Huang, W., Z., Cui, Y., Li, H., Pei, W., Li, X., Li, Z., Wang, M., and Wang, Z.: 3.1 W mid—infrared fiber laser at 4.16 μm based on HBr—filled hollow—core silica fibers. Opt. Lett. **47**, 5785–5788 (2022)
35. Cui, Y., Wang, Z., Zhou, Z., Huang, W., Li, H., Wang, M., Gao, S., Wang, Y.: Towards high—power densely step—tunable mid—infrared fiber source from 4.27 to 4.43 μm in CO_2—filled anti—resonant hollow—core silica fibers. J. Light Technol. **40**, 2503–2510 (2022)
36. Song, W., Yao, J., Zhang, X., Zhang, Q., Hou, Y., Wu, J., Wang, P.: 4.3 μm high—power amplified spontaneous emission fiber source based on CO_2—filled nested hollow—core anti—resonant fiber. Opt. Express. **32**, 14532–14540 (2024)
37. Song, W., Zhang, X., Zhang, Q., Hou, Y., Wang, P.: 21.8 W acetylene—filled hollow—core anti—resonant fiber amplified sponataneous emission source at 3.1 μm. Opt. Lett. **49**, 3636–3639 (2024)

38. Zhou, Z., Tang, N., Li, Z., Huang, W., Wang, Z., Wu, W., Hua, W.: High—power tunable mid—infrared fiber gas laser source by acetylene—filled hollow—core fibers. Opt. Express. **26**, 19144–19153 (2018)
39. Aghbolagh, F.B.A., Nampoothiri, V., Debord, B., Gerome, F., Vincetti, L., Benabid, F., Rudolph, W.: Mid IR hollow core fiber gas laser emitting at 4.6 µm. Opt. Lett. **44**, 383–386 (2019)
40. Wang, Z., Yu, F., Wadsworth, W.J., Knight, J.C.: Efficient 1.9 µm emission in H_2—filled hollow core fiber by pure stimulated vibrational Raman scattering. Laser Phys. Lett. **11**, 105807 (2014)
41. Gladyshev, A.V., Kolyadin, A.N., Kosolapov, A.F., Yatsenko, Y.P., Pryamikov, A.D., Biriukov, A.S., Bufetov, I.A., Dianov, E.M.: Efficient 1.9 - µm Raman generation in a hydrogen—filled hollow—core fibre. Quantum Electron. **45**, 807–812 (2015)
42. Wang, Z., Gu, B., Chen, Y., Li, Z., Xi, X.: Demonstration of a 150—kW—peak—power, 2—GHz—linewidth, 1.9 µm fiber gas Raman source. Appl. Opt. **56**, 7657–7661 (2017)
43. Adamu, A.I., Wang, Y., Habib, M.S., Dasa, M.K., Antonio – Lopez, J. E., Amezcua – Correa, R., Bang, O., and Markos, C.: Multi—wavelength high—energy gas—filled fiber Raman laser spanning from 1.53 µm to 2.4 µm. Opt. Lett. **46**, 452–455 (2021)
44. Gladyshev, A.V., Kosolapov, A.F., Khudyakov, M.M., Yatsenko, Y.P., Kolyadin, A.N., Krylov, A.A., Pryamikov, A.D., Biriukov, A.S., Likhachev, M.E., Bufetov, I.A., Dianov, E.M.: 2.9, 3.3 and 3.5 µm Raman lasers based on revolver hollow—core silica fiber filled by $1H_2/D_2$ gas mixture. *IEEE J. Sel. Top*. Quantum Electron. **24**, 1–8 (2018)
45. Huang, W., Wang, Z., Zhou, Z., Pei, W., Cui, Y., Li, X., Li, Z., Chen, J.: Tunable fiber gas Raman laser of 6 W at 2.9 µm by deuterium—filled hollow—core fiber. IEEE J. Sel. Top. Quantum Electron. **30**, 1–7 (2024)
46. Gladyshev, A.V., Kosolapov, A.F., Khudyakov, M.M., Yatsenko, Y.P., Kolyadin, A.N., Krylov, A.A., Pryamikov, A.D., Biriukov, A.S., Likhachev, M.E., Bufetov, I.A.: 4.4 - µm Raman laser based on hollow—core silica fibre. Quantum Electron. **47**, 491–494 (2017)
47. Astapovich, M.S., Gladyshev, A.V., Khudyakov, M.M., Kosolapov, A.F., Likhachev, M.E., Bufetov, I.A.: Watt—level nanosecond 4.42 - µm Raman laser based on silica fiber. IEEE Photon. Technol. Lett. **31**, 78–81 (2019)
48. Huang, W., Cui, Y., Li, Z., Zhou, Z., Chen, Y., Wang, Z.: Diode—pump single—pass tunable mid—infrared gas Raman source by methane—filled hollow—core fiber. Laser Phys. Lett. **16**, 085107 (2019)
49. Zhan, X., Peng, Z., Dong, Z., Yao, P., Hou, Y., Wang, P.: High- power mid—infrared 2.8 - µm ultrafast Raman laser based on methane—filled anti—resonant fiber. IEEE Photon. Technol. Lett. **34**, 1007–1010 (2022)
50. Pei, W., Lei, L., Shi, J., Li, X., Huang, W., Zhou, Z., Li, Z., Wang, Z.: 2W, 1.5 µm single—mode fiber methane Raman laser pumped by a Yb—doped fiber amplifier. Opt. Express. **32**, 12419–12427 (2024)
51. Gladyshev, A.V., Komissarov, D.G., Nefedov, S.M., Kosolapov, A.F., Velmiskin, V.V., Mineev, A.P., Bufetov, I.A.: Gas—discharge he—Xe fiber laser. IEEE J. Sel. Top. Quantum Electron. **30**, 1–7 (2024)

Chapter 9
Distributed Feedback Fiber Lasers Based on Regular and Random Refractive Index Structures

S. A. Babin, M. I. Skvortsov, S. R. Abdullina, E. V. Podivilov, and D. A. Shapiro

Abstract This chapter covers recent advances in fiber lasers with distributed feedback (DFB) based on either regular or random refractive index structures. Unlike conventional lasers with Fabry-Perot cavities consisting of two mirrors (fiber Bragg gratings in the case of fiber lasers) at both ends of a gain medium, distributed feedback involves long reflective structures that form effective cavities along the gain medium. Experiments with passive and active fibers, in which regular or random reflective structures are formed, including π-shifted gratings inscribed by femtosecond (fs) pulses, gratings with random amplitude and phase variations, or random arrays of multiple gratings, are reviewed. The results are preceded by a theoretical background on Bragg reflectors and Bragg gratings with π-shift. When the cavity feedback and the amplification of light are distributed along the active fiber, the coupled wave equations become nonlinear, for which some analytical solutions are given. On the basis of this background, a record short DFB laser based on π-shifted grating fs-inscribed in a highly doped erbium fiber, high-efficiency single-frequency ytterbium and Raman lasers based on the random grating and the random grating array, respectively, as well as extreme line narrowing (by 4 orders) in a hybrid configuration involving DFB laser and fs-inscribed artificial Rayleigh reflector are demonstrated. The results are followed by a discussion of further developments and potential applications of DFB fiber lasers.

Keywords Fiber laser · Distributed feedback · Regular · Random · Femtosecond writing

---

S. A. Babin · M. I. Skvortsov · S. R. Abdullina (✉) · E. V. Podivilov · D. A. Shapiro
Institute of Automation and Electrometry, Siberian Branch, Russian Academy of Sciences, Novosibirsk, Russia

Novosibirsk State University, Novosibirsk, Russia
e-mail: abdullinasr@iae.nsk.su

M. F. S. Ferreira, M. C. Paul (eds.), *Optical Fiber Lasers*, Springer Series in Optical Sciences 250, https://doi.org/10.1007/978-3-032-04110-4_9

9.1 Introduction

It is well known that the first laser was demonstrated in 1960 by Theodore Maiman [1]: it was a solid-state laser in which a ruby crystal provides light amplification by pumping with a flash lamp, whereas a Fabry-Perot cavity formed by silver mirrors on the crystal ends provides feedback, which is necessary for laser generation. Since that time, in the laser community, the idea of the waveguide structure of the active element has been discussed, and it was demonstrated in 1961 by Elias Snitzer [2], whereas in 1964, the term "fiber laser" was introduced [3]. Despite the clear advantages of a fiber waveguide cavity structure in which a long active waveguide may greatly amplify the light without thermal lensing and other effects deteriorating the beam quality, whereas cavity mirrors may be integrated with the waveguide and do not require adjustment etc., the fiber lasers were considerably inferior to conventional solid-state lasers because of low waveguide quality and high losses as a result.

Glass fibers as optical media have revolutionized optical signal transmission since the pioneering work of Kao and Hockham in 1966 [4], where the authors proposed the use of silica fibers for optical communications. Since the 1980s, the development of fiber optics has been driven mainly by the demands of the optical communication industry, exhibiting especially rapid growth in the 1990s with the beginning of the internet era. During this period, key fiber optic components have been implemented and developed, namely, efficient single-mode diode lasers operating in the 900–1600 nm range [5] used as transmitter sources, fast photodiodes [6] used as a key part of the receiver, diode-pumped Er-doped fiber amplifiers [7] used for in-fiber compensation of losses, and Bragg gratings [8] used for in-fiber spectral filtering in multichannel transmission systems with wavelength division multiplexing (WDM) [9].

Fiber Bragg grating (FBG) is a periodic refractive index structure that can be formed in a fiber core by exposure to UV laser light with an interference pattern [10]. This periodic structure reflects the light at a resonant (Bragg) wavelength proportional to the grating period, whereas its reflection spectrum is defined by the FBG shape along the fiber (length, amplitude, phase shifts, etc.). They may vary over a broad range with typical values of 1–99% and 0.01–10 nm for the reflection coefficient and bandwidth, respectively [8]. Owing to the unique diversity of parameters and features, FBGs have also been adopted as key elements of fiber lasers and fiber sensors, which became drivers of fiber optic development in the twenty-first century.

The sensitivity of the Bragg wavelength to the environment (in particular, temperature and strain) enables remote monitoring of temperature, deformation, vibrations, pressure, and other physical parameters through the fiber optic link with in-fiber FBGs. They may be multiplexed into an array of FBGs operating at different wavelengths [11], similar to WDM communication systems. Together with fire safety, immunity to EMI interference and aggressive chemicals typical for all fiber sensors, FBG-based sensor systems are also featured by extreme sensitivity and spatial resolution, and are applied for structural health monitoring, smart composite materials, the aerospace industry, and many other practical needs [12–14].

In fiber lasers, FBGs are usually used as in-fiber cavity reflectors, both highly reflective end mirrors and highly-transmissive output couplers, which are able to operate at the kW power level [15, 16]. As high as a 10 kW diffraction-limited output beam is generated in a MOPA (master oscillator-power amplifier) Yb-doped fiber laser with tandem pumping [17]. Another unique feature secured by FBGs is the narrow bandwidth of fiber laser radiation. Moreover, a long FBG consisting of two π-shifted parts inscribed directly into an active fiber, thus providing distributed feedback (DFB), is able to generate single-frequency radiation with an extremely narrow linewidth: 10 Hz for the Er-doped fiber DFB laser with active frequency locking [18]. In addition to Er [19, 20], Er/Yb [21], Yb [22], and Tm [23] fiber DFB lasers with single-frequency operation in the 1.5, 1, and 2 μm spectral ranges, respectively, have also been implemented and applied in research and practice. For example, a Yb-doped DFB laser is used as a seed source in high power/high-energy systems on the basis of multichannel amplification and subsequent coherent combining; one such system includes a mega joule-class laser at the National Ignition Facility [24]. Phase-shifted FBGs may be inscribed not only in active fibers but also in passive fibers, thus enabling single-frequency lasing with Raman gain and DFB in long ($\geq$20 cm) FBGs [25]. The development of point-by-point femtosecond pulse direct writing makes it possible to inscribe gratings of arbitrary shape in almost any fiber, including non-photosensitive ones [26]. This technique also offers new opportunities for DFB fiber lasers. In addition, it appears possible to easily fabricate a random grating structure with variable amplitude and phase along the grating or a purely random refractive index variation similar to natural silica glass in homogeneities, leading to Rayleigh scattering [27] but with a much higher amplitude [28].

In this chapter, we review the basics of the FBG and DFB fiber laser theory and some recent developments in new schemes and operation regimes of DFB fiber lasers, including those based on fs-inscribed FBGs in active fibers (including non-photosensitive ones) in comparison with UV-inscribed FBGs, resulting in a record short DFB cavity in a highly Er-doped fiber. A new type of DFB fiber lasers based on random FBGs (with random amplitude and phase variation) or random array of FBGs, both in active and passive fibers, will be treated. The latter development resulted in the first demonstration of a single-frequency single-polarization Raman fiber laser with a random DFB cavity. Purely random refractive index variations in a hybrid configuration with a regular DFB laser cavity are shown to greatly narrow the single-frequency laser linewidth either for a long fiber with weak natural inhomogeneities or for a short fiber with strong fs-inscribed refractive index modifications. The experimental results are compared with the theory, followed by a discussion of further developments and potential applications of DFB fiber lasers.

9.2 Fiber Bragg Grating (FBG) and FBG Cavity

As mentioned above, fiber Bragg grating is a key element in all fiber-optic applications, especially in optical communications. For example, FBGs are exploited as narrow filters [29] in wave division multiplexing transmission, where different frequency bands carry separate information. The selective optical filters based on FBGs are applied as mirrors in fiber or diode laser cavities. FBG chromatic dispersion compensator [30, 31] is widely used in long-distance optical communication links [9].

The periodic modulation of the refractive index in the fiber core is shown in Fig. 9.1. The Bragg reflector can be schematically represented as a periodic multilayer structure with refractive indices of layers $n_1 = n$ and $n_2 = n + \delta n$, as shown in Fig. 9.1. If the phase difference between reflected waves is a multiple of 2π, the interference becomes constructive. The condition of reflected wave amplification can be written as $2k\Lambda = 2\pi m$, where k is the wavevector, $k = n\omega/c$, ω, c is the frequency and speed of light, n is the average refractive index, and m is an integer number, the order of resonance. The first-order Bragg wavelength is:

$$\lambda_B = 2n\Lambda \tag{9.1}$$

For silica fibers with $n \approx 1.5$ at the telecommunication wavelength $\lambda_B \approx 1.5$ μm, Eq. (9.1) gives an estimation $\Lambda \approx 0.5$ μm.

To estimate the reflection coefficient, let us consider the Fresnel formula for normal incidence. The relationship between electric E and magnetic field H vectors is given by Maxwell's equation. The magnetic field vector is shown by circles in Fig. 9.2. Its direction is perpendicular to the plane of the scheme. Let us write the continuity conditions for tangential projections of electric and magnetic fields:

$$E_1 - E_1' = E_2, \quad k_1\left(E_1 + E_1'\right) = k_2 E_2 \tag{9.2}$$

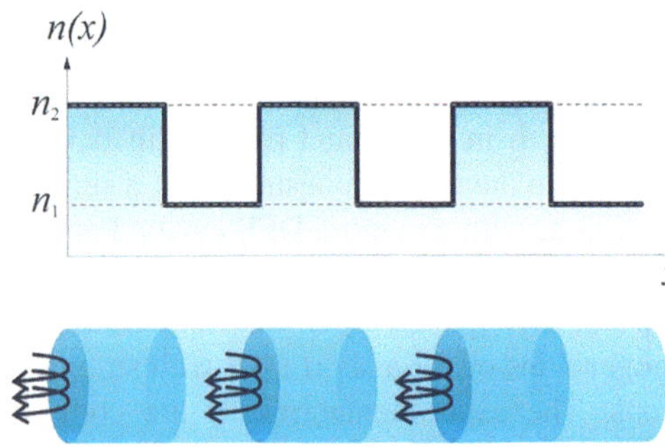

Fig. 9.1 Scheme of a Bragg reflector with period Λ: (**a**) stripes of higher refractive index $n_2 = n_1 + \delta n$ written in the fiber core with index $n_1 \simeq n$, (**b**) incident and reflected waves

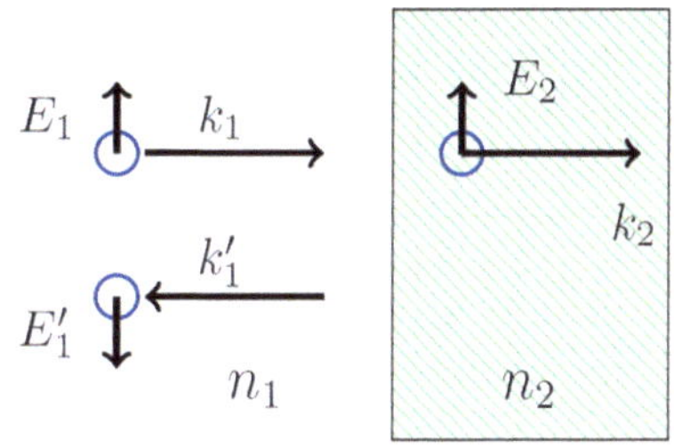

Fig. 9.2 Fresnel reflection: a wave with electric field amplitude E_1 incident from the first medium with refractive index n_1 to the second dielectric n_2, where its amplitude is E_2. The reflected wave with wavenumber k_1' and amplitude E_1' is directed opposite to the incident wave

Then, the complex reflection coefficient is

$$r = \frac{E_1'}{E_1} = \frac{n_2 - n_1}{n_2 + n_1} \tag{9.3}$$

For weak modulation of the index $|n_2 - n_1| = \delta n \ll n$, Eq. (9.3) gives $r \approx \delta n/2n$. The characteristic modulation depth of the FBG is $\delta n \sim 10^{-3}$, and the reflection from one stripe is negligible. The effect is amplified because of the coherent summation from a number of equidistant stripes, for a typical length $L \sim 1$ cm, it is $N \sim L/\Lambda \sim 10^3$. When $\zeta = \delta nN/2n \gg 1$, the reflection is practically full: $1 - |r|^2 \ll 1$, and then the grating is considered as strong. The parameter ζ controlling the reflectivity is the optical density or strength of the grating.

Since the discovery of FBGs [32, 33], the mechanisms of photosensitivity, inscription techniques, special types of FBGs (tilted and long-period gratings), and applications in telecommunications and sensors have been considered in detail in books [8, 34], reviews [31, 35, 36], and references therein.

The theory of the Bragg reflector is based on coupled wave equations (CWEs), see, e.g., the books [8, 34]. The simplest exactly solvable model is a model of a uniform grating, which allows for Bragg bandwidth estimation.

The FBG reflection is the square of the absolute value of the amplitude reflection:

$$R(q) = |r|^2 = \frac{|\kappa|^2 \sinh^2 \mu L}{\mu^2 \cosh^2 \mu L + q^2 \sinh^2 \mu L} = \frac{|\kappa|^2}{\mu^2 \coth^2 \mu L + q^2} \tag{9.4}$$

where $q = k - \beta$ is the detuning from resonance and κ is the coupling coefficient that is equal to the grating strength per unit length: $\kappa = \zeta/L$, $\mu = \sqrt{|\kappa|^2 - q^2}$.

The reflection is maximal at the line center $q = 0$. The maximum value

$$R_{\max} = \tanh^2 \zeta \tag{9.5}$$

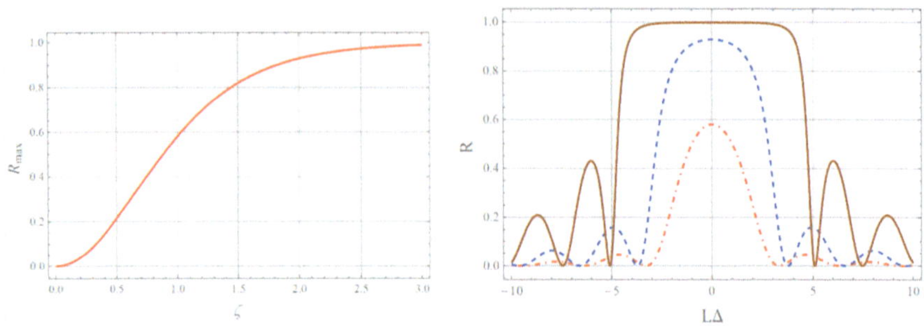

Fig. 9.3 Reflection of a uniform grating: (**a**) the maximum reflectivity as a function of strength $\zeta = |\kappa|L$; (**b**) FBG spectrum $R(q)$ at $\zeta = 1$ (dot-dash), 2 (dash), and 4 (solid curve)

is shown in Fig. 9.3(a). We observe quadratic dependence for a weak grating $\zeta \ll 1$ and saturation for a strong grating at $\zeta \gtrsim 1$. The reflection spectra at $\zeta = 1, 2, 4$ are presented in Fig. 9.3(b). One can see that the central lobe grows in amplitude with ζ, then broadens and saturates. Beyond the stop zone, at $|q| > |\kappa|$, the argument μL of the cotangent in Eq. (9.4) becomes purely imaginary, and then the hyperbolic functions turn into trigonometric: coth →-i·cot. Consequently, the zeros of spectra are placed at $\sqrt{q^2 - |\kappa|^2} = \pi m / L$, m = 1, 2, …, where the cotangent tends to infinity. The first part of zeros at $m = 1$ are at detuning $q_1 = \sqrt{|\kappa|^2 + \pi^2 / L^2}$. Let us choose this position as the definition of the characteristic spectrum half-width

$$\omega = \sqrt{|\kappa|^2 + \pi^2 / L^2} \tag{9.6}$$

For a strong grating $\omega \approx |\kappa| = \zeta / L$. At a subsequent increase in the detuning, the next zeros appear in the spectrum: $q_m = \sqrt{|\kappa|^2 + \pi^2 \mathrm{m}^2 / L^2}$. This effect, undesirable for applications, is called "ripple."

An analytical description of more complicated gratings, including apodized (ripple free), chirped (phase-modulated), etc., is presented in papers [37–39] and citations therein. An inverse problem for the synthesis of FBGs with desired characteristics is solved in articles [40–52].

Next, let us consider the use of FBGs in fiber lasers. Two types of laser cavities involving FBGs are schematically shown in Fig. 9.4. To obtain narrowband generation (single-frequency regime as a limiting case) of a fiber laser, an active fiber should be placed in a Fabry-Perot resonator made of two reflective mirrors. If FBGs are used as such mirrors, the laser has a robust and reliable all-fiber structures free of adjustments, thermal lensing, and other effects inherent to laser cavities made of bulk elements. If the cavity is made of two FBGs separated by a section of an active optical fiber, the FBGs themselves are usually inscribed in passive fibers. To obtain a narrow reflection spectrum, long and weak FBGs should be used according to (9.6); such a type is called a distributed Bragg reflector (DBR) cavity, as schematically shown in Fig. 9.4(a). In this case, the frequency interval between the longitudinal modes is $\delta\nu = c/2nL_c$, where c is the speed of light, n is the refractive index of

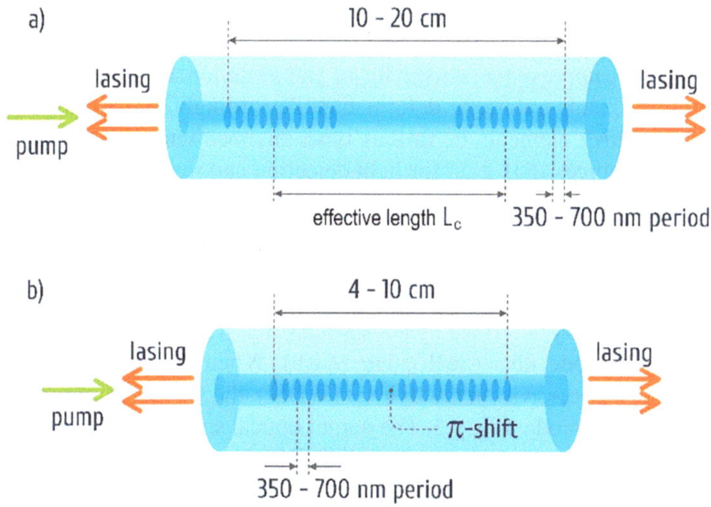

Fig. 9.4 Scheme of DBR and DFB laser cavities: (**a**) Distributed Bragg reflector (DBR) cavity of two long FBGs; (**b**) distributed feedback (DFB) cavity of a π-shifted FBG

the fiber core, and L_c is the effective cavity length, including the active fiber as well as the sections of FBGs in which the radiation attenuates exponentially. By shortening the effective length, the intermode interval becomes larger than the FBG reflection bandwidth, thus selecting a single longitudinal mode; however, the single-frequency regime in this case is highly sensitive to relative FBG phases, which are inherent to environmental disturbances leading to mode hops, etc.

In another configuration known as the distributed feedback (DFB) laser, see Fig. 9.4(b), a point phase shift of approximately π is introduced into a long FBG inscribed in an active fiber. As a result, two FBG parts reflect in the anti-phase for the central frequency, which corresponds to the single longitudinal mode selected automatically. As a result, the problem of mode hopping during heating or stretching/compression of the resonator is absent for a DFB laser, unlike the DBR configuration. The effective length of the cavity in this case is defined by the FBG strength $L_{DFB} \sim 1/\kappa$.

A more detailed theoretical consideration of the DFB laser physics is presented in the next section.

9.3 DFB Laser Scheme and Model

Unlike solid-state lasers, fiber lasers are not limited to a set of certain wavelengths because fiber Bragg gratings act as wavelength selectors within the wide luminescence spectrum of rare-earth ions in silica glass. FBG temperature control provides slow wavelength tuning in the range of 1 nm. In addition, fast frequency adjustment has been achieved by FBG stretching with a piezoelectric transducer [53]. A piece of fiber with a DFB fiber laser (FL) cavity can be packaged to significantly enhance or dramatically suppress the sensitivity of the emission frequency to acoustical noise, vibration, temperature, or deformation. Accordingly, DFB FLs can serve as extremely sensitive elements in various types of fiber sensors [54] or as highly coherent light sources [53], e.g., for light detection and ranging. In both cases, the device can be implemented in a compact and reliable all-fiber design. As examples of scientific applications of DFB FLs, we can mention frequency doubling for iodine spectroscopy [55, 56] and fourth harmonic generation for Mg^{+} ion cooling [57].

The DFB FL cavity is formed by a π-shifted FBG, which is written in the core of an active single-mode fiber (SMF), Fig. 9.4(b). When pumped by a laser diode through a wavelength-division multiplexer (WDM), the fiber DFB laser generates a single longitudinal mode, Fig. 9.5. The output optical spectrum and power are usually measured by an optical spectrum analyzer (OSA).

The π phase shift in the central region of the FBG breaks it into a pair of distributed mirrors reflecting the anti-phase. This cavity structure leads to strong intensity variation along the Bragg grating and to a highly non-uniform gain saturation profile. This is why numerical models are widely used to describe power-related quantities for DFB FLs [58–61]; however, great advances in the analytical description of DFB FLs have been achieved by Foster [62–65]. The agreement of analytical models with numerical calculations for an erbium DFB FL has been demonstrated [66].

Here, we focus on the analytical treatment. The narrow cavity resonance, resolved in the FBG transmission (T) and reflection spectra ($R \approx 1 - T$), see the inset in Fig. 9.5, allowed us to extract the parameters of the cavity required for the theoretical description of the laser. The analytical model is based on Foster's results for erbium DFB FLs [62, 63]. To achieve better agreement with the experimental data, it is necessary to consider the pump absorption along the cavity, which is significant for ytterbium DFB FLs [67].

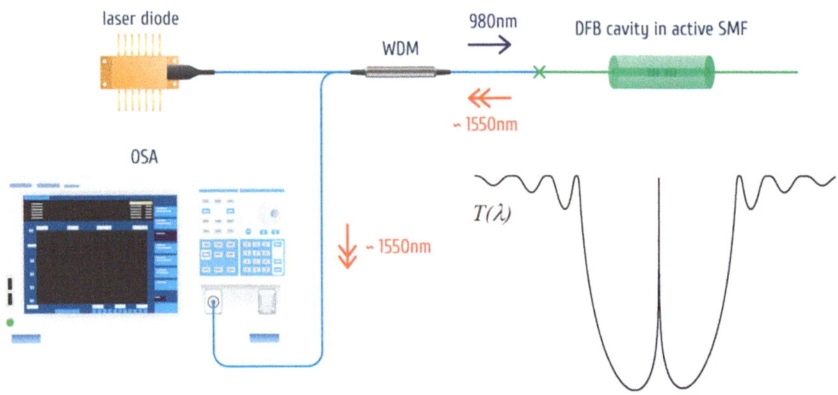

Fig. 9.5 DFB laser scheme in the case of Er-doped active fibers. Inset: transmission spectrum of the π-shifted FBG

9.3.1 Coupled Wave Equations, Intensity Profile, and Output Power in the DFB Cavity

The refractive index of the grating with Bragg wavelength $\lambda = 2\pi/\beta$ can be written as

$$n(z) = n_0 - \frac{\delta n}{2}\sin\left(2\beta|z|\right) \tag{9.7}$$

A π-shifted FBG consists of two simple gratings, where the phase of the right grating ($0 < z < L_R$) is shifted by $\phi = \pi$ with respect to the left grating ($-L_L < z < 0$). The lengths of the left and right gratings L_L and L_R may differ, whereas the amplitude modulation δn is the same.

The one-dimensional scalar Maxwell equation (for a single transverse mode amplitude E normalized by the power) has the form

$$\frac{d^2E}{dz^2} = -k^2E + 2k\left(\kappa\sin\left(2\beta|z|\right)\right) - i\frac{g-\alpha}{2}E \tag{9.8}$$

where $k = n_0\omega/c$ is the wavenumber of light, $\kappa = \delta nk/2n_0$ is the coupling coefficient, g is the amplification, and α is the loss per unit length. Near the Bragg resonance ($k - \beta = q \ll k$), when the field is a sum of counter-propagating waves running in the positive and negative directions

$$E(z,t) = \frac{\exp(i\omega t)}{\sqrt{2}}\left[a^+(z)\exp(-i\beta z) + a^-(z)\exp(i\beta z)\right] + c.c.$$

we obtain coupled wave equations for amplitudes of rapidly oscillating components $a^{\pm}$:

$$\frac{da^+}{dz} = (iq + g/2 - \alpha/2)a^+ - sign(z)\kappa a^- \tag{9.9}$$

$$\frac{da^-}{dz} = -(iq + g/2 - \alpha/2)a^- - sign(z)\kappa a^+$$

Here, $\text{sign}(z) = z/|z|$, $z \neq 0$. Since the fiber Bragg grating is relatively short, $L \sim 1$ cm, the gain per pass is small. Thus, the optical density of the grating must be high. Inequality α, $g(z) \ll \kappa$ is the key condition in the theory below.

The solution to Eq. (9.9) at constant gain $g(z) = g$ (unsaturated mode near the lasing threshold) at $z > 0$ with conditions $a^-(L_R) = 0$ has the form

$$a^+(z) = \left[\cosh\left(\mu(z - L_R)\right) + \frac{iq + g/2 - \alpha/2}{\mu}\sinh\left(\mu(z - L_R)\right)\right]a^+(L_R) \tag{9.10}$$

$$a^{-}(z) = -\frac{\kappa}{\mu}\sinh\left(\mu\left(z - L_R\right)\right)a^{+}\left(L_R\right)$$

where $\mu = \sqrt{\kappa^2 - (q - ig/2 + i\alpha/2)^2}$. For the left grating ($z < 0$), the solution is

$$a^{-}(z) = \left[\cosh\left(\mu\left(z + L_L\right)\right) + \frac{iq + g/2 - \alpha/2}{\mu}\sinh\left(\mu\left(z + L_L\right)\right)\right]a^{-}\left(-L_L\right) \quad (9.11)$$

$$a^{+}(z) = \frac{\kappa}{\mu}\sinh\left(\mu\left(z + L_L\right)\right)a^{-}\left(-L_L\right)$$

provided $a^{+}(-L_L) = 0$. The lasing threshold can be found by equating the expressions (9.10) and (9.11) at $z = 0$:

$$g_{th} = \alpha + \frac{2\mu\cosh\left(\mu\left(L_R - L_L\right)\right)}{\sinh\left(\mu L_R\right)\sinh\left(\mu L_L\right)} = \alpha + 4\kappa\exp\left(-\kappa\left(L_R + L_L\right)\right) \ll \kappa$$

For an active fiber pumped by power P, the gain coefficient in the quasi two-level approximation is

$$g(z) = \frac{g_l P(z)/P_s}{1 + P(z)/P_s + I(z)/I_s}$$

where $g_l = 0.023$ cm^{-1} is the unsaturated amplification, $P_s = 0.5$ mW is the saturation parameter in pumping, and $I_s \simeq 25P_s = 13$ mW is the saturation parameter in laser radiation for Yb-doped fiber [67]. Near the threshold $I(z) = 0$, the pump depletion is negligibly small, then $P(z) = P_{in}\exp(-\alpha_p(z + L_L))$, where $\alpha_p \simeq 0.03$ cm^{-1} represents passive (non-saturated) losses and the pump is coupled to the left port $P(-L_L) = P_{in}$.

Integrating $g(z)$ along the cavity and setting the gain equal to losses $\int_{L_L}^{L_R} dz\, g(z) = g_{th}\left(L_R + L_L\right)$, we find the threshold pump power:

$$P_{th} = P_s \frac{1 - \exp\left(-g_{th}\alpha_p\left(L_L + L_R\right)/g_l\right)}{\exp\left(-g_{th}\alpha_p\left(L_L + L_R\right)/g_l\right) - \exp\left(-\alpha_p\left(L_L + L_R\right)\right)} \simeq P_s \frac{g_{th}}{g_l - g_{th}}$$

The approximate expression is derived in the limit of small passive losses. The threshold occurs on the order of the pump saturation parameter: $g_{th} \simeq g_l/2$.

The threshold is achieved at $q = 0$, and then, we may consider $\mu \simeq \kappa$. For the output power, we obtain from (9.10) and (9.11)

$$I_{out}^{+} = \left|a^{+}\left(L_R\right)\right|^2 = I^{+}(0)/\cosh^2\left(\kappa L_R\right) = T_R I^{+}(0)$$

$$I_{out}^{-} = \left|a^{-}\left(-L_L\right)\right|^2 = I^{-}(0)/\cosh^2\left(\kappa L_L\right) = T_L I^{-}(0)$$

where $I^{+}(0) = I^{-}(0) = I_0/2$ and $T_{L,R} = 1/\cosh^2(\kappa L_{L,R})$ are the transmission coefficients of the left or right grating. The obtained solutions for the amplitude inside the cavity can be rewritten in simple approximate form:

$$a^{+}(z) \simeq a^{-}(z) \simeq \exp(-\kappa|z|)a^{\pm}(0) \tag{9.12}$$

This form describes the power distribution in the DFB π-shifted grating except in the vicinity of its edges, namely, the standing wave:

$$I(z) = |E|^2 = I_0 \exp(-2\kappa|z|) 2\cos^2(\beta z) \tag{9.13}$$

The longitudinal distribution of intensity (9.13) is shown in Fig. 9.6. The exponential decay of the envelope is controlled by the constant κ. The approximate expression for the saturated gain coefficient

$$g(z) \simeq \frac{g_l}{1 + I(z)P_s / P(z)I_s}$$

is valid well above the threshold $P_{in} \gg P_s$.

To find the generation power above the threshold, we use energy conservation. The losses consist of two parts: active $I_{out} = I_{out}^{-} + I_{out}^{+} = I_0(T_L + T_R)/2$ and passive $\int_{-L_L}^{L_R} dz\alpha I(z) = \alpha I_0 / \kappa$. Consequently, the value $L_{DFB} = \int dz I(z)/I_0 = 1/\kappa$ is the length of the effective Fabry-Perot resonator.

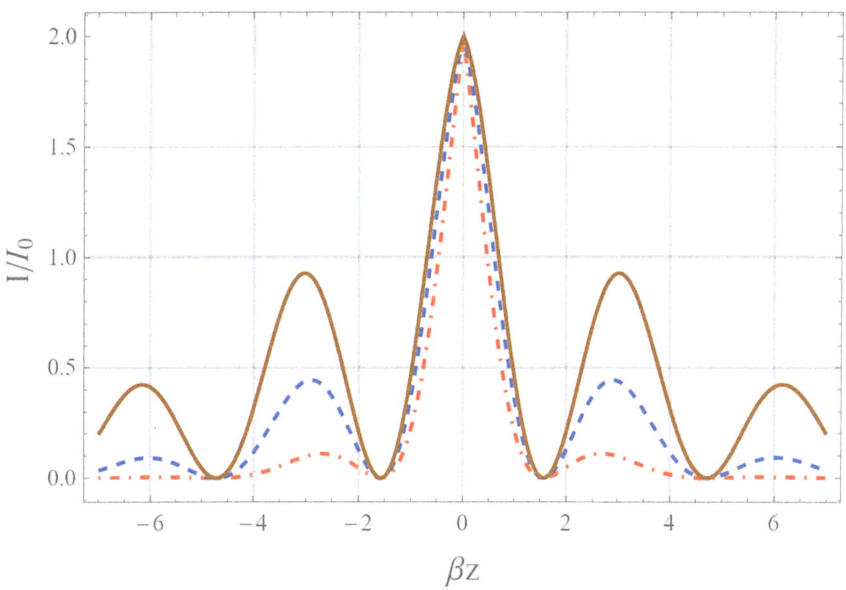

Fig. 9.6 Intensity distributions along the DFB FL at $\kappa/\beta = 1$ (dot-dashed), 0.5 (dashed), and 0.25 (solid line)

The energy transformed into the lasing mode per period is defined by the integral

$$\int_{-L_L}^{L_R} dz \frac{g_l}{1+I(z)P_s/PI_s} I(z) = I_0 \frac{2g_l P_0 I_s}{\kappa P_s I_0} \ln\left(1/2+\sqrt{1/4+P_s I_0/2P_0 I_s}\right) \quad (9.14)$$

For its calculation, we first average over a period of standing wave and then integrate the envelope exponent while taking into account the hole burning in the gain spectra. The equations for the intensities are as follows from Eq. (9.9):

$$\pm\frac{dI^{\pm}}{dz} = \left(g(I)-\alpha\right)I$$

can be integrated from $-L_L$ to L_R. Taking the pump power $P(z)$ at $z = 0$, where the main contribution is accumulated, and equating the gain and losses, we derive the transcendental equation for the self-consistent generation power

$$\frac{4g_l}{\kappa X}\ln\left(\frac{1}{2}+\frac{\sqrt{1+X}}{2}\right) = \frac{T_L+T_R}{2}+\frac{\alpha}{\kappa} \quad (9.15)$$

Here, $X = 2I_0P_s/P_0I_s$ is the normalized generation power (saturation parameter). The left side of Eq. (9.15) decreases monotonically with the dimensionless parameter X; its limit at $X \to 0$ is g_l/κ. Therefore, lasing is possible if the unsaturated gain must exceed the losses.

$$\frac{g_l}{\kappa} > \frac{T_L+T_R}{2}+\frac{\alpha}{\kappa}$$

The generation power obtained from (9.15) via numerical iterations is shown in Fig. 9.7 as a function of losses.

The differential efficiency of the DFB FL can be found as follows:

$$\eta = \frac{I_{out}}{P(-L_L)} = \left(1-\alpha_p L_L - F/2\right)X\left(T_L+T_R\right)\frac{I_s}{4P_s} \quad (9.16)$$

Here, X is the solution of the transcendental Eq. (9.15), and the first factor takes into account the absorption and depletion $P_0 \simeq P(-L_L)(1-\alpha_p L_L - F/2)$, where

$$F \simeq \frac{2g_p}{\kappa}\ln\left(1/2+\sqrt{1+X}/2\right) < 1$$

is the fraction of energy transferred from the pump. The quantity $g_p \simeq 0.6\ \text{cm}^{-1}$ is the absorption of pumping power, including saturation by lasing, and the value F can be found by integrating the equation for the pump $dP/dz = -g_p P\left(I(z)/I_s+1\right)/\left(1+I(z)/I_s+P/P_s\right)-\alpha_p P$ with a known $I(z)$.

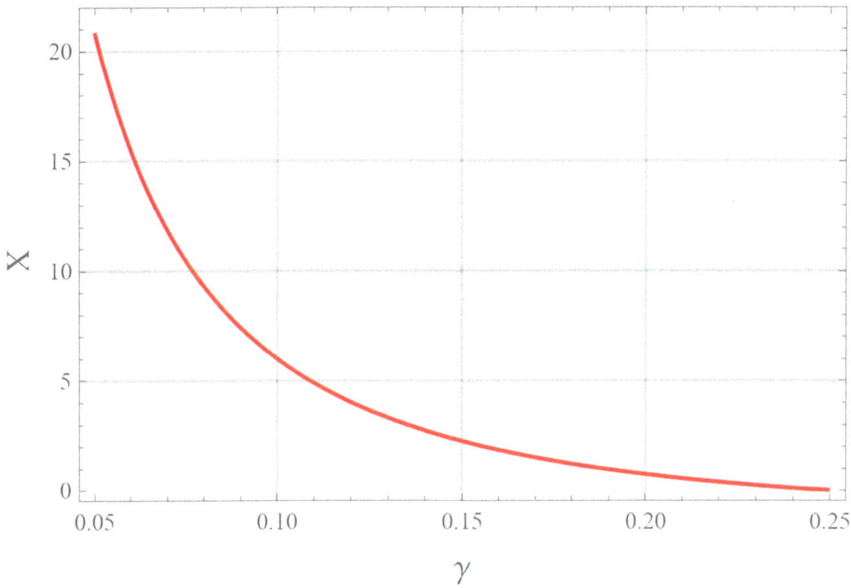

Fig. 9.7 Normalized generation power $X = 2I_0P_s/P_0I_s$ as a function of dimensionless losses $\gamma = \left(\kappa\left(T_L + T_R\right)/2 + \alpha\right)/g_l$

It is also significant that the single-frequency DFB FL has a low output power in principle, because of its high cavity Q-factor. The limitation of power $I_0 < 10$ W at the π-shift point (due to the distortion at the high-temperature FBG being glued to a radiator) yields the restriction for the maximum output power $I_{out} < I_0T_R/2 \ll I_0$.

9.3.2 DFB Fiber Laser Linewidth

The DFB cavity can be characterized by a standard method used for a Fabry-Perot cavity [67]. Under the conditions of high FBG reflectivity, constant gain of the active fiber with a relatively low coefficient $g_s \ll \delta nk_B$ and weak detuning $|q| < \delta nk_B$ from the resonance $\lambda = \lambda_B$ (δnk_B characterizes the width of the central dip in the reflection spectrum), the amplitude reflection coefficient of the FBGs is approximately equal to

$$r_1 = r_2 = r\left(k_B + q - ig_s\right) \simeq \tanh\left(\delta nk_BL_j/2n\right)\left(1 + \frac{g_s + iq}{\delta nk_B/2n}\right) \tag{9.17}$$

For an intracavity roundtrip of a monochromatic wave with resonant frequency $\omega = 2\pi c/\lambda_B$, the following condition is accomplished: $A = r_1r_2A + A_{sp}$, where A is the wave amplitude and A_{sp} is the amplitude of spontaneous emission. Hence, at detuning $\Omega = qc/n$ relative to the resonance frequency, the ratio of the intracavity spectral intensity $I_q(\Omega) = |A|^2$ to the spontaneous emission noise spectral density takes the following form:

$$\frac{I_q(\Omega)}{\langle I_{sp}(\Omega)\rangle} = \frac{1}{\left|1 - r^2(k_B + q - ig_s)\right|^2}$$

Taking into account the approximate expression for the FBG reflectivity, Eq. (9.17), we obtain the Lorentz form of the Schawlow-Townes linewidth [68].

$$\begin{aligned}\frac{I_q(\Omega)}{I_{sp}(\Omega)} &= \frac{1}{\left|1 - r^2(k_B)\left(1 + 2\dfrac{g_s + iq}{\delta n k_B / 2n}\right)\right|^2} = \frac{1}{\delta^2 + \left(2q / (\delta n k_B / 2n)\right)^2} \\ &= \frac{1}{(\Delta T_{rt})^2 + (\Omega T_{rt})^2}\end{aligned} \tag{9.18}$$

where $\delta = \Delta T_{rt} = 1 - r^2(k_B) - 4ng_s / (\delta n k_B) \ll 1$ is related to the intensity through the gain saturation $g_s(I)$ and effective roundtrip time T_{rt} for the DFB cavity. $I_{sp}(\Omega)$ is the average intensity spectral density of spontaneous emission in the defined longitudinal mode (which is constant within the resonance width because its linewidth is much greater than the width of the resonances).

The linewidth broadening model for the DFB laser differs from the linear Fabry-Perot cavity model in that the effective cavity length $L_{DFB} = 2n/(\delta n k_B)$ instead of its physical length; thus, the L_{DFB} is defined as the point where the intracavity field amplitude decreases by a factor *e*. Therefore, the effective roundtrip time in the DFB laser is $T_{rt} = 2L_{DFB}n/c$.

For a 6-cm-long DFB structure with transmissivity $T \approx (T_1 + T_2)/2 \approx 10^{-3}$, its effective length is estimated as $L_{DFB} \approx 14$ mm.

Calculating the integral of the spectral density over frequency

$$I = \int \frac{d\Omega I(\Omega)}{2\pi} = \int \frac{d\Omega}{2\pi} \frac{I_{sp}(\Omega)}{\left(\Delta^2 + \Omega^2\right) T_{rt}^2} = \frac{I_{sp}(\Omega)}{2\Delta T_{rt}^2}$$

we obtain the dependence of the laser linewidth (half-width at half maximum) on the total intensity of laser generation I and roundtrip time T_{rt}:

$$\Delta = \frac{I_{sp}(\Omega)}{2T_{rt}^2 I} = \frac{I_{sp}(\Omega)}{2I} \cdot \left(\frac{c}{2L_{DFB} n}\right)^2 \tag{9.19}$$

For spontaneous emission power $I_{sp} = 10$ pW at 20 pm interval measured by an optical spectrum analyzer (OSA) and a laser power level of $I = 1$ mW, the half-width of the instantaneous (< 1 μs) laser line is estimated as 15 rad·sec^{-1} in frequency units (or ~ 2 Hz), but at long scales (~ 1 ms), the typical linewidth is ~ 1 kHz for Er- and ~ 100 kHz for the Yb-fiber DFB laser and is defined by thermal and acoustic noise.

9.4 DFB Lasers Based on π-Shifted FBGs and Random Refractive Index Structures

The main feature of DFB lasers is efficient and stable single-longitudinal-mode (single-frequency) generation. The influence of the FBG parameters on the DFB laser output has been considered in several studies [58, 60, 63]. The DFB laser output power reaches several tens of milliwatts for the most efficient Yb-doped fiber laser, while its efficiency varies from several to tens of percent with special measures [69, 70].

Erbium-doped fiber DFB lasers [20, 21] are the most attractive for applications because they emit in the 1.5 μm spectral range, featuring the highest transparency of the SiO_2 fiber. The characteristic values of the lasing power are rather low: 0.1–1 mW, since erbium ions have a small pump absorption cross-section at 980 and 1480 nm; therefore, only a small part of the pump power is absorbed in the DFB FL resonator. Despite their relatively low lasing power, these lasers have high temporal stability. The characteristic value of the relative intensity noise (RIN) lies in the range between −80 and −120 dB/Hz, and the use of electronic feedback can further reduce these values by a factor of −20 dB [71], which is important for the development of commercial devices. A detailed theoretical description of the characteristics of Er-doped fiber DFB lasers is given by Foster [62, 63]. For Yb-doped fiber DFB lasers, the theory has been slightly modified [67]; the details are described in the previous section.

A phase-shifted fiber Bragg grating (PSFBG), which forms the resonator of the DFB laser, is usually inscribed in photosensitive fibers using UV laser sources and a phase mask [70]. Burgmeier [72] demonstrated the possibility of fabricating PSFBGs using the femtosecond point-by-point (PbP) inscription technique [26], which allows one to tune the FBG reflection wavelength over a wide range and offers the FBG inscription in non-photosensitive fibers. The results of the formation of PbP PSFBGs with lengths of up to 8.6 mm have been reported. However, this length is less than the characteristic length required for the DFB laser. A recently developed technique of long FBG PbP inscription [73] made it possible to significantly reduce the number of errors in the grating structure, thereby achieving a high grating quality at lengths up to 50 mm. On the basis of this technique, a method to introduce phase shifts during the FBG inscription was proposed [74]. As a result, the fabrication of a 32-mm π-phase-shifted FBG in a photosensitive PM fiber (CorActive ER35-7-PM) and the use of the PSFBG as a resonator of a DFB fiber laser have been demonstrated [75]. It becomes possible to implement this type of PSFBG in the DFB laser described below.

9.4.1 DFB Fiber Laser Based on a Fs-Inscribed π-Shifted FBG in Er-Doped Fibers

To create PSFBGs, we used the modified femtosecond PbP inscription technique [74], where a phase shift is introduced in the FBG structure due to the rapid piezo-displacement of the optical fiber during its translation with a constant velocity (through the ferrule). Figure 9.8 shows the transmission spectra of the 32-mm π-phase-shifted FBG inscribed in the CorActive EDF-L 1500 fiber measured by an Apex AP2050 optical spectrum analyzer (OSA) in orthogonal linear polarizations with a resolution of 40 fm [75]. The wavelength detuning of the PSFBG for the fast and slow polarization axes is equal to 14.9 pm, corresponding to an induced birefringence value of $\Delta n \approx 1.4 \times 10^{-5}$. This birefringence is an intrinsic characteristic of PbP-inscribed FBGs because of the anisotropy of the induced refractive index change in the focal spot of femtosecond laser pulses inside the fiber core. The effect is also known for FBGs inscribed by CW UV radiation [70], but typical birefringence values ($\sim 10^{-6}$) are lower in that case.

The fs-inscribed π-phase-shifted FBG is pumped with a single-mode laser diode at a wavelength of 976 nm through the 980/1550 nm wavelength division multiplexer, as shown in Fig. 9.5. The output lasing spectra and power are measured with the help of a Yokogawa AQ6370D optical spectrum analyzer, and the noise characteristics are measured via a 5 GHz Thorlab DET08CFC photodiode and an Agilent N9010A radio-frequency signal analyzer. The special self-heterodyne [76] scheme,

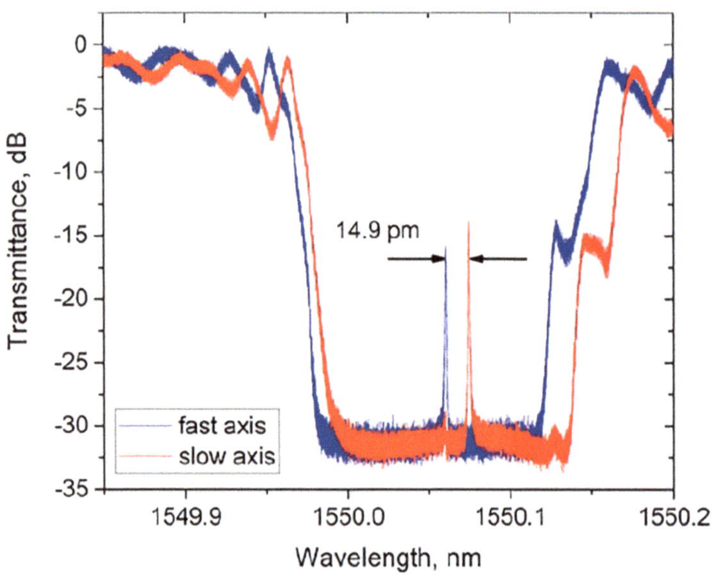

Fig. 9.8 Transmission spectra of the 32-mm π-phase-shifted FBGs measured for the fast (blue) and slow (red) polarization axes [75]

which is based on a ~56 km delay line, enables high-resolution linewidth measurements. Figure 9.9(a) shows the experimental data for the lasing power of the DFB laser based on the fs-inscribed PSFBG as a function of pump power (points), whereas the inset shows the lasing spectrum of the DFB laser at the maximum pump power. The threshold pump power is 8.5 mW. The measured differential efficiency is approximately 0.13%. Over the entire power range, single-frequency generation is observed in the spectrum in only one polarization mode, indicating linear polarization of the output laser radiation. Since the spectral resolution of the OSA is limited to 20 pm, we used a self-heterodyne measurement technique to resolve the spectral linewidth more precisely. Figure 9.9(b) shows the beat spectrum in the case of signal heterodyning. The full width at half maximum of the beat signal spectrum is 33.4 kHz for the laser operating at a power of ~ 200 μW, which corresponds to a spectral linewidth of approximately 17 kHz. The radio-frequency spectrum of the laser relative intensity noise is shown in the inset of Fig. 9.9(b). At the maximum pump power, the peak value amounts to 85 dB/Hz at a frequency of ~660 Hz.

Let us compare the obtained experimental results with the theory. The threshold condition [77] for the generation of a DFB laser is given by the following expression (see the previous section for details of theory):

$$g_s\left(P_{pump}\right) \approx 4\kappa \cdot \exp\left(-\kappa L\right) + \alpha^* + \alpha_{PbP} \tag{9.20}$$

where $g_s(P_{pump})$ is the signal gain per unit length in the fiber due to the population inversion of Er ions, which increases with increasing pump power P_{pump}, $\kappa = \pi\delta n/\lambda$ and L are the coupling coefficient of the PSFBG and its length; and α^* is the coefficient of unsaturated losses (~50 dB/km). Owing to the presence of scattering losses in the FBG induced by point-by-point fs pulse inscription, they are also taken into account (the term α_{PbP}). The measured scattering loss value is $\alpha_{PbP}L \approx 5\%$. In contrast to other work [72], where the inscription is performed in a passive fiber (with femtosecond PbP technique), in our case, the PSFBG is inscribed in the active fiber, where the transmission spectrum turned out to be wider for the slow

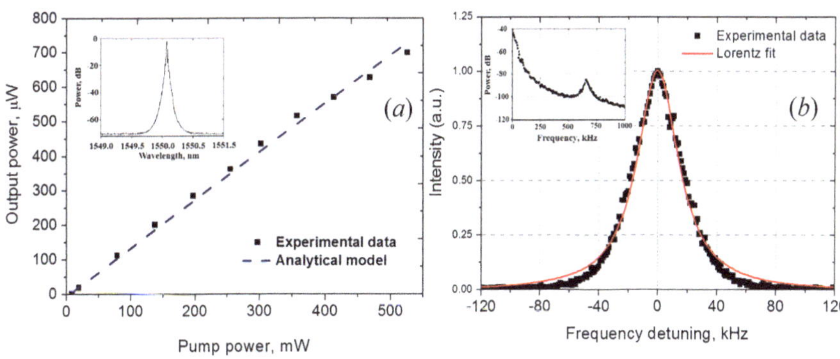

Fig. 9.9 (**a**) Output power versus pump power, inset: lasing spectrum at the maximum pump power; (**b**) Experimental self-heterodyne RF spectrum with its Lorentzian fit, inset: RIN spectrum

polarization axis (see Fig. 9.8), thus indicating a higher effective value of the refractive index change for this axis. Because of the induced birefringence, polarization modes with $\kappa_{fast} = \pi\delta n/\lambda$ and $\kappa_{slow} = \pi(\delta n + \Delta n)/\lambda$ correspond to gratings with different strengths ($\kappa_{slow} > \kappa_{fast}$). Taking into account Eq. (9.20), we see that the generation threshold can be reached for only one polarization component.

Two orthogonal polarization modes are generally observed in the generation of a DBF laser on the basis of UV inscription in non-PM fibers. For a ytterbium DFB laser in a single-mode non-PM fiber, a spatial chirp of FBGs appeared duc to the inhomogeneous heating of the resonator at relatively high pump powers [70]. This effect led to the generation threshold for one of the polarization axes not being reached, which made it possible to obtain a single-polarization generation regime. A method of mode suppression based on fiber twisting along the longitudinal axis [78] is proposed, but optimization of the laser parameters is rather complicated with this technique. Because the birefringence of structures inscribed with femtosecond pulses is an order of magnitude greater than that in the case of UV inscription, the difference between the strengths of the resonator gratings is significant, whereas for the UV technique, the difference is practically indistinguishable. Thus, when the PbP technique is used for phase-shifted FBGs inscription, one polarization mode of a DFB laser fabricated in non-PM fibers can be selected.

In addition, we measured the RF spectra of beat signals for DFB lasers inscribed with both the femtosecond PbP technique (Fig. 9.10(a)) and the UV-irradiation technique (Fig. 9.10(b)) in the active CorActive EDF-L 1500 fiber [75]. For the PbP technique, the difference in the central FBG wavelengths for orthogonal linear polarizations is $\Delta\lambda \approx 14.9$ pm, which corresponds to a beating frequency of $\Delta\nu \approx 1.86$ GHz. The absence of a beating peak at a given frequency in a radio frequency analyzer, Fig. 9.10(a), shows a generation regime with a single-polarization mode. The measured beat frequency of the polarization modes for UV-inscribed 32-mm DFB fiber laser radiation is 560 MHz, which corresponds to a birefringence value of $\Delta n \approx 4.2 \times 10^{-6}$.

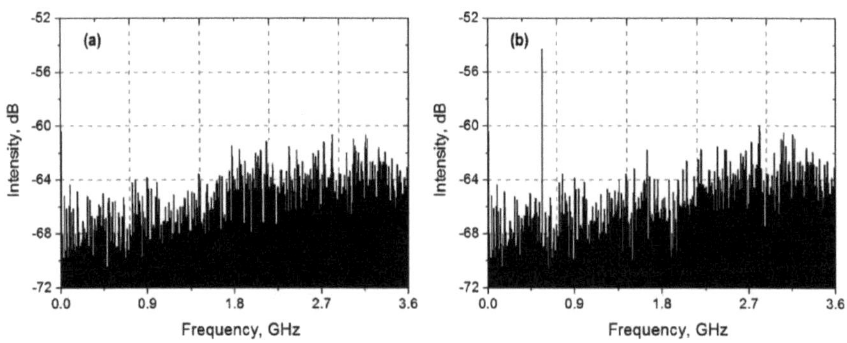

Fig. 9.10 RF spectrum of beating signals of the DFB laser inscribed with the femtosecond PbP technique (**a**) and UV-irradiation technique (**b**) [75]

As we mentioned above, Foster's theory [62] allows one to calculate the differential efficiency of an Er-doped fiber DFB laser via the following formula (see (9.16) in Sect. 9.3.1):

$$\eta = \left(\frac{\lambda_p}{\lambda_l}\right) \cdot \frac{4\kappa \cdot \exp(-\kappa L)}{4\kappa \cdot \exp(-\kappa L) + \alpha^* + \alpha_{PbP}} \alpha_p l_c \cdot \frac{g_s \ln\left[2g_s / \left(4\kappa \exp(-\kappa L) + \alpha^* + \alpha_{PbP}\right)\right]}{\alpha_l + g_s} \quad (9.21)$$

where α_p is the absorption coefficient at the pump wavelength λ_p, $l_c \equiv L_{DFB}$ is the effective length of the laser cavity ($L_{DFB} \approx 1/\kappa$), λ_l is the wavelength of laser generation, and α_l is the loss coefficient at the generation wavelength in the absence of the inverse population. The insertion loss coefficient α_{PbP} related to the PbP inscription process is added to the total losses to correctly determine the efficiency of the laser, similar to the case for threshold estimation. The efficiency calculated in this way is compared with that of the experiment in Fig. 9.9(a), which shows good agreement between the experimental data and the model.

Thus, a linearly polarized erbium-doped DFB fiber laser based on PSFBG inscribed by a femtosecond PbP technique in non-PM fibers is demonstrated for the first time [75]. The maximum lasing power obtained for this laser is approximately 700 μW, with a pumping of 525 mW. The measured linewidth of the laser is 17 kHz, which is an order of magnitude larger than that for the UV-inscribed PSFBG. However, in contrast to UV-based inscription technology, the proposed method allows the creation of DFB fiber lasers that generate radiation in a single-polarization mode. They also feature high grating stability. Therefore, DFB fiber lasers based on FBGs inscribed by femtosecond pulses can be used for sensors [77] in harsh environments, such as high temperature, radiation, and high hydrogen concentration. This method is also attractive for the fabrication of a variety of DFB lasers in various active fibers owing to its flexibility.

9.4.2 *Ultrashort DFB Er-Doped Fiber Laser*

The typical length of DFB and DBR fiber laser cavities is usually several centimeters, which may be too large for some specific applications where they are used as active point-action sensors [77], which are much more sensitive and precise than conventional passive FBG sensors. Moreover, they may be easily assembled into an array of sensors providing new capabilities such as three-element vector hydrophones with high sensitivity [79], which have extensive applications in underwater acoustic monitoring for defense, oil and gas exploration, etc. For such applications, fiber lasers with cavity lengths of ≤1 cm are in great demand. In addition, such short DFB lasers are attractive for their fabrication (reduced errors) and packaging (compact footprint and lower sensitivity to the environment).

Development in this direction resulted in the demonstration of Er/Yb-codoped fiber lasers with an 8.4-mm-long DBR cavity, providing single-frequency generation near 1.54 μm with an ≈ 3 kHz linewidth [80]. Considering that DBR lasers have significant drawbacks, such as mode hopping, the development of short DFB lasers is more attractive. As known from the literature, the shortest DFB fiber lasers have 16 mm long cavity in Yb^{3+}-doped fiber [81] and 17–20 mm cavity in Er^{3+}-doped fiber [82, 83] which is attributed mainly to the absence of active fibers with high gain at low concentration of clusters (ion pairs), limiting the CW single-frequency regime of generation in DFB lasers. The highest level of fiber core doping by rare-earth ions at the minimum level of clustering in the case of silica fibers has been obtained by implementing the preform manufacturing technology of surface plasma chemical vapor deposition of silica glass at reduced pressure (SPCVD) [81, 83, 84] featured by the absence of the glass-melting stage during the synthesis process.

Another possibility is to explore FBG inscription in phosphate fibers, which enables even higher doping concentrations at weaker cluster formations [85]. However, it is difficult to directly inscribe a short in-core Bragg grating using traditional inscription technologies because the photosensitivity of phosphate fibers is sufficiently lower than that of silica fibers. In Er/Yb-codoped phosphate fibers, efficient DFB lasing at direct UV inscription of an ~5 cm long phase-shifted FBG has been shown to be possible, see [86] and citation therein.

We investigated a DFB laser fabricated from a composite fiber with a high Er^{3+} concentration of about 3 wt% (weight percent). The composite fiber was fabricated by melting rare-earth-doped phosphate glass in a silica tube [87, 88]. Since the presence of phosphorus oxide reduces photosensitivity, femtosecond technology enabling FBG fabrication in non-photosensitive fibers has been applied. As a result, a CW DFB laser operating at 1550 nm with a record short cavity length of 5.3 mm has been demonstrated [89].

The main parameters that determine the minimum possible cavity length of a DFB laser are the gain of the active fiber g_s, the passive optical fiber loss α, and the mode coupling coefficient $\kappa = m\pi\delta n/n_{eff}\,\Lambda$, where m is the FBG order, δn is the amplitude of the refractive index modulation, n_{eff} is the effective refractive index of the core mode, and Λ is the FBG period. The relation that determines the lasing threshold can be expressed as [77]: $g_s \approx \alpha + 4\kappa \cdot \exp(-\kappa L)$.

To evaluate the gain, we used a 2-cm-long segment of an active, highly Er-doped fiber and a pump laser diode with a wavelength of 980 nm and an output power of up to 600 mW. The measured gain coefficient at a wavelength of 1550 nm for the maximum pump power is $g_s \approx 2$ dB/cm. The value of passive optical loss estimated from the transmission spectrum of a fiber waveguide in the wavelength region of 1300 nm is $\alpha \approx 4$–5 dB/m. For a highly reflective FBG in the wavelength region of 1550 nm at a coupling coefficient $\kappa = 450$ m^{-1} and $g_s = 2$ dB/cm, the minimum DFB cavity length is estimated to be approximately 8 mm. For $\kappa = 1170$ m^{-1} and $g_s = 2$ dB/cm, the minimum cavity length is approximately 4 mm.

Inscription of an FBG with a phase shift in the center of the structure was carried out by the femtosecond point-by-point method in the fiber workpiece, which was a section of an active fiber with a length of 8 mm to which the passive tails of the

Nufern 1060-XP fiber were spliced. Femtosecond laser pulses (λ = 1026 nm, Δt_p = 232 fs) were focused into the core region of the active fiber via a Mitutoyo 50× Plan Apo NIR HR objective (NA = 0.6), and the fiber was moved using the Aerotech ABL1000 high-precision air-bearing linear stage. A phase shift in the center of the FBG section was provided by amplitude modulation of a 1 kHz sequence of fs laser pulses, as proposed in paper [72]. The total length of the inscribed 2nd-order FBG is 5.3 mm, and the resonance wavelength is approximately 1550 nm.

The inset of Fig. 9.11 shows the transmission spectra of the inscribed phase-shifted FBG, which were measured for different polarization axes using an Apex AP2050 optical spectrum analyzer with a resolution of 1 pm, as well as the spectra obtained via numerical simulation [89]. From their comparison, one can see that the fast axis of the FBG has a higher coupling coefficient than the slow axis does, which is associated with the peculiarity of the method of femtosecond point-by-point inscription [90, 91]. Numerical simulation with coupling coefficients $\kappa_{\mathrm{fast}} = 1357\ \mathrm{m}^{-1}$ and $\kappa_{\mathrm{slow}} = 1115\ \mathrm{m}^{-1}$ for the fast and slow axes, respectively, agrees well with the corresponding experimental spectra, despite the limited dynamic range at which the transmission spectra of the phase-shifted FBG are measured. The simulation also allows us to estimate the phase shift value, which is $\Delta\Lambda = 0.37\ \mu\mathrm{m}$ (or 0.7π) for the FBG period of about Λ = 1.07 μm, and the birefringence value, which is $\Delta n = \delta n_{\mathrm{fast}} - \delta n_{\mathrm{slow}} \approx 1.1 \cdot 10^{-4}$ at the refractive index modulation $\delta n_{\mathrm{fast}} \approx 6.7 \cdot 10^{-4}$.

The experimental scheme of the DFB laser based on the phase-shifted FBG inscribed in the active fiber is presented in Fig. 9.11. The DFB cavity is pumped with a single-mode laser diode with a wavelength of 980 nm and an output power of up to 600 mW through a 980/1550 nm wavelength division multiplexer (WDM). The edge of the FBG coincides with the splicing point between the active fiber segment and the passive fiber tail through which pump radiation is coupled. Thus, the

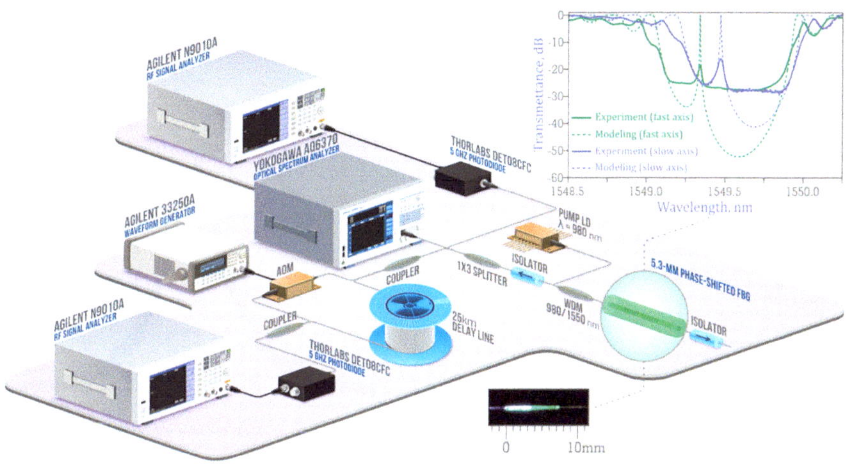

Fig. 9.11 Scheme of the DFB laser and related measuring equipment. The inset at the top shows the transmission spectra of the FBG with a phase shift for the fast and slow polarization axes; the inset at the bottom is a photograph of the DFB laser cavity while pumping [89]

part of an active fiber that does not contain an FBG did not affect the lasing threshold or other characteristics, since we measured the laser output radiation propagating in the opposite direction to the pump radiation.

The laser output radiation counter propagating with the pump radiation was divided into three measuring channels by a 1 × 3 splitter. The measurements of relative intensity noise (RIN) were carried out via an Agilent N9010A radio frequency (RF) analyzer and a 5 GHz Thorlabs DET08CFC photodiode, and the laser wavelength and output power were measured via a Yokogawa AQ6370 optical spectrum analyzer. We used a self-heterodyne technique based on the Mach-Zehnder interferometer (MZI) to measure the laser linewidth with high precision [92]. One of the MZI arms contains a 25-km delay line corresponding to the limiting spectral resolution of ≈ 8.3 kHz. Another arm contains an acousto-optic modulator (AOM) driven by an Agilent 33250A waveform generator at a carrier frequency of 80 MHz, while the RF signal is analyzed by an Agilent N9010A/5-GHz Thorlabs DET08CFC set.

Figure 9.12a shows the dependence of the output radiation power on the pump power. The lasing threshold is observed at a pump power of ≈ 75 mW. We assume that the relatively high threshold power is due to the short ion lifetime at the metastable level $\tau \approx 3.5$ ms, since for a high-Q cavity $P_{th} \sim 1/\tau^3$. For comparison, in silicate glass, Er^{3+} active fibers have $\tau \approx 10$ ms. The obtained differential efficiency ($\eta \approx 0.1\%$) is comparable with typical values for erbium DFB lasers, which have an order of magnitude longer cavity [20, 93], see also Sect. 9.4.1. Lasing was observed only for the fast polarization axis at 1549.4 nm over a broad range of powers from the threshold to maximum power, see the optical spectrum in the inset of Fig. 9.12a. This behavior is typical for resonators inscribed by femtosecond laser radiation and the point-by-point writing technique [91], where the FBG coupling coefficient is different for different polarization axes. More precise self-heterodyne measurements yield a beat signal spectral width of approximately 70 kHz at a level of −20 dB, which corresponds to a laser linewidth of 3.5 kHz (Fig. 9.12b).

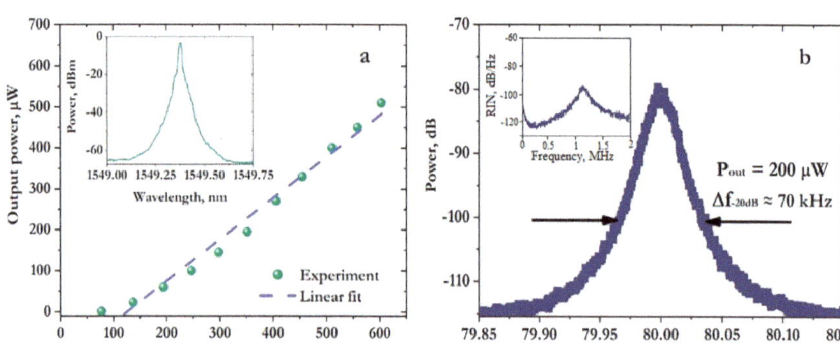

Fig. 9.12 (**a**) Dependence of the output power on the pump power; the inset shows the DFB laser spectrum at the maximum output power. (**b**) RF spectrum of beat signals measured via the self-heterodyne technique. The spectrum was averaged over 100 sweeps, and the measurement time was 5 s. Inset: Relative intensity noise measured at the maximum output power of the laser

The relative intensity noise (RIN) at the maximum laser power is shown in the inset of Fig. 9.12b. For an erbium DFB laser, the typical frequency of the relaxation oscillation peak does not exceed 0.6 MHz [77]; in our case, this value is 1.15 MHz at an output power of 520 μW. This is because the frequency at which the RIN peak is located also depends on the lifetime of active ions τ at a metastable level, as $\nu \sim \sqrt{1/\tau}$ [94]. As mentioned earlier, the lifetime at the metastable level is almost three times less than that for standard commercially available fibers because of the presence of hydroxyl groups in the phosphate fiber. The measured level of the RIN is −95 dB/Hz at 1.15 MHz, which is a typical value for this type of laser without external active stabilization.

Thus, the developed technology of phase-shifted FBG inscription in specially designed Er-doped composite fibers by means of femtosecond pulses enables the fabrication of an efficient DFB laser with a cavity length of 5.3 mm, which is the shortest reported for fiber lasers based only on erbium-doped active media, to the best of our knowledge, see [89] for more details. The power and spectral characteristics of such lasers are comparable to those of conventional Er-doped fiber DFB lasers with much longer cavities. In the whole power range from the threshold to a maximum power of ≈0.5 mW, single-frequency generation with a narrow linewidth (≈ 3.5 kHz) is observed in one linear polarization mode with a peak relative intensity noise of −95 dB/Hz at 1.15 MHz. Such short Er-doped fiber DFB lasers are very attractive for applications as point active sensors, which may be easily integrated into a sensor chain with great potential for high-sensitivity acoustic sensing and similar applications [79].

9.4.3 Random DFB Lasers in Active Fibers

Together with regular DFB fiber lasers, a great deal of attention has recently been given to fiber lasers with randomly distributed feedback (RDFBs) based on non-regular refractive index structures [95, 96]. In the literature, various types of RDFB fiber lasers have been reported. First, lasers based on an array of FBGs are inscribed in an active or passive fiber with random intervals, which form random phase shifts between the FBGs, have been described. A number of sub-gratings counts up to several tens. The number of generated longitudinal modes depends on the length of the random grating and the pump power. To the utmost, it is principally possible to obtain the single-frequency regime.

Therefore, a random Er-doped fiber laser [95] generated at a wavelength of ∼ 1.535 μm was demonstrated. In 150-cm-long fibers, an array of 7–31 gratings was inscribed, each of which has a length of 5 mm and a reflection coefficient of 7–8%. The distance between gratings was randomly chosen from 4.2 to 5.8 mm intervals. The generation spectrum consists of several competing longitudinal modes whose number varies from 2 to 7, depending on the grating number, whereas the power distribution between them changes as the pump power varies. Another random Er-doped fiber laser [96] generated at a wavelength of ∼ 1.534 μm. In contrast to

[95], instead of an array with rather large intervals between gratings, a continuous grating with random phase shifts between sub-gratings of variable length is explored. Polarization-maintaining (PM) active fibers of 30 and 20 cm in length were tested. The lasing regime appears to depend on the fiber length and pump power, which varies from the multimode regime, with an average linewidth of 0.5 pm, to single-longitudinal-mode operation, as demonstrated for random grating for the first time. A rather short random DFB fiber laser [97] was demonstrated. A 25-mm-long FBG was inscribed in an Er/Yb-codoped fiber by UV beam scanning through the phase mask. During scanning, 20 random phase shifts are induced between sub-gratings of equal length (1.19 mm). Single- or double-longitudinal modes are generated depending on the pump direction and power. A tunable Er-doped fiber laser [98] generated near 1.55 μm is also demonstrated. A random DFB is provided by a 10-cm piece of passive SMF-28 fiber, in which a random grating consisting of 100 parts with random intervals is inscribed by a CO_2 laser. In this configuration, a linewidth as low as 2.4 kHz was obtained with central wavelength tuning within 0.5 nm. A Raman fiber laser [99] generated at 1.576 μm due to the random DFB provided by a non-regular grating with 1500 random phase shifts was inscribed in a 1-m-long passive fiber. Multiple- or single-mode generation is observed depending on the pump power. The single-frequency linewidth is 430 kHz. Compared with Raman fiber lasers operating via RDFBs arising from Rayleigh backscattering on natural refractive index fluctuations [68, 100], the implementation of random FBGs allows for greatly shortening the length of passive fibers and obtaining the single-frequency regime of such a random DFB Raman fiber laser.

Thus, the most promising type of RDFB laser is based on the implementation of multiple random phase shifts into long homogeneous gratings. We proposed a technique to introduce additional random variations of the grating amplitude to obtain an entirely random FBG [101]. As a result, the output characteristics of the designed RDFB laser based on an FBG with a randomly changing phase and amplitude formed in a short (4 cm) Yb-doped fiber appeared to be even better than those of the regular DFB Yb-doped fiber laser of the same length [102].

Random FBGs were inscribed in a commercial polarization-maintaining (PM) Yb-doped optical fiber using the experimental setup based on a reliable phase-mask writing scheme [103]. Variation in the UV laser power (in the interval of 40–100 mW) simultaneously with the piezo shift of the phase mask (in the interval of 0-2π) enabled variations in the subgrating phases and amplitudes to be controlled according to a predefined random pattern. To characterize the inscribed random FBGs, we used a linearly polarized tunable DFB Yb-fiber laser [70]. Its lasing wavelength was tuned by temperature variation. The interferometric method was used for wavelength monitoring: one of the fiber coupler arms was connected to the fiber Mach-Zehnder interferometer with a fiber arm difference of $\Delta z = 1.4$ m, which corresponds to a free spectral range of approximately 140 MHz. As a result of data processing, the FBG reflection coefficient as a function of wavelength was measured. Figure 9.13(a) shows the measured reflection spectrum of the random FBG consisting of 10 sub-gratings.

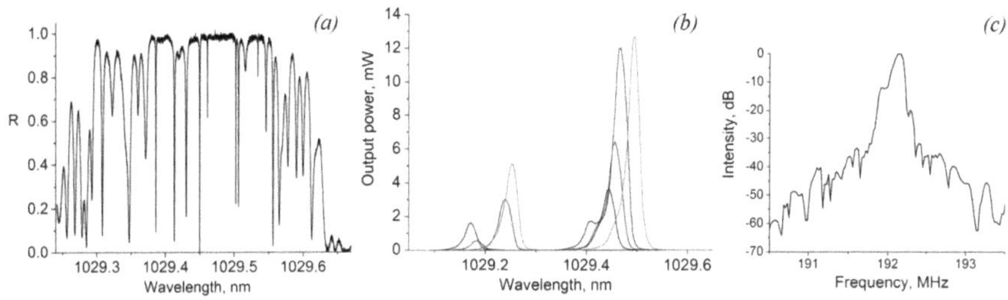

Fig. 9.13 (**a**) Experimental spectrum of the random FBG; (**b**) random DFB laser spectra at a pump power of ~ 145 mW (black line), 255 mW (red line), 310 mW (blue line), and 470 mW (green dashed-dot line); (**c**) RDFB and DFB lasers beat the spectrum at a power of the milliwatt level

Testing of a random FBG with laser diode pumping has demonstrated the single-frequency operation of such an RDFB fiber laser. Figure 9.13(b) shows the output spectra of the random laser measured by an optical spectral analyzer (OSA) at different pump powers. Two separated peaks correspond to two orthogonal polarization modes. At low pump powers, the spectrum contained the addition of another longitudinal mode, which disappeared as the pump power increased. The laser spectrum shifts to longer wavelengths with increasing pump power because of fiber heating. The measured linewidth corresponds to the OSA resolution (0.02 nm). For precise laser linewidth measurements, a frequency beating method was employed. Figure 9.13(c) shows the radio-frequency spectra of the beating RDFB laser with the single-frequency DFB laser whose frequency difference is about 190 MHz. The measured linewidth of the beat spectra at the −3 dB level was about 120 kHz, which is comparable to the linewidth of the beat spectra of two identical DFB lasers in Yb-doped PM fibers, which is approximately 100 kHz [102]. Therefore, the realized RDBF laser is single-frequency with a linewidth close to that of the regular DFB laser inscribed in the PANDA-type PM fiber [102].

To investigate the intensity distribution inside the laser cavity at the pump and lasing wavelengths, the sideward scattered radiation was measured along the fiber by a micro objective mounted on a translation stage, and its spectrum was registered by an OSA. Analogous measurements were produced for a DFB laser with a length of 4 cm and a phase shift equal to π located 2.1 cm from the left end of the grating. The modulation amplitude of the refractive index was close to the average value of the RDFB laser. For both lasers, the longitudinal intensity distribution at the lasing wavelength measured in the same setup has its maximum in the central region of the grating because of the light localization effect, which is known to occur in a regular grating with one π-shift, as well as Anderson-type localization in random gratings [95, 96, 104]. The lasing intensity is reduced at points ±10 mm from the central maximum by 15–20 dB in the DFB laser (for different sides) and by 12–15 dB in the RDFB laser, whereas the pump power is depleted in the center. The DFB and RDFB laser output power as a function of the pump power is close to linear, and the slope value is 3.5 times greater for a random DFB laser (and, accordingly, the maximum power). The random DFB laser power reaches ~25 mW at a pump power of

250 mW, i.e., the absolute efficiency is approximately 10%, while the slope efficiency exceeds this value.

To compare the experimental results with theory, a numerical simulation of random FBG spectra and longitudinal distributions of the light intensity inside the FBG at wavelengths corresponding to its transmission peaks was performed. We used software developed by Belai on the basis of the T-matrix technique for regular FBGs [38]. The program was modified to calculate the longitudinal intensity distribution inside the grating, which consists of several sub-gratings with definite phases and amplitudes of the harmonic modulation [101]. Numerical simulations of the FBG spectrum with a modulation amplitude of $\Delta n = 10^{-4}$ and a π-shift in the center (DFB laser) and the calculated intensity distributions are shown in Fig. 9.14(a) and (b). At the wavelength corresponding to the π-shift-induced transmission peak, the intensity is exponentially reduced by more than 25 dB at the points ±10 mm from the central maximum. The simulation results agree with the experimental data, Fig. 9.14(c), although the absolute value of the intensity decrease in the experiment is lower than the calculated value, which could be explained by the imperfections of the real grating.

A numerical simulation of the reflection spectrum for a random FBG with experimental parameters is shown in Fig. 9.15(a). According to the calculation, narrow transmission peaks caused by phase shifts are located mainly in the left part of the broad reflection peak, whereas in the experimental spectrum, the transmission peaks have larger amplitudes and are distributed more uniformly throughout the main peak, as shown in Fig. 9.13(a). Figure 9.15(b) shows the calculated intensity

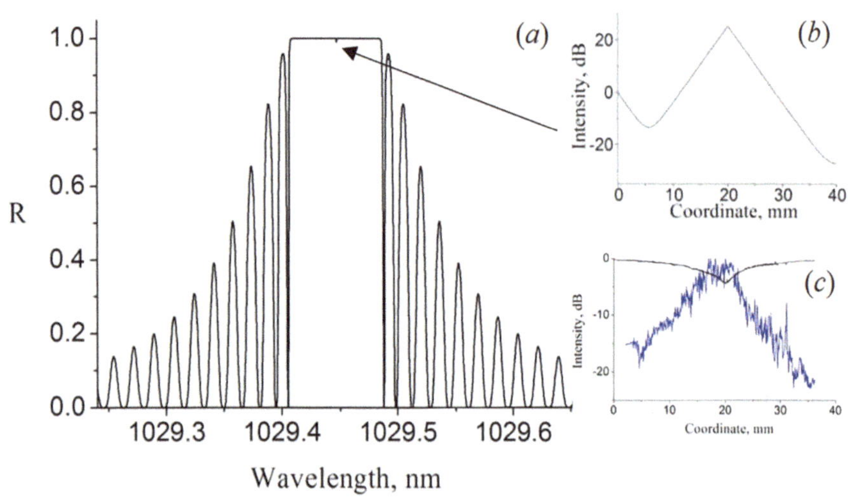

Fig. 9.14 (**a**) Numerical simulation of single π-shift FBG reflection spectra; (**b**) calculated intensity distribution within the grating at the wavelength of the transmission peak; (**c**) experimental data for the pump (black) and generation (blue) wavelength

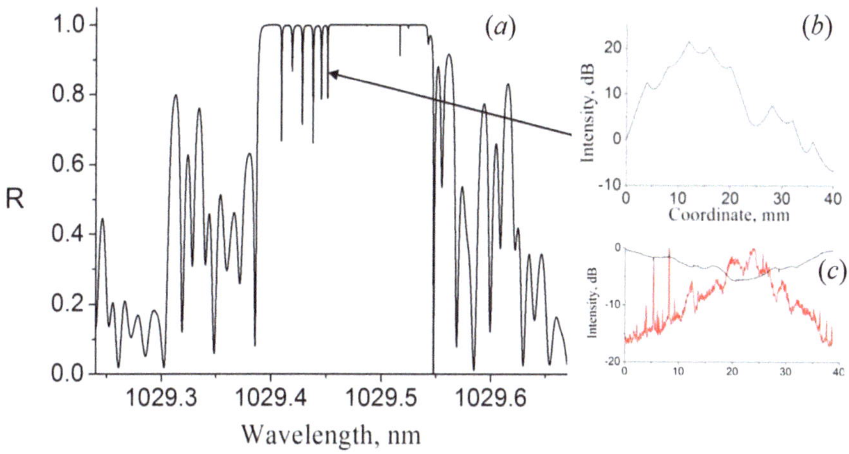

Fig. 9.15 (**a**) Numerical simulation of reflection spectra for the random FBG with experimental parameters of subgrating phases and amplitudes; (**b**) calculated intensity distribution within the random FBG at the wavelength of one of the transmission peaks; (**c**) experimental data for the pump (black) and generation (red) wavelength

distribution along the grating at the wavelength of the transmission peak corresponding to the maximum intensity.

According to calculations, the radiation is localized in the central part of the grating. The decrease in intensity at points ±10 mm from the maximum amounts to 15–20 dB. The simulation results are in qualitative agreement with the experimental data [101], Fig. 9.15(c). Such a distribution with good localization does not occur for other transmission peaks, while their distributions also have much lower intensities.

Therefore, the comparison of intensity distributions for different random gratings shows that for the FBG in which both the phases and amplitudes of the segments are random, the most pronounced localization with maximum intensity in the center is obtained, resulting in single-frequency generation of the random DFB laser. A comparison of regular and random DFB lasers realized in Yb-doped PM fibers reveals that the output power and generation efficiency of random DFB lasers are 3.5–5 times (depending on the pump power) greater than those of DFB lasers, whereas DFB and random DFB lasers have close generation linewidths. The increase in the efficiency could be connected with a wider localization area of the generated radiation for the random grating in comparison to the grating with one π-shift, herewith the dip in the saturated pump radiation distribution is also wider and deeper, which indicates a more effective conversion of the pump to laser radiation in the case of a random DFB laser. This principal differences between regular and random DFB fiber lasers are confirmed by the numerical simulation performed for these two configurations.

9.4.4 Random DFB Raman Lasers in Passive Fibers

Raman fiber lasers operating via randomly distributed feedback (RDFB) arising from Rayleigh backscattering on natural refractive index fluctuations in passive fibers have been proposed and developed in the last few years [68, 105–110]. Due to the weakness of Rayleigh backscattering per unit length ($\sim 10^{-5}$ km^{-1}), such lasers have rather long fiber cavities, from hundreds of meters to several tens of kilometers. In addition to standard telecommunication fibers operating in the 1.5 micron range [68, 100], Rayleigh backscattering in passive fibers in the presence of optical pumping enables the generation of Stokes components at <1 micron [105], 1.1 micron [106], 1.2 micron [107], 1.4 micron [108]. The value of the Stokes shift is approximately 13 THz in silica fibers [100]. In addition to the opportunity to tune within the broad (~ 10 THz) Raman gain peak, a cascaded generation of higher Stokes orders enables almost continuous wavelength tuning of the RDFB Raman fiber laser in the ultra-broad range of 1.1–2 microns [109, 110]. Herewith, the utilization of PM fibers results in linearly polarized Raman generation with a high efficiency of pump-to-Stokes conversion (70–80%), which is almost independent of the Stokes order [109]. Without additional spectral filtering, the generated linewidth is defined by the Raman gain bandwidth (and its Schawlow-Townes narrowing near the threshold combined with nonlinear broadening well above the threshold [68, 111]), amounting to ~1 nm and increasing to ~3 nm at high powers and high Stokes orders [100, 105–110]. To reduce the linewidth, intracavity spectral filters of different types are usually inserted. Therefore, using a tunable acousto-optic filter, it is possible to tune the wavelength continuously [112] with extremely low power variation (0.1 dB) in the ~40 nm range. Using multi-wavelength filters, it is possible to obtain multi-wavelength generation free of line competition [113]. Raman fiber lasers with random DFB based on Rayleigh backscattering have already been applied in various fields, from distributed sensing [114, 115] to fiber optics communications [115, 116]. A more detailed survey of basic schemes and characteristics of RDFB fiber lasers, as well as their applications, may be found in review papers [68, 117].

As mentioned above, there is another opportunity to create RDFB Raman fiber lasers, namely, to use random FBGs with much higher refractive index modulation. Compared with Rayleigh backscattering-based RDFB fiber lasers, the implementation of a random FBG allows one to greatly shorten the length of the passive fiber and obtain the single-frequency regime of the RDFB Raman fiber laser [99]. A regular phase-shifted FBG [25] of 12.4 cm length provided single-frequency Raman lasing at 1584 nm under pumping at 1480 nm. In the case of commercial single-mode fiber, the maximum power and linewidth are 150 mW and 7.5 MHz, respectively, with a 39 W threshold pump power.

Following these results, we have proposed and realized [118, 119] a new scheme of an RDFB fiber laser based on a random FBG array inscribed in a longer passive fiber (13 m, Fujikura SM98-PS-U25D). It consists of 57 sub-gratings with equal Bragg wavelengths of 1092.3 nm and lengths of 4 cm, but different phases and amplitudes. The distance between adjacent sub-gratings varies around 19 cm, whereas a continuous 1-m grating of constant amplitude and random phase modulation was employed earlier, demonstrating random lasing with low efficiency (< 1%) [99]. As a result, we obtain a lower generation threshold (1.3 W) and much higher maximum output power and efficiency (approximately 6 W and 40%, respectively), see Fig. 9.16(a), in comparison with [25, 99]. The linewidth was characterized at a pump power of ~ 1.5 W via a self-heterodyne technique [76]; the FWHM value was measured for the backward wave. The beating spectrum width at half maximum amounts to approximately 370 kHz, thus confirming single-frequency operation. At lower powers (near the threshold), the beating spectrum width decreases to 140 kHz, so the linewidth is below 100 kHz, Fig. 9.16(b). The power for the backward Stokes wave is much greater than that for the forward Stokes wave, which mainly propagates in the region of the depleted pump. This power behavior is similar to that of a Raman fiber laser with RDFB based on natural Rayleigh backscattering [68]. The line broadens sufficiently as the pump power increases to 2 W, whereas a further increase in pumping does not lead to a significant effect, and the linewidth increases to 60–80 pm.

To explain such behavior, the following theoretical model has been developed. The FBG array with a length of $L \sim 13$ m consists of $n = 57$ uniform slightly reflective FBGs with grating strength $f_j = \delta n_j s_j \ll 1$, where δn_j is the modulation of the FBG refractive index, s_j is the number of periods, and k_0 is the frequency of parametric resonance, $\Delta\text{FBG} \approx 30$ GHz. The distances between adjacent gratings (center-to-center spacing) are random, $L/n \approx 0.2$ m; therefore, the phases of the gratings depend on detuning with respect to k_0: $q = k - k_0$. As a result, for different resonator modes with $q_m = \pi m/L$, any pair of FBGs provides constructive reflection $r = r_i + r_j$ for $q_m(x_j - x_i) = 2\pi l$ or destructive reflection $r = r_j - r_i$ for $q_m(x_j - x_i) = 2\pi(l + 1/2)$ and all intermediate cases.

The effective Raman gain coefficient is $g = g_R P - \alpha$, where α represents passive losses, including distributed and point losses, P is the pump power. The lowest threshold is expected for mode q_n with a π-shift in the center of the grating, i.e., the

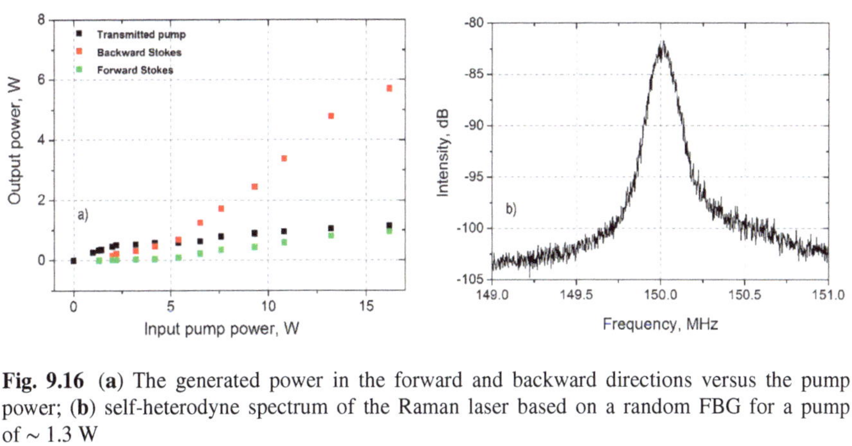

Fig. 9.16 (**a**) The generated power in the forward and backward directions versus the pump power; (**b**) self-heterodyne spectrum of the Raman laser based on a random FBG for a pump of ~ 1.3 W

left and right sides of the grating have opposite phases. If the FBG's half strengths are $f_L \approx f_R \approx f$, then the probability of the existence of such a mode is

$$W = \frac{\Delta FBG}{\Delta MM} \exp\left(-2\frac{f^2}{\frac{n}{2}\langle f_j^2 \rangle}\right),$$

where $\Delta MM = c/(2nL) \approx 8$ MHz.

Presuming $W \approx 1$, we obtain $f = \sqrt{\frac{10}{4} n \langle f_j^2 \rangle}$, then $\ln(\Delta FBG/\Delta MM) = 10$. The left and right side transmission coefficients are $T_R = T_L = 1/\cosh^2 f \approx 4e^{-2f}$. The gain-loss balance condition gives the expression $g_R P_{th} L_{eff} I_0 = \alpha L_{eff} I_0 + 2TI_0$, where $L_{eff} = L/f$. Then $P_{th} = \left(\alpha L + 8fe^{-2f}\right)/g_R L \sim 2W$, since $f_{exp} \approx 4$, $4e^{-2f} \approx 10^{-3}$, $\alpha L \approx 0.03$.

Therefore, near the threshold, Raman generation with a random FBG array may be single-frequency, whereas at a high gain-to-loss ratio, it becomes multi-frequency, with descriptions based on a balance model similar to that used for Rayleigh-based RDFB Raman fiber lasers [68]. In this case, each grating reflects $R_j = f_j^2 \ll 1$ part of the incident power, and then the balance model is reduced to

$$\frac{\partial I_+}{\partial z} = gI_+ - \frac{\langle R_j \rangle}{\langle L_j \rangle}(I_+ - I_-), \frac{\partial I_-}{\partial z} = -gI_- - \frac{\langle R_j \rangle}{\langle L_j \rangle}(I_+ - I_-),$$

where $\langle L_j \rangle = L/n$. Taking.

$$I = I_+ + I_-, V = I_+ - I_-, I_Z = \frac{\partial I}{\partial z} = gV - 2\frac{R}{L}V, V_Z = \frac{\partial V}{\partial z} = gI, g \ll \frac{\langle R \rangle}{\langle L \rangle},$$

in the symmetric case, we obtain.

$$I = I_0 \cos(pz), V = I_0 \frac{g}{p} \sin(pz)$$

where $p = \sqrt{g\left(\frac{2R}{L} - g\right)} \simeq \sqrt{\frac{2Rg}{L}}$

Applying the boundary condition $I_+(-L/2) = I_-(L/2) = 0$, we obtain

$$\sqrt{\frac{gL}{2R}} \tan\left(\sqrt{\frac{gLR}{2}}\right) = 1$$

hence $gL \approx 0.8$, $P_{th} \approx 10$–30 W.

This treatment shows that the incoherent Rayleigh-like limit is realized at the maximum pumping, whereas coherent few-mode generation is realized near the threshold. In qualitative correspondence with the model, the experimental data have similar behavior, as shown in Fig. 9.16(a): there are two different regimes with different slopes for power and different spectra: a coherent regime from 1.3 to 3–5 W pump power with relatively narrow bandwidth and lower efficiency, and incoherent lasing in the 5–16 W pump power range with higher efficiency and a relatively broad spectrum consisting of many random modes with a total bandwidth of ~80 pm.

Thus, a new type of linearly polarized Raman fiber laser with random distributed feedback has been demonstrated on the basis of an array of relatively short FBGs with random amplitudes and phases easily fabricated without the rather complicated procedure used earlier for inscription of long regular gratings with phase shift [25] or long gratings with multiple phase shifts [99]. As a result, the maximum power has been increased to ~6 W at a pump power of approximately 16 W, whereas the linewidth at maximum power is as small as ~80 pm. Compared to Gagn'e and Kashyap [99], the threshold is much lower, and the efficiency is much greater because of the much larger effective amplification length in our case. The linewidth near the threshold pump power (1.3 W) is narrower than 100 kHz in the single-frequency regime. The developed model qualitatively describes the obtained results.

Optimized RDFB Raman fiber lasers can find a variety of applications: having a narrow spectral linewidth and rather high power, such lasers can be used for remote FBG sensor interrogation and as light sources in distributed sensor systems. Moreover, this configuration can be adapted to the 1.5 mm range, which will allow such lasers to be employed in long-haul telecommunication and sensor systems. Their high power, narrowband output can be efficiently converted to the visible range via frequency doubling in a nonlinear crystal or periodically poled fiber [120].

9.4.5 DFB Laser Line Narrowing by a Random Rayleigh Reflector

Random distributed feedback due to Rayleigh backscattering on naturally present refractive index fluctuations in fibers has been actively explored since the first demonstration of stable narrowband random lasing in a cavity-free passive fiber with Raman gain [100]. It is also implemented in active-passive fiber configurations with an open or half-open cavity in which conventional gain in a short active fiber is combined with the random DFB through Rayleigh backscattering in a long passive fiber. As the Rayleigh backscattering is broadband, the generated line may be easily tuned within the corresponding gain bandwidth. For example, a tunable erbium-doped fiber (EDF) laser with a 20-km-long single-mode passive fiber demonstrates a broad tuning range (1525–1565 nm) with a narrow linewidth of ~ 0.04 nm provided by a tunable intracavity Fabry-Perot interferometer (FPI) filter [121].

Moreover, Rayleigh backscattering in a piecc of passive fiber may also be used for line narrowing of existing lasers, including single-frequency DFB fiber lasers. Such an opportunity has been actively studied in recent years with the use of long conventional passive fibers with natural Rayleigh scattering, as well as with short fibers in which artificial random structures are formed, enhancing Rayleigh backscattering. In such a hybrid configuration, the implementation of an additional scattering fiber with a random reflection spectrum leads to additional spectral filtering and line narrowing of the generated light, correspondingly. Therefore, in early work [122], linewidth narrowing of a semiconductor laser generating at 1.58 μm by ~2000 times (from tens of MHz to 20 kHz) was achieved with the use of an external fiber ring interferometer with Rayleigh backscattering. The fiber interferometer is supposed to increase the effective scattering length because of the large number of roundtrips. In [123], a single-frequency Er-doped fiber ring laser with a tunable operating wavelength was proposed and demonstrated. The Rayleigh backscattering feedback in a 660-m SMF-28e fiber piece acts as an element that suppresses side modes, thus ensuring single-frequency laser operation. The laser linewidth did not exceed 3 kHz throughout the entire tuning range (1549.7–1550.18 nm). In [124], a low-loss fiber ring resonator with a total length of 4 m comprising a conventional SMF-28 fiber and directional couplers was utilized as a high-finesse filter for the self-injection locking of a semiconductor DFB laser. By varying the coupling coefficient, the authors define the regime providing the best locking and significant narrowing of the laser linewidth to the level of 2.5 kHz. The same group has demonstrated self-stabilization of a semiconductor DFB laser accompanied by linewidth narrowing to 2.8 kHz with the use of an external 4-m fiber ring cavity in conjunction with simple active optoelectronic feedback, ensuring stable mode hopping-free laser operation in the SLM regime [125]. In [126], a narrow-linewidth dual-wavelength random Er-doped fiber ring laser with a semiconductor optical amplifier (SOA) was demonstrated. The randomly distributed feedback from a 5-km non-uniform fiber with enhanced Rayleigh backscattering (−34 dB/km) results in narrow linewidth (~1 kHz) generation at both wavelengths. In [127], a tunable single-frequency Er-doped fiber ring laser based on the so-called longitude-purification induced by randomly distributed feedback was demonstrated. An external distributed reflector with enhanced Rayleigh backscattering in a single-mode special fiber (HSF UHNA3 by Thorlabs with a length of ~ 50 m, providing an integral reflection level of −38 dB) functions as a mode selector to ensure robust side mode suppression of ~ 67 dB. The laser linewidth did not exceed 1.2 kHz within the whole tuning range (approximately 40 nm). In [128], random lasing was obtained in an LD-pumped 5-m-long Er-doped fiber comprising a distributed array of weak FBGs with an individual peak reflectivity of ~0.00003% at ~1547.6 nm. Owing to the distributed feedback arising at the reflection from multiple FBGs and the dynamic population inversion grating enabling laser line filtering, the laser linewidth was <300 Hz. In [129], a compact random fiber laser with a half-open cavity based on a heavily doped Er fiber and a short (10 cm) artificial Rayleigh reflector based on random refractive index structures inscribed by femtosecond pulses in a single-mode fiber with a mean scattering level of +41.3 dB/mm relative to the natural scattering has been demonstrated. A single-frequency regime with a ~10 kHz linewidth was observed at an output power of up to ~ 3 mW. Although the use of artificial random reflectors enables compact design, random DFB via natural Rayleigh backscattering in a long conventional fiber does not require any inscription of in-fiber artificial structures, which simplifies experimental studies of basic effects. Therefore, random feedback in conventional single-mode fibers such as SMF-28 has been well characterized, and it is random indeed, which is why it is the most suitable for a proof of principle.

Here, we present a model and experimental test for linewidth narrowing in a hybrid configuration consisting of a 6-cm Er-doped fiber DFB laser and a spool of 25-km standard SMF-28 fiber. It is shown that a large narrowing (up to four orders of magnitude) of generation linewidth may be achieved so that the linewidth can reach extreme sub-Hertz values (0.1–0.001 Hz) [130]. The details of the developed model and its comparison with the experimental data are presented below, together with further opportunities for the method with the use of natural as well as artificial Rayleigh reflectors.

First, let us take the DFB laser linewidth as a starting point, see Sect. 9.3.2. As a next step, let us consider the model of the DFB fiber laser in a hybrid configuration with an additional weak random Rayleigh reflector based on a conventional single-mode fiber. Let us consider Rayleigh backscattering in a conventional fiber due to naturally presented refractive index fluctuations, which are 7–8 orders of magnitude lower than the induced modulation of the refractive index in highly reflective FBGs, $\delta n \sim 10^{-3}$–10^{-4}, see Fig. 9.17. The effective fiber length of a Rayleigh reflector is defined by its physical length and losses:

$$L_{eff} = \frac{1 - \exp(-2\alpha L)}{2\alpha}$$

For the 25-km fiber coil, the effective length is about 10 km at typical losses of 0.2 dB/km. Although local reflection is weak and strongly fluctuating along the fiber (Fig. 9.17), the average reflectivity R_c of the whole coil is the product of its effective length and average Rayleigh backscattering coefficient: $R_c = \varepsilon L_{\text{eff}}$. For a typical

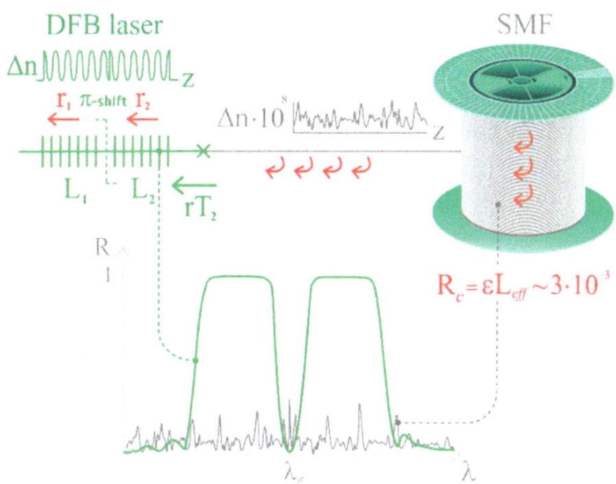

Fig. 9.17 Basic scheme of the hybrid configuration with a DFB fiber laser and a 25-km SMF spool

value of $\varepsilon = 3\cdot10^{-4}$ 1/km, we can estimate $R_c = 3\cdot10^{-3}$, so the characteristic absolute value of the random reflectivity amplitude is $|r_c| = \sqrt{R_c} \approx 0.05$. In general, the spectral dependence of the reflectivity $r_c(k)$ is a quickly varying complex parameter.

When the Rayleigh reflector based on the 25-km SMF fiber is spliced to one of the FBG mirrors of the DFB laser cavity (e.g., to the right, see Fig. 9.17), this mirror becomes complex with reflectivity

$$\tilde{r}_2 = \frac{r_2(k)+r_c(k)}{1+r_2^*(k)r_c(k)} \simeq r_2(k)+r_c(k)T_2$$

Here, T_2 is the transmissivity of the right DFB laser mirror, which can be estimated from the total transmissivity of the π-shifted FBG structure. In our case, $T_2 \approx 5\%$.

At a narrow generation spectral range when $|q|L_{DFB} < \delta$, $\delta = \Delta T_{rt} = \Delta\frac{2L_{DFB}n_0}{c}$ (or $|\Omega| < \Delta$), it could be taken that $r_2(k) = r_2(k_B - ig_s) = const$. If the phase of $r_c(k_j)$ is approximately equal to the phase of r_2 for a definite wavelength λ_j corresponding to one of the random spectral peaks near resonance (see Fig. 9.17), the reflected beams add up constructively, and the absolute value of the reflectivity increases, thus increasing the resonator quality factor. Generation is achieved for the mode with the maximum absolute reflectivity of the complex mirror. For simplicity, we can replace the distributed Rayleigh reflector with the point one with $r_c = \sqrt{\varepsilon L_{eff}}$ placed at distance $z = L_{eff}$ from the FBG. Then, following the model presented in Sect. 9.3.2, we obtain

$$\frac{I_q(\Omega)}{\langle I_{sp}(\Omega)\rangle} = \frac{1}{\left|1-r_1(k_B-ig_s)\left(r_2(k_B-ig_s)+T_2r_c\cdot\exp(2iqL_{eff})\right)\right|^2} = \frac{1}{\delta^2+\left(\Omega T_{rt}^{eff}\right)^2},$$

where $T_{rt}^{eff} = T_{rt}\frac{L_{eff}}{L_{DFB}}T_2r_c$.

The shape of the laser line is described by Eq. (9.18), but its half-width Δ is defined by the characteristic width of random reflector peaks, not by the bandwidth of the DFB laser cavity consisting of a π-shifted FBG, so that

$$\Delta = \left(\frac{c}{2T_2\sqrt{\varepsilon L_{eff}}L_{eff}n}\right)^2\frac{I_{sp}(\Omega)}{2I}$$

As a result, the hybrid laser linewidth is narrowed by $\left(\frac{T_2\sqrt{\varepsilon L_{eff}}L_{eff}}{L_{DFB}}\right)^2 = \left(\frac{T_{rt}^{eff}}{T_{rt}}\right)^2$ times as compared with DFB laser, which is estimated to be ~5•10^{-6} in our case. This large factor is defined by the ratio of the effective cavity lengths of the hybrid and DFB lasers (≈ 10 km and ≈ 1 cm, respectively), which in turn are inversely proportional to the corresponding width of the spectral resonances.

To implement a single-frequency laser scheme, a DFB cavity consisting of π-shifted FBGs was fabricated using the UV laser inscription technique [70].

The experimental scheme for characterization of the DFB and hybrid laser configurations [130] is presented in Fig. 9.18. The DFB cavity is pumped by a single-mode laser diode with a wavelength of 980 nm and an output power of up to 600 mW through a 980/1550 nm wavelength division multiplexer (WDM). For the hybrid configuration, a 25-km SMF-28 fiber (with a natural Rayleigh scattering level of −105 dB/mm) was spliced to the right end of the phase-shifted FBG, providing additional randomly distributed feedback (RDFB). The power characteristics of the DFB laser are comparable to those of conventional Er-doped fiber DFB lasers: the differential efficiency is 0.1%, and the peak value of the relative intensity noise (RIN) is −100 dB/Hz. The deviation of the power from the average value at time intervals of ~ 10 s does not exceed 1%.

To increase the power of the measured signal, an optical amplifier (EDFA) was added to the setup, consisting of a 1.5-m active fiber, Fibercore I-25 (980/125), which was also pumped by a 980 nm laser diode with a power of up to 500 mW through a WDM 980/1550. To suppress the influence of back reflection from the components, an isolator was added between the WDMs, Fig. 9.18. To measure the spectral characteristics of the amplified signal, a Yokogawa AQ6370 optical spectrum analyzer (OSA) was used. The measured signal-to-noise ratio is ≈ 60 dB, and the generation linewidth corresponds to the resolution of the OSA (20 pm). For a more precise measurement of the linewidth, the signal was launched to the

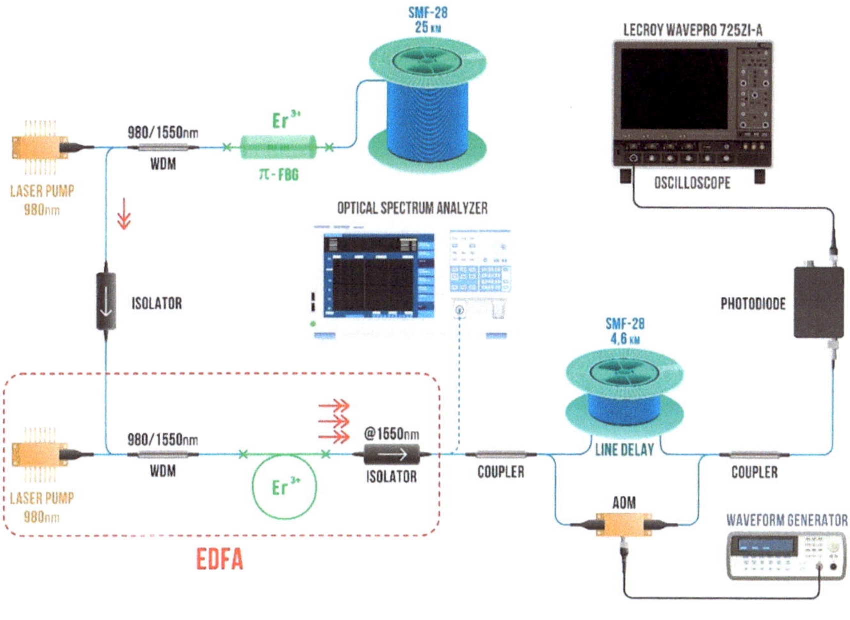

Fig. 9.18 Scheme of the DFB and hybrid laser configurations and related measuring equipment [130]

Mach-Zehnder interferometer (MZI) through an optical connector. The signal power at the input of the interferometer was ~ 30 mW. One of the MZI arms contains a 4.6-km fiber, whereas the other arm contains an acousto-optic modulator (AOM) driven by an Agilent 33250A waveform generator at a carrier frequency of 80 MHz. The beat signal was measured and analyzed by a Thorlabs DET08CFC photodiode and a LeCroy WavePro 725Zi-A oscilloscope/5-GHz with an FFT function, which allows the measurement of the waveform and RF spectra of the signal. Using the technique of processing the waveform of the beat signal, as described previously [131, 132], the spectra of phase and frequency noises were obtained.

A comparison of the phase noise and frequency fluctuation power spectral density (PSD) obtained for a DFB laser in conventional and hybrid configurations is shown in Fig. 9.19(a) and (b). The measurements were carried out for a frequency range of more than 10 kHz, since the influence of the environmental and temperature noise is significantly reduced.

The frequency separation between the peaks in the spectra corresponds to the intermodal separation for the 4.6-km-long Mach-Zehnder interferometer, $\Delta\nu \approx 45$ MHz. The dotted lines in Fig. 9.19(a) correspond to the theoretically calculated phase-noise values for the laser with Lorentzian linewidths varying in the range of 0.1 Hz-10 kHz. For the frequency range above 100 kHz, only white noise exists, characterized by S_0 (the first term of random frequency PSD expansion into a frequency power series), which determines the lower limit of the laser linewidth [133]: $\Delta\nu = S_0\pi$. For a DFB laser without additional random DFB from the SMF, the value of the instant generation linewidth determined by white noise is $\approx$ 15 Hz (Fig. 9.19(b), black color), which is in good agreement with the theoretical estimate. For the hybrid resonator, the linewidth for frequencies above 100 kHz tends to be constant, which is determined by the electrical noise. The upper limit estimation of the generation linewidth, determined from the "bursts" of the signal above the level of electrical noise in the frequency range of 10–100 kHz, is approximately 0.1 Hz. Since the RIN peak is completely "sinks" in the electrical noise (Fig. 9.19(b)), it can

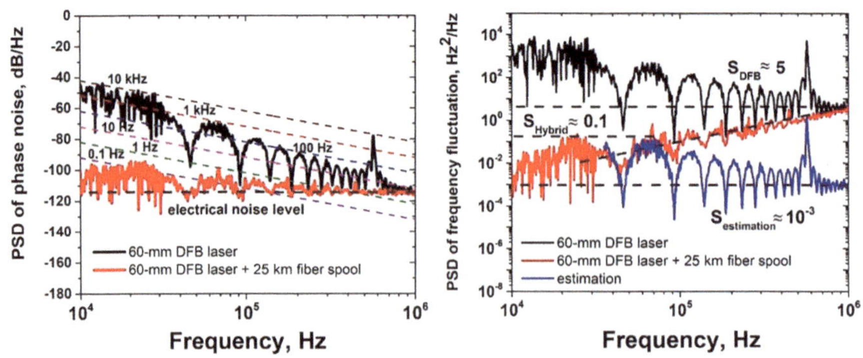

Fig. 9.19 (**a**) Phase-noise PSD obtained for the DFB laser in the conventional (black) and hybrid (red) configurations; (**b**) frequency fluctuation PSD obtained for the DFB laser in the conventional (black) and hybrid (red) configurations and estimation above 100 kHz (blue) [130]

be assumed that the white noise level (S_0) for the composite resonator is much lower, but its variation with frequency is similar. Taking this into account, we can estimate the instantaneous linewidth from the corresponding correction of the noise curve above 100 kHz, which gives an estimated value of 10^{-3} Hz.

From the frequency fluctuation PSD spectrum, one can estimate the linewidth as a function of the observation time [134]. The linewidth is determined as follows:

$$FWHM = \sqrt{8\ln(2)\cdot A}$$

where $A = \int_{1/T}^{\infty} S_v(f)df$. The linewidth increases as the lower limit of the integral decreases, or in other words, as the observation time increases. This is due to the presence of $1/f^{\alpha}$ noise in the frequency fluctuation PSD [77]. Thus, the theory makes it possible to estimate the linewidth of the generation line at long times. Nevertheless, in [135], this model was applied at times of ~ 0.1 ms.

The value of ~ 100 kHz was taken as the upper limit of integration, since white noise S_0 prevails in the range above 100 kHz. Similarly, for a delay line of 4.6 km (observation time is 23 μs), the width of the DFB laser line measured using the technique described in [134] is 5.7 kHz, and the generation linewidth obtained using the self-delay heterodyne (SDH) technique [92] amounts to 120 kHz at −20 dB or 6 kHz at −3 dB. For the hybrid configuration, the linewidth measured using the technique [134] is 150 Hz, whereas the values obtained via the SDH technique amount to 3.2 kHz at the level of −20 dB or 160 Hz at the level of −3 dB.

Thus, narrowing of the generation line of a DFB laser in a hybrid configuration with a random DFB from a 25-km SMF was demonstrated, see [130] for more details. For instantaneous widths, the values obtained from the estimation based on phase-noise data processing differ by more than four orders of magnitude (from 15 Hz for the conventional configuration to 10^{-3} Hz for the hybrid configuration). This extreme narrowing could not be observed directly from the experimental phase-noise data because of the limitation caused by electrical noise, but may be estimated from the fitting of their frequency dependence. Moreover, the estimated limit obtained from the theoretical model for the instantaneous linewidth narrowing factor is approximately $5\cdot10^6$, which leaves room for further narrowing with the use of more precise measurement techniques. Significant narrowing was also observed at longer-term scales by the self-delay heterodyne technique with delay line times of ≥10 μs: the linewidths for the DFB laser in the conventional and hybrid configurations are approximately 6 kHz and 160 Hz, respectively. In practice, the linewidth narrowing of the Er-doped DFB fiber laser generating in the telecommunication window around 1.55 μm allows expansion of the application range of this source in areas such as coherent reflectometry, sensing and telecommunications due to an increase in the coherence length, as well as for high-resolution spectroscopy, metrology and frequency standards in this and other spectral ranges.

As an alternative to natural Rayleigh backscattering in a long SMF-28 fiber, it is possible to use an artificial refractive index structure randomly inscribed in SMF-28 by the point-by-point fs technique [136] as random DFB, Fig. 9.20.

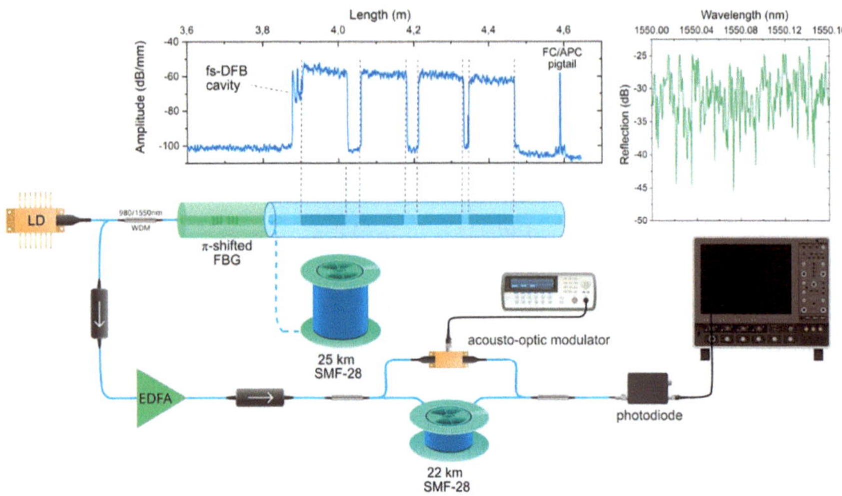

Fig. 9.20 Experimental scheme of the DFB Er fiber laser combined with an artificial random reflector [136]

At the same integral reflection coefficient, an artificial random reflector (ARR) will have a length several orders of magnitude smaller than that of the natural refractive index in homogeneities. Four random reflectors fs-inscribed in SMF-28 fibers (10 cm long each) are spliced into a 3-cm DFB fiber laser, which is also fabricated via the P-b-P fs technique in the commercially available Er-doped fiber Fibercore I-25 (980/125). The top inset shows a reflectogram of a hybrid resonator obtained using a high-resolution backscattering reflectometer, LUNA OBR4600. The scattering level of the ARR is +40 dB to the natural Rayleigh scattering level; thus, the integral reflection from the ARR is ~0.05%, which corresponds to the Rayleigh backscattering of the ~10 km SMF-28 spool. The optical spectrum of the reflector, shown in Fig. 9.20, demonstrates broadband reflection with narrow peaks and random amplitudes due to the interference of reflected radiation from laser-induced inhomogeneities. In the experiment, the linewidth was determined via the self-heterodyne technique with a delay line of 22 km. The optical spectrum of the signal amplified to ~30 mW by the EDFA is shown in Fig. 9.21(a). The measured RF beat spectrum of the amplified signal is shown in Fig. 9.21(b) in comparison with those for the DFB laser, as well as for the hybrid configuration based on a 25-km SMF-28 spool. The measured linewidth at the −20 dB level is 16 kHz, which is sufficiently narrower than that for the DFB laser (120 kHz) but broader than that for the configuration with long SMF-28 (6 kHz), which corresponds with the model predicting the linewidth reduction with increasing length.

Thus, the narrowing of the DFB laser line is demonstrated with artificial reflectors fabricated via the fs technique, which indicates the common nature of the observed effect with the scheme involving natural Rayleigh backscattering.

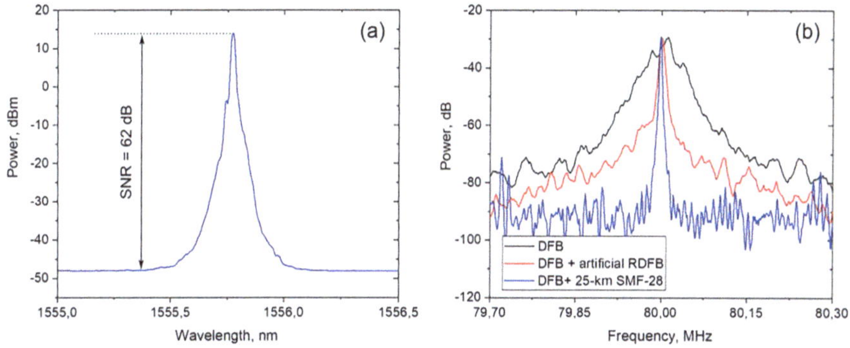

Fig. 9.21 (**a**) Optical spectrum of the amplified signal at the input of the Mach-Zehnder interferometer; (**b**) the radio frequency spectrum of the beat signal obtained using the self-heterodyne technique for different configurations: DFB laser (black color), with added ARR (red color) or a 25-km SMF-28 fiber (blue color) [136]

9.5 Conclusions

Thus, we reviewed the basics of the FBG and DFB fiber laser theory and compared it with experiments on DFB lasers inscribed in active fibers by traditional UV and new femtosecond techniques, which demonstrated good agreement between the theoretical and experimental data. At the same time, a linearly polarized erbium-doped DFB fiber based on a 32-mm-long FBG inscribed by a femtosecond point-by-point inscription technique in a non-PM fiber is demonstrated for the first time. In contrast to UV-based inscription technology, the proposed method allows the creation of DFB fiber lasers that generate radiation in a single-polarization mode. They are also characterized by a high grating stability and may be written without the removal of plastic coatings, which is important for applications. The flexibility of the method also makes it attractive for the fabrication of a variety of DFB lasers in various active fibers with either regular or random refractive index structures, for example, in holmium-doped fibers [137]. The great potential of femtosecond technology has also been demonstrated by the fabrication of a record short (5.3 mm) DFB cavity inscribed in highly Er-doped fiber, which has quite good laser characteristics, both for power and spectrum (linewidth) comparable to those in conventional Er-doped fiber DFB lasers with a one-order longer cavity.

Another new type of DFB laser was realized in an active fiber based on an FBG with random amplitude and phase variations simultaneously. The theory of such a laser has also been developed. A comparison of intensity distributions for different random gratings shows that for the FBG in which both the phases and amplitudes of the segments are random, Anderson-type localization with maximum intensity in the center is obtained, resulting in single-frequency generation of the random DFB laser. A comparison of regular and random DFB lasers realized in Yb-doped PM fiber reveals that the output power and generation efficiency of random DFB lasers are 3.5–5 times greater than those of regular DFB lasers, whereas DFB and random DFB lasers have close generation linewidths. The increase in efficiency is explained by the wider localization area of the generated radiation for the random grating in comparison to the grating with one π-shift; herewith, the dip in the saturated pump radiation distribution is also wider and deeper, which indicates a more effective conversion of the pump to laser radiation in the case of a random DFB laser.

A random DFB fiber laser based on an array of relatively short FBGs with random amplitudes and phases inscribed in a PM passive fiber has resulted in the first demonstration of a linearly polarized Raman fiber laser operating in the single-frequency regime at the ten mW level. Moreover, an output power of up to ~6 W is obtained while the linewidth at maximum power is as small as ~80 pm. A theoretical model has been developed that qualitatively describes the obtained results.

A hybrid configuration consisting of a regular DFB laser and purely random refractive index variations is shown to greatly narrow the single-frequency laser linewidth either for a long fiber with weak natural inhomogeneities or for a short fiber with strong fs-inscribed refractive index modifications. The possibility of narrowing the instantaneous linewidth of the Er-doped fiber DFB laser in this configuration to the 0.1–0.001 Hz level has been demonstrated.

Together with new knowledge, the developed DFB fiber lasers have great potential for various applications. DFB fiber lasers based on FBGs inscribed by femtosecond pulses can be used for sensors in harsh environments: high temperature, radiation, and high hydrogen concentrations. At that, the DFB laser can be used as a sensor itself, securing easy interrogation and high precision. An array of many DFB lasers in one fiber may be used for multipoint distributed sensing systems, for example, up to 32 DFB lasers have been used recently for distributed hydrophone applications [138].

Optimized random DFB Raman fiber lasers can also be used in a variety of applications: having a narrow spectral linewidth and rather high power, such lasers can be used for remote FBG sensor interrogation and as light sources in distributed sensor systems. Moreover, this configuration can be adapted to the 1.5 μm range, which will allow such lasers to be employed in long-haul telecommunication systems. Their high power, narrowband output can be efficiently converted to the visible range via frequency doubling in a nonlinear crystal or periodically poled fibers with potential applications in biomedicine.

Femtosecond writing technology also offers new types of lasers in multicore and multimode fibers. For example, fundamental mode selection by an FBG with a special 3D structure inscribed in a graded-index fiber has been demonstrated in the multimode Raman fiber laser directly pumped by multimode laser diodes [139]. Interference between partially reflected light from FBGs in a random 3D array, which is fs-inscribed either in passive multimode fiber [140] or in passive [141] and active [142] multicore fibers, is shown to result in significant line narrowing in all these cases. Therefore, DFB lasers in multimode and multicore fibers seem to also be feasible, but new theoretical and experimental techniques should be developed for that purpose.

References

1. Maiman, T.H.: Stimulated optical radiation in ruby. Nature. **187**(4736), 493–494 (1960)
2. Snitzer, E.: Optial MASER action of Nd^{+3} in a barium crown glass. Phys. Rev. Lett. **7**(12), 444–446 (1961)
3. Koester, C.J., Snitzer, E.: Amplification in a fiber laser. Appl. Opt. **3**(10), 1182–1186 (1964)
4. Kao, K.C., Hockham, G.A.: Dielectric-fibre surface waveguides for optical frequencies. Proc. Instit. Electr. Eng. **113**, 1151–1158 (1966)
5. Carroll, J.E., Whiteaway, J., Plumb, D.: Distributed Feedback Semiconductor Lasers. IET, London (1998)
6. Alexander, S.B.: Optical communication receiver design. SPIE Optical Engineering Press, London (1997)
7. Becker, P.M., Olsson, A.A., Simpson, S.J.R.: Erbium-Doped Fiber Amplifiers: Fundamentals and Technology. Academic Press, San Diego (1999)
8. Kashyap, R.: Fiber Bragg Gratings. Academic Press, New York (1999)
9. Agrawal, G.P.: Fiber-Optic Communication Systems. Wiley, New York/Toronto (2002)
10. Meltz, G., Morey, W.W., Glenn, W.H.: Formation of Bragg gratings in optical fibers by a transverse holographic method. Opt. Lett. **14**(15), 823–825 (1989)
11. L'opez-Amo, M., L'opez-Higuera, J.M.: Multiplexing techniques for FBG sensors. In: Cusano, A., Cutolo, A., Albert, J. (eds.) Fiber Bragg Grating Sensors: Recent Advancements, Industrial Applications and Market Exploitation, pp. 99–115. Bentham Science, Sharjah (2011)
12. Grattan, K.T.V., Meggitt, B.T.: Optical Fiber Sensor Technology: Fundamentals. Kluwer Academic, Boston (2000)
13. L'opez-Higuera, J.M., Cobo, R.L., Incera, A.Q., Cobo, A.: Fiber optic sensors in structural health monitoring. J. Lightwave Technol. **29**(4), 587–608 (2011)
14. Ramakrishnan, M.: Optical fiber sensors for smart composite materials structures. In: Iniewski, K., Rajan, G. (eds.) Optical Fiber Sensors: Advanced Techniques and Applications, pp. 491–520. CRC Press, New York (2015)
15. Limpert, J., Roser, F., Klingebiel, S., Schreiber, T., Wirth, C., Peschel, T., Eberhardt, R., Tunnermann, A.: The rising power of fiber lasers and amplifiers. IEEE J. Select. Top. Quant. Electron. **13**(3), 537–545 (2007)
16. Richardson, D.J., Nilsson, J., Clarkson, W.A.: High power fiber lasers: current status and future perspectives. JOSA B. **27**(11), B63–B92 (2010)
17. Gapontsev, V., Fomin, V., Ferin, A., Abramov, M.: Diffraction limited ultra-highpower fiber lasers. In: Lasers, Sources and Related Photonic Devices, OSA Technical Digest Series (CD), paper AWA1 (2010)
18. Cliche, J.-F., Allard, M., Tetu, M.: Ultra-narrow linewidth and high frequency stability laser sources. In: Optical Amplifiers and Their Applications/Coherent Optical Technologies and Applications, Technical Digest (CD). IEEE (2006)
19. Loh, W., Laming, R.: 1.55 μm phase-shifted distributed feedback fibre laser. Electron. Lett. **31**(17), 1440–1442 (1995)
20. Sejka, M., Varming, P., Hubner, J., Kristensen, M.: Distributed feedback Er^{3+}-doped fibre laser. Electron. Lett. **31**(17), 1445–1446 (1995)
21. Kringlebotn, J.T., Archambault, J.-L., Reekie, L., Payne, D.N.: Er^{3+}: Yb^{3+}- codoped fiber distributed-feedback laser. Opt. Lett. **19**(24), 2101–2103 (1994)
22. Asseh, A., Storoy, H., Kringlebotn, J., Margulis, W., Sahlgren, B., Sandgren, S., Stubbe, R., Edwall, G.: 10 cm Yb^{3+} DFB fibre laser with permanent phase shifted grating. Electron. Lett. **31**(12), 969–970 (1995)
23. Agger, S., Povlsen, J.H., Varming, P.: Single-frequency thulium-doped distributed feedback fiber laser. Opt. Lett. **29**(13), 1503–1505 (2004)
24. Bowers, M., Burkhart, S., Cohen, S., Erbert, G., Heebner, J., Hermann, M., Jedlovec, D.: The injection laser system on the national ignition facility. In: Proceedings SPIE 6451, Solid State Lasers XVI: Technology and Devices, vol. 64511. IEEE (2007)

25. Westbrook, P.S., Abedin, K.S., Nicholson, J.W., Kremp, T., Porque, J.: Raman fiber distributed feedback lasers. Opt. Lett. **36**(15), 2895–2897 (2011)
26. Martinez, A., Dubov, M., Khrushchev, I., Bennion, I.: Direct writing of fibre Bragg gratings by femtosecond laser. Electron. Lett. **40**(19), 4697–4702 (2004)
27. Rayleigh, F.R.S.: Lord: on the transmission of light through an atmosphere containing small particles in suspension and on the origin of the blue sky. Philos. Mag. **47**(287), 375–384 (1899)
28. Lu, P., Mikhailov, S.J., Coulas, D., Ding, H., Bao, X.: Low loss random fiber gratings made with a fs-IR laser for distributed fiber sensing. J. Lightw. Techn. **37**(18), 2895–2897 (2011)
29. Madsen, C.K., Zhao, J.H.: Optical Filter Design and Analysis. Wiley, New York/ Toronto (2004)
30. Litchinitser, N.M., Eggleton, B.J., Patterson, D.B.: Fiber Bragg gratings for dispersion compensation in transmission: theoretical model and design criteria for nearly ideal pulse recompression. J. Lightwave Technol. **15**(8), 1303–1313 (1997)
31. Sumetsky, M., Eggleton, B.J.: Fiber Bragg gratings for dispersion compensation in optical communication systems. J. Opt. Fiber. Commun. Rep. **2**, 256–278 (2005)
32. Kawasaki, B.S., Hill, K.O., Johnson, D.C., Fujii, Y.: Narrow-band Bragg reflectors in optical fibers. Opt. Lett. **3**(2), 66–68 (1978)
33. Hill, K.O., Fujii, Y., Johnson, D.C., Kawasaki, B.S.: Photosensitivity in optical fiber waveguides: application to reftection filter fabrication. Appl. Phys. Lett. **32**(10), 647–649 (1978)
34. Othonos, A., Kalli, K.: Fiber Bragg Gratings: Fundamentals and Applications in Telecommunications and Sensing. Artech House, Norwood, MA (1999)
35. Hill, K.O., Meltz, G.: Fiber Bragg grating technology fundamentals and overview. J. Lightw. Technol. **15**(8), 1263–1276 (1997)
36. Ivanov, O.V., Nikitov, S.A., Gulyaev, Y.V.: Cladding modes of optical fibers: properties and applications. Physics-Uspekhi. **49**, 167–191 (2006)
37. Shapiro, D.A.: Family of exact solutions for reflection spectrum of Bragg grating. Opt. Commun. **215**(4–6), 295–301 (2003)
38. Belai, O.V., Podivilov, E.V., Shapiro, D.A.: Group delay in Bragg grating with linear chirp. Opt. Commun. **266**(2), 512–520 (2006)
39. Podivilov, E.V., Shapiro, D.A., Trubitsyn, D.A.: Exactly solvable profiles of quasi-rectangular Bragg filter with dispersion compensation. J. Opt. A Pure Appl. Opt. **8**(9), 788–795 (2006)
40. Peral, E., Capmany, J., Marti, J.: Iterative solution to the Gel'fand–Levitan–Marchenko coupled equations and application to synthesis of fiber gratings. IEEE J. Quant. Electr. **32**(12), 2078–2084 (1996)
41. Feced, R., Zervas, M.N., Muriel, M.A.: An efficient inverse scattering algorithm for the design of nonuniform Bragg gratings. IEEE J. Quant. Electr. **35**, 1105–1115 (1999)
42. Poladian, L.: Simple grating synthesis algorithm. Opt. Lett. **25**(11), 787–789 (2000) [Errata: Opt. Lett. 25 (18), 1400 (2000)]
43. Skaar, J., Wang, L., Erdogan, T.: On the synthesis of fiber Bragg gratings by layer peeling. IEEE J. Quant. Electr. **37**(2), 165–173 (2001)
44. Frangos, P., Jaggard, D.: A numerical solution to the Zakharov–Shabat inverse scattering problem. IEEE Trans. Antenn. Propag. **39**(1), 74–79 (1991)
45. Xiao, G.B., Yashiro, K.: An efficient algorithm for solving Zakharov-Shabat inverse scattering problem. IEEE Trans. Antennas Propag. **50**(6), 807–811 (2002)
46. Papachristos, C., Frangos, P.: Design of corrugated optical waveguide filters through a direct numerical solution of the coupled Gel'fand–Levitan–Marchenko integral equations. J. Opt. Soc. Am. A. **19**(5), 1005–1012 (2002)
47. Papachristos, C., Frangos, P.: Synthesis of single- and multi-mode planar optical waveguides by a direct numerical solution of the Gel'fand–Levitan–Marchenko integral equations. Opt. Commun. **203**, 27–37 (2002)
48. Rosenthal, A., Horowitz, M.: Inverse scattering algorithm for reconstructing strongly reflecting fiber Bragg gratings. IEEE J. Quant. Electr. **39**(8), 1018–1026 (2003)

49. Belai, O.V., Frumin, L.L., Podivilov, E.V., Shapiro, D.A.: Efficient numerical method of the fiber Bragg grating synthesis. J. Opt. Soc. Am. B. **24**(7), 1451–1457 (2007)
50. Belai, O.V., Podivilov, E.V., Frumin, L.L., Shapiro, D.A.: Numerical reconstruction stability of fiber Bragg gratings. Opt. Spectrosc. **105**(1), 103–110 (2008)
51. Belai, O.V., Frumin, L.L., Podivilov, E.V., Shapiro, D.A.: Inverse scattering for the one-dimensional Helmholtz equation: fast numerical method. Opt. Lett. **33**(18), 2101–2103 (2008)
52. Frumin, L.L., Belai, O.V., Podivilov, E.V., Shapiro, D.A.: Efficient numerical method for solving the direct Zakharov–Shabat scattering problem. J. Opt. Soc. Am. B. **32**(2), 290–296 (2015)
53. Simonsen, H., Henningsen, J., Søgaard, S.: DFB fiber lasers as optical wavelength standards in the 1.5-μm region. IEEE Trans. Instrum. Meas. **50**(2), 482–485 (2001)
54. Hill, D.J., Hoddera, B., De Freitasa, J., Thomasa, S.D., Hickeyb, L.: DFB fibre-laser sensor developments. Proc. SPIE. **5855**, 904–907 (2005)
55. Wallerand, J.-P., Robertsson, L., Ma, L.-S., Zucco, M.: Absolute frequency measurement of molecular iodine lines at 514.7 nm, interrogated by a frequency-doubled yb-doped fibre laser. Metrologia. **43**(3), 294 (2006)
56. Markert, F., Scheid, M., Kolbe, D., Walz, J.: 4W continuous-wave narrow-linewidth tunable solid-state laser source at 546 nm by externally frequency doubling a ytterbium-doped single-mode fiber laser system. Opt. Express. **15**(22), 14476–14481 (2007)
57. Friedenauer, A., Markert, F., Schmitz, H., Petersen, L., Kahra, S., Herrmann, M., Udem, T., H¨ansch, T., Sch¨atz, T.: High power all solid state laser system near 280 nm. Appl. Phys. B Lasers Opt. **84**(3), 371–373 (2006)
58. Jaskorzynska, B., Vanin, E., Helmfrid, S., Asseh, A.: Gain saturation and pump depletion in high-efficiency distributed-feedback rare-earth-doped lasers. Opt. Lett. **21**(17), 1366–1368 (1996)
59. Lauridsen, V.C., Povlsen, J.H., Varming, P.: Design of DFB fibre lasers. Electron. Lett. **34**(21), 2028–2030 (1998)
60. Lauridsen, V.C., Povlsen, J.H., Varming, P.: Optimising erbium-doped DFB fibre laser length with respect to maximum output power. Electron. Lett. **35**(4), 300–302 (1999)
61. Yelen, K., Hickey, L.M., Zervas, M.N.: Experimentally verified modeling of erbium-ytterbium co-doped DFB fiber lasers. J. Lightwave Technol. **23**(3), 1380–1392 (2005)
62. Foster, S.: Spatial mode structure of the distributed feedback fiber laser. IEEE J. Quantum Electron. **40**(7), 884–892 (2004)
63. Foster, S.: Dynamical noise in single-mode distributed feedback fiber lasers. IEEE J. Quantum Electron. **40**(9), 1283–1293 (2004)
64. Foster, S.: A new derivation of the fundamental mode equations for low gain distributed feedback lasers. IEEE J. Quantum Electron. **43**(1), 4–5 (2007)
65. Foster, S.: Fundamental limits on 1/f frequency noise in rare-earth-metal-doped fiber lasers due to spontaneous emission. Phys. Rev. A. **78**(1), 013820 (2008)
66. Agger, S.D., Povlsen, J.H.: Comments on "dynamical noise in single-mode distributed feedback fiber lasers". IEEE J. Quantum Electron. **42**(7), 733–734 (2006)
67. Nikulin, M.A., Churin, D.E., Vlasov, A.A., Podivilov, E.V.: Distributed feedback ytterbium fiber laser: experiment and analytical model. J. Opt. Soc. Am. B. **27**(7), 1414–1420 (2010)
68. Turitsyn, S.K., Babin, S.A., Churkin, D.V., Vatnik, I.D., Nikulin, M., Podivilov, E.V.: Random distributed feedback fibre lasers. Phys. Rep. **542**(2), 133–193 (2014)
69. Yelen, K., Zervas, M.N., Hickey, L.M.B.: Fiber DFB lasers with ultimate efficiency. J. Lightwave Technol. **23**(1), 32–43 (2005)
70. Babin, S.A., Churkin, D.V., Ismagulov, A.E., Kablukov, S.I., Nikulin, M.A.: Single frequency single polarization DFB fiber laser. Laser Phys. Lett. **4**(6), 428 (2007)
71. Cranch, G.A.: Frequency noise reduction in erbium-doped fiber distributed-feedback lasers by electronic feedback. Opt. Lett. **27**(13), 1114–1116 (2002)

72. Burgmeier, J., Waltermann, C., Flachenecker, G., Schade, W.: Point-by-point inscription of phase-shifted fiber Bragg gratings with electro-optic amplitude modulated femtosecond laser pulses. Opt. Lett. **39**(3), 540–543 (2014)
73. Dostovalov, A., Wolf, A., Parygin, A., Zyubin, V., Babin, S.: Femtosecond point-by-point inscription of Bragg gratings by drawing a coated fiber through ferrule. Opt. Express. **24**(15), 16232–16237 (2016)
74. Wolf, A.A., Dostovalov, A.V., Skvortsov, M.I., Raspopin, K.S., Parygin, A.V., Babin, S.A.: Femtosecond-pulse inscription of fiber Bragg gratings with single or multiple phase-shifts in the structure. Opt. Laser Technol. **101**, 202–207 (2018)
75. Skvortsov, M.I., Wolf, A.A., Dostovalov, A.V., Vlasov, A.A., Akulov, V.A., Babin, S.A.: Distributed feedback laser based on fiber Bragg grating inscribed using the femtosecond point-by-point technique. Laser Phys. Lett. **15**(3) (2018)
76. Derickson, D.: Fiber Optic Test and Measurement. Pearson Education, Upper Saddle River (1997)
77. Cranch, G.A., Flockhart, G.M.H., Kirkendall, C.K.: Distributed feedback fiber laser strain sensors. IEEE Sensors J. **8**(7), 1161–1172 (2008)
78. Harutjunian, Z., Loh, W., Laming, R., Payne, D.: Single polarisation twisted distributed feedback fibre laser. Electron. Lett. **32**(4), 346–348 (1996)
79. Zhang, X., Zhang, F., Jiang, S., et al.: Short cavity DFB fiber laser based vector hydrophone for low frequency signal detection. Photon. Sens., 7325–7328 (2017)
80. Zhang, Y., Guan, B.-O., Tam, H.-Y.: Ultra-short distributed Bragg reflector fiber laser for sensing applications. Opt. Express. **17**(12), 10050–10055 (2009)
81. Butov, O.V., Rybaltovsky, A.A., Bazakutsa, A.P., Golant, K.M., Vyatkin, M.Y., Popov, S.M., Chamorovskiy, Y.K.: 1030 nm Yb^{3+} distributed feedback short cavity silica-based fiber laser. J. Opt. Soc. Am. B. **34**(3), A43–A48 (2017)
82. Wong, C.L., et al.: Extremely short distributed Bragg reflector fibre lasers with subkilohertz linewidth and ultra-low polarization beat frequency for sensing applications. Meas. Sci. Technol. **22**, 045202 (2011)
83. Butov, O.V., et al.: Short-cavity DFB fiber lasers. In: Progress in Electromagnetics Research Symposium - Spring (PIERS), pp. 1594–1597 (2017)
84. Smirnov, A.M., et al.: Thermal switching of lasing regimes in heavily doped Er^{3+} fiber lasers. ACS Photon. **5**, 5038–5046 (2018)
85. Albert, J., Schulzgen, A., Temyanko, V.L., Honkanen, S., Peyghambarian, N.: Strong Bragg gratings in phosphate glass single mode fiber. Appl. Phys. Lett. **89**, 14–17 (2006)
86. Jiang, M., Zhou, P., Gu, X.: Ultralong π-phase shift fiber Bragg grating empowered single-longitudinal mode DFB phosphate fiber laser with low-threshold and high efficiency. Sci. Rep. **8**, 3–8 (2018)
87. Egorova, O.N., et al.: Phosphate-core silica-clad Er/Yb-doped optical fiber and cladding pumped laser. Opt. Express. **22**, 7632–7637 (2014)
88. Egorova, O.N., et al.: Single-frequency fibre laser with a cavity formed by Bragg gratings written in the core of an active composite fibre using KrF laser radiation (248 nm). Quant. Electron. **49**, 2–7 (2019)
89. Skvortsov, M.I., Wolf, A.A., Vlasov, A.A., et al.: Advanced distributed feedback lasers based on composite fiber heavily doped with erbium ions. Sci. Rep. **10**, Article number: 14487 (2020)
90. Jovanovic, N., et al.: Polarization-dependent effects in point-by-point fiber Bragg gratings enable simple, linearly polarized fiber lasers. Opt. Express. **17**, 6082 (2009)
91. Skvortsov, M.I., et al.: Distributed feedback fiber laser based on a fiber Bragg grating inscribed using the femtosecond point-by-point technique. Laser Phys. Lett. **15**, 35103 (2018)
92. Okoshi, T., Kikuchi, K., Nakayama, A.: Novel method for high resolution measurement of laser output spectrum. Electron. Lett. **16**, 630–631 (1980)
93. Loh, W.H., Laming, R.I.: 1.55 mkm phase-shifted distributed feedback fibre laser. Electron. Lett. **31**, 1440–1442 (1995)

94. Svelto, O.: Principles of Lasers, 5th edn. Springer, New York/Dordrecht/Heidelberg/London (2018)
95. Liz'arraga, N., Puente, N.P., Chaikina, E.I., Leskova, T.A., M'endez, E.R.: Single-mode Er-doped fiber random laser with distributed Bragg grating feedback. Opt. Express. **17**(2), 395–404 (2009)
96. Gagn'e, M., Kashyap, R.: Demonstration of a 3 mW threshold Er-doped random fiber laser based on a unique fiber Bragg grating. Opt. Express. **17**(21), 19067–19074 (2009)
97. Wang, L., Dong, X., Shum, P.P., Liu, X., Su, H.: Random laser with multiphase-shifted Bragg grating in Er/Yb-codoped fiber. J. Lightwave Technol. **33**(1), 95–99 (2015)
98. Li, Y., Lu, P., Bao, X., Ou, Z.: Random spaced index modulation for a narrow linewidth tunable fiber laser with low intensity noise. Opt. Lett. **39**(8), 2294–2297 (2014)
99. Gagn'e, M., Kashyap, R.: Random fiber Bragg grating Raman fiber laser. Opt. Lett. **39**(9), 2755–2758 (2014)
100. Turitsyn, S.K., Babin, S.A., El-Taher, A.E., Harper, P., Churkin, D.V., Kablukov, S.I., Ania-Castan'on, J.D., Karalekas, V., Podivilov, E.V.: Random distributed feedback fibre laser. Nat. Photon. **4**(9), 231 (2010)
101. Abdullina, S.R., Vlasov, A.A., Lobach, I.A., Belai, O.V., Shapiro, D.A., Babin, S.A.: Single-frequency Yb-doped fiber laser with distributed feedback based on a random FBG. Laser Phys. Lett. **13**(7), 075104 (2016)
102. Vlasov, A.A., Churin, D.E., Babin, S.A.: Specifics of Bragg gratings inscription and characterization in polarization-maintaining Yb-doped fiber for DFB lasers. Laser Phys. **20**(12), 2045–2049 (2010)
103. Abdullina, S.R., Babin, S.A., Vlasov, A.A., Kablukov, S.I.: Simple method for apodization of fibre Bragg gratings written by a Gaussian beam. Quant. Electron. **36**(10), 966 (2006)
104. Shapira, O., Fischer, B.: Localization of light in a random-grating array in a single-mode fiber. J. Opt. Soc. Am. B. **22**(12), 2542–2552 (2005)
105. Babin, S.A., Dontsova, E.I., Kablukov, S.I.: Random fiber laser directly pumped by a high-power laser diode. Opt. Lett. **38**(17), 3301–3303 (2013)
106. Teng, R., Ding, Y., Chen, L.: Random fiber laser operating at 1,115 nm. Appl. Phys. B. **111**(2), 169–172 (2013)
107. Churkin, D.V., Vatnik, I.D., Turitsyn, S.K., Babin, S.A.: Random distributed feedback Raman fiber laser operating in a 1.2 μm wavelength range. Laser Phys. **21**(8), 1525–1529 (2011)
108. Lobach, I.A., Kablukov, S.I., Skvortsov, M.I., Podivilov, E.V., Melkumov, M.A., Babin, S.A., Dianov, E.M.: Narrowband random lasing in a bismuth-doped active fiber. Sci. Rep. **6**, 30083 (2016)
109. Babin, S.A., Zlobina, E.A., Kablukov, S.I., Podivilov, E.V.: High-order random Raman lasing in a PM fiber with ultimate efficiency and narrow bandwidth. Sci. Rep. **6**, 22625 (2016)
110. Zhang, L., Jiang, H., Yang, X., Pan, W., Cui, S., Feng, Y.: Nearly-octave wavelength tuning of a continuous wave fiber laser. Sci. Rep. **7**, 42611 (2017)
111. Churkin, D.V., Kolokolov, I.V., Podivilov, E.V., Vatnik, I.D., Nikulin, M.A., Vergeles, S.S., Terekhov, I.S., Lebedev, V.V., Falkovich, G., Babin, S.A., Turitsyn, S.K.: Wave kinetics of random fibre lasers. Nature Comm. **6**, 6214 (2015)
112. Babin, S.A., El-Taher, A.E., Harper, P., Podivilov, E.V., Turitsyn, S.K.: Tunable random fiber laser. Phys. Rev. A. **84**, 021805 (2011)
113. El-Taher, A.E., Harper, P., Babin, S.A., Churkin, D.V., Podivilov, E.V., Ania-Castanon, J.D., Turitsyn, S.K.: Effect of Rayleigh-scattering distributed feedback on multiwavelength Raman fiber laser generation. Opt. Lett. **36**(2), 130–132 (2011)
114. Jia, X.-H., Rao, Y.-J., Yuan, C.-X., Li, J., Yan, X.-D., Wang, Z.-N., Zhang, W.-L., Wu, H., Zhu, Y.-Y., Peng, F.: Hybrid distributed Raman amplification combining random fiber laser based 2nd-order and low-noise LD based 1st-order pumping. Opt. Express. **21**(21), 24611–24619 (2013)
115. Martins, H., Marques, M.B., Frazao, O.: 300 km-ultralong Raman fiber lasers using a distributed mirror for sensing applications. Opt. Express. **19**(19), 18149–18154 (2011)

116. Jia, X.-H., Rao, Y.-J., Peng, F., Wang, Z.-N., Zhang, W.-L., Wu, H.-J., Jiang, Y.: Random-lasing-based distributed fiber-optic amplification. Opt. Express. **21**(5), 6572–6577 (2013)
117. Churkin, D.V., Sugavanam, S., Vatnik, I.D., Wang, Z., Podivilov, E.V., Babin, S.A., Rao, Y., Turitsyn, S.K.: Recent advances in fundamentals and applications of random fiber lasers. Adv. Opt. Photon. **7**(3), 516–569 (2015)
118. Skvortsov, M., Abdullina, S., Vlasov, A., Zlobina, E., Lobach, I., Terentyev, V., Babin, S.: FBG array-based random distributed feedback Raman fibre laser. Quant. Electron. **47**(8), 696 (2017)
119. Abdullina, S.R., Skvortsov, M.I., Vlasov, A.A., Podivilov, E.V., Babin, S.A.: Coherent Raman lasing in a short polarization-maintaining fiber with a random fiber Bragg grating array. Laser Phys. Lett. **16**(10), 105001 (2019)
120. Dontsova, E., Kablukov, S., Lobach, I., Dostovalov, A., Babin, S., Gladyshev, A., Dianov, E., Corbary, C., Ibsen, M., Kazansky, P.: Extracavity and external cavity second-harmonic generation in a periodically poled silica fibre. Quant. Electron. **46**(11), 989 (2016)
121. Wang, L., Dong, X., Shum, P.P., Su, H.: Tunable Erbium-doped fiber laser based on random distributed feedback. IEEE Photon. J. **6**, 1–5 (2014)
122. Bulushev, A.G., Dianov, E.M., Kuznetsov, A.V., Okhotnikov, O.G.: Reduction in the line width of an injection laser by resonant Rayleigh scattering in a fiber ring interferometer. Sov. J. Quantum Electron. **19**, 479–481 (1989)
123. Yin, G., Saxena, B., Bao, X.: Tunable Er-doped fiber ring laser with single longitudinal mode operation based on Rayleigh backscattering in single mode fiber. Opt. Express. **19**, 25981–25989 (2011)
124. L'opez-Mercado, C.A., Spirin, V.V., Bueno Escobedo, J.L., M'arquez Lùcero, A., M'egret, P., Zolotovskii, I.O., Fotiadi, A.A.: Locking of the DFB laser through fiber optic resonator on different coupling regimes. Opt. Commun. **359**, 195–199 (2016)
125. Spirin, V.V., Bueno Escobedo, J.L., Korobko, D.A., M'egret, P., Fotiadi, A.A.: Stabilizing DFB laser injection-locked to an external fiber-optic ring resonator. Opt. Express. **28**, 478–484 (2020)
126. Xu, Y., Zhang, L., Chen, L., Bao, X.: Single-mode SOA-based 1 kHz-linewidth dual-wavelength random fiber laser. Opt. Express. **25**, 15828–15837 (2017)
127. Dang, L., Huang, L., Li, Y., Cao, Y., Lan, T., Iroegbu, P.I., Cao, Z., Mei, K., Liang, L., Fu, S., et al.: A longitude-purification mechanism for tunable fiber laser based on distributed feedback. Light. Technol. **40**, 206–214 (2022)
128. Popov, S.M., Butov, O.V., Bazakutsa, A.P., Vyatkin, M.Y., Chamorovskii, Y.K., Fotiadi, A.A.: Random lasing in a short Er-doped artificial Rayleigh fiber. Resul. Phys. **16**, 102868 (2020)
129. Skvortsov, M.I., Wolf, A.A., Dostovalov, A.V., Egorova, O.N., Semjonov, S.L., Babin, S.A.: Narrow-linewidth Er-doped fiber lasers with random distributed feedback provided by artificial Rayleigh scattering. J. Light. Technol. **40**, 1829–1835 (2022)
130. Skvortsov, M.I., et al.: Extreme narrowing of the distributed feedback fiber laser linewidth due to the Rayleigh backscattering in a single-mode fiber: model and experimental test. Photonics. **9**(8), 590 (2022)
131. Fomiryakov, E., Kharasov, D., Nikitin, S., Nanii, O., Treshchikov, V.: New approach to laser characterization using delayed self-heterodyne interferometry. J. Light. Technol. **39**, 5191–5196 (2021)
132. Von Bandel, N., Myara, M., Sellahi, M., Souici, T., Dardaillon, R., Signoret, P.: Time-dependent laser linewidth: beat-note digital acquisition and numerical analysis. Opt. Exp. **24**, 401–407 (2016)
133. Owens, D.K., Weiss, R.: Measurement of the phase fluctuations in a He–Ne Zeeman Laser. Rev. Sci. Instrum. **45**, 1060–1062 (1974)
134. Di Domenico, G., Schilt, S., Thomann, P.: Simple approach to the relation between laser frequency noise and laser line shape. Appl. Opt. **49**, 4801–4807 (2010)

135. Xu, D., Yang, F., Chen, D., Wei, F., Cai, H., Fang, Z., Qu, R.: Laser phase and frequency noise measurement by Michelson interferometer composed of a 3x3 optical fiber coupler. Opt. Express. **23**, 22386–22393 (2015)
136. Dostovalov, A., Wolf, A., Munkueva, Z., Skvortsov, M., Abdullina, S., Kuznetsov, A., Babin, S.: Continuous and discrete-point Rayleigh reflectors inscribed by femtosecond pulses in singlemode and multimode fibers. Opt. Laser Technol. **167**, 109692 (2023)
137. Abdullina, S.R., Skvortsov, M.I., Dostovalov, A.V., Proskurina, K.V., Munkueva, Z.E., Babin, S.A.: A narrow-linewidth holmium fiber laser with random distributed feedback based on artificial Rayleigh backscattering. Bull. Lebedev Phys. Inst. **50**(Suppl. 13), S1425–S1430 (2023)
138. Harrison, J., Foster, S.: Wavelength division multiplexing of 32 distributed-feedback fiber lasers. In: Advanced Photonics 2017 (IPR, NOMA, Sensors, Networks, SPPCom, PS), OSA Technical Digest (online) (2017)
139. Babin, S.A., Zlobina, E.A., Kablukov, S.I.: Multimode fiber Raman lasers directly pumped by laser diodes. IEEE J. Select. Top. Quant. Electron. **24**(3), 1–10 (2018)
140. Kuznetsov, A.G., Wolf, A.A., Munkueva, Z.E., Dostovalov, A.V., Babin, S.A.: Multimode graded-index fiber with random array of Bragg gratings and its Raman lasing properties. Fibers. **11**, 48 (2023)
141. Wolf, A., Dostovalov, A., Bronnikov, K., Skvortsov, M., Wabnitz, S., Babin, S.: Advances in femtosecond direct writing of fiber Bragg gratings in multicore fibers: technology, sensor and laser applications. Opto-Electron. Adv. **5**(4), 210055 (2022)
142. Kuznetsov, A.G., Wolf, A.A., Egorova, O.N., Semjonov, S.L., Dostovalov, A.V., Podivilov, E.V., Babin, S.A.: Spectrum collapse in a 7-core Yb-doped fiber laser with an array of fs-inscribed fiber Bragg gratings. Opt. Lett. **48**(13), 3603–3606 (2023)

Chapter 10
Random Fiber Laser: From Demonstration to Application

Jiangming Xu, Tianfu Yao, Jun Ye, Yanzhao Ke, Xiaoya Ma, Chenchen Fan, Hanwei Zhang, and Pu Zhou

Abstract Random fiber laser (RFL), whose lasing is based on the remarkable random distributed feedback without a conventional defined laser resonator, is nowadays the object of intensive research for its unique features such as cavity-free, mode-free, and structural simplicity. In this chapter, we will review the multidisciplinary demonstrations and applications of RFLs in the past decade. The power of single-stage structured RFL has been scaled to 2 kW level. By employing fiber MOPA scheme, it was amplified to 10 kW successfully. In the temporal domain, pulsed random laser has been obtained with pulsation width ranging from milliseconds to picoseconds, which can enrich the laser physics and related dynamics occurring in the interaction between light and disordered media. As to the spatial dimension, advancements on linearly polarized operation, higher order mode generation, cylindrical vector beam and vortex beam carrying orbital angular momentum, and beam cleanup-induced brightness enhancement indicated that RFLs have tremendous potential in controlling and diversifying output spatial modes. Furthermore, spectral flexibility of RFLs, such as wide-range tunability, high spectral purity, multiwavelength operation, narrowband and supercontinuum generation, has been demonstrated. Benefitting from the fast performance developments, RFLs have been versatile tools for wide applications, such as pumping, sensing, imaging, random bits generation, and new laser source construction. Furthermore, the potential new horizon of RFLs will be proposed preliminarily.

Keywords Random fiber laser · Random distributed feedback · Multidisciplinary demonstration · Temporal-spatial-spectral domain · Power scalability · Application exploration

---

J. Xu · T. Yao · J. Ye · Y. Ke · X. Ma · C. Fan · H. Zhang · P. Zhou (✉)
College of Advanced Interdisciplinary Studies, National University of Defense Technology, Changsha, Hunan, China

M. F. S. Ferreira, M. C. Paul (eds.), *Optical Fiber Lasers*, Springer Series in Optical Sciences 250, https://doi.org/10.1007/978-3-032-04110-4_10

10.1 Introduction

The concept of random laser employing multiple scattering of photons in an amplifying disordered material to achieve coherent light has attracted increasing attention due to its special features, such as cavity-free, mode-free, and structural simplicity [1–3]. Meanwhile, because of the lack of sharp resonances, it also has presented many challenges to conventional laser theory, such as light localization phenomenon in the disordered amplifying media [4–6]. Moreover, random fiber laser (RFL), whose operation is based on the remarkable random distributed feedback (RDFB) in a piece of passive fiber, can trap the random laser radiation in one-dimensional waveguide structure with efficiency and performance comparable to conventional fiber lasers [7–9]. It presents rich physical and flexible manipulation properties in temporal, spatial, and spectral domains. Furthermore, promising applications in many fields, such as new pump source, speck-free imaging, long-distance sensing, and other scientific or industrial fields, are also attempted in the past decade. In this chapter, the fast performance development and application exploration of RFLs will be reviewed and discussed in Sects. 10.2 and 10.3, and the potential new horizon will be drawn preliminarily in Sect. 10.4.

10.2 Experimental Demonstration

The performance developments of RFLs can provide unprecedented opportunity for the application exploration. In this section, we will focus on the demonstration of RFLs in recent years, including power scalability, temporal modulation, spatial manipulation, and spectral flexibility.

10.2.1 Power Scalability

Early implementations of RFLs usually utilized long passive fibers (on the order of tens of kilometers) to obtain sufficient random distributed feedback [7, 10], resulting in a low threshold for second-order SRS and output powers typically in the hundreds of milliwatts. In 2014, both experimental [11] and theoretical [12] investigations demonstrated that shortening the length of passive fibers could significantly enhance output power and efficiency. By utilizing a 1-km-long passive fiber with a core diameter of 6 μm to provide random distributed feedback, an output power of 11.8 W was achieved [13]. Moreover, increasing the mode area of the fiber core can also suppress the second-order SRS effect, thereby augmenting the output power of RFLs. In the same year, by increasing the core diameter of the passive fiber to 8.2 μm and reducing the fiber length to 300 m, the maximum output power reached 73.7 W [11]. In 2015, further increasing the core diameter of the passive fiber to 10 μm while maintaining a relatively constant fiber length of 320 m resulted in a maximum output power of 124 W [14].

It is important to note that the aforementioned RFLs are all in a full-opened structure, meaning there are no point feedback devices. Consequently, random lasing outputs from both ends, and the stated power represents the sum of the output power in both forward and backward directions. To achieve unidirectional high-power output, RFLs are progressively shifting toward a half-opened structure. By replacing the WDM devices in the system with highly reflective (HR) fiber Bragg gratings (FBGs), a unidirectional high-power output of 112 W was realized [14]. Subsequently, by continuously shortening the length of the passive fiber and increasing the core diameter, the output power of the RFLs was consistently improved. The typical results of high-power single-stage RFLs are summarized in Table 10.1. For instance, in 2016, our research group reduced the length of the passive fiber to 120 m, achieving a high-power output of 193.5 W [15]. In 2017, passive fiber with a larger core (20 μm) was first utilized to provide random distributed feedback [16]. Theoretical research indicated that the maximum output power could reach 1250 W when using 130 m of this passive fiber. However, the highest power achieved in the experiment was only 418 W, nearly one-third of the theoretical prediction. This discrepancy can be attributed to the Raman-assisted four-wave mixing (FWM) process. To further investigate the limiting physical mechanisms, the uniform core passive fibers were replaced with tapered fiber [17], which has a wide-side diameter of 20 μm, a decreasing slope of 0.013 μm/m, and a numerical aperture (NA) of 0.08. This change suppressed the FWM effect, resulting in a maximum output power of 491 W. The phase matching required in the FWM effect can be satisfied by modal dispersion; thus, reducing the number of transverse modes can also mitigate the FWM effect. A straightforward method to achieve this is by decreasing the NA of the passive fiber. For example, utilizing 90 m of 20/400 μm passive fiber with a NA of 0.06 achieved a maximum random lasing output of 919 W in the forward direction [18]. However, because the broadened spectrum of random lasing exceeds the

Table 10.1 Research progress of high-power single-stage RFLs in recent years

Year	Power	Efficiency	Central wavelength	Reference
2014	73.7 W	74.7%	1184 nm	[11]
2015	7 W	70%	1140 nm	[12]
	226 W (theoretical simulation)	94%		
2015	124 W	79%	1150 nm	[14]
2016	193.5 W	~89%	1173 nm	[15]
2017	418 W	71%	1120 nm	[16]
2018	491 W	68.2%	1120 nm	[17]
2019	206.7 W	59.7%	1238 nm	[19]
2019	985 W	78.9%	1150 nm	[18]
2021	1570 W	77.5%	1120 nm	[20]
2022	1277 W	67.1%	1134 nm	[21]
2023	1972 W	68.4%	1130 nm	[22]

reflective bandwidth of HR-FBG, some power leaks into the pump source and is further amplified. In the aforementioned work, the backward output power reached 66 W.

To suppress spectral broadening and high-power backward random lasing, temporally stable pump sources were introduced into RFLs. In 2016, an incoherent broadband amplified spontaneous emission (ASE) source [23, 24], often considered more stable than fiber oscillators, was first employed to pump RFLs. And in 2021, a phase-modulated single-frequency fiber amplifier was utilized as another stable pump source [20], which further increased the output power to 1.57 kW. However, the challenge of high-power backward random lasing remains inadequately addressed. Recent research indicated that spectral broadening and power leakage are nearly impossible to eliminate [22], suggesting that a new approach is necessary. The proposed Yb-Raman hybrid gain structure, combined with the temporally stable pump source, further improved the output power of the RFL to 1972 W.

Undoubtedly, the high-power operation of RFLs has made significant progress in recent years, becoming one of the most important research topics. In addition to core-pumped and half-opened structures, RFLs have also been developed in full-opened configurations [21, 25], as well as laser diode (LD) direct pumping [26] and brightness enhancement [27, 28]. Therefore, it is reasonable to believe that RFLs will make substantial contributions to high-power fiber lasers, particularly at specialized wavelengths, in the future.

Furthermore, RFLs exhibit a more stable temporal characteristic compared to traditional fiber oscillators with the aid of the unique operating principle, making them suitable as seed sources for high-power master oscillator power amplifiers (MOPAs), capable of achieving power levels in the kilowatt to tens of kilowatts range [29–32]. In 2015, a 1070-nm RFL was first utilized in a MOPA configuration [29], achieving a maximum power of 1030 W with a one-stage power amplifier, limited only by pump brightness and the power handling of the pump combiner. Subsequently, by optimizing the RFL seed [31, 32], introducing tandem pumping technique [31, 33], and utilizing band-pass filters [34] or chirped tilted fiber Bragg gratings [35] to eliminate spectral noise, the randomly polarized MOPA amplifier seeded by RFLs has effectively suppressed nonlinear effects such as stimulated Raman scattering (SRS) and transverse mode instability (TMI), achieving a maximum output power of over 10 kW. Additionally, by controlling the optical spectrum of RFL seed [36] and introducing polarization-maintaining (PM) structures [37], the MOPA amplifier seeded by RFLs has achieved not only kilowatt-level output power but also PM and narrow-linewidth operation [30, 36, 38, 39].

10.2.2 Temporal Modulation

Most reported RFLs operate in a continuous-wave state. In fact, RFLs can also achieve pulsed output, with the main implementation schemes including active modulation and passive modulation. High-peak-power pulsed RFLs can induce various nonlinear responses when experiencing multiple scattering in passive fibers, enriching the laser physics and related dynamics occurring in the interaction between light and disordered media.

Active modulation techniques in pulsed RFLs are generally based on various optical modulation devices such as electro-optic modulators (EOM) [40], acousto-optic modulators (AOM) [41], and polarization switches [42], and can also be directly achieved by pulsed pumping [43], offering notable advantages such as high pulse energy and flexible control. Additionally, integrating external signal modulation with spectral control techniques allows for simultaneous modulation of RFLs in both the time and spectral domains [44–47].

Passive modulation techniques in pulsed RFLs can be realized through mode locking and passive Q-switching based on saturable absorbers (SAs), characterized by their simplicity, compactness, and cost-effectiveness, without the need for external triggers or complex modulators [48–51]. Another passive modulation scheme involves utilizing nonlinear effects such as stimulated Brillouin scattering (SBS) and Raman scattering to induce self-Q-switching mechanisms, facilitating the generation of pulses and multiwavelength outputs in RFLs [52–57]. For instance, Xu et al. [54] achieved stable pulses with a period of 37.42 μs and pulse energy of 79.7 μJ via Raman gain self-modulation induced by mutual modulation between the bidirectional pump light field and the signal light field. Under backward-pumping action, passive spatial-temporal gain modulation can cause self-modulation of the Raman gain, thereby achieving pulsed random laser output.

In addition to individual active or passive modulation techniques, integrating both methods enables multidimensional control of pulsed RFLs. Liu et al. [58] combined EOM and SBS to control the pulse amplitude, width, and repetition rate simultaneously. However, the pulsed RFLs produced by the aforementioned methods generally have pulse widths in the nanosecond range or longer. Notably, W. Pan et al. [43] achieved an ultrafast RFL by employing synchronous pulse pumping within a compound cavity. They successfully generated a stable picosecond pulsed RFL, delivering a single pulse energy of approximately 6 nJ and an average power of around 170 mW.

Additionally, by studying the time-domain noise characteristics of RFLs, researchers have found that the relative intensity noise (RIN) of RFLs has transmission characteristics that are distinct from those of traditional Raman oscillators [59, 60]. In the study of the time-domain statistical properties of RFLs, many novel physical phenomena have been discovered, including optical rogue waves [61], hierarchical structures of optical turbulence [62], replica symmetry breaking phenomena [63], and spectral intensity distributions that satisfy Lévy statistics [64].

10.2.3 Spatial Manipulation

Linearly polarized laser is crucial for some practical application fields such as sensing and communication, and the linearly polarized RFLs have been widely investigated [42, 50, 65–67]. Furthermore, the demonstration of high-power RFL with linearly polarized output was also attempted. Du et al. [66] obtained linearly polarized random lasing via linearly polarized passive fiber and nonlinearly polarized pump source by introducing a certain fiber coiling technique. Zlobina et al. [67] and Babin et al. [68] presented single-order and high-order linearly polarized RFL via linearly polarized passive fiber and linearly polarized pump source with maximal output power of about ten-watt level and polarization extinction ratio (PER) of >22 dB. In 2017, Lou et al. [69] reported the linearly polarized RFLs with 86.3 W at 1120 nm referring to the first-order Raman Stokes. At the same year, Xu et al. scaled the power of 1st-order RFLs to 1 kW with PER of 17 dB [36], and that of 2nd-order RFLs up to over 100 W with PER of 23.5 dB [24]. In 2018, Huang et al. [37] achieved the power scaling of linearly polarized RFLs based on MOPA configuration. A 442 W linearly polarized output with a narrow 3 dB linewidth of 0.28 nm is realized. Meanwhile, a 621 W linearly polarized output is also realized with a relatively wider 3 dB linewidth of 2.7 nm (Table 10.2).

In addition, modulation on optical beam field has been investigated in RFLs to extend the application recently. Until now, both beam cleanup with fundamental mode optimization and higher order mode generation including LP_{11} mode, cylindrical vector beam (CVB), and vortex beam carrying orbital angular momentum have been demonstrated in RFLs, indicating the ability on spatial modulation.

Conventional fiber lasers operate at fundamental mode LP_{01}. For the RFLs, due to the weak feedback provided by Rayleigh scattering, to reduce the laser threshold, the aforementioned studies all used conventional single-mode fiber with single-mode pumping. In that case, the injected pump power and thus the output power are limited, as well as the pump wavelength. To scale the injected pump power and extend the pump wavelength, pump beam with low brightness is expected to be utilized, while the output laser brightness can be enhanced by the beam cleanup effect. In 2018, researchers [71] in Russian Academy of Sciences firstly to build a multimode laser diode pumped RFL based on graded-index fiber (GRIN), achieving a 27 W output with a central wavelength of 996 nm. In that demonstration, GRIN

Table 10.2 Recent progress of RFLs with linear polarized output

Central wavelength	Power	Polarization extinction ratio	Reference
1106/1169/1232 nm	4.4 W/7.4 W/ 9.1 W	25 dB/25 dB/22 dB	[68]
1120 nm	86.3 W	22 dB	[69]
1178 nm	100.7 W	23.5 dB	[24]
1081.4 nm	381.1 W	11 dB	[70]
1262/1515 nm	8 W/9 W	27 dB	[67]
1080 nm	1010 W	17 dB	[36]
1081.39 nm	621 W	11.9 dB	[37]

fiber indicates great potential on brightness enhancement from low pump brightness to high signal brightness with a beam quality M^2 factor of 1.6. In 2021, researchers [27] in National University of Defense Technology built a fully fiber-structured RFL based on a 120-m-long GRIN fiber with a core diameter of 62.5 μm, achieving an output of 306 W, with the beam quality M^2 factor being purified from 9.25 of the pump light to 2.35, corresponding to a brightness enhancement factor of 6.1. In 2024, the same group [72] reported the power scaling on GRIN fiber RFLs with brightness enhancement. The output power reaches up to 656 W with beam quality M^2 factor improving from >20 to 2.9, corresponding to the brightness enhancement factor of 14.9. This result is also the current highest output power level for GRIN fiber RFLs.

In addition, RFLs with different output optical field profiles have made significant research progress, including the LP_{11} mode [73–75], cylindrical vector beam [76, 77], and vortex beam [78–80] carrying orbital angular momentum, as shown in Table 10.3. Recently, there are several methods to generate LP_{11} mode in RFLs, such as the lateral offset splicing [73], long-period fiber gratings [74],

Table 10.3 Research progress in RFLs with higher order modes

Generation method	Output beam	Output power	Central wavelength	Reference
Few-mode fiber Bragg grating (FMFBG)	LP_{11}	~10 mW	1070 nm	[73]
Mode-selective coupler (MSC)	TE_{01} and TM_{01}	~4 mW	1060 nm	[76]
Long-period fiber gratings	LP_{11}	2.78 W	1554.5 nm	[74]
Few-mode fiber Bragg grating (FMFBG)	TE_{01} and TM_{01}	~17 mW	1055 nm	[77]
Helical long-period fiber grating	$OAM_{\pm 1}$	2.09 W	1690 nm	[79]
Digital micromirror device (DMD)	LG 1 ~ 100	1.703 W	1044 ~ 1403 nm	[78]
Acoustically induced fiber grating	LP_{11} and $OAM_{\pm 1}$	23.6 W	1133.9/1197.6/1260.5/1331.8/1414.5/1513.7 nm	[80]
Acoustically induced fiber grating	LP_{11}	93.8 W	1134 nm	[75]

mode-selective couplers [76], acoustically induced fiber grating [75, 80], while the power scaling up to 93.8 W has been achieved [75]. Furthermore, cylindrical vector beams (CVBs), such as azimuthally and radially polarized beams (TE_{01} and TM_{01}), were obtained by adjustment of the polarization controllers on LP_{11} mode [76, 77]. In 2023, researchers reported the wavelength-tunable LG modes to 100 order spanning from 1044 to 1403 nm [78]. Besides, the vortex beam carrying orbital angular momentum in the 1.1–1.5 μm band was generated, another result indicating the ultra-wide wavelength tuning capability of RFLs [80]. These research advancements indicate that RFLs have tremendous potential in controlling and diversifying output spatial modes, offering new possibilities for future applications in optical communications, imaging, and industrial applications.

10.2.4 Spectral Flexibility

Herein, we review the recent advancements in spectral manipulation of RFLs. RFLs have demonstrated the flexible ability to produce wide-range tunability, high spectral purity, multiwavelength operation, narrowband and supercontinuum generation.

10.2.4.1 Wide-Range Tunability

Broad wavelength tunability is a prominent advantage of RFLs, and the operating wavelength can be altered through various schemes. Leveraging a spectrum of gain mechanisms, including nonlinear effects like SRS and SBS, as well as active gains from rare-earth ion doping in optical fibers, specifically ytterbium, erbium, erbium/ytterbium codoped, bismuth, and thulium, RFLs are capable of spanning an extensive operating wavelength range [9]. By integrating tunable filters, wavelength-dependent point reflectors, or shapers in modeless RFLs, or by using a fixed pump to change the center wavelength of the filter, the operating wavelength can be tuned over a wide range with a unique flat output power and high efficiency, as presented in Table 10.4.

10.2.4.2 High Spectral Purity

Cascaded RFLs are wavelength versatile. Nevertheless, residual low-order Stokes waves lead to low spectral purity of cascaded RFLs, which is defined by the ratio of the target lasing power to the total power. The residual pump or low-order Stokes waves lead to low spectral purity in cascaded RFLs. In cascaded RFLs, pump intensity fluctuations are transferred to the SRS process, with femtosecond response times, inducing larger power variations in the generated RFLs [94]. To improve the spectral purity of cascade RFLs, a temporally stable pump source such as the incoherent broadband ASE light source used in ytterbium-doped fiber lasers and the

Table 10.4 Advancements of the wavelength-tunable RFLs

Gain mechanism	Realization method	Typical wavelength tuning range	Reference
SRS	Tuning the pump wavelength of the half-open structure RFL with a broadband mirror	1060–1900 nm	[81–83]
SRS	Changing the central wavelength of the filter in the point reflector with a wavelength-fixed pump	1535–1570 nm, 1114–1132 nm, 1138–1152 nm, 1193–1222 nm	[84–86]
SRS	Simultaneously tuning the pump wavelength and point reflector wavelength	1105–1160 nm and 1217–1280 nm dual-wavelength tunable	[87]
SBS	Tuning the pump wavelength of the RFL	1543–1565 nm	[88, 89]
Ytterbium-doped	Changing the central wavelength of the filter in the point reflector	1040–1090 nm	[86, 90]
Erbium-doped	Tuning the wavelength of the fiber Fabry-Perot interferometer filter	1525–1565 nm	[91]
Erbium/ytterbium codoped	Changing the central wavelength of the filter in the point reflector	1531–1553 nm	[92]
Bismuth-doped	Tensioning the FBG in the laser structure	1352.03–1356.03 nm/ 1419.20–1467.19 nm	[93]

ytterbium-Raman hybrid laser with better timing characteristics is preferred [95]. In 2019, Zhang et al. [96] developed a 1.7-μm eighth-order RFL, achieving a spectral purity of 98.27% through pumping by an amplified ASE source. In 2022, Ye et al. [97] increased the spectral purity of the first-order RFL to over 99.9% by utilizing the amplified ASE as the pumping source. In 2023, Qi et al. [95] demonstrated that by forming a backward-pumped fully open common-structure ytterbium/Raman RFL to suppress forward Raman scattering, the spectral purity can reach as high as 99.4% and 99.5% for continuous-wave and pulsed first-order Raman random fiber lasing, respectively.

10.2.4.3 Multiwavelength Operation

Multiwavelength fiber lasers (MWFLs) can generally be categorized into two types: one utilizes nonlinear optical effects, such as FWM and SBS, to achieve multiwavelength lasing directly. MWFLs based on these nonlinear optical effects show advantages in having narrower laser linewidths and increasing the number of laser lines. However, achieving flexible control of spectral envelopes and wavelength intervals is challenging because the characteristics of pump lasers and fibers greatly influence the nonlinear conversion process. At the same time, the other method involves using spectral filters, such as Fabry-Perot interferometers [98], Mach-Zehnder interferometers [99], Lyot filter [100], and Sagnac interferometers [101].

10.2.4.4 Narrowband and Supercontinuum Generation

At high pump powers, due to nonlinear effects such as FWM and cross-phase modulation in the fiber, the bandwidth of RFLs broadens to the nanometer level [7, 102]. However, for applications such as efficient frequency doubling, high-precision metrology, fiber sensing, and coherent fiber communication, narrowband laser emission is necessary. To achieve narrowband random laser radiation, one can choose to insert point reflectors with narrow-linewidth filtering bandwidths in semi-open RFLs [103–106]. Another effective way to narrow the bandwidth is to use different gain mechanisms or fibers in RFLs, including SBS gain, polarization-maintaining fiber (PMF), and high-scattering fibers [107]. Additionally, compared to forward-pumping structures, backward-pumping structures can produce laser outputs with a narrower spectral width. Furthermore, enhanced random feedback can be another option for achieving narrowband random laser emission.

It's interesting that RFLs also have been extensively studied as a platform for generating supercontinuum (SC) generation, as the random temporal fluctuations of RFLs can promote the formation of Raman solitons induced by modulation instability, thereby stimulating the SC [108, 109]. Compared to traditional laser-pumped SC spectrum generation, the RFL seed method integrates the generation of pump lasers and the spectral broadening process in the same transmission nonlinear optical fiber, enhancing the compactness and robustness of the structure. Short/open-cavity structures endow RFLs with high-power/efficiency characteristics, which are also conducive to exciting kilowatt-level high-power SC. Compared to traditional laser pumping, low-coherence RFL pumping improves the relative intensity noise (RIN) and the smoothness of the obtained SC spectrum. In 2022, employing a photonic crystal fiber-based RFL, He et al. [110] presented 50.7 W SC output with spectral range of 500–2400 nm and a 20-dB bandwidth of 1900 nm. At the same year, Cheng et al. [111] adopted standard telecom fiber-based RFL seeded by the oscillator as the pump source to increase the output power of the SC to 134 W, with an 20-dB spectrum ranging from 852 nm to 2055 nm. In 2023, Jiang et al. [112] reported a single-stage RFL-pumped kilowatt-level SC output. With the maximum pump power of 1847 W, the output power of SC achieved 1300 W with a spectrum covering 887–1920 nm and conversion efficiency of 66%.

10.3 Application Exploration

10.3.1 Pumping

Using RFLs as pump sources for nonlinear frequency conversion can greatly expand the operating wavelength of RFLs (from visible to mid-infrared). The resulting laser can inherit the advantages of RFLs such as high temporal stability, low noise, and low coherence. Commonly used methods of RFL-based nonlinear frequency conversion include second-harmonic generation (SHG) [113–115], sum-frequency generation (SFG) [87, 116], and difference-frequency generation (DFG) [117, 118]. In 2022, using silica-related Raman gain peak with the frequency shifts of 13.2 THz and phosphorus-related Raman gain peak with the frequency shifts of 39.9 THz in phosphosilicate fiber, Wu et al. [119] built a double-wavelength-tunable RFL, which achieved single-wavelength or double-wavelength output at 1120 and 1238 nm by adjusting pump wavelength. The generated random laser was incident on a periodically poled lithium niobate (PPLN) crystal array. Through single-wavelength SHG process and a dual-wavelength SFG process, the tunable wavelength output of 560-nm green light, 588-nm yellow light, and 619-nm red light was realized, respectively. In 2024, Hu et al. [120] reported the DFG process pumped by 1.3 μm tunable RFL and a 1.5 μm erbium-doped fiber amplifier. By using a $ZnGeP_2$ crystal, they obtained the long-wavelength infrared laser with the wavelength tuning range of 9.5 to 11.5 μm. A stable and tunable mid-infrared band of 3–5 μm can be realized by using RFL-based optical parametric oscillator (OPO). In 2018, Wu et al. [121] demonstrated a wavelength-tunable RFL-pumped OPO system. Using a MgO:PPLN crystal as the nonlinear medium, they achieved 2-W tunable mid-infrared laser with the wavelength tuned from 3977.34 to 4059.65 nm.

10.3.2 Sensing

The long-distance single-mode fiber structure gives RFL excellent environmental robustness, enabling it to work stably in a variety of complex environments, thus widely used in high sensitivity sensing and ultra-long-distance sensing. On the one hand, by integrating advanced sensing elements such as random fiber gratings, phase-shifted FBG, and shifted Fabry-Perot interferometers, RFL can realize highly sensitive sensing of multiple physical parameters, such as temperature, stress, and displacement [122–124]. In 2021, Deng et al. [125] reported a partial reflection random fiber grating for high temperature sensing, which showed a single reflection peak with a very high signal-to-noise ratio. Applying the random fiber grating in RFL greatly improved the sensing stability; they achieved temperature sensing ranging from 25 °C to 500 °C, with an optical signal-to-noise ratio of up to 70 dB. In 2022, Lv et al. [126] developed a narrow-linewidth erbium-doped RFL based on three-grating ring resonator, with a linewidth of ~205 Hz. The three-grating ring resonator not only suppressed mode competition and optimized system stability and frequency noise, but also improved the resolution of strain sensing by using double-cavity FBG-FP interferometer embedded in the resonator. Applying the narrow-linewidth RFL for strain measurement, they obtained a high dynamic strain measuring resolution of 35 femto-strain (fε)/$\sqrt{\text{Hz}}$, with the frequency noise of ~2 Hz/$\sqrt{\text{Hz}}$.

On the other hand, by combining with optical time-domain reflectometry (OTDR), Brillouin optical time-domain analysis (BOTDA), and other sensing systems, RFL can realize ultra-long-distance multiplex detection, with the sensing distance exceeding hundreds of kilometers, which has a wide application range in long-distance communication, industrial monitoring, and geological exploration [127–129]. In addition, distributed Raman amplifiers (DRAs) are a commonly used signal amplification system in optical fiber communication and sensing. The application of RFL in DRAs can significantly reduce the relative intensity noise and nonlinear distortion, and improve signal power stability, which is widely used in WDM/MDM transmission systems and BOTDAs [130–132].

In recent years, RFL has been widely used in Doppler velocity detection [133], acoustic sensing [134], and liquid refractive index sensing [135] as well, providing more accurate and reliable solutions for ultrasonic detection and composite material damage monitoring applications [136, 137].

10.3.3 Imaging

Due to the special feedback and optical amplification mechanism, RFLs present low coherence at high-power output, causing a more significant speckle suppression effect than the fiber resonator, which has obvious advantages in high-power multimode output and speckle-free imaging. In recent years, via the application of MOPA amplifiers and the optimization of loss and heat generation, the output power of multimode RFLs has exceeded hundreds of watts, which greatly improved the imaging quality of speckle-free imaging and caries imaging [138–140]. Speckle correlation imaging can reconstruct the object image from the camera speckles, which shows important application significance in many fields such as opaque media imaging, multimode fiber image transmission, and encrypted communication. The low coherence, high temporal stability, and high spectral density of RFLs can suppress the extra speckle noise and improve the quality of reconstructed image [141]. Wang et al. [142] compared the imaging effects of five different lighting sources in a single-arm multimode fiber image transmission system. By the optimized inverse transmission matrix to recover object images, they found that narrow-linewidth RFL showed the best image quality with structural similarity 0.673 and contrast noise ratio close to 8.5. Furthermore, the unique physical properties of RFLs in the spectral and time domains give other imaging techniques many advantages. For example, the random intensity fluctuation of RFL in time sequence can suppress the artifacts of ghost imaging, which can reconstruct the original time sequence with high accuracy [143]. RFL based on randomly distributed Fabry-Perot cavity has unstable and sparsely distributed peaks in the spectrum, and dynamically changing self-reconstructing speckle patterns can be generated by passing through multimode fiber, which can improve the contrast of super-resolution imaging and spatial ghost imaging [144].

10.3.4 Random Bits Generation

Random bits generation (RBG) is of great significance in the fields of secure communication and data security. With the rapid development of information technology, the demand for data security is increasing day by day; however, the traditional pseudorandom number generators are often difficult to meet the strict requirements of randomness. In this context, true random bit generation based on RFLs provides a new solution. The generation of RFL involves unpredictable physical processes, resulted from the noise and nonlinear effects inside the laser, which caused random fluctuations of the output signal and made the generated RBG completely unpredictable, thus effectively improving data security. In recent years, RBG technology has made significant progress in many types of RFL, including Brillouin RFL [145, 146], semiconductor optical amplifier-pumped RFL [147, 148], single-mode fiber-based RFL [149]. In 2021, Monet et al. [150] demonstrated high-speed RBG generation using half-opened cavity-based RFL. Due to the intense mode competition caused by nonlinear effects, the RBG generation rate was achieved as high as 540 Gbps with minimal postprocessing. Moreover, by the NIST statistical test suite, bit sequences generated by this random laser passed all the statistical tests with a significance level of 0.01.

10.3.5 New Laser Source

RFL, which integrates unique physical properties and flexible regulatory capabilities, provides new ideas and directions for high energy physics and precision spectral analysis. RFL has shown significant application potential in laser-driven inertial confinement fusion (ICF). The low coherence allows RFL to effectively suppress laser-plasma interactions, resulting in a more uniform energy distribution and improved compression symmetry during fusion. In addition, by precisely controlling the shape, intensity, and time series of the RFL pulses, it is possible to achieve a uniform distribution of laser energy on the fusion target. This fine-tuning not only improves the compression effect, but also optimizes the output characteristics of the laser under different experimental conditions to suit different fusion target configurations. By adjusting the time delay and intensity of the pulse, the optimal energy transfer can be achieved, thus significantly improving the efficiency of the fusion reaction [151, 152].

The stochastic optical reconstruction spectroscopy (STORS) can use a low-resolution spectrometer to obtain high-resolution spectra, which requires the sparse emission characteristics of the light source, so as to avoid the overlap of wavelengths that cause the spectrometer to be unable to resolve. RFL based on disordered multipoint feedback mechanism exhibits random and sparsely distributed unimodal characteristics in the spectrum. This property can be used to achieve faster wavelength sparse sampling in a shorter sampling time, thus improving the speed and accuracy of spectral reconstruction. In addition, the spectral reconstruction technique based on chaotic RFL is not affected by convolution, and its resolution is as high as several pm, far exceeding the resolution of commercial spectrometers [153].

10.4 Conclusion and Prospect

Over the past decade, the performances of RFL have gained significant development. The power of single-stage structured RFL has been scaled to 2 kW level. By employing fiber MOPA scheme, the random laser was amplified to 10 kW successfully. In the temporal domain, pulsed random laser has been obtained with pulsation width ranging from milliseconds to picoseconds, which can enrich the laser physics and related dynamics occurring in the interaction between light and disordered media. As to the spatial dimension, research advancements on linearly polarized operation, higher order mode generation, cylindrical vector beam and vortex beam carrying orbital angular momentum, and beam cleanup-induced brightness enhancement indicated that RFLs have tremendous potential in controlling and diversifying output spatial modes. Furthermore, spectral flexibility of RFLs, such as wide-range tunability, high spectral purity, multiwavelength operation, narrowband and SC generation, has been demonstrated. The spectral coverage has been extended from visible to the mid-infrared region, which opens a new horizon for seeking for light source in the spectrum region. The spectral purity has been scaled to over 99.9% by utilizing novel pump source. Benefitting from the fast performance developments, RFLs have been versatile tools for wide applications, such as pumping, sensing, imaging, random bits generation, and new laser source construction. For the future endeavors, RFL might be more expected in the case that there is high-power operation with flexible editing characteristics in temporal-spatial-frequency domains. Furthermore, as a powerful tool to reveal the statistical feature and fundamental physics in different laser scenarios, RFL can be developed as an optimal platform for complex system investigation. The temporal-spatial-frequency flexibility and intrinsic unique coherence might help RFL to be a powerful tool for both scientific research and practical applications.

References

1. Ambartsumyan, R., Basov, N., Kryukov, P., Letokhov, V.: A laser with a nonresonant feedback [J]. IEEE J. Quantum Electron. **2**(9), 442–446 (1966)
2. Lawandy, N.M., Balachandran, R.M., Gomes, A.S.L., Sauvain, E.: Laser action in strongly scattering media [J]. Nature. **368**(6470), 436–438 (1994)
3. Wiersma, D.S.: The physics and applications of random lasers [J]. Nat. Phys. **4**(5), 359–367 (2008)

4. Milner, V., Genack, A.Z.: Photon localization laser: low-threshold lasing in a random amplifying layered medium via wave localization [J]. Phys. Rev. Lett. **94**(7), 073901 (2005)
5. Türeci, H.E., Ge, L., Rotter, S., Stone, A.D.: Strong interactions in multimode random lasers [J]. Science. **320**(5876), 643–646 (2008)
6. Abaie, B., Mobini, E., Karbasi, S., Ballato, J., Mafi, A.: Random lasing in an Anderson localizing optical fiber [J]. Light Sci Appl. **6**(8), e17041 (2017)
7. Turitsyn, S.K., Babin, S.A., El-Taher, A.E., Harper, P., Churkin, D.V., Kablukov, S.I., Ania-Castañón, J.D., Karalekas, V., Podivilov, E.V.: Random distributed feedback fibre laser [J]. Nat. Photonics. **4**(4), 231–235 (2010)
8. Churkin, D.V., Podivilov, E.V., Vatnik, I.D., Turitsyn, S.K., Babin, S.A., Sugavanam, S., Rao, Y., Wang, Z.: Recent advances in fundamentals and applications of random fiber lasers [J]. Adv. Optics Photon. **7**(3), 516 (2015)
9. Han, B., Cheng, Q., Tao, Y., Ma, Y., Liang, H., Ma, R., Qi, Y., Zhao, Y., Wang, Z., Wu, H.: Spectral manipulations of random fiber lasers: principles, characteristics, and applications [J]. Laser Photonics Rev. **18**(7), 2400122 (2024)
10. Churkin, D.V., Babin, S.A., El-Taher, A.E., Harper, P., Kablukov, S.I., Karalekas, V., Ania Castañón, J.D., Podivilov, E.V., Turitsyn, S.K.: Raman fiber lasers with a random distributed feedback based on Rayleigh scattering [J]. Phys. Rev. A. **82**(3), 033828 (2010)
11. Zhang, H., Zhou, P., Xiao, H., Xu, X.: Efficient Raman fiber laser based on random Rayleigh distributed feedback with record high power [J]. Laser Phys. Lett. **11**(7), 075104 (2014)
12. Wang, Z., Wu, H., Fan, M., Zhang, L., Rao, Y., Zhang, W., Jia, X.: High power random fiber laser with short cavity length: theoretical and experimental investigations [J]. IEEE J. Select. Topics Quant. Electron. **21**(1), 10–15 (2015)
13. Zhang, H., Xiao, H., Zhou, P., Wang, X., Xu, X.: High-power random distributed feedback Raman fiber laser operating at 1.2-μm [J]. Chin. Opt. Lett. **12**(s2), 021410 (2014)
14. Zhang, H., Zhou, P., Wang, X., Du, X., Xiao, H., Xu, X.: Hundred-watt-level high power random distributed feedback Raman fiber laser at 1150 nm and its application in mid-infrared laser generation [J]. Opt. Express. **23**(13), 17138–17144 (2015)
15. Du, X., Zhang, H., Wang, X., Zhou, P., Liu, Z.: Short cavity-length random fiber laser with record power and ultrahigh efficiency [J]. Opt. Lett. **41**(3), 571–574 (2016)
16. Zhang, H., Huang, L., Zhou, P., Wang, X., Xu, J., Xu, X.: More than 400 W random fiber laser with excellent beam quality [J]. Opt. Lett. **42**(17), 3347–3350 (2017)
17. Zhang, H., Ye, J., Zhou, P., Wang, X., Leng, J., Xu, J., Wu, J., Xu, X.: Tapered-fiber-enabled high-power, high-spectral-purity random fiber lasing [J]. Opt. Lett. **43**(17), 4152–4155 (2018)
18. Zhang, H., Huang, L., Song, J., Wu, H., Zhou, P., Wang, X., Wu, J., Xu, J., Wang, Z., Xu, X., Rao, Y.: Quasi-kilowatt random fiber laser [J]. Opt. Lett. **44**(11), 2613–2616 (2019)
19. Dong, J., Zhang, L., Zhou, J., Pan, W., Gu, X., Feng, Y.: More than 200 W random Raman fiber laser with ultra-short cavity length based on phosphosilicate fiber [J]. Opt. Lett. **44**(7), 1801–1804 (2019)
20. Song, J., Ren, S., Liu, W., Li, W., Wu, H., Ma, P., Zhang, H., Zhou, P.: Temporally stable fiber amplifier pumped random distributed feedback Raman fiber laser with record output power [J]. Opt. Lett. **46**(19), 5031–5034 (2021)
21. Zhang, H., Wu, J., Wan, Y., Wang, P., Yang, B., Xi, X., Wang, X., Zhou, P.: Kilowatt random Raman fiber laser with full-open cavity [J]. Opt. Lett. **47**(3), 493–496 (2022)
22. Ye, J., Zhang, Y., Liang, J., Ma, X., Xu, J., Yao, T., Leng, J., Zhou, P.: 2 kW random fiber laser based on hybrid Yb-Raman gain [invited] [J]. Chin. Opt. Lett. **21**(9), 090004 (2023)
23. Xu, J., Zhou, P., Leng, J., Wu, J., Zhang, H.: Powerful linearly-polarized high-order random fiber laser pumped by broadband amplified spontaneous emission source [J]. Sci. Rep. **6**, 35213 (2016)
24. Xu, J., Lou, Z., Ye, J., Wu, J., Leng, J., Xiao, H., Zhang, H., Zhou, P.: Incoherently pumped high-power linearly-polarized single-mode random fiber laser: experimental investigations and theoretical prospects [J]. Opt. Express. **25**(5), 5609–5617 (2017)

25. Wu, J., Wang, P., Zhang, H., Wu, H., Xi, X., Shi, C., Yang, B., Wang, X., Han, K., Chen, J.: Power optimization of high-power random Raman laser with a full-open cavity [J]. Opt. Express. **32**(16), 27723 (2024)
26. Babin, S.A., Dontsova, E.I., Kablukov, S.I.: Random fiber laser directly pumped by a high-power laser diode [J]. Opt. Lett. **38**(17), 3301–3303 (2013)
27. Chen, Y., Fan, C., Yao, T., Xiao, H., Leng, J., Zhou, P., Nemov, I.N., Kuznetsov, A.G., Babin, S.A.: Brightness enhancement in random Raman fiber laser based on a graded-index fiber with high-power multimode pumping [J]. Opt. Lett. **46**(5), 1185–1188 (2021)
28. Fan, C., An, Y., Li, Y., Hao, X., Yao, T., Xiao, H., Huang, L., Xu, J., Leng, J., Zhou, P.: Modal dynamics in kilowatt cladding-pumped random distributed feedback Raman fiber laser with brightness enhancement [J]. J. Lightwave Technol. **40**(19), 6486–6492 (2022)
29. Du, X., Zhang, H., Ma, P., Xiao, H., Wang, X., Zhou, P., Liu, Z.: Kilowatt-level fiber amplifier with spectral-broadening-free property, seeded by a random fiber laser [J]. Opt. Lett. **40**(22), 5311–5314 (2015)
30. Li, T., Li, Y., Ke, W., Zha, C., Peng, W., Sun, Y., Ma, Y.: Power scaling of narrow-linewidth fiber amplifier seeded by Yb-doped random fiber laser [J]. IEEE J. Select. Topics Quant. Electron. **24**(5), 0903208 (2018)
31. Xu, J., Ye, J., Zhou, P., Leng, J., Xiao, H., Zhang, H., Wu, J., Chen, J.: Tandem pumping architecture enabled high power random fiber laser with near-diffraction-limited beam quality [J]. SCIENCE CHINA Technol. Sci. **62**(1), 80–86 (2019)
32. Qi, T., Li, D., Fu, G., Yang, Y., Li, G., Wang, L., Du, S., Yan, P., Gong, M., Xiao, Q.: Amplification of random lasing enables a 10-kW-level high-spectral-purity Yb–Raman fiber laser [J]. Opt. Lett. **48**(7), 1794–1797 (2023)
33. Wang, Z., Yan, P., Huang, Y., Tian, J., Cai, C., Li, D., Yi, Y., Xiao, Q., Gong, M.: An efficient 4-kW level random fiber laser based on a tandem-pumping scheme [J]. IEEE Photon. Technol. Lett. **31**(11), 817–820 (2019)
34. Wang, Z., Yu, W., Tian, J., Qi, T., Li, D., Xiao, Q., Yan, P., Gong, M.: 5.1 kW tandem-pumped fiber amplifier seeded by random fiber laser with high suppression of stimulated Raman scattering [J]. IEEE J. Quantum Electron. **57**(2), 1–9 (2021)
35. Du, S., Qi, T., Li, D., Yan, P., Gong, M., Xiao, Q.: 10 kW fiber amplifier seeded by random fiber laser with suppression of spectral broadening and SRS [J]. IEEE Photon. Technol. Lett. **34**(14), 721–724 (2022)
36. Xu, J., Huang, L., Jiang, M., Ye, J., Ma, P., Leng, J., Wu, J., Zhang, H., Zhou, P.: Near-diffraction-limited linearly polarized narrow-linewidth random fiber laser with record kilowatt output [J]. Photon. Res. **5**(4), 350–354 (2017)
37. Huang, L., Xu, J., Ye, J., Liu, X., Zhang, H., Wang, X., Zhou, P.: Power scaling of linearly polarized random fiber laser [J]. IEEE J. Select. Topics Quant. Electron. **24**(3), 1–8 (2018)
38. Li, Y., Li, T., Peng, W., Su, H., Wang, X.: Narrow spectrum kilowatt-level MOPA seeded by Yb-doped random fiber laser [J]. IEEE Photon. Technol. Lett. **29**(21), 1844–1847 (2017)
39. Zhang, J., Bai, G., Li, X., Yang, Y., He, B., Zhou, J.: 1.36-kW Spectral-Narrowing Fiber Laser Seeded by Random Fiber Laser [J]. IEEE Photon. Technol. Lett. **31**(16), 1343–1346 (2019)
40. Bravo, M., Fernandez-Vallejo, M., Lopez-Amo, M.: Internal modulation of a random fiber laser [J]. Opt. Lett. **38**(9), 1542–1544 (2013)
41. Xu, J., Ye, J., Xiao, H., Leng, J., Wu, J., Zhang, H., Zhou, P.: Narrow-linewidth Q-switched random distributed feedback fiber laser [J]. Opt. Express. **24**(17), 19203–19210 (2016)
42. Wu, H., Wang, Z., He, Q., Fan, M., Li, Y., Sun, W., Zhang, L., Li, Y., Rao, Y.: Polarization-modulated random fiber laser [J]. Laser Phys. Lett. **13**(5), 055101 (2016)
43. Pan, W., Zhang, L., Jiang, H., Yang, X., Cui, S., Feng, Y.: Ultrafast raman fiber laser with random distributed feedback [J]. Laser Photonics Rev. **12**(4), 1700326 (2018)
44. Zhang, B., Pan, H., Zhu, Y., Wang, D., Chen, C., Li, R.: Wavelength injection locking actively Q-switched with wide locking range random fiber laser based on single-mode fiber [J]. Opt. Fiber Technol. **84**, 103732 (2024)

45. Chen, C., Li, R., Pan, H., Zhang, B., Chen, Z., Zhao, Z.: Tunable and switchable multi-wavelength bright and dark pulse actively Q-switched random fiber laser based on NPR and EOM [J]. Infrared Phys. Technol. **136**, 105038 (2024)
46. Liu, Z., Pan, H., Zhang, A., Peng, Y., Li, S., Zhang, B., Chen, Z.: Tunable and switchable multi-wavelength actively Q-switched random fiber laser based on an electro-optic modulator and sagnac loop filter [J]. Appl. Opt. **62**(17), 4563 (2023)
47. Zhang, B., Pan, H., Chen, Z., Chen, C., Li, R., Wang, D.: Multi-pulse random fiber laser with wide tuning range in L-band based on tunable band-pass filter and electro-optic modulator [J]. Infr. Phys. Technol. **136**, 105061 (2024)
48. Zhang, A., Zhen, C., Pan, H., Chen, H., Liu, F., Chang, P., Wang, J.: A passively Q-switched random fiber laser with cascaded doped fiber segments and random phase-shift bragg grating [J]. Infr. Phys. Technol. **131**, 104710 (2023)
49. Hu, B., Cui, H., Zhang, Y.L., Ma, R., Xiao, Y.C., Qu, P.F., Zhang, W.L.: Mode locking of a coherent random fiber laser with selectable repetition rates [J]. Opt. Express. **28**(24), 36380 (2020)
50. Yao, B.C., Rao, Y.J., Wang, Z.N., Wu, Y., Zhou, J.H., Wu, H., Fan, M.Q., Cao, X.L., Zhang, W.L., Chen, Y.F., Li, Y.R., Churkin, D., Turitsyn, S., Wong, C.W.: Graphene based widely-tunable and singly-polarized pulse generation with random fiber lasers [J]. Sci. Rep. **5**, 18526 (2015)
51. Ma, R., Zhang, W.L., Zeng, X.P., Yang, Z.J., Rao, Y.J., Yao, B.C., Yu, C.B., Wu, Y., Yu, S.F.: Quasi mode-locking of coherent feedback random fiber laser [J]. Sci. Rep. **6**, 39703 (2016)
52. Tang, Y., Xu, J.: A random Q-switched fiber laser [J]. Sci. Rep. **5**, 9338 (2015)
53. Zeng, X.P., Zhang, W.L., Ma, R., Yang, Z.J., Zeng, X., Dong, X., Rao, Y.J.: Regulation of a pulsed random fiber laser in the Q-switched regime [J]. Laser Phys. Lett. **13**(11), 115105 (2016)
54. Xu, J., Ye, J., Liu, W., Wu, J., Zhang, H., Leng, J., Zhou, P.: Passively spatiotemporal gain-modulation-induced stable pulsing operation of a random fiber laser [J]. Photon. Res. **5**(6), 598–603 (2017)
55. Tarasov, N., Melnikov, L.A., Vatnik, I.D., Mazhirina, Y.A., Churkin, D.V.: Self-gain-modulation random distributed feedback raman fiber laser with switchable repetition rate [J]. Opt. Express. **29**(19), 29857 (2021)
56. Liu, X., Hao, W., Yang, Z., Tang, Y.: High-peak-power random Yb-fiber laser with intracavity Raman-frequency comb generation [J]. High Power Laser Sci. Eng. **11**, e11 (2022)
57. Guo, T., Zhang, A., Pan, H., Sui, P., Peng, Y., Liu, Z.: Tunable multi-wavelength SBS Q-switched random fiber laser based on SMF [J]. Appl. Phys. B. **128**(5), 99 (2022)
58. Liu, C., Zhang, A., Pan, H., Zhen, C.: Pulse regulation by additional attenuation in a random erbium-doped fiber laser utilizing EOM and stimulated Brillouin scattering [J]. Optik. **248**, 168022 (2021)
59. Nuño, J., Alcon-Camas, M., Ania-Castañón, J.D.: RIN transfer in random distributed feedback fiber lasers [J]. Opt. Express. **20**(24), 27376 (2012)
60. Sergio, R.-R., Giuseppe, R., Daniel, L., Nuño, J., Manuel, L.-A., Giorgio, S., Juan Diego, A.-C.: Anomalous relative intensity noise transfer in ultralong random fiber lasers [J]. Opt. Express. **28**(19), 28234 (2020)
61. Xu, J., Wu, J., Ye, J., Song, J., Yao, B., Zhang, H., Leng, J., Zhang, W., Zhou, P., Rao, Y.: Optical rogue wave in random fiber laser [J]. Photon. Res. **8**(1), 1–7 (2020)
62. Iván, R.R.G., Bismarck, C.L., Pablo, I.R.P., Arthur, A.B., Antônio, M.S.M., Giovani, L.V., Leonardo De, S.M., Ernesto, P.R., Anderson, S.L.G., Raman, K.: Turbulence hierarchy in a random fibre laser [J]. Nat. Commun. **8**, 15731 (2017)
63. Edwin, C., Avishek, D., Iván, R.R.G., Anderson, S.L.G., Walter, M., Weid, J.P.V.D., Ernesto, P.R.: Evaluation of Pearson correlation coefficient and Parisi parameter of replica symmetry breaking in a hybrid electronically addressable random fiber laser [J]. Opt. Express. **29**(15), 24422 (2021)

64. Li, J., Wu, H., Wang, Z., Lin, S., Lu, C.P., Raposo, E., Gomes, A., Rao, Y.: Lévy spectral intensity statistics in a Raman random fiber laser [J]. Opt. Lett. **44**(11), 2799–2802 (2019)
65. Hu, Z., Liang, Y., Qian, X., Gao, P., Xie, K., Jiang, H.: Polarized random laser emission from an oriented disorder polymer optical fiber [J]. Opt. Lett. **41**(11), 2584–2587 (2016)
66. Du, X., Zhang, H., Wang, X., Zhou, P., Liu, Z.: Investigation on random distributed feedback Raman fiber laser with linear polarized output [J]. Photon. Res. **3**(2), 28–31 (2015)
67. Zlobina, E.A., Kablukov, S.I., Babin, S.A.: Linearly polarized random fiber laser with ultimate efficiency [J]. Opt. Lett. **40**(17), 4074–4077 (2015)
68. Babin, S.A., Zlobina, E.A., Kablukov, S.I., Podivilov, E.V.: High-order random Raman lasing in a PM fiber with ultimate efficiency and narrow bandwidth [J]. Sci. Rep. **6**, 22625 (2016)
69. Lou, Z., Xu, J., Huang, L., Zhang, H., Leng, J., Xiao, H., Zhou, P.: Linearly-polarized random distributed feedback Raman fiber laser with record power [J]. Laser Phys. Lett. **14**(5), 055102 (2017)
70. Xu J, Huang L, Ye J, Wu J, Zhang H, Xiao H, Leng J, and Zhou P. Power Scalability of a Linearly-Polarized Narrowband Random Fiber Laser in an all-Fiber MOPA Structure with 0.1 Nm Linewidth [J]. Laser Phys. Lett., 2017, 14(9): 095101
71. Evmenova, E.A., Kuznetsov, A.G., Nemov, I.N., Wolf, A.A., Dostovalov, A.V., Kablukov, S.I., Babin, S.A.: 2nd-order random lasing in a multimode diode-pumped graded-index fiber [J]. Sci. Rep. **8**, 17495 (2018)
72. Hao, X., Fan, C., Li, Y., Pan, Z., Leng, J., Yao, T., Lei, B., Zhou, P.: Brightness enhancement on random-distributed-feedback Raman fiber lasers pumped by multimode diodes [J]. High Power Laser Sci. Eng. **12**, e29 (2024)
73. Du, X., Zhang, H., Ma, P., Wang, X., Zhou, P., Liu, Z.: Spatial mode switchable fiber laser based on FM-FBG and random distributed feedback [J]. Laser Phys. **25**(9), 095102 (2015)
74. Zulkifli, M.Z., Lau, K.Y., Muhammad, F.D., Yasin, M.: Dual-mode output in half open cavity random fibre laser [J]. Opt. Commun. **430**, 273–277 (2019)
75. Li, Y., Hao, X., An, Y., Zhu, Y., Yao, T., Zeng, X., Zhou, P.: Hundred-watt level high-order mode all-fiberized random distributed feedback Raman fiber laser with high mode purity [J]. Chin. Opt. Lett. **21**(2), 021406 (2023)
76. Wang, J., Chen, R., Yao, J., Ming, H., Wang, A., Zhan, Q.: Random distributed feedback fiber laser generating cylindrical vector beams [J]. Phys. Rev. Appl. **11**(4), 044051 (2019)
77. Lv, J., Li, H., Zhang, Y., Tao, R., Dong, Z., Gu, C., Yao, P., Zhu, Y., Chen, W., Zhan, Q., Xu, L.: Few-mode random fiber laser with a switchable oscillating spatial mode [J]. Opt. Express. **28**(26), 38973–38982 (2020)
78. Luo, K.H., Ma, R., Wu, H., Li, X.Y., Zhang, C.X., Fan, D.Y., Liu, J.: Flexible generation of broadly wavelength- and OAM-tunable Laguerre–Gaussian (LG) modes from a random fiber laser [J]. Opt. Express. **31**(9), 30639–30649 (2023)
79. Guanyu, Y., Chunxiang, Z., Zheng, H., Rui, L., Rui, M., Zhiyong, B., Dianyuan, F., Jun, L.: 1.7 μm high-power vortex beam generation based on random fiber lasers [J]. Acta Opt. Sin. **43**(22), 2214003 (2023)
80. Li, Y., Yao, T., Fan, C., Hao, X., Ma, X., Xu, J., Zhang, Q., Zeng, X., Zhou, P.: 1.1-1.5 μm waveband high power random vortex beams based on acoustically-induced fiber grating [J]. Acta Opt. Sin. **44**(10), 1026031 (2024)
81. Balaswamy, V., Aparanji, S., Arun, S., Ramachandran, S., Supradeepa, V.R.: High-power, widely wavelength tunable, grating-free raman fiber laser based on filtered feedback [J]. Opt. Lett. **44**(2), 279–282 (2019)
82. Zhang, L., Jiang, H., Yang, X., Pan, W., Feng, Y.: Ultra-wide wavelength tuning of a cascaded Raman random fiber laser [J]. Opt. Lett. **41**(2), 215–218 (2016)
83. Zhang, L., Jiang, H., Yang, X., Pan, W., Cui, S., Feng, Y.: Nearly-octave wavelength tuning of a continuous wave fiber laser [J]. Sci. Rep. **7**, 42611 (2017)
84. Bravo, M., De Miguel, S.V., Ortigosa, A., Lopez-Amo, M.: Fully switchable multi-wavelength fiber lasers based on random distributed feedback for sensors interrogation [J]. J. Lightwave Technol. **33**(12), 2598–2604 (2015)

85. Zhang, Y., Wang, S., She, M., Rao, Y., Zhang, W.: Spectrally programmable Raman fiber laser with adaptive wavefront shaping [J]. Photon. Res. **11**(1), 20–26 (2023)
86. Li, S., Xu, J., Liang, J., Ye, J., Zhang, Y., Ma, X., Leng, J., Zhou, P.: Multi-wavelength random fiber laser with a spectral-flexible characteristic [J]. Photon. Res. **11**(2), 159–164 (2023)
87. Wu, H., Wang, W., Hu, B., Li, Y., Tian, K., Ma, R., Li, C., Liu, J., Yao, J., Liang, H.: Widely tunable continuous-wave visible and mid-infrared light generation based on a dual-wavelength switchable and tunable random Raman fiber laser [J]. Photon. Res. **11**(5), 808–816 (2023)
88. Xu, J., Wang, L., Zhang, L., Teng, L., Zhang, Z., Berko, R.R., Zhang, L., Pang, F., Zeng, X.: Dynamic mode-switchable and wavelength-tunable Brillouin random fiber laser by a high-order mode pump [J]. Opt. Express. **29**(21), 34109–34117 (2021)
89. Xiang, D., Lu, P., Xu, Y., Chen, L., Bao, X.: Random Brillouin fiber laser for tunable ultra-narrow linewidth microwave generation [J]. Opt. Lett. **41**(20), 4839–4842 (2016)
90. Du, X., Zhang, H., Wang, X., Zhou, P.: Tunable random distributed feedback fiber laser operating at 1 μm [J]. Appl. Opt. **54**(4), 908–911 (2015)
91. Wang, L., Dong, X., Shum, P.P., Su, H.: Tunable erbium-doped fiber laser based on random distributed feedback [J]. IEEE Photon. J. **6**(5), 1–5 (2014)
92. Ma, R., Quan, X., Wu, H., Gao, W.C., Huang, D.J., Wang, X.C., Xu, S.X., Fan, D.Y., Liu, J.: 20 watt-level single transverse mode narrow linewidth and tunable random fiber laser at 1.5 μm band [J]. Opt. Express. **30**(16), 28795 (2022)
93. Wang, H., Jia, W., Yao, Y., Yang, X., Melkumov, M., Firstov, S., Lobanov, A., Dong, Z., Luo, Z.: Generation of 1.3/1.4 μm random fiber laser by bismuth-doped phosphosilicate fiber [J]. Chin. Opt. Lett. **21**(7), 071401 (2023)
94. Dong, J., Zhang, L., Jiang, H., Yang, X., Pan, W., Cui, S., Gu, X., Feng, Y.: High order cascaded raman random fiber laser with high spectral purity [J]. Opt. Express. **26**(5), 5275 (2018)
95. Qi, T., Tian, J., Yang, Y., Du, S., Fu, G., Wu, Y., Li, D., Yan, P., Gong, M., Xiao, Q.: High-spectral-purity random raman fiber laser oscillator with full-open cavity counter-pumped by random lasing [J]. Opt. Laser Technol. **161**, 109145 (2023)
96. Zhang, Y., Song, J., Ye, J., Xu, J., Yao, T., Zhou, P.: Tunable random raman fiber laser at 1.7 μm region with high spectral purity [J]. Opt. Express. **27**(20), 28800 (2019)
97. Ye, J., Ma, X., Zhang, Y., Xu, J., Zhang, H., Yao, T., Leng, J., Zhou, P.: Revealing the dynamics of intensity fluctuation transfer in a random Raman fiber laser [J]. Photon. Res. **10**(3), 618 (2022)
98. Bing, L., Wentao, Z., Wenzhu, H., Fang, L.: Switchable and compact dual-wavelength random fiber laser based on random Bragg grating array [J]. Opt. Fiber Technol. **70**, 102858 (2022)
99. Wu, H., Meng, Q., Li, J., Han, B., Wang, Z., Rao, Y.: Spectral tailoring of random fiber laser based on the multimode interference filter [J]. IEEE Access. **6**, 39435–39441 (2018)
100. Ye, J., Zhang, Y., Xu, J., Song, J., Yao, T., Xiao, H., Leng, J., Zhou, P.: Broadband pumping enabled flat-amplitude multi-wavelength random Raman fiber laser [J]. Opt. Lett. **45**(7), 1786 (2020)
101. Min, T., Youchao, J., Haisu, L., Qi, Z., Min, C., Yuean, M., Wei, J., Wenhua, R., Guobin, R.: Multi-wavelength fiber laser based on dual-sagnac comb filter for LP_{11} modes output [J]. J. Lightwave Technol. **38**(14), 3745–3750 (2020)
102. Churkin, D.V., Kolokolov, I.V., Podivilov, E.V., Vatnik, I.D., Nikulin, M.A., Vergeles, S.S., Terekhov, I.S., Lebedev, V.V., Falkovich, G., Babin, S.A., Turitsyn, S.K.: Wave kinetics of random fibre lasers [J]. Nat. Commun. **6**, 6214 (2015)
103. Sugavanam, S., Tarasov, N., Shu, X., Churkin, D.V.: Narrow-band generation in random distributed feedback fiber laser [J]. Opt. Express. **21**(14), 16466 (2013)
104. Ye, J., Xu, J., Zhang, H., Zhou, P.: Powerful narrow linewidth random fiber laser [J]. Photon. Sens. **7**(1), 82–87 (2017)
105. Lv, B., Zhang, W., Huang, W., Li, F.: Low frequency-noise ring random fiber laser with a dual-cavity FBG Fabry-Perot filter [J]. J. Lightwave Technol. **40**(15), 5286–5293 (2022)
106. Wu, H., Han, B., Liu, Y.: Tunable narrowband cascaded random raman fiber laser [J]. Opt. Express. **29**(14), 21539 (2021)

107. Zhou, W., Ou, Z., Fan, X., Zhang, L., Liu, S., Liu, Y.: Ultra-narrow linewidth and a low-noise cascading Brillouin random fiber laser with a dual pump [J]. Opt. Lett. **48**(20), 5415 (2023)
108. Ma, R., Rao, Y.J., Zhang, W.L., Zeng, X., Dong, X., Wu, H., Wang, Z.N., Zeng, X.P.: Backward supercontinuum generation excited by random lasing [J]. IEEE J. Select. Top. Quant. Electron. **24**(3), 1–5 (2018)
109. Qi, T., Yang, Y., Li, D., Yan, P., Gong, M., Xiao, Q.: Kilowatt-level supercontinuum generation in random Raman fiber laser oscillator with full-open cavity [J]. J. Lightwave Technol. **40**(21), 7159–7166 (2022)
110. He, J., Song, R., Jiang, L., Hou, J.: Ultra-compact 50 W flat supercontinuum generation in single-stage self-Q-switched fiber laser with photonic crystal fiber [J]. IEEE Photon. J. **14**(2), 1–6 (2022)
111. Cheng, X., Dong, J., Zeng, X., Zhou, J., Cui, S., Qi, W., Lin, Z., Jiang, H., Feng, Y.: 130 W continuous-wave supercontinuum generation within a random Raman fiber laser [J]. Opt. Fiber Technol. **68**, 102825 (2022)
112. Jiang, L., Wu, J., Song, R., Chen, Z., Zhu, X., Li, F., Li, K., Zhang, H., Hou, J.: Kilowatt-level supercontinuum generation in a single-stage random fiber laser with a half-open cavity [J]. High Power Laser Sci. Eng. **11**, e80 (2023)
113. Dontsova, E.I., Kablukov, S.I., Vatnik, I.D., Babin, S.A.: Frequency doubling of Raman fiber lasers with random distributed feedback [J]. Opt. Lett. **41**(7), 1439–1442 (2016)
114. Rota-Rodrigo, S., Gouhier, B., Dixneuf, C., Antoni-Micollier, L., Guiraud, G., Leandro, D., Lopez-Amo, M., Traynor, N., Santarelli, G.: Watt-level green random laser at 532 nm by SHG of a Yb-doped fiber laser [J]. Opt. Lett. **43**(17), 4284–4287 (2018)
115. Cui, S., Qian, J., Zeng, X., Cheng, X., Gu, X., Feng, Y.: A watt-level yellow random laser via single-pass frequency doubling of a random Raman fiber laser [J]. Opt. Fiber Technol. **64**, 102552 (2021)
116. Liu, X., Hao, W., Yang, Z., Mao, M., Tang, Y.: Ultra-broadband visible random laser generation via hybrid SHG and SFG processes [J]. J. Lightwave Technol. **42**(6), 2131–2136 (2024)
117. Wu, H., Wang, W., Li, Y., Li, C., Yao, J., Wang, Z., Liang, H.: Difference-frequency generation of random fiber lasers for broadly tunable mid-infrared continuous-wave random lasing generation [J]. J. Lightwave Technol. **40**(9), 2965–2970 (2022)
118. Hu, B., Wu, H., Tian, K., Liang, H.: Continuous-wave 2.9–3.8 μm random lasing via temperature-tunning free difference-frequency generation of random fiber lasers in PPLN crystal [J]. Sci. China Inf. Sci. **66**(8), 189401 (2023)
119. Wu, H., Wang, W., Hu, B., Ma, R., Liu, J., Liang, H.: Multi-color switchable visible light source generated via nonlinear frequency conversion of a random fiber laser [J]. Opt. Express. **30**(25), 44785–44797 (2022)
120. Hu, B., Chen, L., Yang, X., He, L., Li, Y., Li, W., Wu, H., Liang, H.: Continuous-wave long-wavelength infrared difference-frequency generation in ZGP driven by near-infrared fiber lasers [J]. Opt. Lett. **49**(4), 1101–1104 (2024)
121. Wu, H., Wang, P., Song, J., Ye, J., Xu, J., Li, X., Zhou, P.: High power tunable mid-infrared optical parametric oscillator enabled by random fiber laser [J]. Opt. Express. **26**(5), 6446–6455 (2018)
122. Lu, P., Mihailov, S.J., Coulas, D., Ding, H., Bao, X.: Low-loss random fiber gratings made with an fs-IR laser for distributed fiber sensing [J]. J. Lightwave Technol. **37**(18), 4697–4702 (2019)
123. Miao, S., Zhang, W., Huang, W., Song, Y.: High-resolution static strain sensor based on random fiber laser and beat frequency interrogation [J]. IEEE Photon. Technol. Lett. **31**(18), 1530–1533 (2019)
124. Qi, Y., Lin, S., Zhang, J., Wang, P., Wang, Z.: Impact of feedback bandwidth on Raman random fiber laser remote-sensing [J]. Opt. Express. **30**(12), 21268–21275 (2022)
125. Deng, J., Churkin, D.V., Xu, Z., Shu, X.: Random fiber laser based on a partial-reflection random fiber grating for high temperature sensing [J]. Opt. Lett. **46**(5), 957–960 (2021)

126. Lv, B., Zhang, W., Huang, W., Li, F., Li, Y.: Narrow linewidth half-open-cavity random laser assisted by a three-grating ring resonator for strain detection [J]. Sensors. **22**(20), 7882 (2022)
127. Leandro, D., deMiguel, S.V., Perez-Herrera, R.A., Bravo Acha, M., López-Amo, M.: Random DFB fiber laser for remote (200 km) sensor monitoring using hybrid WDM/TDM [J]. J. Lightwave Technol. **34**(19), 4430–4436 (2016)
128. DeMiguel-Soto, V., Leandro, D., Lopez-Amo, M.: Ultra-long (290 km) remote interrogation sensor network based on a random distributed feedback fiber laser [J]. Opt. Express. **26**(21), 27189–27200 (2018)
129. Ran, Z., Dong, S., Liu, Y., Deng, K., Wang, J., Rao, Y.: Ultralong Raman fiber laser and sensor with optimized remotely optical pump amplification [J]. Opt. Laser Technol. **175**, 110717 (2024)
130. Chen Y, Du J, Li J, Shen L, Luo J, and He Z. Time-wavelength-mode equalization by PSO for random fiber laser based FMF Raman amplifier [C] 2020 Optical Fiber Communications Conference and Exhibition (OFC). San Diego, CA: IEEE, 2020:W4B.1
131. Nuño, J., Ania-Castañón, J.D.: Cavity and random ultralong fibre laser amplification in BOTDAs: a comparison [J]. Laser Phys. **24**(6), 065107 (2014)
132. Tan, M., Rosa, P., Le, S.T., Dvoyrin, V.V., Iqbal, M.A., Sugavanam, S., Turitsyn, S.K., Harper, P.: RIN mitigation and transmission performance enhancement with forward broadband pump [J]. IEEE Photon. Technol. Lett. **30**(3), 254–257 (2018)
133. Hu, J., Li, R., Hu, Z., Li, H., Yang, Y., Li, H., Lv, J., Yu, Q., Zhao, Y., Yu, B., Lu, L.: Implementation of Er-doped random fiber laser self-mixing sensor with ultra-limit sensitivity [J]. APL Photon. **9**(3), 036113 (2024)
134. Pang, Y., Ma, S., Zhao, X., Qin, Z., Liu, Z., Xu, Y.: Broadband high-sensitivity acoustic sensing based on Brillouin random fiber laser [J]. Opt. Laser Technol. **161**, 109195 (2023)
135. Han, B., Ma, Y., Wu, H., Zhao, Y.: Random Raman fiber laser as a liquid refractive index sensor [J]. Photon. Sens. **14**(1), 240121 (2023)
136. Zhang, L., Lu, P., Zhou, Z., Wang, Y., Mihailov, S., Chen, L., Bao, X.: High-efficiency random fiber laser based on strong random fiber grating for MHz ultrasonic sensing [J]. IEEE Sensors J. **20**(11), 5885–5892 (2020)
137. Lv, B., Tian, M., Zhang, W., Yan, K., Huang, W., Li, F., Li, Y.: Embedded random fiber laser sensor for in situ AE monitoring inside buoyancy material [J]. IEEE Sensors J. **23**(19), 22564–22571 (2023)
138. Wang, S., Zhang, W., Yang, N., Ma, R., Zhang, Y., Wang, Z., Zhang, J., Rao, Y.: High-power multimode random fiber laser for speckle-free imaging [J]. Ann. Phys. **533**(12), 2100390 (2021)
139. Guo, J.Y., Zhang, W.L., Rao, Y.J., Zhang, H.H., Ma, R., Lopes, D.S., Lins, I.C.X., Gomes, A.S.L.: High contrast dental imaging using a random fiber laser in backscattering configuration [J]. OSA Contin. **3**(4), 759–766 (2020)
140. Guo, J., Rao, Y., Zhang, W., Cui, Z., Liu, A., Yan, Y.: Dental imaging with near-infrared Transillumination using random fiber laser [J]. Photon. Sens. **10**(4), 333–339 (2020)
141. Ma, R., Wang, Z., Hui Zhang, H., Zhang, W.L., Jiang, R.Y.: Imaging through opacity using a near-infrared low-spatial-coherence fiber light source [J]. Opt. Lett. **45**(13), 3816–3819 (2020)
142. Wang, L., Qi, T., Liu, Z., Meng, Y., Li, D., Yan, P., Gong, M., Xiao, Q.: Complex pattern transmission through multimode fiber under diverse light sources [J]. APL Photon. **7**(10), 106104 (2022)
143. Wu, H., Han, B., Wang, Z., Genty, G., Feng, G., Liang, H.: Temporal ghost imaging with random fiber lasers [J]. Opt. Express. **28**(7), 9957–9964 (2020)
144. Wang, Z., Wang, S., Ma, R., Liu, Y., Zhu, H., Zhang, Y., Liu, J., Mu, Y., Rao, Y., Zhang, W.: Near-infrared speckle-illumination imaging based on a multidimensionally disordered fiber laser [J]. Phys. Rev. Appl. **18**(2), 024031 (2022)
145. Xiang, D., Lu, P., Xu, Y., Gao, S., Chen, L., Bao, X.: Truly random bit generation based on a novel random Brillouin fiber laser [J]. Opt. Lett. **40**(22), 5415–5418 (2015)

146. Gao, S., Zhang, L., Xu, Y., Chen, L., Bao, X.: High-speed random bit generation via Brillouin random fiber laser with non-uniform fibers [J]. IEEE Photon. Technol. Lett. **29**(16), 1352–1355 (2017)
147. Xu, Y., Lu, P., Mihailov, S., Bao, X.: Real-time physical random bit generation at Gbps based on random fiber lasers [J]. Opt. Lett. **42**(23), 4796–4799 (2017)
148. Lima B C and Weid J P v d. Fast True Random Bit Generation with an SOA-Based Random Fiber Laser [J]. IEEE Photonics Technology Letters, 2023, 35(4): 191–194
149. Wu, H., Xiong, J., Han, B., Wang, Z., Zhang, W., Jia, X., Liang, H.: Ultra-high speed random bit generation based on Rayleigh feedback assisted ytterbium-doped random fiber laser [J]. SCIENCE CHINA Technol. Sci. **64**(6), 1295–1301 (2021)
150. Monet, F., Boisvert, J.-S., Kashyap, R.: A simple high-speed random number generator with minimal post-processing using a random Raman fiber laser [J]. Sci. Rep. **11**, 13182 (2021)
151. Fan, M., Lin, S., Yao, K., Qi, Y., Zhang, J., Zheng, J., Wang, P., Ni, L., Bao, X., Zhou, D., Zhang, B., Xiao, K., Xia, H., Zhang, R., Li, P., Zheng, W., Wang, Z.: Spectrum-tailored random fiber laser towards ICF laser facility [J]. Matter Radiat. Extr. **8**(2), 025902 (2023)
152. Ma, R., Luo, K.H., He, J.S., Zhang, W.L., Fan, D.Y., Gomes, A.S.L., Liu, J.: Digital generation of super-Gaussian perfect vortex beams via wavefront shaping with globally adaptive feedback [J]. High Power Laser Sci. Eng. **12**, e5 (2024)
153. Zhang, Y., Wang, S., Zheng, C., Zhang, W.: Coherent random fiber laser-enabled super-resolution spectroscopy [J]. ACS Photon. **10**(8), 2670–2678 (2023)

Chapter 11
Self-Sweeping Fiber Lasers

Alexander V. Kir'yanov

Abstract A review of the articles dedicated to a novel for fiber laser effect of self-induced laser-line sweeping (SLLS), since its opening to nowadays' detailed exploration, is given. Due to its unusual appearance in general and already demonstrated versatile applications, SLLS ought to present interest for the community involved in studying fiber lasers. Here, a historical insight into the theme with a resume of the meaningful details of SLLS provided.

Keywords Fiber lasers · Self-sweeping effect · Laser dynamics · Rare-earth fibers · Spatial hole burning

11.1 Introduction

The 'self-induced laser-line sweeping' (SLLS) is recently opened for fiber lasers (FL) effect. Its essence is in its very name. Meanwhile, to avoid misunderstanding, it must be defined more precisely. As for me, it consists of spontaneous, periodical, rather slow in time (seconds) movement of a lasing wavelength in an FL, occurring in between some initial and some final value (a span reaching tens of nanometers), followed by a virtually immediate bounce backward to its start value. The spontaneous nature of the SLLS phenomenon, on one hand, and its robustness and reproducibility, on the other hand, are to be accented. It is very different from the other opportunities of making laser wavelength swept, as the SLLS effect happens in the absence of any optical or electrical units commonly used to force a laser line to tune.

Since its first observation and primary explanation in 2011, for an Ytterbium-doped fiber laser (YDFL) [1, 2], in the following years it was uncovered and explored in detail in virtually all known types of FLs, nowadays covering the spectral range from ~1.0 to ~2.1 μm [3–92].

A. V. Kir'yanov (✉)
Centro de Investigaciones en Óptica, León, Guanajuato, Mexico
e-mail: kiryanov@cio.mx

M. F. S. Ferreira, M. C. Paul (eds.), *Optical Fiber Lasers*, Springer Series in Optical Sciences 250, https://doi.org/10.1007/978-3-032-04110-4_11

Because of its beauty in general, simplicity in realization, non-trivial physics standing behind its appearance, and apparent usefulness for potential applications, the SLLS phenomenon attracted a great deal of attention from a few research groups working in the field of FLs. As was revealed just at the early stages of its exploration, SLLS was found to relate to spatial hole-burning (SHB) in a pumped 'active' (capable of lasing) fiber by standing waves formed via superposition of narrow-bandwidth contra-propagating beams inside the laser cavity. As a result of this, dynamic self-induced grating in the gain fiber is formed by the cavity's longitudinal modes that initially get lasing after passing threshold. {Note here that, generally, nodes and antinodes of the standing waves can 'draw' both types of gratings—of metastable level's population along the active fiber and of refractive index; these two play the role of internal, spontaneously induced, spectral modulation.} In turn, such gratings become themselves a source for perturbations in the competition for gain for the cavity's other longitudinal modes. Hence, the primary lasing mode dies out while the adjacent ones—because of spatially more favorable gain/loss conditions owing to the SHB effect—enter lasing. Then, the patterns of inversion and refractive index formed by the new lasing modes become a source for forming the next gratings. Such a process is repeated and repeated, and so new and new longitudinal modes, supported by the cavity, are getting in laser action. Eventually, these repeatable (periodic) changes in the oscillating frequency, or wavelength, lead to what is known as the SLLS phenomenon. Furthermore, as was recently found, the scenario of establishing self-sweeping in an FL, briefly described above for an intracavity gain fiber, can be applied, too, to an intracavity passive fiber that plays the role of saturable absorber (SA).

One needs to notice that the linearity and relative slowness (seconds) of changing the laser wavelength in the time domain, along with a relatively large absolute difference (span) between the start and stop wavelengths inherent in the process (sweeping span that can reach tens of nm), are the key attractivities of the SLLS phenomenon. Owing to these factors, and again bearing in mind the simplicity of implementation of this regime in an FL, it seems invaluable for such applications as sensors' interrogation, optical frequency domain reflectometry, high-resolution spectroscopy, etc.

In the current review, I provide a historical 'snapshot' of the SLLS phenomenon's exploration. All the papers dedicated to the topic till nowadays are referred to and briefly described and discussed herewith.

11.2 A Historical Snapshot of how Knowledge about SLLS Was Raising

11.2.1 2011

The SLLS, a new FL physics effect, was first reported in 2011 [1]. It was uncovered as one of the regimes inherent in a continuous-wave (CW) pumped YDFL, implemented in the simplest (Fabry–Perot) cavity's design, where cleaved ends of the active fiber play the role of non-resonant Fresnel reflectors. The laser's setup is shown in Fig. 11.1; the GTWave (a kind of double-clad) YDF was pumped at ~976 nm by a laser diode (LD).

The effect's main (and spectacular) characteristic is 'adiabatic' movement (during seconds) of the laser line from shorter to longer wavelengths, spanning over a spectral interval as large as ~7 nm, followed by its instantaneous bounce backward; see Fig. 11.2.

The SLLS regime was revealed to occur nearly above the YFDL threshold and to hold for pump powers increased up to ~1.7…1.8 thresholds, while for higher pump powers it gets replaced by intensive bursts of power in the time domain. Authors hypothesized, just at this very primary stage of the SLLS regime's realization, that it appears owing to nonlinear wave-mixing and SHB, i.e. the effects closely related to the intracavity modes' interference in highly uniform spectral and spatial circumstances of low Q-factor cavity, provided by the chosen YDFL's design. Meanwhile, it was mentioned in [1] that a likely behavior in a ring YDFL [2] was registered earlier, in 2009, but without an adequate assessment that time.

Furthermore, and this deserves mentioning, too, a couple of months after publication [1], a similar phenomenon (observed in a YDFL based on standard double-clad YDF) was reported in [3], with a more complicated setup design realized. The two kinds of cavity were under study: one based on a fiber Bragg grating (FBG) and another—on a fiber loop mirror (FLM). Self-sweeping of a narrow laser line (with bandwidth <1 pm) over a broad spectral range, even broader than in [1] (with a span of ~16 nm), was observed. Authors of [3] realized that the SLLS effect in their case originates from the laser's self-sustained relaxation oscillations. Furthermore, they revealed that the sweeping rate increases as a square root of the intracavity laser

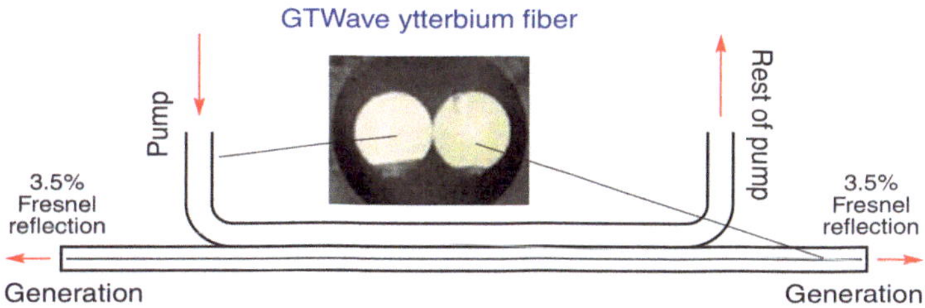

Fig. 11.1 Schematic of the first YDFL showcasing the SLLS effect. [Adapted with permission from [1]; Astro Ltd. ©2011]

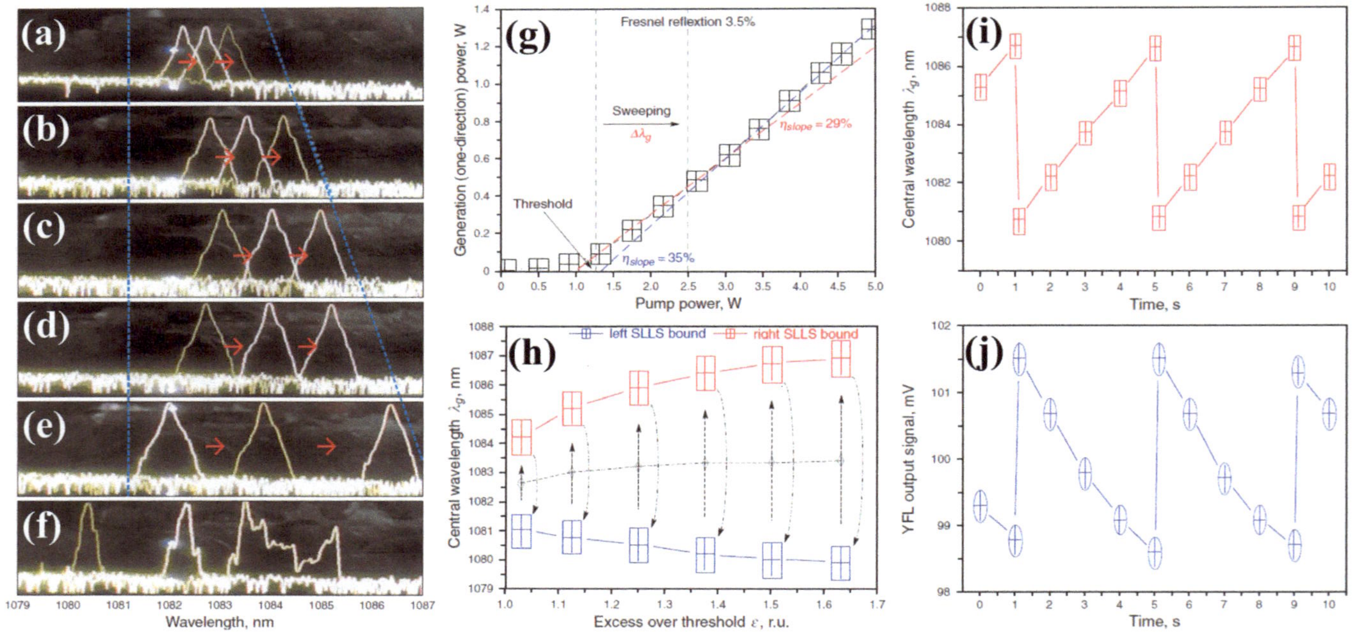

Fig. 11.2 (**a–f**) Laser line's temporal dynamics vs. pump power. The SLLS regime is revealed by spectrograms (**a**)–(**e**) while the replacing one (stochastic spectral bursts)—by spectrogram (**f**); pump powers at which the spectrograms were obtained are 1.4 (**a**), 1.7 (**b**), 2.0 (**c**), 2.3 (**d**), 2.6 (**e**), and 2.9 W (**f**). Each spectrogram is a photo that covers three snapshots in time (counting 200 ms), taken from optical spectrum analyzer (OSA) and recorded with a camera at the exposition time of 0.7 s. (**g**) YDFL output power vs. pump power (the dashed lines and arrow mark correspondingly the range and direction of the SLLS process). (**h**) Dependences of spectral positions of laser line λ_g at the left and right SLLS regimes' limits vs. excess of pump power over threshold; the dotted line shows the central wavelength around which SLLS exists. (**i, j**) Dependences of (**i**) spectral position of laser line λ_g and (**j**) YDFL output signal against time (the data were obtained for pump power 1.8 W). [Adapted with permission from [1]; Astro Ltd. ©2011]

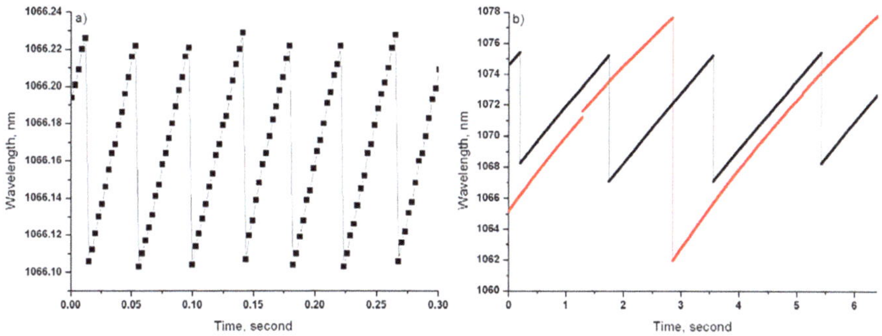

Fig. 11.3 Temporal dynamics of wavelength for (**a**) FBG and (**b**) FLM-based cavities. [Adapted with permission from [3]; Optica (OSA) ©2011]

power while decreasing with enlarging the cavity length. They also proposed a qualitative model that accounts for the effects of gain saturation at the laser transition and spatial SHB, explaining well the laser frequency dynamics.

Examples of temporal dynamics of the laser wavelength in [3] are shown in Fig. 11.3. It was found that the range of its periodic sweeping is limited by the spectral width of the cavity mirror: for FBG, ~0.1 nm and for FLM, ≥10 nm. The laser wavelength was revealed to increase linearly with time at a sweeping rate of 0.5–16 nm/s for different pump power and cavity lengths.

11.2.2 2012–2015

In 2012–2015, the SLLS effect became the subject of a more detailed examination. Particularly, in more detail it was studied in [4] on an example of a YDFL laser (based on double-clad YDF) with a cavity formed by a broadband FLM on one side of the YDF and Fresnel reflection on the other side.

The sweeping range in this scheme was limited by the width of the broadband FLM-reflector, reaching ~16 nm (as in [3]). It was also revealed that the SLLS effect is interrelated with self-pulsations in the form of relaxation oscillations. Furthermore, it was found that the sweeping rate and average frequency of relaxation oscillations both increase with increasing pump (and laser) power (see Fig. 11.4) while decreasing cavity length. The RF and optical spectra (with linewidth ≤100 MHz) show that, during the evolution of a single pulse, only a very few longitudinal modes (or just a single mode) take part. Based on these results, the authors proposed a model describing the dynamics of the laser frequency, treating the SHB effect, and gaining saturation in YDF.

In [5], an analytical model for self-phase modulation in an YDFL with FBGs was developed, describing the output spectrum and its broadening with increasing laser power. Among plenty of data presented in [5], the authors showed that self-sustained pulses are observed in the laser above the threshold but below 0.8 W; in this range,

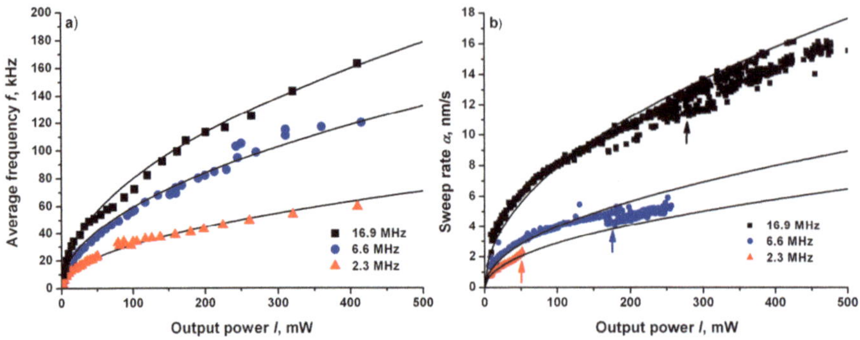

Fig. 11.4 Average frequency of relaxation oscillations (**a**) and sweeping rate (**b**) vs. YDFL's output power at different cavity-mode spacings (lines indicate square root approximation and arrows indicate the instability thresholds of SLLS regime). [Adapted with permission from [4]; SPIE ©2012]

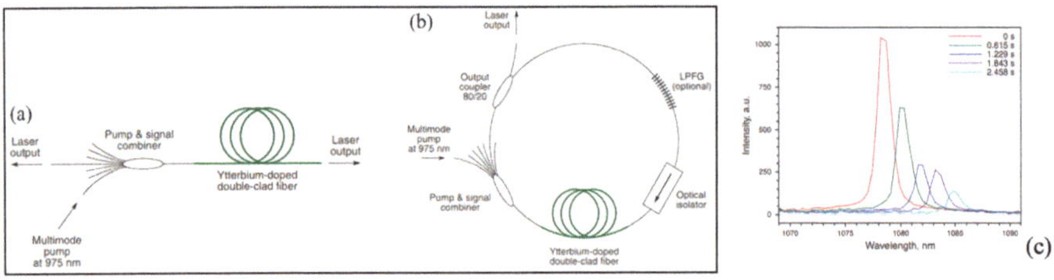

Fig. 11.5 Setups of YDFL in (**a**) Fabry–Perot and (**b**) ring configuration. (**c**) A few records of the laser line's time drift in the YDFL with Fabry–Perot cavity at 976-nm pump power of 0.8 W. [Adapted with permission from [6]; Astro Ltd. ©2012]

a typical SLLS regime arises where the operational wavelength sweeps near the reflection maximum of the output FBG (in a very similar manner as in [1, 3, 4] where SLLS is inherent in an FL with a low-Q cavity). The sweeping range was estimated in [5] to be ~10 pm, whereas the measured 'instant' linewidth was defined by the intracavity interferometer's resolution. In turn, at setting output power above 0.8 W but below 1 W, the SLLS regime got intermitted with common CW lasing, while at further increments of output power, the YDFL turned to pure CW mode.

In 2012, a couple of studies on SLLS was published by a Czeck group of authors [6–8] where were compared the phenomenon's features in YDFLs with ring and Fabry-Perot configurations (importantly, in all realizations the same gain fiber, i.e. double-clad YDF was used); see Fig. 11.5a, b [6]. In [6], authors confirmed the basic appearances of the SLLS effect, uncovered in [1, 3–5], i.e. drift of the laser wavelength in the time domain with a speed of the order of nm/s from shorter to longer wavelengths in the range of several nm, terminated by its fast bounce backward. An example of how authors registered the SLLS effect in an 'instant' way (using a spectrometer based on a CCD camera) is shown in Fig. 11.5c. As seen, output laser power varies (drops) during a single sweeping run.

In the meantime, and this deserves stress, they for the first time used a standard (grating-monochromator-based) OSA to detect the SLLS operation (in this case, the OSA was turned into a mode of high (0.2 nm) resolution and long-time (around 40 sec) acquisition). The two examples of registering SLLS regime, for YDFLs with Fabry–Perot and ring cavities, are demonstrated in Fig. 11.6a, b. An overview of what the fine differences are between these two types of lasers in the sense of pump-power limits of SLLS operation is provided by Fig. 11.6c.

One more novelty in [6] was that, in this paper, the authors discussed some discrepancies between the qualitative model of SLLS published in [1, 3–5] and their own experimental observations, made for YDFLs with Fabry–Perot and ring cavities.

In [7], the Czeck group of authors reported further details of the SLLS regime in the two YDFL versions (with Fabry–Perot and ring cavities). Among these note a kind of it, featured by placing in the ring cavity a couple of long-period fiber gratings (LPFGs). Such an action can be useful for precisely selecting a central wavelength, around which the sweeping is to be realized. In Fig. 11.7, the spectral patterns characteristic to SLLS (refer to the upper spectra) and SQS operation, replacing it at higher powers (see the lower spectra), obtained for the YDFLs assembled in the three mentioned cavity types.

Another point addressed in [7] for the first time is the demonstration of SLLS with 'reverse' directionality, namely, a drift of the laser wavelength from longer towards shorter wavelengths; in the latter case, it was done using a YDFL forced to lase at ~1080 nm (Fig. 11.8).

In [7], as well for the first time while exploring the SLLS effect, it was registered also in an Erbium-doped FL (EDFL) where the laser line was self-sweeping around 1560 nm.

In one more work published in 2012 [8] by the same group of authors, the SLLS phenomenon in EDFL was studied more comprehensively. In this case, the active

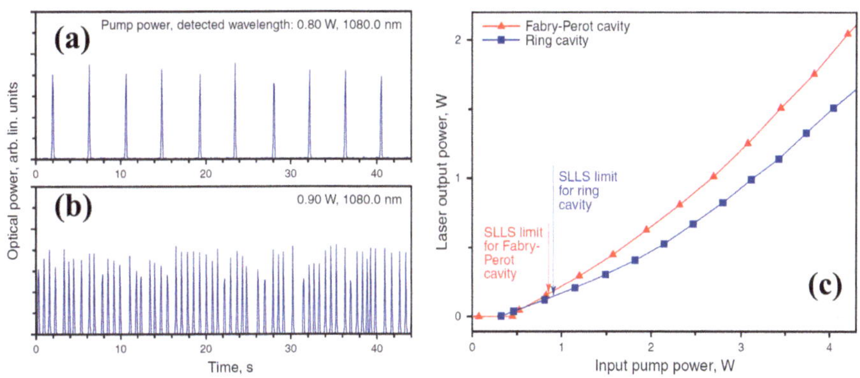

Fig. 11.6 (**a, b**) Output signals from YDFL measured using monochromator-based OSA for the cases of: (**a**) Fabry–Perot and (**b**) ring cavities, respectively. (**c**) Output power vs. 975-nm pump power for the two cavity types: the SLLS regime holds in both cases up to pump levels marked by arrows. [Adapted with permission from [6]; Astro Ltd. ©2012]

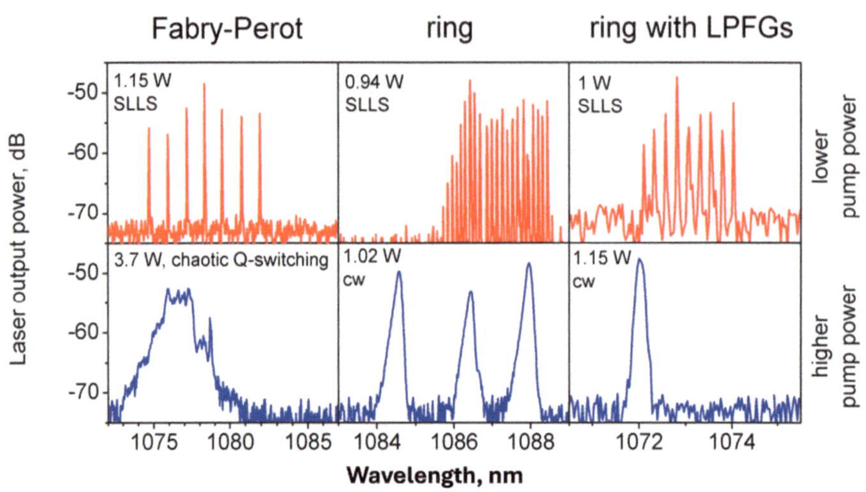

Fig. 11.7 Laser output spectra, measured for three types of fiber laser cavities: Fabry–Perot (left), ring (center), and ring with LPFGs (right). The upper spectra were taken for the pump power where the SLLS regime occurs, while the lower ones were taken for the pump power outside the SLLS regime. [Adapted with permission from [7]; SPIE ©2012]

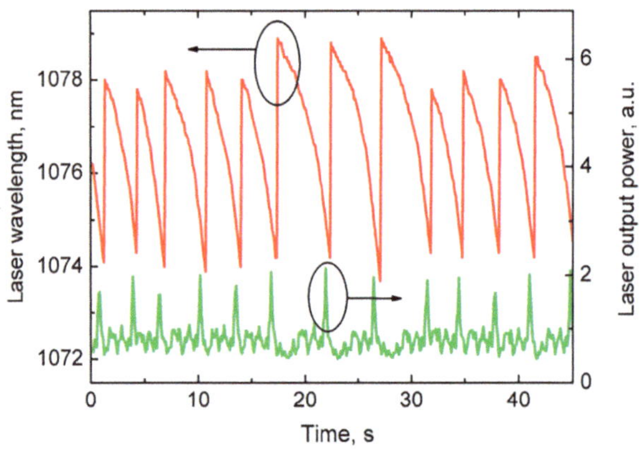

Fig. 11.8 Temporal dynamics of the laser and peak power at SLLS in the reverse (towards shorter wavelengths) version; pump-diode temperature in both cases was 40 °C and pump powers were 0.28 and 0.38 W, respectively. [Adapted with permission from [7]; SPIE ©2012]

EDF was pumped through a WDM by an LD at 980 nm (a piece of a Thulium-doped fiber laser (TDFL) added to the linear cavity played the role of saturable absorber—for another application as the authors claimed). The laser cavity in this case was assembled by an FLM on one side and cleaved fiber end on the other side; it also contained a Koshin–Kogaku 3-nm bandpass filter and polarization controller. As a result, authors obtained lasing delimited by the tunable filter; above the threshold they were able to get a sweeping regime of the EDFL within a 0.5 nm interval, defined by the filter. In contrast to YDFLs, the sweeping period in EDFL is measured by 100–150 ms and is tunable within the spectral range 1550–1565 nm.

In 2013, just 2 years after the first observation of the SLLS phenomenon, it was reported the first application of an FL with self-sweeping, namely, its use for high-resolution spectral analysis of π-shifted FBGs [9].

It was done using a YDFL with SLLS in the cavity design comprising a cleaved cut and FLM; the self-sweeping range was measured at around 8 nm. As the self-sweeping laser generates nearly single-frequency radiation during a single pulse, this fact permitted spectral selection of a shift-induced dip of 14-MHz width in the reflection spectrum of the FBG under test. The sampling interval during the scan is shown to be limited by pulse-to-pulse frequency shift, equal to the cavity mode spacing of 6 MHz. The main feature of the applied technique is its simplicity due to the absence of special elements for frequency scanning, inherent at SLLS.

Furthermore, in [10], the same group of authors attempted a more profound insight into the dynamics of YDFL with wavelength self-scanning. The laser scheme was very alike to that used in [3] with the main difference being handling polarization maintaining (PM) fibers and other units inside the cavity to omit polarization artifacts in the laser; see Fig. 11.9.

The laser under study [10], in all its examined versions (different lengths of intra-cavity passive fibers and lengths of active YDF), demonstrated SLLS in the range of ~20 nm. The other important details of this regime, revealed by Fig. 11.10, point to variation of the number of longitudinal modes involved at SLLS depending on a cavity version (see above) used. In addition to the experimental pattern inherent in this laser with SLLS, the authors of [10] developed a theoretical model that reveals

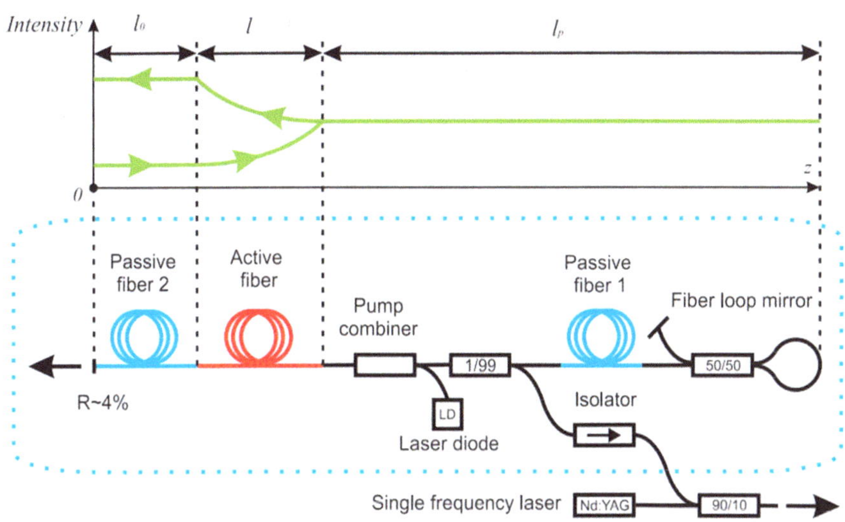

Fig. 11.9 Experimental setup of the YDFL with PM components (the lower dotted part) and the heterodyne scheme used for spectral analysis. Longitudinal intensity distribution for two running waves is schematized on the figure's top. [Adapted with permission from [10]; Astro Ltd. ©2014]

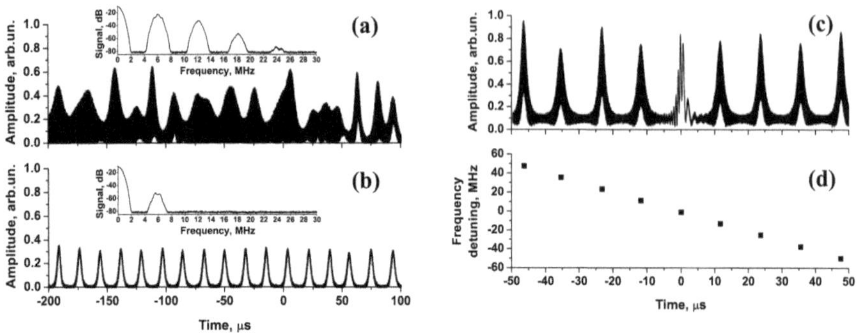

Fig. 11.10 (**a**, **b**) Dynamics of SLLS at pump power of 1.5 W, 14 m of passive fiber '1' and 0 m of passive fiber '2' (**a**) and 1.3 m (**b**); it is seen (from the insights where the correspondent RF-spectra are plotted) that 5 and 1 to 2 longitudinal modes are involved at lasing. (**c**, **d**) Heterodyne signal at pump power of 1.8 W (**c**) and modulation frequency of individual pulses (**d**). [Adapted with permission from [10]; Astro Ltd. ©2014]

the critical impact of self-induced dynamic phase gratings (DPGs) formed in such lasers: in fact, both the frequency and intensity evolutions in them were demonstrated to be driven by the interplay of the DPGs and gain saturation in the active YDF. Authors also claimed that the theory can provide a way to fine optimization of FLs in SLLS mode.

Note that subsequently an almost identical to [10] single-frequency YDFL, self-sweeping within a 20 nm range (1045–1065 nm), was successfully used for another application, viz. exploiting of the SLLS phenomenon for characterization of polarization properties of LPFGs [11].

In [12], the Czeck group of authors attempted more research to characterize a self-swept YDFL (with double-clad YDF and a Fabry–Perot cavity with perpendicularly cleaved fiber ends; its setup was almost the same as that shown in Fig. 11.5 but now with an aim to reveal a possible effect of variation of the LD pump wavelength on SLLS. The pump wavelength was changed in two ways: either by changing the LD's case temperature or current through it; at both measures, the pump wavelength grew (from ~965 to ~977 nm) while increasing these two LD's parameters. Using this laser design, the authors found that the SLLS wavelength's range was always 5–7 nm (around 1080 nm) for a wide span of pump-LD power and temperature. Meanwhile, the other signatures of the SLLS mode (its rate and period) were changed—while not largely—as the result of the undertaken modifications.

The first ~2-μm self-sweeping FL based on thulium-holmium co-doped fiber was reported in [13]. The laser cavity in this case was formed by a flat cleaved end of a fiber coupler and FLM consisting of a PC and coupler. The pump wavelength was 1570 nm. As seen from Fig. 11.11, the laser's central wavelength self-sweeps periodically around 1.9 μm within the spectral range, increasing from 4 to 17 nm; it was also found that the sweeping rate decreases from 0.4 to 1.5 nm/s at increasing pump power (the SLLS operation was observed for pump power varied from 260 to 340 mW).

In [14], the authors presented a study of self-pulsed regimes in Fabry–Perot-cavity YDFL, where the role of the SLLS effect in triggering self-Q-switched (SQS) regime was discussed.

As shown in [14], the SLLS regime is observed at up to 0.5 W of output power; at further power growth, it is replaced by SQS. At SLLS, as seen, a train of ~μs narrow-linewidth pulses is generated, with single or a few longitudinal modes to be generated. The latter is justified by the presence of a sinusoidal beat signal under the pulse's envelope. From the latter figure, it also follows a gradual transition of the sinusoidal signal to the 10-ns giant pulses.

In [15], it was announced, for the first time, that dynamic gratings and SBS constraints are behind the SLLS regime and its transformation to SQS operation; these matters have been later discussed in more detail in the subsequent work of the same authors; see below [19, 30]. For example, for the SLLS regime, one can note that just above the laser threshold (0.14 W of output power), the laser wavelength drifted slowly from 1060 to 1064.5 nm (the repetition rate of power bursts is low), whereas at increasing pump power, the drift is accelerated, and so the bursts appear at an increased repetition rate. At a further increase of pump power, the SLLS regime is replaced by SQS operation, where bursts' sequencing is replaced by a 'common' spectrum, registered for an output power of 1.4 W.

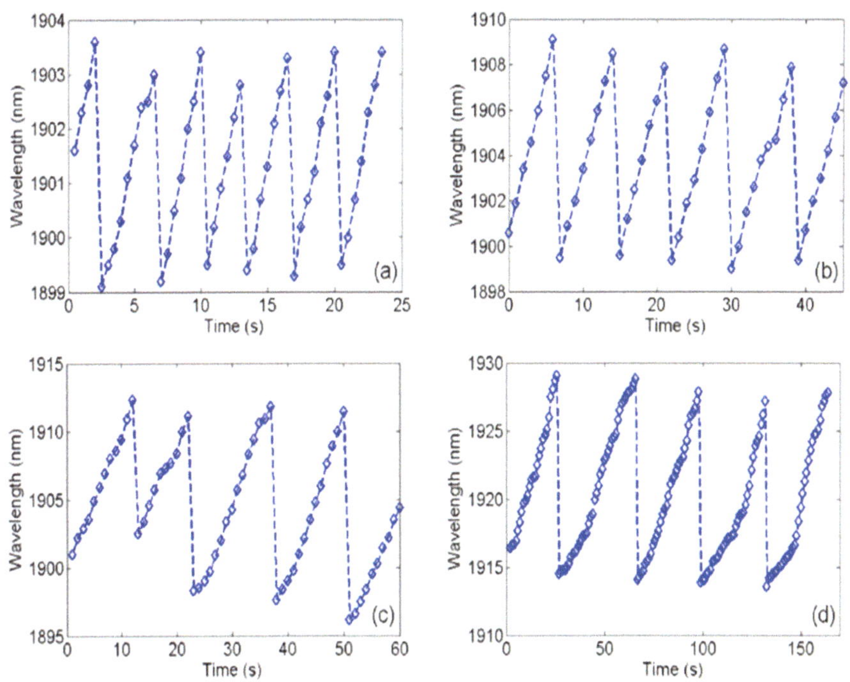

Fig. 11.11 Temporal dynamics of self-sweeping in Tm-Ho FL at pump powers of 275 (**a**), 291 (**b**), 310 (**c**), and 333 (**d**) mW. [Adapted with permission from [13]; Optica (OSA) ©2013]

Furthermore, in [15], a method for estimating the reflection spectra of transient gratings in doped fiber lasers with SLLS was presented. It was shown that such gratings can be highly reflective (up to several tens of percent), a firstly made conclusion for FLs with SLLS. A demonstration of this achievement was done by authors on an example of the YDFL sketched above [14]. It was particularly shown that the gratings spectra are highly asymmetric due to the apodization of the refractive index to be self-modulated at SLLS.

In 2014, a study [16] was published where, among the other regimes of self-pulsing operation established in large-mode-area, end-pumped, double-clad YDFL with non-resonant cavities, formed either by using one or two perpendicularly cleaved YDF-ends, SLLS was registered (again just above the laser threshold), thus once more justifying its 'universality' for YDFLs of such type. Note that the way to register SLLS in [16] was like that used in [6].

In 2015, a group of authors from Russia reported, for the first time, the generation of linearly polarized pulses at SLLS, characterized by ~10-nm sweeping of laser wavelength around 1.46 μm in one more type of FL, i.e. in an all-fiber bismuth-doped FL (BDFL) [17]. It was shown that these pulses—relatively long (~10 μs) and so overlapping in the time domain—are single-frequency with a small frequency chirp, while the pulse-to-pulse frequency shift (~1 MHz) corresponds to the inter-mode interval of the laser cavity (~100 m). Thus, the data presented in [17] reveal that the SLLS regime is nearly steady-state in both frequency and time domains. This work has value as BDFL can be an alternative to common 1.5…1.6-μm EDFL; particularly, it is the case regarding the SLLS effect, shown to be barely expressed (with self-sweeping range not exceeding 1 nm: refer to [8]).

In [17], the BDFL's low-Q cavity was formed by a FLM based on a 50/50 fused coupler on the right side and Fresnel reflection on the left side, which provides a base for effective SLLS to develop. The laser was pumped at 1310 nm, a wavelength resonant to Bismuth active centers having absorption at this wavelength of ~1 dB/m. The BDFL's threshold was rather low (~20 mW) and had high differential efficiency (33%). Interestingly, SLLS was observed in the laser at pump powers widely ranging from 50 to 300 mW; since a rather slow sweeping rate was inherent in this laser implementation, it was straightforwardly registered using a standard OSA. Furthermore, the authors of [17] revealed that the BDFL's temporal characteristics nearly fit those of a classic quasi-CW regime; hence, the SLLS process was quasi-continuous, too, with the scan rate increasing with pump power owing to the decreasing period of pulses.

In the same year, a collective of authors, also from Russia, reported a study of one more version of YDFL with SLLS, characterized by a kind of 'regular' dynamics where the phases of modes generated within different pulses are not in fact random [18], thus, an issue regarding their coherent properties arose. The laser was built using an analogous to [17] configuration where the roles of the cavity's couplers were played by an FLM (with 80/20 coupling) and Fresnel reflection. The active fiber employed was a 2.6-m piece of double-clad YDF; it was pumped by an LD in the band of Yb^{3+} ions. The YDF's and passive fibers' lengths were chosen to ensure single-frequency SLLS operation in the form of periodic pulses. Another part of the scheme under study was an external ring fiber cavity that, for the summation of consecutive modes, was generated by the YDFL. A superposition of 18 consequent laser pulses inside an external ring fiber cavity highlights their coherent combining into a train of nanosecond pulses.

Analysis of laser pulse dynamics and their superposition in the external cavity revealed that the initial phase of the 'mode-pulse' radiation is transmitted over the pulse train indicating that information about the phase is kept in the laser cavity for a long time. This fact allowed authors to conclude that an FL with self-sweeping represents a new kind of operational regimes, different from classic Q-switched and mode-locked lasers.

In [19], the Czeck group of authors provided a more comprehensive analysis of transient FBGs at SLLS in an YDFL with the simplest (Fabry–Perot) cavity; this research was somewhat a generalization of the approach suggested by them in [15].

It is claimed in [19] that as the SLLS mode is caused by the presence of standing waves in the laser cavity, such a series of waves ought to create an interferometric pattern, which in turn is a source of a highly reflective grating to form through resonant index change in the active fiber's core. An analysis of the spectra of such transient (and superimposed: in difference with the work [15], the case of a few longitudinal modes in the cavity was considered in [19], too) FBGs stands behind the results of a numerical modeling that the authors developed.

In Fig. 11.12, an example of the results obtained for the longitudinal distributions of the pump and signal powers' evolutions along the active YDF length (left Y-scale). Furthermore, in the same figure, the authors provide an example of calculations for the visibility of the interference pattern formed in the cavity (right Y-scale). The latter is noticeable, showcasing that the fringes' visibility is maximal near the point where pump light is launched into the cavity.

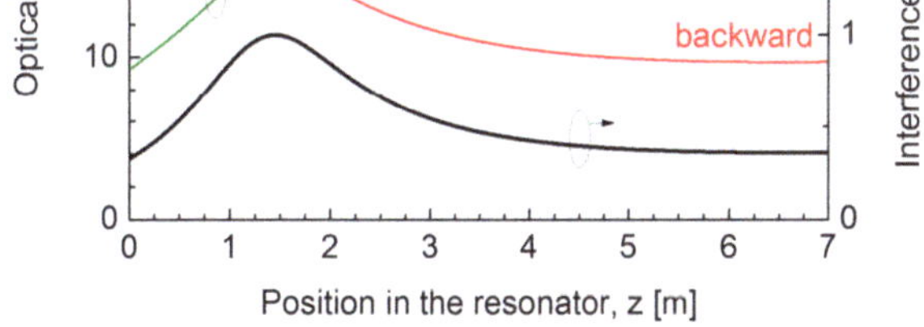

Fig. 11.12 Results of modeling of the pump and signal powers evolutions, as well as the interference fringe visibility for the case of single-longitudinal-mode lasing, along the YDF fiber. [Adapted from [19]; SPIE ©2015]

Starting from the results shown in Fig. 11.12, the authors modeled the longitudinal dependencies of the refractive index (RI) modulation; see Fig. 11.13a, with the three cases being addressed: the single-frequency regime (top) and the regimes of 10 (middle) and 20 (bottom) modes, respectively (it was implied that the Δn-value is given by the metastable level population). As seen, apparently, the refractive-index modulation strength vs. coordinate is coherent with that demonstrated by the visibility of the interference pattern. Accordingly, the corresponding reflection spectra of the FBGs to be formed in the cavity (shown in Fig. 11.13b) because of the index modulation take the shapes inherent in the regimes of a single (top) and 10 (middle) or 20 (bottom) modes established. Importantly, despite the small RI-modulation (of around 10^{-6}; see Fig. 11.13a), the narrowband (~1 pm) FBGs with reflectivity of almost 100% (Fig. 11.13b). Certainly, the stream of data presented in Fig. 11.13 is one of the clue laws underlying the SLLS effect in FLs; moreover, as authors of [19] correctly note, the transient FBGs should severely change the Q-factor of the cavity (very low initially) and, along with the narrow-linewidth nature of the SLLS regime itself, can stand behind triggering SQS operation in the laser.

Note that later, in [20], an erratum was published regarding the reflection spectra of transient FBGs spontaneously built-up in the YDFL with SLLS [15]. This was based on the newer, more correct, results [19] showing that reflectivity of such FBGs can approach 100% (see above).

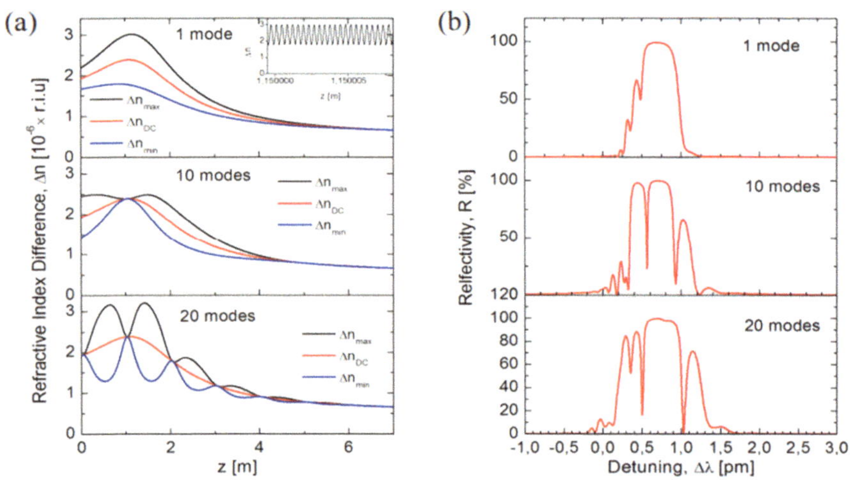

Fig. 11.13 Modulation of RI along the YDF (**a**) and corresponding reflectivity of the dynamic FBG (**b**) for 1, 10, and 20 longitudinal modes. The modulation depth of RI is given by Δn_{max} and Δn_{min}, see inset in (**a**). [Adapted from [19]; SPIE ©2015]

11.2.3 2016–2020

In 2016–2020, the SLLS phenomenon in FLs of different types was mainly explored by the two teams of authors—from the Czeck Republic and Russia. As a result, the knowledge about it and its use in the practice of FLs were steadily enriched, in a somewhat competitive manner.

Say, in [21], the Russian group examined the effect of laser cavity parameters (an active fiber length and output coupling losses) and the temperature of the elements used (active fiber and pump LD) on the sweeping range in an YDFL with SLLS; the correspondent setup arrangements are shown in Fig. 11.14a, b. The experimental data set was compared with the modeling one, and good agreement between the two was revealed.

Particularly, the results obtained showed that the sweeping region gets spectrally shifted to shorter wavelengths for shorter active fibers and with increasing absorbed pump power. This allows one to obtain a self-sweeping operation in such a type of FL in a broad range within an Yb^{3+} gain bandwidth, coverable from 1028 to 1080 nm (i.e. over ~50 nm but, certainly, not for a fixed set of parameters); meanwhile, in its optimal realization, the YDFL provided a maximal sweeping span of ~21 nm. Two examples of the data [21] are demonstrated in Fig. 11.15.

In [22, 23], the same group developed a single-frequency, all-fiber, temperature-controllable YDFL with self-sweeping covering an 18-nm range (1058–1076 nm) and output power exceeding 50 mW. They also proposed a roadmap towards a fibered interrogator based on the developed laser and tested it on the example of a line of six FBGs as sensors; the tests revealed that FBG reflection can be measured with 2-pm accuracy.

The Russian group also published a study [24] where a single-frequency, large-mode-area YDFL with SLLS was developed and comprehensively investigated. The main motivation of this research was to compare details of the SLLS regime in two versions of the YDFL, viz. based on double-clad YDFs with core areas of 11 μm (length, 9.1 m) and 5 μm (length, 7.9 m), respectively. It was found that in the latter (standard) version of YDFL, the SLLS range reached ~21 nm, while in the former, it was limited to ~13 nm (at the same pump power of around 5 W). In the meantime, the average and peak powers (and overall efficiencies) of the two laser schemes differed nearly twice (viz. YDF with the 11-μm core being a better case in this sense). Thus, according to the authors' claim, the use of a YDF with a bigger or smaller

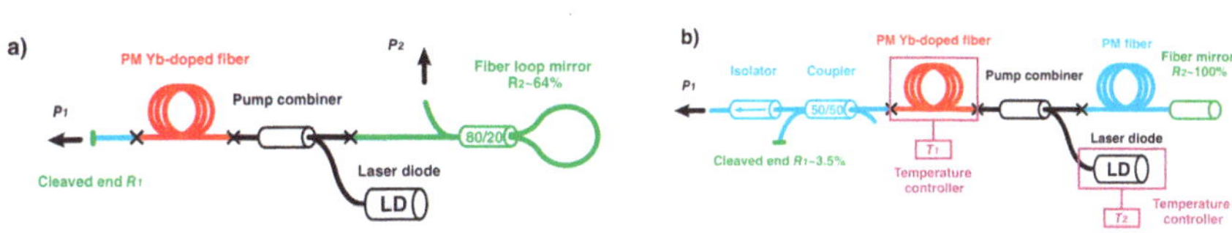

Fig. 11.14 Setups of the YDFLs with SLLS, used for: (**a**) for output coupling and active fiber length optimization and (**b**) for temperature control. [Adapted with permission from [21]; Astro Ltd. ©2016]

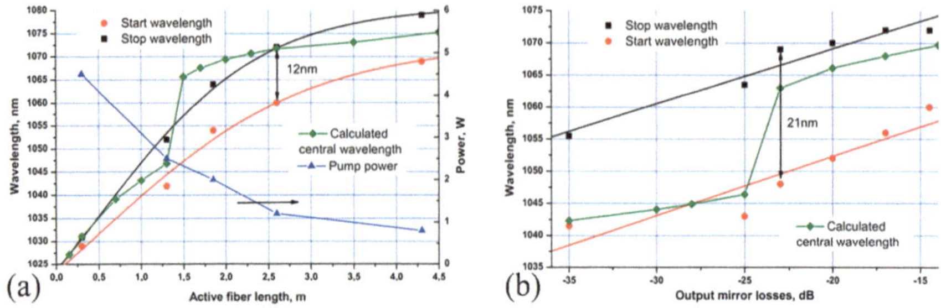

Fig. 11.15 (**a**) Dependence of the SLLS range and calculated central wavelength at SLLS (left axis) on the YDF length and pump powers at which observed the maximal SLLS span (right axis). (**b**) Dependence of the SLLS and calculated central wavelength at SLLS on losses of output mirror. [Adapted with permission from [21]; Astro Ltd. ©2016]

core ought to be compromised for a particular application; meanwhile, it is found that the instability threshold for the SLLS regime in either laser version is associated with the route to multi-frequency lasing.

In turn, the Czeck group of authors, deeply involved in the examination of the SLLS phenomenon since its first publication, visited the issue in 2016–2019 a few times, releasing a line of interesting reports [25–32] that are briefly enlightened in the following piece of text.

In [25–27], SLLS was obtained (presumably, for the first time) and comprehensively examined in a Holmium-doped FL (HDFL) near ~2.1 μm. The sweeping was shown to occur within ~3…5-nm interval with a rate of ~0.7…0.9 nm/s, from longer towards shorter wavelengths (reverse SLLS). Interestingly, the effect was registered using an FTIR spectrometer. Thus, this work is a breakthrough, being the first demonstration of the SLLS effect in the mid-IR range.

The HDFL was made in this case in an all-fiber Fabry-Perot design, with the output couplers being a cleaved end of the HDF and a highly reflective FLM based on a 50/50 broadband coupler. The HDFL was pumped by a thulium-doped FL (TDFL) at ~2030 nm. It was shown in [25–27] that, at low output powers, the output laser light at ~2.1 μm presents either quasiperiodic bursts when it is measured behind an optical circulator and Fabry–Perot grating (FBG) (case I) or merely relaxation oscillations when measured directly (case II), the typical appearances of the SLLS effect. Furthermore, when recorded with the FTIR spectrometer, the effect gets visualized more firmly (see Fig. 11.16).

Furthermore, it was found that changing the pump TDFL operating at 2030 nm to a similar one operating at 1940 nm led to the disappearance of SLLS in the 2.1-μm HDFL. Also note that in [25–27] possible origins of the 'normal' (shorter to longer wavelengths) and 'reverse' (longer to shorter wavelengths) SLLS were discussed. Meanwhile, authors mentioned, too, that stability of the SLLS regime in their HDFL was short-lived (minutes, becoming then gradually unstable), hypothesizing that thermal effects might be a reason for that.

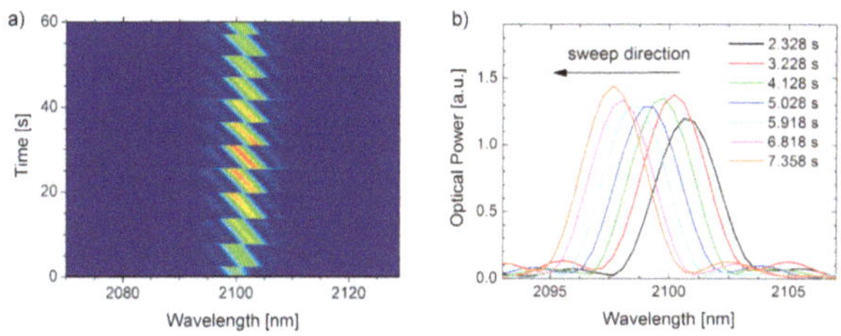

Fig. 11.16 (**a**) Sweeping of the laser wavelength recorded by the FTIR spectrometer in a color contour plot and (**b**) selected spectra slices of the time series recorded within a single sweep; the resolution of the spectrometer was set to 8 cm^{-1} or ~ 3 nm. [Adapted with permission from [26]; Optica (OSA) ©2017]

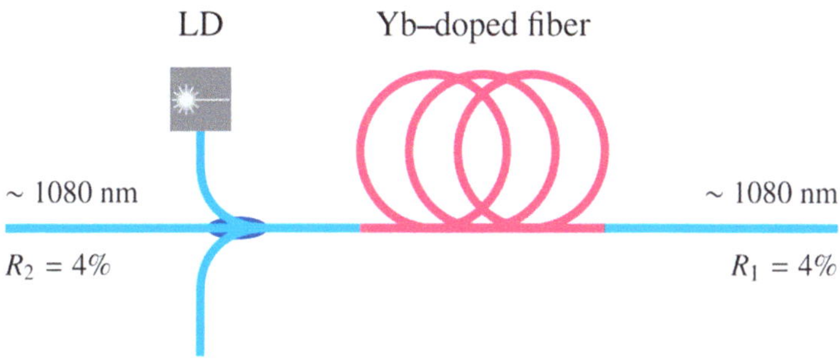

Fig. 11.17 Experimental setup of the YDFLs with SLLS. [Adapted with permission from [28]; Astro Ltd. ©2017]

In the meantime, a study was published reporting various directionalities (normal, reverse, and mixed) of self-sweeping in an YDFL based on double-clad YDF [28]; Fig. 11.17. An interesting detail of the study was that the directionality of sweeping at SLLS (towards Stokes, 1076 nm → 1083 nm, or anti-Stokes, 1079 nm → 1073 nm, side) could be selected merely through changing the pump wavelength. The authors provided, too, an explanation for this behavior, based on a semi-empirical model. The experimental schematic of the YDFL was a standard Fabry–Perot design, the simplest for getting the SLLS regime. In turn, the diversity of the laser wavelength drifting with time is exemplified in Fig. 11.18 for a few pump powers, at which the three different scenarios of self-sweeping (normal, reverse, and mixed) are supported. In [28], authors also presented a theory clarifying the sweep direction based on the gain curve profile and its spectral inclinations.

In [30], the Czeck group of authors reported a summarizing study dedicated to modeling reflectivity of transient FBGs self-created in active fiber of an FL at SLLS. The research was conducted using the coupled-mode theory, allowing a theoretical model of superimposed FBGs in this case, eventually defined by the fiber's RI modulation inside the active fiber through the Kramers–Kronig relations. Authors considered an example of an YDFL where RI change around 1.06 μm is directly proportional to the metastable level population in the system of Yb^{3+} ions.

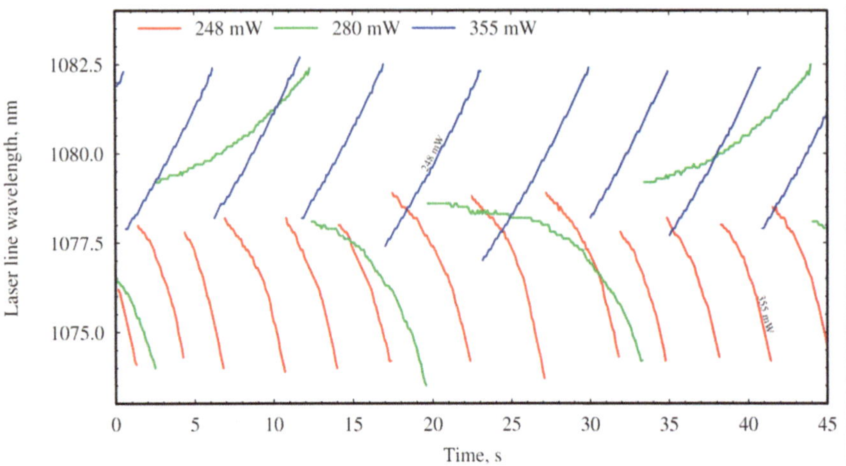

Fig. 11.18 Time dependence of laser line wavelength for pump laser diode case temperature – 40 °C and pump power 248, 280, and 355 mW. [Adapted with permission from [28]; Astro Ltd. ©2017]

Considering realistic values of temporal damping of the gratings, significant reflectivity of the order of tens of percent was obtained; see Fig. 11.19. It was revealed that the reflectivity of transient FBGs can be controlled by setting the SLLS laser parameters, while the effect of the cavity length was shown as an example.

It should be emphasized that the model published in [30] (being in turn a generalization of studies [19, 20]) is an important step towards the understanding of the origins of SLLS in FLs.

In [31], the Czech group of authors presented research, generalizing and giving more detailed data results on the SLLS phenomenon in a 1.56-μm EDFL, presented earlier [8]. Both laser-line sweeping directions, normal (redshift) and reverse (blueshift), were addressed in more detail experimentally and analytically. Note in this sense that, experimentally, these two directions of sweeping were found to depend on settings of a few intracavity elements (as the SLLS regime in EDFLs is hard to accomplish, and so additional measures ought to be applied to fulfill the task). Regarding the analytical section of [32], the authors discussed the two presumable origins for different sweeping paths, namely, those caused by spectral 'inclinations' of the gain and refractive-index gratings (FBGs) inside the active fiber. Particularly, authors developed here, too, a semi-empirical model based on the gain curve spectra that can explain predetermination of the sweep direction in an EDFL.

Also note that the same group presented in [29, 32, 58] a couple of resuming reviews on the problematics of FLs oscillating in SLLS mode. It provided a snapshot of the basic appearances and laws inherent in FLs with SLLS; they also provided clear physical explanations for them, allowing the reader a better understanding of what happens behind SLLS. The SLLS mode in FLs was considered as a kind of longitudinal-mode instability, resulting in, on one hand, self-pulsing in the form of

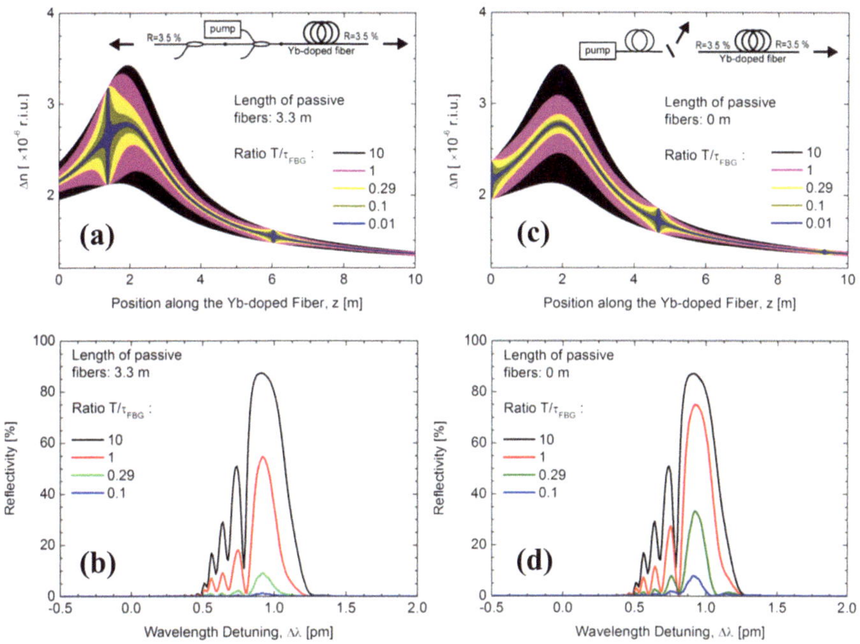

Fig. 11.19 (**a, c**) Refractive index modulation along the active fiber for various values of T/τ_{FBG}, modeled for the length of passive fiber being 3.3 (**a**) and 0 (**c**) m. (**b, d**) Reflectivity of transient FBGs formed by the standing waves of the consequential hopping longitudinal modes for the same values of T/τ_{FBG} and the same lengths of intracavity passive fiber: 3.3 (**b**) and 0 (**d**) m. [Adapted with permission from [30]; IEEE ©2018]

self-sustained relaxation oscillations and a spectacular laser-line drift with time, on the other hand. Particularly, these works were focused on SLLS as a factor igniting SQS (e.g. revealing the role of spontaneously induced FBGs in the active fiber at SLLS). The reviews highlighted that its origin is presumed to be determined by the type and strength of the dynamic FBGs created along the active fiber in the cavity as well as drawing potential usefulness of the SLLS effect for applications in photonics and sensing.

It is worth noting that, starting from 2017, i.e. just 6 years after the first report on the SLLS effect, there has arisen a trend to use it for applications. Since then, a dozen applications have been reported, e.g. the primary ones authored by the Russian group [33–36].

Say, in [33, 34], authors proposed and demonstrated an all-fiber scheme for precise optical spectrum analyzer based on stimulated Brillouin scattering (SBS) and the SLLS effect in a YDFL, as said, without external drivers and frequency-tunable elements. Certainly, it operated in the spectral range of ~1 μm in the single-frequency regime. Noticeably, its scheme was built on PM components. The resolution and measuring range of the analyzer were measured to be 23 MHz (defined by the Brillouin gain window's bandwidth) and 5 THz (defined by the limits of the sweeping range of SLLS), respectively.

Using the invented SLLS/SBS-based OSA, the authors tested two different type lasers, one being an YDFL and another based on YAG:Nd^{3+} bulk crystal, both operating at ~1064 nm; moreover, they compared the spectra obtained with the ones measured using 'standard' (grating-based) OSA. As was shown, OSAs of both types demonstrate very similar optical spectra; in the meantime, it deserves noting that the line shape measured with the SLLS/SBS-based OSA was more symmetrical (due to its narrower instrumental function), while its noise level exceeded that inherent in the standard OSA.

One more application of SLLS was proposed by the same group of authors in [35, 36, 59], where a coherent optical frequency-domain reflectometer (C-OFDR) is demonstrated. In this case, the authors dealt with a kind of YDFL with self-sweeping in a purely single-frequency regime of lasing. Again, see Fig. 11.20, it was realized in the simplest cavity's configuration comprising an FLM-based coupler and a 90°-cleaved fiber end; the spectral range covered at SLLS was ~20 nm. Interestingly, on one hand, each μsec-range pulse in the train inherent in SLLS developed on only one longitudinal mode (with linewidth <1 MHz), while, on the other hand, the pulse frequency linearly changed (decreased) with pulse number (Fig. 11.21). This provided a high linearity of frequency tuning, invaluable for making a C-OFDR without any additional spectral correction and active elements. The authors claimed that the C-OFDR they performed demonstrates attainment of spatial sampling of ~200 μm and reflectance sensitivity of down to −80 dB at a test line length of up to ~9 m.

In [37], authors reported an open-cavity YDFL studied from different points of view, including the one where distributed feedback gets provided by a dynamic grating written in the cavity in different ways, e.g. due to weak (~0.1%) reflection from an angle-cleaved fiber end. As mentioned in [37], a meter-long dynamic grating can be formed at these conditions, causing an initially single-frequency operation at SLLS while subsequently—via the hole-burning effects—self-switching from one longitudinal mode to another.

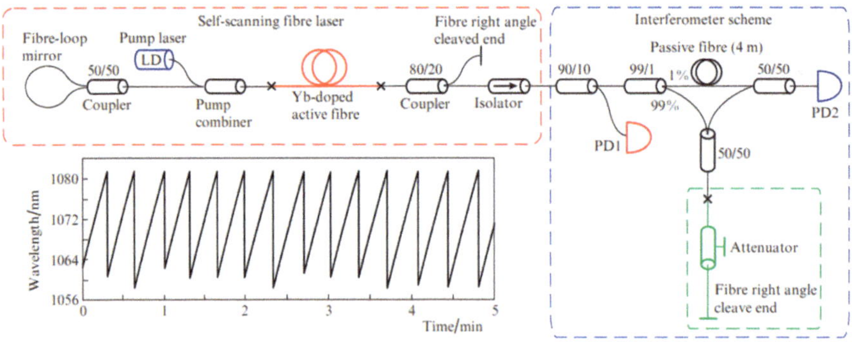

Fig. 11.20 Schematic of C-OFDR based on YDFL with SLLS. The typical laser spectral dynamics at SLLS is exemplified in the inset. [Adapted with permission from [36]; IOP ©2019]

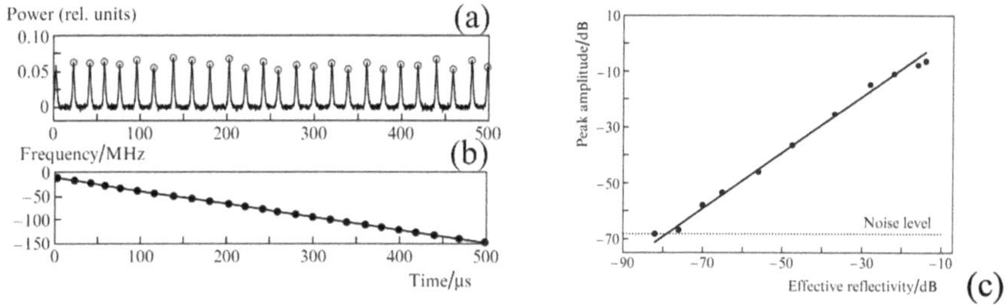

Fig. 11.21 (**a**, **b**) Snapshots of (**a**) pulse intensity and (**b**) optical frequency. (**c**) Dependence of peak amplitude vs. effective reflectivity of a unit under study. [Adapted with permission from [36]; IOP ©2019]

In the meantime, in 2018–2020, the same Russian group, deeply involved in exploration of the SLLS phenomenon, suggested a few methodologies towards stabilizing boundaries of a spectral range of its occurrence; the first results on the theme were published in [38–41]. The motivation for such studies stemmed from the fact of the matter that spectral positions of the SLLS regime's boundaries (i.e. the start and stop wavelengths between which self-sweeping exists) frequently—if not always—suffer from certain instabilities, presumably inevitable due to self-induced nature of the sweeping process. Certainly, finding ways to improve the basic sides of the SLLS phenomenon was ultimately demanded by the requirement of exploiting it for practical needs.

In this sense, one ought to refer to a cycle of research [38–42], where authors reported—for the first time—the ideas towards stabilization of the SLLS range. Say, in [38, 39], this was done by means of adding a narrowband FBG to the output mirror in the Michelson-interferometer configuration. The influence of FBG reflectivity and optical path difference in the interferometer upon the laser spectral dynamics was under study. The optimization allowed authors to get broadband (>16 nm) self-sweeping and diminish the fluctuations (down to a few pm) in its start and stop wavelengths by ~100 and ~ 15 times, respectively; some of the results reported in [38, 39] are presented in Fig. 11.22. Certainly, this ought to largely improve the SLLS-regime stability overall, which is valuable for its exploitation.

Then, in [40–42], the same group of authors undertook further steps towards stabilization of or control over the start and stop boundaries of SLLS in FLs using intracavity-placed FBG (or FBGs) or/and spectrally selective filters of other types. In all cases, they dealt with YDFLs. In fact, these works continued the primary research opened in [38, 39]. The idea behind the suggested scheme of stabilization (control) was the creating—in one or another way—the selective loss and/or reflection in the cavity of an FL with SLLS, which might help in the intention. Authors used in these experiments the FBGs as spectral filters inside the Michelson-interferometer cavity, being interested in how their reflection coefficients co-influence with those of spontaneous FBGs self-established at SLLS in the active fiber and result in control over the SLLS regime's range [40, 41]. An example revealing the suitability of the measures is demonstrated in Fig. 11.23. Also note

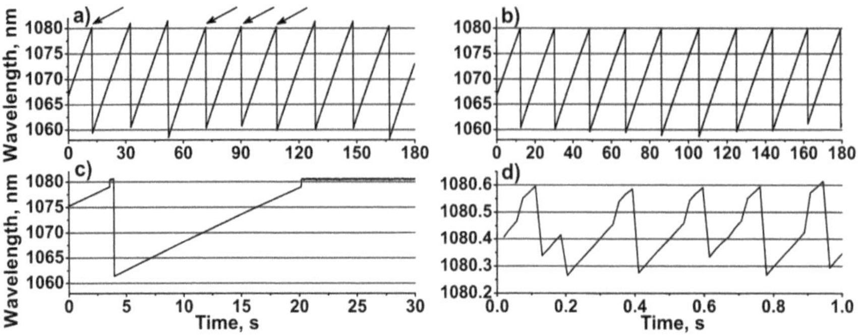

Fig. 11.22 Spectral dynamics at different FBG's reflectivity at ~1080 nm (it increases from (**a**) to (**d**)). Arrows in (**a**) correspond to the times of the stop wavelength's stabilization. [Adapted with permission from [39]; Optica (OSA) ©2018]

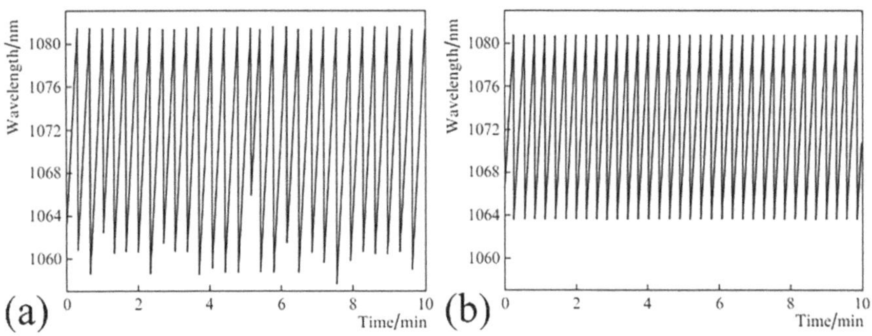

Fig. 11.23 Temporal dynamics of the radiation wavelength of a self-sweeping laser (**a**) without an FBG and (**b**) with two FBGs. [Adapted with permission from [40]; IOP ©2018]

that the effect of the temperature of the active FL upon the performance of the self-sweeping characteristics (in the presence of additional FBGs in the cavity) was under study, too, in [40].

Furthermore, in [42], the authors attempted another way towards control over and enlarging the spectral range covered at SLLS using a broadly tunable Lyot filter in the cavity of an FL with SLLS. The idea relied on the vision that an FBG that is intrinsically formed in active fiber at self-sweeping ought to be affected by the bandpass of the filter, in turn controlled by temperature biased to it; the laser scheme is shown in Fig. 11.24. Thus, the interplay of the two (of the Lyot filter and the spontaneously induced grating) were thought to positively affect upon broadening of the tuning range. As was demonstrated in [42], this measure led indeed to expanding the SLLS range from 20 to 42 nm (see Figs. 11.25 and 11.26) while keeping the linewidth of lasing in this regime below the level of 20 MHz.

Resuming, the results reported in [38–42] serve to conclude that the approaches used were very fruitful in the sense of enhancing predictability of the lasing

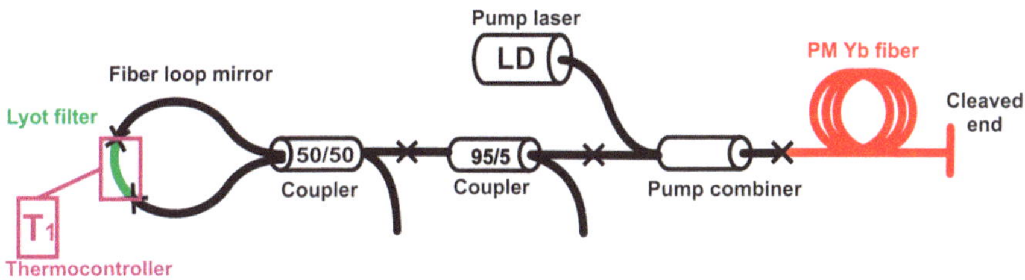

Fig. 11.24 The scheme of a tunable YDFL with SLLS using a Lyot filter. [Adapted with permission from [42]; Astro Ltd. ©2019]

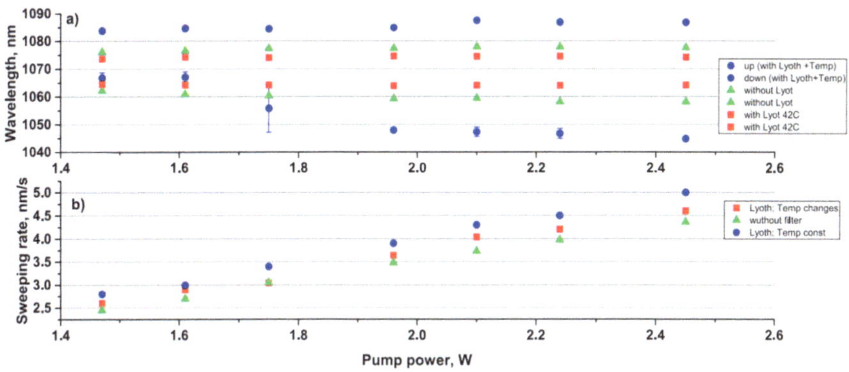

Fig. 11.25 Dependences of (**a**) SLLS spectral bounds and (**b**) sweeping rate vs. pump power for the laser without the Lyot filter (green triangles), with the filter at fixed temperature (42 °C) (red squares), and with the filter at tuning temperature (blue circles). [Adapted with permission from [42]; Astro Ltd. ©2019]

wavelength tuning in FLs with SLLS and enlarging the boundaries (and control over them) of the regime.

In the meantime, a line of research devoted to searching for and exploring the SLLS phenomenon in one more type of FLs, viz. a 2-µm TDFL, was published in 2018–2019. The first achievements in the area were done, nearly simultaneously, by the two research groups: one from Japan [43, 44] and the other—from Russia [45–50].

In [43, 44], the Japanese group presented—seemingly for the first time—a TDFL with SLLS. In its primary [43] and following [44] versions, the TDFLs in SLLS mode were turned to oscillate at central wavelengths of ~1.95 and 1.97 µm, respectively, with maximal sweeping ranges being ~12 to ~15 nm. The sweeping rate was measured to be around 0.2 nm/s. In both cases, the TDFLs were pumped at 1.57 µm (i.e. directly into the absorption band of Tm^{3+} ions) by a tunable semiconductor laser through an EDF amplifier; the laser schemes [43, 44] were assembled in bi-directional ring cavities, containing, apart from the active TDF, a polarization controller, pump-to-signal WDM, and output coupler. It was noted by authors that SLLS operation was, as in other FLs, inherently associated with self-pulsing

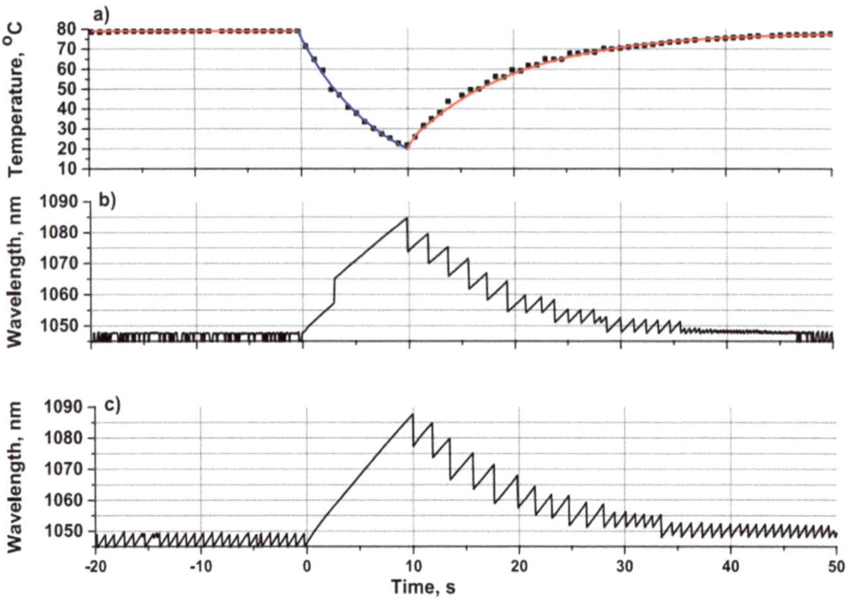

Fig. 11.26 Temporal snapshots of (**a**) temperature and the laser's wavelength dynamics at the corresponding temperatures of the Lyot filter at pump powers of 1.6 (**b**) and 2.1 (**c**) W, respectively. [Adapted with permission from [42]; Astro Ltd. ©2019]

(namely, with generation of trains of quasi-periodic μsec pulses, each containing a virtually single longitudinal mode) and that the sweeping rate was controllable (10–15 nm) by changing the polarization state using PC. Interestingly, the sweeping direction in both its realizations was of the 'reverse' type, i.e. the laser wavelength changed smoothly in time (seconds) from a longer to a shorter wavelength with subsequent immediate bouncing to the start (longer) one. Authors also noted in [43, 44] that the regime of SLLS could be transformed to quasi-CW or CW at a 'static' wavelength by means of simple manipulations with intracavity PC.

In turn, the Russian team published—almost simultaneously with [43, 44]—a few reports on attaining the SLLS regime in TDFLs [45–50].

Say, in [45, 46], authors released a primary study on the basic appearances of SLLS in a TDFL at ~2 μm (the scheme of which was nearly the same in [45, 46]). Namely, a linearly polarized radiation in the SLLS mode around 1.92 μm with a spectral span of up to ~20…25 nm with a sweeping rate up to 10 nm/s from the TDFL at 1.54-μm pumping (an EDFL) was realized experimentally. This was done using a 'standard' cavity design comprising an FLP and right-angle cleaved fiber end and PM-fiber-based intracavity components. In the time domain, the laser generated a train of periodic μs-scale pulses, each containing highly coherent single-mode radiation with a linewidth of ~1 MHz. Interestingly, each subsequent pulse had a laser frequency that differed from the preceding one by one intermodal beating frequency (~8 MHz). The laser-line movement at SLLS occurred in a 'normal' regime, i.e. towards the Stokes side of the spectrum. The TDFL's average output power at SLLS exceeded 400 mW. It is also demonstrated (in [46]) a successful use of the developed TDFL with SLLS for measuring the absorption spectrum of water in air.

As a next step, the same group of authors reported in [47, 48] a 1.9-μm TDFL (with analogous to [45, 46] schematics) where the SLLS spectral dynamics could be 'manually' controlled by means of simple pump power adjustment. It was shown that the laser operates in three qualitatively different self-sweeping regimes, obtainable depending on pump rate. That is, the two 'opposite' sub-regimes—one with a 'normal' scanning direction (from shorter to longer wavelengths) with a high sweeping rate (~5 nm/s) and the other with a 'reverse' one (from longer to shorter wavelengths) at a much lower sweeping rate (~0.1 nm/s)—were the limited ones; the overall spectral range where these two kinds of laser-line self-sweeping was limited to 15 nm. It is curious that whereas at normal sweeping the laser emitted highly coherent light (almost in a single-frequency regime), at the reverse one it was essentially multi-frequency.

Furthermore, one more intermediate between the two sub-regimes was found for the TDFL, named in [47, 48] as 'wavelength stopping,' at which a lasing wavelength could be stopped at an arbitrary value (but depending upon a prehistory of the laser dynamics) within the range of 1.912–1.923 μm with a high tolerance of less than 50 pm (in fact, the spectral resolution of the OSA used), for a few (at least 5) min. The most meaningful sets of data enlightened by the features drawn above are resumed in Figs. 11.27 and 11.28. It is claimed by authors of [47, 48] that the spectacular result for SLLS in a TDFL can be a base for a row of practical applications.

Note that the second and third sub-regimes, framed by SLLS in a TDFL, viz. the 'slow' reverse laser-line sweeping and its stopping (see above), were explored in [49] in more detail with the use of a heterodyne technique. Say, in the latter case, the experiments revealed the wavelength stopping is nothing more than a series of normal-scans' SLLS drifts (covering 5–7 consecutive μs-pulses) without a drift of

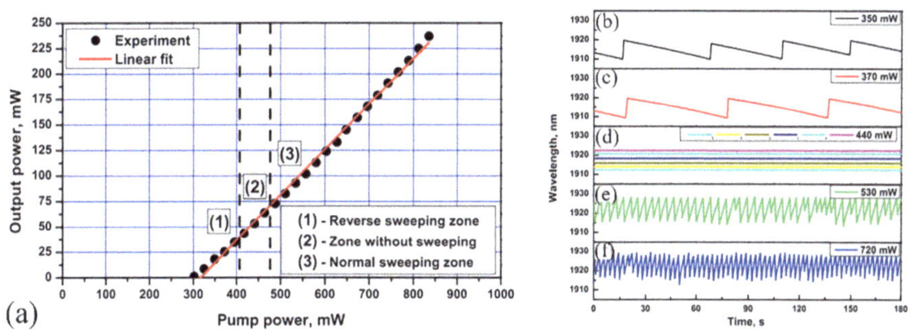

Fig. 11.27 (**a**) Output vs. pump power (the dashed vertical lines delimit the ranges where sub-regimes of normal, reverse, and no sweepings are observed); (**b–f**) Typical laser-line dynamics obtained for different pump levels and corresponding to (**b**, **c**) reverse sweeping, (**d**) wavelength stopping, and (**e**, **f**) normal sweeping, respectively. [Adapted with permission from [48]; Astro Ltd. ©2019]

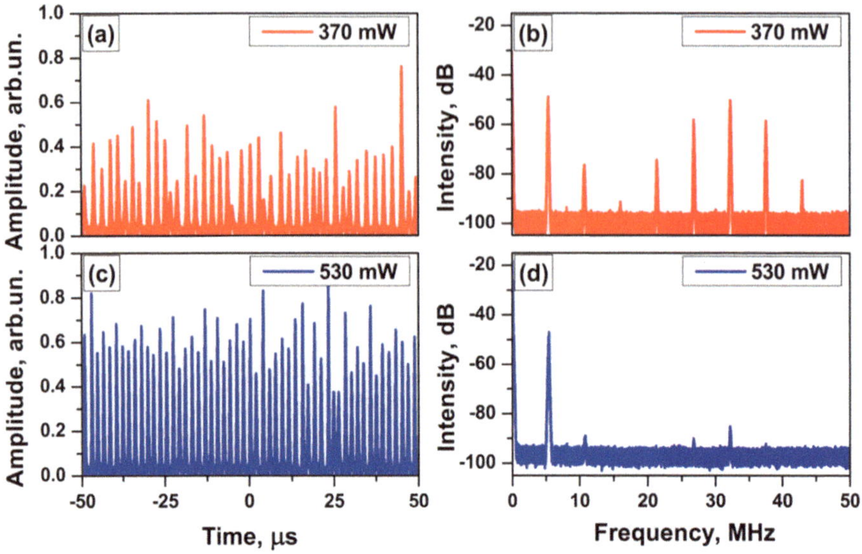

Fig. 11.28 (**a**, **b**) Intensity and (**c**, **d**) RF spectrum for (**a**, **c**) reverse and (**b**, **d**) normal sweeping directions. [Adapted with permission from [48]; Astro Ltd. ©2019]

central wavelength. In other words, the wavelength stopping is fictive (being in fact an oviposition of such unresolved snapshots).

This pattern gets clearer from Fig. 11.29. Note that, additionally, authors paid more attention in [49] to the influence of absorption and gain as well as to dynamic FBGs in the TDFL upon the effect of wavelength stopping. Some further detailed information about the laser-line sweeping in TDFLs, e.g. the effects of TDF length and pump wavelength, is considered in [50], aiming optimization of output sweeping characteristics and spectral dynamics in such kinds of FLs.

One more type of FL, viz. neodymium-doped FL (NDFL), became a subject of research towards the same aim, i.e. getting SLLS. A unique positive result in this sense was reported in 2019 [51, 52]; it is curious that from that time till now no data on SLLS using NDFLs has been published. Seemingly, this is caused by the fact that an NDFL represents a four-level laser system where—in contrast to a three-level one—this phenomenon is hard to attain, most probably because of inhomogeneous gain saturation inherent in the former laser system.

Pumped by a couple of multimode LDs at 805 nm (with a total of 6 W launched), the NDFL [51, 52] was built using the cavity comprising a highly reflective FLM from one side and a normally cleaved fiber end from another side; furthermore, it was added with a couple of fibered Lyot filters (temperature-controllable) and a 10-nm bandpass filter.

It deserves to be noticed, first, that the laser scheme as a whole was built on PM fibers, facilitating access to a highly coherent regime of generation, and, second, that the set of narrowband filters used allowed to reduce the spectral bandwidth of

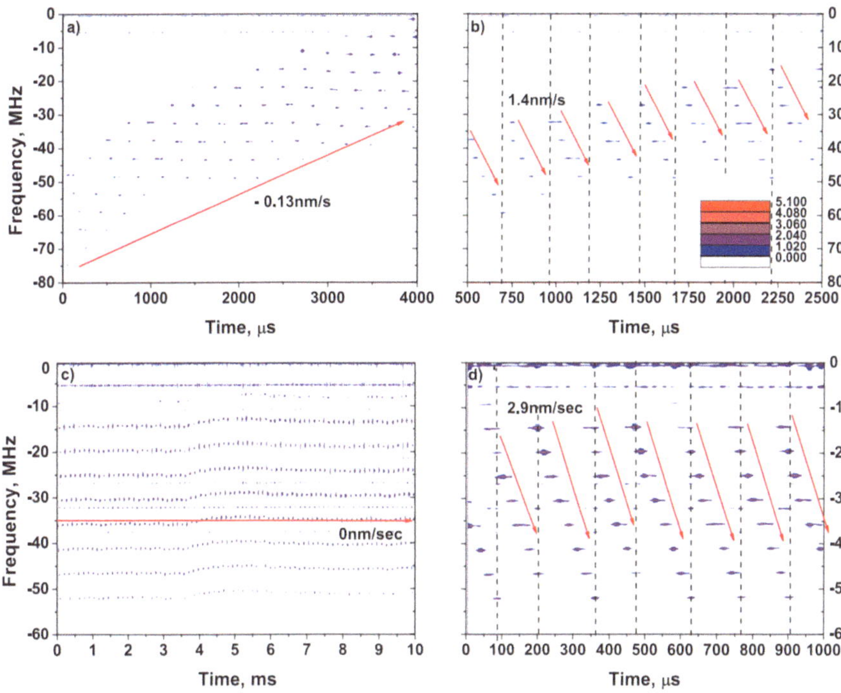

Fig. 11.29 Spectral dynamics for (**a**, **b**) reverse sweeping and (**c**, **d**) wavelength stopping sub-regimes at pump powers of 325 and 440 mW, respectively; (**b**, **d**) are zooms of (**a**, **c**). Arrows and dashed vertical lines designate sweeping directions and borders between them, respectively. [Adapted with permission from [49]; Astro Ltd. ©2019]

the multicomponent effective reflector to be restricted to ~2.2 nm near 1.06 μm. As a result, the NDFL was demonstrated to be turned into an SLLS operation, stable at pump powers varying from 1.5 to 4.5 W and characterized by a rather narrow spectral domain of self-sweeping; see its snapshot for pump power of 3 W in Fig. 11.30. As it sometimes happens at attaining the SLLS operation, the self-sweeping appeared in the NDFL in the form of periodic μs pulses, each bearing a single longitudinal mode (linewidth of ~1 MHz) while spectrally shifted from the adjacent pulses by a single intermodal interval (~7.5 MHz, corresponding to the cavity length used). Note that the latter was confirmed in [51, 52] using a heterodyne method where the output of the NDFL with SLLS was probed by an outside single-frequency probe radiation. The SLLS range was restricted to ~1.8 nm around 1.06 μm, and the sweeping rate was revealed to grow (up to 10 nm/s) with increasing pump level; the output power at SLLS exceeded 600 mW. Authors also claimed that such an NDFL may be turned to SLLS-mode at a shorter (~0.93 μm) wavelength instead of the 'standard' one (~1.06 μm).

A Chinese group of authors reported in 2019–2020 a study [53, 54] where SLLS of reverse type (viz. sweeping from longer to shorter wavelengths) was observed in a bidirectional PM YDFL (it was the first observation of reverse self-sweeping in

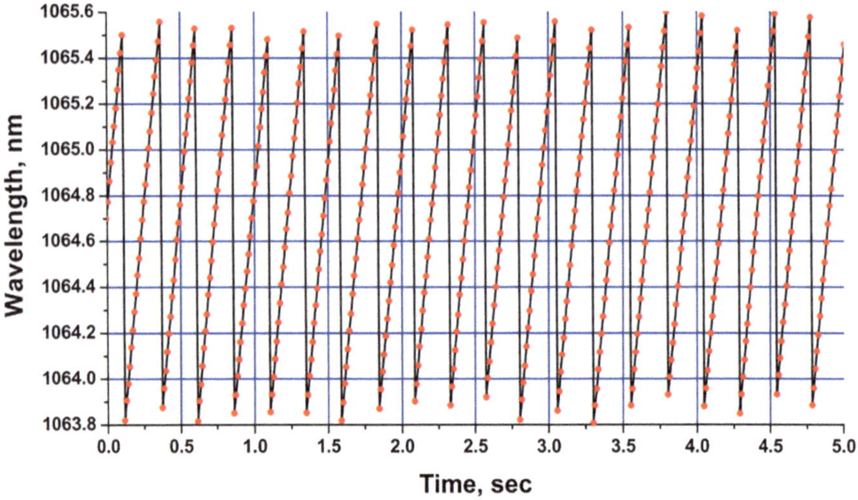

Fig. 11.30 Wavelength dynamics at SLLS in NDFL, obtained at pump power of 3 W. [Adapted with permission from [52]; Optica (OSA) ©2019]

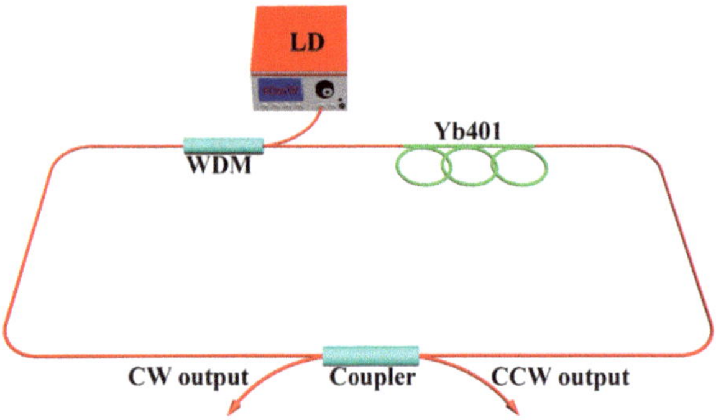

Fig. 11.31 Experimental setup of bidirectional YDFL with SLLS. [Adapted with permission from [54]; Optica (OSA) ©2020]

such type of FL). The laser's scheme was nearly the same as in [43, 44]; see Fig. 11.31.

In this work, the YDFL was forced to operate at SLLS at ~1.037 μm with the maximal range being around 3 nm and the sweeping rate up to 0.38 nm/s; the regime in the time domain presented a kind of self-pulsing in the form of μs-pulses with intermodal beating at 116–128 kHz (see Figs. 11.32 and 11.33). Another novelty reported in [53, 54] was that decoupling of output radiation in the bidirectional design of the YDFL was realized in two versions using couplers with ratios 50/50 and 10/90; interestingly, in the latter case, a kind of SLLS with ultra-slow (with a period measured by hours) variation of laser wavelength was registered.

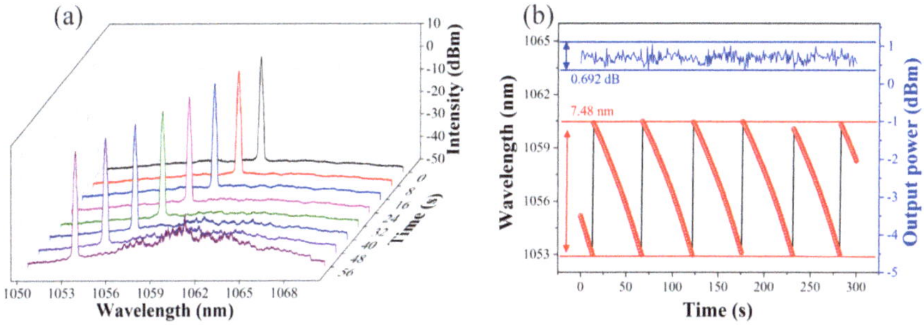

Fig. 11.32 Spectral performance of YDFL with SLLS: (**a**) temporal snapshots of lasing spectra; (**b**) an example of spectral appearance of SLLS and output power variation at pump power of 65 mW. [Adapted with permission from [54]; Optica (OSA) ©2020]

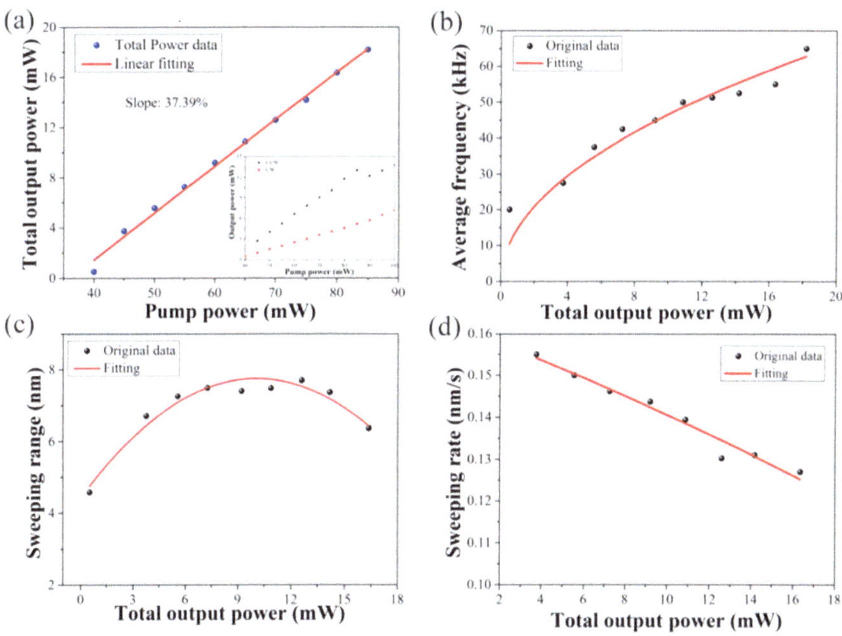

Fig. 11.33 Performance of YDFL using 50/50 output coupler: (**a**) output laser power vs. pump power; (**b**) pulse repetition rate, (**c**) SLLS range, and (**d**) sweeping rate vs. total output power. [Adapted with permission from [54]; Optica (OSA) ©2020]

11.2.4 2021–2024

In 2020, when worldwide scientific activities were largely suspended because of the COVID-19 pandemia, the studies towards further exploration of the SLLS phenomenon were slowing down, too. In the meantime, a couple of research ought to be mentioned where authors drew the ways to its practical applications near 2 μm, one dealing with the use of a 1.91-μm TDFL with SLLS as a base of a NIR spectrometer for analyzing the water nuclear spin isomer abundance [55] and another utilizing a 2.06-μm HDFL as a spectroscopic tool for studying the absorptive features of carbon dioxide (its 12 and 13 isotopes) [56]. In the first case [55], the characteristic NIR lines of water, corresponding to its vibrational $\nu2 + \nu3$ band, were the target, with the result being an accurate estimate for the ortho/para ratio of water nuclear spin isomers (3.06 ± 0.02), whereas in the second one [56], the two main isotopes of CO_2 were claimed to be a possible target. Say, in [56], authors presented an improved version of an HDFL pumped by a specially designed ~5-W 1.125 μm YDFL; the reverse sweeping's range in the HDFL was better performed in this case (it was more stable in time and enlarged up to ~4 nm) due to the use of a specially oriented PC. Such a laser source as claimed can be an effective tool for sensing the isotopic structure of CO_2 molecules.

A more fundamental and interesting study was reported in 2020 in [57] where under study was a kind of SLLS regime in EDFL based on PM fibers and a cavity built using an FLM and a right-cut fiber end; a Mach–Zehnder interferometer for homodyne measurements was added at the laser output. Under certain conditions, a 'dual mode' CW version of reverse SLLS was uncovered in the laser, turned to operate at 1.605 μm with a sweeping range of 2.8 nm, at 1.55-μm pumping; the reached sweeping range was significantly wider than the maximal one, ~0.5 nm, assessed earlier for EDFLs with SLLS in [8, 31]). However, sweeping range largely fluctuated in time. It was recognized by authors that partially this is explained by the simultaneous generation of two adjacent longitudinal modes of equal intensity, one of which (at lower frequency) at some moment faded while the other (at higher frequency) grew. Thus, the regime obtained in [57] is a kind of 'intermitted quasi-CW' operation because of the SLLS phenomenon.

Meanwhile, in [60], the Russian group reported the first demonstration of SLLS in an EDFL with a ring cavity and entirely based on PM fibers. In this case, a double-clad ytterbium-erbium co-doped fiber (YEDF) was used as an active medium (mostly as an amplifier); it was pumped by a 976-nm, ~10-W LD into the Yb^{3+} ion band, from which excitation was transferred to Er^{3+} ions. Furthermore, the authors used un-pumped EDF and a 1.6-μm FBG (85%) in a linear bench of their setup, partially coupled with the ring cavity. Namely, in this part the radiation in the form of SLLS lasing was formed due to the combined effect of reflection from the FBG and the SHB effect, which was amplified inside the YEDF in the ring cavity. The authors carefully experimented with different parameters of their laser scheme, and one of the regimes found was a kind of SLLS mode. It was concluded that, in the presented scheme, self-sweeping of reverse type was observed, ranged by ~0.15 nm at ~1.6 μm (defined by the reflection peak of the FBG) and characterized by an extremely slow movement of the laser line (of ~5 pm/s: it is an order of value slower than in other EDFLs with SLLS). As in earlier realized linear-cavity EDFLs, SLLS operation is gone in a dual-longitudinal-mode scenario in the form of very long pulses (of ~10 ms in duration), sequenced by a single intermodal interval. It was also demonstrated that pulses at SLLS were elongating (for up to 500 ms) and sweeping themselves decelerating at increasing pump power. The authors claimed that an FL of this kind can be a good choice for interrogator applications.

The Russian group continued in 2021 the research on SLLS in YDFLs, too [61, 62]. Again, as in [60], a ring cavity of similar design was under study (see the upper part of Fig. 11.34a). However, in this case, the YDF placed in the ring acted as an active medium, while the YDF in the linear bench (terminated by an FBG) played the role of SA. The laser was pumped in-core of the active YDF by a fibered 975-nm LD. With this scheme, authors accessed a curious (as they termed it themselves) kind of SLLS (ranged by ~100 pm) to be attributed as quasi-CW operation, where a train of single frequency, very long, almost rectangular in shape pulses was generated, but in some moments (namely, those when the frequency of operation was stepwise changed) intermitted by regular bursts of intensity; see Fig. 11.34b, c and 11.35. Each pulse in the train was spaced from the adjacent ones by one intermodal space (of ~10 MHz) whereas pulses, ~1 ms in duration, were oscillated at a single longitudinal mode with a linewidth of around 0.7 MHz). All these characteristics were measured using a special heterodyne high-resolution scheme developed on the base of a highly coherent Nd:YAG laser, the output of which was a probe for the output of the YDFL laser with SLLS (see the lower part of Fig. 11.34a).

One more feature standing behind such uncommon oscillation was, as the authors discussed, a mixture of running and standing waves in different sections of the cavity. It was also noticed that an 'SA-mirror' (refer to the bench on the right side of the laser scheme in Fig. 11.34a) can be used for control over the self-sweeping directionality (see the examples shown in Fig. 11.35a, b) and the value of the laser frequency jumps.

A set of new results on SLLS at ~2 μm was presented in 2021 in [63–65], where a stable self-sweeping lasing within a spectral range as large as ~10 nm around 2.1 μm in an HDFL (maximal value obtained so far) was reported; the released radiation was rather narrowband and linearly polarized, at output powers of up to 200 mW. The clue novelties in the experimental technique utilized to get such impressive results were, first, employing essentially PM fiber components in the cavity; second, optimization of the HDF length and intracavity losses; and third, water cooling applied to the active HDF (refer to Fig. 11.36a). Meanwhile, it is

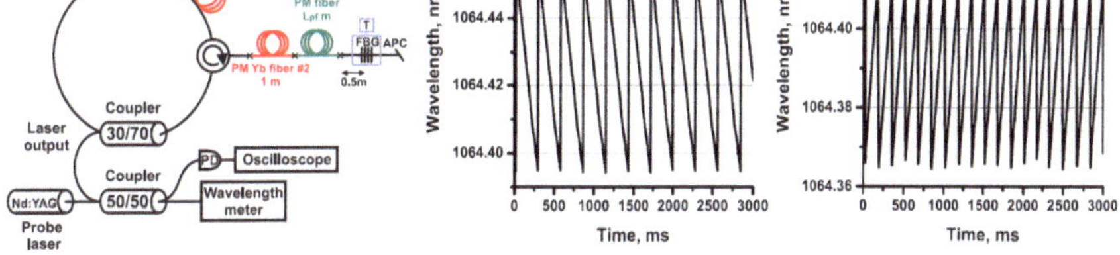

Fig. 11.34 (**a**) Setup of YDFL with ring cavity; (**b**, **c**) two examples showcasing the opposite directionalities of the SLLS process, accessible in the laser by means of manipulations with its parameters. [Adapted with permission from [62]; Astro Ltd. ©2021]

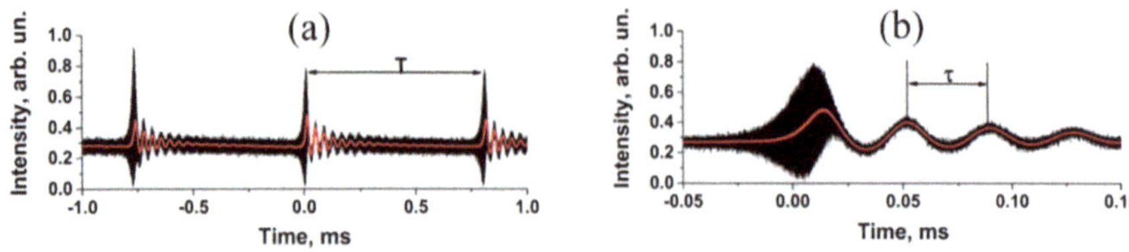

Fig. 11.35 (**a**, **b**) Examples of the laser's intensity dynamics at a pump power of 120 mW and a length of the bench containing the 'SA-mirror' of 2 m; the presented graphs are made at different scales. [Adapted with permission from [62]; Astro Ltd. ©2021]

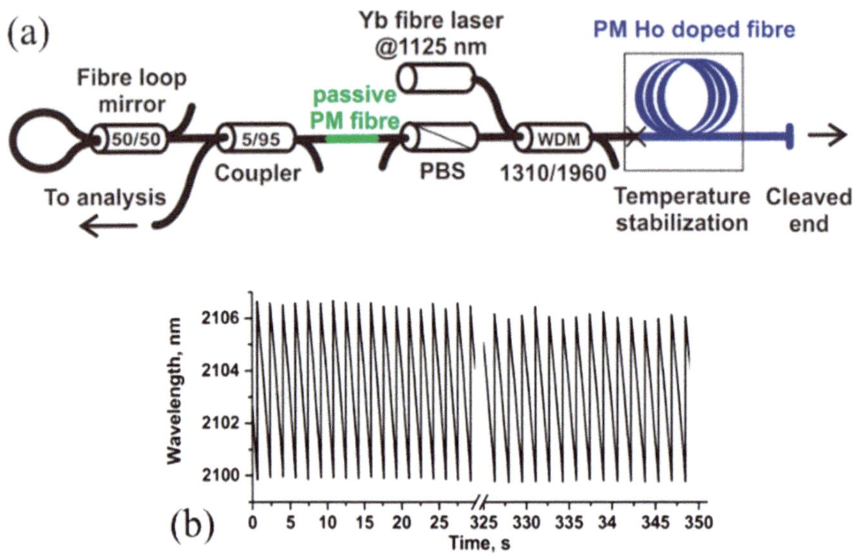

Fig. 11.36 (**a**) Schematic of PM HDFL with SLLS; (**b**) long-term spectral dynamics of self-sweeping at output power of 180 mW. [Adapted with permission from [65]; Astro Ltd. ©2021]

necessary to note that the HDF used in these studies was heavily doped, which, according to the discussion, might be a reason for getting such a broadly swept laser line. All the above measures helped authors to obtain a more robust operation of the HDFL with SLLS as compared to the earlier approaches (see [25–27]); the snapshots of the SLLS regime attained using this system are shown in Fig. 11.36b.

Furthermore, authors revealed—seemingly for the first time for a 2-μm FL with self-sweeping—regular intermittence of quasi-CW operation by self-pulsing, occurring at a certain moment during each self-sweeping period. The dependences of sweeping range and rate at two wavelengths and frequency of relaxation oscillations vs. output power are shown in Fig. 11.37.

As was mentioned in [65], the developed 2-μm self-swept source based on HDFL can be applied for spectroscopy of nitrous oxide, the absorption lines of which lie in the spectral range covered by the SLLS regime.

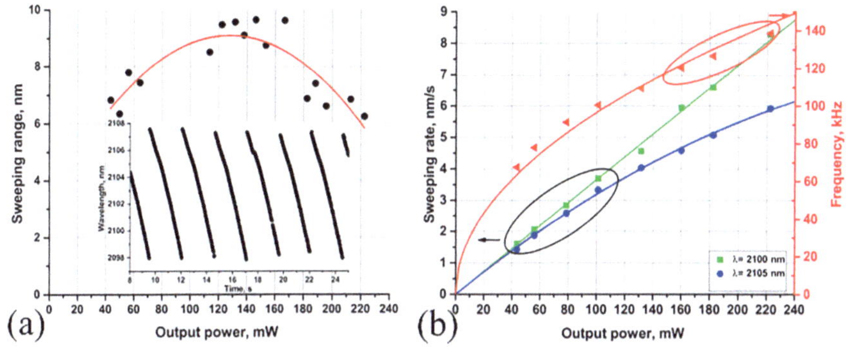

Fig. 11.37 (**a**) Dependence of sweeping range vs. laser output power (inset shows the laser-line dynamics at output power of 150 mW). (**b**) Dependencies of sweeping rate at two wavelengths and frequency of relaxation oscillations vs. output power. The blue and green lines show the fits of the sweeping rate, specified for wavelengths 2.100 and 2.105 μm, respectively, while the red one shows the fit for the relaxation oscillations frequency. [Adapted with permission from [65]; Astro Ltd. ©2021]

Also note that in [66] the Russian group reported on self-sweeping in a TDFL under, as said, an automatic control over the tuning parameters (sweeping rate and wavelength). The approach consisted of measuring the laser wavelength and fast data processing for revealing the sweeping directionality and rate, with posterior adjustment of the pump power using a feedback loop. It was claimed that the suggested technique will offer a possibility to, e.g. execute more delicate measurements in gas absorption spectroscopy.

In 2021–2022, a Chinese group reported on novel features regarding the SLLS phenomenon [67–69]. In these works, authors dealt with YDFLs oscillating in SLLS mode where their clue characteristics arisen due to special assembling of laser cavities embracing the active fiber; note that in all cases under study, the experiments were performed at direct core-pumping YDF (using fiberized low-power 976-nm LDs), while the whole of fibers forming the cavities were of PM type. Whereas in [67] the ring cavity was added through a circulator by a piece of un-pumped YDF, and a loop with a polarization controller inside served as a kind of saturable absorber (SA), in [68] a passive fiber inserting controllable bending loss was added (see Fig. 11.38).

As stems from the results reported in [67], see Fig. 11.39, directionality of SLLS was normal, that is, the lasing wavelength smoothly moved from shorter to longer wavelengths; the sweeping rate and range (of up to 5 nm around 1068 nm) altered against pump power by a common for SLLS way; the regime was observed in the YDFL at output power of up to 5.5 mW. In the meantime, a novelty provided by the presence of the fibered SA was pronounced changes in the pulsing dynamics at SLLS, defined—as authors claimed—by 'interplay' between the FBG spontaneously drawn in the cavity and the bleaching effect in the SA.

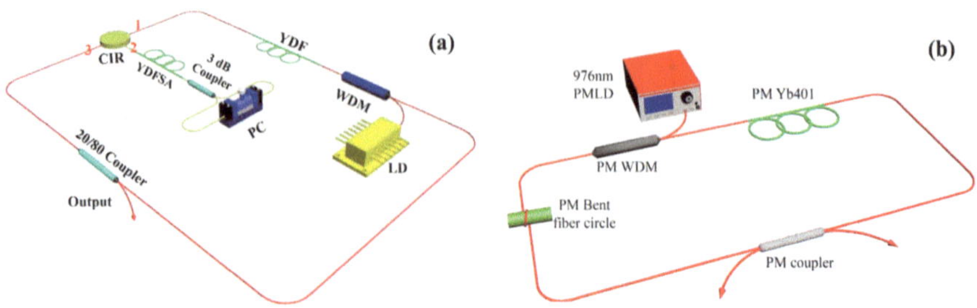

Fig. 11.38 Experimental setups of YDFLs with SLLS, studied in (**a**) [67] and (**b**) [68]. [Adapted with permission from [67, 68], The Japan Society of Applied Physics ©2021 & Chinese Optics Letters ©2021]

In turn, in [68], authors focused attention on what happens to the SLLS regime when changing the cavity loss via the bending effect in a section of intracavity fiber. It was established that at the descending diameter of the bent fiber, the sweeping range moved to the anti-Stokes part of the spectrum; see Figs. 11.40 and 11.41. Moreover, both the start wavelength of SLLS and the laser threshold showed an exponential correlation of change vs. bending diameter.

One more study undertaken by the Chinese group was focused on a possibility to rule over the basic parameters of self-sweeping by means of employing a length of space-path to achieve intracavity ranging [69]. In that case, authors got single-frequency SLLS with a few-longitudinal-modes range, as claimed. They used in experiments a setup like the one shown in Fig. 11.38, while adding, through a circulator, by a collimator and a bulk mirror (the distance between the two units was variable). Based on this YDFL design, a regime of a few (seven) longitudinal modes at self-sweeping was obtained. By analyzing the RF spectra of lasing, it was demonstrated that information about the collimator-to-mirror distance can be found, which seems to be promising for laser ranging.

In the meantime, in [72], the Russian group examined an experimental setup (Fig. 11.42a) recalling that [67], where—instead of active YDF and SA composed of passive YDF and FBG [67]—authors employed to fulfill analogous demands a YEDF as the active medium and a set comprising EDF and 99% FBG (for 1.56 μm), respectively. Based on this design, they conducted comprehensive research aimed at a detailed examination of the modal properties of the regimes, supported by this design of FL. Note that the laser scheme [72] was built on PM fibers.

In this case, double-clad YEDF was pumped, through a combiner, by a 976-nm LD; so, the excitation stored in the band of Yb^{3+} subsystem was transferred to Er^{3+} ions, providing effective gain at ~1.56 μm of the latter subsystem. The SA was composed of un-pumped EDF piece, and the FBG was spliced with the ring cavity through a circulator.

In experiments, the main attention was paid to changes in laser dynamics at varying both pump power and length of the EDF-SA. One of the interesting results obtained in [72] is shown in Fig. 11.42b, c, i.e. the bistable character of oscillation above the threshold. Among the other results reported in [72], notice a transition

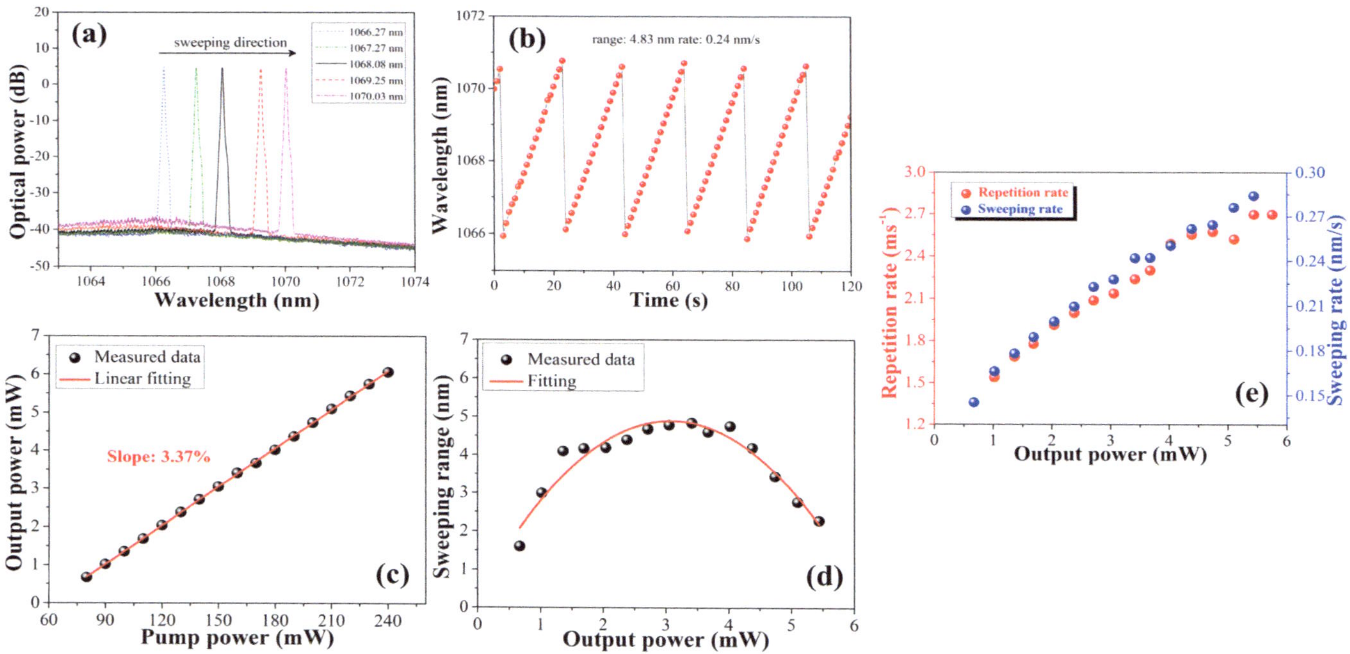

Fig. 11.39 (**a**) The spectral pattern of YDFL with SLLS at a pump power of 180 mW. (**b**) Wavelength dynamics. (**c**) The slope efficiency of the self-sweeping fiber laser. (**d**) The sweeping range changes with the increasing output power. (**e**) The change of repetition rate and sweeping rate as the increased output power. [Adapted with permission from [67]; The Japan Society of Applied Physics ©2021]

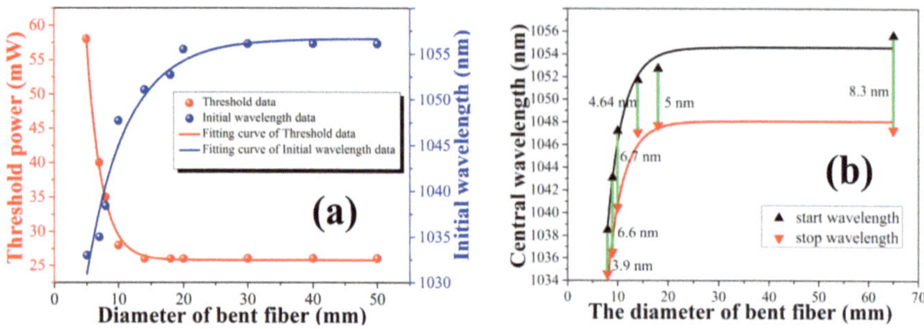

Fig. 11.40 Relationships between (**a**) diameter of the bent fiber and both threshold power and initial wavelength, (**b**) start and stop wavelengths vs. diameter of the bent fiber. [Adapted with permission from [68]; Chinese Optics Letters ©2021]

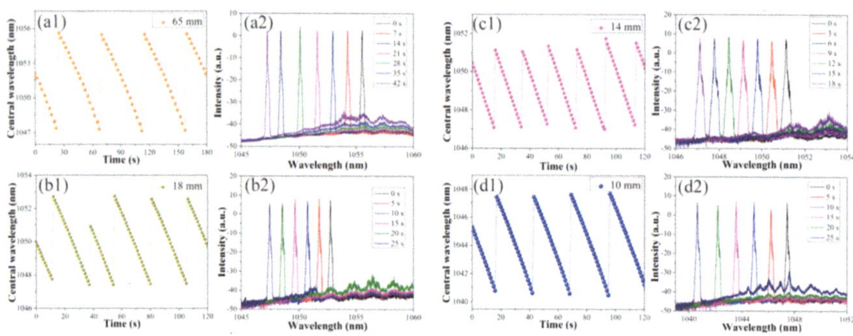

Fig. 11.41 Spectral dynamics of SLLS, recorded for different bending diameters: (**a1**) and (**a2**) 65 mm; (**b1**) and (**b2**) 18 mm; (**c1**) and (**c2**) 14 mm; (**d1**) and (**d2**) 10 mm. [Adapted with permission from [68]; Chinese Optics Letters ©2021]

from operation at a single longitudinal mode to regular mode-hopping. It was, e.g. shown that the laser could be switched—by means of increasing pump power—from the sub-regime of wavelength stabilization (supported for up to tens of seconds; Fig. 11.43a) through the one of stronger mode-hopping (Fig. 11.43b) to the sub-regime of SLLS (within the range of ~20 pm around 1.56 μm; Fig. 11.43c, d), terminated by quasi-CW operation at further increasing pump power. Thus, as seen from Fig. 11.43, the EDFL demonstrates a diversity of operational 'modes' through simple control over pump power. These phenomena were discussed in detail and eventually explained by an interplay of amplitude and phase dynamic gratings in the cavity.

Note that the results [72] referred to above were successfully applied for generating tunable THz-range radiation in [73]; this was done by means of combining the output of the EDFL with SLLS and CW radiation delivered from a DFB laser with a fixed wavelength. Moreover, authors discussed in [73] the measures for building a spectrometer with a < 50-MHz spectral resolution and a tuning range of ~5 GHz around the central frequency of 100 GHz.

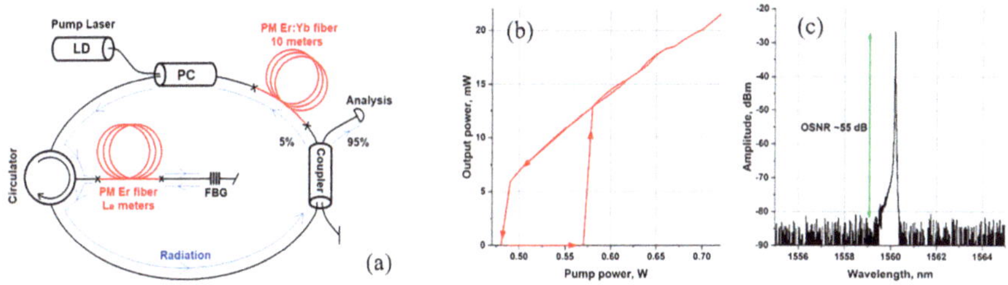

Fig. 11.42 (**a**) Schematic of ring EDFL with fibered SA; (**b**) dependence of output power on pump power for EDF length of 8 m; (**c**) optical spectrum at pump power of 0.67 W. [Adapted with permission from [72]; Elsevier Ltd. ©2022]

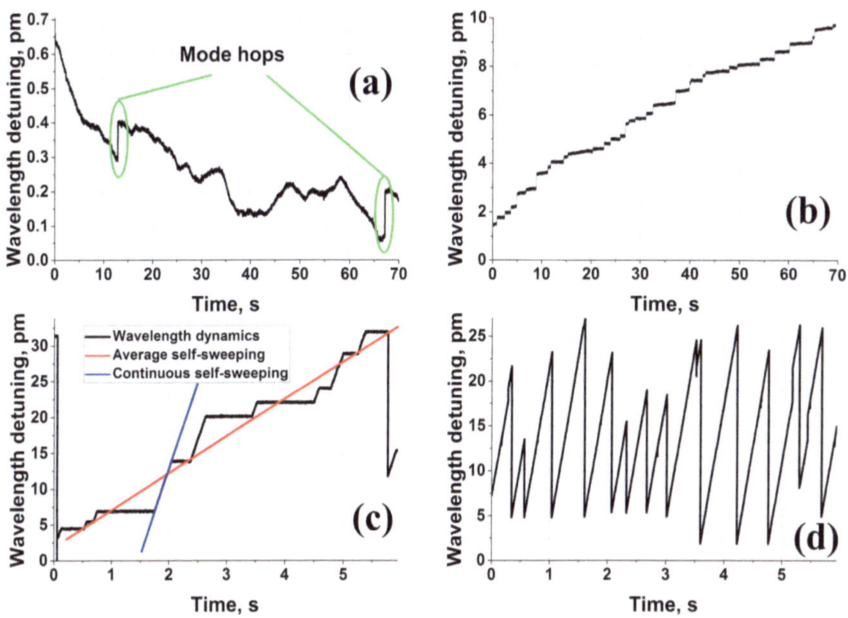

Fig. 11.43 Typical wavelength dynamics for certain laser configurations (L = 8 m0 and pump powers of (**a**) 0.51, (**b**) 0.6, (**c**) 0.65, and (**d**) 0.68 W (slopes of the red and blue lines in (**c**) correspond to average and CW self-sweeping rates, respectively). [Adapted with permission from [72]; Elsevier Ltd. ©2022]

The same group of authors presented in [74] a more detailed study of the modal pattern inherent in the EDFL with a ring cavity, accessed by means of homodyne (interferometric) and heterodyne measurements; see Fig. 11.44. The typical snapshots of spectral and intensity behavior of the laser at SLLS are demonstrated in Fig. 11.45.

As measured using the homodyne technique (see setup (A) in Fig. 11.45), the laser line at SLLS did not exceed 4 kHz (at each 'step' of nearly single-mode

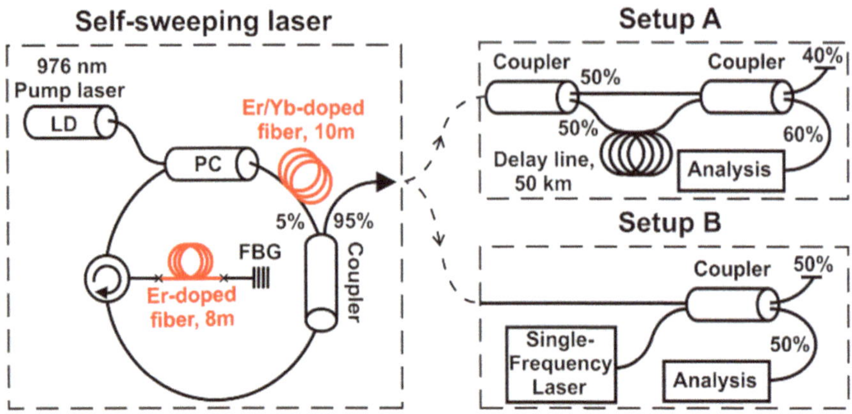

Fig. 11.44 Experimental schemes of: (**a**) Er, Er/Yb FL with SLLS and (**b**) interferometric (top (A)) and heterodyne (bottom (B)) measurements. [Adapted with permission from [74]; Astro Ltd. ©2022]

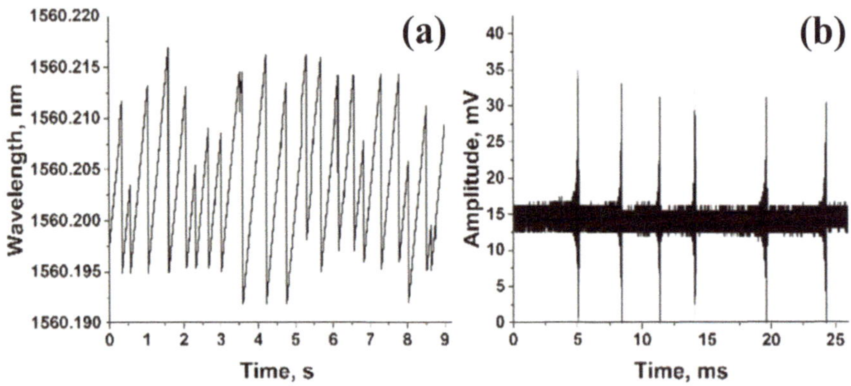

Fig. 11.45 Temporal behavior of the laser: (**a**) wavelength and (**b**) intensity dynamics. [Adapted with permission from [74]; Astro Ltd. ©2022]

operation), while the corresponding chirp was around 40 kHz. On the other hand, by applying the heterodyne measurements (see setup (B) in Fig. 11.45), the authors detected that the laser frequency changes in a stepwise manner by about 12 MHz (Fig. 11.46). By analyzing the latter results, they arrived at the conclusion that the laser output is composed of weakly overlapping packets of longitudinal modes, which are shifted by the frequency corresponding to two free spectral ranges of the cavity.

Furthermore, based on an analysis of the results obtained with the heterodyne technique, the authors inferred the presence of quite strong influence of parametric processes in the laser system under study.

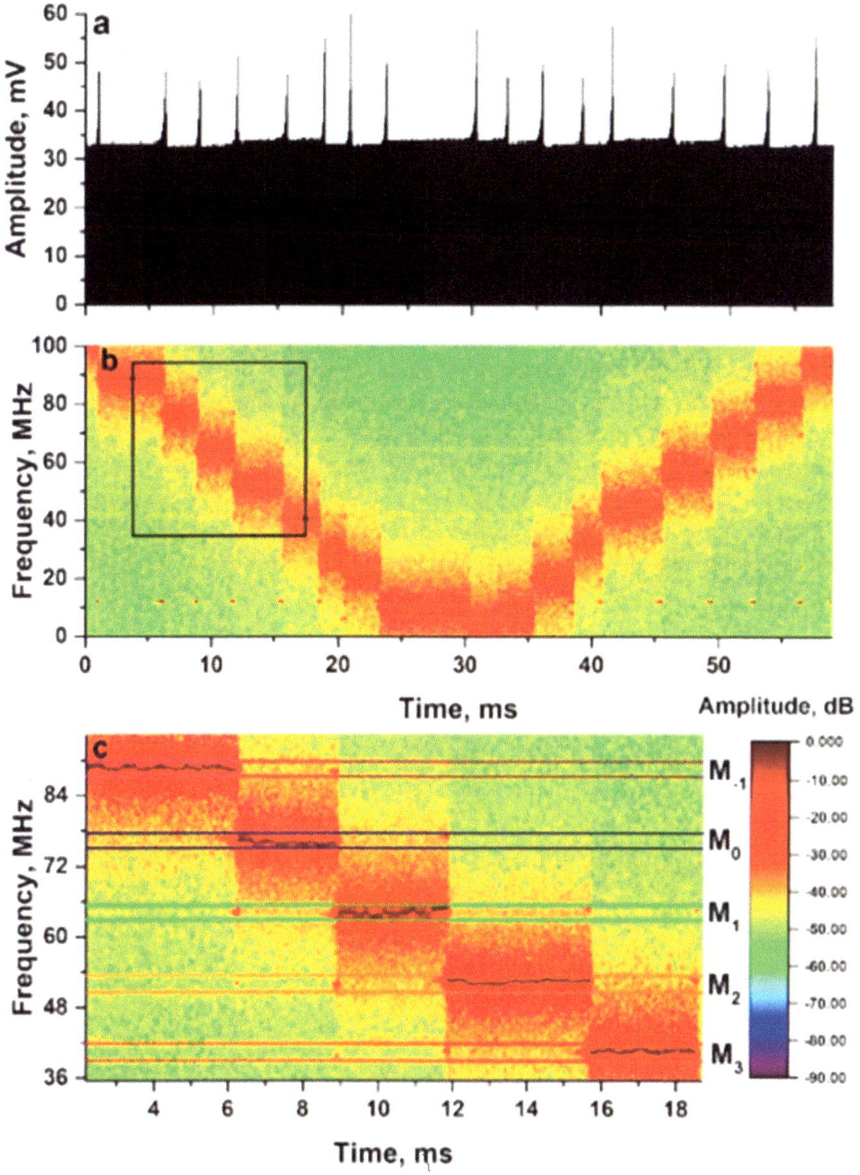

Fig. 11.46 Temporal dependences of (**a**) heterodyne signal, (**b**) its spectrogram, and (**c**) a zoom of the latter (the colored horizontal lines indicate the borders of the band-pass filter for corresponding longitudinal modes). [Adapted with permission from [74]; Astro Ltd. ©2022]

Meanwhile, the Chinese group reported in 2022–2024 a cycle of studies dedicated to the SLLS effect in various FLs; in fact, it was a continuation of the research undertaken in the earlier works [67–69].

Say, in [75], the authors constructed a single-frequency YDFL capable of oscillating in both normal and reverse self-sweeping sub-regimes, where the directionality of SLLS was realized by varying the length of SA placed in the cavity. Here, a kind of ring-cavity design was employed (earlier used in [67, 69, 72, 73]), where the main ring cavity was added through a circulator with a linear bench composed of a piece of passive YDF, a polarization controller, and an FLM reflector, playing the role of SA. The active 1.5-m YDF was in-core pumped by a 975-µm 600-mW LD; two lengths of the passive YDF (in SA) were examined: 1.5 and 2 m. The SLLS sub-regimes assessed were thoroughly documented from the viewpoints of optical spectra and pulse dynamics; particularly, the maximal ranges covered at these two kinds of self-sweeping were measured to be ~3.5 (normal SLLS) and ~ 5 nm (reverse SLLS), respectively. The featured properties of reverse self-sweeping obtained in [75] are resumed in Fig. 11.47. Likewise, in some preceding works, dealing with other FLs with SLLS, it was revealed in [75] that the interval of frequency change during SLLS is twice the spacing of longitudinal modes supported

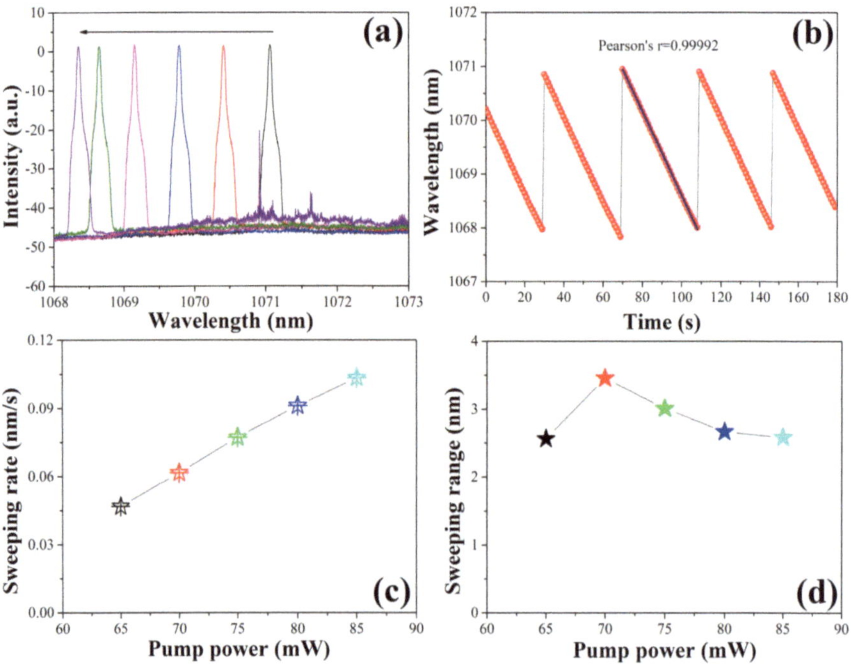

Fig. 11.47 Spectral patterns of the SLLS sub-regimes with reverse directionality of self-sweeping, obtained at the use of 2- and 1.5-m passive YDF SA, respectively: (**a**) static spectra, (**b**) dynamical spectra, (**c**) self-sweeping rate, and (**d**) self-sweeping range. [Adapted with permission from [75]; IEEE ©2023]

in the cavity. Modeling [75], too, allowed authors to explain the main trends uncovered experimentally.

In [76, 77], the same group of authors presented an ultranarrow-bandwidth ring-cavity YDFL with self-sweeping, the cavity design of which was featured, first, by the use of a bench (spliced with the main ring through a circulator) that contains a reflector being a set comprising a collimator and bulk mirror and, second, by the presence—as in the preceding works of this group—of a 2-m fibered SA; see Fig. 11.48.

The SLLS regime in this case was realized at the central wavelength of ~1065.382 nm and ranged by 2 pm only; as claimed by the authors, such characteristics can find applications in laser ranging and tunable microwave signal generation (note that in the subsequent paper [78], a ranging tool based on the narrowband SLLS was practically implemented). Furthermore, in [76, 77], it was found—both experimentally and through modeling—that the RF spectrum at this kind of lasing contains both the contributions from relaxation oscillations and beating frequencies (see Fig. 11.49).

Another group from China, also involved in studies of the SLLS phenomenon, reported a study [79] where they detected three qualitatively different sub-regimes of self-sweeping in a single-mode bidirectional ring-cavity YDFL, one-others' replacing at increasing pump power. The active 1.3-m YDF was pumped in-core by a 100-mW 976-nm LD; output radiation of the YDFL was monitored at two ends of a coupler; see Fig. 11.50.

The three sub-regimes observed were, correspondingly, the normal self-sweeping (from shorter to longer wavelengths), the wavelength's 'stationary state,' and the reverse self-sweeping (from longer to shorter wavelengths); the characteristic snapshots of the dynamics of laser wavelength and output intensity in these sub-regimes are demonstrated in Figs. 11.51 and 11.52. {Note here that a similar SLLS pattern, comprising three distinct operational modes, but for a ~2-μm TDFL, was earlier observed and examined in detail in [47–49].}

As seen from the figures, the intensity dynamics of both the normal and reverse SLLS had a dual-longitudinal-mode quasi-CW character where there were

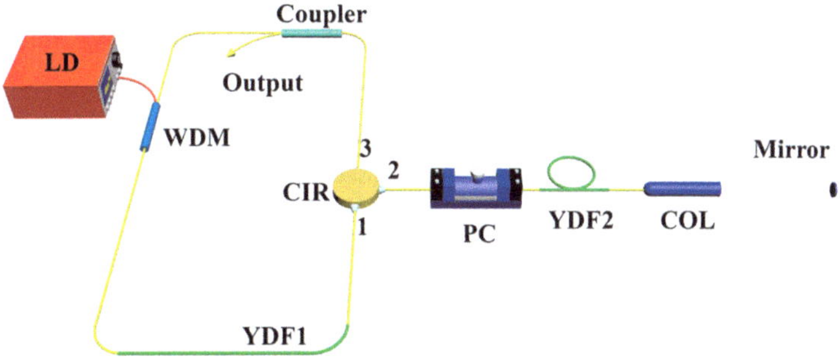

Fig. 11.48 (**a**) Schematic of ring-cavity YDFL with fibered SA. [Adapted with permission from [77]; IEEE ©2023]

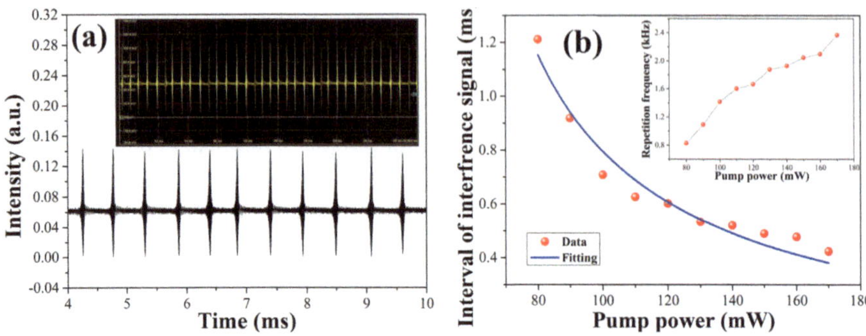

Fig. 11.49 (**a**) An example of a pulse train at SLLS in YDFL at a pump power of 130 mW (the inset shows the regime on a larger (30 ms) time scale); (**b**) dependence of intervals of signals at beat frequency vs. pump power (the inset shows the average pulse repetition against pump power). [Adapted with permission from [77]; IEEE ©2023]

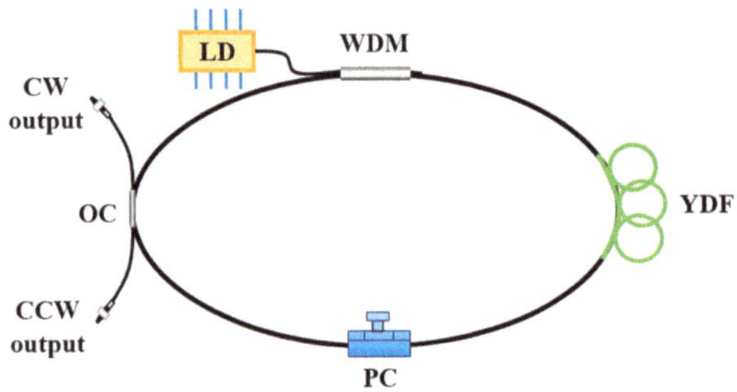

Fig. 11.50 The schematic of the single-mode ring-cavity YDFL capable of self-sweeping in three distinct sub-regimes. [Adapted with permission from [79]; Elsevier Ltd. ©2024]

alternating flat and bursts' phases; the former occurs in the range of 1069.5...1072.3 (at pump powers between 33 and 40 mW) while the latter—in the range of 1067.5...1071.5 nm (at pump powers between 45 and 95 mW). On the other hand, the wavelength's stationary state arises at any wavelength between 1067.5 and 1072.3 nm at pump powers varied from 40 to 45 mW (depending on the historical process of its approaching); this sub-regime had intensity dynamics, where intermitted chaotic pulses and periodic bursts.

One more aspect deserving attention in study [79] is that the authors observed pronounceable inequality in laser powers, measured at the two opposite YDFL's outputs, a circumstance not yet fully clear whilst resembling the one encountered in [80] for a linear-cavity YDFL with SLLS (but see below).

Let us now stop on a block of data that relates to SLLS operation in a more powerful YDFL with a Fabry–Perot cavity formed by two perpendicularly cut fiber ends (the case of low-Q-factor). The laser was based on double-clad YDF, pumped at 976 nm by two 25-W LDs (see Fig. 11.53a) [80]. In this work, recently published by a Mexican group, the authors reported on a curious behavior of the laser, being

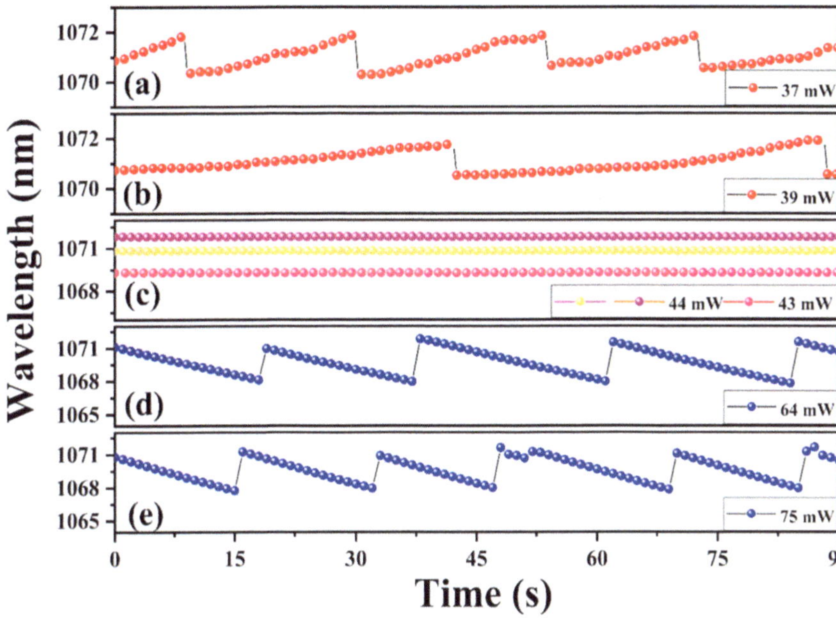

Fig. 11.51 (**a–e**) Spectral characteristics of the FL at different pump power levels, being: (**a**) 37, (**b**) 39, (**c**) 43…44, (**d**) 64, and (**e**) 75 mW. [Adapted with permission from [79]; Elsevier Ltd. ©2024]

the occurrence of a few radically different sub-regimes arising at increasing pump power; these were amplified spontaneous emission (ASE), SLLS, self-Q-switching (SQS) and, finally (at highest pumps), narrowband 'thermal-light' (noise-pulsing, NP) emission. Special attention in [80] was paid to a discussion of the statistical and coherent properties of light, corresponding to the operation of the YDFL in each of the aforementioned sub-regimes. As concerns the SLLS sub-regime, as seen from Fig. 11.53b, c, it followed ASE lighting and appeared just after passing the laser's threshold and was characterized by notable inequality in laser powers at the opposite ends of the YDFL.

As further seen from Fig. 11.54a, b, where laser powers—total and partial ones measured at the two outputs—are plotted as a function of pump power, this inequality is mostly expressed at SLLS (at lowest powers of 0.2…0.4 W) while it steadily vanishes with power's growth (i.e. in sub-regimes of SQS and noise-like thermal emission). In other words, the nearly unidirectional path of laser radiation to the right side in Fig. 11.53a ($P_{out1} \gg P_{out2}$) is one of the SLLS sub-regime's signatures. As authors of [80] claimed, this can be explained by a recently uncovered feature [19, 30] that the highest reflectivity (of up to ~100%) of the dynamical FBGs self-establishing in an YDF at SLLS is observed close to a point of pump-light launching (where inversion of Yb^{3+} ions is highest), whereas it steadily drops toward an opposite side of the fiber. Thus, this circumstance stands behind the breaking of the

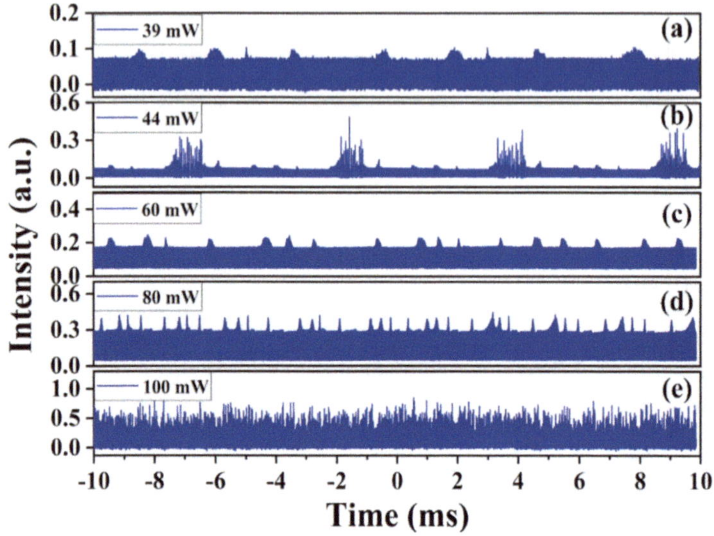

Fig. 11.52 (**a–e**) The intensity dynamics of the fiber laser at different pump power levels, being: (**a**) 39, (**b**) 44, (**c**) 60, (**d**) 80, and (**e**) 100 mW. [Adapted with permission from [79]; Elsevier Ltd. ©2024]

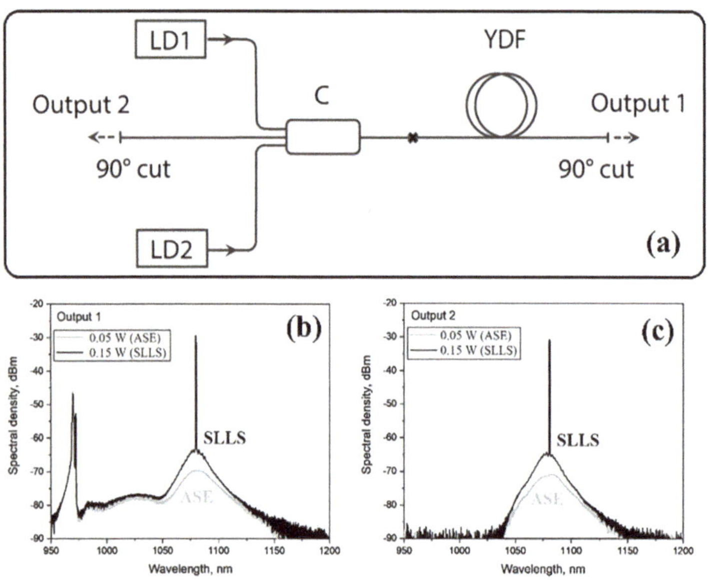

Fig. 11.53 (**a**) Experimental setup of YDFL; (**b, c**) optical spectra measured near threshold at the YDFL's outputs 1 (**b**) and 2 (**c**). [Adapted with permission from [80]; Astro Ltd. ©2023]

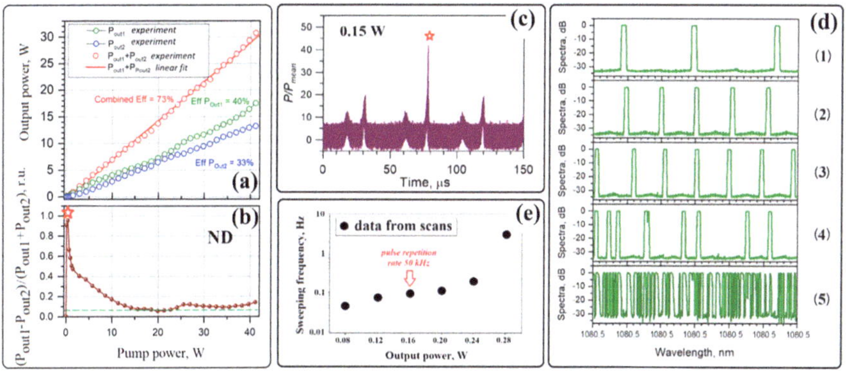

Fig. 11.54 (**a**) Dependencies of laser power at YDFL outputs vs. pump power (the green, blue, and red curves stand for outputs 1 and 2 and their sum, respectively); (**b**) normalized difference (ND) between the output powers at ports 1 and 2, vs. pump power; (**c**) typical snapshot of pulsing at SLLS; (**d**) typical OSA's scans, obtained in SLLS sub-regime: traces 1 to 5 were measured at smooth increase of output power (from 0.1 to 0.3 W); (**e**) dependence of sweeping frequency at SLLS vs. output power. [Adapted with permission from [80]; Astro Ltd. ©2023]

presumed symmetry of Fabry–Perot cavity and eventually leads to the inequality of laser powers emerging at the opposite outputs of the laser.

The SLLS sub-regime appeared as shown in Fig. 11.54c–e). Spectrally, it was manifested as smooth periodic laser wavelength movement to the Stokes side (from 1077 to 1085 nm), culminating in instantaneous bounce backward to the start wavelength (like it happens in all other realizations of SLLS with normal directionality); the sweeping frequency at SLLS is in a sub-Hz range (Fig. 11.54e). To visualize it in [80], the authors used the approach suggested in [6], i.e. they reduced the OSA's wavelength span to zero at 1080.5 nm (the central wavelength of self-sweeping) and kept its scanning rate and sensitivity fixed. The result is shown in Fig. 11.54d. Meanwhile, the SLLS operation gets destroyed by the SBS process when the peak magnitude of more powerful pulses (see Fig. 11.54c an asterisked pulse) reaches the SBS threshold, and, as a result, the laser rapidly enters a completely different operational mode, i.e. SQS.

As mentioned above, one of the aims of study [80] was assessing the statistical properties of all sub-regimes of the YDFL with low-Q cavity. As e.g. seen from Fig. 11.55, at SLLS, the photon correlation function of the second order (CFSO) $g^{(2)}(\tau = 0)$ is ~1, revealing its 'full' coherence; meanwhile, at increasing laser power, when SLLS operation is replaced by SQS (where powerful bursts of pulses are produced) and, then, by noisy thermal light, the system's coherence dies out (when $g^{(2)}(\tau = 0)$ gets more, or much more, than 1), eventually taken fully incoherent character; for more details, see the remainder of work [80].

Let us briefly consider research [81] where authors from Russia suggested a novel cavity design to realize SLLS in FL where they separated the two types of spontaneously formed laser cavity gratings, namely, dynamic gain grating (DGG) and dynamic absorption grating (DAG). These two are accordingly responsible for

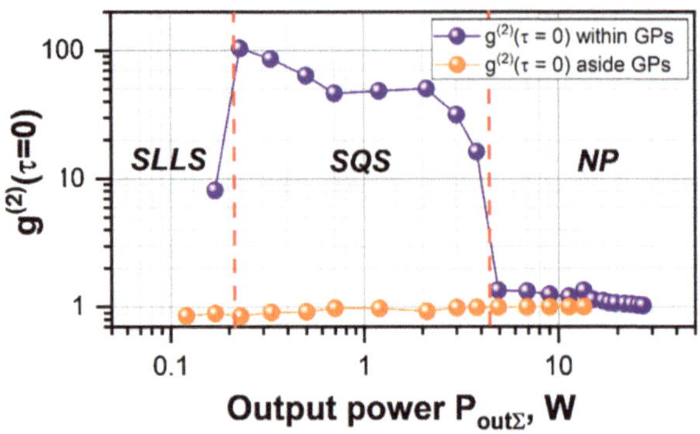

Fig. 11.55 The upper graph shows the dependencies of CFSO vs. YDFL's total output power, measured in SLLS, SQS, and NP sub-regimes (delimited by the vertical dashed red lines). The CFSO values in pulses and outside them are shown by violet and orange circles, respectively. The lower graph schematizes the hierarchy of the ASE, SLLS, SQS, and NP sub-regimes, all inherent in the YDFL with a linear low-Q-factor cavity. [Adapted with permission from [80]; Astro Ltd. ©2023]

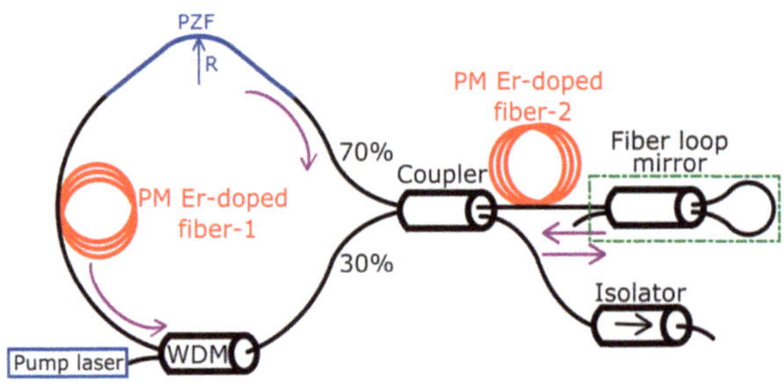

Fig. 11.56 The schematic of a ring-cavity EDFL with SLLS, based on both DGG and DAG phenomena. [Adapted with permission from [81]; the authors ©2023]

a particular intensity dynamic, inherent in SLLS: pulsed or quasi-CW. In [81], an introduction is given to how the DGG and DAG, separately or by a combined manner, are established and act at SLLS in a FL—while authors in [81] surprisingly re-named the effect as 'wavelength self-sweeping effect (WSSE)—and, then, an approach is presented for using this knowledge for an experimental realization of the effect on an example of an EDFL oscillating in CW / SLLS regime at ~1.6 μm; the setup is shown in Fig. 11.56.

The scheme comprised a ring cavity containing a piece of EDF and a linear bench, also containing a piece of EDF plus FLM; the two parts were coupled by a coupler. This design allowed them to separate the DGG and DAG in the same laser. Also note that the EDFL was in-core pumped through 1540/1610 WDM by a 1548-nm LD; the use of all-PM fibers and special polarizing fiber (PZF) permitted establishing a stable polarization in the laser and additional losses (by bending the PZF).

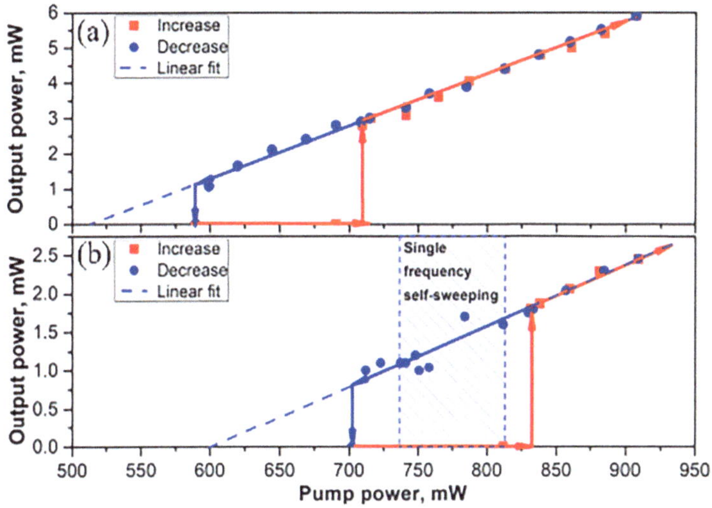

Fig. 11.57 Dependences of EDFL's output power on pump power at (**a**) no additional loss and (**b**) maximum bending loss. [Adapted with permission from [81]; the authors ©2023]

In this design, two standing waves were formed: one in the linear part of the cavity (forming mostly DAG) and the other—in the ring part (forming mostly DGG).

Not entering into the whole experimental details presented in [81], only note here a possibility of control over the SLLS regime's limits on pump power by means of inserting additional loss in the ring part of the EDFL cavity (Fig. 11.57). Also note that the measures to optimize the entire set of parameters of the laser scheme allowed authors to reach a record (for EDFLs) self-sweeping range of ~2.6 nm (at single-frequency operation); see Fig. 11.58.

It is worth noting, too, that a big piece of research on SLLS, made in the recent (2022–2024) years, was dedicated to versatile applications of the SLLS phenomenon [82–90]. This has become natural, as nowadays almost the whole of its state-of-the-art (see above) has been addressed with a high degree of relevance, and so its use in practice has encountered a call for. Not stopping on details of any of this research (it is not the current review's aim), annotate here their basic directions and achievements. Specifically, in [82], authors discussed the way to use a ~2-μm TDFL with SLLS with optimized characteristics for LIDAR applications; in [83], a similar TDFL/SLLS source was used as a tunable prober for remote gas analysis; in [84–87], an ~1.06-μm YDFL with SLLS was optimized to proceed with C-OFDR measurements; in [88], a similar YDFL with SLLS was optimized for the FBG-interrogation technique; in [89] an ~1.6-μm EDFL (see [81]) was investigated and employed for the C-OFDR purposes; at last, in [90], a similar ~1.56-μm EDFL was used, for the first time, for Brillouin optical time domain analysis (BOTDA).

Finally, we ought to mention a few review articles devoted to versatile aspects of the SLLS phenomenon [70, 71, 91, 92]; all these were published in recent years (2021–2024). The reader is advised to pay attention to these reviews for a more

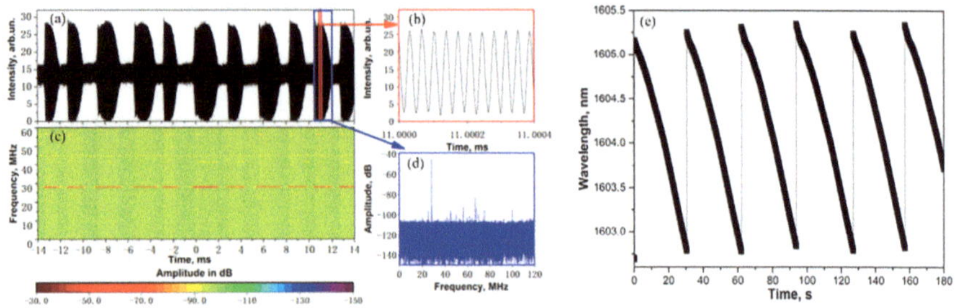

Fig. 11.58 (**a**) Intensity dynamics at pump power of 740 mW (inset (**b**) shows a closer view of the area, highlighted by the red rectangle); (**c**) the corresponding Fourier spectrogram (inset (**d**) shows a segment in the burst region, highlighted by the blue rectangle); (**e**) laser line's dynamics (SLLS) at the same pump. [Adapted with permission from [81]; the authors ©2023]

comprehensive understanding of the effect's background and fine details, not covered above.

11.3 Concluding Remarks

As one can reveal from the above-referred results, SLLS is a common operational regime of a fiber laser. It is inherent in fiber-laser cavities of both linear (Fabry-Perot) and ring types with spectrally unselective feedback but where the role of a spectral selector, dynamic in nature, is played by spontaneously establishing gratings of either amplitude (in the inverted system of doping ions or phototropic centers) or phase type. One of the key factors that ignites an SLLS-process is the formation of standing waves of intensity inside the cavity just above threshold, with the interference pattern produced being of high visibility and so notable inequality of gain/loss in its minimum and maximum. Another essential factor behind the wavelength dynamics at SLLS (namely, self-sweeping) is the spatial hole-burning effect, responsible for fading initially established longitudinal modes that overcame the threshold first and the simultaneous arising of neighboring longitudinal modes that get encountered with more favorable gain-to-loss conditions. Eventually, the above-revealed set of processes defines the enrolling and the following dynamics of the SLLS regime in a fiber laser. In turn, disappearance of the regime is associated with growth of pump power, which leads to reducing of the interference pattern's visibility (because of a tremendous growth of longitudinal modes participating in lasing) and so diminishing of impact of the intracavity gratings. As a rule, the SLLS regime is replaced in a fiber laser first by the one that generated powerful bursts of intensity, inherent in various kinds of self-Q-switching and eventually by noise-like thermal emission. Furthermore, during exploration of the SLLS phenomenon, plenty of its details were addressed; among these, note that it can be realized in two directions, normal (from shorter to longer wavelengths) and reverse (from longer to shorter wavelengths); it can be regular, very fast, and very slow (with its cycles to be measured from milliseconds to hours); it can occur in the forms of irregular pulsing or quasi-continuous-wave; etc.

The dynamics of publishing activity on the SLLS problematics is showcased on Fig. 11.59. As seen, it primarily grew very fast, while subsequently the growth got saturating and is currently slowing down. Seemingly, this happens owing to one's understanding of the phenomenon overall and aiming for the right paths for its practical applications.

Another important point to mention is how the SLLS spectral range grew historically. In Fig. 11.60, I provide this trend in two aspects: one showing growth with the year of the SLLS-range in YDFLs (in which the phenomenon has been explored

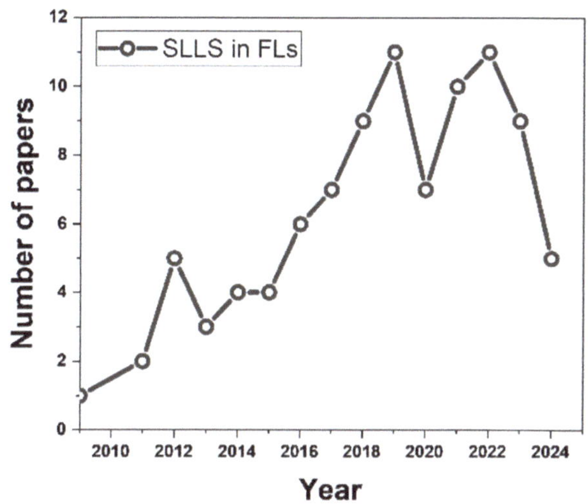

Fig. 11.59 Dependence of number of papers dedicated to SLLS phenomenon in FLs vs. publication year

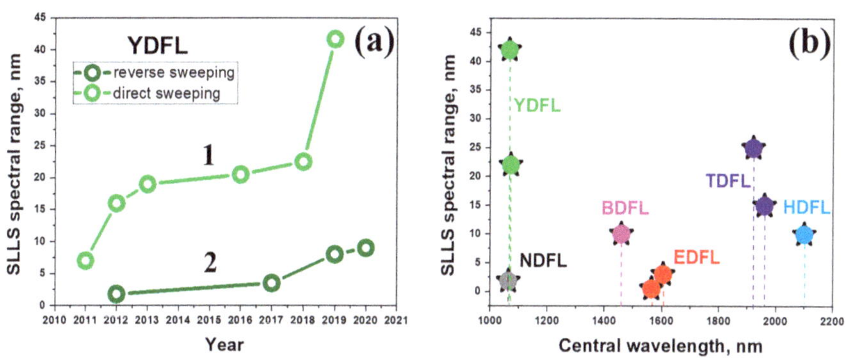

Fig. 11.60 (**a**) Growth with years of SLLS spectral range for YDFLs with direct (curve 1) and reverse (curve 2) wavelength's self-sweeping; (**b**) maximal SLLS spectral ranges reached in FLs of different types nowadays

thoroughly), as an example, see Fig. 11.60a, and another demonstrating the maximal SLLS-ranges reached till now in all FLs where SLLS was realized, see Fig. 11.60b. As seen, progress in the area is noticeable.

Certainly, the present review provides a 'snapshot' of the SLLS phenomenon's state-of-the-art for nowadays. Meanwhile, it is the author's feeling that the current knowledge and applications of FLs with SLLS are still limited; so more detailed experimental research, on the one hand, and developing a more adequate model of the effect, on the other hand, are being awaited in the near future.

References

1. Kir'yanov, A.V., Il'ichev, N.N.: Self-induced laser line sweeping in an ytterbium fiber laser with non-resonant Fabry-Perot cavity. Laser Phys. Lett. **8**(4), 305–312 (2011)
2. Peterka, P., Maria, J., Dussardier, B., Slavik, R., Honzatko, P., Kubecek, K.: Long-period fiber grating as wavelength selective element in double-clad Yb-doped fiber-ring lasers. Laser Phys. Lett. **6**(10), 732–736 (2009)
3. Lobach, I.A., Kablukov, S.I., Podivilov, E.V., Babin, S.A.: Broad-range self-sweeping of a narrow-line self-pulsing Yb-doped fiber laser. Opt. Express. **19**(18), 17632–17640 (2011)
4. Lobach, I.A., Babin, S.I., Podivilov, E.V., Kablukov, S.A.: All-fiber broad-range self-sweeping Yb-doped fiber laser. Proc. SPIE-The Int. Soc. Opt. Eng. **8237**, 30 (2012)
5. Kablukov, S.I., Zlobin, E.A., Podivilov, E.V., Babin, S.A.: Output spectrum of Yb-doped fiber lasers. Opt. Lett. **37**(13), 2508–2510 (2012)
6. Peterka, P., Navratil, P., Maria, J., Dussardier, B., Slavık, R., Honzatko, P., Kubecek, V.: Self-induced laser line sweeping in double-clad Yb-doped fiber-ring lasers. Laser Phys. Lett. **9**(6), 445–450 (2012)
7. Navratil, P., Vojtisek, P., Peterka, P., Honzatko, P., Kubecek, V.: Self-induced laser line sweeping and self-pulsing in rare-earth doped fiber lasers. Proc. SPIE-The Int. Soc. Opt. Eng. **8697**, 86971M (2012)
8. Peterka, P., Navrátil, P., Dussardier, B., Slavík, R., Honzátko, P., Kubeček, V.: Self-induced laser line sweeping and self-pulsing in double-clad fiber lasers in Fabry-Perot and unidirectional ring cavities. Proc. SPIE Int. Soc. Opt. Eng. **8433**, 843309 (2012)
9. Lobach, I.A., Kablukov, S.I.: Application of a self-sweeping Yb-doped fiber laser for high-resolution characterization of phase-shifted FBGs. J. Lightwave Technol. **31**(18), 2982–2987 (2013)
10. Lobach, I.A., Kablukov, S.A., Podivilov, E.V., Babin, S.I.: Self-scanned single-frequency operation of a fiber laser driven by a self-induced phase grating. Laser Phys. Lett. **11**(4), 045103 (2014)
11. Wolf, A.A., Dostovalov, A.V., Lobach, I.A., Babin, S.A.: Femtosecond laser inscription of long-period fiber gratings in a polarization-maintaining fiber. IEEE J. Lightwave Technol. **33**(24), 5178–5183 (2015)
12. Navratil, P., Peterka, P., Kubecek, V., Kalli, K., Kanka, J., Mendez, A.: Effect of pump wavelength on self-induced laser line sweeping in Yb-doped fiber laser. Micro-Struct. Special. Opt. Fibres. **8775**, 87750D (2013)
13. Wang, X., Zhou, P., Wang, X., Xiao, H., Si, L.: Tm-ho co-doped all-fiber broad-range self-sweeping laser around 1.9 μm. Opt. Express. **21**(14), 16290 (2013)
14. Peterka, P., Honzátko, P., Todorov, F., Kašík, I., Aubrecht, J., Podrazký, O.: Self-Q-switched regime of fiber lasers as a transition from self-induced laser line sweeping. Adv. Photon., SoTh2B.6 (2014)

15. Peterka, P., Honzátko, P., Koška, P., Todorov, F., Aubrecht, J., Podrazký, O.: Reflectivity of transient Bragg reflection gratings in fiber laser with laser-wavelength self-sweeping. Opt. Express. **22**(24), 30024 (2014)
16. Toral-Acosta, D., Martinez-Rios, A., Selvas-Aguilar, R., Kir'yanov, A.V., Anzueto-Sanchez, G., Duran-Ramirez, V.M.: Self-pulsing in a large mode area, end-pumped, double-clad ytterbium-doped fiber laser. Laser Phys. **24**(10), 10107 (2014)
17. Lobach, I.A., Kablukov, S.I., Melkumov, M.A., Khopin, V.F., Babin, S.A., Dianov, E.M.: Single-frequency bismuth-doped fiber laser with quasi-continuous self-sweeping. Opt. Express. **23**(19), 24833–24842 (2015)
18. Lobach, I.A., Kablukov, S.I., Podivilov, E.V., Fotiadi, A.A., Babin, S.A.: Pulse coherence in self-sweeping fiber laser. CLEO, STh4L.4 (2015)
19. Peterka, P., Honzátko, P., Koška, P., Podrazký, O., Kašík, I.: Transient-fiber-Bragg grating spectra in self-swept Fabry-Perot fiber lasers proc SPIE-the Int Soc. Opt. Eng. **9344**, 934423 (2015)
20. Peterka, P., Honzátko, P., Koška, P., Todorov, F., Aubrecht, J., Podrazký, O.: Reflectivity of transient Bragg reflection gratings in fiber laser with laser-wavelength self-sweeping: erratum. Opt. Express. **24**(14), 16221 (2016)
21. Lobach, I.A., Yu Tkachenko, A., Kablukov, S.I.: Optimization and control of the sweeping range in an Yb-doped self-sweeping fiber laser. Laser Phys. Lett. **13**(4), 045104 (2016)
22. Tkachenko, A.Y., Lobach, I.A., Guskov, L.N.: Fiber Sensor Interrogator Based on Self-Sweeping Fiber Laser. In: Proceeding of the 2016 17th international conference of young specialists on micro/nanotechnologies and electron devices (EDM), pp. 348–351. IEEE (2016)
23. Tkachenko, A.I., Lobach, I.A.: Fiber sensor sensor interrogator based on self-sweeping fiber laser. Appl. Photon. **3**(1), 37–49 (2016)
24. Drobyshev, R.V., Lobach, I.A., Kablukov, S.I.: Self-sweeping Yb-doped fiber laser based on large-mode area fiber. Appl. Photonics. **3**(4), 413–426 (2016)
25. Aubrecht, J., Peterka, P., Honzátko, P., Koška, P., Podrazký, O., Kašík, I.: Self-swept holmium-doped fiber laser near 2100 nm. Lasers Congress 2016 (ASSL, LSC, LAC): JTu2A.7 (2016)
26. Aubrecht, J., Peterka, P., Koška, P., Podrazký, O., Todorov, F., Honzátko, P.: Self-swept holmium fiber laser near 2100 nm. Opt. Express. **25**(4), 4120–4125 (2017)
27. Aubrecht, J., Peterka, P., Koška, P., Honzátko, P., Jelínek, M., Kamrádek, M.: Spontaneous laser-line sweeping in ho-doped fiber laser. Fiber Lasers XIV Tech. Syst. **10083**, 100831V (2017)
28. Navratil, P., Peterka, P., Honzatko, P., Kubecek, V.: Reverse spontaneous laser line sweeping in ytterbium fiber laser. Laser Phys. Lett. **14**(3), 035102 (2017)
29. Peterka, P., Honzátko, P., Aubrecht, J., Navrátil, P., Koška, P., Todorov, F., Podrazký, O., Čtyroký, J., Kašík, I.: Self-sweeping of laser wavelength and associated mode instabilities in fiber lasers. ICTON, Tu.B6.2 (2017)
30. Peterka, P., Koska, P., Ctyroky, J.: Reflectivity of superimposed Bragg gratings induced by longitudinal mode instabilities in fiber lasers. IEEE J. Select. Top. Quant. Electron. **24**, 1–8 (2018)
31. Navratil, P., Peterka, P., Vojtisek, P., Kasik, I., Aubrecht, J., Honzatko, P.: Self-swept erbium fiber laser around 1.56 µm. Opto-Electron Rev. **26**(1), 29–34 (2018)
32. Peterka, P., Aubrecht, J., Cajzl, J., Podrazký, O., Todorov, F., Kašík, I., Honzátko, P.: Dynamic gratings induced by mode instabilities in fiber lasers. Proc. SPIE Int. Soc. Opt. Eng. **10976**, 109760H (2018)
33. Tkachenko, A.Y., Lobach, I.A., Kablukov, S.I.: Brillouin optical Spectrum analyzer based on self-sweeping fiber laser. In: PIERS—Progress in Electromagnetics Research Symposium, pp. 2316–2320 (2017)
34. Tkachenko, A.Y., Lobach, I.A., Kablukov, S.I.: All-fiber Brillouin optical spectrum analyzer based on self-sweeping fiber laser. Opt. Express. **25**(15), 17600 (2017)
35. Tkachenko, A.Y., Lobach, I.A., Kablukov, S.I.: Optical frequency domain reflectometry based on self-sweeping fiber laser. Proc. SPIE Int. Soc. Opt. Eng. **11199**, 111992H (2019)
36. Tkachenko, A.Y., Lobach, I.A., Kablukov, S.I.: Coherent optical frequency-domain reflectometer based on a fibre laser with frequency self-scanning. Quantum Electron. **49**(12), 1121–1126 (2019)

37. Lobach, I.A., Drobyshev, R.V., Fotiadi, A.A., Podivilov, E.V., Kablukov, S.I., Babin, S.A.: Open-cavity fiber laser with distributed feedback based on externally or self-induced dynamic gratings. Opt. Lett. **42**(20), 4207–4210 (2017)
38. Tkachenko, A.Y., Vladimirskaya, A.D., Lobach, I.A., Kablukov, S.I.: Michelson reflector for spectral range stabilization in a self-sweeping fiber laser. In: International Conference "Laser Optics" (2018). https://doi.org/10.1109/LO.2018.8435584
39. Tkachenko, A.Y., Vladimirskaya, A.D., Lobach, I.A., Kablukov, S.I.: Michelson mode selector for spectral range stabilization in a self-sweeping fiber laser. Opt. Lett. **43**(7), 1558–1561 (2018)
40. Tkachenko, A.Y., Lobach, I.A., Podivilov, E.V., Kablukov, S.I.: Fibre loop mirror with FBGs for the scanning range stabilisation in a fibre self-sweeping laser. Quantum Electron. **48**(12), 1132–1137 (2018)
41. Tkachenko, A.Y., Vladimirskaya, A.D., Lobach, I.A., Kablukov, S.I.: Sweeping range control in a self-sweeping laser with selective mirrors. Proc. SPIE Int. Soc. Opt. Eng. **10821**, 108210V (2018)
42. Drobyshev, R.V., Lobach, I.A., Kablukov, S.I.: Narrow-linewidth self-sweeping fiber laser with scanning range control by a tunable Lyot filter. Laser Phys. **29**(10), 105104 (2019)
43. Jiang, H., Lei, J., Set, S.Y., Yamashita, S.: Spontaneous laser line sweeping in Tm doped fiber laser. Laser Congress 2018 (ASSL), OSA Technical Digest: AM6A.34
44. Jiang, H., Zhao, Z., Jin, L., Set, S.Y., Yamashita, S.: Self-wavelength-sweeping in a bidirectional thulium-doped fiber ring laser. Appl. Phys. Express. **12**(4), 042006 (2019)
45. Budarnykh, A.E., Vladimirskaya, A.D., Lobach, I.A., Kablukov, S.I.: Broad-range self-sweeping single-frequency tm-doped fiber laser for sensing applications. Proc. SPIE Int. Soc. Opt. Eng. **10814**, 108140R (2018)
46. Budarnykh, A.E., Vladimirskaya, A.D., Lobach, I.A., Kablukov, S.I.: Broad-range self sweeping single-frequency linearly polarized tm-doped fiber laser. Opt. Lett. **43**(21), 5307–5310 (2018)
47. Budarnykh, A.E., Vladimirskaya, A.D., Skvortsov, M.I., Wolf, A.A., Lobach, I.A., Kablukov, S.I.: Tm-doped fiber laser with control of spectral dynamics. Proc. SPIE Int. Soc. Opt. Eng. **11028**, 110282T (2019)
48. Budarnykh, A.E., Lobach, I.A., Kablukov, S.I.: Self-sweeping tm-doped fiber laser with wavelength stopping. Laser Phys. Lett. **16**(2), 025108 (2019)
49. Budarnykh, A.E., Vladimirskaya, A.D., Lobach, I.A., Kablukov, S.I.: High-resolution analysis of the wavelength stopping effect in a self-sweeping fiber laser. Laser Phys. Lett. **16**(2), 085104 (2019)
50. Budarnykh, A.E., Lobach, I.A., Kablukov, S.I.: Sweeping regimes in a Tm-doped self-sweeping fiber laser. Intern. Conf. Laser Optics (2020) https://doi.org/10.1109/ICLO48556.2020.9285829
51. Kashirina, E.K., Lobach, I.A., Kablukov, S.I.: Self-sweeping operation of Nd-doped fiber laser. 2019 Conference on Lasers and Electro-Optics Europe & European Quantum Electronics Conference (CLEO/Europe-EQEC) https://doi.org/10.1109/CLEOE-EQEC.2019.8871796
52. Kashirina, E.K., Lobach, I.A., Kablukov, S.I.: Single-frequency self-sweeping Nd-doped fiber laser. Opt. Lett. **44**(9), 2252–2255 (2019)
53. Wang, K., Wen, Z., Lu, B., Qi, X., Bai, J.: Wavelength self-sweeping Yb-doped bidirectional fiber ring laser[C]. Proc. SPIE Int. Soc. Opt. Eng. **11209**, 112092I (2019)
54. Wang, K., Wen, Z., Chen, H., Qi, X., Lu, B., Bai, J.: Observation of reverse self-sweeping effect in an all-polarization-maintaining bidirectional ytterbium doped fiber laser. Opt. Express. **28**(9), 13913–13920 (2020)
55. Budarnykh, A.E., Lobach, I.A., Kablukov, S.I., Chapovsky, P.L.: Detection of nuclear spin isomers of water molecules using self-scanning tm-doped fiber laser. Laser Phys. Lett. **17**(6), 065702 (2020)

56. Vladimirskaya, A.D., Skvortsov, M.I., Wolf, A.A., Kaminin, V.A., Lobach, I.A., Kablukov, S.I.: Ho-doped fiber laser with wavelength self-sweeping near 2.06 μm for CO_2 spectroscopy. Intern. Conf. Laser Optics. (2020). https://doi.org/10.1109/ICLO48556.2020.9285573
57. Kashirina, E.K., Lobach, I.A., Kablukov, S.I.: Dual-longitudinal-mode CW self-sweeping operation in Er-doped fiber laser. Opt. Lett. **45**(24), 6659–6662 (2020)
58. Todorov, F., Aubrecht, J., Peterka, P., Schreiber, O., Jasim, A.A., Mrázek, J., Podrazký, O., Kamrádek, M., Kanagaraj, N., Grábner, M., Baravets, Y., Cajzl, J., Koška, P., Fišar, A., Kašík, I., Honzátko, P.: Active optical fibers and components for fiber lasers emitting in the 2-μm spectral range. Materials. **13**(22), 5177 (2020)
59. Tkachenko, A.Y., Smolyaninov, N.N., Skvortsov, M.I., Lobach, I.A., Kablukov, S.I.: A coherent optical frequency-domain reflectometer based on a self-sweeping fiber laser for sensing applications. Instrum. Exp. Tech. **63**(4), 532–537 (2020)
60. Poddubrovskii, N.R., Lobach, I.A., Kablukov, S.I.: Self-sweeping Er-doped ring fiber laser. Proc. SPIE Int. Soc. Opt. Eng. **11890**, 118900F (2021)
61. Drobyshev, R.V., Poddubrovskii, N.R., Lobach, I.A., Kablukov, S.I.: Self-sweeping Yb-doped fiber laser generating ultra-long single-frequency pulse. Proc. SPIE Int. Soc. Opt. Eng. **11890**, 118900Z (2021)
62. Drobyshev, R.V., Poddubrovskii, N.R., Lobach, I.A., Kablukov, S.I.: High-resolution spectral analysis of long single-frequency pulses generated by a self-sweeping Yb-doped fiber laser. Laser Phys. Lett. **18**(8), 085102 (2021)
63. Vladimirskaya, A.D., Lobach, I.A., Kablukov, S.I., Kamynin, V.A.: Broad-range self-sweeping ho-doped fiber laser for N_2O spectroscopy tasks. Proc. SPIE Int. Soc. Opt. Eng. **11901**, 119010H (2021)
64. Vladimirskaya, A.D., Lobach, I.A., Kablukov, S.I.: Broad-range self-sweeping linearly polarized ho-doped fiber laser. CLEO, JW1A.121 (2021)
65. Vladimirskaya, A.D., Kamynin, V.A., Lobach, I.A., Kablukov, S.I.: Robust operation of linearly polarized broad-range self-sweeping ho-doped fiber laser. Laser Phys. Lett. **18**(7), 075101 (2021)
66. Budarnykh, A.E., Lobach, I.A., Smolyaninov, N.N.: Automatic control of the tuning parameters of a self-sweeping tm-doped fiber laser. Proc. SPIE Int. Soc. Opt. Eng. **11901**, 119010M (2021)
67. Wen, Z., Wang, K., Chen, H., Lu, B., Bai, J.: Self-sweeping ytterbium-doped fiber laser based on a fiber saturable absorber. Appl. Phys. Express. **14**(1), 012005 (2021)
68. Wang, K., Wen, Z., Chen, H., Lu, B., Bai, J.: Wavelength-flexible all-polarization maintaining self-sweeping fiber laser based on intracavity loss tuning. Chin. Opt. Lett. **19**(4), 041401 (2021)
69. Wang, K., Wen, Z., Wang, P.: Intracavity ranging enabled by a single-frequency self-sweeping fiber laser with a few-longitudinal-mode range. Opt. Express. **30**(26), 47115–47123 (2022)
70. Zuo, Z., Wang, K., Chen, H., Lu, B., Bai, J.: Recent progress and development trend of self-sweeping fiber laser. Front. Phys. **9**, 809049 (2021)
71. Poddubrovskii, N.R., Drobyshev, R.V., Lobach, I.A., Kablukov, S.I.: Fiber lasers based on dynamic population gratings in rare-earth-doped optical fibers. Photonics. **9**(9), 613 (2022)
72. Poddubrovskii, N.R., Lobach, I.A., Kablukov, S.I.: Regular mode-hopping dynamics in erbium-doped ring fiber laser with saturable absorber. Opt. Laser Technol. **156**(12), 108568 (2022)
73. Poddubrovskii, N.R., Lobach, I.A., Nikolaeva, N.A., Kablukov, S.I.: Generation of tunable THz-radiation with self-sweeping Er-doped fiber laser. Proc. SPIE Int. Soc. Opt. Eng. **12324**, 123240F (2022)
74. Poddubrovskii, N.R., Lobach, I.A., Podivilov, E.V., Kablukov, S.I.: Time-resolved mode analysis in Er-doped self-sweeping ring fiber laser. Laser Phys. Lett. **19**(12), 125102 (2022)
75. Wang, K., Wen, Z., Wang, P.: Observation of optical Spectrum dynamics and longitudinal-mode feature in a single frequency self-sweeping fiber laser. J. Lightwave Technol. **41**(14), 4786–4791 (2023)
76. Wang, K., Wen, Z., Wang, P.: Direct observation of beat-frequency switching in a self-sweeping fiber laser using variable space length. In: Opto-Electronics and Communications Conference (OECC) (2023). https://doi.org/10.1109/OECC56963.2023.10209841

77. Wang, K., Wen, Z., Wang, P.: Ultra-narrow-bandwidth single-frequency self-sweeping ytterbium-doped fiber laser. IEEE Photon. Technol. Lett. **35**(7), 357–360 (2023)
78. Wang, K., Wen, Z., Guo, W., Xiong, Y., Yang, L., Wang, P.: Amplitude modulation ranging achieved by self-sweeping fiber laser in a short distance. Infrared Phys. Technol. **134**(11), 104895 (2023)
79. Huang, X., Lu, C., Zuo, Z., Chen, H., Lu, B., Bai, J.: Self-sweeping regimes control in a bi-directional Yb-doped ring fiber laser. Opt. Laser Technol. **168**(8), 109943 (2024)
80. Kir'yanov, A.V., Muniz-Cánovas, P., Barmenkov, Y.O.: The diversity of operational regimes in an ytterbium-doped fiber laser with a low-Q-factor cavity. Laser Phys. Lett. **20**(11), 115101 (2023)
81. Kashirina, E.K., Lobach, I.A., Kablukov, S.I.: Single-frequency continuous-wave self-sweeping fiber laser based on separated gain and absorption dynamics gratings. Photonics. **10**(7), 843 (2023)
82. Budarnykh, A.E., Lobach, I.A., Kablukov, S.I.: 2 μm self-sweeping fiber laser oscillator for LIDAR applications. Proc. SPIE Int. Soc. Opt. Eng. **12321**, 123210P (2022)
83. Budarnykh, A.E., Lobach, I.A., Kablukov, S.I.: Optical frequency domain reflectometry based on thulium-doped self-sweeping fiber laser for gas analysis applications. Russ. Fiber Lasers. **2**(1), 67–68 (2022)
84. Volikova, A.M., Lobach, I.A., Kablukov, S.I.: Modeling of laser vibrometer-rangefinder based on self-sweeping Yb-doped fiber laser. In: IEEE 23-rd International Conference of Young Professionals in Electron Devices and Materials (EDM) (2022). https://doi.org/10.1109/ICLO54117.2022.9840263
85. Volikova, A.M., Lobach, I.A., Kablukov, S.I.: Laser vibrometer-rangefinder based on self-sweeping fiber laser. Opt. Express. **30**(12), 22025–22039 (2022)
86. Krivosheina, D.A., Tkachenko, A.Y., Lobach, I.A., Kablukov, S.I.: Sensitivity optimization for a coherent optical frequency-domain reflectometer based on a self-sweeping fiber laser. In: IEEE 23-rd International Conference of Young Professionals in Electron Devices and Materials (EDM) (2022). https://doi.org/10.1109/EDM55285.2022.9855153
87. Krivosheina, D.A., Tkachenko, A.Y., Lobach, I.A., Kablukov, S.I.: Coherent optical frequency reflectometer based on a self-sweeping fiber laser for distributed measurements. Bull. Lebedev Phys. Inst. **50**(Suppl 3), S305–S313 (2023)
88. Drobyshev, R.V., Tkachenko, A.Y., Lobach, I.A., Kablukov, S.I.: Fiber Bragg grating interrogation based on self-sweeping Yb-doped ring fiber laser. In: IEEE 23-rd International Conference of Young Professionals in Electron Devices and Materials (EDM) (2022). https://doi.org/10.1109/ICLO54117.2022.9839971
89. Poddubrovskii, N.R., Lobach, I.A., Kablukov, S.I.: Signal processing in optical frequency domain reflectometry systems based on self-sweeping fiber laser with continuous-wave intensity dynamics. Algorithms. **16**(5), 260 (2023)
90. Poddubrovskii, N.R., Lobach, I.A., Kablukov, S.I.: Microwave-free BOTDA based on a continuous-wave self-sweeping laser. Opt. Lett. **49**(2), 282–285 (2024)
91. Tkachenko, A.Y., Lobach, I.A., Kablukov, S.I.: Coherent optical frequency reflectometry based on a fiber self-scanning laser: current status and development prospects (review). Instrum. Exp. Tech. **66**, 730–736 (2023)
92. Lobach, I.A., Fotiadi, A.A., Yatseev, V.Y., Konstantinov, Y.A., Barkov, F.L., Claude, D., Kambur, D.A., Belokrylov, M.E., Turov, A.T., Korobko, D.A.: Newest methods and approaches to enhance the performance of optical frequency-domain reflectometers. Sensors. **24**, 5432 (2024)

Chapter 12
Wavelength-Tunable Mode-Locked Polarization-Maintaining Fiber Laser

Xuanyi Liu, Maolin Dai, and Guanyu Liu

Abstract Wavelength-tunable mode-locked fiber lasers are highly desirable in diverse applications, such as optical parametric generation (OPG), optical frequency comb (OFC) generation, and bio-imaging. Wide operating bandwidth and high-quality mode-lockers are essential for enhancing the spectral tuning range. Wavelength tuning can be accomplished either through intra-cavity effects or by external modulation. Moreover, implementing an all-polarization-maintaining (all-PM) laser cavity configuration helps to improve environmental robustness. This chapter introduces the optical characteristics of wavelength-tunable mode-locked PM fiber oscillators using various pulse-picking techniques across different bands. It focuses on the tunable range and tuning methods of the central wavelength in the mode-locked spectrum, which is critical for applications such as coherent anti-Stokes Raman scattering (CARS) microscopy.

Keywords Ultrafast laser · Wavelength tunable · Polarization-maintaining fiber · Mode-locking · Coherent anti-Stokes Raman scattering

X. Liu (✉)
State Key Laboratory of Optoelectronic Materials and Technologies, School of Physics, Sun Yat-sen University, Guangzhou, China
e-mail: liuxy728@mail.sysu.edu.cn

M. Dai
Department of Electrical Engineering and Information Systems, The University of Tokyo, Bunkyo-ku, Tokyo, Japan

Research Center for Advanced Science and Technology, The University of Tokyo, Meguro-ku, Tokyo, Japan

G. Liu
School of Physics & Photoelectric Engineering, Guangdong University of Technology, Guangzhou, China

M. F. S. Ferreira, M. C. Paul (eds.), *Optical Fiber Lasers*, Springer Series in Optical Sciences 250, https://doi.org/10.1007/978-3-032-04110-4_12

12.1 Introduction

The ultrashort pulses produced by ultrafast lasers are characterized by high peak power and short pulse duration, making them invaluable for scientific and industrial applications in fields such as optical metrology [1] and high-precision micromachining [2]. When the central wavelength of the mode-locked pulse becomes tunable, the application scope of ultrafast lasers is significantly expanded, particularly for biological imaging. Coherent anti-Stokes Raman scattering (CARS) microscopy, which employs wavelength-tunable pulsed lasers as the light source, is a preferred choice among various imaging techniques. It offers several advantages, such as label-free detection, the ability to detect small molecules, and the capability to capture hyperspectral images. It is critical for wavelength-tunable ultrafast lasers to meet multiple requirements to be effectively applied in complex practical scenarios. Using optical fiber as the working medium of the ultrafast laser provides several advantages. It enhances the compactness and simplicity of the laser system, while the use of polarization-maintaining (PM) fiber increases robustness against environmental factors such as temperature fluctuations and mechanical vibrations. Another important aspect is the wide wavelength tuning range, or the absorption band for a particular sample, which depends on the working bandwidth of the mode-locker and gain medium used. A fast-tuning speed is beneficial for improving sampling speed, which relies on various intra-cavity and extra-cavity wavelength switching methods. Moreover, picosecond pulses are regarded as a more efficient light source for achieving high signal-to-noise ratio imaging.

To meet the specified requirements, it is important to design and fabricate stable and reliable wavelength-tunable mode-locked PM fiber lasers. Two crucial factors in this process are the utilization of an effective saturable absorber (SA) and the implementation of various wavelength tuning approaches. Choosing an appropriate SA requires considering its compatibility with PM fiber. Unlike wavelength-tunable non-PM fiber lasers, which can utilize mode-locking mechanisms including nonlinear polarization evolution (NPE) and most two-dimensional (2D) materials for pulse picking [3–5], options for the all-PM laser configuration are more restricted. Real SAs employed in wavelength-tunable ultrafast fiber lasers need to accommodate the relevant spectral bands. Among them, single-wall carbon nanotubes (SWCNTs) and semiconductor saturable absorber mirrors (SESAMs) are favored as excellent mode-lockers because of their broad operating bandwidth and perfect compatibility with PM fibers [6, 7]. Additionally, ultrafast fiber lasers based on these real SAs benefit from their good self-starting capabilities. In contrast to NPE in non-PM fibers, the nonlinear optical/amplifying loop mirror (NOLM/NALM) mode-locking mechanism, which is also based on the nonlinear Kerr effect, is versatile for use in all-PM fiber oscillators [8, 9]. It is particularly employed in constructing wavelength-tunable mode-locked PM fiber lasers. NALM exhibits numerous benefits, including a high damage threshold and adaptability to various gain media across different operational bands. Thus, the operating bandwidth of

NALM-based ultrafast lasers is primarily constrained by the fiber-pigtailed and free-space devices employed, as well as the gain bandwidth.

The technical schemes for achieving a tunable central wavelength of the mode-locked spectrum can be categorized into two types: intra-cavity effects and extra-cavity modulation. To implement efficient intra-cavity wavelength tunability, mechanical, pressure-sensitive, and heat-sensitive spectral filters are employed for rapid intra-cavity wavelength tuning. A few laser examples have demonstrated that variations in pump power can modify the central wavelength of the generated pulses, owing to their unique cavity structures and mode-locking mechanisms. Besides, adjusting the intra-cavity loss is also believed to influence the central wavelength of the mode-locked spectrum. For extra-cavity spectral modulation, fast spectral shift can also be achieved through mechanical spectral filtering. In addition, nonlinear effects such as soliton self-frequency shift (SSFS) can be utilized for spectral tuning [10].

The current book chapter provides an overview of the design, fabrication, performance, and applications of wavelength-tunable mode-locked PM fiber lasers across various wavelength bands. The chapter is structured as follows: Section 2 introduces the main SAs used in the design and fabrication of these ultrafast lasers. Sections 3 and 4 discuss different schemes for tuning the central wavelength of the mode-locked spectrum, classified based on intra-cavity and extra-cavity methods. Section 5 summarizes the key points and provides a forward-looking perspective on the book chapter.

12.2 Mode-Lockers in Polarization-Maintaining Fiber Oscillators

Real SAs such as SWCNT and SESAM, along with artificial SAs like NALM and NPE based on the Kerr nonlinear effect, are employed as mode-lockers in wavelength-tunable mode-locked polarization-maintaining fiber lasers.

SWCNTs are cylindrical nanostructures with diameters typically ranging from 1 to 2 nanometers. They possess unique electrical, thermal, and optical properties due to their one-dimensional structure and quantum confinement effects. The band structure of SWCNTs allows for efficient light interaction, resulting in desirable saturable absorption behavior. Due to their nanoscale size and high carrier mobility, SWCNTs exhibit very fast response times, making them ideal for ultrafast laser applications. Moreover, SWCNTs can be engineered to cover a wide range of wavelengths, enhancing their versatility for various laser systems. However, SWCNTs cannot withstand high intensities without damage, making them less suitable for high-power laser systems. Besides, they exhibit poor thermal stability, which can lead to inconsistent performance over a range of operating temperatures. As shown in Fig. 12.1a, the simplest way to fabricate SWCNT-based SAs is to sandwich CNTs

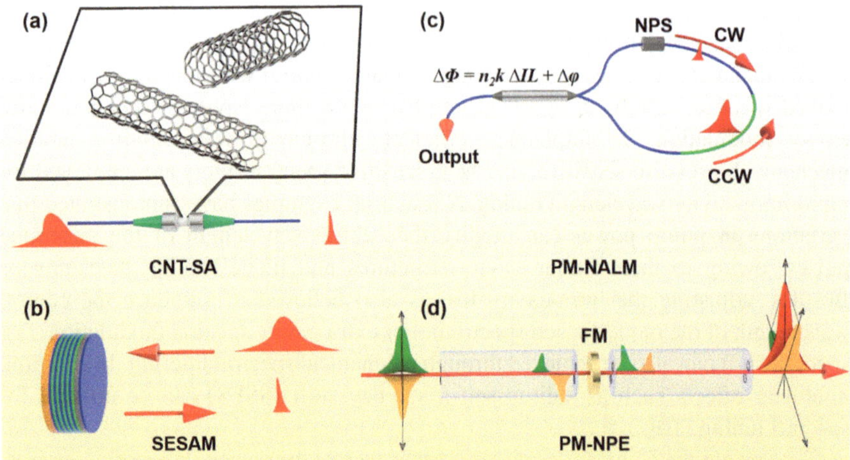

Fig. 12.1 Various mode-locking devices. (**a**) CNT-SA; (**b**) SESAM; (**c**) PM-NALM, *NPS* non-reciprocal phase shifter, *CW* clockwise, *CCW* counter-clockwise; (**d**) PM-NPE, *FM* Faraday mirror

between fiber connectors using techniques such as mechanical exfoliation, high-pressure carbon monoxide (HiPCO), or chemical vapor deposition (CVD) [6].

SESAMs are structured to optimize their function as mode-locking devices for ultrafast solid-state or fiber lasers. The core components of an SESAM depicted in Fig. 12.1b include a substrate, a distributed Bragg reflector (DBR), a spacer layer, an SA layer, and optionally an anti-reflection coating. The substrate, typically made from materials like sapphire, silicon, or gallium arsenide (GaAs), provides mechanical support and thermal management. The DBR, composed of multiple pairs of quarter-wavelength-thick layers of alternating high and low refractive index materials, acts as a highly reflective mirror, typically reflecting more than 99% of incident light. The spacer layer ensures the correct optical thickness to position the SA at an anti-node of the standing wave formed by the DBR, enhancing light interaction. The SA layer, often made from materials like indium gallium arsenide (InGaAs), indium phosphide (InP), or GaAs, absorbs low-intensity light but becomes transparent at high intensity, facilitating passive mode-locking. An optional anti-reflection coating reduces unwanted reflections and enhances efficiency. The design and functionality of SESAM are crucial for initiating and stabilizing ultrashort pulse generation in ultrafast lasers. The DBR ensures high reflectivity, maintaining the necessary feedback for continuous wave mode-locking, while the modulation depth and recovery time of the SA layer are tailored by its thickness, material, and doping levels. Advanced SESAM designs may incorporate features like quantum wells for precise absorption control, bandgap engineering for tailored wavelength or pulse duration response, and thermal management layers to dissipate heat, improving durability and performance under high-power operation.

The NALM mode-locking mechanism, based on Kerr nonlinear effects, can be implemented in ultrafast fiber lasers to generate ultrashort pulses. The basic working principle is illustrated in Fig. 12.1c. It operates by splitting an incoming pulse into two counter-propagating pulses within a fiber loop, where the intensity-dependent refractive index induces nonlinear phase shifts. These shifts cause constructive or destructive interference upon recombination at the loop mirror, typically a fiber coupler. An amplifying medium like an erbium-doped fiber amplifier (EDFA) within the loop boosts pulse intensity, enhancing the difference in the accumulation of nonlinear phase shifts between the clockwise (CW) and counter-clockwise (CCW) pulses [11]. NALM can be engineered in an all-PM configuration and can help reduce amplitude noise and timing jitter, improving the overall performance of ultrafast lasers. Increasing the splitting ratio of the fiber coupler or extending the fiber length can partially address the issue of the difficulty in self-starting at low power, but it limits performance factors such as pulse repetition rate. Therefore, a non-reciprocal phase shifter (NPS) is typically introduced into the fiber ring to enhance the self-starting ability of NALM-based lasers [12].

The NPE mode-locking mechanism exploits intensity-dependent changes in the polarization state of light as it propagates through an optical fiber, where the Kerr nonlinearity induces phase shifts that alter pulse polarization. A polarizer in the laser cavity selectively transmits high-intensity pulses with the desired polarization, effectively discriminating against low-intensity pulses. NPE employs wave plates or polarization controllers to control polarization states. To enhance environmental stability, the upgraded version of NPE implements the NPE mechanism in PM fibers [13]. Figure 12.1d illustrates the propagation of orthogonally polarized pulses along the fast and slow axes of the PM fiber. The group velocity mismatch (GVM) is compensated by PM fiber off-axis splicing or by using a Faraday mirror (FM) to exchange the axes [14–16].

12.3 Wavelength Tunability Controlled by Intra-cavity Effects

In SWCNT-mode-locked wavelength-tunable PM fiber lasers, wavelength tuning is primarily achieved by adjusting inserted filters through mechanical adjustments, temperature changes, or strain variations in the laser cavity. The first spectrally tunable pulsed fiber laser effectively exploits the broad working bandwidth of SWCNT-SA in the 1.55 μm band, combined with a dual-stage wavelength filter. The pass bandwidth of the utilized filter is set to 13 nm, allowing the center wavelength of the mode-locked spectrum to be continuously tunable between 1532 nm and 1562 nm [17]. The mechanical adjustment scheme used in the proposed laser lacks the characteristic of rapid wavelength tunability, but it demonstrates the nonlinear saturable absorption behavior of SWCNT-SA within the EDF gain bandwidth for effective pulse picking. The transmission characteristics of a Lyot filter made from PM fibers are sensitive to external influences. Specifically, temperature changes induce differential thermal expansion between the stress-applying parts (SAPs) and the cladding, causing unequal deformation of the fast and slow axes of the PM fibers. This disparity alters the birefringence and consequently affects the free spectral range (FSR) of the Lyot filter. Therefore, a temperature-controlled, wavelength-tunable PM fiber laser utilizing a transmissive Lyot filter has been proposed and characterized. The central wavelength of the soliton pulses generated by the CNT-mode-locked fiber oscillator can be rapidly tuned between 1546 nm and 1571 nm at a rate of 0.602 nm/°C [18]. Furthermore, a wavelength-tunable PM fiber laser featuring a reflective Lyot filter has been designed, offering a simpler laser configuration. The proposed filter, illustrated in Fig. 12.2a, includes a reflector with an 8.2 cm PM fiber pigtail fused at a 45-degree angle and is incorporated into the laser system via a slow-axis working PM circulator. The designed thermal-sensitive filter supports an FSR of 31.32 nm, enabling the laser to achieve a tuning efficiency of 0.589 nm/°C. As depicted in Fig. 12.2b, increasing the temperature from 22 to 70 degrees in 8-degree increments results in the central wavelength of the mode-locked spectrum linearly decreasing from 1572.71 nm to 1544.46 nm, spanning the 28.25 nm tuning range of the communication band [19]. Strain-controlled central wavelength scanning operates on the same principle of manipulating the

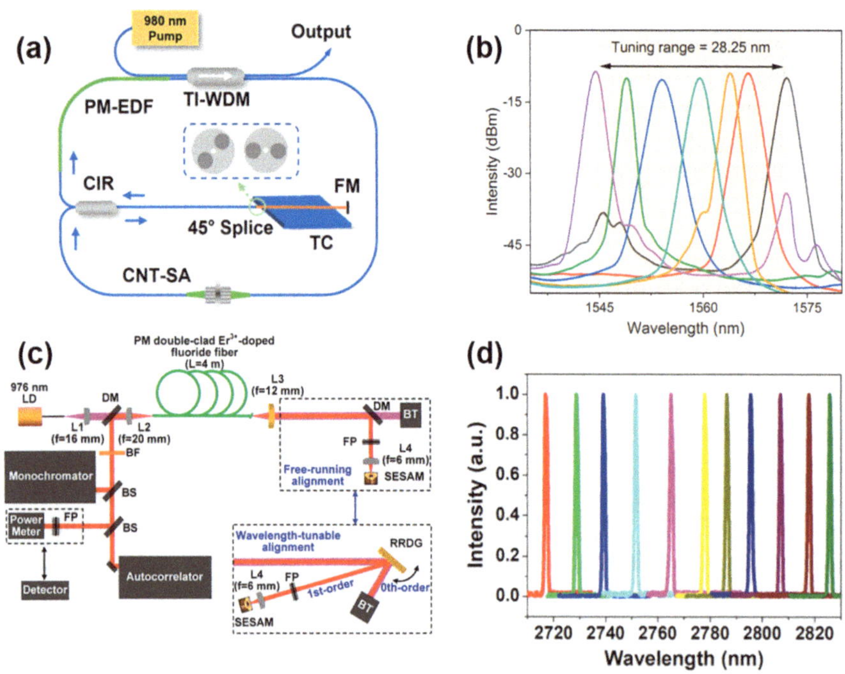

Fig. 12.2 (**a**) SWCNT-mode-locked wavelength-tunable PM fiber laser with a reflective Lyot filter. (**b**) Thermally controlled mode-locked spectra with a range of 28.25 nm [19]; Copyright 2023, with permission from Optica Publishing Group. (**c**) SESAM-mode-locked wavelength-tunable PM fiber laser. (**d**) Mechanically controlled mode-locked spectra with a range of 110 nm [27]. Copyright 2021, with permission from Optica Publishing Group

birefringence of PM fibers. Thus, a piezoelectric lead zirconate titanate (PZT) fiber stretcher has been employed in a transmission-type Lyot filter, achieving a strain sensitivity of 0.052 nm/με. The designed filter has been introduced into a CNT-mode-locked fiber laser, enabling the central wavelength of the emitted ultrashort pulses to achieve a fast wavelength-sweeping speed of 13,000 nm/s [20]. As excellent mid-infrared mode-lockers, CNT-SAs can be extended to the 2-μm band and are intensively investigated in thulium-doped or holmium-doped fiber lasers. A recent study has demonstrated a 300 nm wavelength tuning range in an all-PM thulium-doped fiber laser, thanks to the broad gain bandwidth of the thulium-doped fiber and a custom mechanical tunable filter. The homemade filter comprises a diffraction grating, a lens, a mechanical slit, and a gold mirror. Adjusting the horizontal position and width of the slit allows for tunability of the spectral central wavelength and filter bandwidth, respectively [21].

As commercially available mode-locking devices, SESAMs have been applied to pulsed fiber lasers in both the near-infrared and mid-infrared bands. The ferrule connector/physical contact (FC/PC) connector can be butt-coupled to the SESAM, allowing for the construction of an all-fiber laser configuration. Alternatively, the laser beam can be focused onto the SESAM using a free-space lens. In a neodymium-doped fiber laser mode-locked by a SESAM, a bandpass filter positioned in free space is employed for dual functionality, suppressing unwanted parasitic emissions near 1.06 μm while enabling wavelength tuning between 903 and 912 nm [22]. An in-fiber bandpass filter with a full width at half maximum (FWHM) bandwidth of 100 pm and a resolution of 0.02 nm is employed in a SESAM-based seed oscillator to achieve wavelength tunability. The proposed ytterbium-doped all-PM fiber laser with a tunable wavelength range from 1025 nm to 1055 nm is designed as a pump source for optical parametric generation (OPG). A segment of PM photonic crystal fiber provides parametric gain, allowing the all-fiber system to generate tunable signal radiation from 770 to 940 nm and tunable idler radiation from 1225 nm to 1510 nm. The generated signal and pump pulses are spatially and temporally overlapped and are utilized for coherent anti-Stokes Raman scattering (CARS) imaging [23]. To achieve fast wavelength tuning, a SESAM mode-locked EDF laser is equipped with an intra-cavity filter that contains a galvanometer mirror and a diffraction grating. The wavelength can be swiftly adjusted by 30 nm in under 5 ms and by 18 nm in just 1 ms. By shifting the position of the grating, the variation in pulse repetition rate is reduced to less than 200 Hz. These characteristics make the proposed wavelength-tunable picosecond pulsed fiber laser ideal for hyperspectral CARS imaging [24]. Incorporating high pulse repetition rate into wavelength-tunable mode-locked lasers could broaden their application range to include optical metrology and broadband absorption spectroscopy. By optimizing the laser configuration to be simple and compact, wavelength-tunable ultrashort pulses with a repetition rate of 126.5 MHz are generated in a SESAM mode-locked linear-type fiber laser. The center wavelength is adjusted from 1533.7 nm to 1565.6 nm by tuning the intra-cavity bandpass filter. Since all components in the laser cavity are equipped with fiber connectors, the pulse repetition rate can be customized by tailoring the pigtail length [25]. Moreover, using a free-space bandpass filter, the repetition rate of pulses with a spectral center wavelength adjustable from 1505 nm to 1561 nm increases to 250 MHz [26]. Mid-infrared pulsed light sources are valuable tools for applications including laser surgery, strong-field physics, and spectroscopy. The use of PM double-clad erbium-doped fluoride fiber gain medium extends the central wavelength of ultrashort pulses to the 2.8 μm band. The SESAM mode-locked fiber laser shown in Fig. 12.2c features wavelength tunability. The wavelength tuning module employs a ruled reflection diffraction grating (RRDG) to filter out the zeroth-order diffraction, allowing only the first-order diffraction to be resonated by the SESAM. As shown in Fig. 12.2d, the rotation of the RRDG makes the central wavelength of the emitted pulses continuously tunable within the 110 nm range from 2716.8 nm to 2826.6 nm [27].

Unlike directly inserting bandpass filters in fiber lasers mode-locked by real SAs for wavelength tunability, designing and fabricating wavelength-tunable pulsed fiber lasers with artificial SAs based on nonlinear effects requires careful consideration of the cavity structure and pulse-shaping mechanism. The self-consistency of the 1-μm all-normal-dispersion (ANDi) fiber laser requires bandpass filters to tailor the highly chirped pulses both spectrally and temporally. The utilization of various bandpass filters also provides the possibility of wavelength tuning. In an NALM-mode-locked figure-8 fiber laser shown in Fig. 12.3a, a grating pair combined with a collimator (Col-2) serves as a tunable spectral filter to select different lasing wavelengths. While the homemade filter partially compensates for the net-normal dispersion, spectral filtering remains the dominant factor, resulting in the formation of dissipative solitons. The horizontal position of the collimator controls the shift in the central wavelength of the generated pulses, varying between 1020.8 nm and 1059.5 nm, as illustrated in Fig. 12.3b [28]. A recently developed figure-8 fiber laser relies on a liquid-crystal-on-silicon-based electrically programmable filter to engineer the pulse-shaping mechanism, delivering ultrashort pulses with tunable central wavelengths and pulse widths [29]. Furthermore, high peak power dissipative soliton resonance (DSR) is achieved in a wavelength-tunable figure-8 fiber laser. The generated seed pulses are boosted to an average power of 20 W after passing through a two-stage fiber amplifier [30]. When the dispersion element compensates for the intra-cavity normal dispersion to varying degrees, the fiber laser operates in different dispersion regimes, producing ultrashort pulses with distinct temporal and spectral characteristics without requiring the bandpass filter. In a dispersion-managed figure-9 fiber laser mode-locked by NALM, tilting the silver mirror at the laser end allows for a 90 nm wavelength tuning range, which is attributed to the spatial dispersion and limited aperture. Additionally, since different central wavelengths correspond to different fiber dispersion coefficients, the net-cavity dispersion gradually evolves from anomalous at short wavelengths to normal at long wavelengths. The corresponding pulse-shaping mechanism transitions the generated pulses from

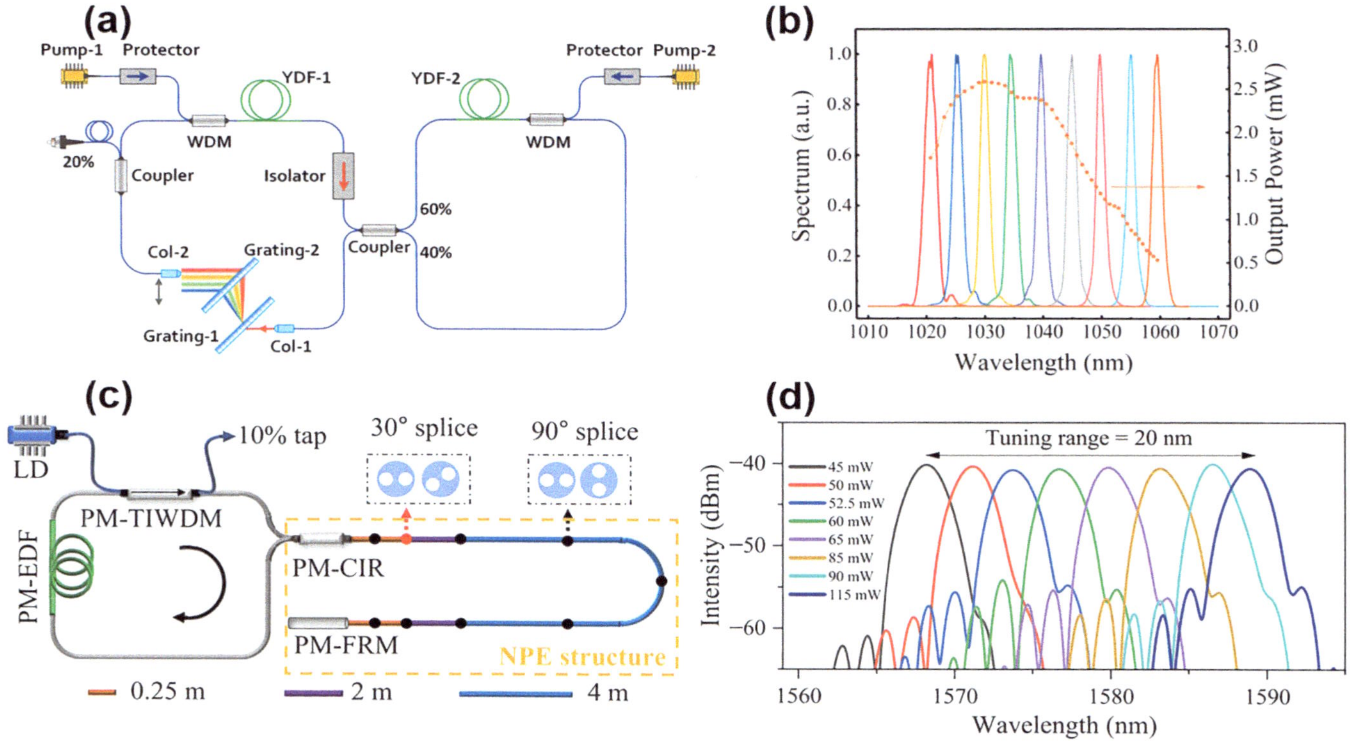

Fig. 12.3 (**a**) NALM-mode-locked wavelength-tunable PM fiber laser with a homemade spectral filter. (**b**) Mode-locked spectra with a range of 38.7 nm [28]; Copyright 2021, with permission from Elsevier. (**c**) NPE-mode-locked wavelength-tunable PM fiber laser. (**d**) Pump-power-controlled mode-locked spectra with a range of 20.7 nm [37]. Copyright 2024, with permission from Optica Publishing Group

dissipative solitons to stretched pulses and ultimately to conventional solitons [31]. Another impressive research effort has extended the length of the gain fiber in a Ytterbium-doped figure-9 fiber oscillator to enhance long-wavelength stimulated emission. By adjusting the intra-cavity loss, the central wavelength of the emitted stretched pulses can be shifted from 1034 to 1104 nm [32].

For the 1.55- and 2-μm bands, ultrafast fiber lasers typically exhibit net-anomalous dispersion, which, in balance with nonlinearities, promotes the formation of conventional solitons approaching the Fourier transform limit. The combination of a reflective diffraction grating and a collimator can serve as a spectral filter for wavelength tuning in a figure-9 erbium-doped fiber laser [33]. With the assistance of a galvanometer scanner, the designed NALM-based fiber laser can achieve rapid wavelength switching in under 10 ms [34]. Certain components of a mode-locked fiber laser can naturally provide filtering effects under specific conditions. In an all-PM thulium-doped fiber laser mode-locked by NOLM, a temperature-sensitive high-birefringence fiber optical loop filter tunes the central wavelength of the generated noise-like pulses (NLPs) in the 2-μm band, spanning from 1910.4 nm to 1942.8 nm [35]. Even in the absence of artificially introduced filters or inherent filtering effects, the mode-locked fiber laser may experience gain-related frequency shifts. For instance, the spectrum of solitons produced by an NALM-locked figure-9 fiber laser exhibits a significant redshift as the pump power increases. Wavelength tuning occurs as increased pump power modifies the nonlinear phase shift, resulting in a redshift in the transmittance curve at the laser output [36]. Another case where changes in pump power induce frequency shifts has been demonstrated and explained in a PM-NPE mode-locked fiber laser. The artificial SA shown in Fig. 12.3c adopts off-axis splicing of multiple sections of PM fiber. At the end of the laser, the FM switches the fast and slow-axis polarization states, effectively compensating for the walk-off effect. As illustrated in Fig. 12.3d, the central wavelength of the mode-locked spectrum shifts red by 20.7 nm with increased pump power, a change attributed to the PM-NPE mode-locking mechanism. The phase difference between the orthogonally polarized pulses at the original and shifted wavelengths needs to be preserved. Changes in pump power directly affect the phase difference, which in turn indirectly modulates the PM-NPE mode-locking wavelength [37].

To evaluate the lasing performance of wavelength-tunable PM fiber lasers employing different mode-locking techniques and pulse-shaping mechanisms, Table 12.1 provides a comprehensive overview across various bands. The focus is on wavelength tuning capabilities, including range, efficiency, and methods.

Table 12.1 A comparison of the lasing performance of wavelength-tunable PM fiber lasers

SA	Pulse-shaping	Gain fiber	Wavelength tuning range	Tuning efficiency	Wavelength tuning methods	Ref.
Graphene	Soliton	Er	1556–1560 nm	–	Intra-cavity loss	[38]
SWCNT	Soliton	Er	1532–1562 nm	–	Mechanically tunable filter	[17]
SWCNT	Soliton	Er	1546–1571 nm	0.602 nm/°C	Thermally controlled transmissive Lyot filter	[18]
SWCNT	Soliton	Er	1544.46–1572.71 nm	0.589 nm/°C	Thermally controlled reflective Lyot filter	[19]
SWCNT	Soliton	Er	1540–1567 nm	0.052 nm/με	Strain-controlled transmissive Lyot filter	[20]
SWCNT	Soliton	Tm	1733–2033 nm	–	Grating-based filter	[21]
SESAM	Dissipative soliton	Nd	903–912 nm	–	Mechanically tunable filter	[22]
SESAM	Dissipative soliton	Yb	1025–1055 nm	–	Mechanically tunable filter	[23]
SESAM	Soliton	Er	1541–1574 nm	<5 ms/30 nm	Galvanometer-driven intra-cavity filter	[24]
SESAM	Soliton	Er	1533.7–1565.6 nm	–	Mechanically tunable filter	[25]
SESAM	Soliton	Er	1505–1561 nm	–	Mechanically tunable filter	[26]
SESAM	Soliton	Er	2716.8–2826.6 nm	–	RRDG-based filter	[27]
NOLM	NLP	Tm	1910.4–1942.8 nm	–	Thermally controlled birefringence filter	[35]
NALM	Dissipative soliton	Yb	1020.8–1059.5 nm	–	Mechanically tunable filter	[28]
NALM	Soliton/ Stretched pulse/ Dissipative soliton	Yb	1015–1105 nm	–	Spatial dispersion/ Limited aperture	[31]
NALM	Stretched pulse	Yb	1034–1104 nm	–	Intra-cavity loss	[32]
NALM	NLP/ Dissipative soliton	Yb	1018–1065 nm	–	Liquid-crystal-on-silicon-based electrically programmable filter	[29]

(continued)

Table 12.1 (continued)

SA	Pulse-shaping	Gain fiber	Wavelength tuning range	Tuning efficiency	Wavelength tuning methods	Ref.
NALM	DSR	Yb	1017–1042 nm	–	Mechanically tunable filter	[30]
NALM	Soliton	Er	1525–1565 nm	–	Grating-based filter	[33]
NALM	Soliton	Er	1527–1559 nm	<10 ms	Galvanometer-driven intra-cavity filter	[34]
NALM	Soliton	Er	1572–1577 nm	–	Pump power	[36]
NPE	Soliton	Er	1568.2–1588.9 nm	–	Pump power	[37]

12.4 Wavelength Tunability Controlled by Extra-cavity Effects

In the aforementioned studies, the filtering effects, losses, or pump power responsible for wavelength tuning all originate from the intra-cavity components of the laser. The wavelength manipulation by intra-cavity effects depends to some extent on the mode-locking mechanism employed and the laser structure adopted. A more direct approach is to place the spectral filter outside the laser cavity, allowing wavelength tuning to be independent of the mode-locking mechanism and enabling the use of any pulsed laser as a seed source. In a recently developed high-speed wavelength-tunable laser system, amplified and spectrally broadened pulses are seeded by a figure-9 fiber laser with a phase-biased NALM. A wavelength tunability of >280 cm^{-1} is implemented by a fast wavelength-tunable filter comprising a galvanometric scanner, a lens pair, and a diffraction grating. Video-rate hyperspectral stimulated Raman imaging has been achieved by synchronizing the designed laser with a picosecond Ti:sapphire laser [39]. Synchronizing pulses from two separate ultrafast lasers complicates the imaging process. In contrast, using a single laser to generate dual-color synchronized pulses enhances system stability and practicality. As illustrated in Fig. 12.4a, the CARS imaging system employs a figure-9 NALM mode-locked fiber laser as a seed oscillator. The output pulses are split into two branches: the Stokes branch and the pump branch. The Stokes branch forms a supercontinuum after passing through two EDFAs and a segment of highly nonlinear fiber and is subsequently wavelength-tuned using a filter based on a high-speed galvanometric scanner. The pump branch produces picosecond pulses with a central wavelength of 790 nm through second harmonic generation (SHG) following two-stage amplification. The two obtained beams are injected into the microscope system for CARS detection [40]. SSFS exploits stimulated Raman scattering in optical fibers to achieve wavelength tuning in the mode-locked spectrum. In the experimental demonstration shown in Fig. 12.4b, a NALM-based figure-9 fiber laser generates

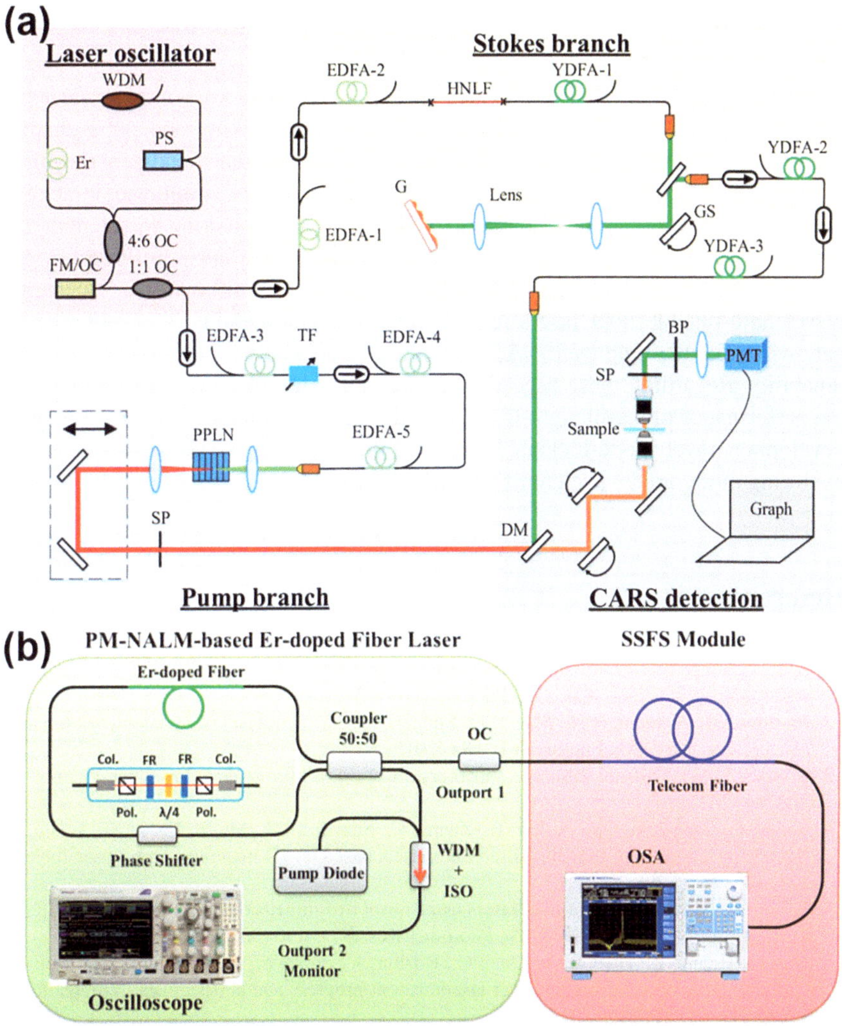

Fig. 12.4 (**a**) Experimental setup of a wavelength-tunable fiber laser system for CARS microscopy [40]; Copyright 2021, with permission from IOP Publishing. (**b**) Experimental setup of a 1.7-μm wavelength-tunable PM fiber laser based on SSFS [41]. Copyright 2019, with permission from IOP Publishing

180 fs, 0.52 nJ soliton pulses, which are then coupled to a 460-m PM fiber, shifting the spectral center wavelength to 1.7 μm [41]. Additionally, the SSFS technology enables the dual-comb pulses obtained by SWCNT-SA to be tunable in wavelength from 1625 nm to 1700 nm. The constructed dual-comb spectroscopy system is then used to measure the absorption spectra of CH_4 [42].

12.5 Summary

This book chapter summarizes a category of pulsed lasers known as wavelength-tunable mode-locked PM fiber lasers. Different pulse-picking and pulse-shaping techniques are essential for the laser to achieve self-consistency and generate ultra-short pulses. Various intra- and extra-cavity effects can be employed to achieve tunable center wavelengths of mode-locked spectra. Incorporating wavelength tuning into mode-locked fiber lasers can greatly broaden their application range, particularly in biological imaging. To adapt to multifunctional scenarios, wavelength-tunable fiber lasers require environmental stability, a wide operating bandwidth for the employed SA, robust laser architecture, and rapid wavelength tunability. Regarding these aspects, there remains significant room for enhancing the performance of wavelength-tunable mode-locked PM fiber lasers. This includes developing novel mode-lockers, extending the operating wavelength range, and exploring more rapid wavelength tuning schemes.

References

1. Kim, J., Song, Y.: Ultralow-noise mode-locked fiber lasers and frequency combs: principles, status, and applications. Adv. Opt. Photon. **8**(3), 465 (2016)
2. Phillips, K.C., Gandhi, H.H., Mazur, E., Sundaram, S.K.: Ultrafast laser processing of materials: a review. Adv. Opt. Photon. **7**(4), 684 (2015)
3. Kobtsev, S.M.: Artificial saturable absorbers for ultrafast fibre lasers. Opt. Fiber Technol. **68**, 102764 (2022)
4. Shang, C., Zhang, Y., Qin, H., He, B., Zhang, C., Sun, J., Li, J., Ma, J., Ji, X., Xu, L., Fu, B.: Review on wavelength-tunable pulsed fiber lasers based on 2D materials. Opt. Laser Technol. **131**, 106375 (2020)
5. Sobon, G.: Mode-locking of fiber lasers using novel two-dimensional nanomaterials: graphene and topological insulators [Invited]. Photonics Res. **3**(2), A56 (2015)
6. Dai, L., Huang, Z., Huang, Q., Zhao, C., Rozhin, A., Sergeyev, S., Al Araimi, M., Mou, C.: Carbon nanotube mode-locked fiber lasers: recent progress and perspectives. Nanophotonics **10**(2), 749–775 (2020)
7. de Souza, E.A., Soccolich, C., Pleibel, W., Stolen, R.H., Islam, M.N., Simpson, J.R., DiGiovanni, D.J.: Saturable absorber modelocked polarisation maintaining erbium-doped fibre laser. Electron. Lett. **29**(5), 447 (1993)
8. Szczepanek, J., Kardaś, T.M., Michalska, M., Radzewicz, C., Stepanenko, Y.: Simple all-PM-fiber laser mode-locked with a nonlinear loop mirror. Opt. Lett. **40**(15), 3500 (2015)
9. Jiang, T., Cui, Y., Lu, P., Li, C., Wang, A., Zhang, Z.: All PM Fiber Laser Mode Locked With a Compact Phase Biased Amplifier Loop Mirror. IEEE Photon. Technol. Lett. **28**(16), 1786–1789 (2016)
10. Lee, J.H., van Howe, J., Xu, C., Liu, X.: Soliton Self-Frequency Shift: Experimental Demonstrations and Applications. IEEE J. Sel. Top. Quantum Electron. **14**(3), 713–723 (2008)
11. Fermann, M.E., Haberl, F., Hofer, M., Hochreiter, H.: Nonlinear amplifying loop mirror. Opt. Lett. **15**(13), 752 (1990)
12. Hänsel, W., Hoogland, H., Giunta, M., Schmid, S., Steinmetz, T., Doubek, R., Mayer, P., Dobner, S., Cleff, C., Fischer, M., Holzwarth, R.: All polarization-maintaining fiber laser architecture for robust femtosecond pulse generation. Appl. Phys. B Lasers Opt. **123**(1), 41 (2017)

13. Szczepanek, J., Kardaś, T.M., Radzewicz, C., Stepanenko, Y.: Nonlinear polarization evolution of ultrashort pulses in polarization maintaining fibers. Opt. Express. **26**(10), 13590 (2018)
14. Nielsen, C.K., Keiding, S.R.: All-fiber mode-locked fiber laser. Opt. Lett. **32**(11), 1474 (2007)
15. Liu, X., Ye, F., Zhao, M., Malomed, B.A., Fu, H.Y., Li, Q.: All-Polarization-Maintaining Linear Cavity Fiber Lasers Mode-Locked by Nonlinear Polarization Evolution in Stretched Pulse Regime. J. Light. Technol. **41**(15), 5107–5115 (2023)
16. Liu, X., Zhou, R., Pan, D., Li, Q., Fu, H.Y.: 115-MHz Linear NPE Fiber Laser Using All Polarization-Maintaining Fibers. IEEE Photon. Technol. Lett. **33**(2), 81–84 (2021)
17. Senoo, Y., Nishizawa, N., Sakakibara, Y., Sumimura, K., Itoga, E., Kataura, H., Itoh, K.: Polarization-maintaining, high-energy, wavelength-tunable, Er-doped ultrashort pulse fiber laser using carbon-nanotube polyimide film. Opt. Express. **17**(22), 20233 (2009)
18. Sun, X., Zhu, Y., Jin, L., Yamashita, S., Set, S.Y.: Polarization-maintaining all-fiber tunable mode-locked laser based on a thermally controlled Lyot filter. Opt. Lett. **47**(19), 4913 (2022)
19. Dai, M., Liu, B., Shirahata, T., Sun, X., Set, S.Y., Yamashita, S.: All-polarization-maintaining, efficiently wavelength-tunable, Er-doped mode-locked fiber laser with a compact reflective Lyot filter. Opt. Express. **31**(17), 27810 (2023)
20. Sun, X., Yamashita, S., Set, S.: Fast wavelength-swept polarization maintaining all-fiber mode-locked laser based on a piezo-stretched fiber Lyot filter. Opt. Express. **31**(8), 12837 (2023)
21. Dai, R., Meng, Y., Li, Y., Qin, J., Zhu, S., Wang, F.: Nanotube mode-locked, wavelength and pulsewidth tunable thulium fiber laser. Opt. Express. **27**(3), 3518 (2019)
22. Le Corre, K., Robin, T., Cadier, B., Becheker, R., Godin, T., Hideur, A., Gilles, H., Girard, S., Laroche, M.: Mode-locked all-PM Nd-doped fiber laser near 910 nm. Opt. Lett. **46**(15), 3564 (2021)
23. Aporta, I., Quintela, M.A., Lopez-Higuera, J.M.: Broadband Continuously Tunable All-Fiber Laser Based on OPG for CARS Imaging. J. Light. Technol. **39**(8), 2489–2496 (2021)
24. Ozeki, Y., Tashiro, D.: Fast wavelength-tunable picosecond pulses from a passively mode-locked Er fiber laser using a galvanometer-driven intracavity filter. Opt. Express. **23**(12), 15186 (2015)
25. Jang, Y.-S., Park, J., Jin, J.: Linear-cavity Er-doped fiber mode-locked laser with large wavelength tunability. Appl. Opt. **61**(34), 10116 (2022)
26. Jang, Y.-S., Park, J., Jin, J.: Full C-band wavelength-tunable, 250 MHz repetition rate mode-locked polarization-maintaining fiber laser. Sci. Rep. **13**(1), 3623 (2023)
27. Luo, H., Wang, Y., Li, J., Liu, Y.: High-stability, linearly polarized mode-locking generation from a polarization-maintaining fiber oscillator around 2.8 μm. Opt. Lett. **46**(18), 4550 (2021)
28. Zhao, K., Yan, W., Liu, M., Wang, L., Li, H., Zhang, M., Jia, Z., Zhai, R., Liu, M.: Tunable ytterbium fiber laser mode-locked with a nonlinear amplifying loop mirror. Opt. Laser Technol. **148**, 107764 (2022)
29. Nishiura, M., Shioda, T.: Wavelength and pulse width programmable mode-locked Yb fiber laser. Opt. Express. **31**(4), 5347 (2023)
30. Li, J., Wu, J., Hao, Q., Huang, K., Yang, K., Zeng, H.: Wavelength and repetition rate tunable high peak power dissipative soliton resonance in an all polarization maintaining Yb-doped fiber laser. Opt. Laser Technol. **162**, 109204 (2023)
31. Gao, G., Wang, S., Zhao, Q., Cong, Z., Liu, Z., Zhao, Z.: Consecutive 1015–1105-nm wavelength tunable "figure-of-9" mode-locked Yb:fiber oscillator. Opt. Lett. **47**(22), 5869 (2022)
32. Gao, K., Liu, Y., Qiao, W., Song, Y., Zhao, X., Wang, A., Li, T.: Wavelength-tunable 1104 nm nonlinear amplifier loop mirror laser based on a polarization-maintaining double-cladding fiber. Opt. Lett. **47**(1), 5 (2022)
33. Ozeki, Y., Fukazu, T.: A wavelength-tunable, polarization-maintaining picosecond figure-nine fiber laser. In: Conference on Lasers and Electro-Optics, p. JTu5A.119. OSA (2016)
34. Fujita, T., Ozeki, Y.: Fast wavelength-switchable figure-nine Er fiber laser using a galvanometer-driven intracavity filter. In: Conference on Lasers and Electro-Optics, p. SM4L.6. OSA (2017)
35. Zhang, Y., Zheng, Y., Su, X., Zhang, H., Sun, T., Cui, M., Peng, J.: Wavelength tunable all-polarization-maintaining fiber laser at 2 μm. Opt. Laser Technol. **168**, 109850 (2024)

36. Zhang, H., Xia, H., Fan, M., Zheng, J., Li, J., Tian, X., Zhou, D., Huang, Z., Zhang, F., Zhang, R., Peng, Z., Zhu, Q.: Observation of wavelength tuning in a mode-locked Figure-9 fiber laser. Photonics. **10**, 2 (2023)
37. Ye, G., Liu, B., Dai, M., Ma, Y., Shirahata, T., Yamashita, S., Set, S.Y.: Pump-power-controlled L-band wavelength-tunable mode-locked fiber laser utilizing nonlinear polarization evolution in all-polarization-maintaining fibers. Opt. Lett. **49**(9), 2433 (2024)
38. Sobon, G., Sotor, J., Pasternak, I., Krzempek, K., Strupinski, W., Abramski, K.M.: A tunable, linearly polarized Er-fiber laser mode-locked by graphene/PMMA composite. Laser Phys. **23**(12), 125101 (2013)
39. Asai, T., Yoshimi, H., Shou, J., Fujita, T., Ozeki, Y.: Generation of Wavelength-Tunable Picosecond Pulses with Polarization-Maintaining Yb Figure-Nine Laser for Stimulated Raman Scattering Microscopy. In: Conference on Lasers and Electro-Optics, p. SW4K.2. OSA (2018)
40. Yang, K., Huo, L., Ao, J., Wang, Q., Hao, Q., Yan, M., Huang, K., Ji, M., Zeng, H.: Fast tunable all-polarization-maintaining supercontinuum fiber laser for CARS microscopy. Appl. Phys. Express. **14**(6), 062004 (2021)
41. He, X., Lin, Q., Guo, H., Sun, J., Bai, J., Hou, L., Wang, K.: Robust 1.7- μ m, all-polarization-maintaining femtosecond fiber laser source based on standard telecom fibers. Appl. Phys. Express. **12**(7), 072007 (2019)
42. Nishizawa, N., Kanehara, Y., Yamanaka, M., Terabayashi, R., Omoda, E., Kataura, H., Sakakibara, Y.: 1.6-1.7 um Wavelength Tunable, All-Polarization Maintaining, Dual-Comb Fiber Laser with Single Wall Carbon Nanotube for Dual Comb Spectroscopy. In: Conference on Lasers and Electro-Optics, p. SW4R.5. OSA (2020)

Chapter 13
Frequency Conversion of Fiber Lasers

Bing Han and Han Wu

Abstract Visible and mid-infrared (MIR) lasers are crucial in a wide range of practical applications. To develop lasers with high beam quality, excellent wavelength tunability, and high output power in the visible and MIR regions, nonlinear frequency conversion of fiber lasers has emerged as a highly promising approach. Here, the operation principles and recent progress in the nonlinear frequency conversion of various types of fiber lasers toward visible and MIR ranges are overviewed.

Keywords Frequency conversion · Fiber laser · Mid-infrared region · Random fiber lasers · Rare-earth-doped fiber lasers · Raman fiber lasers

13.1 Introduction

In practical applications, lasers with special wavelengths in the visible and mid-infrared (MIR) ranges have found extensive use across diverse fields such as display technology, biomedicine, molecular detection, material processing, laser-guide-star systems, military applications, and nonlinear optics [1–3]. For instance, in color display systems, red, green, and blue are the three primary colors. Due to the strong

B. Han
College of Information Science and Engineering, Northeastern University, Shenyang, China

Hebei Key Laboratory of Micro-Nano Precision Optical Sensing and Measurement Technology, Northeastern University at Qinhuangdao, Qinhuangdao, China
e-mail: hanbing@neuq.edu.cn

H. Wu (✉)
College of Electronics and Information Engineering, Sichuan University, Chengdu, China
e-mail: hanwu@scu.edu.cn

M. F. S. Ferreira, M. C. Paul (eds.), *Optical Fiber Lasers*, Springer Series in Optical Sciences 250, https://doi.org/10.1007/978-3-032-04110-4_13

absorption lines of CO, CO_2, and NO_2 at 2.4, 2.8, and 2.9 μm, respectively, molecular detection based on MIR lasers is feasible. However, the high transmission loss of silica fibers in the visible and MIR spectral regions poses a significant challenge in extending the operating wavelengths of fiber lasers. To address this issue, nonlinear frequency conversion technology emerges as an effective solution for constructing visible and MIR fiber lasers.

Conventionally, solid-state lasers have been employed as light sources for visible light generation in display applications. However, the lasing wavelengths of the solid-state lasers are determined by the materials used, which makes it difficult to extend the wavelength range. In addition, the brightness of solid-state lasers tends to decrease as their emission area increases. On the other hand, achieving direct emission in the 2–10 μm wavelength range for fiber lasers is challenging due to the high transmission loss in that band and the lack of suitable active fibers. Hence, fiber lasers, which offer a compact structure, good wavelength flexibility, high output power, high spatial beam quality, efficient heat dissipation, and excellent temporal stability, are favored as a powerful alternative for visible and MIR lasing generation by nonlinear frequency conversion.

With the growing interest in new scientific and industrial applications, frequency-converted fiber laser systems have been developed rapidly. In recent years, the performances of frequency-converted fiber lasers, including output power, conversion efficiency, wavelength tuning range, and reliability, have been improved via optimizing the pumping schemes, nonlinear media, and other related technologies. In this chapter, we will look into the general principles and recent progress in the nonlinear frequency conversion of fiber lasers.

13.2 Visible Light Based on Frequency Conversion of Fiber Lasers

With the significant advancements in fiber laser technology and the development of quasi-phase-matched (QPM) and birefringent nonlinear materials, the realization of visible fiber lasers has become possible. To enable the wavelength range extension of near-infrared fiber lasers toward shorter visible wavelengths, several nonlinear frequency conversion techniques, such as frequency doubling for second-harmonic generation (SHG), sum-frequency generation (SFG), and optical parametric oscillator (OPO), have emerged as promising solutions. The fundamental sources of these nonlinear frequency conversion processes can be rare-earth-doped fiber lasers, Raman fiber lasers, incoherent fiber lasers, and random Raman fiber lasers, in which the availability of specific lasing wavelengths in these sources ultimately determines the achievable emission wavelengths of the visible light.

13.2.1 *Frequency Conversion of Conventional Cavity-Based Fiber Lasers*

13.2.1.1 Frequency Doubling of Fiber Lasers

In the aspect of frequency conversion of fiber lasers, SHG is a widely used method. To achieve frequency doubling of fiber lasers, two primary approaches are employed: single-pass SHG and resonantly enhanced SHG.

In a single-pass SHG system for fiber lasers, a near-infrared fiber laser is utilized as the fundamental laser source, which is frequency doubled in a nonlinear crystal. Conventionally, periodically poled nonlinear crystals are employed to achieve efficient and compact wavelength conversion of fiber lasers. However, the power scaling of frequency-converted fiber lasers using periodically poled nonlinear crystals is limited by thermal dephasing. Then, QPM nonlinear crystals, including PP-MgLN and PP-MgSLT crystals, have been developed and successfully applied in high-power visible light generation by frequency doubling of fiber lasers.

Referring to the fundamental sources for frequency doubling, rare-earth-doped fiber lasers and Raman fiber lasers are the basic ones. Among these, ytterbium-doped fiber lasers (YFLs) have been widely used in green lasing generation due to their efficient radiation and broad wavelength tunability without excited state absorption or concentration quenching effects. The frequency doubling occurring in a KTP nonlinear crystal with a YFL as a fundamental source in a single-pass configuration has been experimentally verified [4]. As shown in Fig. 13.1, the output

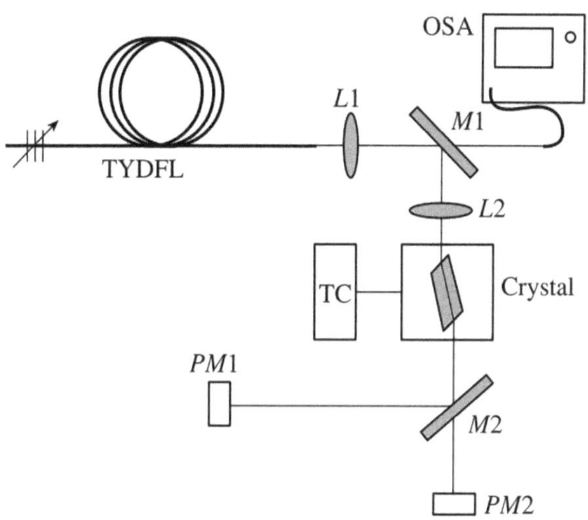

Fig. 13.1 Schematic diagram for the frequency doubling of the YFL [4]

beam of the YFL is collimated by an aspheric lens $L1$. Then, the collimated light beam is reflected by a highly reflective mirror $M1$ and focused by another lens $L2$ inside the KTP nonlinear crystal, which is put inside a thermostat to maintain the temperature at a fixed level. In such a frequency-doubling laser system, the parameters of the lens should be optimized to meet the system requirements. The generated second-harmonic radiation and the fundamental lasing are separated by a selective mirror $M2$ and injected into $PM1$ and $PM2$, respectively. In addition, by tuning the wavelength of the YFL from 1080 to 1120 nm, the generated visible fiber lasing shifts from 540 nm to 560 nm, correspondingly. Since the operation of YFL at short wavelengths is also known to be possible with relatively short active fibers, S. I. Kablukov et al. demonstrated the 515 nm SHG of a 1030 nm YFL by optimizing the length of the ytterbium-doped fiber [5]. Additionally, a tunable green fiber laser in the range of 510–520 nm was realized using a 1017–1040 nm tunable YFL as the fundamental source and frequency doubling in a KTP crystal [6]. Furthermore, bidirectional blue lasers at 490 nm were obtained in a single-pass configuration using two highly nonlinear BIBO crystals with a quasi-three-level YFL emitting at 980 nm, achieving a total output power of 15 mW at 490.8 nm [7].

To achieve frequency doubling in the pure blue spectral range, fiber laser sources emitting at wavelengths shorter than 920 nm are essential. For this purpose, Nd-doped fiber lasers (NFLs) are particularly suitable. Operating on a three-level transition, these lasers enable continuous lasing radiation in the 872–936 nm range. By frequency doubling a 900 nm NFL within an LBO crystal, a 7.5 W blue laser at 450 nm was demonstrated [8]. Concerning the power of the 900 nm laser source coupled into the resonant cavity, the conversion efficiency achieved was 55%.

Raman fiber lasers possess the unique advantage of good wavelength flexibility, which is available at arbitrary wavelengths across the transparency window of the fiber. Leveraging their wide wavelength range (1120–1550 nm) and high output power, Raman fiber lasers can be used for SHG to produce visible fiber lasers with longer wavelengths. For example, by employing a YFL-pumped high-power Raman fiber laser, emitting at 1178 nm, as the fundamental source, the single-pass frequency doubling to 589 nm yellow lasing in a PP-MgLN crystal with an output power of 3 W was presented [9]. Moreover, a Raman fiber laser with an operating wavelength of 1310 nm has been utilized to generate the second harmonic of 655 nm red light with a power of >60 mW in a single-pass configuration based on a PP-MgLN crystal [10].

To further scale the power of the laser systems, coherent beam combining (CBC) of ultrafast fiber lasers is a powerful approach. Compared to the CBC of the single-core fibers, the multicore fiber-based CBC can be achieved with no need for individual beam alignment, enabling more efficient integration, due to the cores being inherently arranged into a dense array. In [11], the frequency doubling of a Q-switched oscillator-amplifier system based on a 4 × 4 Yb-doped tapered MCF was described to generate a high-repetition-rate green laser. A 49 mJ total energy in 30-ns pulses was achieved, which promises not only for CBC but also for applications that can be driven by a high-energy incoherent superposition of laser light.

Directly converting the frequency of the fiber laser in a single-pass way requires a high incident pump power and a crystal with a high nonlinear coefficient, which restricts the output power of the frequency-doubling system. Another method for SHG is employing an external resonant cavity, in which a single-frequency fundamental source and an active control of the cavity length are needed. In addition, LBO crystals with a high damage threshold and a good thermal performance are adopted in the external resonantly enhanced cavities to achieve high-power SHG. X. Zeng et al. demonstrated a high-power and high-efficiency cavity-enhanced SHG based on a 1064 nm periodically phase-modulated fundamental laser [12]. By employing periodic phase modulation on a single-frequency laser, it was converted to a multiple-frequency laser with a spacing equal to the fundamental modulation period. By adjusting the frequency spacing of the modulated multiple-frequency laser to equal or be an integer multiple of the free spectral range of the SHG cavity, the resonance can be enhanced after cavity locking, which is helpful for power scaling of the frequency-doubled lasers. As a result, a 532 nm laser with a high output power of 30 W and a high conversion efficiency of 80% is realized.

13.2.1.2 Sum-Frequency Generation of Fiber Lasers

In the SHG process of the fiber lasers, the spectral range of the generated lasing is determined by the doubled frequency of the fundamental laser sources. To extend the emitting wavelength of the visible fiber lasers, SFG of two fiber lasers or a fiber laser with two operating wavelengths in a nonlinear crystal is an alternative method.

Via employing the 1.5 μm Er-doped or Er/Yb co-doped and 1 μm Yb-doped lasing for SFG, red fiber lasers can be generated. Based on a dual-wavelength EYFL with the operating wavelength at both 1064 and 1550 nm, the 631 nm red fiber lasing can be generated by the SFG of the two lightwaves [13]. In the EYFL, the Yb ions can both stimulate 1 μm lasing emission and transfer the energy to the Er ions for 1.5 μm lasing generation. The schematic diagram of the dual-wavelength fiber laser is shown in Fig. 13.2a. The dual-wavelength laser is fabricated in a 5 m-long Er/Yb co-doped fiber. A broadband metallic mirror is used as a point reflector at the pump side of the laser. The lasing beam is collimated by a lens and reflected by a diffraction grating at the output end. By rotating the mirrors M_3 and M_4 independently, the emitting wavelength can be tuned. By tuning the wavelength of the ytterbium-doped fiber lasing in the range of 1054–1074.5 nm, maintaining the wavelength of the erbium-doped fiber lasing at 1550 nm, and adjusting the temperature of the PPLN crystal from 25 °C to 140 °C, the red fiber lasing wavelength generated in the PPLN crystal was successfully tuned from 627.5 to 634.5 nm, which can be seen in Fig. 13.2b.

Moreover, hybrid frequency conversion methods can be adopted in one laser system to achieve wavelength extension. In [14], a 1560 nm laser source was divided into two portions, with one portion of the laser being frequency doubled in an MgO:PPLN crystal to 780 nm. Then, the generated 780 nm fiber lasing and the other

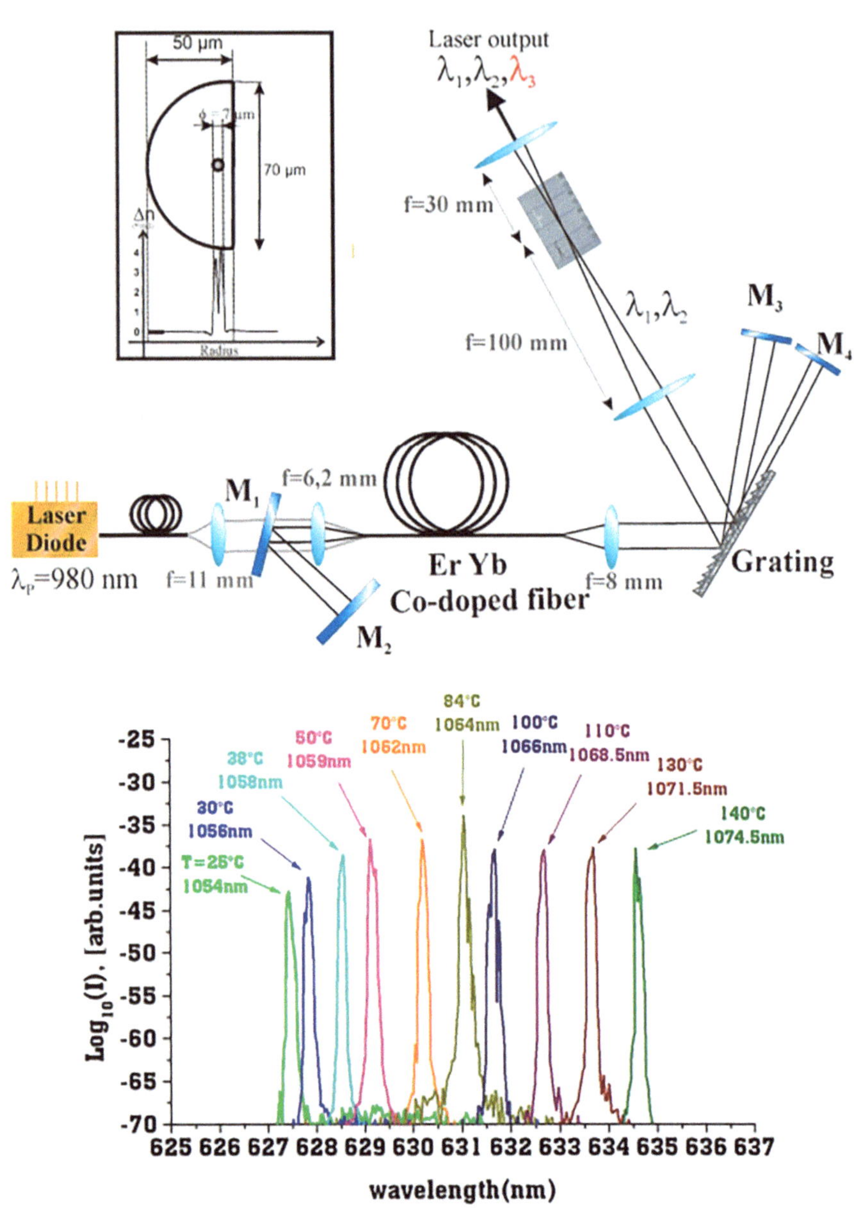

Fig. 13.2 (**a**) Experimental setup of the dual-wavelength EYFL. (**b**) Tuning spectra of the red fiber lasing [13]

portion of the original 1560 nm fiber lasing are employed for SFG. As a result, a 520 nm CW lasing is achieved in a KTP crystal with an output power of 268 mW. In addition, by SFG of a 1064 nm fiber laser and a 775 nm fiber laser, which is produced by an SHG process with a 1550 nm fundamental laser, the 448 nm blue fiber laser was realized [15].

13.2.1.3 OPO of Fiber Lasers

To extend the working wavelength of fiber lasers, an optical parametric oscillator (OPO) is a promising choice for generating coherent light at every wavelength in a broad spectral range from the visible to the mid-infrared range. In an OPO, a signal and an idler lasing are generated in the nonlinear conversion process via pumping by a high-power laser source. By phase matching, wavelength tuning of the signal and idler wave can be achieved. In order to generate visible lasing in the fiber laser-pumped OPO, a combination of nonlinear optical processes, including intracavity frequency doubling of a resonant infrared idler [16] or SFG of the generated signal or idler lasing with the pump lasing [17], is required.

By utilizing the OPO with the SFG in the same resonator, a 605–616 nm tunable laser has been demonstrated [18]. The schematic diagram of a visible fiber laser system based on the combination of OPO and SFG by a 1064 nm fiber laser is shown in Fig. 13.3. The pump source is an amplified 1064 nm fiber laser controlled by a piezoelectric transducer with the output power of 16 W and a frequency jitter of <100 kHz. To enable both the OPO and SFG processes, the pump laser is injected into a matched MgO: PPLN nonlinear crystal. Two dichroic mirrors behind the system are used to separate the lasing beam with different wavelengths. In the MgO: PPLN crystal, the OPO process occurs first. The poling period changes linearly from 27.2 to 28.9 μm perpendicular to the beam propagation direction, enabling tuning the signal wavelength from 1386 to 1445 nm via shifting the position of the nonlinear crystal in the resonator. In the second stage of the MgO: PPLN crystal, the

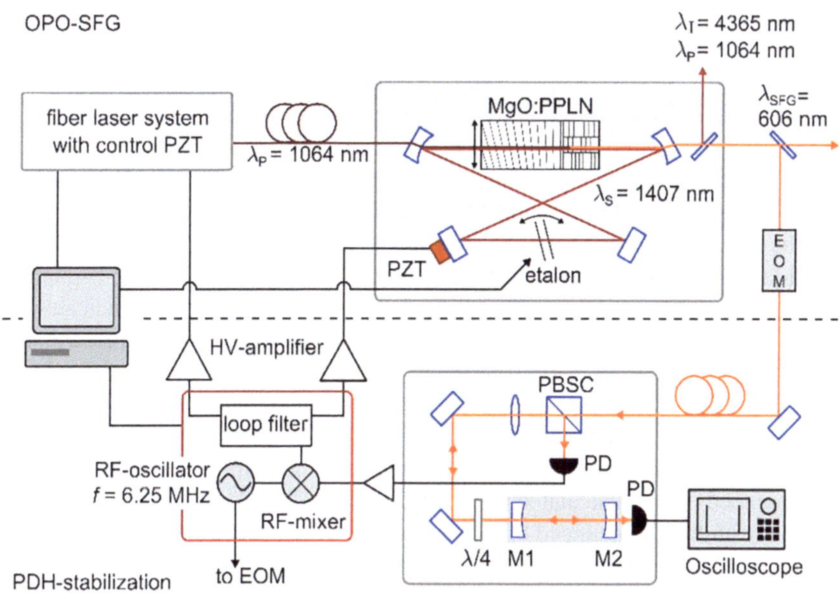

Fig. 13.3 Experimental setup of the visible laser system based on the combination of OPO and SFG [18]

SFG process takes place with the poling periods well designed to obtain a large tuning range around the wavelength of the lasing beam generated by the SFG of the 1064 nm pump and the 1.4 μm signal. In this configuration, the operating wavelength of the lasing generated by SFG can be varied by changing the crystal position. Thus, the emitting wavelength of the OPO can be tuned by continuously changing the poling periods. Besides, the SFG-generated lasing wavelength can be changed via adjusting the crystal temperature with simultaneous phase matching of both OPO and SFG processes. The methods to enhance the efficiency of the visible OPO with the help of SFG are further investigated [19]. The results indicate that to obtain high SFG power, the resonating signal beam waist should be decreased.

The OPO pumped by a YFL can also be used to stimulate a visible frequency comb for cavity-enhanced ultrafast transient absorption spectroscopy [20]. The 1070 nm pump comb with pulse duration, pulse energy, and repetition rate of 120 fs, 115 nJ, and 100 MHz, respectively, is firstly frequency doubled in an LBO crystal to generate the 535 nm pump lasing for the OPO. In the OPO cavity, another LBO crystal is used for parametric gain, and a BBO crystal is employed for intracavity frequency doubling of the signal and idler. By changing the temperature of the LBO and the cavity length of the OPO, the signal and idler wavelengths can be tuned from 835 to 990 nm and from 1490 to 1160 nm, respectively. The OPO cavity is also designed to intracavity double both the signal and idler beams using a single BBO crystal. Hence, a 420–700 nm continuously tuning frequency comb with a linewidth of 100 kHz was obtained.

13.2.2 Frequency Doubling of Incoherent Amplified Spontaneous Emission Sources

The amplified spontaneous emission (ASE) source can be considered a chaotic light source with low temporal coherence, which has wide applications in low-coherence interferometry, imaging, low-coherence tomography, etc. In [21], the enhancement of the ASE-pumped SHG in a PPLN crystal was investigated. The results revealed that the conversion efficiency of the ASE-pumped SHG is twice that of the conventional laser-pumped one. In addition, the visible lasing power of the ASE-pumped SHG remains almost unchanged with the temperature tuning, which indicates the relaxation of the phase-matching requirements in the nonlinear optical process. The influences of the pulsed ASE source on SHG have then been demonstrated by N. Valero et al. [22]. Compared to the laser pump with the same parameters, the ASE-pumped SHG power with a quadratic behavior is higher, which can be seen in Fig. 13.4a. The output power ratio of two between the ASE- and single-frequency laser-pumped SHG is calculated and plotted in Fig. 13.4b. The results show that the efficiency of the ASE-pumped SHG is twice that of the laser-pumped SHG as well. Nevertheless, the intensity probability density function of the ASE source is fixed with the exponential distribution [22–24], which means the enhancement factor of the ASE-pumped SHG efficiency is limited to some extent.

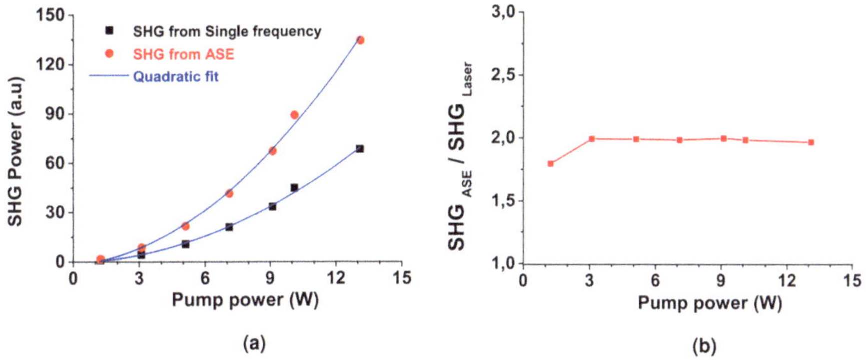

Fig. 13.4 (**a**) Comparison of the output power of the SHG pumped by an ASE source and a single-frequency laser. (**b**) The ratio of the SHG power of the ASE and laser pump [22]

13.2.3 Visible Light Based on Frequency Conversion of Random Fiber Lasers

It has been verified that the frequency doubling efficiency depends on both the bandwidth and the temporal dynamics and statistics of the pump laser [22, 25, 26]. In addition, the intensity fluctuations of the visible lasing are proportional to the square of those of the fundamental source for SHG. Considering these reasons, random fiber lasers based on randomly distributed feedback rather than well-defined resonant cavities, which have unique advantages including good wavelength flexibility, high output power, low temporal noise, etc. [27–29], can be good candidates as the fundamental source for frequency conversion.

Due to the wide wavelength tunability of the random Raman fiber laser based on the stimulated Raman scattering gain, visible lasing emissions in different wavelength ranges are made possible. A 654 nm red random fiber laser was realized by employing a 1308 nm random Raman fiber laser as the fundamental source and utilizing second harmonic generation (SHG) in a PPLN crystal through a single-pass configuration [25]. By adopting a 1178 nm random Raman fiber laser as the fundamental source, a 589 nm yellow random fiber laser with output power and conversion efficiency of 1.09 W and 10.8%, respectively, was demonstrated [30].

Taking advantage of the large Raman frequency shift of the phosphosilicate fiber, a dual-wavelength switchable and tunable random fiber laser was fabricated, in which the spectrum can be manipulated to meet different frequency conversion requirements [31]. As depicted in Fig. 13.5a, a 1 km-long phosphosilicate fiber with both the silica-related and phosphorus-related Raman peaks was utilized to provide both random Rayleigh backscattering feedback and Raman gain. The point feedback unit consists of a fiber collimator, a beam splitter, and two gratings in Littrow configuration to provide wavelength-selective reflection separately. By rotating the angle of the two gratings, the reflection wavelengths can be varied individually. Based on such a point feedback unit and a 1040–1095 nm tunable ytterbium-doped

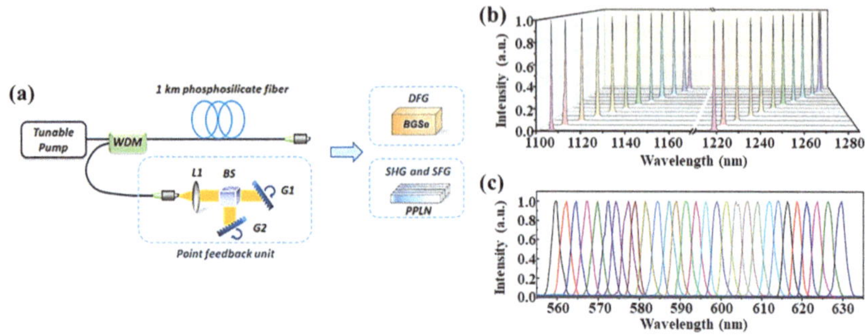

Fig. 13.5 (**a**) Experimental setup of the dual-wavelength switchable and tunable random Raman fiber laser. (**b**) Output spectra of the dual-wavelength-tunable random fiber lasing. (**c**) Tunable spectra of the visible random fiber lasing in the range of 560–630 nm [31]

random fiber laser pump, it is flexible to achieve single- or dual-wavelength-tunable random fiber lasing emission corresponding to the silica- and/or phosphorus-related peaks. By simultaneously changing the pump wavelength and the two gratings' angles, the tunable spectra of the dual-wavelength random fiber lasing in the range of 1106–1160 nm and 1217–1280 nm are measured and shown in Fig. 13.5b. Benefiting from the wavelength flexibility of such a switchable and tunable random Raman fiber laser as the fundamental source, the visible random fiber lasing with a broad tuning range by combining SHG and SFG in a PPLN crystal was realized. By combining the SHG of the tunable single-wavelength random Raman fiber lasing and the SFG of the tunable dual-wavelength random Raman fiber lasing, the visible fiber lasing with a 560–630 nm gap-free tuning range was generated, as illustrated in Fig. 13.5c. These results reveal that the random Raman fiber laser can provide an excellent platform for performing multifunctional nonlinear frequency conversion processes and generating widely tunable visible light.

The ytterbium-doped random fiber laser can be employed to realize frequency doubling for green light stimulation as well [26, 32]. Based on a 1064 nm ytterbium-doped random fiber laser as the fundamental source, a 900 mW green random fiber laser was achieved by frequency doubling in a single-pass SHG [26]. Then, a wavelength-tunable ytterbium-doped random fiber laser was utilized for 529.9–537.3 nm tunable green random fiber lasing generation [32]. Particularly, it is found that based on the statistics property-tailored random fiber laser, by filtering different spectral components, the SHG conversion efficiency can be nearly doubled.

13.3 Mid-Infrared Light Generation Based on Frequency Conversion of Fiber Lasers

In order to construct a MIR fiber laser, one way is to pump active fibers with dopants including Tm^{3+}, Ho^{3+}, etc., which can emit fiber lasing in the 2–3 μm region. However, MIR light sources with longer wavelengths are desired for spectroscopy, free-space communications, imaging, etc. [33–35]. Fiber lasers with unique superiorities are promising to generate MIR radiation via nonlinear optical processes of difference frequency generation (DFG) and OPO in the nonlinear crystals.

13.3.1 DFG of Fiber Lasers

Compared to the quantum cascaded lasers, MIR fiber lasing based on DFG in nonlinear crystals has a broad wavelength tunability. Various nonlinear crystals can be employed for DFG with different operating wavelengths. For the DFG for fiber lasing with a wavelength shorter than 5 μm, a widely available commercial PPLN crystal with a large nonlinear coefficient (~15 pm/V) can be adopted [36, 37]. In the longer wavelength region, MIR lasing enabled by the DFG process has been achieved in orientation-patterned (OP)-GaAs [38] and OP-GaP [39] nonlinear crystals.

The DFG system requires a pump source and an additional driving signal source to radiate an idler lasing via nonlinear wave-mixing in a single-pass process. In particular, by using the 1 μm YFLs and 1.5 μm fiber lasers, such as ErFLs or Raman fiber lasers, as the pump and signal sources for DFG in PPLN crystals, respectively, 3 μm fiber lasers can be easily achieved [36, 40]. The wavelength tuning of the DFG can be achieved by changing the nonlinear crystal temperature and angle and/or varying the pump and signal wavelengths. Figure 13.6a shows the experimental setup of the DFG system based on a YFL and an ErFL [36]. The pump and signal lasings are amplified in two master oscillator-power amplifier systems and then coupled into the PP-MgLN crystal for DFG to stimulate idler lasing emission. The pump beam (the green lines in Fig. 13.6a) is collimated by an aspheric lens L1 and then injected into the dichroic mirror M1, which separates the 1 μm pump lasing and the unabsorbed 960 nm lasing, a half-wave plate, and a high-power isolator. A polarizing beam splitter is used to ensure the transmission of only the e-polarized light to pump the PP-MgLN crystal. Then, the pump beam is reflected by a mirror M2 and combined with the signal beam in a dichroic mirror M3. The signal beam (the yellow lines in Fig. 13.6a) is collimated in a lens L2 and coupled into M3. A lens L3 is employed to couple both the pump and the signal beams into the PP-MgLN crystal, which is mounted on a clip and temperature-controlled. At the output end of the nonlinear crystal, two CaF_2 mirrors, M4 and M5, are utilized to block the unconverted pump and signal lasing, respectively. In such a system configuration, an idler lasing can be stimulated via the DFG process between the pump and the signal beams in the nonlinear crystal. The generated idler lasing (the red lines in Fig. 13.6a) is height-adjusted and collimated by mirrors M6 and M7 and a lens L4 sequentially. With the pump wavelength tuning from 1040 to 1084.6 nm and the signal wavelength changing between 1545.2 and 1561.4 nm for phase matching, the generated idler wavelength can be shifted from 3117.2 to 3598.8 nm, as depicted in Fig. 13.6b.

The DFG scheme can also be used for pulsed MIR fiber lasing radiation. In 2015, P. Belden et al. proposed a 3.5 μm pulsed MIR laser, which was generated by DFG using a 1064 nm pulsed YFL and a 1525 nm CW laser diode in a PPLN crystal [41]. As a result, the pulsed MIR lasing was achieved with an average power of >1 W and

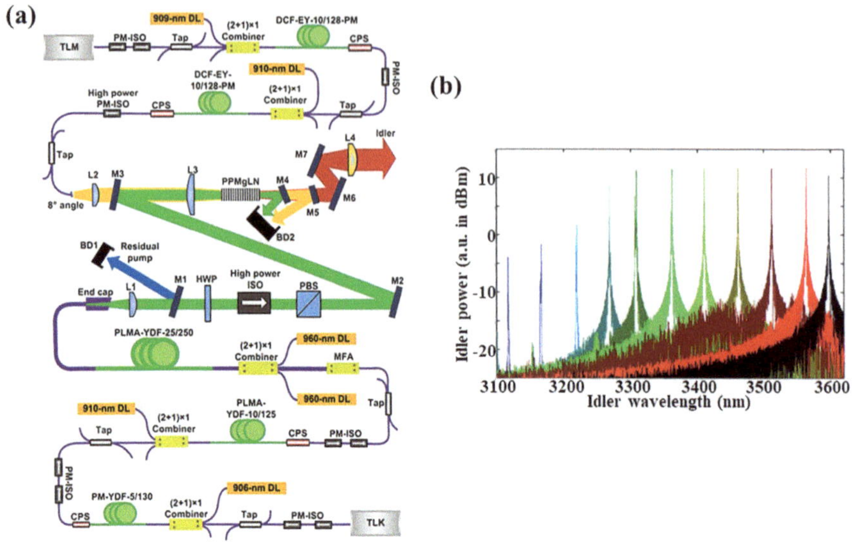

Fig. 13.6 (**a**) Schematic diagram of the wavelength-tunable DFG system based on a YFL and an ErFL. (**b**) Generated tunable spectra of the idler lasing [36]

a pulse energy of 5 μJ. In addition, frequency combs can also be the signal and pump lasers to achieve MIR light by DFG. In [42], a 2.9 μm frequency comb with the average power, repetition rate, and pulse duration of 6.7 W, 100 MHz, and 100 fs, respectively, was generated by the DFG based on a 1035 nm pump frequency comb and a 1500–1650 nm signal frequency comb from an Er-fiber oscillator.

13.3.2 Fiber Lasers Pumped OPO

Since the first demonstration of the fiber laser-pumped OPO based on optical parametric conversion in a PPLN crystal [43], it has been widely used in MIR light generation with broad bandwidth, high power, high efficiency, and a wide tuning range [44–46]. Different from the DFG process, the OPO is driven by only one pump source with no need for high-power 1.5 μm laser sources. To obtain high conversion efficiency, phase matching between the pump and the parametric waves is required by temperature phase matching, angular phase matching, and quasi-phase matching.

Based on a YFL pump, a MIR laser can be generated in an OPO with an MgO: PPLN crystal [47]. The schematic diagram of the OPO system pumped by a 1064 nm high-power linearly polarized YFL is illustrated in Fig. 13.7a. The isolator

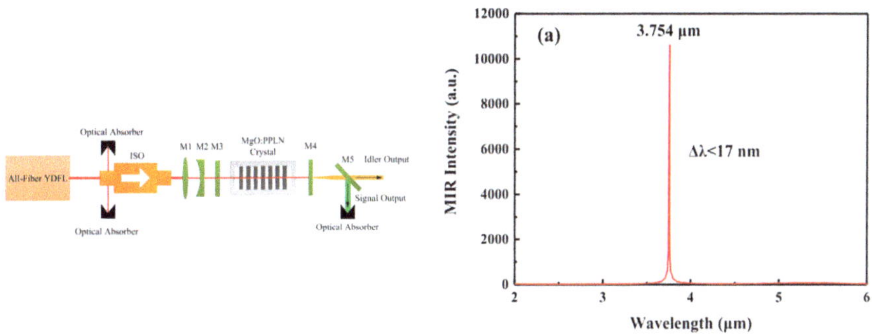

Fig. 13.7 (**a**) Schematic diagram of the YFL-pumped OPO system. (**b**) Output spectrum of the generated MIR lasing in the OPO system [47]

connected to the YFL pump is employed to avoid unwanted reflections. The polarization state of the pump is aligned along the principal axis of the nonlinear crystal to ensure high conversion efficiency. The output pump lasing is collimated by the lenses M1 and M2 and then coupled into the MgO: PPLN crystal. The two ends of the MgO: PPLN crystal are both antireflection-coated for wavelengths of 1.06, 1.48, and 3.8 μm. The OPO cavity is constructed by the nonlinear crystal and two mirrors, M3 and M4. In addition, a dichroic mirror M5 is used to separate the signal lasing and the MIR idler lasing. As shown in Fig. 13.7b, the 3.754 μm idler lasing was stimulated in the OPO process with the output power of 10.82 W and the linewidth of <17 nm by optimizing the repetition rate, pulse width, and waist diameter of the YFL pump. Moreover, a thulium-doped fiber laser operating at 1.95 μm can also be utilized as the pump source for an OP-GaAs crystal-based OPO system, which can stimulate idler lasing with the wavelength, average output power, and conversion efficiency of ~10.6 μm, 800 mW, and 36.8%, respectively [48].

To achieve wavelength-tunable OPO systems, several technologies can be employed, including crystal temperature adjustment, crystal angle changing, and pump wavelength tuning. Due to the fact that the temperature of the nonlinear crystal cannot change rapidly and needs several minutes to become stable, fast wavelength tuning of the OPO in this way is difficult [36]. It is hard to obtain continuous wavelength tuning of the OPO system by changing the crystal angle. Thus, pump wavelength tuning is a promising method to achieve a tunable OPO with fast and continuous tunability. Benefiting from a 1060 nm and 1080 nm dual-wavelength YFL pump for the OPO system, a three-wavelength MIR lasing with a total output power of 7.38 W and the wavelengths of 3127, 3303, and 3346 nm, respectively, was presented [49]. On the other hand, with a dual-wavelength fiber laser pump, in which one wavelength is fixed at 1060 nm and the other wavelength is tuned from 1070 nm to 1090 nm, two idler emissions with one wavelength at 3140 nm and the other wavelength tuning from 3227 nm to 3413 nm in the OPO were realized [50].

13.3.3 Mid-Infrared Light Sources Based on Frequency Conversion of Random Fiber Lasers

Furthermore, random fiber lasers can also be promising candidates as the seed sources for MIR lasing generation based on nonlinear frequency conversion. Via the single-pass nonlinear frequency conversion process, the stimulated MIR lasing can inherit the properties of a modeless frequency spectrum, temporal stability, and low coherence of the random fiber lasers.

Taking advantage of the good wavelength flexibility of the random fiber lasers, tunable MIR random fiber lasing in different wavelength regions can be obtained via DFG, which breaks the limits of the tunable range for DFG based on conventional YFLs and ErFLs [51, 52]. For example, based on a YRFL and a cascaded random Raman fiber laser as the pump and signal sources, respectively, the DFG process can be enabled in a BGSe crystal in a single-pass configuration, as shown in [51]. Both the YRFL and the cascaded random Raman fiber laser are wavelength tunable in the range of 1038–1091 nm and 1213–1281 nm, respectively. Thus, a 5.5–9.5 μm continuously tunable MIR random fiber lasing was realized. Besides the MIR random fiber lasing radiated by DFG of two random fiber lasers with different types, it can be achieved based on cascaded random Raman fiber lasers with different wavelengths. By utilizing the dual-wavelength-tunable cascaded random Raman fiber laser as the source for DFG, an experimental demonstration of tunable MIR random fiber lasing radiation in the 10.7–12.3 μm range was successfully realized [31].

On the other hand, random fiber lasers are suitable for OPO pumping as well [53, 54]. By utilizing a 1089.78–1103.25 nm tunable random Raman fiber laser as the pump source for the OPO system with an MgO: PPLN nonlinear crystal, an MIR idler lasing was obtained, which can be continuously tuned in the range of 3977.34–4059.65 nm with an output power of ~2 W, which can be seen in Figs. 13.8a and 13.8b, respectively [54].

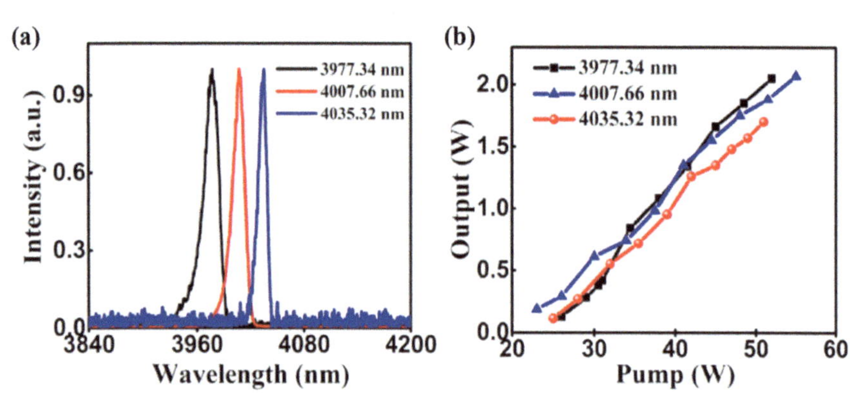

Fig. 13.8 (**a**) Tunable output spectra of the idler light. (**b**) Output power of the idler light versus the pump power [54]

13.4 Summary

In this chapter, we introduced the technology and recent progress of nonlinear frequency conversion of fiber lasers to visible and MIR spectral regions in the aspects of SHG, SFG, OPO, and DFG schemes of rare-earth-doped fiber lasers, Raman fiber lasers, and random fiber lasers. It is anticipated that the special wavelength fiber lasers generated by frequency-converted fiber lasers will further enhance the importance of fiber lasers in the coming years.

References

1. Zhang, L., Jiang, H., Cui, S., et al.: Versatile Raman fiber laser for sodium laser guide star. Laser Photonics Rev. **8**(6), 889–895 (2014)
2. Qiao, S., He, Y., Sun, H., et al.: Ultra-highly sensitive dual gases detection based on photoacoustic spectroscopy by exploiting a long-wave, high-power, wide-tunable, single-longitudinal-mode solid-state laser. Light Sci. Appl. **13**(1), 100 (2024)
3. Yao, Y., Hoffman, A.J., Gmachl, C.F.: Mid-infrared quantum cascade lasers. Nat. Photonics. **6**(7), 432–439 (2012)
4. Akulov, V.A., Babin, S.A., Kablukov, S.I., et al.: Fiber lasers with a tunable green output. Laser Phys. **18**, 1225–1229 (2008)
5. Kablukov, S.I., Dontsova, E.I., Akulov, V.A., et al.: Frequency doubling of Yb-doped fiber laser to 515 nm. Laser Phys. **20**, 360–364 (2010)
6. Dontsova, E.I., Kablukov, S.I., Babin, S.A.: Ytterbium-doped fibre laser tunable in the range 1017–1040 nm with second-harmonic generation. Quantum Electron. **43**(5), 467–471 (2013)
7. Zou, S., Li, P., Wang, L., et al.: 980 nm Yb-doped single-mode fiber laser and its frequency-doubling with BIBO. Appl. Phys. B. **95**, 685–690 (2009)
8. Leconte, B., Gilles, H., Robin, T., et al.: 7.5 W blue light generation at 452 nm by internal frequency doubling of a continuous-wave Nd-doped fiber laser. Opt. Express. **26**(8), 10000–10006 (2018)
9. Georgiev, D., Gapontsev, V.P., Dronov, A.G., et al.: Watts-level frequency doubling of a narrow line linearly polarized Raman fiber laser to 589 nm. Opt. Express. **13**(18), 6772–6776 (2005)
10. Kablukov, S.I., Babin, S.A., Churkin, D.V., et al.: Frequency doubling of a broadband Raman fiber laser to 655 nm. Opt. Express. **17**(8), 5980–5986 (2009)
11. Aleshire, C., Eichner, T., Steinkopff, A., et al.: Frequency-doubled q-switched 4 × 4 multicore fiber laser system. Opt. Lett. **48**(8), 2198–2201 (2023)
12. Zeng, X., Cui, S., Cheng, X., et al.: Resonant frequency doubling of phase-modulation-generated few-frequency fiber laser. Opt. Lett. **45**(17), 4944–4947 (2020)
13. Boullet, J., Lavoute, L., Berthelemot, A.D., et al.: Tunable red-light source by frequency mixing from dual band Er/Yb co-doped fiber laser. Opt. Express. **14**(9), 3936–3941 (2006)
14. Guo, S., Ge, Y., He, J., et al.: Singly resonant sum-frequency generation of 520-nm laser via a variable input-coupling transmission cavity. J. Mod. Opt. **62**(19), 1583–1590 (2015)
15. Anderegg, J.P., Chernysheva, T.A., Elkins, D.F., et al.: RGB laser generation from fiber MOPAs coupled to external enhancement cavities, fiber lasers VII: technology, systems, and applications. SPIE. **7580**, 130–144 (2010)
16. Mhibik, O., Paboeuf, D., Drag, C., et al.: Sub-kHz-level relative stabilization of an intracavity doubled continuous wave optical parametric oscillator using Pound-Drever-Hall scheme. Opt. Express. **19**(19), 18049–18057 (2011)

17. Lin, S.T., Lin, Y.Y., Tu, R.Y., et al.: Fiber-laser-pumped CW OPO for red, green, blue laser generation. Opt. Express. **18**(3), 2361–2367 (2010)
18. Mieth, S., Henderson, A., Halfmann, T.: Tunable, continuous-wave optical parametric oscillator with more than 1W output power in the orange visible spectrum. Opt. Express. **22**(9), 11182–11191 (2014)
19. Ghavami Sabouri, S., Khorsandi, A.: Effects of resonator design on the efficiency of singly-resonant optical parametric oscillator with intracavity sum-frequency generation. Opt. Quant. Electron. **47**, 3349–3363 (2015)
20. Chen, Y., Silfies, M.C., Kowzan, G., et al.: Tunable visible frequency combs from a Yb-fiber-laser-pumped optical parametric oscillator. Appl. Phys. B. **125**, 81 (2019)
21. Arahira, S., Murai, H.: Wavelength conversion of incoherent light by sum-frequency generation. Opt. Express. **22**(11), 12944–12961 (2014)
22. Valero, N., Marion, D., Lhermite, J., et al.: High-power amplified spontaneous emission pulses with tunable coherence for efficient non-linear processes. Sci. Rep. **11**(1), 4844 (2021)
23. Liu, W., Ma, P., Lv, H., et al.: Investigation of stimulated Raman scattering effect in high-power fiber amplifiers seeded by narrow-band filtered superfluorescent source. Opt. Express. **24**(8), 8708–8717 (2016)
24. Gorbunov, O.A., Sugavanam, S., Vatnik, I.D., et al.: Poisson distribution of extreme events in radiation of random distributed feedback fiber laser. Opt. Lett. **45**(8), 2375–2378 (2020)
25. Dontsova, E.I., Kablukov, S.I., Vatnik, I.D., et al.: Frequency doubling of Raman fiber lasers with random distributed feedback. Opt. Lett. **41**(7), 1439–1442 (2016)
26. Rota-Rodrigo, S., Gouhier, B., Dixneuf, C., et al.: Watt-level green random laser at 532 nm by SHG of a Yb-doped fiber laser. Opt. Lett. **43**(17), 4284–4287 (2018)
27. Turitsyn, S.K., Babin, S.A., El-Taher, A.E., et al.: Random distributed feedback fibre laser. Nat. Photonics. **4**(4), 231–235 (2010)
28. Han, B., Cheng, Q., Tao, Y., et al.: Spectral manipulations of random fiber lasers: principles, characteristics, and applications. Laser Photonics Rev., 2400122 (2024)
29. Han, B., Ma, Y., Zhao, Y., et al.: The applications of random fiber lasers in optical fiber communication and sensing systems: a review. IEEE Trans. Instrum. Meas. **7001017** (2024)
30. Cui, S., Qian, J., Zeng, X., et al.: A watt-level yellow random laser via single-pass frequency doubling of a random Raman fiber laser. Opt. Fiber Technol. **64**, 102552 (2021)
31. Wu, H., Wang, W., Hu, B., et al.: Widely tunable continuous-wave visible and mid-infrared light generation based on a dual-wavelength switchable and tunable random Raman fiber laser. Photonics Res. **11**(5), 808–816 (2023)
32. Wu, H., Liu, H., Wang, W., et al.: Tailoring the efficiency and spectrum of a green random laser generated by frequency doubling of random fiber lasers. Opt. Express. **29**(14), 21521–21529 (2021)
33. Ghimire, S., Reis, D.A.: High-harmonic generation from solids. Nat. Phys. **15**(1), 10–16 (2019)
34. Vasilyev, S., Moskalev, I.S., Smolski, V.O., et al.: Super-octave longwave mid-infrared coherent transients produced by optical rectification of few-cycle 2.5-μm pulses. Optica. **6**(1), 111–114 (2019)
35. Seidel, M., Xiao, X., Hussain, S.A., et al.: Multi-watt, multi-octave, mid-infrared femtosecond source. Sci. Adv. **4**(4), eaaq1526 (2018)
36. Zhao, J., Jia, F., Feng, Y., et al.: Continuous-wave 3.1–3.6 μm difference-frequency generation of dual wavelength-tunable fiber sources in PPMgLN-based rapid-tuning design. IEEE J. Sel. Top. Quantum Electron. **24**(3), 3100208 (2017)
37. Krzempek, K., Sobon, G., Abramski, K.M.: DFG-based mid-IR generation using a compact dual-wavelength all-fiber amplifier for laser spectroscopy applications. Opt. Express. **21**(17), 20023–20031 (2013)
38. Devi, K., Schunemann, P.G., Ebrahim-Zadeh, M.: Continuous-wave, multimilliwatt, mid-infrared source tunable across 6.4–7.5 μm based on orientation-patterned GaAs. Opt. Lett. **39**(23), 6751–6754 (2014)

39. Devi, K., Padhye, A., Schunemann, P.G., et al.: Multimilliwatt, tunable, continuous-wave, mid-infrared generation across 4.6–4.7 μm based on orientation-patterned gallium phosphide. Opt. Lett. **43**(10), 2284–2287 (2018)
40. Guha, S., Barnes, J.O., Gonzalez, L.P.: Multiwatt-level continuous-wave midwave infrared generation using difference frequency mixing in periodically poled MgO-doped lithium niobate. Opt. Lett. **39**(17), 5018–5021 (2014)
41. Belden, P., Chen, D.W., Teodoro, F.D.: Watt-level, gigahertz-linewidth difference-frequency generation in PPLN pumped by an nanosecond-pulse fiber laser source. Opt. Lett. **40**(6), 958–961 (2015)
42. Catanese, A., Rutledge, J., Silfies, M.C., et al.: Mid-infrared frequency comb with 6.7 W average power based on difference frequency generation. Opt. Lett. **45**(5), 1248–1251 (2020)
43. Gross, P., Klein, M.E., Walde, T., et al.: Fiber-laser-pumped continuous-wave singly resonant optical parametric oscillator. Opt. Lett. **27**(6), 418–420 (2002)
44. Mo, Q., Li, S., Liu, Y., et al.: Widely tunable optical parametric oscillator in periodically poled congruently grown lithium tantalite whispering gallery mode resonators. Chin. Opt. Lett. **14**(9), 091902 (2016)
45. Xu, L., Chan, H.Y., Alam, S.U., et al.: Fiber-laser-pumped, high-energy, mid-IR, picosecond optical parametric oscillator with a high-harmonic cavity. Opt. Lett. **40**(14), 3288–3291 (2015)
46. Shukla, M.K., Das, R.: High-power single-frequency source in the mid-infrared using a singly resonant optical parametric oscillator pumped by Yb-fiber laser. IEEE J. Sel. Top. Quantum Electron. **24**(5), 5100206 (2018)
47. He, Y., Ji, Y., Wan, H., et al.: High-power mid-infrared pulse MgO: PPLN optical parametric oscillator pumped by linearly polarized Yb-doped all-fiber laser. Opt. Laser Technol. **146**, 107545 (2022)
48. Wueppen, J., Nyga, S., Jungbluth, B., et al.: 1.95 μm-pumped OP-GaAs optical parametric oscillator with 10.6 μm idler wavelength. Opt. Lett. **41**(18), 4225–4228 (2016)
49. Wang, P., Shang, Y., Li, X., et al.: Multiwavelength mid-infrared laser generation based on optical parametric oscillation and intracavity difference frequency generation. IEEE Photonics J. **9**(1), 1500107 (2017)
50. Wang, P., Cheng, X., Li, X., et al.: Tunable dual-wavelength, continuous-wave, mid-infrared generation using intracavity difference frequency mixing in PPLN-based optical parametric oscillator. IEEE J. Sel. Top. Quantum Electron. **24**(5), 0902807 (2018)
51. Wu, H., Wang, W., Li, Y., et al.: Difference-frequency generation of random fiber lasers for broadly tunable mid-infrared continuous-wave random lasing generation. J. Lightwave Technol. **40**(9), 2965–2970 (2022)
52. Hu, B., Wu, H., Tian, K., et al.: Continuous-wave 2.9–3.8 μm random lasing via temperature-tunning free difference-frequency generation of random fiber lasers in PPLN crystal. Sci. China Inf. Sci. **66**(8), 189401 (2023)
53. Shang, Y., Shen, M., Wang, P., et al.: Amplified random fiber laser-pumped mid-infrared optical parametric oscillator. Chin. Opt. Lett. **14**(12), 121901 (2016)
54. Wu, H., Wang, P., Song, J., et al.: High power tunable mid-infrared optical parametric oscillator enabled by random fiber laser. Opt. Express. **26**(5), 6446–6455 (2018)

Chapter 14
Development of Pulse Fiber Laser Based on Two-Dimensional Materials

Cong Wang

Abstract Two-dimensional materials (2DMs) are crystals with one to a few layers of atoms and are being used in many fields such as optical modulator, photodetector, optical switch, and ultrafast lasers. Their exceptional opto-electronic and nonlinear optical properties make them as a suitable saturable absorber for laser cavities. This chapter focuses on the recent progress in ultrafast laser use 2DM as a saturable absorber (SA). 2DMs traditionally include graphene, topological insulators (TIs), transition metal dichalcogenides (TMDs), as well as new materials such as black phosphorus (BP), bismuthene, antimonene, and MXenes. Material characteristics, fabrication techniques, and nonlinear properties are also introduced. Finally, future perspectives of ultrafast lasers based on 2D materials are also addressed.

Keywords Ultrafast laser · Two-dimensional materials · Graphene · Transitional metal dichalcogenides · Topological insulator · Black phosphorus · MXenes

14.1 Introduction

In the fields of ultrafast photonics, two-dimensional materials (2DMs) are characterized by their broadband saturable absorptions, ultrafast recovery, large nonlinear refractive indices, and potential as outstanding mode-lockers for ultrafast fiber laser. What follows is a brief overview of 2DMs' atomic structures, bandgap structures, and recovery times. Table 14.1 and Fig. 14.1 show a brief comparison of the 2DM family. Later in this section, corresponding material transfer and fiber integration techniques are also introduced.

C. Wang (✉)
School of Microelectronics, Shenzhen Institute of Information Technology, Shenzhen, P. R. China

Tsinghua Shenzhen International Graduate School, & Center of Double Helix, Tsinghua University, Shenzhen, P. R. China

M. F. S. Ferreira, M. C. Paul (eds.), *Optical Fiber Lasers*, Springer Series in Optical Sciences 250, https://doi.org/10.1007/978-3-032-04110-4_14

Table 14.1 2D material family for ultrafast fiber laser

Type	Graphene [68, 69]	TIs [70–72]	TMDs [73–75]	BP [76, 77]	MXenes [40]	Bismuthene [78–80]
Atomic structure	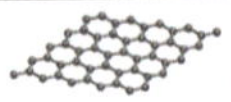	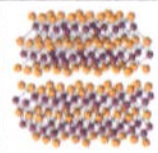	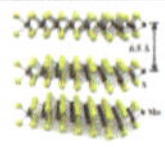	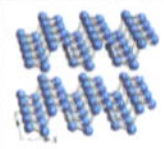	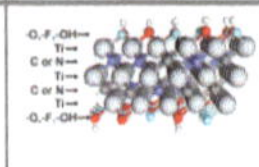	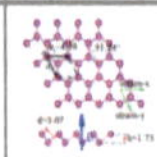
Band structure	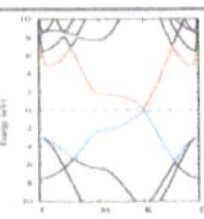	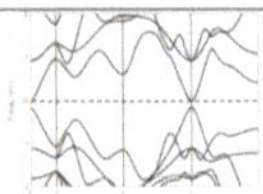	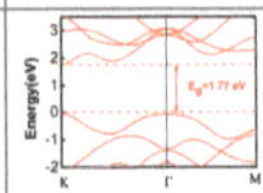	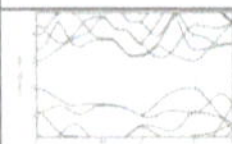	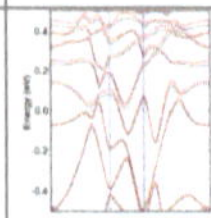	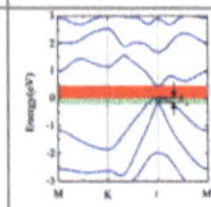
Bandgap	0 eV (monolayer) 0.25 eV (bilayer)	0 ~ 0.7 eV	1 ~ 2.5 eV	0.35 ~ 2 eV	<0.2 eV	0.5 eV (monolayer)
Carrier lifetime	Fast: <200 fs Slow: ~1 ps	Fast: 0.2 ~ 2 ps Slow: 3 ~ 23 ps	Fast: ~1–3 ps Slow: 70 ~ 400 ps	Fast: 360 fs Slow: ~1.36 ps	–	Fast: 3 ps Slow: 420 ps

Graphene, the firstly discovered 2DM, is a single layer of carbon atoms arranged in a 2D honeycomb lattice. Benefiting from its gapless Dirac cone, monolayer graphene is calculated to possess about 2.3% absorption of incident visible to infrared (IR) light. Graphene is special in that it enjoys ultrashort recovery time (<200 fs), low saturable absorption (~10 MW/cm^2), great relative modulation depth (>60% per layer), and wavelength-independent operation (ranging from the visible to the terahertz), allowing it to operate efficiently for the generation of broadband ultrafast laser pulses.

TIs define a new kind of material with nontrivial, symmetry-protected topological order that behaves as insulators in their interior but whose surfaces contain gapless conducting states. A small indirect bulk bandgap of 0.2–0.3 eV gives TIs a strong graphene-like broadband nonlinear response from the visible to the mid-IR. The lifetime of their phonon-induced carriers is short of several picoseconds, also making them useful for ultrafast light modulators. Three kinds of TIs, namely Sb_2Te_3, Bi_2Se_3, and Bi_2Te_3, are the most widely used TI saturable absorbers (SAs).

TMDs are a class of more than 40 different semiconductors that share a formula of MX_2, where M stands for a transition metal (e.g., Mo, W, Ti, Nb) and X stands for a chalcogen (e.g., S, Se, or Te). In a TMD monolayer, the single transition metal layer is sandwiched between the two chalcogen layers, showing graphene-like layered structure. As for different monolayer TMDs, they have energy bandgaps varying from 1 to 2.5 eV. Surprisingly, the sub-bandgaps created by the edge states of TMDs allow efficient absorptions of light with photon energy much lower than their normal bandgaps. Meanwhile, the short recovery times of TMDs are several picoseconds, which are also fast enough for ultrafast light modulation. For example, MoS_2, $MoTe_2$, WSe_2, and WS_2 have been widely used for ultrafast fiber laser in 1.5- and 2-μm wavelength regions.

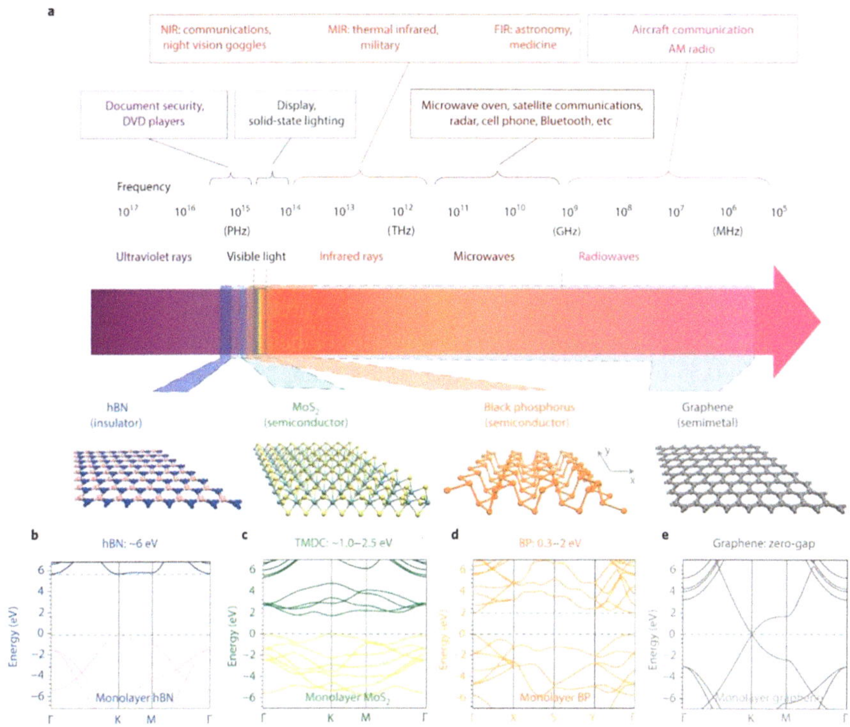

Fig. 14.1 (**a**) Electromagnetic spectrum. Applications that utilize the different spectral ranges are presented in the top portion of the panel. NIR, MIR, and FIR indicate near-, mid-, and far-infrared, respectively. The atomic structures of hBN, MoS_2, BP, and graphene are shown at the bottom of the panel, left to right. The crystalline directions (*x* and *y*) of anisotropic BP are indicated. The possible spectral ranges covered by different materials are indicated using colored polygons. (**b–e**) Band structure of single-layer hBN (**b**), MoS_2 (**c**), BP (**d**), and graphene (**e**). Reproduced with permission [1]

BP is a thermodynamically stable allotrope of phosphorus at room temperature. Like graphene, each phosphorus atom in BP is connected to three adjacent phosphorus atoms, forming a stable six-atom linked ring structure. The difference is that the BP structure is puckered, which reduces its symmetry and brings an angle-dependent nonlinearity. It has tunable direct bandgaps varying from 0.35 eV (bulk) to 2 eV (monolayer), indicating its broadband nonlinear response deep into the mid-IR, which has been widely used for ultrafast fiber laser. Research shows that nanosheet BP has a wavelength-dependent recovery time of 0.36–1.36 ps excited with photon energies from 1.55 to 0.61 eV. It should be noted that BP would be oxidized when exposed in ambient conditions and needs encapsulation for scaling its long-term stability.

MXenes represent a new class of 2D transition metal carbides, carbonitrides, or nitrides whose chemical formula is $M_{n+1}X_nT_x$ (n = 1–3). Here, M stands for transition metals (Sc, Ti, Zr, Hf, V, Nb, Ta, Cr, Mo, etc.), X is carbide and/or nitride, and

T_x represents surface terminations (F, O, OH, etc.). Few-layer $Ti_3C_2T_x$ has an indirect energy bandgap of <0.2 eV and a low absorption of ~1%/nm. Stacking of 2D MXene materials generally occurs through van der Waals interaction without internal surface termination, as in the cases of graphene, phosphorene, and TMDs. A recent study showed that the main features observed in MXene monolayers were well conserved in stacked ones, indicating that well-functioning fiber SAs could be made using MXenes, thus avoiding the troublesome processes of monolayer dispersion.

Due to its distinct electronic and mechanical properties, as well as its strong stability, bismuthene has attracted tremendous research interest. Recent research shows that the layer-dependent optical bandgap of beta-bismuthene ranges from almost 0 to 0.55 eV, suggesting that it is a promising broadband optical material from the near-IR to the terahertz regions. The short recovery time of bismuthene is 2.8 ps, which also indicates it is a potential ultrafast SA.

The aforementioned 2DM offer distinct, yet complementary properties and hence, new opportunities for optical applications in ultrafast fiber laser. But these are not enough, since better SAs with enhanced optical properties, such as much shorter carrier lifetimes, higher damage thresholds, and larger modulation depths, are always desired. On the one hand, explorations for new 2DM will never stop. On the other hand, modification of existing materials also provides an opportunity. The possibility of combining different 2DM to form van der Waals (vdW) heterostructures offers an exciting prospect for a wide range of new engineerable photonic devices. This overcomes the intrinsic drawbacks of single materials, while enhancing performance greatly. Recently, several ultrafast fiber mode-locked lasers by heterostructure SAs have been reported.

14.2 Transfer Method of 2DM

2DM transfer methods include wet transfer, dry transfer, and nonatmospheric transfer [2]. Among them, wet transfer is a common transfer method for constructing large-area complex heterojunctions. According to the different transfer media and exfoliation method, it can be divided into polyvinyl alcohol sorption transfer method [3], polymethyl methacrylate-assisted transfer method [4], poly(L-lactic acid) rapid transfer method [5], small molecule doped polystyrene (PS) transfer method [6], cellulose film transfer method [7], chemical etching transfer method [8], electrochemical stripping transfer method [9], sacrifice layer transfer method [10], metal-assisted stripping transfer method [11]. Dry transfer method includes polydimethylsiloxane peel transfer method [12] and vdWs interaction transfer method [13]. Transfer in nonatmospheric environment includes inert atmosphere transfer method and vacuum transfer method.

For fiber lasers, there are many methods of material and cavity coupling as shown in Fig. 14.2. When 2DMs were placed on a substrate, it reflects/transmits the incoming light in a way called reflective/transmittance in Figure 14.2a, b. 2DMs can be

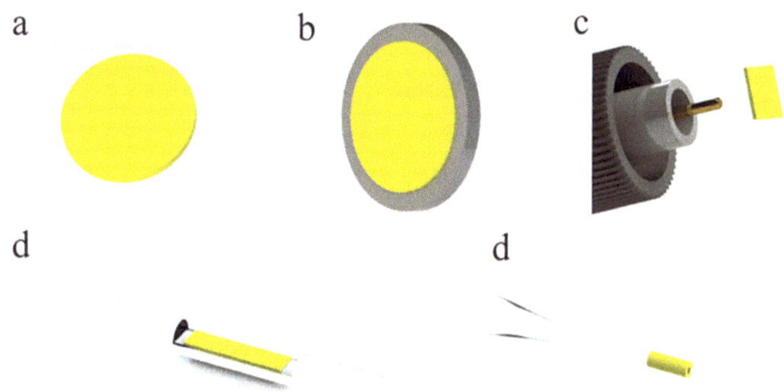

Fig. 14.2 Fiber integration with 2DM: (**a**) Transmission coupling; (**b–d**) evanescent wave coupling; 2DM are deposited or transferred on (**b**) reflection mirror, (**c**) fiber ends, (**d**) D-shaped fiber, (**e**) tapered fiber

mixed with the high polymer to make a film, which is placed in the middle of the two jump fiber joints, which is called the sandwich structure in Figure 14.2c. Using the sandwich structure composed of the film, the thermal damage and thermal stability of the film are two factors affecting the laser performance. In addition, 2DMs grown by mechanical exfoliation or chemical vapor deposition can be transferred by wet or dry transfer, transferring the material to the end face of the fiber connector, or forming a sandwich structure. In addition, the nanosheet solution prepared by liquid phase exfoliation can deposit two-dimensional materials on the surface of D-shaped and tapered fiber by light deposition in Figure 14.2d, e.

14.3 Nonlinear Optics in 2D Materials

Nonlinear optics is an important branch of optics which describes the interaction between light and nonlinear optics materials. It plays a significant role in modern nanophotonics, and therefore arouses great interest in design, preparation, and characterization of new nonlinear optics nanomaterials. We will introduce the two well-known techniques for characterization of nonlinear optics nanomaterials response of 2D layered materials: ultrafast pump–probe and Z-scan measurements, which can give the key parameters of saturable absorber, for example, ultrafast carrier relaxation time, saturation intensity, modulation depth, and so on.

Ultrafast Pump-Probe Measurement

The pump-probe measurement is an important scientific technique that has been widely used for investigating the carrier dynamics of materials by using ultrafast lasers, and has been utilized in the fields of spintronics, valleytronics, photocatalysis, photonics, optoelectronics, etc. The schematic experimental setup is shown in Fig. 14.3. In the experiment, an ultrafast pulsed laser is split into two parts: the

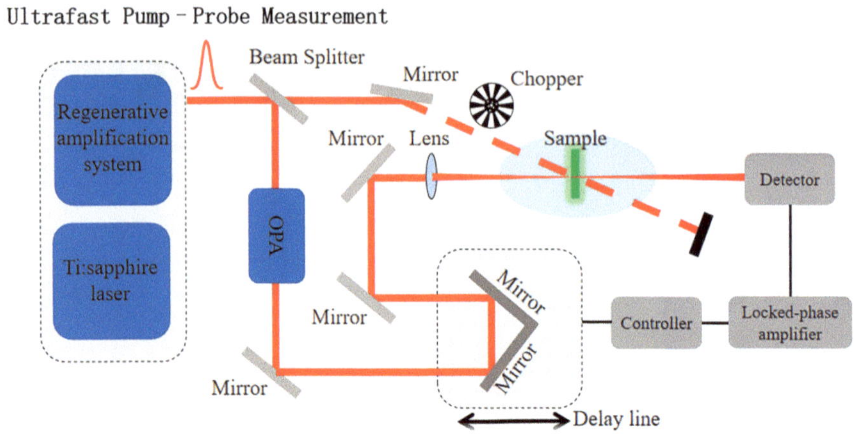

Fig. 14.3 Experimental setup of the ultra-fast pump-probe measurement. Reproduced with permission [14]

stronger (pump) beam used to excite the sample and generate the excited carriers; the weaker (probe) beam is utilized to detect the pump-induced changes in the optical constants (e.g., reflectivity or transmission) of the sample. When the laser wavelengths of the excited laser and probe laser are the same, it is named degenerate pump-probe, or it is called as nondegenerate pump-probe. Measuring the modulation of the probe laser pulse is the key to deduce the physical states of the sample as the generated excited carrier decays back to the equilibrium state after the excitation by the pump pulse. If most of the generated carriers are still in the excited state by the time the probe pulse arrives, the probe laser pulse will transmit rather than be absorbed. Otherwise, if most of the generated carriers have already decayed to the equilibrium state by the time the probe laser pulse arrives, most of the probe pulse will be absorbed since many electrons reside again in the lower energy state. Thus, measuring the intensity modulation of the probe pulse as a function of delay will describe the population dynamics of the electron energy levels.

The relaxation process of the generated excited carriers is generally assumed to follow the exponential decays, of which the double exponential decays with two characteristic times are commonly used. Take the pump-probe measurement of graphene, for example. The time-dependent normalized transmission ($\Delta T/T_0$) changes of graphene at room temperature are detected, where T_0 is the transmittance of the probe laser beam before the pump laser excitation, and ΔT is the corresponding variation in the probe transmission after being excited by the pump laser. The measured $\Delta T/T_0$ delay kinetics can be fitted to an exponential decay function:

$$\frac{\Delta T}{T_0} = A_1 \exp\left(-\frac{t}{\tau_1}\right) + A_2 \exp\left(-\frac{t}{\tau_2}\right) \tag{14.1}$$

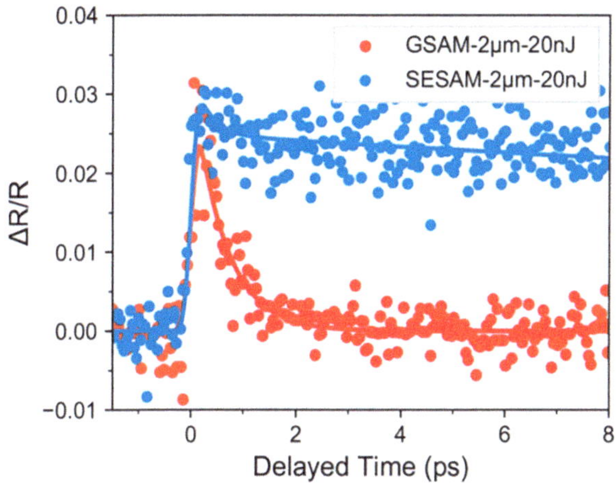

Fig. 14.4 The carrier lifetime comparison of our graphene saturable absorption mirror (GSAM) and the semiconductor saturable absorption mirror (SESAM). Reproduced with permission [68]

Here, A_1 and A_2 are the relative amplitudes of the double temporal components, τ_1 is attributed to intraband carrier scattering (hot carrier distribution), and τ_2 is attributed to the interband carrier-phonon scattering (possibly electron-hole recombination). Table 14.1 summarized the ultrafast relaxation times of the 2D materials mentioned. In Fig. 14.4, the carrier relaxation of GSAM is determined to be 168 fs, which was shorter than 1.109 ps of SESAM at 2-μm wavelength, indicating that GSAM is a much faster saturable absorber than the commercial SESAM.

14.3.1 Z-Scan Measurement

Z-scan measurement, based on the spatial beam distortion principle, is a simply and highly sensitive method for studying the third-order nonlinear optics properties of optical materials. As shown in Fig. 14.5, in Z-scan measurement, the sample is translated along the axis of a tightly focused Gaussian laser beam and experiences with different intensities depending on the position of the sample (relative to focus $z = 0$). The power transmitted through the sample is measured as the function of the sample position. Using Z-scan technique, the magnitude of nonlinear absorption and the sign and magnitude of nonlinear refraction can be determined simultaneously. When a high-intensity laser beam propagates through a material, induced refractive index changes lead to self-focusing or defocusing of the laser beam. This enables to determine the third-order nonlinear optical properties of the materials in liquid, thin films, or crystal forms. It contains open (Figure 14.5) and closed (Figure 14.5) Z-scan measurements, determined by whether there is an aperture before the detector. The nonlinear refractive index n_2 and nonlinear absorptive coefficient β_{eff} are two important parameters required to determine the third-order

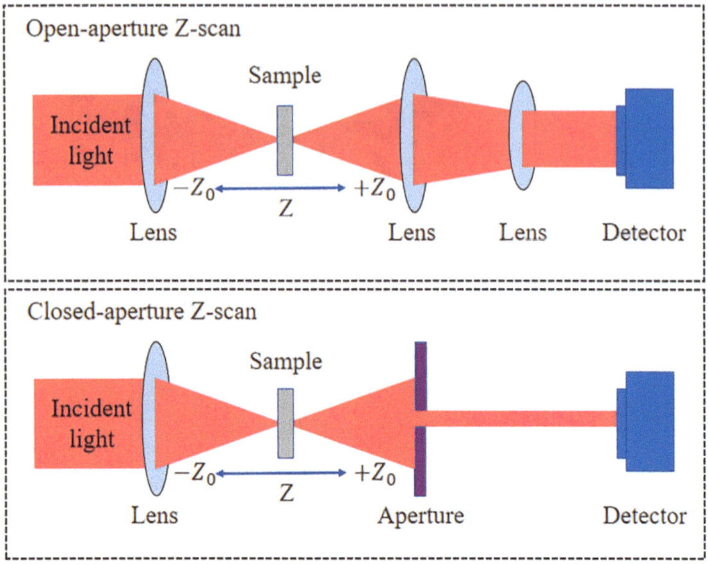

Fig. 14.5 Experimental setup of the Z-scan measurement

susceptibility $\chi^{(3)}$. The nonlinear refractive index n_2 can be determined by performing closed aperture Z-scan experiments and similarly the nonlinear absorptive coefficient β_{eff} can be determined by performing open aperture Z-scan experiments.

Figure 14.5 shows the open Z-scan measurement of several 2DM. According to the nonlinear optics theory, light attenuation of propagating through a nonlinear optics material can be described as follows [15]:

$$\frac{dI}{dz'} = -\alpha(I)I \tag{14.2}$$

where I is the incident laser intensity, z' is the propagating distance in the material, and $\alpha(I)$ is the total absorption coefficient, which includes linear and nonlinear absorption processes and can be described as [15, 16]:

$$\alpha(I) = \alpha_0 + \beta_{\mathrm{eff}} I \tag{14.3}$$

where α_0 and β_{eff} are the linear and nonlinear absorption coefficients, respectively. Therefore, β_{eff} can be achieved by fitting the open aperture Z-scan measurement data with Eqs. (14.2) and (14.3). After that, the imaginary part of $\chi(3)$ can be obtained from the formula [15]:

$$Im\left(\chi^{(3)}\right) = \frac{2n_0^2\varepsilon_0 c^2}{3\omega}\beta_{\mathrm{eff}} \tag{14.4}$$

where n_0 is the linear refractive index, ε_0 is the vacuum permittivity, c is the vacuum light speed, and ω is the angular frequency. Considering the saturable absorption, the total nonlinear absorption coefficient $\alpha(I)$ can be written as:

$$\alpha(I) = \frac{\alpha_0}{1 + I / I_S} \tag{14.5}$$

where I_s is the saturable intensity. Combining Eqs. (14.3) and (14.5), we can get the following expression:

$$\alpha_0 + \beta_{\text{eff}} I = \frac{\alpha_0}{1 + I / I_s} \tag{14.6}$$

For saturable absorption, the sign of β_{eff} is negative. To obtain the saturable absorption parameters, the Z-scan data can be transferred to nonlinear transmission curve and need to be fitted by the following equation [17]:

$$T = \left[1 - \frac{\Delta R \times I_s}{I_s + \frac{I_0}{\left(1 + \frac{Z^2}{Z_0^2}\right)}} \right] / (1 - \Delta R) \tag{14.7}$$

where T is the transmittance, ΔR is the modulation depth, I_0 is the incident pulse intensity, and Z_0 is the diffraction length of the beam. The nonlinear optics parameters of the commonly used two-dimensional materials have been summarized [18].

In Fig. 14.6, the real and imaginary parts of the complex nonlinear refractive index of graphene were measured by Z-scan. Graphene possesses a giant nonlinear refractive index $n_2 \simeq 10^{-7}$ cm^2 W^{-1}, almost 9 orders of magnitude larger than bulk dielectrics. We find that the nonlinear refractive index decreases with increasing excitation flux but slower than the absorption. This suggests that graphene may be a very promising nonlinear medium, paving the way for graphene-based nonlinear photonics.

14.3.2 Ultrafast Lasers Using Different 2D Materials as an SA

Since graphene was successfully developed to mode-lock a fiber laser in 2009, more and more 2DMs are being found with potential use for ultrafast lasers. Graphene, TMDs, TIs, black phosphorus, and recently newer 2DMs are being used in

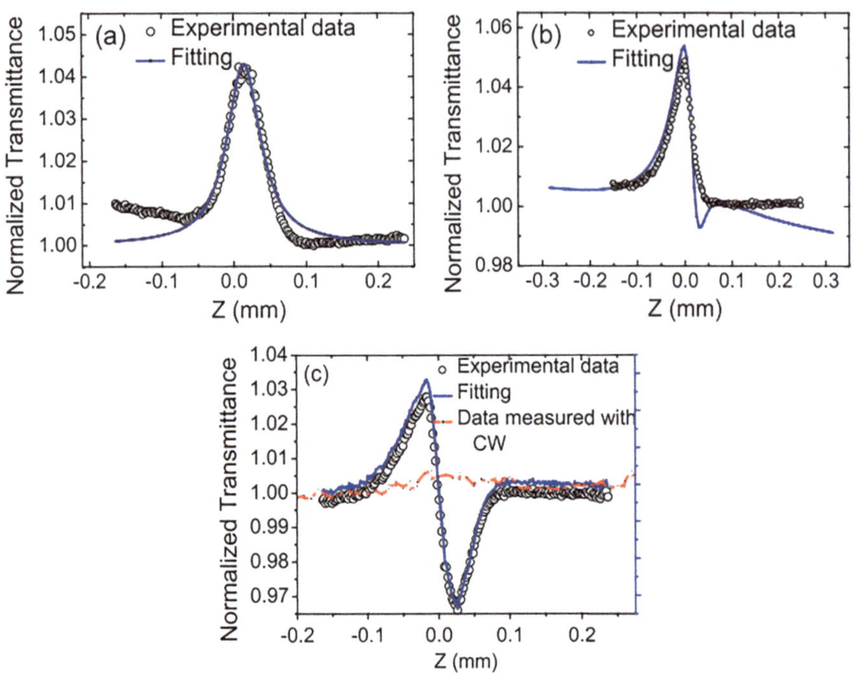

Fig. 14.6 Z-scan traces for three-layer graphene. (**a**) Near field (open aperture). (**b**) Far field (closed aperture). Upon dividing by the near-field curve, one obtains the data of panel (**c**), which exhibits the typical shape of a Z-scan curve with positive nonlinear phase shift. Reproduced with permission [19]

passively mode-locked (PML) fiber lasers in different wavelengths. Herein, we briefly introduce these mode-locked fiber lasers.

Yb-doped lasers can generate ultrashort pulses with high energy and are used in medical and biological research. Table 14.2 lists recent years' progress in PML fiber lasers based on 2DM as an SA in 1-μm regime. For example, dissipative soliton operation can be obtained from graphene as an SA [20, 21], with a maximum pulse energy of 159.4 nJ and a shortest pulse duration of 580 ps [20]. As for TIs, a 46-ps pulse was demonstrated with a 0.756-nJ pulse energy in dispersion-managed soliton fiber laser [22]. The shortest pulse duration using TIs as a saturable absorber is Sb2Te3, which achieved a 5.9-ps pulse duration with an 8.28-nm bandwidth in PML fiber lasers [23]. TMDs using as an SA also can be achieved in a mode-locked fiber laser at the 1-μm regime [24–26]. Zhang et al. generated an 800-ps pulse with a 2.7-nm bandwidth centered at 1054.3 nm using MoS2 as an SA [25]. The longest wavelength (1085.58 nm) can be achieved using BP as an SA with a 7.54-ps pulse duration and a 5.93-nJ pulse energy [27]. Bismuthene as a new 2D material found in recent years shows tremendous interest in nonlinear optics; Chai et al. first demonstrated Yb-doped PML fiber lasers based on bismuthene as an SA in Fig. 14.7 [28].

Table 14.2 Performance summary of mode-locked laser based on 2D materials as an SA in a 1-μm regime

Material type		Fabrication method	Incorporation method	Nonlinear characterization		Laser type	Laser performance				Ref.
				α_s (%)	I_{sat} (MW/cm^2)		λ (nm)	Time	Energy	Repetition frequency (MHz)	
Graphene		CVD	Sandwiched	8	…	Yb	1069.8	590 ps	0.41 nJ	0.9	[20]
		…	Sandwiched	1.5	…	Yb	1064.9	60.8 ns	159.4 nJ	0.927	[21]
TIs	Bi_2Se_3	Polyol method	Fiber end facet	5.2	70 μJ/cm^2	Yb	1031.7	46 ps	0.756 nJ	44.6	[22]
	Bi_2Te_3	ME	Side-polished	1.8	…	Yb	1057.82	230 ps	0.599 nJ	1.44	[81]
	Bi_2Te_3	Hydrothermal exfoliation	TI-PCE Photonic crystal fiber	19.1	14.9	Yb	1064.47	960 ps	…	1.11	[82]
	Sb_2Te_3	PMS	D-shaped	3.1	…	Yb	1065.3	5.9 ps	4 mW	19.28	[23]
TMDs	WS_2	Hydrothermal exfoliation	Fiber end facet	5.8	406	Yb	1052.45	713 ps	30 mW	23.26	[24]
	MoS_2	Hydrothermal exfoliation	Quartz	9.3	15.9	Yb	1054.3	800 ps	9.3 mW	7	[25]
	SnS_2	Solution process	Fiber end facet	0.76	78	Yb	1062.66	656 ps	2.23 mW	39.33	[26]
BP		ME	Sandwiched	8	0.35	Yb	1085.5	7.54 ps	5.93 nJ	13.5	[27]
		LPE	Fiber end facet	8.5	2.95	Yb	1030.6	400 ps	0.7 nJ	46.3	[83]
Bismuthene		Sonochemical exfoliation	Microfiber	2.2	13	Yb	1034.4	30.25 ps	…	21.74	[28]

CVD chemical vapor deposition, *ME* mechanical exfoliation, *PMS* pulsed magnetron sputtering; *LPE* liquid phase stripping

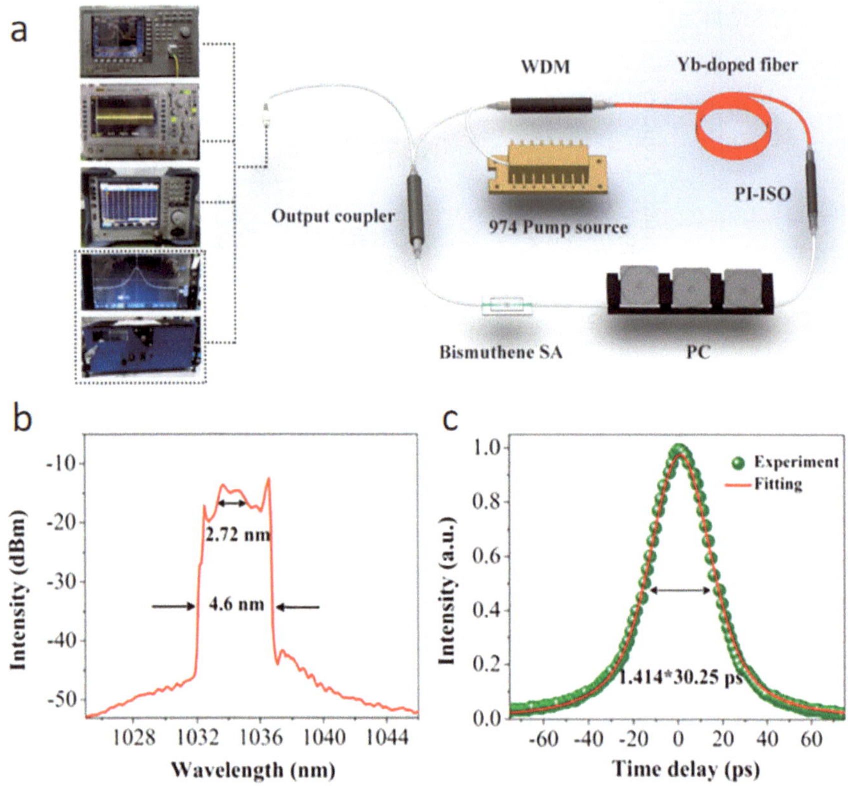

Fig. 14.7 (**a**) Experimental setup of the fiber laser with a microfiber-based bismuthene SA. (**b**) Typical spectrum of the dissipative solitons with a center wavelength of 1034.4 nm. (**c**) Autocorrelation trace, with a pulse width of 30.25 ps. Reproduced with permission [28]

Performance in 1.5-μm regime PML fiber lasers using 2D materials is shown in Table 14.3. As we can see, most of the achievements in 2D materials as an SA are in 1.5-μm regime. The highest repetition rate obtained is 101 MHz using graphene SA [29]. The shortest pulse duration (88 fs) with a 48-nm bandwidth can be generated directly from PML fiber laser based on graphene as an SA in Fig. 14.8c [30]. Sobon et al. demonstrated by scaling the number of graphene layers that they can control the SA parameters; the shortest pulse duration can be obtained using a 37-layer graphene as an SA [31]. Purdie et al. combined a graphene mode-locked oscillator with an external compressor and achieved few cycle pulses (263 fs directly from oscillator and 29 fs after compressor); this is the shortest pulse generated from PML fiber laser based on 2D materials as an SA so far [32]. Sotor et al. used Sb2Te3 as an SA in Er-doped fiber laser that can generate pulses in different laser cavities: all-anomalous, all-normal, and balanced dispersion, the shortest pulse of 125 fs was obtained in a dispersion-managed cavity. A 70-fs pulse was generated used hybrid mode-locked Er-doped fiber laser with Sb2Te3 as an SA [33]. It proved that by

Table 14.3 Performance summary of mode-locked laser based on 2D materials as an SA in a 1.5-μm regime

Material type	Fabrication method	Incorporation method	Nonlinear characterization		Laser type	Laser performance				References
			α_s (%)	I_{sat} (MW/cm²)		λ (nm)	Time	Energy	Repetition frequency (MHz)	
Graphene	CVD	Sandwiched	2.93%	53.25	Er	1559.1	432.47 fs	0.09 nJ	25.51	[84]
	CVD	Sandwiched	7.5%	750 μJ/cm²	Er	1565.3	148 fs	15 pJ	101	[29]
	CVD	Sandwiched	11%	460 μJ/cm²	Er	1545	88 fs	71 pJ	21.15	[30]
	CVD	Microfiber	4.37%	1.51 GW/cm²	Er	1531.3	1.21 ps	…	1.88	[85]
	CVD	Sandwiched	7.5%	450 μJ/cm²	Er	1559.3	345 fs	2.1 mW	54.28	[31]
	CVD	Fiber end facet	65.9%	0.53	Er	1561	1.23 ps	…	2.54	[86]
	CVD	Fiber end facet	66.5%	0.71 μJ/cm²	Er	1576.3	415 fs	7.3 pJ	6.84	[87]
Graphene-GO	…	Etched fiber	5.25%	…	Er	1556.5	615	…	17.09	[88]
Graphite	ME	D-shaped	3.3%	53	Er	1567	220	83 pJ	15.7	[32]
Graphene	Solution process	Sandwiched	…	…	Er	1550	29	2.8 nJ	18.67	[89]
	Marcano method	Sandwiched	1.4%	96.2	Er	1565.9	613	22.3 pJ	37.2	[90]
	…	Sandwiched	2.6%	…	Er	1554	168	55 pJ	63	[91]

(continued)

Table 14.3 (continued)

Material type		Fabrication method	Incorporation method	Nonlinear characterization		Laser type	Laser performance				References
				α_s (%)	I_{sat} (MW/cm^2)		λ (nm)	Time	Energy	Repetition frequency (MHz)	
TIs	Bi_2Se_3	Polyol method	Sandwiched	5.2%	12	Er	1600	360	0.86 mW	35.45	[92]
		Polyol method	Fiber end facet	3.9%	12	Er	1557.5	660	0.144 nJ	12.5	[93]
		LPE	Fiber end facet	88%	3.7	Er	1571	579	1.59 mW	12.54	[94]
		LPE	TI-PCF	5%	…	Er	1559	245	75 mW	20.27	[95]
		Solution processing	Fiber end facet	36%	3.7	Er	1571	579	1.59 mW	12.54	[94]
	Bi_2Te_3	ME	Side-polished	15.7%	44 W	Er	1547	543	0.8 mW	15.11	[96]
		PLD	Microfiber	5.7%	26.7	Er	1560.8	286	0.027 nJ	18.55	[97]
		Hydrothermal exfoliation	Fiber end facet	2%	180	Er	1557	1100	0.25 mW	8.635	[98]
	Sb_2Te_3	LPE	Side-polished	3.9%	106	Er	1556	449	39.6 pJ	22.13	[99]
		ME	Side-polished	6%	31.6	Er	1561	270	0.029 pJ	34.58	[100]
		Direct fusion technique	Side-polished	4.5	129 μJ/cm^2	Er	1568.6	195	0.27 nJ	33	[101]
		ME	Side-polished	6	31	Er	1565	128	0.0448 nJ	23.32	[102]
TIs		PMS	Side-polished	7.3	280 μJ/cm^2	Er	1558	167	0.21 nJ	25.38	[103]
		PLD	Tapered fiber	7.42	175	Er	1542	70	63 mW	95.4	[33]
	$CoSb_3$	…	Side-polished	5	8.7 W	Er	1557.9	833	0.1 mW	14.48	[34]

TMDs	WS_2	Magnetron-sputtering deposition	Tapered fiber	14.79	3.273	Er	1560	288	0.044 pJ	41.4	[104]
		Solution process	D-shaped	11	5	Er	1557	660	…	10.2	[105]
		LPE	Sandwiched	2.9	370	Er	1550	595	…	…	[106]
		PLD	Tapered fiber	6.5	20	Er	1560	220	…	…	[107]
		PLD	Tapered fiber	35.1	22.8	Er	1540	67	…	135	[108]
		LPE	Tapered fiber	0.6	…	Er	1561	369	1.93 mW	24.93	[109]
		CVD	Side-polished	0.14	…	Er	1565	332	0.43 mW	31.11	[110]
		PLD	Tapered fiber	17.2	34.02	Er	1561	246	18 mW	101.4	[111]
	MoS_2	CVD	Side-polished	2.5	…	Er	1568	4980	…	26.02	[112]
		CVD	Side-polished	2.5	…	Er	1568	637	…	33.48	[112]
		LPE	Polyvinyl alcohol film	4.3	34	Er	1569.5	710	22 mW	12.09	[113]
		LPE	D-shaped	0.2	…	Er	1550	125	…	14.53	[114]
TMDs		LPE	Polyvinyl alcohol film	…	…	Er	1573.7	630	0.141 nJ	27.1	[115]
	WTe_2	LPE	Side-polished	2.85	64.6 W	Er	1556.2	770	…	13.98	[116]
	$MoSe_2$	CVD	Tapered fiber	22.57	7.747	Er	1552	207	…	64.56	[117]
	ReS_2	CVD	Side-polished	1.2	62 μ J/cm^2	Er	1564	1900	…	3.43	[118]
		LPE	Polyvinyl alcohol film	0.12	74	Er	1558.6	1600	0.4 mW	5.48	[119]
	WSe_2	CVD	Tapered fiber	21.89	15.43	Er	1557.4	163.5	28.5 mW	63.13	[120]
	SnS_2	Solution process	Polyvinyl alcohol film	4.6	125	Er	1562	623	1.2 mW	29.33	[35]

(continued)

Table 14.3 (continued)

Material type	Fabrication method	Incorporation method	Nonlinear characterization		Laser type	Laser performance				References
			α_s (%)	I_{sat} (MW/cm²)		λ (nm)	Time	Energy	Repetition frequency (MHz)	
BP	LPE	Sandwiched	10.03	14.98	Er	1555	102	0.071 nJ	23.9	[36]
	LPE	Polyvinyl alcohol film	0.3	…	Er	1562	1236	…	5.42	[121]
	ME	Sandwiched	7	0.25	Er	1560.7	570	0.74 nJ	6.88	[122]
	LPE	Fiber end facet	21	12	Er	1559.5	670	…	8.77	[123]
	ME	Fiber end facet	…	…	Er	1558.7	786	0.11 nJ	14.7	[37]
	ME	Fiber end facet	8.1	6.55	Er	1571.4	946	…	5.96	[124]
	ME	Fiber end facet	4.6	…	Er	1560.5	272	65 mW	28.2	[125]
	Solution process	Fiber end facet	12.3	6 GW/cm²	Er	1562	635	…	12.5	[126]
BP quantum dots	LPE	Microfiber	9	1.5	Er	1562.8	291	…	10.36	[127]
	LPE	Fiber end facet	8.1	1.69 mW	Er	1561.7	882	…	5.47	[128]
Bismuthene	Sonochemical exfoliation	Microfiber	2.03	30	Er	1559.1	652	1.15 mW	8.83	[129]
	Sonochemical exfoliation	Microfiber	5.6	48.2	Er	1561	193	5.6 mW	8.85	[39]
Antimonene	LPE	Microfiber	19.72	15.10 GW/cm²	Er	1557.6	552	10.8 mW	10.27	[40]
Metallic MXene (Ti_3CNT_x)	Drop-casting method	Side-polished	1.7	45	Er	1557	660	0.05 mW	15.4	[130]

PLD pulsed laser deposition

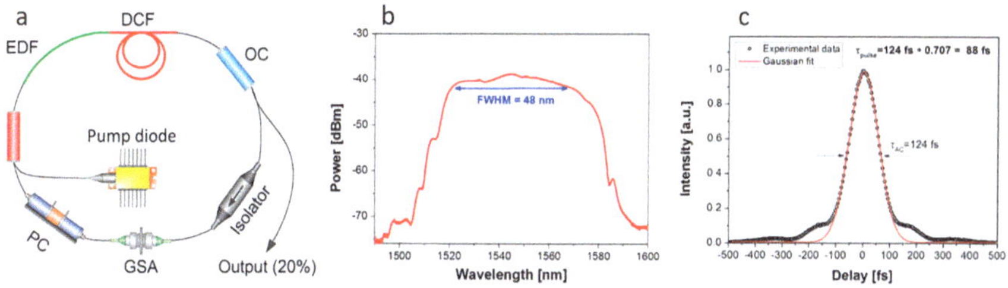

Fig. 14.8 Ultrafast laser using graphene (**a**–**c**) saturable absorber. (**a**) Experimental setup of the pulsed fiber laser. WDM, wavelength-division multiplexer. (**b**) Optical spectrum. (**c**) Autocorrelation trace with a Gaussian fit. Reproduced with permission [30, 36]

using TI, the performance is better than using them separately. CoSb3, which is also known to be a sort of TI, has been proved to be a good candidate as an SA using in mode-locked fiber lasers in 2018. Lee et al. demonstrated an ultrafast mode locker based on CoSb3 achieving an 833-fs pulse duration at the 1.5-μm region [34]. IV–VI group TMDs, SnS2, which has held certain attention in the fields of ultrafast optics due to their suitable layer independent bandgap value and anisotropic crystal structure, was successfully prepared and employed as an SA using in Er-doped mode-locked fiber laser [35]. BP as an expected 2D material also achieved significant progresses in 1.5 μm. The shortest pulse duration generated from BP as an SA in PML fiber laser was achieved in 2018. The pulse duration and spectrum are 102 fs and 40 nm, respectively [36]. Song et al. demonstrated that they can generate soliton pulses down to 670 fs by incorporating the BP nanoflake-based saturable absorber into Er-doped fiber laser and particularly generate vector solitons in fiber lasers [37]. In the past 2 years, novel 2DMs, including antimonene and bismuthene, have been used in mode-locked fiber lasers. Guo et al. demonstrated a 193-fs pulse duration Er-doped fiber laser using a microfiber-based bismuthene SA for the first time [38]. Antimonene, a new type of 2DM, was also demonstrated in 2017 [39]. A new family of 2DM called MXene was also achieved in recent years; Jhon et al. discovered MXene's potential application in pulsed lasers; PML fiber lasers with 660 fs at a repetition rate of 15.4 MHz and the central wavelength of 1557 nm using Ti3CN as an SA were reported [40]. Though these 2D materials are still in the initial experiment, future development promises good results for passively mode-locked (PML) fiber lasers.

Table 14.4 shows a summary of PML fiber lasers based on 2D materials as an SA in the 2-μm regime. The longest wavelength achieved in a 2-μm regime is 2082.5 nm with a 6.1-nm bandwidth using graphene as an SA in a mode-locked fiber laser [41]. The first demonstration of dispersion-managed soliton operation of a graphene-based fiber laser was achieved by Sotor et al.; a 205-fs pulse duration and a 27.5-nm bandwidth are shown in Figure 14.9a–c, respectively [42]. The first dispersion-managed fiber laser based on graphene was achieved in 2018, which generated a 190-fs pulse duration and a 2.55-nJ pulse energy [43]. A 795-fs pulse duration can be generated from Bi2Te3 as an SA at a wavelength of 1935 nm from a Tm/Ho

Table 14.4 Performance summary of mode-locked laser based on 2D materials as an SA in a 2-μm regime

Material type		Fabrication method	Incorporation method	Nonlinear characterization		Laser type	Laser performance				Reference
				α_s (%)	I_{sat} (MW/cm^2)		λ (nm)	Time	Energy	Repetition frequency (MHz)	
Graphene		CVD	Sandwiched	7.5	450 μJ/cm^2	Tm	1923.3	737	1.21 mW	28.25	[31]
		Marcano method	Sandwiched	1.4	96.2	Tm	1961.6	1360	…	27.37	[90]
		CVD	Sandwiched	3.7	110 μJ/cm^2	Tm	1876	603	1.5 mW	41.46	[130]
		ME	Sandwiched	2.84	110 μJ/cm^2	Ho	2082.5	811	1.3 nJ	34	[41]
		CVD	Sandwiched	13.9	250 μJ/cm^2	Tm	1945	205	220 pJ	58.87	[42]
		…	Sandwiched	12	70 μJ/cm^2	Ho	2060	190	2.55 nJ	20.98	[43]
TIs	Bi_2Te_3	ME	Side-polished	20.6	29 W	Tm/ho	1935	795	1 mW	27.9	[44]
		LPE	Tapered fiber	9.8	…	Tm/ho	1909.5	1260	…	21.5	[131]
TMDs	WS_2	LPE	Side-polished	10.9	3.2 W	Tm/ho	1941	1300	17.2 pJ	34.8	[132]
	$MoSe_2$	LPE	Side-polished	4.4	25.7	Tm/ho	1912	920	4.3 mW	18.21	[47]
	WTe_2	MSD	Microfiber	31	7.6	Tm	1915.5	1250	39.9 mW	18.72	[45]
	WSe_2	ME	Microfiber	3.7	3.7	Tm	1863.9	1120	32.5 mW	11.36	[46]
BP		ME	Fiber end facet	4.1	10–550 μJ/cm^2	Tm	1910	739	0.047 nJ	36.8	[48]
		Solution processing	Tapered fiber	9.8	…	Tm	1898	1580	8.45 mW	19.2	[133]

MSD magnetron sputtering deposition

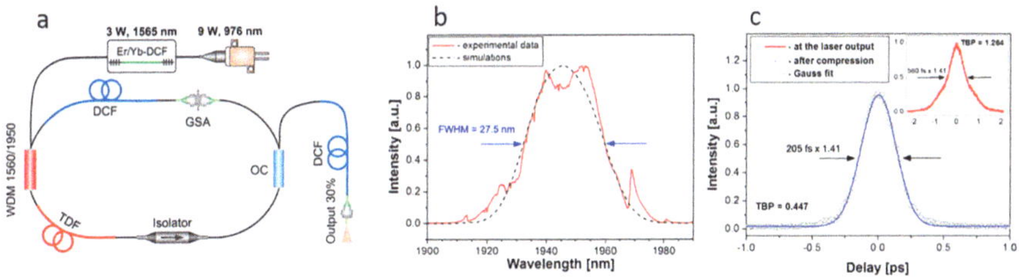

Fig. 14.9 (**a**) Experimental setup of the all-PM fiber stretched-pulse oscillator. (**b**) Generated and calculated optical spectrum of the laser. (**c**) Autocorrelation trace measured after the DCF, while the inset graph presets the uncompressed pulse directly at the laser output. Reproduced with permission [42]

co-doped fiber laser cavity [44]. The first demonstration of WTe2 and WSe2 as an SA was reported in 2017 [45] and 2018 [46], respectively. A 920-fs pulse could be produced at 1912 nm from a Tm/Ho co-doped fiber laser using MoSe2 as an SA [47]. The first femtosecond pulse in Tm-doped fiber laser using BP as an SA was obtained in 2015 by Sotor et al. using mechanically exfoliated BP deposited on a fiber connector tip, generating 739 fs pulses centered at 1910 nm [48].

Lasers in mid-infrared region have seen a rapid development driven by many important applications such as environmental monitor, disease diagnosis, spectroscopy, etc. The continuous improvement of infrared glass (e.g., fluoride, chalcogenide, fluorotellurite, etc.) fiber drawing technology has also spurred the development of mid-infrared fiber lasers. Table 14.5 shows the performance of mode-locked fiber laser based on 2D materials as an SA in the 3-μm region. By using ZBLAN fiber, a mode-locked capability of graphene in the 3-μm region has been demonstrated [49]. By inserting BP SA into specifically designed rare earth ions doped with fluoride fiber lasers, an 8.6-ps pulse duration can be obtained [50]. Qin et al. achieved the longest wavelength, near 3.5 μm [51], obtained by a BP SA (Fig. 14.10).

In addition to using 2D material SAs in mode-locked fiber lasers for different wavelengths, we now discuss mode-locked lasers with harmonic pulse generation. Harmonic mode-locked fiber lasers have many applications such as optical communication, frequency comb, and spectroscopy. Table 14.6 shows the recent progress on a harmonic mode-locked fiber laser using 2D materials as an SA. Sobon et al. achieved a 2.22-GHz (21st harmonic) repetition rate with 900 fs in a mode-locked fiber laser with multilayer graphene as an SA [52]. Meng et al. first reported Bi2Se3 as an SA using Er-doped fiber laser operating above 1.6 μm with a 640.9-MHz repetition rate and received the highest output power reported in TI-SA-based PML fiber laser [53]. For TMDs, Koo et al. demonstrated using MoSe2 as an SA in a mode-locked fiber laser and achieved a 751-fs pulse duration with 3.27 GHz (Fig. 14.11), which is the highest frequency (212th harmonic) demonstrated regarding the production of femtosecond pulses from fiber lasers [54]. Pawliszeska et al. reported a 290-MHz (tenth harmonic) repetition rate with 1.3 ps using BP as an SA in mode-locked fiber lasers at a 2094-nm central wavelength [55].

Table 14.5 Performance summary of mode-locked laser based on 2D materials as an SA in a 3-μm regime

Material type	Fabrication method	Incorporation method	Nonlinear characterization		Laser type	Laser performance				Reference
			α_s (%)	I_{sat} (MW/cm^2)		λ (nm)	Time	Energy	Repetition frequency (MHz)	
Graphene	CVD	Gold mirror	10	2	Er:ZBLAN	2780	42 ps	0.7 nJ	25.4	[49]
BP	LPE	SA mirror	41.2	3.767	Ho^{3+}/Pr^{3+}	2866.7	8.6 ps	6.28 nJ	13.987	[50]
	Sonication	SA mirror	7.7	0.36 μJ/cm²	Er:ZBLAN	3489	…	40 mW	28.91	[51]

ZBLAN ZrF_4-BaF_2-LaF_3-AlF_3-NaF

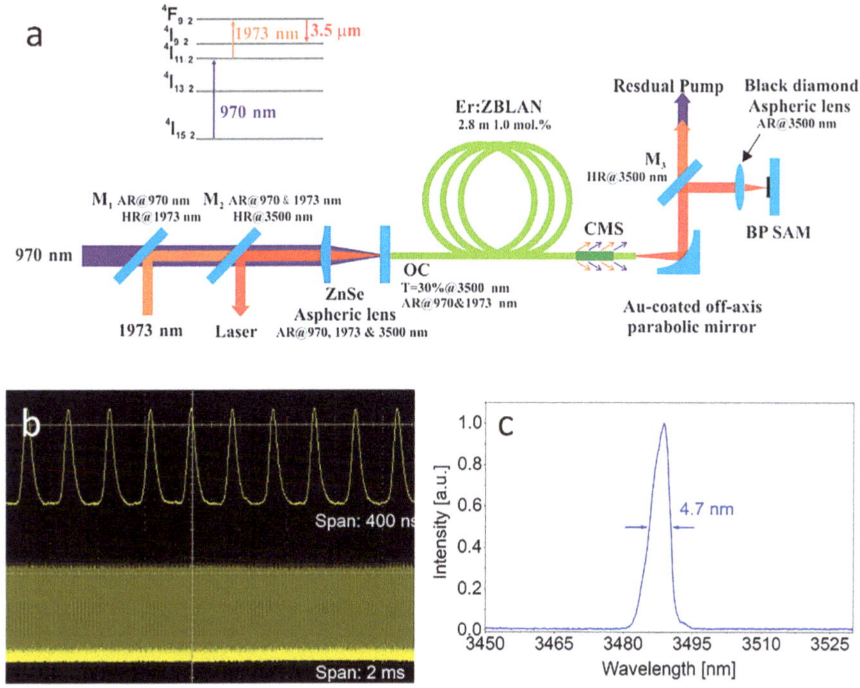

Fig. 14.10 (**a**) Schematic of the BP Q-switched and mode-locked Er:ZBLAN fiber lasers. OC, output coupler; CMS, cladding mode stripper. (**b**) Mode-locked pulse train in microsecond and millisecond time scales. (**c**) Mode-locked pulse spectrum. Reproduced with permission [51]

The broadband saturable absorption characteristics of the 2D materials ensure they can realize the mode-locking in different wavelength regions. The wavelength-tuning mechanisms based on 2D materials in pulsed fiber lasers include cavity birefringence filter, tunable filter, Mach-Zehnder filter, and Sagnac interferometer filter. Generally, there are two main ways to realize the tunable output in fiber lasers. One is achieved by inserting additional filter components, and the other one is obtained by using fiber birefringence or various structures in laser cavities. As an artificial filter, tuning or squeezing the polarization controller can change the birefringence in laser cavity, resulting in the change of transmittance curve of the light in cavity. Both the transmittance curve of the filter and the gain curve of the gain fiber determine the gain properties of the laser. Therefore, the lasing wavelength can be changed by tuning the polarization controller in cavity.

Tunable mode-locked fiber lasers and dual wavelength mode-locked fiber lasers can also be achieved by using 2DM as an SA (Table 14.7). As we can see, all 2DM can achieve tunable PML fiber lasers. For example, Sun et al. achieved a 1-ps pulse duration, tunable range from 1525 to 1559 nm, with stable mode-locking, insensitive to environmental perturbations by using graphene as an SA [56]. Using graphene as an SA can achieve the tunable wavelength at 2 μm in mode-locked fiber lasers [57]. A high modulation depth up to 98% by using Bi2Se3 is obtained by

Table 14.6 Performance summary of harmonic mode-locked laser based on 2D materials as an SA

Material type		Fabrication method	Incorporation method	Nonlinear characterization		Laser type	Laser performance				Reference
				α_s (%)	I_{sat} (MW/cm^2)		λ (nm)	Time (fs)	Energy	Repetition frequency	
Graphene		ME	Sandwiched	8.4	0.1	Er	1560.5	900	…	2.22 GHz	[52]
		…	Fiber end facet	…	…	Er	1560	992	…	0.49 GHz	[134]
TIs	Bi_2Te_3	PLD	Microfiber	6.2	28	Er	1562.4	320	…	2.95 GHz	[135]
	Bi_2Se_3	Hydrothermal	Tapered fiber	4.3	…	Er	1610	0.7 ns	308 mW	640.9 MHz	[53]
		LPE	TI-PCF	5	…	Er	1559	245	75 mW	202.7 MHz	[95]
TMDs	WS_2	PLD	Polyvinyl alcohol film	8	200	Er	1559.7	452	11.3 mW	1.04 GHz	[136]
	MoS_2	Hydrothermal	Microfiber	2.82	…	Er	1556.8	3 ps	5.39 mW	2.5 GHz	[137]
		Solution process	D-shaped	1.2	158	Er	1530.4	1.2 ps	43 mW	125 MHz	[138]
		Solution process	Side-polished	5.4	23	Er	1555.6	737	6.7 pJ	3.27 GHz	[54]
BP		Solution process	Sandwiched	0.67	194 μJ/cm^2	Tm	2094	1300	0.397	290 MHz	[55]

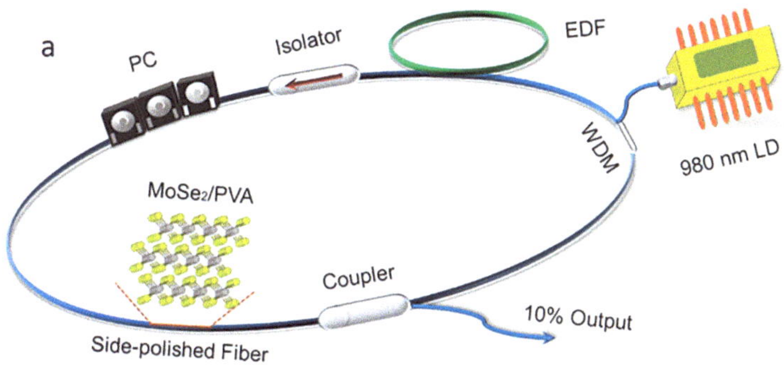

Fig. 14.11 Schematic of the laser used in the experiment to achieve the 212th-harmonic-output optical pulses. Reproduced with permission [54]

Zhao et al., and a 1.57-ps pulse duration with an 8-nm tunable range is achieved in soliton fiber laser [58]. The largest tunable range was achieved by using BP as an SA in mode-locked fiber lasers with a tuning range of 1529–1592 nm (Fig. 14.12) [59]. Multiple wavelength mode-locked fiber lasers are suitable for environmental sensing and microwave radiation. Zhao et al. achieved dual wavelength centered at 1061.8 and 1068.8 nm by using graphene as an SA [60]. Femtosecond level pulse duration can also be generated by graphene as an SA, at approximately 1530 and 1560 nm with pulse durations of 900 fs and 940 fs, respectively [61]. Dual wavelength polarization vector soliton can be generated from the mode-locked fiber laser based on BP as an SA, with the center at 1533 and 1558 nm and with bandwidths of 3.7 and 6.9 nm simultaneously [62].

The pursuit of higher repetition rates is another important aspect. The ultrafast laser based on 2DM SA was summarized in Table 14.8. Lasers with repetition rates from several to hundreds of gigahertz are important for high-speed optical communication systems and microwave generation. The short recovery times of 2DM make them good candidates for gigahertz pulse generation. Table 14.3 shows the representative results of high repetition rate UFLs mode-locked by 2DM. For fundamental mode-locking operation, the repetition rate of the pulse train is the same as the longitudinal mode spacing, whose increment can only be achieved by reducing cavity length. In 2012, Martinez et al. reported a graphene mode-locked EDFL with a ~ 1-cm linear cavity, whose pulse repetition rate was 9.67 GHz [63]. In 2015, Wu et al. also demonstrated a MoS2-based short-cavity EDFL at a fundamental repetition rate of 463 MHz [64]. However, due to the fiber gain limitation, short-cavity configuration is difficult to improve the repetition rate further (e.g., >10 GHz). By inserting a comb filter like Fabry–Perot (FP) filter or microknot filter into a fiber laser cavity to form a composite cavity, it is also possible to increase output repetition rate by filtering out some longitudinal modes. In 2015, Qi et al. reported the generation of a 100-GHz pulse train in an Er-doped fiber laser (EDFL) by using a graphene tapered fiber SA and an FP filter [65]. Figure 14.13 shows a recent work by Liu et al. in which a graphene-deposited microfiber knot filter was used in ring

Table 14.7 Performance summary of tunable mode-locked lasers and dual wavelength based on 2D materials as an SA

Material type		Fabrication method	Incorporation method	Nonlinear characterization		Laser type	Laser performance				Reference
				α_s (%)	I_{sat} (MW/cm^2)		λ (nm)	Time	Energy	Repetition frequency (MHz)	
Graphene		…	Fiber end facet	…	0.71	Er	1570–1600	140–40 ps	2.3 nJ	…	[139]
		LPE	Fiber end facet	4.5	266	Er	1525–1559	1 ps	125 pJ	8	[56]
		CVD	Microfiber	19	…	Tm	1880–1940	1.9 ps	1.96 mW	19.7	[57]
TIs	Bi_2Se_3	Polyol method	Quartz	98	0.49	Er	1557–1565	1.57 ps	…	1.21	[58]
TMDs	MoS_2	Solution process	Sandwiched	10.7	2.02	Er	1535–1565	0.96–7.1 ps	…	12.99	[140]
BP		LPE	Microfiber	9	25	Er	1532–1570	940 fs	5.6 mW	4.96	[141]
		LPE	Tapered fiber	10.1	9.27	Er	1549–1575	280 fs	…	60.5	[142]
		LPE	Sandwiched	0.76	5.1	Er	1529–1592	0.8–1.3 ps	…	5.6	[59]
Graphene		Optical deposition	Microfiber	…	…	Yb	1061.8, 1068.8	1.41 ns, 4.23 ns	1.712 nJ	1.78	[60]
		…	Sandwiched	6	3.4	Er	1530, 1560	900 fs 940 fs	1.57 nJ, 1.38 nJ	…	[61]
TMDs	WS_2	LPE	Tapered fiber	0.6	…	Er	1561, 1563	369 fs 563 fs	1.93 mW 2.77 mW	24.93,20.39	[109]
		PLD	Tapered fiber	6.5	370	Er	1558.54, 1565.99	585 fs 605 fs	1.14 nJ	8.83	[107]
TIs	Bi_2Te_3	…	Microfiber	16.3	…	Er	1557.4, 1559.4	3.42 ps, 2.02 ps	…	388, 239	[143]
	Bi_2Se_3	LPE	Sandwiched	3.8	25	Er	1567.2, 1568	22 ps	1.1 nJ	8.83	[144]
BP		LPE	Fiber end facet	0.35	…	Er	1533, 1558	…	0.07 nJ	20.82	[62]

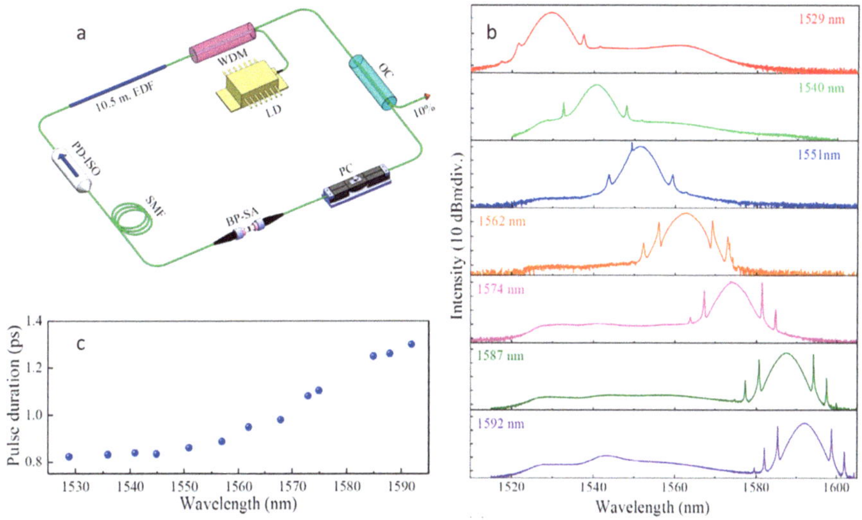

Fig. 14.12 (**a**) Experimental setup of the wavelength-tunable mode-locked fiber laser. EDF: erbium-doped fiber; WDM: wavelength-division multiplexer; *LD* laser diode, *OC* output coupler, *PC* polarization controller, *BP-SA* black phosphorus-based saturable absorber, *SMF* single-mode fiber, *PD-ISO* polarization-dependent isolator. (**b**) Mode-locking operation with wideband tunable output spectra from 1529 to 1592 nm; (**c**) output pulse duration at different wavelengths. Reproduced with permission [59]

Table 14.8 Mode-locked fiber laser by 2D materials with high repetition rates

Laser configuration	Saturable absorber	Repetition rate (GHz)	Reference
~1 cm short linear-cavity EDFL	LPE-prepared graphene	9.67	[63]
21st-harmonic mode-locked ring-cavity EDFL	ME-prepared multilayer graphene	2.22	[52]
Ring-cavity EDFL with FP filter	LPE-prepared graphene	100	[65]
Ring-cavity EDFL with microfiber knot filter	LPE-prepared graphene	106.7	[66]
Ring-cavity YDFL with microfiber knot filter	LPE-prepared graphene	162	[66]
418th-harmonic mode-locked ring-cavity EDFL	LPE-prepared TI – Bi2Te3	2.04	[67]
170th-harmonic mode-locked ring-cavity EDFL	PLD-prepared TI – Bi2Te3	2.95	[135]
~22 cm short linear-cavity EDFL	LPE-prepared TMD – MoS2	495 MHz	[64]
212th-harmonic mode-locked ring-cavity EDFL	LPE-prepared TMD – MoSe2	3.27	[54]
10th-harmonic mode-locked ring-cavity HDFL	LPE-prepared BP	290 MHz	[55]

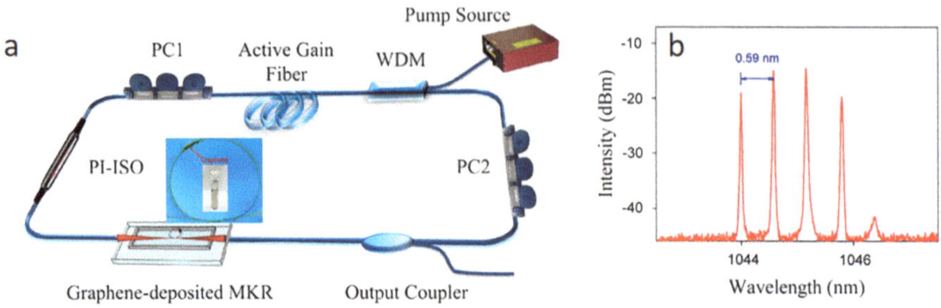

Fig. 14.13 (**a**) Schematic of the ultrafast fiber laser used in the experiment. (**b**) FD-FWM mode-locking operation in the YDF laser based on graphene-deposited MKR: Output spectrum; Reproduced with permission [66]

cavity fiber lasers, providing the spectral filtering and saturable absorption effect [66]. When the filter was implemented into a Yb-doped fiber laser (YDFL), a pulse train at 1 μm with a 162-GHz repetition rate was generated. When it was inserted into an EDFL, a pulse train in the 1.5-μm region with a 106.7-GHz repetition rate was addressed. The harmonic mode-locking technique is an alternative important approach to realize high repetition rate outputs. In 2012, a graphene-based 21st-harmonic mode-locked fiber laser was reported by Sobon et al., with a pulse repetition rate of 2.22 GHz [52]. Other kinds of 2D materials like TIs, TMDs, and BP are also efficient for achieving harmonic mode-locking operations. With a TI-Bi_2Te_3 fiber SA implemented in a ring-cavity EDFL, Luo et al. raised the repetition rate to 2.04 GHz, corresponding to a high harmonic order of 418 [67]. Another work by Yan et al. also utilized a TI-Bi_2Se_3 fiber SA in a ring-cavity EDFL, but with a higher fundamental repetition rate of 200 MHz. In this case, a harmonic mode locking repetition rate of 2.95 GHz was obtained at a 170th harmonic order. To date, the highest repetition rate of a TMD mode-locked UFL was accomplished by Koo et al. in 2016 [54]. In their work, a MoSe2 /PVA composite depositing side-polished fiber was incorporated as an SA within a ring-cavity EDFL, where a repetition rate up to 3.27 GHz at a harmonic order of 212 was obtained. Considering various applications of high repetition rate pulses, we anticipate that high repetition rate mode-locked by 2D materials will continue to be a hot topic in the future.

References

1. Xia, F., Wang, H., Xiao, D., Dubey, M., Ramasubramaniam, A.: Two-dimensional material nanophotonics. Nat. Photon. **8**, 899–907 (2014)
2. Liao, J.Y., Wu, J.X., Dang, C.H., Xie, L.M.: Methods of transferring two-dimensional materials. Acta Phys. Sin. **70**, 028201 (2021)
3. Huang, Z., et al.: Versatile construction of van der Waals heterostructures using a dual-function polymeric film. Nat. Commun. **11**, 3029 (2020)

4. Dean, C.R., et al.: Boron nitride substrates for high-quality graphene electronics. Nat. Nanotechnol. **5**, 722–726 (2010)
5. Li, H., et al.: A universal, rapid method for clean transfer of nanostructures onto various substrates. ACS Nano. **8**, 6563–6570 (2014)
6. Wang, P., et al.: High-fidelity transfer of chemical vapor deposition grown 2D transition metal dichalcogenides via substrate decoupling and polymer/small molecule composite. ACS Nano. **14**, 7370–7379 (2020)
7. Schneider, G.F., Calado, V.E., Zandbergen, H., Vandersypen, L.M., Dekker, C.: Wedging transfer of nanostructures. Nano Lett. **10**, 1912–1916 (2010)
8. Lee, Y., et al.: Wafer-scale synthesis and transfer of graphene films. Nano Lett. **10**, 490–493 (2010)
9. Wang, Y., et al.: Electrochemical delamination of CVD grown graphene film: toward the recyclable use of copper catalyst. ACS Nano. **5**, 9927–9933 (2011)
10. Fan, S., Vu, Q.A., Tran, M.D., Adhikari, S., Lee, Y.H.: Transfer assembly for two-dimensional van der Waals heterostructures. 2D Mater. **7**, 022005 (2020)
11. Liu, F., et al.: Disassembling 2D van der Waals crystals into macroscopic monolayers and reassembling into artificial lattices. Science. **367**, 903–906 (2020)
12. Castellanos-Gomez, A., et al.: Deterministic transfer of two-dimensional materials by all-dry viscoelastic stamping. 2D Mater. **1**, 011002 (2014)
13. Pizzocchero, F., et al.: The hot pick-up technique for batch assembly of van der Waals heterostructures. Nat. Commun. **7**, 11894 (2016)
14. Dawlaty, J.M., Shivaraman, S., Chandrashekhar, M., Rana, F., Spencer, M.G.: Measurement of ultrafast carrier dynamics in epitaxial graphene. Appl. Phys. Lett. **92**, 042116 (2008)
15. Wang, K., et al.: Ultrafast nonlinear excitation dynamics of black phosphorus nanosheets from visible to mid-infrared. ACS Nano. **10**, 6923–6932 (2016)
16. Feng, Y., et al.: Saturable absorption behavior of free-standing graphene polymer composite films over broad wavelength and time ranges. Opt. Express. **23**, 559–569 (2015)
17. Jiang, X., et al.: Broadband nonlinear photonics in few-layer MXene $Ti_3C_2T_x$ (T = F, O, or OH). Laser Photon. Rev. **12**, 1700229 (2017)
18. Guo, B.: 2D noncarbon materials-based nonlinear optical devices for ultrafast photonics [invited]. Chinese Opt. Lett. **16**, 020004 (2018)
19. Zhang, H., et al.: Z-scan measurement of the nonlinear refractive index of graphene. Opt. Lett. **37**, 1856–1858 (2012)
20. Zhao, L.M., et al.: Dissipative soliton operation of an ytterbium-doped fiber laser mode locked with atomic multilayer graphene. Opt. Lett. **35**, 3622 (2010)
21. Cheng, Z., et al.: Dissipative soliton resonance and reverse saturable absorption in graphene oxide mode-locked all-normal-dispersion Yb-doped fiber laser. Opt. Express. **23**, 7000–7006 (2015)
22. Dou, Z., et al.: Mode-locked ytterbium-doped fiber laser based on topological insulator: Bi_2Se_3. Opt. Express. **22**, 24055–24061 (2014)
23. Kowalczyk, M., et al.: Sb_2Te_3-deposited D-shaped fiber as a saturable absorber for mode-locked Yb-doped fiber lasers. Opt Mater Express. **6**, 2273–2282 (2016)
24. Li, L., et al.: WS_2/fluorine mica (FM) saturable absorbers for all-normal-dispersion mode-locked fiber laser. Opt. Express. **23**, 28698–28706 (2015)
25. Zhang, H., et al.: Molybdenum disulfide (MoS_2) as a broadband saturable absorber for ultrafast photonics. Opt. Express. **22**, 7249–7260 (2014)
26. Li, J., et al.: Passively mode-locked ytterbium-doped fiber laser based on SnS_2 as saturable absorber. IEEE Photon. J. **9**, 1506707 (2017)
27. Hisyam, M.B., Rusdi, M.F.M., Latiff, A.A., Harun, S.W.: Generation of mode-locked ytterbium doped fiber ring laser using few-layer black phosphorus as a saturable absorber. IEEE J. Sel. Top. Quant. **23**, 39–43 (2017)
28. Chai, T., et al.: Few-layer bismuthene for ultrashort pulse generation in a dissipative system based on an evanescent field. Nanoscale. **10**, 17617–17622 (2018)

29. Tarka, J., et al.: Power scaling of an all-PM fiber Er-doped mode-locked laser based on graphene saturable absorber. IEEE J. Sel. Top. Quant. **23**, 1100506 (2017)
30. Sotor, J., Pasternak, I., Krajewska, A., Strupinski, W., Sobon, G.: Sub-90 fs a stretched-pulse mode-locked fiber laser based on a graphene saturable absorber. Opt. Express. **23**, 27503–27508 (2015)
31. Sobon, G., et al.: Multilayer graphene-based saturable absorbers with scalable modulation depth for mode-locked Er- and Tm-doped fiber lasers. Opt. Mater. Express. **5**, 2884–2894 (2015)
32. Purdie, D.G., et al.: Few-cycle pulses from a graphene mode-locked all-fiber laser. Appl. Phys. Lett. **106**, 253101 (2015)
33. Liu, W., et al.: 70-fs mode-locked erbium-doped fiber laser with topological insulator. Sci. Rep. **6**, 19997 (2016)
34. Liu, M., et al.: Ultrashort pulse generation in mode-locked erbium-doped fiber lasers with tungsten disulfide saturable absorber. Opt. Commun. **406**, 72–75 (2018)
35. Niu, K., Sun, R., Chen, Q., Man, B., Zhang, H.: Passively mode-locked Er-doped fiber laser based on SnS_2 nanosheets as a saturable absorber. Photon. Res. **6**, 72–76 (2018)
36. Jin, X., et al.: 102 fs pulse generation from a long-term stable, inkjet-printed black phosphorus-mode-locked fiber laser. Opt. Express. **26**, 12506–12513 (2018)
37. Li, D., et al.: Polarization and thickness dependent absorption properties of black phosphorus: new saturable absorber for ultrafast pulse generation. Sci. Rep. **5**, 15899 (2015)
38. Guo, B., et al.: Sub-200 fs soliton mode-locked fiber laser based on bismuthene saturable absorber. Opt. Express. **26**, 22750–22760 (2018)
39. Song, Y., et al.: Few-layer antimonene decorated microfiber: ultra-short pulse generation and all-optical thresholding with enhanced long term stability. 2D Mater. **4**, 2017
40. Jhon, Y.I., et al.: Metallic MXene Saturable absorber for femtosecond mode-locked lasers. Adv. Mater. **29**, 1702496 (2017)
41. Sotor, J., et al.: All-fiber Ho-doped mode-locked oscillator based on a graphene saturable absorber. Opt. Lett. **41**, 2592–2595 (2016)
42. Sotor, J., et al.: All-polarization-maintaining, stretched-pulse Tm-doped fiber laser, mode-locked by a graphene saturable absorber. Opt. Lett. **42**, 1592–1595 (2017)
43. Pawliszewska, M., Martynkien, T., Przewloka, A., Sotor, J.: Dispersion-managed Ho-doped fiber laser mode-locked with a graphene saturable absorber. Opt. Lett. **43**, 38–41 (2018)
44. Jung, M., et al.: A femtosecond pulse fiber laser at 1935 nm using a bulk-structured Bi_2Te_3 topological insulator. Opt. Express. **22**, 7865–7874 (2014)
45. Wang, J., et al.: Magnetron-sputtering deposited WTe_2 for an ultrafast thulium-doped fiber laser. Opt. Lett. **42**, 5010–5013 (2017)
46. Liu, W., et al.: Tungsten diselenide for mode-locked erbium-doped fiber lasers with short pulse duration. Nanotechnology. **29**, 174002 (2018)
47. Lee, J., Koo, J., Lee, J., Jhon, Y.M., Lee, J.H.: All-fiberized, femtosecond laser at 1912 nm using a bulk-like $MoSe_2$ saturable absorber. Opt. Mater. Express. **7**, 2968–2979 (2017)
48. Sotor, J., et al.: Ultrafast thulium-doped fiber laser mode locked with black phosphorus. Opt. Lett. **40**, 3885–3888 (2015)
49. Zhu, G., et al.: Graphene mode-locked fiber laser at 2.8 um. IEEE Photon. Technol. Lett. **28**, 7–10 (2016)
50. Li, J., et al.: Black phosphorus: a two-dimension saturable absorption material for mid-infrared Q-switched and mode-locked fiber lasers. Sci. Rep. **6**, 30361 (2016)
51. Qin, Z., et al.: Black phosphorus Q-switched and mode-locked mid-infrared Er:ZBLAN fiber laser at 3.5 um wavelength. Opt. Express. **26**, 8224–8231 (2018)
52. Sobon, G., Sotor, J., Abramski, K.M.: Passive harmonic mode-locking in Er-doped fiber laser based on graphene saturable absorber with repetition rates scalable to 2.22 GHz. Appl. Phys. Lett. **100**, 161109 (2012)
53. Meng, Y., et al.: High power L-band mode-locked fiber laser based on topological insulator saturable absorber. Opt. Express. **23**, 23053–23058 (2015)

54. Koo, J., Park, J., Lee, J., Jhon, Y.M., Lee, J.H.: Femtosecond harmonic mode-locking of a fiber laser at 3.27 GHz using a bulk-like, $MoSe_2$-based saturable absorber. Opt. Express. **24**, 10575–10589 (2016)
55. Pawliszewska, M., Ge, Y., Li, Z., Zhang, H., Sotor, J.: Fundamental and harmonic mode-locking at 2.1 um with black phosphorus saturable absorber. Opt. Express. **25**, 16916–16921 (2017)
56. Sun, Z., et al.: A stable, wideband tunable, near transform-limited, graphene-mode-locked, ultrafast laser. Nano Res. **3**, 653–660 (2010)
57. Yang, G., et al.: Broadband wavelength tunable mode-locked thulium-doped fiber laser operating in the 2 μm region by using a graphene saturable absorber on microfiber. Laser Phys. Lett. **13**, 065105 (2016)
58. Zhao, C., et al.: Wavelength-tunable picosecond soliton fiber laser with topological insulator: Bi_2Se_3 as a mode locker. Opt. Express. **20**, 27888–27895 (2012)
59. Li, W., Chen, G., Wang, G., Zeng, C., Zhao, W.: Wideband wavelength-tunable ultrafast fiber laser based on black phosphorus saturable absorber. Laser Phys. Lett. **15**, 125102 (2018)
60. Zhao, N., et al.: Dual-wavelength rectangular pulse Yb-doped fiber laser using a microfiber-based graphene saturable absorber. Opt. Express. **22**, 10906–10913 (2014)
61. Lau, K.Y., et al.: Dual-wavelength, mode-locked erbium-doped fiber laser employing a graphene/polymethyl-methacrylate saturable absorber. Opt. Express. **26**, 12790–12800 (2018)
62. Yun, L.: Black phosphorus saturable absorber for dual-wavelength polarization-locked vector soliton generation. Opt. Express. **25**, 32380–32385 (2017)
63. Martinez, A., Yamashita, S.: 10 GHz fundamental mode fiber laser using a graphene saturable absorber. Appl. Phys. Lett. **101**, 041118 (2012)
64. Wu, K., Zhang, X., Wang, J., Chen, J.: 463-MHz fundamental mode-locked fiber laser based on few-layer MoS_2 saturable absorber. Opt. Lett. **40**, 1374–1377 (2015)
65. Qi, Y.L., et al.: Graphene-deposited microfiber photonic device for ultrahigh-repetition rate pulse generation in a fiber laser. Opt. Express. **23**, 17720–17726 (2015)
66. Liu, M., Tang, R., Luo, A.P., Xu, W.C., Luo, Z.C.: Graphene-decorated microfiber knot as a broadband resonator for ultrahigh-repetition-rate pulse fiber lasers. Photon. Res. **6**, C1–C7 (2018)
67. Luo, Z.C., et al.: 2 GHz passively harmonic mode-locked fiber laser by a microfiber-based topological insulator saturable absorber. Opt. Lett. **38**, 5212–5215 (2013)
68. Wang, G., et al.: Ultrafast nonlinear optical properties of a graphene saturable mirror in the 2 μm wavelength region. Laser Photon. Rev. **11**, 1700166 (2017)
69. Miro, P., Audiffred, M., Heine, T.: An atlas of two-dimensional materials. Chem. Soc. Rev. **43**, 6537–6554 (2014)
70. Zhao, J., et al.: Thickness-dependent carrier and phonon dynamics of topological insulator Bi_2Te_3 thin films. Opt. Express. **25**, 14635–14643 (2017)
71. Koski, K.J., et al.: Chemical intercalation of zerovalent metals into 2D layered Bi_2Se_3 nanoribbons. J. Am. Chem. Soc. **134**, 13773–13779 (2012)
72. Zhang, H., et al.: Topological insulators in Bi_2Se_3, Bi_2Te_3 and Sb_2Te_3 with a single Dirac cone on the surface. Nat. Phys. **5**, 438–442 (2009)
73. Lu, S.C., Leburton, J.P.: Electronic structures of defects and magnetic impurities in MoS_2 monolayers. Nanoscale Res. Lett. **9**, 676 (2014)
74. Xu, M., Liang, T., Shi, M., Chen, H.: Graphene-like two-dimensional materials. Chem. Rev. **113**, 3766–3798 (2013)
75. Splendiani, A., et al.: Emerging photoluminescence in monolayer MoS_2. Nano Lett. **10**, 1271–1275 (2010)
76. Tran, V., Fei, R., Yang, L.: Quasiparticle energies, excitons, and optical spectra of few-layer black phosphorus. 2D Mater. **2**, 044014 (2015)
77. Tran, V., Soklaski, R., Liang, Y., Yang, L.: Layer-controlled band gap and anisotropic excitons in few-layer black phosphorus. Phys. Rev. B. **89**, 235319 (2014)

78. Yang, Q.Q., et al.: 2D bismuthene fabricated via acid-intercalated exfoliation showing strong nonlinear near-infrared responses for mode-locking lasers. Nanoscale. **10**, 21106–21115 (2018)
79. Liu, M.Y., et al.: Strain and electric field tunable electronic structure of buckled bismuthene. RSC Adv. **7**, 39546–39555 (2017)
80. Pumera, M., Sofer, Z.: 2D monoelemental arsenene, antimonene, and bismuthene: beyond black phosphorus. Adv. Mater. **29**, 1605299 (2017)
81. Chi, C., Lee, J., Koo, J., Han Lee, J.: All-normal-dispersion dissipative-soliton fiber laser at 1.06 μm using a bulk-structured Bi_2Te_3 topological insulator-deposited side-polished fiber. Laser Phys. **24**, 105106 (2014)
82. Yan, P., et al.: Topological insulator solution filled in photonic crystal fiber for passive mode-locked fiber laser. IEEE Photon. Technol. Lett. **27**, 264–267 (2015)
83. Song, H.S., Wang, Q., Zhang, Y., Li, L.: Mode-locked ytterbium-doped all-fiber lasers based on few-layer black phosphorus saturable absorbers. Opt. Commun. **394**, 157–160 (2017)
84. Huang, P.L., et al.: Stable mode-locked fiber laser based on CVD fabricated graphene saturable absorber. Opt. Express. **20**, 2460–2465 (2012)
85. Liu, X.M., et al.: Graphene-clad microfibre saturable absorber for ultrafast fibre lasers. Sci. Rep. **6**, 26024 (2016)
86. Bao, Q., et al.: Monolayer graphene as a saturable absorber in a mode-locked laser. Nano Res. **4**, 297–307 (2010)
87. Bao, Q., et al.: Atomic-layer graphene as a saturable absorber for ultrafast pulsed lasers. Adv. Funct. Mater. **19**, 3077–3083 (2009)
88. Ko, S., et al.: Chemical wet etching of an optical fiber using a hydrogen fluoride-free solution for a saturable absorber based on the evanescent field interaction. J. Light. Technol. **34**, 3776–3784 (2016)
89. Steinberg, D., Zapata, J.D., Thoroh, D.S., Eunezio, A.S., Lucia, A.M.: Mechanically exfoliated graphite onto D-shaped optical fiber for femtosecond mode-locked erbium-doped fiber laser. J. Lightw. Technol. **36**, 1868–1874 (2018)
90. Boguslawski, J., et al.: Graphene oxide paper as a saturable absorber for Er- and tm-doped fiber lasers. Photon. Res. **3**, 119–124 (2015)
91. Tarka, J., et al.: 168 fs pulse generation from graphene-chitosan mode-locked fiber laser. Opt. Mater. Express. **4**, 1981–1986 (2014)
92. Li, K., et al.: L-band femtosecond fibre laser based on Bi_2Se_3 topological insulator. Laser Phys. Lett. **12**, 105103 (2015)
93. Liu, H., et al.: Femtosecond pulse generation from a topological insulator mode-locked fiber laser. Opt. Express. **22**, 6868–6873 (2014)
94. Xu, Y., et al.: Bilayer bismuth selenide nanoplatelets based saturable absorber for ultra-short pulse generation (invited). Opt. Commun. **395**, 55–60 (2017)
95. Gao, L., Zhu, T., Huang, W., Luo, Z.: Stable, ultrafast pulse mode-locked by topological insulator Bi_2Se_3 nanosheets interacting with photonic crystal fiber: from anomalous dispersion to normal dispersion. IEEE Photonics J. **7**, 3300108 (2015)
96. Lee, J., Koo, J., Jhon, Y.M., Lee, J.H.: A femtosecond pulse erbium fiber laser incorporating a saturable absorber based on bulk-structured Bi_2Te_3 topological insulator. Opt. Express. **22**, 6165–6173 (2014)
97. Yan, P., et al.: A practical topological insulator saturable absorber for mode-locked fiber laser. Sci. Rep. **5**, 8690 (2015)
98. Mao, D., et al.: Soliton fiber laser mode locked with two types of film-based Bi_2Te_3 saturable absorbers. Photon. Res. **3**, A43–A46 (2015)
99. Boguslawski, J., et al.: Mode-locked Er-doped fiber laser based on liquid phase exfoliated Sb_2Te_3 topological insulator. Laser Phys. **24**, 105111 (2014)
100. Sotor, J., Sobon, G., Grodecki, K., Abramski, K.M.: Mode-locked erbium-doped fiber laser based on evanescent field interaction with Sb_2Te_3 topological insulator. Appl. Phys. Lett. **104**, 251112 (2014)

101. Boguslawski, J., et al.: Investigation on pulse shaping in fiber laser hybrid mode-locked by Sb_2Te_3 saturable absorber. Opt. Express. **23**, 29014–29023 (2015)
102. Sotor, J., Sobon, G., Abramski, K.M.: Sub-130 fs mode-locked Er-doped fiber laser based on topological insulator. Opt. Express. **22**, 13244–13249 (2014)
103. Boguslawski, J., Sobon, G., Zybala, R., Sotor, J.: Dissipative soliton generation in Er-doped fiber laser mode-locked by Sb_2Te_3 topological insulator. Opt. Lett. **40**, 2786–2789 (2015)
104. Lee, J., Kim, Y., Lee, K., Lee, J.H.: Femtosecond mode-locking of a fiber laser using a $CoSb_3$-skutterudite-based saturable absorber. Photon. Res. **6**, C36–C43 (2018)
105. Li, L., et al.: Femtosecond passively Er-doped mode-locked fiber laser with WS_2 solution saturable absorber. IEEE J. Sel. Top. Quant. **23**, 44–49 (2017)
106. Wu, K., Zhang, X., Wang, J., Li, X., Chen, J.: WS_2 as a saturable absorber for ultrafast photonic applications of mode-locked and Q-switched lasers. Opt. Express. **23**, 11453–11461 (2015)
107. Guo, B., et al.: Dual-wavelength soliton mode-locked fiber laser with a WS_2-based fiber taper. IEEE Photon. Technol. Lett. **28**, 323–326 (2016)
108. Liu, W., et al.: Tungsten disulfide saturable absorbers for 67 fs mode-locked erbium-doped fiber lasers. Opt. Express. **25**, 2950–2959 (2017)
109. Khazaeinezhad, R., et al.: Ultrafast pulsed all-fiber laser based on tapered fiber enclosed by few-layer WS_2 nanosheets. IEEE Photon. Technol. Lett. **27**, 1581–1584 (2015)
110. Khazaeinezhad, R., Hosseinzadeh Kassani, S., Jeong, H., Yeom, D.I., Oh, K.: Femtosecond soliton pulse generation using evanescent field interaction through tungsten disulfide (WS_2) film. J. Lightw. Technol. **33**, 3550–3557 (2015)
111. Liu, W., et al.: Tungsten disulphide for ultrashort pulse generation in all-fiber lasers. Nanoscale. **9**, 5806–5811 (2017)
112. Khazaeizhad, R., Kassani, S.H., Jeong, H., Yeom, D.I., Oh, K.: Mode-locking of Er-doped fiber laser using a multilayer MoS_2 thin film as a saturable absorber in both anomalous and normal dispersion regimes. Opt. Express. **22**, 23732–23742 (2014)
113. Liu, H., et al.: Femtosecond pulse erbium-doped fiber laser by a few-layer MoS_2 saturable absorber. Opt. Lett. **39**, 4591–4594 (2014)
114. Aiub, E.J., Steinberg, D., Thoroh de Souza, E.A., Saito, L.A.M.: 200-fs mode-locked erbium-doped fiber laser by using mechanically exfoliated MoS_2 saturable absorber onto D-shaped optical fiber. Opt. Express. **25**, 10546–10552 (2017)
115. Ahmed, M.H.M., Latiff, A.A., Arof, H., Ahmad, H., Harun, S.W.: Femtosecond mode-locked erbium-doped fiber laser based on MoS_2–PVA saturable absorber. Opt. Laser Technol. **82**, 145–149 (2016)
116. Koo, J., et al.: Near-infrared saturable absorption of defective bulk-structured WTe_2 for femtosecond laser mode-locking. Adv. Funct. Mater. **26**, 7454–7461 (2016)
117. Liu, W., et al.: CVD-grown $MoSe_2$ with high modulation depth for ultrafast mode-locked erbium-doped fiber laser. Nanotechnology. **29**, 394002 (2018)
118. Cui, Y., Lu, F., Liu, X.: Nonlinear saturable and polarization-induced absorption of rhenium disulfide. Sci. Rep. **7**, 40080 (2017)
119. Mao, D., et al.: Passively Q-switched and mode-locked fiber laser based on an ReS_2 saturable absorber. IEEE J. Sel. Top. Quant. **24**, 1100406 (2018)
120. Wang, J., et al.: Ultrafast thulium-doped fiber laser mode locked by monolayer WSe_2. IEEE J. Sel. Top. Quant. **24**, 1100706 (2018)
121. Mao, D., et al.: Stable high-power saturable absorber based on polymer-black-phosphorus films. Opt. Commun. **406**, 254–259 (2018)
122. Ahmed, M.H.M., Latiff, A.A., Arof, H., Harun, S.W.: Ultrafast erbium-doped fiber laser mode-locked with a black phosphorus saturable absorber. Laser Phys. Lett. **13** (2016)
123. Song, Y., et al.: Vector soliton fiber laser passively mode locked by few layer black phosphorus-based optical saturable absorber. Opt. Express. **24**, 25933–25942 (2016)
124. Chen, Y., et al.: Mechanically exfoliated black phosphorus as a new saturable absorber for both Q-switching and mode-locking laser operation. Opt. Express. **23**, 12823–12833 (2015)

125. Sotor, J., Sobon, G., Macherzynski, W., Paletko, P., Abramski, K.M.: Black phosphorus saturable absorber for ultrashort pulse generation. Appl. Phys. Lett. **107**, 051108 (2015)
126. Xu, Y., et al.: Size-dependent nonlinear optical properties of black phosphorus nanosheets and their applications in ultrafast photonics. J. Mater. Chem. C. **5**, 3007–3013 (2017)
127. Liu, M., et al.: Black phosphorus quantum dots for femtosecond laser photonics. Opt. Commun. **406**, 85–90 (2018)
128. Du, J., et al.: Phosphorene quantum dot saturable absorbers for ultrafast fiber lasers. Sci. Rep. **7**, 42357 (2017)
129. Lu, L., et al.: Few-layer bismuthene: sonochemical exfoliation, nonlinear optics and applications for ultrafast photonics with enhanced stability. Laser Photon. Rev. **12**, 1700221 (2017)
130. Sobon, G., et al.: All-polarization maintaining, graphene-based femtosecond Tm-doped all-fiber laser. Opt. Express. **23**, 9339–9346 (2015)
131. Yin, K., et al.: Soliton mode-locked fiber laser based on topological insulator Bi_2Te_3 nanosheets at 2 μm. Photon. Res. **3**, 72–76 (2015)
132. Jung, M., et al.: Mode-locked, 1.94 um, all-fiberized laser using WS_2 based evanescent field interaction. Opt. Express. **23**, 19996–20006 (2015)
133. Yu, H., Zheng, X., Yin, K., Cheng, X., Jiang, T.: Thulium/holmium-doped fiber laser passively mode locked by black phosphorus nanoplatelets-based saturable absorber. Appl. Opt. **54**, 10290–10294 (2015)
134. Meng, Y.C., et al.: Passive harmonically mode-locked fiber laser with low pumping power based on a graphene saturable absorber. Laser Phys. Lett. **9**, 537–541 (2012)
135. Yan, P., Lin, R., Ruan, S., Liu, A., Chen, H.: A 2.95 GHz, femtosecond passive harmonic mode-locked fiber laser based on evanescent field interaction with topological insulator film. Opt. Express. **23**, 154–164 (2015)
136. Ferrari, A.C., et al.: Science and technology roadmap for graphene, related two-dimensional crystals, and hybrid systems. Nanoscale. **7**, 4598–4810 (2015)
137. Liu, M., et al.: Microfiber-based few-layer MoS_2 saturable absorber for 2.5 GHz passively harmonic mode-locked fiber laser. Opt. Express. **22**, 22841–22846 (2014)
138. Wang, Y., et al.: Harmonic mode locking of bound-state solitons fiber laser based on MoS_2 saturable absorber. Opt. Express. **23**, 205–210 (2015)
139. Zhang, H., et al.: Graphene mode locked, wavelength-tunable, dissipative soliton fiber laser. Appl. Phys. Lett. **96**, 111112 (2010)
140. Zhang, M., et al.: Solution processed MoS_2-PVA composite for sub-bandgap mode-locking of a wideband tunable ultrafast Er:fiber laser. Nano Res. **8**, 1522–1534 (2015)
141. Luo, Z.C., et al.: Microfiber-based few-layer black phosphorus saturable absorber for ultrafast fiber laser. Opt. Express. **23**, 20030–20039 (2015)
142. Chen, Y., Chen, S., Liu, J., Gao, Y., Zhang, W.: Sub-300 femtosecond soliton tunable fiber laser with all-anomalous dispersion passively mode locked by black phosphorus. Opt. Express. **24**, 13316–13324 (2016)
143. Zheng, X.W., et al.: Dual-wavelength harmonically mode-locked fiber laser with topological insulator saturable absorber. IEEE Photon. Technol. Lett. **26**, 983–986 (2014)
144. Guo, B., et al.: Topological insulator: Bi_2Se_3/polyvinyl alcohol film-assisted multi-wavelength ultrafast erbium-doped fiber laser. J. Appl. Phys. **117**, 063108 (2015)

Chapter 15
Large-Mode-Area (LMA) Fibers for High-Power Lasers

Wang Pengfei, Wan Rui, Chen Chao, Li Xianda, Guo Chen, and Pan Jiapeng

Abstract High-power fiber lasers have experienced dramatic development over the past few decades. Further upscaling of the laser power greatly depends on the development of laser fibers with an enlarged large-mode area (LMA) while maintaining a single-mode (SM) output, which can effectively reduce the power density and mitigate the influence of nonlinear and thermo-optic effects that induce transverse mode instability (TMI). The development of fiber lasers is inseparable from the advancement of laser fibers from conventional step-index fibers to maturing large mode area (LMA) double-clad fibers and microstructured fibers (MOFs). This chapter begins with a brief history and overview of LMA fibers and then summarizes the most recent research progress on the structural design principles and fabrication of solid-core LMA fibers, which are made mainly of silica-based glasses, as well as their applications in high-power fiber lasers.

Keywords Large mode area fibers · Microstructured fibers · Photonic crystal fibers (PCFs) · All-solid photonic bandgap fibers (AS-PBGFs) · Single-mode operation · Mode field diameter (MFD) · Transverse mode instability · High-power fiber lasers

---

W. Pengfei (✉) · W. Rui · L. Xianda · G. Chen · P. Jiapeng
State Key Laboratory of Ultrafast Optical Science and Technology, Xi'an Institute of Optics and Precision Mechanics, Chinese Academy of Sciences (CAS), Xi'an, Shaanxi, China

Center of Materials Science and Optoelectronics Engineering, University of Chinese Academy of Sciences (UCAS), Beijing, China
e-mail: pfwang@opt.ac.cn

C. Chao
State Key Laboratory of Transient Optics and Photonics, Xi'an Institute of Optics and Precision Mechanics, Chinese Academy of Sciences (CAS), Xi'an, Shaanxi, China

M. F. S. Ferreira, M. C. Paul (eds.), *Optical Fiber Lasers*, Springer Series in Optical Sciences 250, https://doi.org/10.1007/978-3-032-04110-4_15

15.1 Fundamentals of LMA Fibers

Interest in LMA fibers for realizing high-power fiber laser sources has continuously increased over the last few decades due to the multiple application fields. For most of those applications, the main requirement regarding the amplifying medium, in addition to the output power, is to ensure high beam quality, generally a Gaussian-like beam profile, while delaying the appearance of unwanted nonlinear effects (Raman, Kerr, etc.) when the power is increased. However, the improvement in the output power of fiber lasers is limited by two main aspects: the laser-induced damage to the fiber end face and the nonlinear effect on the fibers. To achieve laser output with high beam quality and overcome the limitation of power scaling-up caused by laser damage and nonlinear effects, the number aperture (NA) of the fiber core should be reduced as much as possible when designing and selecting the laser fibers, and the core diameter should be increased accordingly [1]. These demands have led to LMA optical fibers, which greatly promote the development of high-power fiber laser technologies. Many LMA optical fiber designs have been proposed on the basis of both conventional and microstructured fiber geometries, for which light confinement is ensured by total internal reflection (TIR), modified total internal reflection (M-TIR), the photonic band gap (PBG), or inhibited coupling (IC).

The main problem with increasing core size is that fibers tend to support the propagation of different transverse modes, which usually results in a decrease in beam quality and pointing stability. Furthermore, in extreme cases, this may even lead to the appearance of transverse mode instabilities (TMIs) [2]. Thus, the main challenge when increasing the core size is to keep the fiber core effectively single-mode. During the development of LMA fibers, initial efforts focused on increasing the core diameter while reducing the NA accordingly to maintain pure single-mode operation [3]. However, as the NA decreases, the guidance becomes progressively weaker, and the guided laser light is more readily lost from the core as the fiber is bent, which limits the minimum NA that can be used in practice to 0.04–0.05 [4]. The technically achievable minimum NA of the fiber core through the modified chemical vapor deposition (MCVD) method is approximately 0.05–0.06, but the corresponding maximum single-mode fiber (SMF) core diameter does not exceed 35 μm. This limits further application of the MCVD method for the fabrication of SMFs with larger core diameters with stricter requirements of tiny refractive index differences between the core and clad to ensure high stability of the mode field of the fiber laser during high-power operation. Therefore, conventional fiber designs based on step-index cores have to be largely abandoned and substituted by more sophisticated approaches based on MOFs [5–7].

In this case, MOFs, or so-called photonic-crystal fibers (PCFs), have emerged as alternative solutions for realizing LMA fibers. These fibers have a microstructured geometry in the cross section that is repeated over the entire fiber's length. MOFs are considered to be the most effective way to achieve large-mode field single-mode fibers. An increase in the core area increases the threshold of laser damage and nonlinear effects. On the other hand, increasing the ratio of the cross-sectional area of the core and the inner cladding improves the pumping efficiency so that much shorter fibers can be used to achieve high-power laser output, which is conducive to overcoming the nonlinear effect of the limit on the output power [1]. MOFs have attracted a great deal of interest because of their unusual properties, including a wide range of single-mode operations [8, 9], tailorable dispersion, reduced or enhanced effective nonlinearity, and easily attainable high birefringence [10], breaking through traditional step-index fibers that cannot cross their own limits, indicating attractive application prospects [11]. MOFs are promising for nonlinear optics applications such as supercontinuum generation [12], ultrashort fiber lasers, and amplifiers, as well as transmission or sensing applications, including high-power light transmission, evanescent sensing devices, and fiber grating devices [13].

The development of MOFs dates far more back to 1987, when S. John [14] and E. Yablonovitch et al. [15] first proposed the conception of photonic crystals, which were introduced in optical fibers by Philip Russell in 1992 [16], by introducing two-dimensional photonic crystals in the fiber cladding that are close in size to the wavelength of light, where the fiber core has a defective structure missing an air hole and where light is restricted to propagate in the fiber core. In 1996, J. C. Knight and P. St. J. Russell et al. successfully fabricated the first refractive index-guided TIR-PCF with a solid core and confirmed its endless single mode characteristic [8]. These works aroused widespread interest, setting off a wave of research on MOFs.

There are two categories of MOFs: solid-core MOFs (SC-MOFs) and hollow-core MOFs (HC-MOFs) [17]. HC-MOFs were realized for the first time in 1999 [18] and present a central region containing air. The photonic crystal that constitutes the cladding only allows the propagation of light in certain frequency bands. Light guidance in the fiber core occurs at frequencies within the photonic bandgap of the cladding periodic structure. Extremely low losses are expected from these types of fibers since the light travels predominantly in empty spaces. On the other hand, the nonlinear effects are strongly reduced, and the dispersion becomes negligible in air-core MOFs [19]. In contrast, SC-MOF has a solid core, in which the guidance of light is generally due to the modified total internal reflection effect, similar to the guidance mechanism of conventional optical fibers. In fact, the periodic nature of air holes in the cladding is not crucial in SC-MOFs, as long as they effectively reduce the refractive index below that of the fiber core [13]. Moreover, when an asymmetric structure is used, SC-MOFs also exhibit high birefringence, which provides for the maintenance of the polarization of the propagated light field.

According to the different fiber structures, in this chapter, we classify the discussed LMA fibers into three main types, i.e., cladding structure fibers, MOFs, and others, such as multicore fibers (MCFs) and segmented cladding fibers (SCFs). Here, the LMA cladding structure fibers include traditional double-clad fibers, very large-mode-area (VLMA) fibers, and taper double-clad fibers (T-DCFs). LMA MOFs are mainly introduced in terms of rod-type PCFs, photonic bandgap fibers (PBGFs), leakage channel PCFs (LC-PCFs), large-pitch PCFs (LPFs), solid-core antiresonant fibers (SC-ARFs), and polarization-maintaining (PM) LMA fibers. This chapter reviews the history and most recent research progress of the above-mentioned two categories of LMA fibers and introduces the structural design principles of LMA-PCFs and the single-mode determination criterion, the fabrication techniques of solid-core LMA-PCFs, which are made of mainly silica-based and some soft glasses, and their applications in high-power fiber lasers.

15.2 Brief History and Overview of LMA Fibers

Fiber lasers with high-average-power outputs basically adopt a double-cladding structure. Cladding structure fibers have been widely applied in high-power continuous wave (CW) fiber laser systems, and double-cladding structure fiber lasers can easily reach high-average-power outputs of kilowatts or even >10 kW. The development of LMA double-clad fibers has rapidly increased the output power/energy of fiber lasers, both in CW and pulsed modes [4, 20, 21]. It provides greater pump injection to the fibers, which consequently results in multiple kW-level average power and multiple millijoules of energy within the fiber laser and amplifiers. The upper limits for power scaling of fiber lasers have been predicted to be 28 and 52 kW for the diode pumping and tandem pumping techniques, respectively [22].

To accommodate the normalized frequency V-parameter requirements of single-mode transmission, LMA fibers with large-core areas and relatively small-core NAs are required to increase the output power and beam quality. The incredible properties underlying LMA fibers will pave the way for innovative optical fiber devices and technology [23]. Three main strategies can be followed to obtain single-mode operation in an LMA fiber [7]. The first consists of reducing the NA of the core to the point at which only the fundamental mode is guided. Owing to technical limitations, this direct approach is very challenging for cores >40 μm. This limitation can be partially solved by microstructuring the cladding region, as done with index-guiding PCF rods [24]. However, single-mode cores in excess of ~60 μm are still difficult to obtain. On the other hand, if the core supports the propagation of several modes, single-mode operation can still be achieved if the higher order modes (HOMs) experience significantly larger propagation losses than the fundamental mode does. This approach is followed by designs such as leakage channel fibers (LCFs) [25] or chirally coupled core fibers (CCCFs) [26]. Another proven approach to obtain single-mode operation in an active fiber is to exploit the delocalization of the HOMs. The structure of the fiber is such that it acts as a modal sieve [11], ensuring that most of the energy of the HOMs is located outside of the core area. This has two benefits: on the one hand, HOMs are difficult to excite with an incoming Gaussian beam, and they also have a significantly lower gain than the fundamental mode of the fiber [6]. Thus, after amplification, only the fundamental mode survives. Representatives of fiber designs exploiting HOM delocalization are the distributed mode-filtering (DMF) rod [27], the multitrench fiber [28], and the LPF [29].

The progress in mode area scaling appears most impressive in the issue of whether the approaches providing the largest claimed mode area will ever be practical, given that the concerns of robustness, mode stability, manufacturability, and reliability of pump coupling remain a concern. Significant further research into the whole area of mode scaling with a particular focus on improving practicality and high-power operation is needed, and increased mode areas will be achievable by extending the concepts described above to longer-wavelength transitions (i.e., one might anticipate a factor of 2–4 increase moving from 1 to 1.5 and 2 μm, or even longer) [4, 30].

15.2.1 LMA Cladding Structure Fibers

The LMA cladding structure fibers used here are three types, i.e., traditional double-clad fibers, very large mode area (VLMA) fibers and taper double-clad fibers (T-DCFs). LMA double clad fibers (LMA-DCFs) mainly include cores, inner claddings, outer layers, and protective layers. The core is usually a round rare-earth (RE)-doped silica glass, the diameter is generally 20–40 μm, and the minimum NA can be smaller than 0.06. The inner cladding is usually pure silica glass, and the shape can be round, square, rectangular, D-type, hexagonal, or unstable cavity structures. The inner cladding diameter is usually several hundred microns, and its NA is between 0.4 and 0.6. Compared with a traditional SM fiber, a large inner cladding not only increases the area of absorbed pumped light but also reduces the power density per unit area and increases the pumping damage threshold, but a large NA is conducive to receiving more pumped light and improving the coupling efficiency of pumped light. In addition, the absorption coefficient of the double-clad fiber core is related to the ratio of the inner cladding and the cross-sectional area of the fiber core; the thicker the cladding is, the smaller the absorption coefficient. With the development of fiber technology, the structure of LMA-DCFs has become increasingly standardized and matches the pigtail parameters of semiconductor lasers to improve the pumping coupling efficiency as much as possible [31].

15.2.1.1 Large-Mode Area Double-Cladding Fibers (LMA-DCFs)

LMA-DCFs are usually prepared via MCVD and solution doping methods, but it is difficult to obtain DCFs with uniform RE ion doping; the refractive index fluctuation in the core region is on the order of 10^{-4}, the NA is less than 0.06, and the core diameter is greater than 30 μm. To overcome the limitations of the MCVD method in the preparation of RE-doped large-mode field fibers, researchers from Heraeus Quarzglas GmbH & Co. KG and IPHT Jena in Germany have developed an innovative method to prepare Yb-doped silica glass via the powder sintering method. This method was used in the early stage to prepare high-purity silica glass, as well as large RE-doped silica glass mandrels [32], and was successfully applied to the development of LMA silica fibers for high-power fiber lasers [33–35]. This method has been used to produce large-size Yb^{3+}-doped mandrels with a loss of <20 dB/km and refractive index fluctuations of 2×10^{-4} and achieves a >10 kW multimode laser output in a single fiber [36].

The LMA of all-solid step-index fiber (AS-SIF) is realized mainly by lowering the core NA by introducing a highly ytterbium-doped P: Al core and germanium-doped cladding [37]. However, in ultralow NA fibers, the core diameter is limited to less than 50 μm because it is difficult to further reduce the core-clad refractive index difference while ensuring a certain gain [37, 38]. Recently, Kruska Kristopher et al. presented a forward-pumped high-power single-frequency master oscillator power amplifier (MOPA) based on a Yb^{3+}-doped depressed-cladding confined core specialty fiber with a core diameter of 35 μm. The depressed-cladding results in a decreased effective NA for the HOM; thus, the bending losses for these modes increase, whereas the fundamental mode is influenced by the higher refractive index step. With this fiber, an output power of 400 W was achieved with a slope efficiency of 75% and a PER of 15 dB, which further pushed the current power limits for gravitational wave detection lasers toward the 500 W mark [31].

15.2.1.2 Very Large-Mode-Area (VLMA) Fibers

Most approaches struggle with active and effective single-mode operation in SIFs with mode field diameters (MFDs) greater than 50 μm. Therefore, fibers with MFDs in excess of 50 μm represent a new regime, which can be classified as VLMA fibers [39] with outer diameters of more than 1 mm, which are almost rod-type fibers and are used mainly in high-power fiber laser systems. The doping concentration of Yb ions can be reduced, and the thermal effect caused by quantum loss can be reduced while ensuring a sufficiently high gain. However, a low concentration of Yb doping is also conducive to avoiding the photodarkening effect [33].

Andreas Langner et al. designed a fiber with an active core diameter between 35 and 100 μm. The pump cladding had a so-called 4D shape, which optimized the pumping of the fiber core and broke the circular geometry of the pump cladding to avoid helix modes that did not pass through the core. The pump cladding was always surrounded by a second fluorine-doped synthetic fused silica cladding with a depressed refractive index, which was produced by the plasma outside deposition process and featured a depressed refractive index typically in the range of -2×10^{-3} with respect to the undoped silica and corresponded to a pump NA of 0.27. The refractive index variation is typically less than $\pm 1 \times 10^{-4,}$ which is smaller than that of competitive materials produced, for example, by MCVD and solution doping. Using a single laser fiber, a maximum fiber laser output power of 4 kW was achieved in 2011, which is limited by the available laser diode (LD) pump power [33].

15.2.1.3 Taper Double-Clad Fibers (T-DCFs)

In 2008, Filippov et al. introduced a new concept of a double-clad fiber, called T-DCF, which offered a new approach for the amplification of laser signals to high-power/energy levels while maintaining excellent beam quality [40–43]. In T-DCFs, both the core and clad layers vary in diameter along the fiber length. T-DCFs provide varying large-core diameters of up to 200 μm, resulting in the minimum nonlinear phase shift and high pump absorption capability. Since the size of the core gradually increases along the fiber length, the mode content does not change through amplification, which results in excellent beam quality ($M^2 < 1.1$). Indeed, T-DCFs have an inherent capacity to increase the nonlinear effect thresholds [44]. However, for stable high-power/high-energy amplification, in addition to a large MFD, polarization maintenance is needed. In 2018, Fedotov et al. presented the first ultralarge core Yb-doped birefringent active T-DCF, with a core size of 96 μm, an effective MFA of 2000 μm^2, and a beam quality of $M^2 < 1.09$ [45].

T-DCFs have been extensively advanced in recent years because of their enormous potential to deliver outstanding characteristics, resulting in multikW average power and millijoule-level pulse energy. As a case in point, soon after the demonstration of T-DCF in 2008 [40], a 600 W single-mode ($M^2 < 1.08$) laser amplifier was reported by Filippov et al. in 2009, with an efficiency of 63% [46]. One year later, in 2010, they improved their work to a highly efficient 750 W laser amplifier with an efficiency of 80% [47]. Recently, a 4 kW CW single-mode laser via double-tapered double-clad fiber (DT-DCF) with a beam quality of $M^2 < 1.33$ was reported [48]. Later, it was increased to a 5 kW monolithic fiber amplifier, called a spindle-shaped YDF, with an optical efficiency of 66.6% and a beam quality of $M^2 < 1.9$ [49]. With respect to the production of single-frequency lasers, in 2020, a 550 W single-frequency fiber amplifier was demonstrated on the basis of the polarization-maintaining T-DCF [50]. Moreover, 100 ns single-frequency laser pulses with 0.5 mJ energy and a beam quality of $M^2 = 1.08$ were obtained [51]. In the case of energy scaling, 16 and 0.24 mJ energy pulses were achieved via active [52] and passive high-power Q-switched lasers [53], respectively. The latter was the first demonstration of a Q-switched fiber ring laser, which was based on active polarization-maintaining tapered double-clad fibers (APM-T-DCF).

Recently, the development of technology for the fabrication of tapered optical fibers [45, 54–58], which can be bent with a radius of 10 cm or even less, has made it possible to obtain a maximum peak power comparable to that achieved in Yb-doped PCFs. In the case of Er-doped fibers, a tapered Er-doped fiber with a core diameter ranging from 15.8 μm (with an MFD of 14.5 μm) to 93 μm (MFD of 40 μm) along 3.7 m and maintaining single-mode performance at 1555 nm was developed as an amplifier for high-peak-power pulse amplification. The amplification of 0.9 ns pulses with spectral widths below 0.04 nm up to a peak power above 200 kW (limited by self-phase modulation) with a slope pump-to-signal conversion efficiency of 15.6% was demonstrated [57]. This technique allowed the setting of a world record in terms of peak power.

There are several other reports on the use of T-DCFs to simultaneously achieve high peak and average power pulses in picosecond ranges with excellent beam quality, which have resulted in a few hundred watts in average power and megawatts in peak power via the T-DCF [55]. For example, a laser with a peak power of 1.26 MW and an average output power of 200 W was obtained by implementing the T-DCF without the need for additional stretching techniques [59]. These examples indicate that T-DCFs have great potential for high-power amplification of laser beams while maintaining the beam quality, which makes them an ideal gain medium for powerful coherent beam combining (CBC) systems [60–62].

In addition to classic monotonically tapered fibers, different longitudinal profiles in fibers have been manufactured for all-fiber kW-level MOPAs and oscillators. Using the ASE inhibition ability of T-DCFs, Aleshkina et al. reported a monolithic oscillator with a novel saddle-shaped YDF. A record SM output power of 10 W at 976 nm was obtained by good suppression of the ASE at 1030 nm [63]. Taking advantage of the stimulated Raman scattering (SRS) and TMI suppression ability, Zeng et al. achieved a 1.3 kW output at 1080 nm via a saddle-shaped T-DCF oscillator with 915 nm co-pumping [64]. Moreover, 1.8 and 3 kW TMI-free outputs have been achieved via improved bidirectional pumping under an all-fiber spindle-shaped oscillator [65]. Finally, the power was scaled to 5 kW ($M^2 \approx 1.9$) by slightly increasing the fiber size and adopting a MOPA configuration. This is the highest CW power recorded by T-DCF, and the relatively low optical-to-optical conversion efficiency of 66.6% is expected to be further improved by optimizing the longitudinal fiber profile [49].

15.2.2 Large-Mode Area Microstructured Fibers (LMA MOFs)

The MOF or PCF is a glass fiber with a regular array of holes running down its length. A single missing hole in the array forms a region that effectively has a higher refractive index than the surrounding photonic crystal does, which acts as a waveguide core in which light can become trapped, forming a guided mode [8]. J. C. Knight et al. first reported that the PCF remains monomode even for very short wavelengths and explained it via an effective index model [9]. They further demonstrated that it remains monomode even for very large-scale fibers and successfully prepared large-mode field PCFs with an MFD of 22 μm and an MFA of 380 μm^2 [66]. These LMA fibers are useful for generating and propagating high optical powers without limitations due to the onset of intensity-dependent nonlinear effects. However, owing to the complexity of the LMA-PCF structure design and fabrication process, its research progress has been slow in the following years. Until Limpert et al. [24] increased the core diameter by replacing a number of air holes in the microstructure cladding with solid rods and obtained a single-mode photonic-crystal fiber with an MFA of more than 1000 μm^2. After 2005, the design and preparation methods of LMA-PCFs began to diversify, and LMA fiber structures of various shapes, including leakage channel PCFs, rod photonic-crystal fibers, and multicore PCFs, appeared. The MFA of the optical fiber is also greatly improved. In 2007, Dong et al. [67] of the IMRA Company, USA, obtained a PCF with an MFA of 3160 μm^2 by using the leakage channel structure and then increased the MFA of the fiber to 14,000 μm^2 via the same method. In 2009, researchers also designed a new method to obtain the LMA-PCF stress-induced low-index difference method. Fu et al. [68] induced a tiny refractive index difference [$\Delta n \approx (\pm 1) \times 10^{-5}$] between the fiber core and the low refractive index induced by the stress field formed after fiber drawing and cooling. Single-mode low-loss PCFs with an MFA of 17,400 μm^2 (corresponding wavelength of 1.03 μm) and an MFA of 31,600 μm^2 (corresponding wavelength of 1.55 μm) were realized.

In the 2000s, several LMA fibers emerged following the development of photonic crystal fibers, allowing effective single- operation: LPFs [39], AS-PBGs, chirally coupled core fibers [26], LCFs [67], and distributed mode filtering (DMF) fibers [69]. Among them, the chirally coupled core fibers and the LCFs have only achieved mode areas in the range of 1500—3500 μm^2 at 1.06 μm [67], which cannot sustain a further increase in the output power of the fiber laser. AS-PBGs have achieved a single-mode regime near the high-frequency end of the bandgap [70]. In this region, only a narrow spectrum range can support the low-loss transmission of the fundamental mode (FM). This not only increases the difficulty of reproducible fabrication for low-loss and single-mode guidance at the laser wavelength but also has a negative effect on the amplification process of ultrafast lasers. High-power scaling capabilities have been demonstrated by using LPFs, in which single-mode operation is ensured owing to higher-order mode (HOM) delocalization [39]. However, air clad in the cladding structure and index matching between the active core and surrounding material lead to an open structure, in which the core and cladding modes can interact. With increasing cladding diameter, at some point, the effective index of the cladding modes could reach that of the FM in the core, leading to anticrossing and a reduction in the laser reliability [71].

Avoiding nonlinear effects in fibers is one of the most critical issues for many high-power fiber laser systems. SRS is a major limiting factor for current nanosecond master oscillator power amplifiers (MOPAs) and becomes an increasingly limiting factor in CW systems as power levels increase beyond 10 kW. Stimulated Brillouin scattering (SBS) limits the maximum single-frequency power that can be generated in a fiber MOPA and thus is a key issue in power scaling for coherent beam combination schemes. Self-phase modulation (SPM) is currently the primary limiting factor for most ultrashort pulse systems. Techniques to manage nonlinear effects through either fiber design or pulse shape control are thus essential [4].

Several specialized books on PCFs and MOFs have been published. One early book published in 2003, edited by Anders Bjarklev et al. [72], introduces the classification of the various PCFs, theoretical tools for analyzing the fibers, and methods of their production. Another book published in 2005, authored by Frederic Zolla et al., presented a variety of perspectives on PCFs together with a thorough treatment of the theoretical, physical, and mathematical foundations of the optics of PCFs [73]. The recent book edited by Mário Ferreira et al. in 2021 presents a comprehensive account of recent advances and research in optical fiber technology, with a focus on new material composition-based active fibers along with their applications as broadband optical amplifiers, fiber lasers, radiation sensors, etc. [74]

In addition to these books, many review articles emerged alongside the development of PCFs. Among these, Philip St. J. Russell, one of the founders of photonic-crystal fibers, presented several insightful reviews on history, fabrication, theory, numerical modeling, optical properties, guidance mechanisms, and applications of photonic-crystal fibers, published in 2003 [11] and 2006 [75] as well as in 2007 [76] for the special 30th anniversary of PCFs. For others, Yi et al. reviewed the characteristic parameters, design methods, and applications of LMA-PCFs [77] in 2012. In 2013, Li et al. from the Shanghai Institute of Optics and Fine Mechanics (SIOM) summarized several Yb-doped LMA silica glass optical fibers that were successfully used in high-power laser systems and analyzed the development status of LMA-PCFs in China until 2012 [1], when this research group just started their trial of the sol-gel method to prepare Yb-doped silica glass mandrels. In 2019, the research group in SIOM made great progress in the preparation of Yb-doped silica glass mandrels with large diameters and low NAs through sol-gel and vacuum melting methods [78], which helped them achieve breakthroughs in the preparation of a series of Yb-doped silica glass mandrels co-doped with Al, P and F with low loss, a regulable refractive index, and a large diameter (>5 mm), especially a high optical homogeneity as high as $1 \times 10^{-4,}$ which is almost equal to that of pure silica glass made by Heraeus GMBH. This laid the foundation for their rapid development in the preparation of Yb-doped silica LMA-PCFs from 2016 to 2019, which achieved high-beam-quality picosecond pulse amplification with an average power exceeding 100 W and a high beam quality [79, 80], as well as the preparation of a full optical fiber amplifier module on the basis of their 40-μm core diameter Yb-doped PM-LMA PCF. Recently, Hu et al. briefly introduced the progress of research on Yb-doped LMA-PCFs at home and abroad, as well as the design of Yb-doped LMA-PCFs. The effects of thermal history on the refractive index of the Yb/Al/P/F co-doped silica glass and the beam quality of the PCF are demonstrated, especially with respect to the progress of research on Yb-doped LMA-PCFs in the past 10 years at the SIOM [36]. In addition, Ma et al. of the Xi'an Institute of Optics and Precision Mechanics (XIOPM) compile the worldwide development of large-mode area and polarization-maintaining (PM) PCFs and propose possible technical routes to realize the single-mode operation of LMA-PCFs and PM-LMA-PCFs [17].

15.2.2.1 Rod-Type PCFs

The rod-type PCF is a new large-mode field PCF first proposed by Limpert et al. [24, 81] to further improve the output power of PCF lasers and avoid the restriction of nonlinear effects such as stimulated Raman scattering and self-phase modulation caused by high power. The rod-like fiber is composed of a large rod-like outer layer of millimeter magnitude, an inner layer of air hole microstructure, and a large-diameter solid fiber core. One characteristic that most ultra-LMA fibers share is their rod-type geometry, i.e., they are thick, unbendable waveguides that must be kept straight. This is done to minimize bending losses, which increase nonlinearly with the core diameter [7]. There are two important waveguide structures in fibers: one is used to transmit pumped light, and the other can be used to transmit laser light, which has the dual advantages of rods and fiber gain media.

In 2005, Limpert et al. reported Yb^{3+}-doped rod-type PCFs with a diameter of 1.7 mm, an inner cladding diameter of 117 μm, and an MFD of 35 μm. When a length of the above 48 cm fiber was pumped by a laser at 976 nm, a continuous laser output power of 120 W near the diffraction limit was obtained [24]. Subsequently, they prepared active and passive single-mode PCFs with MFDs of 60 and 100 μm, respectively. In 2008, Johan Boullet et al. first reported an 80 μm Yb-doped rod-type PCF laser emitting up to 94 W in the CW regime when operating at 977 nm, reaching the highest output power ever achieved from a single-mode solid-state laser operating at this wavelength at that time [82]. This PCF delivers a nearly diffraction-limited beam from an 80 μm diameter (NA_{core} ~ 0.01), providing a measured MFD of 70 μm, which is significantly higher than that of conventional LMA fibers. The inner cladding has a diameter of 200 μm and a very large NA ($NA_{clad} > 0.7$) defined by the surrounding air-filled microstructure. The extreme optogeometric parameters of the rod-type fiber led to a clad-to-core surface ratio of ~6.25, which is comparable with that of the fiber used in [83], but with a pump clad area of 31,400 μm^2, which is two orders of magnitude larger. For almost the same period of time, more than 710 W was achieved from a 1.2 m long rod-type fiber operating at 1030 nm and pumped at 976 nm, with a record power extraction of approximately 500 W/m. Jes Broeng et al. [84] anticipated that air-clad PCFs hold promise for increasing single-mode power levels beyond the 1 kW limit through the utilization of extremely high NAs, large MFDs, and short fiber lengths. Air-clad PCFs stand as strong candidates to allow single-mode lasing past the 1 kW limit.

LMA fiber technology has achieved great success in near-infrared high-power silica glass fiber lasers. At present, the single-fiber continuous laser output power of large-mode silica fibers has reached more than 10^4 W [85], and the peak output power of single-fiber femtosecond lasers has reached more than 10^9 W [86]. The previously reported maximum MFDs of rare-earth-doped single-mode active silica PCFs and single-mode passive silica PCFs are 135 μm [87] and 205 μm [88], respectively. Compared with ordinary large-mode silica fibers, the near-single-mode VLMA rod-like PCF above 100 μm has a significantly improved ability to withstand high-power and high-energy lasers. For example, femtosecond pulse outputs with pulse energies on the order of millijoules and peak powers above 10^9 W have been achieved in VLMA-active silica PCF laser sources with MFDs above 100 μm [89].

15.2.2.2 All-Solid Photonic Bandgap Fibers (AS-PBGFs)

PBG microstructure fibers are another very important type of MOF. In AS-PBGFs, a mode is guided only when it falls within the photonic bandgap of the cladding lattice. This provides great potential for creating designs that support only the fundamental mode, i.e., selective mode guidance versus selective elimination of mode guidance, as in some other approaches. The robust optical guidance and physical constructs of AS-PBFs enable them to be made and used much like conventional fibers. Double-clad designs and polarization maintenance can be added with ease [90]. Mode area scaling to 20 μm MFD using AS-PBG fibers was reported in 2008 [91]. A detailed theoretical investigation on the limit of mode area scaling with

AS-PBGs was reported in 2010 [92], indicating an upper limit of ~500 μm^2 using a more optimized seven-cell core and operating in the first bandgap. AS-PBG fibers with up to ~700 μm^2 effective mode areas have been demonstrated to operate in the first bandgap [93, 94]. In 2012, Kong et al. reported, for the first time, both theoretical and experimental studies of mode area scaling with a seven-cell core AS-PBGF above 900 μm^2, which has the ability to deliver 30 times more power than regular single-mode fibers without reaching the nonlinear threshold [90].

Among the options to prevent the appearance of unwanted high-order modes (HOMs), AS-PBG fibers have also proven their potential as efficient LMA fibers. Their primary originality lies in their particular guidance mechanism, which permits the filtering out of specific wavelengths and then helps delay the appearance of nonlinear effects such as SRS. AS-PBG fibers with all-solid substrates are also attractive for increasing the length of delivery fibers with enough suppression of the SRS effect. This advanced fiber incorporates the advantages of all-solid fibers, including bending-resistant operation and feasibility for postprocessing, including tapering and splicing. Most importantly, its intrinsic potential superiority in spectral filtering has evolved from photonic bandgap guidance. A mode is guided only when it falls within the photonic bandgap of the cladding lattice; otherwise, a continuous wave band called the photonic band suffers from high confinement loss because of strong resonance leakage. Controllable PBG distributions can be easily adjusted by drawing the fiber to various parameters with a certain refractive index profile (RIP) of the cladding lattice. A further development was the use of high-loss spectral regions between PBGs to selectively suppress undesired gains for some types of laser delivery [95].

Large-core designs are also possible, and several passive and active single-mode LMA AS-PBG fibers have been presented over the past decade [96, 97]. In 2013, Assaad Baz et al. reported on a 19-cell-core LMA PBGF with hetero structured cladding [98], which was previously suggested by Murao et al. [99] and consists of omitting high-index inclusions in the microstructured cladding to reject HOM. By introducing six four-cell diamond-shaped outer cores around a large central core, it has been proven possible to achieve a functional single-mode passive LMA AS-PBGF with an MFD of 44 or 33 μm when operating either in the fourth or third PBG at 1050 nm. This design was also successfully extended to a polarization-maintaining fiber geometry that demonstrated similar modal properties and polarizing behavior [100]. This concept was further developed to realize a double-clad ytterbium-doped and polarization-maintaining fiber geometry used in the last amplification stage of a nanosecond high-power fiber amplifier operating at 1030 nm. Olivier Vanvincq et al. [101] reported on the design, realization, and testing of a double-cladding LMA, polarization-maintaining, solid-core photonic bandgap fiber with a Yb-doped core obtained from a sol-gel-made material in the pulsed amplification regime. A specific heterostructure of the cladding has been applied to efficiently reject the higher-order modes, making the fiber single-mode in practice, have an MFD value of 35 μm. When it was used as the last stage of a high-power fiber amplifier, more than 90 W of average power was extracted at a 1.03 μm wavelength, with a slope efficiency of 75% with respect to the absorbed pump power.

15.2.2.3 Leakage Channel PCFs (LC-PCFs)

To overcome the restrictions of the MFD imposed by the strict single-mode, low loss, and bending resistance requirements in traditional PCFs, LC-PCFs comprising one or several rings of very large air holes provide much greater leakage losses for HOMs while confining the FM in the core by the high-contrast air–glass boundary [102]. LC-PCFs guide light through the total internal reflection mechanism, and by adjusting the size, interval period and refractive index difference of the low refractive index region in the inner cladding, the high-order mode in the core region can be coupled to the cladding region through the resonance effect, and only the fundamental mode propagates in a low-loss mode in the core region. Compared with rod-like PCFs, LCFs have the advantages of maintaining both single-mode large-core transmission and excellent bending characteristics. The all-solid-state structure makes it easy for LCFs to be fused with traditional optical fibers, which is conducive to fiber laser system integration. Dong et al. [67, 102] proposed a new design method of strict single-mode large-mode PCFs with bending resistance, i.e., LCFs. prepared a single-mode low-loss leakage channel PCF with an MFA of 1417 μm^2. The fiber could still work in single mode when the core diameter was increased to 100 μm. The loss of the fundamental mode in LCF is very small (less than 0.0.1 dB/m), and the loss of the high-order mode is very large (greater than 1 dB/m), so it can be strictly single-mode or near-single-mode operation.

In LCFs [102, 103], the core is surrounded by a small number of large holes, making all propagation modes leaky in a very selective way, such that all higher-order modes but not the fundamental mode experience substantial propagation losses. Earlier versions of such fibers have been made as PCFs with air holes, which need to be sealed, and this process can lead to mode distortion, especially in large-core fibers, where small refractive index perturbations can have a significant impact on the mode structure. Solid all-glass designs are also possible [104]. It has been theoretically shown that LCFs can potentially extend the single-mode operation of LMAs to a core diameter of 200 μm via built-in mode filtering [67]. In 2007, Yb^{3+}-doped low-loss single-mode LCFs with a mode diameter of 63.4 μm and an MFA of 3160 μm^2, as well as a polarization-maintaining LCF with an MFA of 1400 μm^2 and excellent bending resistance, were prepared [67]. Dong et al. demonstrated robust single-mode operation in an all-glass fiber with six low-index glass rods inserted into the cladding to define the central core region, for which a core diameter as large as 170 μm and an MFA of 1400 μm^2 were achieved [105]. Although this fiber displays good bending resistance, mode shrinking caused by coiling must be considered for VLMA fibers, which can become the most limiting factor for effective field area (EFA) scaling [106].

15.2.2.4 Large-Pitch PCFs (LPFs)

LPFs were the first design to exploit HOM delocalization to achieve effective single-mode operation at high average powers. They are still among the simplest designs and have an extremely interesting advantage in what is known as proportional scaling [6]. This feature is the ability to arbitrarily scale the structure while preserving intact HOM delocalization. The structure of a typical active LPF is shown in Fig. 15.1a [88]. In contrast, a passive LPF [Fig. 15.1b] has neither the doped core region nor the cladding ring. HOM delocalization, i.e., poor overlap with the core, is a superior novel approach to achieve effective mode discrimination in a VLMA rod-type fiber. LPFs with pitch Λ values greater than 10 times the wavelength are the first representative of this idea [39]. Inside the relatively open structure, HOMs are not easily excited by a Gaussian beam, and their amplification factor is substantially reduced; thus, only FMs propagate with high gain [39].

A single-mode LPF structure with a record large-core diameter of 135 μm and an MFA over 10,000 μm^2 has been proposed and practically applied in fiber amplifiers [87]. With outstanding scalability and robustness, LPFs hold the majority of the representative results for high-performance pulsed fiber laser systems. Typical examples are the 26 mJ pulse energy [87] and 2 kW average power [107] directly obtained from the LPF-based MOPA system. In addition, the simple nonresonant structure enables LPFs to employ an index-depressed core to precompensate for the thermal index profile in high-power operation [108]. In reality, fabrication tolerances still limit the scaling potential of active LPFs, but despite this, an MFD in excess of 100 μm has been demonstrated at high average powers (>100 W) [7].

In 2011, Tino Eidam et al. designed a VLMA-LPF with an MFD of 105 μm and obtained 480 fs diffraction-limited compressed pulses at 3.8 GW peak power from the monolithic CPA configuration [86]. The record was not broken until 2017, when L. Lavenu et al. employed nonlinear compression in a xenon-filled hollow capillary

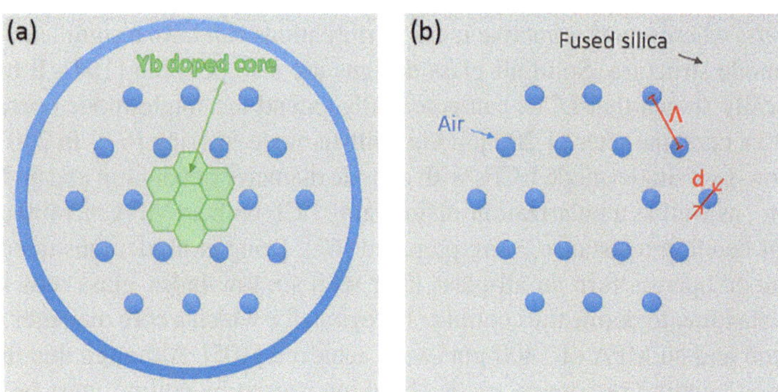

Fig. 15.1 Cross section of (**a**) active and (**b**) passive LPFs. The structure consists of a hexagonal arrangement of equidistant air holes (separated by a pitch Λ) with diameter *d*. An additional cladding and a doped core region are added for the active LPF. [Reprinted/Adapted] with permission from [88] © Optical Society of America

after an LPF-based CPA system to generate 14 fs ultrashort pulses with 5.8 GW peak power [109]. Dia Daewich et al. [110] reported a single-polarization feature out of a passive VLMA fully aperiodic LPF operating at 1942 nm. The core diameter of the fiber core is 140 μm, the corresponding MFA is 14,590 μm^2, and a polarization extinction ratio higher than 16 dB is achieved. Steinkopff Albrecht et al. [88] reported a stable single-transverse mode achieved at a wavelength of 1.03 μm through a straight passive LPF with a pitch of 140 μm, resulting in a measured MFD of 205 μm. The simulations revealed that the propagation losses of both the FM and first HOM decrease with increasing Λ and d/Λ. This result corresponds to a mode area scaling of a factor of ~4 and a factor of ~8, compared with the largest effective single-mode reported until now in an active fiber and a passive fiber, respectively.

15.2.2.5 Solid-Core Ant Resonant Fibers (SC-ARFs)

Recently, solid-core antiresonant fibers (SC-ARFs) have drawn extensive attention for the development of LMA fibers because of their outstanding ability to suppress HOM and ease of fusion splicing for all-fiber laser systems. Zhang et al. reported that a fundamental mode area larger than 5000 μm^2 can be achieved with proper resonant coupling structure design of the SC-ARF, in which single-mode operation was defined with minimum HOM propagation losses >10 dB/m [111]. Afterward, Xing et al. proposed nonuniform [112] and double-layer [113] antiresonant elements to further improve the fundamental mode area of the SC-ARF, where an MFA of up to 7481 μm^2 was simulated on the basis of the defined minimum HOM propagation losses >1 dB/m for single-mode fibers. Zhen Xing [113] presented a double-layer all-solid antiresonant fiber. With a core diameter of 80 μm, the loss ratio for HOMs to the fundamental mode can exceed 3500, and the MFA reaches 3030 μm^2 at a wavelength of 1.064 μm. By further expanding the core size, the MFA can be enlarged to 7481 μm^2 while maintaining single-mode operation. In 2021, Zhang et al. [111] proposed an SC-ARF structure with a large LMA and high HOM suppression. The fiber core size varies from 40 to 100 μm, and single-mode transmission can still be ensured when the core diameter reaches 100 μm with an effective area larger than 5000 μm^2 at 1.06 μm. It can achieve resonant coupling in a broadband range and provide a higher than 20 dB HOM (mainly TE_{01}, TM_{01}, and HE_{21}) suppression ratio while maintaining the fundamental mode loss lower than 1 dB/m.

PM-SC-ARF has also been investigated recently [115], where nested circular tubes used as antiresonant elements resulted in an induced birefringence of 6.93×10^{-5}. However, the fiber designed for the mid-infrared (mid-IR) laser at 3 μm has a fiber core diameter of only 20 μm. Recently, Tian et al. [114] demonstrated a PM-LMA-SC-ARF with an MFA of 3026 μm^2 and birefringence of 1.20×10^{-5}, where all the HOMs exhibit confinement losses (CLs) larger than 10 dB/m. Double-layer antiresonant elements (Fig. 15.2) are used not only for the MFA increment of single-mode operation but also to alleviate the CL degradation of the fundamental core mode during fiber birefringence optimization with parameter adjustment of the cladding tubes. Mode coupling between the fundamental core mode in one

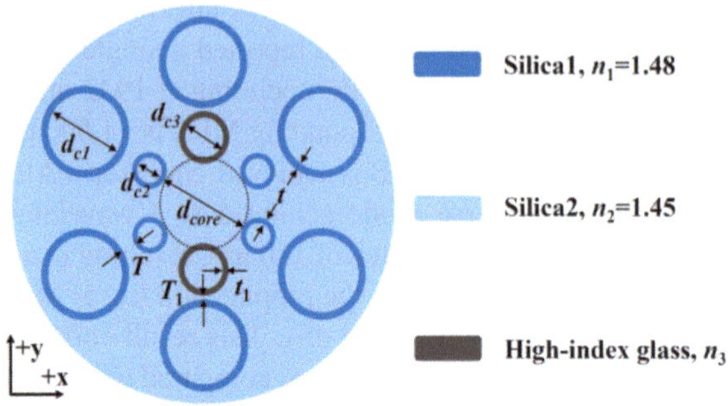

Fig. 15.2 Structure and refractive index arrangement of the designed PM-LMA-SC-ARF [Reprinted/Adapted] with permission from [114] © Elsevier B.V.

polarization direction and the mode on two high-index inner cladding tubes is employed to introduce high birefringence for the designed SC-ARF under the LMA.

15.2.2.6 Polarization-Maintaining Large-Mode-Area (PM-LMA) Fibers

Many experiments and applications in ultrafast optics require polarization-maintaining or single-polarization fibers [116]. The birefringence required for this purpose is usually achieved by geometrical effects (form birefringence, e.g., elliptical cores) or the elasto-optic effect [117]. On the basis of the conventional step-index design with stress-applying parts (SAPs, such as PANDA, bow-tie, and elliptical-stress-layer), commercially available Yb^{3+}-doped PM-LMA fibers can provide an effective MFA of 500–600 μm^2, where the lasers are actually in few-mode operation and reasonable fiber bending is needed for mode filtering in practical applications [118]. Form birefringence has been successfully applied to solid-core PCFs [10]. In addition to the fact that larger values of form birefringence can be obtained compared with step-index fibers, this technique suffers from the disadvantage that birefringence decreases rapidly for larger cores. Thus, to obtain polarization-maintaining LMA fibers, one can use the well-known technique of stress-applying parts (SAP) inside the fiber because the elasto-optical effect birefringence is introduced.

The technique of SAPs was first demonstrated in combination with LMA-PCFs in 2004 [119], and in the same year, impressive results from a Yb-doped large-mode single-mode PCF were obtained by J. Limpert et al. [120], which were only possible by further increasing the core size (~40 μm), but no polarization maintenance mechanism was applied. In 2005, T. Schreiber et al. [121] reported the first numerical investigation of stress-induced birefringence in microstructured solid-core optical fibers and reported that the birefringence induced by stress applying parts is significantly different at values of $d/\Lambda > 0.65$ because the whole holey cladding influences the stress distribution. Compared with LMA SIFs, the birefringence that can be achieved in LMA-PCFs when SAPs are placed outside the holey cladding is only half that value because of the distance of the stress rods to the core. Nevertheless, most birefringence is introduced by the inner SAPs. Very soon after the numerical investigation, they [122] realized the design of PCFs that finally comprise SAPs of 6 boron-doped silica rods on the side of seven missing hole core areas as part of the photonic cladding to simplify the fiber fabrication. This design combines SAPs and inner photonic cladding via index matching for the first time. Using this design, a Yb-doped single-polarization fiber with low nonlinearity and pump light absorption of ~14 dB/m was fabricated, with which a highly polarized laser with an extinction ratio of 15.5 dB and an output power of up to 25 W was demonstrated. PM-LMA fibers have been demonstrated by several types of microstructure fibers, including large-pitch PCFs and LCFs. Yb^{3+}-doped rod-type PCFs with a fundamental mode MFA of 2300 μm^2 were reported with boron-doped SAPs to achieve high birefringence of up to 3.10×10^{-4} and single-polarization laser operation [123]. Benjamin G. Ward et al. [124] predicted through numerical calculations that particular birefringent PCF designs exhibit slightly better performance in a coiled configuration than nonbirefringent SIF designs with respect to HOM suppression for the realization of LMA effectively single transverse mode fibers. The passive losses of the fundamental and first few higher-order modes of a birefringent PCF design with a 41 μm diameter core incorporating SAPs were calculated to have a minimum higher-order mode loss of up to 5.5 dB/m and a fundamental mode loss of only 1.4 dB/km. A structural difference breaking the hexagonal symmetry of the fiber cross section cannot induce large-mode birefringence and MFD, e.g., by incorporating large holes in the central region of the core (Fig. 15.3a), the induced birefringence reaches 1×10^{-4} at 1300 nm and 1.5×10^{-4} at 1500 nm, and the fiber has an MFD of 10 μm [125]. In the same year [123], the University of Jena and Crystal Fiber A/S jointly reported a Yb-doped single-transverse-mode rod-type PCF that combines low nonlinearity and polarization-maintaining properties (Fig. 15.3d), whose MFD reaches 54 μm. In 2013, the NKT developed a fiber (DC-200/40-PZ-Yb) consisting of a 40 μm Yb-doped core, a 200 μm pump cladding, and a stress polarization-maintaining structure (Fig. 15.3b) [69]. The stress zone consists of 13 rods on the left and right sides of the core, and the stress zone is separated from the core by a layer of air hole rings. This PM-LMA-PCF can realize high polarization performance with an extinction ratio of 25–30 dB. In 2015, Clemson University and the Air Force Research Laboratory codeveloped an all-solid Yb-doped polarizing PCF for single-polarization, single-mode operation (Fig. 15.3c). Three low-index boron-doped stress rods on both sides of the fiber core produced birefringence. These boron rods also provided light confinement through total internal reflection. It has an effective mode area of approximately 1150 μm^2 and a polarization extinction ratio of up to 21 dB [126].

In 2019, NKT Photonics reported a double-clad Yb-doped polarization-maintaining fiber called "DC-250/30-PM-Yb-FUD" with a core diameter of approximately 30 μm (MFD, 24 μm) and a 250 μm inner cladding consisting of silica and doped silica [127]. The stress zone consists of hexagons of doped silica rods on both sides of the fiber core, with each hexagon consisting of 30 rods. The polarization is generated by inducing stress-related birefringence in the core, which produces a

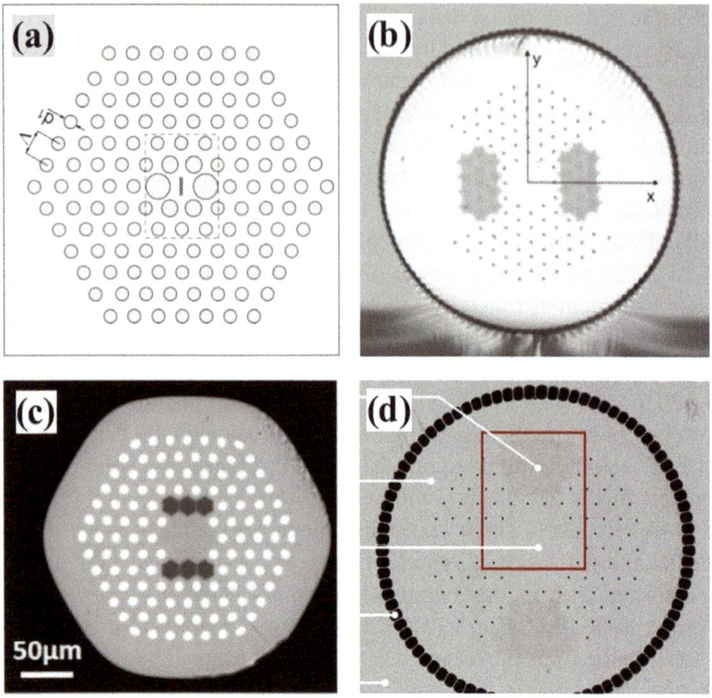

Fig. 15.3 (**a**) (2008) Wroclaw University of Technology introduced small holes in the central region of the core to induce birefringence; (**b**) (2013) NKT-fabricated fiber named DC-200/40-PZ-Yb; (**c**) (2015) Clemson University developed single-polarized, single-mode operating Yb-polarized all-solid-state photonic bandgap fiber [126]; (**d**) (2008) 54 μm mode field diameter reported by Crystal Fiber A/S [Reprinted/Adapted] with permission from [17] © MDPI, Basel, Switzerland

polarization extinction ratio > 18 dB. In the same year, Yangze Soton Laser Co., Ltd. (OYSL) also successfully developed an LMA Yb-doped fiber with polarization-maintaining ability, which has a core diameter of 40 μm, and the fiber structure is very similar to that of NKT's PM-LMA-PCF (Fig. 15.3b). It has excellent single-mode characteristics. Very recently, Yangze Soton Laser Co., Ltd. successfully realized a breakthrough in rod-type PM-LMA-PCF. The fiber's structure and performance are akin to those of the NKT's DC-200/40-PZ-Yb, whose core diameter reaches 85 μm with an MFD of 65 μm and MFA > 3000 μm^2, and it can maintain single-mode operation at the 1030 nm band with a lower beam quality factor (M^2 < 1.1). This fiber's polarization extinction ratio reached 18 dB. In summary, significant breakthroughs in structural design, preform preparation, and fiber fabrication technology for PCFs have been made, which have contributed to overall performance improvement. This is conducive to the development of high-power lasers and femtosecond lasers in the presence of LMA-PCFs and PM-LMA-PCFs, which have broad application potential in scientific lab research, industrial laser processing, national defense, and other fields. To date, the largest MFD that has been achieved was 75 μm in DC-285/100-PM-Yb-Rod rod-type PCFs by NKT in 2011 [128], whose structure is proprietary. Very recently, a 100-μm-core PM-PCF with similar properties was reported by Hu et al. in the SIOM; the stress zone consists of 13 rods on each side of the core, and the stress zone is separated from the core by two layers of air hole rings (see Fig. 15.10d). The PM-PCF has a cladding diameter of 285 μm, an NA > 0.45, and an MFD of 75 μm [36].

High-birefringence LMA fibers have also been investigated with LCFs, where very few air holes are needed to surround the solid core to form sufficient leakage channels for higher-order modes (HOMs). Watt-level fundamental mode laser operation has been demonstrated from a Yb^{3+}-doped PM-LMA LCF with an effective area of 1400 μm^2, in which a high birefringence of 2.1×10^{-4} was achieved from two boron-doped silica stress rods arranged close to the fiber core opposite [129]. Remi et al. [130] reported that ytterbium-doped LMA fibers featured a large-pitch architecture, which demonstrated laser and amplification for high-power laser source applications. A linearly polarized single-mode signal was obtained in both configurations with a 58 μm core diameter fiber, corresponding to a 45 μm MFD with a PER of 17 dB.

15.2.2.7 Brief Summary of LMA-PCFs

Commercial fiber laser markets have experienced tremendous growth in the past few decades. There are still a few technical challenges in improving the performance of high-peak-power pulsed and single-frequency fiber lasers. These systems are currently limited by nonlinear effects. The ultimate solution to this problem is robust single-mode operation in very large effective mode area fibers. Conventional large-mode fibers (LMAs) have an extended core diameter of ~30 μm and operate in fundamental mode in multimode fibers [131]. PCFs with an array of small air holes in the cladding have been operated in single mode with a 100 μm core diameter in a rod with an outer diameter much larger than 1 mm to overcome high bend loss [132].

As shown in Fig. 15.4, early studies focused on increasing the effective MFD and effective field area (EFA) of the fibers, achieving significant progress in EFA from 176 to over 10,000 μm^2 [39, 68]. Unfortunately, VLMA-PCFs with MFD > 50 μm are mostly rod fibers, which experience deadly bending loss that slows the development of compact fiber laser systems. Therefore, many new flexible designs, including the double air–hole diameter [133], double lattice constant [134], triangular core [135], hybrid cladding [136], and circular gradient diameter [137], have been proposed in the past decade. The simulation results prove that these fibers can maintain a bending loss of less than 10^{-2} dBm^{-1} in the fundamental mode [133, 134, 137–139] and that the EFA exceeds 2000 μm^2 [139] at a bending radius close to 10 cm. Nonetheless, the increased difficulty and complexity of fabrication caused by noncircular fiber symmetry and additional codoping to eliminate index mismatching delay their practical application [138].

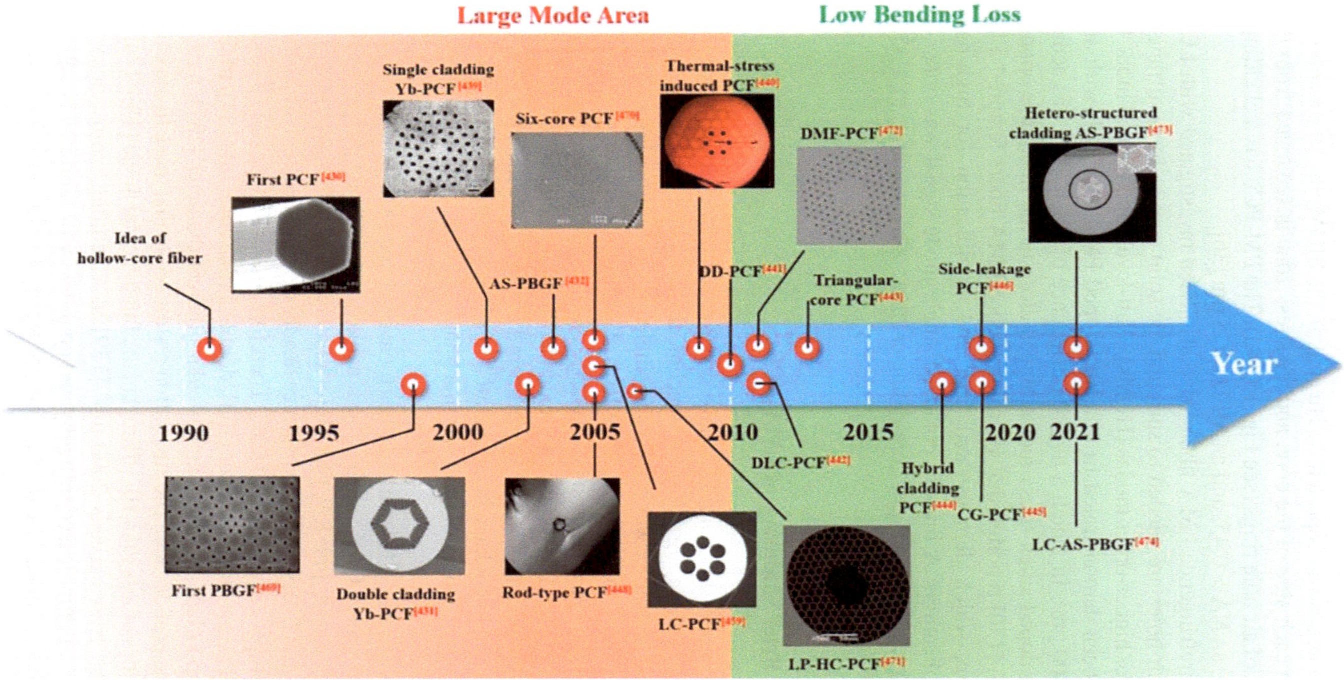

Fig. 15.4 Chronological overview of the leading development of the LMA-PCF. [Reprinted/Adapted] with permission from [62] ©Wiley-VCH GmbH

15.2.3 Other Types of LMA Fibers

15.2.3.1 Multicore Fibers

Despite the effectiveness of HOM delocalization as a means to obtain effective single-mode operation in VLMA fibers, the scalability of the active core is ultimately limited by manufacturing tolerances. Multicore fibers with N smaller cores distributed in a rectangular array that shares a common pump cladding provide a new approach to overcome the above limitation to synthesize the VLMA core because of their flexible design in mode control and simple and reliable fabrication methods compared with those of other methods [140]. Care must be taken during the design so that all the cores are both optically and thermally isolated from one another [7].

Since it was first introduced in 1979, multicore fibers have been extensively researched in both communication and fiber laser fields. Early studies focused on the near-infrared (NIR) laser output of multicore fibers [141–143]. In 2008, a six-core fiber with an MFA of 4200 μm^2 was successfully fabricated, which is capable of outputting a high-power single-pulse laser with a pulse energy of 2.2 mJ, a duration of 26 ns, and a repetition rate of 10 kHz [141]. In 2018, Ji et al. developed a Yb-doped multicore fiber for NIR applications. The MFA of the fiber is 1571 μm^{2}, and a maximum laser output of 115 W was achieved [143]. Jia et al. [144] also proposed a trench-assisted low-bending-loss LMA multicore fiber with an air hole structure capable of dual-mode transmission. The fundamental mode's effective MFA can reach 2003.24 μm^2 at 1550 nm, and the bending loss is as low as 2.57×10^{-3} dB/m when the bending radius is 1 cm. The proposed fiber is feasible with existing fabrication techniques and is applicable to large-capacity optical communication systems and high-power lasers.

Although there have been numerous reports on the realization and transmission of high-power lasers by multicore fibers in the NIR region [140, 145, 146], high-power laser transmission in the mid-IR region based on chalcogenide fibers is rare. In 2014, a six-core fiber based on As_2Se_3 with an MFA of 14.5 μm^2 at 4 μm was designed, which is six times larger than that of single-core fibers of the same size [147]. Most of the studies on chalcogenide multicore fibers for the mid-infrared region are limited to theoretical research [148, 149]. The fabrication of multicore optical fibers based on chalcogenides is a challenge because of the mechanical brittleness of chalcogenide glasses. Although some works have successfully fabricated large MFA multicore fibers [150, 151], the deformation of the fiber is significant, and the output beam quality has not been studied. To overcome the above challenge, very recently, in 2024, Wang et al. [152] designed and fabricated a seven-core chalcogenide fiber with an ultralarge mode field for the mid-infrared range of 2.5–11 μm. A theoretical MFA of 8914 μm^2 can be calculated with a core radius of 24 μm and a pitch of 50 μm. This impressive MFA is realized through the deployment of an improved drilling technique in the fabrication of a Ge-As-Se seven-core fiber. The fiber has a relatively low loss in the wavelength range of 2.5–11 μm, and the minimum loss is 1.4 dB/m at 8.5 μm. The measured MFA of the fiber at 10.6 μm is 7364 μm^2, which is 6.2 times greater than that of traditional step-index single-core fibers.

A further increase in the core number from 7 to 19 was observed: a 19-core all-solid-state single-mode large-mode-area PCF based on chalcogenide glasses was fabricated via the extrusion-stacking method [153], the fiber confinement loss ratio of the high-order mode (LP11) to the fundamental mode (LP01) was greater than 10^2 in the range of 1.5–10 μm, and the effective mode area of the fundamental mode was approximately 5362 μm^2 at 2 μm. A low bending loss of 1 dB is observed in a 1.5 m long fiber with a bending radius of 14 cm, indicating that this AS-PCF has excellent single-mode transmission and bending resistance and has potential for mid-infrared high-power laser applications.

15.2.3.2 Segmented Cladding Fibers (SCFs)

Segmented cladding fibers (SCFs), with cladding in an arrangement of periodic low- and high-refractive-index segments in the azimuth direction, have high potential to simultaneously meet the LMA and SM requirements, similar to PCFs or holey fibers [154]. SCF does not require any air hole arrays; thus, it can provide longitudinal uniformity within a long-length fiber. As a type of leaky fiber, in SCF, all the higher-order modes can be effectively stripped off [155]; therefore, single mode operation is allowed over an extended range of wavelengths while keeping the mode area large [156] to effectively cope with optical nonlinearity issues. As the SCF is also an all-solid fiber, it is potentially less susceptible to geometry-induced birefringence and scattering loss at the core-cladding interfaces [157]. These potentials of SCFs have been investigated for applications in high-power fiber lasers, fiber amplifiers, dense wavelength division multiplexing optical communication systems, and mid-IR sensing [158].

Despite the above potential advantages of SCFs suggested by numerical investigations [157, 159], the fabrication of silica SCFs faces more practical challenges. Few attempts have been made to use powder-in-tube techniques [160] that combine stack-and-draw methods to fabricate preforms. Severe impurity-induced scattering results in index and geometry variations along the SCF and very high losses on the order of 10 dB/m even at 1.310 and 1.550 μm, and the core diameter is only 16.8 μm [38]. To investigate the loss reduction of SCF through design optimization, in 2022, the authors proposed a bend-resistant octo-wing silica segmented cladding fiber that has a uniform circular core and octo-wings of rods (see Fig. 15.5) With a higher refractive index than the silica segments in the cladding, an effective mode area as large as 572 μm^2 (at a bend radius of 10 cm) is obtained at 1.55 μm, with a relatively large loss ratio between the FM and the first higher-order mode of approximately 137 [161]. They further theoretically increased the effective mode area to 715 μm^2 in 2023 through optimization of a segmented cladding fiber via the response surface methodology approach [162].

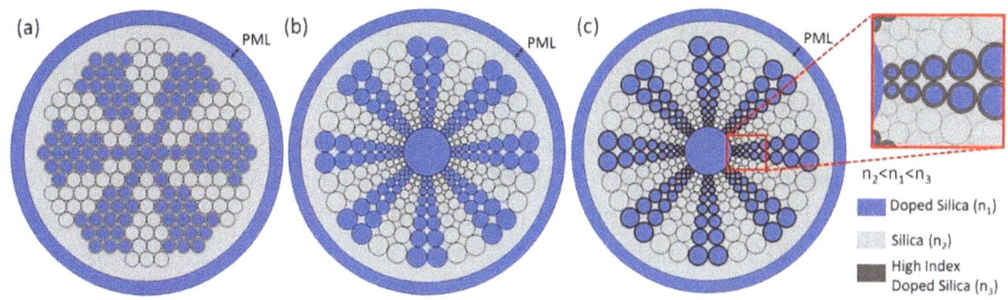

Fig. 15.5 (**a**) Previous studies of hexagonal-wing SCF designs with nonidentical high- and low-index segments [38], (**b**) new octo-wing SCF designs for which the segments are identical, and (**c**) octo-wing SCF designs with high-index rings. PML stands for perfect layer matching. The cross-sectional magnified in (**c**) represents high-index doped rings. [Reprinted/Adapted] with permission from [161] © Elsevier B.V.

Even segmented cladding fibers are thought to have potential in high-performance delivery devices such as amplifiers and high-power lasers because of their excellent SM performance along with a large effective mode area at a tight bending radius. In reality, the potential application of SCFs as a new alternative for LMA fibers to suppress nonlinear optical effects for high-power applications and optical communications still faces great challenges for loss reduction and fabrication technique improvement.

15.2.3.3 Soft-Glass LMA Fibers

Along with the rapid development of silica glass-based LMA fibers, LMA fibers made of soft glasses, including lead silicate [163, 164], silicate [165], phosphate [166], tellurite [167, 168], fluorite [169, 170], bismuth-based [171], chalcogenide [172, 173] glass, etc., have been developed. Soft glasses have their own merits and demerits; the merits include a large adjustable range of refractive indices and high solubility of rare-earth ions, while the disadvantages, such as greater difficulty in achieving very small refractive index differences between the core and cladding and intrinsically lower softening temperatures, make them easily deformed during the PCF fabrication process. This makes it difficult to prepare soft-glass fibers with perfect cross-sectional structures and very large-mode diameters and field areas.

In 2011, Rennes University reported the preparation of ChG (As-Se-Te matrix) single-mode PCFs for mid-infrared regions of 3 ~ 12 μm, and the MFD of single-mode ChG fibers was increased to 40 μm for the first time [172]. In 2013, Yi et al. reported the simulation work of a ChG (Ge-Sb-Se matrix) single-mode VLMA PCF, and the calculated fundamental MFA at a 10.6 μm wavelength was 13,333 μm^2 (the MFD was 130 μm) [174]. In 2019, Ren et al. reported the design and preparation of a PCF, the ChG single mode with the largest MFA thus far [173]. The MFA of the fiber at a wavelength of 4 μm reaches 5200 μm^2 (the corresponding MFD is 80 μm). Owing to the high refractive index of ChG and the low loss of fibers in the mid-infrared band, a mid-infrared supercontinuum spectrum with an average power of

3.5–7.5 μm and an average power of 3.1 mW is generated in the 35 cm long fiber under the pumping of a 4 μm femtosecond laser with an average power of 26 mW. To solve the technical problems in which a single-mode VLMA PCF has high bending loss and the optical fiber needs to be kept straight when used, Yang et al. [175] designed and prepared a ChG few-mode PCF with the largest MFA thus far in 2020. At 2 μm, the MFA of the fundamental mode LP01 reaches 10,400 μm^2 (the corresponding MFD is 115 μm), and the MFA of the LP11 mode reaches 11,300 μm^2 [175].

As a promising alternative approach to the stacking technique for making complex PCF preforms, glass billet extrusion can produce a PCF preform in a single step. The first extruded preform for PCFs with many holes was a 3-ring bismuth glass preform [171]. The special die design, along with advanced extrusion process controls, enabled the fabrication of a 3-ring bismuth PCF with a minimum loss of 1.2 dB/m at 1100 nm. Using the extrusion method to make a cane preform, an innovative mid-infrared polarization-maintaining photonic-crystal fiber (PM-PCF) made of chalcogenide (As_2S_3) glass with an asymmetric orthogonal pattern of longitudinal holes was proposed by Francois Chenard et al. [176]. The PM-PCF fabrication process included extrusion of the PM-PCF structure, cane pulling, and cladding to produce the PM-PCF preform with smaller features and final fiber drawing. The measurements indicate that the new PM-PCF has high birefringence ($\sim 10^{-4}$), low propagation loss (0.2 dB/m), and low insertion loss (<0.1 dB). The measured output beam quality at 2 μm was good and reached $M^2 = 1.282$ along the horizontal axis and $M^2 = 1.581$ along the vertical axis. The LMA enables the transmission of a high-power polarized infrared laser (>10 W CW).

Segmented cladding fibers have also been realized in mid-IR chalcogenides via the extrusion method to obtain an LMA, as well as single-mode conditions. In 2023, Peng et al. [177] fabricated a mid-IR SCF via an extrude-and-stack technique based on chalcogenide glasses for the first time. The mode area of the fabricated SCF is more than 2000 μm^2 in the wavelength range of 3–10 μm. The minimum fiber loss is 3.1 dB/m at 8.6 μm, and the bend loss is 3 dB when the bend radius is 6 mm. Compared with that in traditional step-index single-mode fibers, the mode area in SCFs can be increased by more than 15 times. This finding demonstrates that the chalcogenide SCF possesses the potential for high-power mid-IR laser transmission. A further reduction in fiber loss can be expected via more effective glass purification and elimination of interface defects when the coextrusion and stacking methods are used.

15.2.4 *Transverse Mode Instability (TMI)*

TMI is caused by a power transfer from the FM to some HOMs, which is mediated by the thermo-optic effect. Owing to quantum defects, a significant amount of heat is generated in the fiber core as a side effect of optical amplification, which leads to relevant thermal lensing. As a consequence, the originally single-mode waveguide will support the propagation of HOMs, which will eventually drain power from the FM through the stimulated thermal Rayleigh scattering effect [178, 179]. TMI leads to a mode-field area shrinking or even to a complete change in the guided mode set, and it occurs in all single- and multimode fibers during high-power operation regardless of their inner structure (step index or photonic crystal).

Arlee V. Smith et al. [180] theoretically anticipated for the first time in 2011 that beam quality degradation is caused by transverse thermal gradients induced by the amplification process for powers exceeding a sharp threshold in the vicinity of several hundred watts. Soon after that year, Tino Eidam et al. first reported on their observation and experimental characterization of a threshold-like onset of mode instabilities in high-power rod-type fiber amplifiers [2] in which the Gaussian output beam of a high-power, rod-type fiber laser system suddenly broke apart and started to fluctuate once a certain average power level was reached. This work marks the beginning of research on TMIs worldwide. Considerable effort has since been made worldwide to understand the physical origin of this effect [181–185] and to realize experimental measurements [186–188] of TMIs in mostly active [69, 181, 186] and sometimes passive [188] fibers and to explore mitigation strategies to increase the average power of fiber laser amplifiers. In 2020, Cesar Jauregui et al. reviewed the effects of transverse mode instability in high-power fiber laser systems, and the corresponding investigations led worldwide over the past decade [189]. To date, most TMI mitigation strategies attempt to reduce the strength of the refractive index gradient (RIG) [29, 190], but more recently, several methods have been proposed, which are based on manipulating the phase shift [191, 192]. Two general categories of TMI mitigation strategies, i.e., passive (not controlled by any external element) and active (involving an external control mechanism) mitigation strategies, are specifically presented and discussed in [189].

For fibers with large MFDs, the impact of refractive index changes in the range of 10–5 becomes significant [193] because of two factors: the Kerr effect in high-pulse-energy fiber laser systems and increased popularity due to the thermo-optical effect [194]. Thermal effects are currently the main bottleneck for future power scaling since, beyond a certain average power, they allow guidance of high-order modes and energy transfer to them, causing sudden degradation of the beam quality [187]. The knowledge about this mode shrinking is of utmost importance when low nonlinearity is the reason for mode area scaling [186].

Thermally induced waveguide changes become significant for VLMA fibers. This results in a reduction in the MFD but simultaneously in an improvement in the beam quality. A parabolic thermally induced index profile manifests itself in a decreased MFD during high-power operation. The thermally induced waveguide changes are negligible for fibers with MFDs of approximately ten microns but must be considered for VLMA fibers with MFDs greater than 50 μm, which rely on resonant effects [186], i.e., photonic bandgap fibers, distributed mode filter fibers, or chirally coupled core fibers.

As shown in Fig. 15.6a, the reduction in the MFD shows a characteristic behavior that scales with the core size but that is independent of the particular fiber design for the double-clad LPFs [186]. Figure 15.6b shows the measured evolution of the

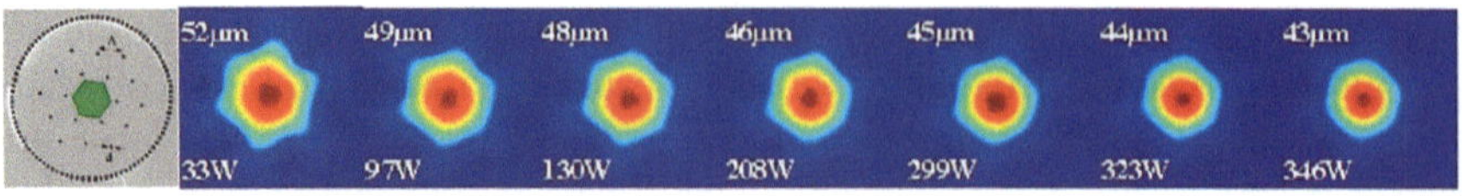

Fig. 15.6 SEM micrograph of a double-clad LPF with a rare-earth doped (green) core (LPF30) and the evolution of the measured output beam mode profile of LPF30 with increasing extracted average output power. [Reprinted/Adapted] with permission from [186] © Optical Society of America

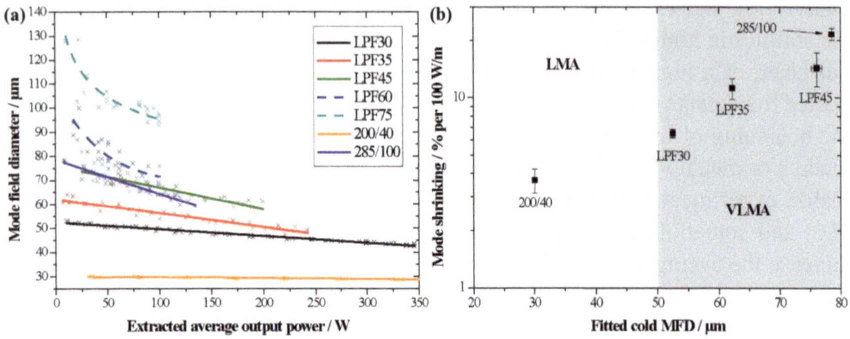

Fig. 15.7 (**a**) Measured mode-field diameter vs. extracted average output power and (**b**) approximate MFD shrinking (slope of the linear regression in Fig. 15.7a) vs. passive MFD (intercept of linear regression in Fig. 15.7a) for all linearly fitted FUTs. Note that MFD shrinking refers to the MFD at the fiber end facet. [Reprinted/Adapted] with permission from [186] © Optical Society of America

fundamental mode of an LPF30 fiber with increasing extracted average output power. This clearly demonstrates the progressive fading of the hexagonal features as the thermal index profile takes over as the main guidance mechanism. Furthermore, a decrease in the fundamental mode diameter by ~20% for the output beam mode profile was measured for this fiber at a 350 W extracted output power compared to the cold setup. The effect of mode shrinking can be seen in all the LMA-LPFs in Fig. 15.7a, but it becomes more pronounced as the initial MFD increases. For LPF30, LPF35, LPF45, and 200/40, the measurements can be approximately fitted via linear regression, which allows for the calculation of an average (linear) mode shrinking per 100 W extracted output power.

Figure 15.7b shows that the slope of the mode shrinks versus the (fitted) cold MFD. The mode shrinking increases exponentially with the MFD. In conclusion, the impact of thermal effects on active high-power fibers exponentially scales with the MFD. Both the indices guiding PCFs and LPFs have similar sensitivities to thermal effects. The core size of VLMA fibers has reached a dimension at which thermal effects must be considered in the fiber design process. Some fiber designs, such as index-guiding PCFs or LPFs, can cope with these thermal effects without experiencing any significant degradation in performance or beam quality. However, thermal effects may become critical for VLMA fiber designs that rely on resonant effects, such as photonic bandgap fibers and fibers with resonant filtering mechanisms, namely, DMF fibers or chirally coupled core fibers. In these fiber designs, a waveguide change during high-power operation can severely affect the resonance condition.

The guiding properties of fibers enable fiber designs that address these thermal effects. This will pave the way for active fibers with MFDs far beyond the 100 μm limit during high-power operation. For example, the thermal index profile may be precompensated, similar to precompensation for bending-induced index changes [195]. Furthermore, the heat load may be tailored by changing the doping geometry. This can be accomplished, for example, with reduced mode overlap fibers, which have been proposed for reduced photodarkening effects [196]. Enrico Coscelli et al. conducted a comprehensive comparison of the thermal resilience of three distinct rod-type DC-PCF designs: the 19-cell core, the DMF PCF, and the LPF. The simulations of the impact of thermally induced refractive index changes on the SM properties of these fibers revealed that while the LPF PCF exhibited a broad SM bandwidth of over 150 nm under low thermal loads, it rapidly transitioned to multimode operation as the temperature gradient across the fiber cross section exceeded 12–14 °C. In contrast, the DMF PCF demonstrated greater thermal resilience, with its SM range undergoing a modest blueshift of approximately 25 nm even at a temperature difference of 25 °C between the core center and the outer fiber edge [197]. One effective way to push the TMI threshold to higher power levels is to ensure that the active fiber provides strong delocalization of the HOMs even under severe heat loading while preserving high pump absorption and a large-core area. Several DC-PCF designs have been developed with this aim, and remarkable progress has been achieved thus far in exploiting the possibility of cladding microstructuration offered by the stack-and-draw fabrication technique [198, 199]. In 2016, Enrico Coscelli et al. systematically studied [187] the modal content of LMA-PCFs under heat loading. The combined effects of the microstructured cladding and thermally induced refractive index change on mode guidance were compared among three DC-PCFs, an LPF [29], a symmetry-free PCF (SF-PCF) [199], a fully periodic LPF (FA-LPF) [200], and a purely thermally guiding fiber (TGF). An obvious decrease in the FM effective area of the four types of LMA fibers was observed as a function of the heat load due to thermal lensing effects. The microstructured cladding of the considered DC-PCFs provides a further source of mode confinement, which sums to the thermal lensing and results in an overall lower effective area. At high heat load values, where the contribution of the thermally induced index gradient to the guidance is stronger, the effective areas of all the PCFs are quite similar.

15.3 Structural Design Principle and Single-Mode Criterion of LMA-PCFs

15.3.1 Structural Design Principle of LMA-PCFs

PCF, as a kind of microstructure fiber usually made of a single material, is distinguished by a periodic air hole array running along the whole length of the fiber. In contrast to conventional SIFs, the arresting feature of PCFs is the endless single-mode (ESM) behavior [9], i.e., remaining monomode at all wavelengths, which can be understood by the modal sieve concept, as shown in Fig. 15.8a, b [75]. The fundamental mode cannot leak through the gaps between the air holes (diameter d, spacing Λ) because of its excessively large effective transverse wavelength component. The high-order modes are the opposite. Numerical results in Fig. 15.8c reveal that proper manipulation of the geometry, typically $d/\Lambda < 0.4$, ensures that only the fundamental mode is supported [75, 201]. Extensive mode area fibers with low nonlinearity and strict single mode characteristics have become possible and can be employed to modify the performance of a high-power fiber system [62].

For an LMA, one requires a large fiber core. Standard single-mode fibers have an effective mode area below 100 μm^2, and LMA fibers reach values of hundreds or even thousands of μm^2. However, large-core fibers do not necessarily have large modes if they are strongly multimode; at least the fundamental mode may then be much smaller than the core. Larger mode areas can be achieved by relaxing the strict single-mode fiber requirements and working with step-index fibers at much higher V values such that the fiber is itself capable of supporting multimode (MM) operation, but engineering a fiber device such that it operates in a single-transverse mode [118, 131, 202]. The lowest-order mode (LP01 mode) is preferred in most instances, but interest in operating on a single HOM instead is increasing [203]. Furthermore, they may be more robust to subsequent mode coupling due to perturbations than the fundamental mode. This, in turn, suggests that larger mode areas can be used. A downside with the use of HOMs is that they often exhibit localized regions of high

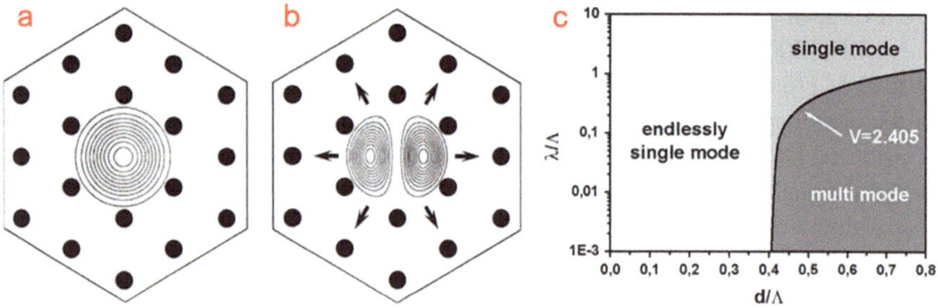

Fig. 15.8 Schematic diagram of the sieve model. (**a**) The fundamental mode is trapped. (**b**) HOM leaks away through the gaps between the air holes. (**c**) Modal characteristics of a one-hole missing PCF. [Reprinted/Adapted] with permission from [62] ©Wiley-VCH GmbH

field intensity across the core that can result in problems with damage at lower values of the effective area than might otherwise be expected. This can potentially be avoided by appropriate engineering of the HOM profile via control of the core refractive index profile and geometry.

Somewhat larger mode areas can be achieved with fiber designs supporting a few propagation modes (i.e., few-mode fibers). It may then still be possible to guide light dominantly in the fundamental mode so that the output, e.g., of a fiber amplifier, is close to diffraction-limited [118, 131]. Stringent limitations arise from the more critical launch conditions and from mode mixing in the fiber, which can spoil the beam quality and lead to beam pointing fluctuations [204]. Various more sophisticated fiber designs (partly based on PCFs) and techniques have been developed. In many cases, one attempts to introduce substantial propagation losses for any higher-order mode, making it easier to maintain robust single-mode propagation in a multimode fiber. Another important aspect is to minimize unwanted mode coupling, essentially by achieving a sufficiently large difference in the propagation constants between the fundamental and higher-order modes. All-glass LCFs have been demonstrated to be promising solutions for achieving large effective mode areas with an all-glass structure very similar to that of conventional fibers [104]. Both experiments and simulations indicate that very low bend loss at a small bend radius can be achieved with a large d/Λ for LCFs, and the best differential mode loss is achieved at a large bend radius [205].

15.3.2 Single-Mode Criterion of LMA-PCFs

PCFs can be designed to have the endless single-mode property, where only the fundamental mode can propagate through the fiber core for all wavelengths [9]. The cutoff wavelength analysis of PCFs is not as simple as that of conventional fibers because all modes propagating in PCFs with a finite number of air hole rings are leaky [67]. An analysis of the single-mode properties of different structures of optical fibers requires careful consideration of the most reasonable analysis method to be used.

Conventional refractive index-guided fibers usually use the V parameter to calculate the normalized frequency. When the calculated $V \leq 2.405$, it can be judged as having a single-mode operation. The refractive indices of the cladding and core in conventional fibers are uniformly distributed, and it is easy to calculate the V-parameter and analyze the single-mode properties. Another single-mode verification method for conventional fibers is when the confinement loss of the fundamental mode is <0.1 dB/m, whereas the confinement loss of all higher-order modes is >1 dB/m. Usually, when the confinement loss of the first higher-order mode is >1 dB/m, it can be judged as having single-mode operation.

Compared with conventional double-clad fibers, PCFs are more flexible because of the greater freedom of their structural design. Single-mode property analysis can be performed by calculating the V-parameter, Q-parameter, effective area, and effective refractive index of the second-order mode, confinement loss, core overlap factor, etc. However, the applicability of each method is different, so it is necessary to analyze different fiber structures and select an appropriate criterion [206].

By calculating the Q-parameters for different normalized wavelengths λ/Λ and estimating the negative minima at different air hole diameter-to-pitch ratios (d/Λ), the single-mode operating intervals of the PCFs can be obtained. The criterion of confinement loss adopted for single-mode operation judgment in traditional optical fibers also applies to PCFs. However, the confinement loss of PCFs strongly correlates with the number of air hole rings. With an increasing number of air hole rings, the confinement loss can be greatly reduced to 10^{-7}–10^{-11} orders of magnitude, which is almost negligible. Hence, when the number of air hole rings in the inner cladding is large, confinement loss is no longer applicable as a single-mode criterion for PCFs. When the V-parameter method is used, it is necessary to calculate the effective refractive indices of the fundamental space-filling mode (n_{FSM}) and the guided fundamental mode (n_{eff}) in the air hole cladding.

For different fiber structures, the values of the effective core radius (ρ) are different. Many different definitions of the effective core radius have been proposed for fiber structures with a triangular lattice replacing an air hole, such as 0.5Λ [207], $\Lambda/\sqrt{3}$ [208], 0.625Λ [209], 0.64Λ [210], and Λ [9, 211]. The calculated $V \leq \pi$ can be used as a judge for single-mode operation. Currently, the control of the laser output mode is no longer strictly limited by the refractive index difference between the fiber core and the cladding. Modal analysis in terms of the V parameter has some technical limitations, such as the fact that the definition of the core diameter seems to be non-unified in different studies [207, 208, 210, 211]. The mode of the fiber can be finely adjusted by the appropriate design of the fiber cross-sectional geometry, so that the introduction of the core to the overlap factor can be used to determine the percentage of fundamental modes. Three main methods for determining whether one PCF can be in single-mode operation are subsequently introduced.

15.3.2.1 Normalized Cutoff Frequency (V) [9, 212–214]

V parameter is easy to calculate for standard fibers because it depends on the core radius and the refractive index difference, which are well defined. For PCFs, the V parameter differs from that of standard fibers, and the normalized frequency V is as expressed in Eq. 15.1 [9, 212–214]:

$$V = \frac{2\pi\rho}{\lambda}\sqrt{n_{eff}^2 - n_{FSM}^2} \tag{15.1}$$

where n_{eff} is the effective refractive index of the fundamental guided mode, and where n_{FSM} is the effective refractive index of the fundamental space-filling mode in the air hole cladding. Typically, Λ is the hole-to-hole distance (also called pitch) of the triangular lattice. For the studied PCFs, a suitable effective core radius needs to be selected, and ρ in Eq. 15.1 represents the effective radius of the fiber core. ρ can be defined as 0.5Λ, $\Lambda/\sqrt{3}$, 0.625Λ, or 0.64Λ, where Λ depends on the fiber structure.

In 2005, Stefano Selleri et al. studied triangular PCFs with different core geometries, with a focus on the cutoff wavelengths of PCFs with one and seven air holes removed [215]. The boundary describing the single-mode and multimode operation regions in the phase diagram of the restless single-mode region was calculated by considering the leakage loss of the second-order mode. For a fiber with a core replaced by one air hole, its operation is in single mode without a cutoff when d/Λ <0.405. For a fiber whose core is replaced by seven air holes, single-mode operation is maintained without a cutoff when d/Λ < 0.035.

The NA of an optical fiber is an essential parameter, which indicates the ability of the end face of the fiber to receive incident light, and the magnitude of its value determines the ability of the fiber to receive light and the effect on mode dispersion. The NA of PCFs is expressed as Eq. 15.2:

$$\mathrm{NA} = \sqrt{n_0^2 - n_{\mathrm{eff}}^2} \tag{15.2}$$

where n_{eff} is the effective refractive index, which is related to the structural parameters of the cladding. When the aperture of the cladding changes, the effective refractive index n_{eff} also changes, so the NA versus d/Λ curve can also be calculated.

15.3.2.2 Q-Parameter Calculation [212, 216]

The analytical method for the cutoff wavelength can be demonstrated by calculating the Q-parameter, which is evaluated by analyzing the turning point of α/k_0 with respect to λ^*/Λ. There is a significant negative minimum at λ^*/Λ in the calculation of the *Q*-parameter, and the expression of the *Q*-parameter is shown in Eq. 15.3 [215], where α is the loss constant and k_0 is the wavenumber [212].

$$Q = \frac{d^2 \log\left[\dfrac{\alpha}{k_0}\right]}{d^2 \log(\Lambda)} \tag{15.3}$$

By fixing the wavelength λ, the *Q*-parameters for different normalized wavelengths λ^*/Λ and estimating the negative minima at different d/Λ conditions, the single-mode and multimode operation intervals of the PCFs can be obtained. As shown in Fig. 15.9a, the larger the number of air hole rings is, the sharper the turning point of the *Q*-parameters [216], and the more reliable this analytical method is. Notably, the Q-parameter method effectively determines the cutoff wavelength only for PCFs with many air apertures.

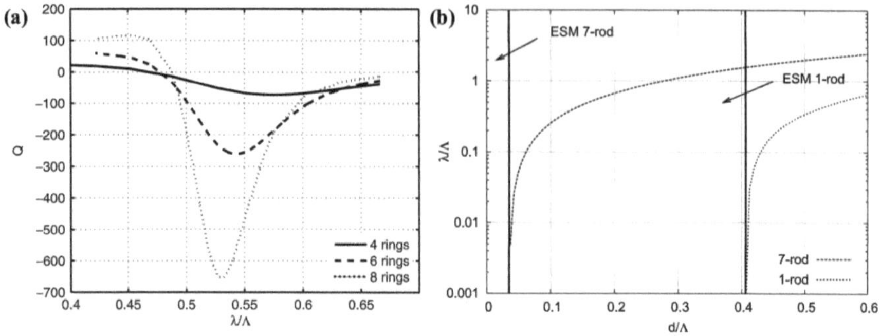

Fig. 15.9 (**a**) Variation in the Q value with normalized wavelength λ*/Λ as the number of air hole rings varies; (**b**) cutoff wavelength of fiber cores replacing 1 and 7 air hole PCFs. [Reprinted/Adapted] with permission from [17] © MDPI, Basel, Switzerland

15.3.2.3 Core Overlap Factor (Γ) [217, 218]

$$\Gamma = \iint_{hex} i(x,y)\,dxdy \tag{15.4}$$

$$i(x,y) = \frac{1}{P}\mathrm{Re}\left[\frac{E \times H^{*}}{2} \bullet \hat{z}\right] \tag{15.5}$$

The expression of the overlap integral is shown in Eq. 15.4, and hex is the cross section of the core's hexagonal fiber core doping. In Eq. 15.5, $i(x, y)$ is the normalized intensity distribution of the guided mode [217, 218], where P is the integral of the intensity over the entire fiber cross section. The fundamental mode intensity distribution is calculated from the Poynting vector definition, which involves the three components of both the electric and the magnetic fields of the guided mode. The overlap integral is an important parameter for describing the single-mode property of Yb-doped PCFs with an LMA. It provides the interaction between the Yb ions and the optical signal and describes how tightly the modes are confined in the fiber core, helping to distinguish whether the modes are guided or not. The concept of normalized intensity is derived from the definition of the Poynting vector, which is the energy flow density vector in the electromagnetic field, so the dual integral over the core and the entire cross section can be calculated via the time-averaged energy density. First, the magnetic field $H = (H_x, Hy, H_z)$ on the fiber cross section is calculated via finite element analysis, and then the electric field $E = (E_x, E_y, E_z)$ is obtained via Maxwell's equations on the basis of the distribution of the magnetic field. Finally, the core overlap factor (Γ) is obtained by double integration of the normalized intensity of the fiber core region.

Mette Marie Jørgensen et al. proposed a single-mode determination criterion when optimizing a distributed mode-filtered fiber amplifier [217]. This determination criterion is based on the evaluation of the mode overlap in the doped core region, defined as the normalized integral of the mode intensity over the core region. To obtain effective SM operation by means of a differential amplification mechanism, the fundamental mode (FM) overlap integral is defined as higher than 0.8, and the overlap difference between the FM and the most detrimental higher-order mode (HOM) should be >0.25, which can be used as a criterion for single-mode operation in optical fibers. This method is more suitable for analyzing the structure of a fiber core that replaces multiple air holes than the *V*-parameter method, and the calculation results are more accurate.

15.4 Fabrication Techniques for LMA-PCFs

The stacking method is a common method for preparing LMA-PCFs. To prepare super LMA-PCFs with core diameters greater than 40 μm or even 100 μm, the ytterbium-doped silica glass core rod (mandrel) required for preparing prefabricated rods must meet the requirements of geometric size and optical (refractive index, uniformity) and spectral (doping concentration, absorption, and emission) properties [79]. The number of guided modes in a PCF is independent of the ratio of the core radius to the optical wavelength. Instead, the number of modes depends only on the ratio of the air hole diameter (d) to the spacing between holes (Λ) [9]. Therefore, no change in the refractive index profile is required to fabricate fibers with different core diameters, and a single preform can be used to fabricate monomode fibers with cores of any size [66].

To suppress the nonlinear effect at peak power and achieve high brightness, high optical homogeneity ytterbium-doped silica glass mandrels with high doping concentrations, refractive indices similar to those of pure silica glass, and diameters greater than 3 mm are required for the preparation of high-brightness LMA-PCFs. A high concentration of rare earth ions leads to an increase in the refractive index and a decrease in the optical uniformity of silica glass, while the refractive index of the core rod required in the preparation of high-brightness ytterbium-doped LMA-PCF is small, and its variation amplitude is also small (high uniformity). Overcoming this major contradiction between the two is the key to the preparation of high-brightness Yb-doped LMA-PCFs [79].

Here, the fabrication methods of silica glass mandrels, which include modified chemical vapor deposition (MCVD) [81, 120, 219], multi-deposition and solution doping [220], direct nanoparticle deposition [221], powder sintering [222–225], glass phase separation [226] and the sol-gel technique [227, 228], are introduced. Among these methods, the sol-gel process for preparing the Yb-doped silica glass mandrel with a large diameter and low NA shows great advantages over the traditional MCVD method, which is reflected in three main aspects [78]:

1. The high dispersion of doped ions effectively reduces the cluster tendency of doped ions, especially rare earth ions, so that it is better than traditional MCVD in controlling the glass refractive index distribution uniformity, photodarkening effect, and concentration quenching effect.

2. In the sol preparation process, it is very easy to introduce a variety of doping elements, which is conducive to regulating the optical and spectral properties of rare-earth-doped silica glass mandrels. For example, it is very easy to introduce co-doped elements such as F and P so that highly doped silica glass mandrels with a low refractive index can be prepared, which has great advantages in the preparation of extremely low-NA fibers.
3. This method can overcome the bottleneck of MCVD process technology, especially the solution immersion doping process technology, to prepare silica glass mandrels with core diameters larger than 3 mm, thus achieving the preparation of rare-earth-doped silica glass core rods with both large diameters (sizes) and small refractive index fluctuations.

The disadvantage of the sol-gel process is its unavoidable container contact-induced relatively high loss compared with that of the MCVD process. Therefore, the sol-gel process is usually used for the preparation of shorter-length LMA-PCFs via the stacking method. Hu et al. of SIOM, China, carried out abundant work on the development of Yb^{3+}-doped LMA-PCFs via the sol-gel method combined with powder sintering techniques. Yb^{3+}/Al^{3+}, $Yb^{3+}/Al^{3+}/P^{5+}$, $Yb^{3+}/Al^{3+}/F^{-}$, and $Yb^{3+}/Al^{3+}/^{5+}/F^{-}$ co-doped silica glasses were prepared as fiber core glasses, and the changes in their optical and spectral properties with composition and thermal history were systematically studied [36]. For example, cross sections of several types of Yb-doped LMA-PCFs are presented in Fig. 15.10. A PCF with a 50 μm core diameter (Fig. 15.10a) is widely used in the laser processing field. A PCF with a 75 μm core diameter was used (Fig. 15.10b), 80 cm in length, an average output power of 102 W, and a peak power > 1 MW were achieved with $M^2 = 2.1$ in a ps pulse system [78]. Picosecond pulse amplification with an average power exceeding one hundred watts and high beam quality ($M^2 < 1.5$) was achieved via 40 μm/200 μm polarization while maintaining the ytterbium-doped PCF (Fig. 15.10c), the polarization degree was greater than 12 dB, and the power fluctuation was less than 1.3% in 2 h under 100 W amplification power operation. Picosecond pulse amplification with a beam quality factor $M^2 < 1.3$ and polarization >95% was realized in 2023 via a custom-made 100-μm core diameter Yb-doped LMA-PCF as the gain fiber (Fig. 15.10d) [36]. The PCF shows a similar performance parameter to that of the DC-285/100-PM-Yb-Rod rod-type PCF developed by NKT Photonics [128].

In the same year, Ma et al. proposed an advanced stress-induced polarization-maintaining Yb-doped PCF structure with an LMA realized by introducing both leakage channels and higher order mode-filtering units. The simulation results indicate that a maximum core diameter of 101 μm and a single-MFD of 76.33 μm at 1064 nm and a birefringence value $>10^{-4}$ orders of magnitude can be achieved [206]. The cross section and refractive index distribution of the LMA microstructured Yb-doped fibers considered in this study are shown in Fig. 15.11. These two fiber structures share several common features, such as an expanded fiber core, by replacing the 19 air hole cells of the triangular lattice with Yb-doped rods. The stress zone consists of 10 boron-doped fused silica rods on each side of the core, and the stress zone is separated from the core by one layer of air hole rings. Figure 15.11a shows a cross-sectional view of a PM-PCF with a stress-induced,

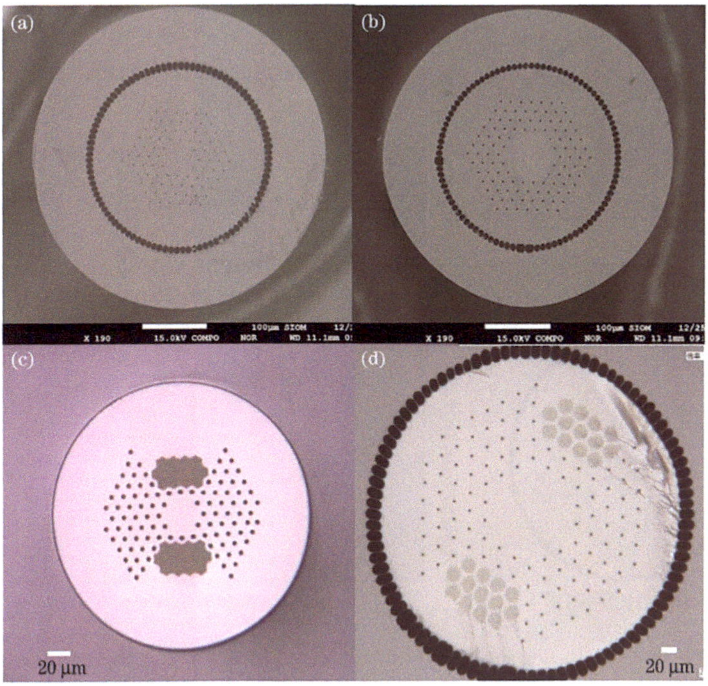

Fig. 15.10 Cross sections of four kinds of Yb^{3+}-doped LMA-PCFs fabricated in the SIOM. (**a**) Cross section of the 50-μm-core PCF; (**b**) cross section of the 75-μm-core PCF; (**c**) cross section of the 40-μm-core PM-PCF; (**d**) cross section of the 100-μm-core PM-PCF. [Reprinted/Adapted] with permission from [36] Chinese Laser Press (CLP)

polarization-maintaining structure with a stomatal diameter of d = 1.2–5.9 μm and a hole spacing of Λ = 20–23.75 μm. The white part is a clad silica substrate with a refractive index of 1.4500, a thermal expansion coefficient of 5.50×10^{-7} K^{-1}, and an inner cladding diameter of 8.5–10 μm. The blue part is set to air, and d/Λ ranges from 0.05–0.25. Boron-doped fused silica rods with a thermal expansion coefficient of 2.54×10^{-6} K^{-1} are introduced as stress elements (yellow section) in the optical waveguide. These stress elements, which have a diameter of 12–15 μm and a refractive index of 1.4412, are distributed in a rectangular shape, replacing the multiple air holes on both sides of the fiber core. Figure 15.11c shows a schematic diagram of the fiber PM-HOF-PCF. A leakage channel and an HOM filter are introduced to achieve single-mode LMA-PM transmission. The HOM filter is a ring structure doped with germanium (green part). To make a fair comparison between these two different structural designs, the fiber–hole spacing Λ for PM-HOF-PCF is set to 20 μm. The HOMs are expected to be doubly filtered out by using Ge ring HOM filters in the high-refractive-index region of the leakage channel. Figure 15.11b, d show the refractive index profiles of the corresponding fibers in Fig. 15.11a, c.

Table 15.1 summarizes the core composition, Yb^{3+} doping concentration, core diameter, laser output power, and mode of Yb-doped LMA-PCFs developed in the

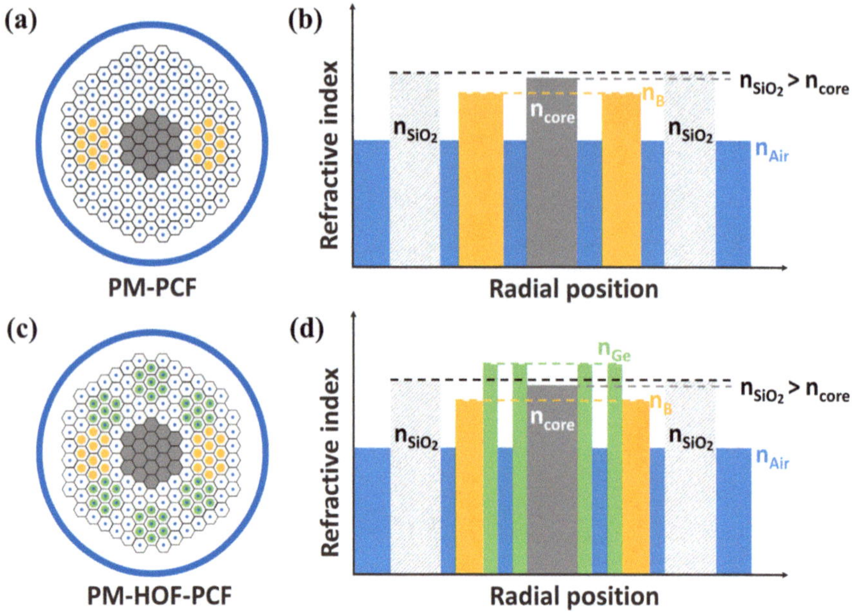

Fig. 15.11 (**a**) Cross section and (**b**) refractive index of PM-PCF with boron rod stress zones and a solid fiber core replacing 19 air holes. (**c**) Cross section and (**d**) refractive index of PM-HOF-PCF with Ge rings as HOM filters, boron rod stress element zones, and a solid fiber core replacing 19 air holes. [Reprinted/Adapted] with permission from [206] © IEEE

Table 15.1 Basic parameters and laser amplification results of the self-developed Yb-doped LMA-PCF in the SIOM. [Reprinted/Adapted] with permission from [36] © Chinese Laser Press (CLP)

Time	Composition	Mole fraction of Yb_2O_3/%	Fiber type	Core diameter/ μm	Laser woke mode	Average power/W	Peak power	M^2
2012	Yb/Al	0.3	PCF	30	CW	6.8	–	–
2013	Yb/Al	0.1	PCF	90	CW	81	–	–
2013	Yb/Al/P	0.05	PCF	35	CW	35	–	–
2015	Yb/Al/P	0.035	DCF PCF	35 60	CW	3.2 46	– –	1.3 –
2016	Yb/Al	0.1	PCF	105	MOPA	255	1.2 MW	>10
2017	Yb/Al	0.1	PCF	100	MOPA	310	1.5 MW	5
2017	Yb/Al/P/F	0.075	PCF	50	MOPA	97	93 kW	1.4
2019	Yb/Al/P/F	0.09	PCF	50	MOPA	272	266 kW	2.2
2019	Yb/Al/P/F	0.15	PCF	75	MOPA	102	1 MW	2.1
2021	Yb/Al/P/F	0.15	PM-PCF	40	MOPA	103	–	1.46
2023	Yb/Al/P/F	0.15	PM-PCF	100	MOPA	~50	–	~1.3

SIOM since 2012. Through 10 years of continuous research, the core composition has evolved from Yb/Al codoping to Yb/Al/P/F tetra-doping, the core diameter varies from 30 to 100 μm, and the laser operating mode includes both CW and main oscillation power amplification (MOPA). The fiber types include nonpolarization-maintaining to polarization-maintaining (PM) fibers, and the beam quality factor M^2 continues to improve (see Fig. 15.12).

The 100 μm core diameter (MFD 75 ± 5 μm) 100/290 PM-LMA PCF in Table 15.1 has a large absorption coefficient, and the cladding absorption coefficient of fibers approximately 80 cm in length can reach 24 dB at 976 nm. The polarization-maintaining structure of the fiber results in a loss difference between the fast axis and the slow axis. The test results show that under the same test conditions, the amplification efficiency of the slow-axis input is more than 30% greater than that of the fast-axis input, and the polarization degree of the amplified output is greater than 95%.

Figure 15.12 shows the research progress of Yb-doped LMA-PCFs developed in the SIOM. Since 2020, the beam quality has greatly improved when the average amplification power of the Yb-doped LMA-PCF is 100 W. Yb-doped LMA-PCFs have expanded from the non-PM to the PM type, and the preparation of a 40 μm core diameter 40/200 PM-PCF all-fiber amplifier module has been realized. The performance of Yb-doped LMA-PCFs amplified with a picosecond pulse laser has steadily improved to >1 MW with a decrease in M^2 from >1.5 to <1.3 [36].

Philippe Roy et al. reviewed fully aperiodic LPFs, including their major advantages and drawbacks, as well as their synthesis methods. This structure was designed

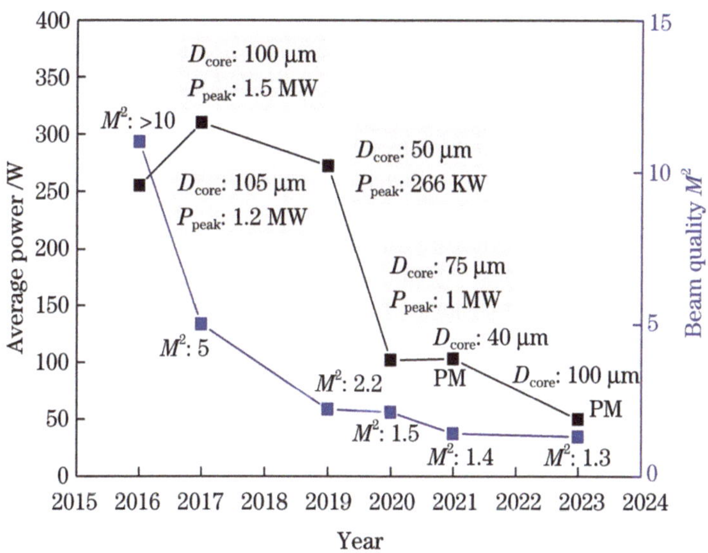

Fig. 15.12 Research progress on Yb-doped LMA-PCFs developed in SIOM. [Reprinted/Adapted] with permission from [36] © Chinese Laser Press (CLP)

to reinforce a single mode, whether the fiber was cold or subjected to significant thermal loading. This approach also significantly simplifies the fabrication process and potentially decreases the production cost of the material, considering the suppression of the multiple-step, stacking, and drawing processes. Finally, promising tests have been performed with an extremely large-core (MFD = 92 μm) fiber specially designed to precompensate for the thermal loading [229]. In 2021, Andreas Langner et al. from Heraeus Quarzglas GmbH & Co. KG discussed the advancements and trends in specialty fiber optic materials. This highlights the unique plasma outside deposition (POD) process used by Heraeus to produce large-core step-index specialty optical fiber preforms, which are capable of withstanding high light intensities and offer excellent transmission performance across a broad wavelength range. They also addressed the challenges in metrology related to these fibers, such as the precise determination of fiber attenuation and refractive index profiling, especially for complex preform designs such as multicadding fibers [230].

PCFs and MOFs made from soft glasses (e.g., heavy-metal oxides, fluoride or chalcogenide glasses) have attracted much interest because of their extended ultraviolet (UV) and mid-IR transmission windows, higher Kerr nonlinearities and refractive indices, and increased rare-earth solubility. However, compared with the fabrication of fused silica, the fabrication of high-quality PCFs from soft glasses remains significantly challenging because of much faster changes in viscosity with temperature, as does glass devitrification during the repeated reheating steps required by the "stack-and-draw" method, which is the most highly developed fabrication technique for such fibers, providing great flexibility for drawing both solid and hollow-core fibers, not only from fused silica but also from many soft glasses. However, drawing soft-glass PCFs is challenging because of a much faster change in viscosity with temperature, and fiber drawing requires much tighter control of temperature since the glasses soften over a much narrower temperature range, which means that the drawing temperature must be controlled within a narrow range (less than ±25 °C), which is much narrower for fused silica (roughly ±150 °C). As a result, even very small temperature fluctuations can result in large distortions in the fiber structure. Another difficulty is the thermal instability of soft glasses, which can cause glass devitrification in fiber preforms during heating. Successful fiber drawing requires both a proper viscosity ($\eta \sim 10^5$–10^6 Pa·s) and a drawing temperature well below the onset of glass crystallization [231]. Not all nanostructures can be drawn from stacked preforms; however, examples include multiple glass nanomembranes or nanospikes. Moreover, when working with soft glasses, many of which are thermally unstable, minimizing the number and duration of reheating steps is important to avoid the risk of glass crystallization. Advanced fiber drawing techniques, such as 3D printing of complex extrusion dies [232], advanced differential etching procedures, and techniques for producing chirally twisted fibers and polymer/glass hybrid structures, have been developed over the last few years. These new techniques enable the development of novel soft-glass fiber structures for a range of different applications.

15.5 Application of LMA Fibers in High-Power Lasers

The high output power and diffraction-limited beam quality achieved by ultrafast fiber lasers make this technique one of the most remarkable and promising laser technologies. For example, several thousand watts of average power [233], up to 26 mJ pulse energy [87], and 22 GW peak power [234] have been reported. Nevertheless, nonlinear effects and mode instabilities have become two main factors hindering the further development of ultrafast fiber lasers. In general, increasing the MFA and reducing the power intensity of the core are the most effective strategies for circumventing the nonlinear effects in the optical fiber. However, fibers with large-mode areas usually support multimode propagation, which may lead to TMIs at high average power.

Compared with traditional fiber lasers, LMA-PCF lasers are expected to increase the output power by an order of magnitude. The excellent properties of LMA-PCF in fiber lasers have attracted much attention from researchers. In 2000, Wadsworth et al. [235] of the University of Bath in the UK used a Ti: sapphire laser (970 nm) to pump an 81-mm-long Yb^{3+}-doped PCF and obtained a laser output of 1040 nm, marking the advent of the first PCF laser. Since then, the LMA-PCF laser has become one of the greatest research hotspots in the field of PCFs. After decades of development, research on LMA-PCF lasers has made great progress.

15.5.1 1 μm High-Power Fiber Lasers

15.5.1.1 Yb-Doped High-Power Fiber Lasers

The 3-level laser transition $^2F_{5/2} \rightarrow {}^2F_{7/2}$ of the Yb^{3+} ion has already been proven to be very efficient for the generation of short IR wavelengths near 980 nm. In 2008, specific fiber laser technology based on rod-type LMA-PCF achieved 94 W near 980 nm with good beam quality [82, 236]. More recently, a Yb-doped double-clad photonic bandgap fiber resulted in 151 W at 978 nm with excellent beam quality ($M^2 = 1.2$) [237]. Moreover, PCFs have been demonstrated to be ideal substrates for multicore structures in coherent beam combining (CBC) systems [238].

To improve PCF technology, the large-pitch photonic-crystal fiber (LP-PCF) concept was introduced for effective single-mode operation in PCF technology for power/energy scaling [39]. The use of LP-PCFs has resulted in 22 GW peak power [234], 23 mJ single pulse [239], and kW-level average power within CBC systems [240, 241]. The best results reported to date on high-peak-power laser systems based on Yb-doped and Tm-doped fibers have been obtained via rod-type PCFs [82, 86, 236, 242, 243] with unprecedentedly low NAs, for which the highest peak power of 4 GW in 2.2 mJ and 500 fs pulses [86] was obtained by Tino Eidam et al. in 2011. The resulting record peak power of 3.8 GW was achieved by combining active phase shaping with an efficient reduction in the acquired nonlinear phase that comprised an LPF with a 105 μm MFD. To date, the highest peak power emitted directly from a fiber amplifier system has been demonstrated via an 85 μm rod-type PCF with a corresponding MFD of 71 μm. This type of LPF with 19 missing holes can be seen as the step-index limit of the LPF design since the rod-type fiber possesses small air holes that can be described by an effective refractive index slightly lower than that of glass. Therefore, the resulting signal core NA can be extremely low.

Recently, Henry Haig et al. [244] demonstrated a simple, power-scalable approach to single-mode regenerative amplification in highly multimode fibers. By using a 35 cm triple-clad multimode fiber (Nufern XLMA-YTF-100/400/480) with a 100 μm core diameter and a fundamental-mode diameter of 71 μm as the main amplifier, they demonstrated a single-stage, high-gain CPA system and generated high-energy, nearly diffraction- and transform-limited pulses. The produced short-pulse fiber source has an unprecedented combination of features, e.g., high gain (>55 dB) with negligible amplified spontaneous emission, high pulse energy (>50 μJ), good beam quality ($M^2 \leq 1.3$), and transform-limited (300 fs) pulses from a single amplification stage. This regenerative amplification technique unlocks the use of multimode fibers in high-performance femtosecond systems and will enable a new class of high-performance fiber lasers. Scaling from the initial results by stretched-pulse duration [86], mode area, and polarization would allow the regenerative fiber amplifier to generate 2 mJ and 300 fs pulses, thus duplicating the performance of a highly complex multistage system in a single low-cost, small-footprint stage.

Double-clad rod-type PCFs make low-brightness pumping possible [39], and the larger outer diameter guarantees the high mechanical stability and heat dissipation capability of PCFs [24], which enables enormous performance improvement of fiber laser systems, particularly for pulsed amplifiers. Nanosecond fiber lasers with a few hundred-kilowatt average power and multi-millijoule pulse energy have been generated from rod-type PCFs with core sizes in the range of 60–100 μm [81, 123, 245–247]. Furthermore, these fibers are also the basis of 100 MW-level to GW-level peak power fs fiber amplifiers, which are routinely operated in state-of-the-art fiber-CPA systems [109, 248–251]. Nevertheless, the challenging fabrication of minute air hole arrays reduces the reproduction of these VLMA fibers [39]. When the core diameter is further increased, rod PCFs are not SM according to strict analysis [39].

Wang et al. [80] reported a ytterbium-doped LMA-PCF with a core NA of 0.034 and a core diameter of 50 μm. Laser power amplification of this LMA-PCF was studied via a seed source with a 21 ps pulse duration and a 48.7 MHz repetition rate at a wavelength of 1030 nm. With a pump power of 520 W, a maximum 272 W (266 kW peak power) quasi-single-mode laser output with an M^2 of 2.2 was achieved in a 4.7 m fiber length bent at a diameter of 47 cm with a slope efficiency of 52%, and no obvious mode instability, stimulated Raman scattering, or thermal damage to the end facet of the fiber was observed. Benjamin Pulford et al. [252] reported a TMI-limited single-mode power of 1.37 kW at 1064 nm from a monolithic fiber amplifier with a 25/400 PBGF, which is the highest output power from a PBG fiber demonstrated to date. This work demonstrates the promise of AS-PBGFs, especially in kW-class monolithic fiber amplifiers, and that AS-PBFs are compatible with kW-class monolithic fiber amplifiers built with commercial components. In addition, with active polarization control, more than 20 dB PER was achieved from this non-PM AS-PBF. In 2022, Zhao et al. demonstrated a high-repetition-rate (500 kHz) high pulse energy (50-μJ-level) 1064-nm, instead of 1035-nm, femtosecond laser, which is based on a commercially available Yb-doped rod-type PCF with a core diameter of 85 μm, an inner cladding diameter of 260 μm, and an MFA of ~3300 μm^2 as the main amplifier [253]. Soon after, on the basis of a similar state-of-the-art single-mode photonic-crystal rod-type ytterbium-doped fiber (improved aeroGAIN-ROD-PM85 from NKT Photonics), a fiber-based CPA power laser amplifier system operating at 248 W of average power and a repetition rate of 750 kHz, with bulk transmission grating compression to a pulse duration of 357 fs, average power of 175 W, and pulse energy of 233 μJ was achieved [254].

Using tapered ytterbium-doped fiber (TYDF) with a core/inner cladding diameter of 30/250 μm in the small-core region and 48/400 μm in the large-core region, Wang et al. established a high-power tandem pumped fiber amplifier that realized up to a 10.28 kW laser free of SRS or TMI, which is significantly improved compared with the 48/400 μm uniform YDF. This is the highest average output power achieved thus far on the basis of the TYDF [255]. Very recently, Carson Shollenbarger et al. proposed a novel architecture for an all-fiber linearly polarized counter-pumped linear amplifier utilizing an 85 μm core diameter rod fiber. Achieving over 95% signal coupling efficiency and 18 dB of gain in an 804 mm long amplification stage is an exciting prospect for high-energy pulsed laser applications. A ns pulsed fiber-coupled 1030 nm source for adequate seeding of the amplifier is either produced or procured [256].

15.5.1.2 Nd-Doped High-Power Fiber Lasers

The 3-level optical transition $^4F_{3/2} \rightarrow {}^4I_{9/2}$ of neodymium-doped silica fibers also has strong potential to generate even shorter IR wavelengths between 880 and 930 nm [257]. However, efficient laser emission based on the 3-level transition of Nd ions is quite challenging because of the strong competition with the 4-level transition of $^4F_{3/2} \rightarrow {}^4I_{11/2}$, which emits near 1050 nm. Several methods have been investigated to overcome this spurious amplified spontaneous emission (ASE) near 1050 nm in Nd-doped fibers. The use of a specific W-type refractive index profile with a finite cutoff wavelength below 1050 nm has resulted in several hundred mW near 910 nm from cladding-pumped Nd-doped fibers (NDFs) [258, 259]. However, this fiber design is based on a small-core diameter (<5 μm), which strongly limits the maximum output power in the single-frequency or pulsed regime [260].

Power scaling to a few tens of watts was achieved via large mode area fibers with either a low clad-to-core diameter ratio to limit the gain near 1050 nm [261] or a PCF structure inducing distributed spectral filtering [262]. To date, the recorded output power from a Nd-doped LMA fiber laser is 113 W near 900 nm with an M^2 of 3.1 by a group from the Shanghai Institute of Optics and Fine Mechanics, Chinese

Academy of Sciences [263], whereas 81 W with a beam quality of $M^2 = 1.5$ was obtained [264] with a 30 μm/130 μm (ratio = 4) fiber geometry. In 2023, Kilian Le Corre et al. reported that a Nd-doped PM-LMA fiber with a core diameter of 30 μm and a cladding diameter of 125 μm was also successfully used as a power amplifier stage in an MOPA, allowing 28 W of average output power [265] to be obtained in the sub-nanosecond pulsed regime. However, despite the very low core NA of 0.045, this fiber is slightly multimodal at 915 nm. Therefore, precise free-space injection of the LP01 mode of the LMA fiber is needed to maintain good output beam quality ($M^2 = 1.25$) [265], which is not suitable for industrial integration.

In all previous reports on LMA-NDFs, the clad-to-core diameter ratio was set to a value close to 4, imposing either a multimode core or a very small inner cladding. Most recently, Raphaël Florentin et al. [266] presented a study combining analytical, experimental, and numerical investigations to determine the most suitable LMA-NDF geometry for high-power operation in an all-fiber configuration while maintaining a near-diffraction-limited beam. They studied a 30/125 μm LMA-NDF with a clad-to-core ratio of 4.2 and explored for the first time the possibility of LMA fibers with higher clad/core ratios and very low core NA to maintain a nearly diffraction-limited output beam in a fully fiberized MOPA system. The 30/125 LMA fiber amplifier with a clad-to-core ratio of 4 allows optimal power extraction (efficiency >31%) at 915 nm at the expense of beam quality. These Nd-doped LMA fibers may therefore pave the way for the realization of an all-fiber MOPA emitting tens of watts at 915 nm, which would be particularly suitable for CW single-frequency or short-pulse amplification [266].

15.5.2 1 μm High-Power Fiber Lasers

A longer emission wavelength also benefits mode area scaling from Er-, Er/Yb-doped fibers to nearly 1.5 μm and Tm-doped fibers to 2 μm [39]. Lee et al. [267] reported high-energy single-frequency optical pulses at 1572 nm with a maximum pulse energy of 1.8 mJ, a peak power of 3.5 kW at 2.5 kHz, as well as 1.3 mJ, a peak power of 2.5 kW at a pulse repetition rate of 7.5 kHz using our proprietary single-mode large-core PM Er-Yb co-doped silicate glass fiber (with a core diameter of ~45 μm) MOPA system pumped at 976 nm, which is the highest pulse energy of a single frequency at 1572 nm from an all-fiber MOPA system by 2018. In their system, they used a 1572 nm performance-improved short-length Er-Yb co-doped silicate glass fiber in a multistage all-fiber MOPA system in combination with pulse shaping to mitigate early pulse energy clamping induced by undesired SBS.

Pumping at 790 nm makes it possible to exploit the cross-relaxation process, leading to two excited Tm^{3+} ions being obtained for one pump photon, enabling Tm^{3+}-doped 2 μm fiber laser systems to achieve a theoretical slope efficiency of ~80% [268]. To overcome the Stokes limit, high doping concentrations as well as Al^{3+} codoping, which reduces unwanted clustering, are critical [269]. For high-power operation, large mode area active fibers, especially those with pedestals, are also needed [270, 271]. Tm-doped PCFs working at approximately 2 μm allow for an MFD twice as large as Yb-doped PCFs under the same NA and cutoff frequency values. Moreover, Tm-doped fiber amplifiers have a much stronger thermal bearing capability and a higher self-focusing threshold [272, 273]. They have excellent power scaling prospects and have already generated fs pulses with 6.9 GW peak power [274] and 1 kW average power [242].

Over the last two decades, the output power in the 2 μm region has increased from 5.4 W [275] to over 1 kW [242]. In 2009, Moulton et al. demonstrated a free-space diode-pumped Tm-doped fiber laser (TDFL) with 300 W of output power at a wavelength of ~2040 nm [276]. The achieved remarkable slope efficiency of 61.8% is still one of the best reported. In 2010, the same group demonstrated that over 1 kW of output power with a slope efficiency of 53.2% was generated from an all-fiber two-stage power amplifier seeded with 50 W at a wavelength of 2045 nm [277]. The output power at the kW-level has been demonstrated only a few times since 2010. In 2018, Gaida et al. developed an ultrafast Tm-doped fiber laser system delivering 1150 W of average power with a diffraction-limited beam quality and a pulse repetition frequency of 80 MHz, spectrally centered at 1960 nm [242]. Recently, Anderson et al. developed an all-fiber narrow linewidth 1950-nm-wavelength TDFA reaching a maximum output power of 1101 W for a launch power of 2165 W, with a slope efficiency of 50.7% and nearly diffraction-limited beam quality [278]. Mathias Lenski et al. [279] investigated the influence of the pump wavelength on the high-power amplification of rod-type (50/250 μm), large-pitch thulium-doped fibers with a large mode area. Pumping at 1692 nm leads to a slope efficiency of 74% with an average output power of 67 W, which paves the way for highly efficient 2 μm fiber-based CPA systems. Very recently, a high-power tunable thulium-doped fiber laser system was reported by Ren et al. [280]. An output power of more than 1 kW with a slope efficiency in the range of 51.5–54.3% over the full wavelength tuning range from 1943 nm to 2050 nm was presented, thereby representing the first demonstration of a widely tunable kW-class 2 μm fiber laser system. A thulium fiber MOPA generating up to 937 W with a slope efficiency of 57.7% and a central wavelength of 2036 nm was also reported by Romano et al. [281]. Using the active fiber in a diode-pumped fiber oscillator configuration, they demonstrated an output power of over 15 W at a wavelength of 1985 nm, reaching a slope efficiency of 74.5%. The solution-doped MCVD was also used to draw a highly doped (>8 wt%, Tm^{3+}) double-clad LMA fiber, on the basis of which Gausmann et al. in 2023 demonstrated an all-fiber single oscillator operating at a wavelength of 1940 nm and featuring a slope efficiency of 69.4% [282]. In 2024, Maria Michalska et al. [283] reported high-power thulium-doped fiber amplifiers built using a nanoparticle-doped double-clad, pedestal-style LMA fiber. The fiber was characterized by a core diameter of 26 μm, a pedestal diameter of 85 μm, and a cladding diameter of ~445 μm. The laser system operating at a wavelength of 2034 nm delivered an output power of up to 441 W with a slope efficiency of 57.4%, and the electro-optical conversion efficiency of the amplifier was 25%. The 1940-nm-wavelength all-fiber amplifier seeded with a narrowband Tm-doped fiber laser provided an output power of up to 273 W with a slope efficiency as high as 61.8%.

15.6 Conclusions

The rapid development of fiber technology in the last two decades, especially LMA special fibers, including LMA double-cladding structure Yb^{3+}-doped silica fibers and VLMA rod-type PM-PCFs, has greatly promoted the power scaling of both high-peak-power and high-average-power fiber lasers. The use of LMA fibers is a direct means to suppress nonlinearity, but this introduces HOMs and triggers the TMI effect. One effective way to push the TMI threshold to higher power levels is to ensure that the active fiber provides strong delocalization of the HOMs even under severe heat loading while preserving high pump absorption and a large-core area. To balance the contradiction between nonlinear suppression and TMI, some LMA fibers, including T-DCFs, AS-PBGFs, LCFs, LPFs, SC-ARFs, MCFs, SCFs, CCCFs, etc., have been designed with their SM operation band strictly verified and fabricated through stack-and-draw or extrude-and-stack techniques, which makes extensive mode area fibers with low nonlinearity and strict single mode characteristics possible. PCFs with an array of small air holes in the cladding have been operated in single mode with a 100 μm core diameter in a rod with an outer diameter much larger than 1 mm to overcome high bend loss, making the LMA-PCFs still maintain the current peak record of the optical fiber. On the other hand, the widely used CCCFs and T-DCFs have realized several kW TMI-free CW outputs and MW-level peak power outputs that are CPA free. With the progress of fiber fabrication and the continuous development of laser technology, compact all-fiber PCF-based laser systems will have greater application prospects in the future.

References

1. Li, W.T., Zhou, Q.L., Chen, D.P., et al.: Research progress of Yb-doped large-mode area silica glass optical fiber and its application in high-power fiber lasers. Laser Optoelectron. Progress. **50**, 020005 (2013)
2. Eidam, T., Wirth, C., Jauregui, C., et al.: Experimental observations of the threshold-like onset of mode instabilities in high power fiber amplifiers. Opt. Express. **19**, 13218–13224 (2011)
3. Taverner, D., Richardson, D.J., Dong, L., et al.: 158-uJ pulses from a single-transverse-mode, large-mode-area erbium-doped fiber amplifier. Opt. Lett. **22**, 378–380 (1997)
4. Richardson, D.J., Nilsson, J., Clarkson, W.A.: High power fiber lasers: current status and future perspectives. J. Opt. Soc. Am. B. **27**, B63–B92 (2010)
5. Jauregui, C., Clarkson, J., Tünnermann, A.: High-power fibre lasers. Nat. Photonics. **7**, 861–867 (2013)
6. Stutzki, F., Jansen, F., Otto, H.J., et al.: Designing advanced very-large-mode-area fibers for power scaling of fiber-laser systems. Optica. **1**, 233–242 (2014)
7. Jauregui, C., Limpert, J., Tünnermann, A.: Ultra-large mode area fibers for high power lasers. In: Optical Fiber Communications Conference and Exposition (OFC) , San Diego, California United States, 11–15 March 2018
8. Knight, J.C., Birks, T.A., Russell, P.S.J., et al.: All-silica single-mode optical fiber with photonic crystal cladding. Opt. Lett. **21**, 1547–1549 (1996)
9. Birks, T.A., Knight, J.C., Russell, P.S.J.: Endlessly single-mode photonic crystal fiber. Opt. Lett. **22**, 961–963 (1997)

10. Ortigosa-Blanch, A., Knight, J.C., Wadsworth, W.J., et al.: Highly birefringent photonic crystal fibers. Opt. Lett. **25**, 1325–1327 (2000)
11. Russell, P.: Photonic crystal fibers. Science. **299**, 358–362 (2003)
12. Ranka, J.K., Windeler, R.S., Stentz, A.J.: Visible continuum generation in air-silica microstructure optical fibers with anomalous dispersion at 800 nm. Opt. Lett. **25**, 25–27 (2000)
13. Eggleton, B.J., Kerbage, C., Westbrook, P.S., et al.: Microstructured optical fiber devices. Opt. Express. **9**, 698–713 (2001)
14. John, S.: Strong localization of photons in certain disordered dielectric superlattices. Phys. Rev. Lett. **58**, 2486–2489 (1987)
15. Yablonovitch, E.: Inhibited spontaneous emission in solid-state physics and electronics. Phys. Rev. Lett. **58**, 2059–2062 (1987)
16. Russell, P.S.J.: Photonic band gaps. Phys. World. **5**, 37–42 (1992)
17. Ma, Y., Wan, R., Li, S.W., et al.: Advances in silica-based large mode area and polarization-maintaining photonic crystal fiber research. Materials. **15**, 23 (2022)
18. Cregan, R.F., Mangan, B.J., Knight, J.C., et al.: Single-mode photonic band gap guidance of light in air. Science. **285**, 1537–1539 (1999)
19. Roberts, P.J., Couny, F., Sabert, H., et al.: Ultimate low loss of hollow-core photonic crystal fibres. Opt. Express. **13**, 236–244 (2005)
20. Zervas, M.N., Codemard, C.A.: High power fiber lasers: a review. IEEE. J. Sel. Top. Quantum. Electron. **20**, 219–241 (2014)
21. Chang, G.Q., Wei, Z.Y.: Ultrafast fiber lasers: an expanding versatile toolbox. iScience. **23**, 101101 (2020)
22. Zervas, M.N.: In: Zervas, M.N. (ed.) Power Scaling Limits in High Power Fiber Amplifiers due to Transverse Mode Instability, Thermal Lensing, and Fiber Mechanical Reliability. SPIE, San Francisco (2018)
23. Tao, Y.G., Wang, F., Shao, W.L., et al.: Large mode-field area multielement silica glass fibers for gigahertz ultrafast lasers. Opt. Express. **31**, 30562–30569 (2023)
24. Limpert, J., Deguil-Robin, N., Manek-Hönninger, I., et al.: High-power rod-type photonic crystal fiber laser. Opt. Express. **13**, 1055–1058 (2005)
25. Gu, G.C., Kong, F.T., Hawkins, T.W., et al.: Impact of fiber outer boundaries on leaky mode losses in leakage channel fibers. Opt. Express. **21**, 24039–24048 (2013)
26. Ma, X.Q., Zhu, C., Hu, I., et al.: Single-mode chirally-coupled-core fibers with larger than 50 μm diameter cores. Opt. Express. **22**, 9206–9219 (2014)
27. Laurila, M., Jørgensen, M.M., Hansen, K.R., et al.: Distributed mode filtering rod fiber amplifier delivering 292 W with improved mode stability. Opt. Express. **20**, 5742–5753 (2012)
28. Jain, D., Baskiotis, C., Sahu, J.K.: Mode area scaling with multi-trench rod-type fibers. Opt. Express. **21**, 1448–1455 (2013)
29. Stutzki, F., Jansen, F., Eidam, T., et al.: High average power large-pitch fiber amplifier with robust single-mode operation. Opt. Lett. **36**, 689–691 (2011)
30. Song, X.Y., Jin, D.Y., Zhou, D.C., et al.: Er^{3+}/Yb^{3+} co-doped bismuthate glass and its large-mode-area double-cladding fiber for 1.53 μm laser. J. Alloys Compd. **853**, 157305 (2021)
31. Kruska, K., Booker, P., Weßels, P., et al.: High-power single-frequency depressed-cladding, confined-doping Yb^{3+} fiber amplifier. In: Conference on Fiber Lasers XXI-Technology and Systems, San Francisco, California, 29 Jan-01 Feb 2024
32. Langner, A., Schötz, G., Such, M., et al.: In: Langner, A. (ed.) A New Material For High-Power Laser Fibers. SPIE, San Jose (2008)
33. Langner, A., Such, M., Schötz, G., et al.: New developments in high power fiber lasers based on alternative materials. In: Conference on Fiber Lasers VIII-Technology, Systems, and Applications, San Francisco, California, 24–27 Jan 2011
34. Leich, M., Just, F., Langner, A., et al.: Highly efficient Yb-doped silica fibers prepared by powder sinter technology. Opt. Lett. **36**, 1557–1559 (2011)
35. Benoît, A., Dauliat, R., Schuster, K., et al.: Optical fiber microstructuration for strengthening single-mode laser operation in high power regime. Opt. Eng. **53**, 071817 (2014)

36. Hu, L.L., Feng, S.Y., Wang, M., et al.: Research progress on Yb-doped large mode field photonic crystal fibers (invited). Chin. J. Lasers. **51**, 288–302 (2024)
37. Sidharthan, R., Ji, J.H., Lim, K.J., et al.: Step-index high-absorption Yb-doped large-mode-area fiber with Ge-doped raised cladding. Opt. Lett. **43**, 5897–5900 (2018)
38. Pournoury, M., Han, S.R., Ghasemi, M., et al.: Silica segmented cladding fiber design and its fabrication using a powder-in-tube technique. J. Lightwave Technol. **39**, 7251–7258 (2021)
39. Limpert, J., Stutzki, F., Jansen, F., et al.: Yb-doped large-pitch fibres: effective single-mode operation based on higher-order mode delocalisation. Light Sci. Appl. **1**, e8 (2012)
40. Filippov, V., Chamorovskii, Y., Kerttula, J., et al.: Double clad tapered fiber for high power applications. Opt. Express. **16**, 1929–1944 (2008)
41. Filippov, V., Chamorovskii, Y.K., Golant, K.M., et al.: Optical amplifiers and lasers based on tapered fiber geometry for power and energy scaling with low signal distortion. In: Conference on Fiber Lasers XIII-Technology, Systems, and Applications, San Francisco, California, 15–18 Feb 2016
42. Huang, L., Ma, P.F., Su, R.T., et al.: Comprehensive investigation on the power scaling of a tapered Yb-doped fiber-based monolithic linearly polarized high-peak-power near-transform-limited nanosecond fiber laser. Opt. Express. **29**, 761–782 (2021)
43. Bobkov, K.K., Aleshkina, S.S., Khudyakov, M.M., et al.: Active tapered fibers for high peak power fiber lasers. In: Conference on Micro-structured and Specialty Optical Fibers VII, Electr Network, 19–23 Apr 2021
44. Ye, Y., Xi, X.M., Shi, C., et al.: Comparative study on transverse mode instability of fiber amplifiers based on long tapered fiber and conventional uniform fiber. Laser Phys. Lett. **16**, 7 (2019)
45. Fedotov, A., Noronen, T., Gumenyuk, R., et al.: Ultra-large core birefringent Yb-doped tapered double clad fiber for high power amplifiers. Opt. Express. **26**, 6581–6592 (2018)
46. Filippov, V., Chamorovskii, Y., Kerttula, J., et al.: 600 W power scalable single transverse mode tapered double-clad fiber laser. Opt. Express. **17**, 1203–1214 (2009)
47. Filippov, V., Kerttula, J., Chamorovskii, Y., et al.: Highly efficient 750 W tapered double-clad ytterbium fiber laser. Opt. Express. **18**, 12499–12512 (2010)
48. An, Y., Yang, H., Chen, X., et al.: 4 kW single-mode laser output assisted by a double-tapered double-clad fiber. In: Conference on Fiber Lasers XVIII-Technology and Systems, Electr Network, 06–11 Mar 2021
49. Zeng, L.F., Pan, Z.Y., Xi, X.M., et al.: 5 kW monolithic fiber amplifier employing homemade spindle-shaped ytterbium-doped fiber. Opt. Lett. **46**, 1393–1396 (2021)
50. Yang, B.L., Zhang, H.W., Shi, C., et al.: High power monolithic tapered ytterbium-doped fiber laser oscillator. Opt. Express. **27**, 7585–7592 (2019)
51. Patokoski, K., Rissanen, J., Noronen, T., et al.: Single-frequency 100 ns/0.5 mJ laser pulses from all-fiber double clad ytterbium doped tapered fiber amplifier. Opt. Express. **27**, 31532–31541 (2019)
52. Kerttula, J., Filippov, V., Chamorovskii, Y., et al.: Actively Q-switched 1.6-mJ tapered double-clad ytterbium-doped fiber laser. Opt. Express. **18**, 18543–18549 (2010)
53. Ustimchik, V.E., Stoliarov, D.A., Korobko, D.A., et al.: Passively Q-switched 0.24 mJ ring laser based on anisotropic tapered fiber. Results Phys. **20**, 103640 (2021)
54. Bobkov, K., Andrianov, A., Koptev, M., et al.: Sub-MW peak power diffraction-limited chirped-pulse monolithic Yb-doped tapered fiber amplifier. Opt. Express. **25**, 26958–26972 (2017)
55. Bobkov, K., Levchenko, A., Kashaykina, T., et al.: Scaling of average power in sub-MW peak power Yb-doped tapered fiber picosecond pulse amplifiers. Opt. Express. **29**, 1722–1735 (2021)
56. Lai, W.C., Ma, P.F., Liu, W., et al.: 550 W single frequency fiber amplifiers emitting at 1030 nm based on a tapered Yb-doped fiber. Opt. Express. **28**, 20908–20919 (2020)
57. Khudyakov, M.M., Levchenko, A.E., Velmiskin, V.V., et al.: Er-doped tapered fiber amplifier for high peak power sub-ns pulse amplification. Photonics. **8**, 523 (2021)
58. Lin, X.F., Zhang, Z.L., Chu, Y.B., et al.: Fabrication and laser performance of cladding uniform core tapered fiber. Opt. Fiber Technol. **64**, 102561 (2021)

59. Petrov, A., Odnoblyudov, M., Gumenyuk, R., et al.: Picosecond Yb-doped tapered fiber laser system with 1.26 MW peak power and 200 W average output power. Sci. Rep. **10**, 17781 (2020)
60. Andrianov, A.V., Kim, A.V., Anashkina, E.A., et al.: Modeling of coherent beam combining from multimillijoule chirped pulse tapered fiber amplifiers. Eur. Phys. J. Spec. Top. **224**, 2579–2583 (2015)
61. Fathi, H., Närhi, M., Gumenyuk, R.: Towards ultimate high-power scaling: coherent beam combining of fiber lasers. Photonics. **8**, 566 (2021)
62. Zuo, J.X., Lin, X.C.: High-power laser systems. Laser. Photonics Rev. **16**, 2100741 (2022)
63. Aleshkina, S.S., Levchenko, A.E., Medvedkov, O.I., et al.: Photodarkening-free Yb-doped saddle-shaped fiber for high power single-mode 976-nm laser. IEEE Photon. Technol. Lett. **30**, 127–130 (2018)
64. Zeng, L.F., Xi, X.M., Ye, Y., et al.: A novel fiber laser oscillator employing saddle-shaped core ytterbium-doped fiber. Appl. Phys. B Lasers Opt. **126**, 185 (2020)
65. Zeng, L.F., Xi, X.M., Ye, Y., et al.: Near-single-mode 3 kW monolithic fiber oscillator based on a longitudinally spindle-shaped Yb-doped fiber. Opt. Lett. **45**, 5792–5795 (2020)
66. Knight, J.C., Birks, T.A., Cregan, R.F., et al.: Large mode area photonic crystal fibre. Electron. Lett. **34**, 1347–1348 (1998)
67. Dong, L., Peng, X., Li, J.: Leakage channel optical fibers with large effective. J. Opt. Soc. Am. B. **24**, 1689–1697 (2007)
68. Fu, L.B., McKay, H.A., Dong, L.: Extremely large mode area optical fibers formed by thermal stress. Opt. Express. **17**, 11782–11793 (2009)
69. Alkeskjold, T.T., Laurila, M., Weirich, J., et al.: Photonic crystal fiber amplifiers for high power ultrafast fiber lasers. Nano. **2**, 369–381 (2013)
70. Gu, G.C., Kong, F.T., Hawkins, T., et al.: Ytterbium-doped large-mode-area all-solid photonic bandgap fiber lasers. Opt. Express. **22**, 13962–13968 (2014)
71. Jansen, F., Stutzki, F., Jauregui, C., et al.: Avoided crossings in photonic crystal fibers. Opt. Express. **19**, 13578–13589 (2011)
72. Bjarklev, A., Broeng, J., Bjarklev, A.S. (eds.): Photonic Crystal Fibres. Springer Science & Business Media, Berlin (2003)
73. Zolla, F.: In: Zolla, F. (ed.) Foundations of Photonic Crystal Fibres. Imperial College Press, London (2005)
74. Ferreira, M.F.S.: In: Ferreira, M.F.D.S. (ed.) Optical Fiber Technology and Applications: Recent Advances. IOP Publishing Limited, London (2021)
75. Russell, P.S.J.: Photonic-crystal fibers. J. Light. Technol. **24**, 4729–4749 (2006)
76. Russell, P.: Photonic crystal fibres: a historical account. IEEE. Leos. Newsletter. **11-15** (2007)
77. Yi, C.S., Zhang, P.Q., Dai, S.X., et al.: Research progress of large-mode area photonic crystal fibers. Laser Optoelectron. Progress. **49**, 5–15 (2012)
78. Yu, C.L., Wang, M., Feng, S.Y., et al.: Research progress on ytterbium-doped large mode area photonic crystal fibers. Laser Optoelectron. Progress. **56**, 20–33 (2019)
79. Wang, M., Wang, F., Yu, C.L., et al.: Ultra-low core numerical aperture large mode area photonic crystal fiber with 1 MW peak power output. Acta Opt. Sin. **39**, 3 (2019)
80. Wang, M., Wang, F., Feng, S.Y., et al.: 272 W quasi-single-mode picosecond pulse laser of ytterbium-doped large-mode-area photonic crystal fiber. Chin. Opt. Lett. **17**, 071401 (2019)
81. Limpert, J., Schmidt, O., Rothhardt, J., et al.: Extended single-mode photonic crystal fiber lasers. Opt. Express. **14**, 2715–2720 (2006)
82. Boullet, J., Zaouter, Y., Desmarchelier, R., et al.: High power ytterbium-doped rod-type three-level photonic crystal fiber laser. Opt. Express. **16**, 17891–17902 (2008)
83. Selvas, R., Sahu, J.K., Fu, L.B., et al.: High-power, low-noise, Yb-doped, cladding-pumped, three-level fiber sources at 980 nm. Opt. Lett. **28**, 1093–1095 (2003)
84. Broeng, J., Vienne, G., Petersson, A., et al.: Air-clad photonic crystal fibers for high-power single-mode lasers. In: Conference on Fiber Lasers, San Jose, California, 26–28 Jan 2004
85. Wang, X.L., Zhang, H.W., Yang, B.L., et al.: High-power ytterbium-doped fiber laser oscillator: current situation and future developments. Chin. J. Lasers. **48**, 59–77 (2021)

86. Eidam, T., Rothhardt, J., Stutzki, F., et al.: Fiber chirped-pulse amplification system emitting 3.8 GW peak power. Opt. Express. **19**, 255–260 (2011)
87. Stutzki, F., Jansen, F., Liem, A., et al.: 26 mJ, 130 W switched fiber-laser system with near-diffraction-limited beam quality. Opt. Lett. **37**, 1073–1075 (2012)
88. Steinkopff, A., Jauregui, C., Stutzki, F., et al.: Transverse single-mode operation in a passive large pitch fiber with more than 200 μm mode-field diameter. Opt. Lett. **44**, 650–653 (2019)
89. Klenke, A., Breitkopf, S., Kienel, M., et al.: 530 W, 1.3 mJ, four-channel coherently combined femtosecond fiber chirped-pulse amplification system. Opt. Lett. **38**, 2283–2285 (2013)
90. Kong, F.T., Saitoh, K., Mcclane, D., et al.: Mode area scaling with all-solid photonic bandgap fibers. Opt. Express. **20**, 26363–26372 (2012)
91. Egorova, O.N., Semjonov, S.L., Kosolapov, A.F., et al.: Single-mode all-silica photonic bandgap fiber with 20-μm mode-field diameter. Opt. Express. **16**, 11735–11740 (2008)
92. Saitoh, K., Murao, T., Rosa, L., et al.: Effective area limit of large-mode-area solid-core photonic bandgap fibers for fiber laser applications. Opt. Fiber Technol. **16**, 409–418 (2010)
93. Kashiwagi, M., Saitoh, K., Takenaga, K., et al.: Low bending loss and effectively single-mode all-solid photonic bandgap fiber with an effective area of 650 μm^2. Opt. Lett. **37**, 1292–1294 (2012)
94. Kashiwagi, M., Saitoh, K., Takenaga, K., et al.: Effectively single-mode all-solid photonic bandgap fiber with large effective area and low bending loss for compact high-power all-fiber lasers. Opt. Express. **20**, 15061–15070 (2012)
95. Chen, X., Yao, T.F., Huang, L.J., et al.: Functional fibers and functional fiber-based components for high-power lasers. Adv. Fiber Mater. **5**, 59–106 (2023)
96. Egorova, O.N., Semjonov, S.L., Kosolapov, A.F., et al.: Large mode area single-mode ytterbium doped all-solid photonic bandgap fiber. In: 35th European Conference on Optical Communication (ECOC), Vienna, Austria, 20–24 Sep 2009
97. Aleshkina, S.S., Likhachev, M.E., Pryamikov, A.D., et al.: Very-large-mode-area photonic bandgap Bragg fiber polarizing in a wide spectral range. Opt. Lett. **36**, 3566–3568 (2011)
98. Baz, A., Bigot, L., Bouwmans, G., et al.: Single-mode, large mode area, solid-core photonic bandgap fiber with hetero-structured cladding. J. Lightwave Technol. **31**, 830–835 (2013)
99. Murao, T., Saitoh, K., Taru, T., et al.: Bend-isensitive and effectively single-moded all-solid photonic bandgap fibers with heterostructured cladding. In: 35th European Conference on Optical Communication (ECOC), Vienna, Austria, 20–24 Sep 2009
100. Baz, A., Bigot, L., Yehouessi, J.P., et al.: Polarizing large mode area photonic bandgap fiber. In: 2015 European Conference on Lasers and Electro-Optics-European Quantum Electronics Conference, Munich, 21–25 June 2015
101. Vanvincq, O., Cassez, A., Habert, R., et al.: Large mode area solid-core photonic bandgap Yb-doped fiber with hetero-structured cladding for compact high-power laser systems. J. Lightwave Technol. **39**, 4809–4813 (2021)
102. Wong, W.S., Peng, X., McLaughlin, J.M., et al.: Breaking the limit of maximum effective area for robust single-mode propagation in optical fibers. Opt. Lett. **30**, 2855–2857 (2005)
103. Dong, L., Li, J., Peng, X.: Bend-resistant fundamental mode operation in ytterbium-doped leakage channel fibers with effective areas up to 3160 μm^2. Opt. Express. **14**, 11512–11519 (2006)
104. Dong, L., Wu, T., McKay, H.A., et al.: All-glass large-core leakage channel fibers. IEEE. J. Sel. Top. Quantum. Electron. **15**, 47–53 (2009)
105. Dong, L., Li, J., McKay, H.A., et al.: Design and applications of large mode area optical fibers. In: Frontiers in Optics 2008/Laser Science XXIV/Plasmonics and Metamaterials/Optical Fabrication and Testing, Rochester, New York, 19–23 October 2008
106. Saitoh, K., Varshney, S., Sasaki, K., et al.: Limitation on effective area of bent large-mode-area leakage channel fibers. J. Lightwave Technol. **29**, 2609–2615 (2011)
107. Otto, H.J., Stutzki, F., Modsching, N., et al.: 2 kW average power from a pulsed Yb-doped rod-type fiber amplifier. Opt. Lett. **39**, 6446–6449 (2014)

108. Malleville, M.A., Leconte, B., Dauliat, R., et al.: Pre-compensation of thermally induced refractive index changes in a depressed core fully aperiodic large-pitch fiber for high average power operation. Opt. Lett. **46**, 2956–2959 (2021)
109. Lavenu, L., Natile, M., Guichard, F., et al.: High-energy few-cycle Yb-doped fiber amplifier source based on a single nonlinear compression stage. Opt. Express. **25**, 7530–7537 (2017)
110. Darwich, D., Sabra, M., Jeu, R.F., et al.: 140 μm single-polarization passive fully aperiodic large-pitch fibers operating near 2 μm. Appl. Opt. **56**, 9221–9224 (2017)
111. Zhang, X., Gao, S.F., Wang, Y.Y., et al.: Design of large mode area all-solid anti-resonant fiber for high-power lasers. High. Power. Laser. Sci. Eng. **9**, e23 (2021)
112. Xing, Z., Wang, X., Lou, S.Q., et al.: Large-mode-area all-solid anti-resonant fiber with single-mode operation for high-power fiber lasers. Opt. Lett. **46**, 1908–1911 (2021)
113. Xing, Z., Wang, X., Gu, S., et al.: Large mode area double-layer all-solid anti-resonant fiber for high-power fiber lasers. Results Phys. **29**, 7 (2021)
114. Tian, H., Fu, S.J., Sheng, Q., et al.: Polarization-maintaining large-mode-area solid-core anti-resonant fiber for high-power fiber lasers. Results Phys. **46**, 7 (2023)
115. Zhang, S., Sun, S., Sheng, Q., et al.: Polarization-maintaining performance of solid-core anti-resonant fiber with nested circular tubes in 3 μm wavelength. J. Lightwave Technol. **40**, 1137–1143 (2022)
116. Noda, J., Okamoto, K., Sasaki, Y.: Polarization-maintaining fibers and their applications. J. Lightwave Technol. **4**, 1071–1089 (1986)
117. Barlow, A.J., Payne, D.N.: The stress-optic effect in optical fibres. IEEE J. Quantum Electron. **19**, 834–839 (1983)
118. Koplow, J.P., Kliner, D.A.V., Goldberg, L.: Single-mode operation of a coiled multimode fiber amplifier. Opt. Lett. **25**, 442–444 (2000)
119. Folkenberg, J.R., Nielsen, M.D., Mortensen, N.A., et al.: Polarization maintaining large mode area photonic crystal fiber. Opt. Express. **12**, 956–960 (2004)
120. Limpert, J., Liem, A., Reich, M., et al.: Low-nonlinearity single-transverse-mode ytterbium-doped photonic crystal fiber amplifier. Opt. Express. **12**, 1313–1319 (2004)
121. Schreiber, T., Schultz, H., Schmidt, O., et al.: Stress-induced birefringence in large-mode-area micro-structured optical fibers. Opt. Express. **13**, 3637–3646 (2005)
122. Schreiber, T., Röser, F., Schmidt, O., et al.: Stress-induced single-polarization single-transverse mode photonic crystal fiber with low nonlinearity. Opt. Express. **13**, 7621–7630 (2005)
123. Schmidt, O., Rothhardt, J., Eidam, T., et al.: Single-polarization ultra-large-mode-area Yb-doped photonic crystal fiber. Opt. Express. **16**, 3918–3923 (2008)
124. Ward, B.G.: Bend performance-enhanced photonic crystal fibers with anisotropic numerical aperture. Opt. Express. **16**, 8532–8548 (2008)
125. Golojuch, G., Urbańczyk, W.: Large mode area photonic crystal fibers with high geometrical birefringence. In: 16th Polish-Slovak-Czech Optical Conference on Wave and Quantum Aspects of Contemporary Optics, Polanica Zdroj, Poland, 08–12 Sep 2008
126. Kong, F.T., Gu, G.C., Hawkins, T.W., et al.: Polarizing ytterbium-doped all-solid photonic bandgap fiber with ~1150 μm^2 effective mode area. Opt. Express. **23**, 4307–4312 (2015)
127. Christensen, S.L., Papior, S.R., Johansen, M.M., et al.: Photonic crystal fiber technology for monolithic single-mode large mode area all-solid amplifier. In: Conference on Fiber Lasers XVI: Technology and Systems, San Francisco, CA, 04–07 Feb 2019
128. Kim, Y.G., Ryu, J.W., Cha, Y.H., et al.: Maintaining the polarization, temperature and amplification characteristics of an ytterbium-doped rod-type photonic crystal fiber (PCF) amplifier. J. Korean Phys. Soc. **59**, 3182–3187 (2011)
129. Peng, X., Dong, L.: Fundamental-mode operation in polarization-maintaining ytterbium-doped fiber with an effective area of 1400 μm^2. Opt. Lett. **32**, 358–360 (2007)
130. du Jeu, R., Dauliat, R., Leconte, B., et al.: Polarization-maintaining Yb-doped large-mode-area fully aperiodic large-pitch fibers. Appl. Opt. **57**, 8582–8585 (2018)
131. Fermann, M.E.: Single-mode excitation of multimode fibers with ultrashort pulses. Opt. Lett. **23**, 52–54 (1998)

132. Di Teodoro, F., Brooks, C.D.: Multi-MW peak power, single transverse mode operation of a 100micron core diameter, Yb-doped photonic crystal rod amplifier. In: 4th Conference on Fiber Laser: Technology, Systems, and Applications, San Jose, California, 22–25 Jan 2007
133. Napierala, M., Nasiłowski, T., Bere's-Pawlik, E., et al.: Extremely large-mode-area photonic crystal fiber with low bending loss. Opt. Express. **18**, 15408–15418 (2010)
134. Napierala, M., Nasiłowski, T., Bere's-Pawlik, E., et al.: Large-mode-area photonic crystal fiber with double lattice constant structure and low bending loss. Opt. Express. **19**, 22628–22636 (2011)
135. Wang, X., Lou, S.Q., Lu, W.L.: Bend-resistant large-mode-area photonic crystal fiber with a triangular-core. Appl. Opt. **52**, 4323–4328 (2013)
136. Qin, Y., Yang, H.J., Jiang, P., et al.: Design and analysis for a bend-resistant and large-mode-area photonic crystal fiber with hybrid cladding. Appl. Opt. **57**, 3976–3982 (2018)
137. Han, J.L., Liu, E.X., Liu, J.J.: Circular gradient-diameter photonic crystal fiber with large mode area and low bending loss. J. Opt. Soc. Am. A-Opt. Image Sci. Vis. **36**, 533–539 (2019)
138. Jain, D., Jung, Y., Nunez-Velazquez, M., et al.: Extending single mode performance of all-solid large-mode-area single trench fiber. Opt. Express. **22**, 31078–31091 (2014)
139. Xing, Z., Wang, X., Lou, S.Q., et al.: Bend-resistant side-leakage photonic crystal fiber with large-mode-area. J. Opt. **21**, 7 (2019)
140. Zhu, B., Taunay, T.F., Yan, M.F., et al.: Seven-core multicore fiber transmissions for passive optical network. Opt. Express. **18**, 11117–11122 (2010)
141. Michaille, L., Taylor, D.M., Bennett, C.R., et al.: Characteristics of a Q-switched multicore photonic crystal fiber laser with a very large mode field area. Opt. Lett. **33**, 71–73 (2008)
142. Zheng, S.W., Ren, G.B., Lin, Z., et al.: A novel four-air-hole multicore dual-mode large-mode-area fiber: proposal and design. Opt. Fiber Technol. **19**, 419–427 (2013)
143. Ji, J.H., Raghuraman, S., Huang, X.S., et al.: 115 W fiber laser with an all solid-structure and a large-mode-area multicore fiber. Opt. Lett. **43**, 3369–3372 (2018)
144. Jia, R.J., Lian, Y.D., Wang, Y.L., et al.: Design and analysis of trench-assisted low-bending-loss large-mode-field-area multi-core fiber with an air hole. Photonics. **8**, 11 (2021)
145. Klenke, A., Jauregui, C., Steinkopff, A., et al.: High-power multicore fiber laser systems. Prog Quantum Electron. **84**, 18 (2022)
146. Vogel, M.M., Abdou-Ahmed, M., Voss, A., et al.: Very-large-mode-area, single-mode multicore fiber. Opt. Lett. **34**, 2876–2878 (2009)
147. Baili, A., Cherif, R., Zghal, M.: New design of multicore nonlinear photonic crystal fiber for mid-IR supercontinuum generation. In: Conference on Micro-Structured and Specialty Optical Fibres III, Brussels, Belgium, 14–15 Apr 2014
148. Anashkina, E.A., Andrianov, A.V.: Design and dispersion control of microstructured multicore tellurite glass fibers with in-phase and out-of-phase supermodes. Photonics. **8**, 113 (2021)
149. Yi, C.S., Zhang, P.Q., Dai, S.X., et al.: Multicore chalcogenide photonic crystal fibers for large mode area and mode shaping. Opt. Commun. **311**, 270–274 (2014)
150. Wang, X.G., Jiao, K., Si, N., et al.: Extruded seven-core tellurium chalcogenide fiber for mid-infrared. Opt Mater Express. **9**, 3863–3870 (2019)
151. Xiao, J., Wang, J.J., Xu, T.S., et al.: Large mode-area chalcogenide multicore fiber prepared by continuous two-stage extrusion. Opt Mater Express. **11**, 791–800 (2021)
152. Wang, X.G., Jiao, K., Liang, X.L., et al.: Mid-infrared seven-core chalcogenide fiber with ultra-large mode field area and high beam quality. J. Phys. D-Appl. Phys. **57**, 225102 (2024)
153. Li, N.Z., Jia, J.S., Zhang, M., et al.: Mid-infrared large mode area single-mode all-solid-state photonic crystal fiber based on chalcogenide glass. Chin. J. Laser. **51**, 1706006 (2024)
154. Chiang, K.S., Rastogi, V.: Ultra-large-core single-mode fiber for optical communications: the segmented cladding fiber. In: Optical Fiber Communication Conference and Exhibit, Anaheim, California, 17–22 Mar 2002
155. Yeung, A., Chiang, K.S., Rastogi, V., et al.: Experimental demonstration of single-mode operation of large-core segmented cladding fiber. In: Optical Fiber Communication Conference, Los Angeles, California, 22 Apr 2004

156. Rastogi, V., Chiang, K.S.: Analysis of segmented-cladding fiber by the radial-effective-index method. J. Opt. Soc. Am. B-Opt. Phys. **21**, 258–265 (2004)
157. Kumar, A., Rastogi, V., Agrawal, A., et al.: Birefringence analysis of segmented cladding fiber. Appl. Opt. **51**, 3104–3108 (2012)
158. Millo, A., Naeh, I., Katzir, A.: Single-mode segmented cladding fibers for the middle infrared. J. Lightwave Technol. **25**, 2115–2121 (2007)
159. Ma, S.S., Ning, T.G., Li, J., et al.: Design and analysis of a modified segmented cladding fiber with large mode area. Opt. Laser Technol. **88**, 172–179 (2017)
160. Pournoury, M., Lee, Y.S., Kim, D., et al.: Granulated silica segmented cladding fiber for optical communication. In: Alexander Wai, P.T.H., Yu, C. (eds.) 26th Opto-electronics and Communications Conference, Hong Kong, 03–07 July 2021
161. Pournoury, M., Kim, D.: Bend-resistant octo-wing silica segmented cladding fiber with high index rings. Results Phys. **36**, 105423 (2022)
162. Pournoury, M., Zamiri, A., Ghasemi, M., et al.: The optimization of a segmented cladding fiber via the response surface methodology approach for a large mode area. Photonics. **10**, 935 (2023)
163. Ghosh, S., Dasgupta, S., Poletti, F., et al.: Soft glass based large mode area photonic bandgap fibre for mid-infrared applications. In: Conference on Optical Fiber Communication (OFC)/National Fiber Optic Engineers Conference (NFOEC), Los Angeles, California, 6–10 Mar 2011
164. Tang, G.W., Zhu, T.T., Lin, W., et al.: Single-mode large mode-field-area Tm^{3+}-doped lead-silicate glass photonic crystal fibers. IEEE Photon. Technol. Lett. **29**, 450–453 (2017)
165. Li, M., Wang, L.F., Han, S., et al.: Large-mode-area neodymium-doped all-solid double-cladding silicate photonic bandgap fiber with a 32 μm core diameter. Opt. Mater. Express. **8**, 1562–1568 (2018)
166. Li, L., Schülzgen, A., Temyanko, V.L., et al.: Short-length microstructured phosphate glass fiber lasers with large mode areas. Opt. Lett. **30**, 1141–1143 (2005)
167. Feng, X., Loh, W.H., Flanagan, J.C., et al.: Single-mode tellurite glass holey fiber with extremely large mode area for infrared nonlinear applications. Opt. Express. **16**, 13651–13656 (2008)
168. Shi, H.X., Feng, X., Tan, F.Z., et al.: Multi-watt mid-infrared supercontinuum generated from a dehydrated large-core tellurite glass fiber. Opt Mater Express. **6**, 3967–3976 (2016)
169. Ebendorff-Heidepriem, H., Foo, T.C., Moore, R.C., et al.: Fluoride glass microstructured optical fiber with large mode area and mid-infrared transmission. Opt. Lett. **33**, 2861–2863 (2008)
170. Yuan, X.Y., Zhou, X.B., Qin, Z.P., et al.: Nearly diffraction-limited picosecond pulse amplification from LMA fluoride fiber at 2.8 μm. Opt. Express. **32**, 1836–1842 (2024)
171. Ebendorff-Heidepriem, H., Monroe, T.M.: Extrusion of complex preforms for microstructured optical fibers. Opt. Express. **15**, 15086–15092 (2007)
172. Conseil, C., Coulombier, Q., Boussard-Plédel, C., et al.: Chalcogenide step index and microstructured single mode fibers. J. Non-Cryst. Solids. **357**, 2480–2483 (2011)
173. Ren, H., Qi, S.S., Hu, Y.S., et al.: All-solid mid-infrared chalcogenide photonic crystal fiber with ultralarge mode area. Opt. Lett. **44**, 5553–5556 (2019)
174. Yi, C.S., Dai, S.X., Zhang, P.Q., et al.: Design of a novel single-mode large mode area infrared chalcogenide glass photonic crystal fibers. Acta Phys. Sin. **62**, 7 (2013)
175. Feng, X., Ren, H., Xu, F., et al.: Few-moded ultralarge mode area chalcogenide photonic crystal fiber for mid-infrared high power applications. Opt. Express. **28**, 16658–16672 (2020)
176. Chenard, F., Alvarez, O., Shartner, E., et al.: Mid-infrared chalcogenide polarization-maintaining single-mode fiber. In: Conference on Optical Components and Materials XIX, San Francisco, California, 22–27 Jan 2022
177. Peng, Q.Q., Sheng, J.K., Yang, K.Y., et al.: Single-mode segmented cladding chalcogenide glass fiber with ultra-large mode area. J. Lightwave Technol. **41**, 5722–5728 (2023)
178. Dong, L.: Stimulated thermal Rayleigh scattering in optical fibers. Opt. Express. **21**, 2642–2656 (2013)

179. Smith, A.V., Smith, J.J.: Overview of a steady-periodic model of modal instability in fiber amplifiers. IEEE. J. Sel. Top. Quantum. Electron. **20**, 12 (2014)
180. Smith, A.V., Smith, J.J.: Mode instability in high power fiber amplifiers. Opt. Express. **19**, 10180–10192 (2011)
181. Ward, B., Robin, C., Dajani, I.: Origin of thermal modal instabilities in large mode area fiber amplifiers. Opt. Express. **20**, 11407–11422 (2012)
182. Hansen, K.R., Alkeskjold, T.T., Broeng, J., et al.: Theoretical analysis of mode instability in high-power fiber amplifiers. Opt. Express. **21**, 1944–1971 (2013)
183. Zervas, M.N.: Transverse mode instability analysis in fibre amplifiers. In: Conference on Fiber Lasers XIV-Technology and Systems, San Francisco, California, 30 Jan-02 Feb 2017
184. Scarnera, V., Ghiringhelli, F., Malinowski, A., et al.: Modal instabilities in high power fiber laser oscillators. Opt. Express. **27**, 4386–4403 (2019)
185. Zervas, M.N.: Transverse-modal-instability gain in high power fiber amplifiers: effect of the perturbation relative phase. APL Phontonics. **4**, 022802 (2019)
186. Jansen, F., Stutzki, F., Otto, H.J., et al.: Thermally induced waveguide changes in active fibers. Opt. Express. **20**, 3997–4008 (2012)
187. Coscelli, E., Dauliat, R., Poli, F., et al.: Analysis of the modal content into large-mode-area photonic crystal fibers under heat load. IEEE. J. Sel. Top. Quantum. Electron. **22**, 323–330 (2016)
188. Distler, V., Möller, F., Strecker, M., et al.: Transverse mode instability in a passive fiber induced by stimulated Raman scattering. Opt. Express. **28**, 22819–22828 (2020)
189. Jauregui, C., Stihler, C., Limpert, J.: Transverse mode instability. Adv. Opt. Photonics. **12**, 429–484 (2020)
190. Huang, L., Tao, R.M., Shi, C., et al.: Towards the enhancement of the TMI threshold in monolithic high-power fiber system by controlling the pump distribution and the seed power. IEEE Photonics J. **10**, 1–12 (2018)
191. Jauregui, C., Stihler, C., Limpert, J., et al.: Transverse mode instabilities in burst operation of high-power fiber laser systems. In: Conference on Fiber Lasers XV-Technology and Systems, San Francisco, California, 29 Jan-01 Feb 2018
192. Jauregui, C., Stihler, C., Tünnermann, A., et al.: Pump-modulation-induced beam stabilization in high-power fiber laser systems above the mode instability threshold. Opt. Express. **26**, 10691–10704 (2018)
193. Jansen, F., Stutzki, F., Otto, H.J., et al.: The influence of index-depressions in core-pumped Yb-doped large pitch fibers. Opt. Express. **18**, 26834–26842 (2010)
194. Hansen, K.R., Alkeskjold, T.T., Broeng, J., et al.: Thermo-optical effects in high-power ytterbium-doped fiber amplifiers. Opt. Express. **19**, 23965–23980 (2011)
195. Fini, J.M.: Large mode area fibers with asymmetric bend compensation. Opt. Express. **19**, 21866–21873 (2011)
196. Mattsson, K.E.: Low photo darkening single mode RMO fiber. Opt. Express. **17**, 17855–17861 (2009)
197. Coscelli, E., Poli, F., Naife, R.W., et al.: Comparison of thermally-induced single-mode regime changes in Yb-doped large mode area photonic crystal fibers. In: Conference on Micro-Structured and Specialty Optical Fibres II, Prague, Czech Republic, 15–17 Apr 2013
198. Dauliat, R., Gaponov, D., Benoit, A., et al.: Inner cladding microstructuration based on symmetry reduction for improvement of singlemode robustness in VLMA fiber. Opt. Express. **21**, 18927–18936 (2013)
199. Coscelli, E., Molardi, C., Masruri, M., et al.: Thermally resilient tm-doped large mode area photonic crystal fiber with symmetry-free cladding. Opt. Express. **22**, 9707–9714 (2014)
200. Benoit, A., Dauliat, R., Jamier, R., et al.: Highly efficient higher-order modes filtering into aperiodic very large mode area fibers for single-mode propagation. Opt. Lett. **39**, 4561–4564 (2014)
201. Limpert, J., Roser, F., Schreiber, T., et al.: High-power ultrafast fiber laser systems. IEEE. J. Sel. Top. Quantum. Electron. **12**, 233–244 (2006)

202. Broderick, N.G.R., Offerhaus, H.L., Richardson, D.J., et al.: Large mode area fibers for high power applications. Opt. Fiber Technol. **5**, 185–196 (1999)
203. Ramachandran, S., Fini, J.M., Mermelstein, M., et al.: Ultra-large effective-area, higher-order mode fibers: a new strategy for high-power lasers. Laser. Photon. Rev. **2**, 429–448 (2008)
204. Wielandy, S.: Implications of higher-order mode content in large mode area fibers with good beam quality. Opt. Express. **15**, 15402–15409 (2007)
205. Wu, T.W., Dong, L., Winful, H.: Bend performance of leakage channel fibers. Opt. Express. **16**, 4278–4285 (2008)
206. Ma, Y., Wan, R., Yang, H.H., et al.: Stress-induced polarization-maintaining large-mode-area photonic crystal fibers with deviation of the single-mode transmission band and delocalization of higher-order modes. IEEE. Photonics J. **16**, 1–11 (2024)
207. Ferrando, A., Silvestre, E., Miret, J.J., et al.: Full-vector analysis of a realistic photonic crystal fiber. Opt. Lett. **24**, 276–278 (1999)
208. Koshiba, M., Saitoh, K.: Applicability of classical optical fiber theories to holey fibers. Opt. Lett. **29**, 1739–1741 (2004)
209. Birkst, T.A., Mogilevtsev, D., Knight, J.C., et al.: The analogy between photonic crystal fibres and step index fibres. In: Optical Fiber Communication Conference and the International Conference on Integrated Optics, San Diego, California, 21–26 Feb 1999
210. Brechet, F., Marcou, J., Pagnoux, D., et al.: Complete analysis of the characteristics of propagation into photonic crystal fibers, by the finite element method. Opt. Fiber Technol. **6**, 181–191 (2000)
211. Saitoh, K., Tsuchida, Y., Koshiba, M., et al.: Endlessly single-mode holey fibers: the influence of core design. Opt. Express. **13**, 10833–10839 (2005)
212. Kuhlmey, B.T., McPhedran, R.C., de Sterke, C.M.: Modal cutoff in microstructured optical. Opt. Lett. **27**, 1684–1686 (2002)
213. Kuhlmey, B.T., McPhedran, R.C., de Sterke, C.M., et al.: Microstructured optical fibers: where's the edge? Opt. Express. **10**, 1285–1290 (2002)
214. Mortensen, N.A., Folkenberg, J.R., Nielsen, M.D., et al.: Modal cutoff and the V parameter in photonic crystal fibers. Opt. Lett. **28**, 1879–1881 (2003)
215. Selleri, S., Cucinotta, A., Foroni, M., et al.: In: Selleri, S. (ed.) New Design of Single-Mode Large-Mode-Area Photonic Crystal Fibers. SPIE, Warsaw (2005)
216. Poli, F., Foroni, M., Bottacini, M., et al.: Single-mode regime of square-lattice photonic crystal fibers. J. Opt. Soc. Am. A. **22**, 1655–1661 (2005)
217. Jorgensen, M.M., Petersen, S.R., Laurila, M., et al.: Optimizing single mode robustness of the distributed modal filtering rod fiber amplifier. Opt. Express. **20**, 7263–7273 (2012)
218. Poli, F., Coscelli, E., Alkeskjold, T.T., et al.: Cut-off analysis of 19-cell Yb-doped double-cladding rod-type photonic crystal fibers. Opt. Express. **19**, 9896–9907 (2011)
219. Chen, G., Jiang, Z.W., Peng, J.G., et al.: Study of air-clad large-mode-area ytterbium doped photonic crystal fiber. Acta. Phys. Sin. **61**, 6 (2012)
220. Pan, R., Wang, D.P., Ning, D., et al.: Research and fabrication of large mode area fiber. Laser Infrared. **40**, 632–634 (2010)
221. Tammela, S., Söderlund, M., Koponen, J., et al.: The potential of direct nanoparticle deposition for the next generation of optical fibers. In: Conference on Optical Components and Materials III, San Jose, California, 23–25 Jan 2006
222. Langner, A., Schötz, G., Such, M., et al.: A new material for high power laser fibers. In: Conference on Fiber Lasers V, San Jose, California, 21–24 Jan 2008
223. Liu, J.T., Zhou, G.Y., Xia, C.M.: Fabrication of Yb^{3+}/Al^{3+} co-doped large-mode-area photonic crystal fiber based on powder sintering technology. Acta. Photonica. Sinica. **42**, 552–554 (2013)
224. Wei, H.F., Chen, K.K., Yang, Y.C., et al.: All-solid very large mode area ytterbium-doped silica microstructured fiber based on accurate control on cladding index. Opt. Express. **24**, 8978–8987 (2016)

225. Chen, Y., Zhao, N., Liu, J.T., et al.: Yb^{3+}-doped large-mode-area photonic crystal fiber for fiber lasers prepared by laser sintering technology. Opt Mater Express. **9**, 1356–1364 (2019)
226. Chu, Y.B., Liu, Y.G., Liao, L., et al.: Yb^{3+} doped gain guided index anti-guided fiber based on the nanoporous silica glass rod. Nanotechnology. **31**, 6 (2020)
227. Pedrazza, U., Romano, V., Lüthy, W.: Yb^{3+}: Al^{3+}: sol-gel silica glass fiber laser. Opt. Mater. **29**, 905–907 (2007)
228. Wang, F., Hu, L.L., Xu, W.B., et al.: Manipulating refractive index, homogeneity and spectroscopy of Yb^{3+}-doped silica-core glass towards high-power large mode area photonic crystal fiber lasers. Opt. Express. **25**, 25960–25969 (2017)
229. Roy, P., Jamier, R., Malleville, M.A., et al.: High peak and average power delivery from fully-aperiodic large-pitch-fiber lasers (invited). In: IEEE 8th International Conference on Advanced Optoelectronics and Lasers (CAOL), Sozopol, Bulgaria, 06–08 Sep 2019
230. Langner, A., Bauera, P., Schönfelda, D., et al.: Trends in specialty fiber optic materials characterization and applications. In: Conference on Optical Fibers and Sensors for Medical Diagnostics, Treatment and Environmental Applications XXI, Electr Network, 06–11 Mar 2021
231. Jiang, X., Babic, F., Joly, N.Y., et al.: Soft-glass photonic crystal fibres: from advanced fabrication techniques to novel applications. In: 2017 Conference on Lasers and Electro-Optics Pacific Rim, 31 July-4 Aug. 2017 (2014)
232. Ebendorff-Heidepriem, H., Warren-Smith, S.C., Monro, T.M.: Suspended nanowires: fabrication, design and characterization of fibers with nanoscale cores. Opt. Express. **17**, 2646–2657 (2009)
233. Müller, M., Klenke, A., Steinkopff, A., et al.: 3.5 kW coherently combined ultrafast fiber laser. Opt. Lett. **43**, 6037–6040 (2018)
234. Klenke, A., Hädrich, S., Eidam, T., et al.: 22 GW peak-power fiber chirped-pulse-amplification system. Opt. Lett. **39**, 6875–6878 (2014)
235. Wadsworth, W.J., Knight, J.C., Reeves, W.H., et al.: Yb^{3+}-doped photonic crystal fiber laser. Electron. Lett. **36**, 1452–1454 (2000)
236. Röser, F., Jauregui, C., Limpert, J., et al.: 94 W 980 nm high brightness Yb-doped fiber laser. Opt. Express. **16**, 17310–17318 (2008)
237. Li, W.S., Matniyaz, T., Gafsi, S., et al.: 151 W monolithic diffraction-limited Yb-doped photonic bandgap fiber laser at ~978 nm. Opt. Express. **27**, 24972–24977 (2019)
238. Kawamura, T., Shirakawa, A., Saito, K.: Phase-locked and mode-locked multicore photonic crystal fiber laser with a saturable absorber. Opt. Express. **29**, 17023–17028 (2021)
239. Stark, H., Buldt, J., Müller, M., et al.: 23 mJ high-power fiber CPA system using electro-optically controlled divided-pulse amplification. Opt. Lett. **44**, 5529–5532 (2019)
240. Müller, M., Kienel, M., Klenke, A., et al.: 1 kW 1 mJ eight-channel ultrafast fiber laser. Opt. Lett. **41**, 3439–3442 (2016)
241. Stark, H., Buldt, J., Müller, M., et al.: 1 kW, 10 mJ, 120 fs coherently combined fiber CPA laser system. Opt. Lett. **46**, 969–972 (2021)
242. Gaida, C., Gebhardt, M., Heuermann, T., et al.: Ultrafast thulium fiber laser system emitting more than 1 kW of average power. Opt. Lett. **43**, 5853–5856 (2018)
243. Stutzki, F., Gaida, C., Gebhardt, M., et al.: Tm-based fiber-laser system with more than 200 MW peak power. Opt. Lett. **40**, 9–12 (2015)
244. Haig, H., Bender, N., Eisenberg, Y., et al.: Single-mode regenerative amplification in multimode fiber. Optica. **10**, 1417–1420 (2023)
245. Brooks, C.D., di Teodoro, F.: Multimegawatt peak-power, single-transverse-mode operation of a 100 μm core diameter, Yb-doped rodlike photonic crystal fiber amplifier. Appl. Phys. Lett. **89**, 111119 (2006)
246. Di Teodoro, F., Brooks, C.D.: Very large-core, single-mode Yb-doped photonic-crystal rod for multi-MW peak power generation. In: Conference on Fiber Lasers III, San Jose, California, 23–26 Jan 2006

247. Schmidt, O., Rothhardt, J., Röser, F., et al.: Millijoule pulse energy Q-switched short-length fiber laser. Opt. Lett. **32**, 1551–1553 (2007)
248. Zhao, Z.G., Kobayashi, Y.: Ytterbium fiber-based, 270 fs, 100 W chirped pulse amplification laser system with 1 MHz repetition rate. Appl. Phys. Express. **9**, 012701 (2016)
249. Yang, P.L., Hao, Z.Q., Fang, S.B., et al.: Highly stable Yb-fiber laser amplifier of delivering 32-μJ, 153-fs pulses at 1-MHz repetition rate. Appl. Phys. B Lasers Opt. **124**, 169 (2018)
250. Manchee, C.P.K., Möller, J., Miller, R.J.D.: Highly stable, 100 W average power from fiber-based ultrafast laser system at 1030 nm based on single-pass photonic-crystal rod amplifier. Opt. Commun. **437**, 6–10 (2019)
251. Bai, Z.N., Bai, Z.X., Sun, X.L., et al.: A 33.2 W high beam quality chirped-pulse amplification-based femtosecond laser for industrial processing. Materials. **13**, 2841 (2020)
252. Pulford, B., Holten, R., Matniyaz, T., et al.: kW-level monolithic single-mode narrow-linewidth all-solid photonic bandgap fiber amplifier. Opt. Lett. **46**, 4458–4461 (2021)
253. Zhao, Q.K., Gao, G.G., Cong, Z.H., et al.: High-repetition-rate, 50-μJ-level, 1064-nm, CPA laser system based on a single-stage double-pass Yb-doped rod-type fiber amplifier. Opt. Express. **30**, 3611–3619 (2022)
254. Pedersen, M.E.V., Johansen, M.M., Olesen, A.S., et al.: 175 W average power from a single-core rod fiber-based chirped-pulse-amplification system. Opt. Lett. **47**, 5172–5175 (2022)
255. Wang, C.W., Xiao, H., Xi, X.M., et al.: 10 kW tandem pumping fiber amplifier with good beam quality based on tapered ytterbium-doped fiber. Opt. Express. **31**, 40980–40990 (2023)
256. Shollenbarger, C., Tafoya, J., Schulz, B., et al.: A monolithic linearly polarized counter-pumped 85 μm core rod fiber amplifier architecture for high-energy pulsed applications. In: Conference on Fiber Lasers XXI-Technology and Systems, San Francisco, California, 29 Jan-01 Feb 2024
257. Leconte, B., Cadier, B., Gilles, H., et al.: Extended tunability of Nd-doped fiber lasers operating at 872-936 nm. Opt. Lett. **40**, 4098–4101 (2015)
258. Bufetov, I.A., Dudin, V.V., Shubin, A.V., et al.: Efficient 0.9-μm neodymium-doped single-mode fibre laser. Quantum. Electron. **33**, 1035–1037 (2003)
259. Bartolacci, C., Laroche, M., Gilles, H., et al.: Generation of picosecond blue light pulses at 464 nm by frequency doubling an Nd-doped fiber based master oscillator power amplifier. Opt. Express. **18**, 5100–5105 (2010)
260. Rota-Rodrigo, S., Gouhier, B., Laroche, M., et al.: Watt-level single-frequency tunable neodymium MOPA fiber laser operating at 915-937 nm. Opt. Lett. **42**, 4557–4560 (2017)
261. Laroche, M., Cadier, B., Gilles, H., et al.: 20 W continuous-wave cladding-pumped Nd-doped fiber laser at 910 nm. Opt. Lett. **38**, 3065–3067 (2013)
262. Pax, P.H., Khitrov, V.V., Drachenberg, D.R., et al.: Scalable waveguide design for three-level operation in neodymium doped fiber laser. Opt. Express. **24**, 28633–28647 (2016)
263. Chen, Y.G., Lin, Z.Q., Sun, H.T., et al.: High-power lasing at ~900 nm in Nd^{3+}-doped fiber: a direct coordination engineering approach to enhance fluorescence. Optica. **10**, 905–912 (2023)
264. Le Corre, K., Gilles, H., Girard, S., et al.: In: Le Corre, K. (ed.) 83 W efficient Nd-doped LMA fiber laser at 910 nm. SPIE, San Francisco (2022)
265. Le Corre, K., Barnini, A., Robin, T., et al.: Watt-level deep-UV subnanosecond laser system based on Nd-doped fiber at 229 nm. Opt. Lett. **48**, 1276–1279 (2023)
266. Florentin, R., Le Corre, K., Robin, T., et al.: Optimization of Nd-doped LMA fibers for high-power laser emission near 915 nm. IEEE Photonics J. **16**, 1–6 (2024)
267. Lee, W.K., Geng, J.H., Jiang, S.B., et al.: 1.8 mJ, 3.5 kW single-frequency optical pulses at 1572 nm generated from an all-fiber MOPA system. Opt. Lett. **43**, 2264–2267 (2018)
268. Jackson, S.D.: Cross relaxation and energy transfer upconversion processes relevant to the functioning of 2 μm Tm^{3+}-doped silica fibre lasers. Opt. Commun. **230**, 197–203 (2004)
269. Sincore, A., Bradford, J.D., Cook, J., et al.: High average power thulium-doped silica fiber lasers: review of systems and concepts. IEEE J. Sel. Top. Quantum. Electron. **24**, 1–8 (2018)
270. Jollivet, C., Farley, K., Conroy, M., et al.: Design optimization of Tm-doped large-mode area fibers for power scaling of 2 μm lasers and amplifiers. In: Conference on Fiber Lasers XIV—Technology and Systems, San Francisco, California, 30 Jan-02 Feb 2017

271. Simakov, N., Hemming, A.V., Carter, A., et al.: Design and experimental demonstration of a large pedestal thulium-doped fibre. Opt. Express. **23**, 3126–3133 (2015)
272. Smith, A.V., Smith, J.J.: Mode instability thresholds for tm-doped fiber amplifiers pumped at 790 nm. Opt. Express. **24**, 975–992 (2016)
273. Fibich, G., Gaeta, A.L.: Critical power for self-focusing in bulk media and in hollow waveguides. Opt. Lett. **25**, 335–337 (2000)
274. Gaida, C., Gebhardt, M., Stutzki, F., et al.: 90 fs pulses with >5 GW peak power from a high repetition rate Tm-doped fiber CPA system. In: Laser Congress 2017 (ASSL, LAC), Nagoya, Aichi, 1–5 October 2017
275. Jackson, S.D., King, T.A.: High-power diode-cladding-pumped Tm-doped silica fiber laser. Opt. Lett. **23**, 1462–1464 (1998)
276. Moulton, P.F., Rines, G.A., Slobodtchikov, E.V., et al.: Tm-doped fiber lasers: fundamentals and power scaling. IEEE J. Sel. Top. Quantum. Electron. **15**, 85–92 (2009)
277. Kokki, T., Koponen, J., Laurila, M., et al.: In: Kokki, T. (ed.) Fiber Amplifier Utilizing an Yb-Doped Large-Mode-Area Fiber with Confined Doping and Tailored Refractive Index Profile. SPIE, San Francisco (2010)
278. Anderson, B.M., Solomon, J., Flores, A.: 1.1 kW, beam-combinable thulium doped all-fiber amplifier. In: Conference on Fiber Lasers XVIII—Technology and Systems, Electr Network, 06–11 Mar 2021
279. Lenski, M., Heuermann, T., Gebhardt, M., et al.: Inband-pumped, high-power thulium-doped fiber amplifiers for an ultrafast pulsed operation. Opt. Express. **30**, 44270–44282 (2022)
280. Ren, C.Y., Shen, Y.Q., Zheng, Y.H., et al.: Widely-tunable all-fiber tm doped MOPA with >1 kW of output power. Opt. Express. **31**, 22733–22739 (2023)
281. Romano, C., Panitzek, D., Lorenz, D., et al.: 937 W thulium: silica fiber MOPA operating at 2036 nm. EPJ. Web. Conf. **267**, 02016 (2022)
282. Gausmann, S., Faugas, B., Bradford, J., et al.: In: Gausmann, S. (ed.) High Concentration Large-Mode-Area Tm-Doped Double-Clad Fiber for High Efficiency Operation. SPIE, San Francisco (2023)
283. Michalska, M., Honzatko, P., Grzes, P., et al.: Thulium-doped 1940-and 2034-nm fiber amplifiers: towards highly efficient, high-power all-fiber laser systems. J. Lightwave Technol. **42**, 339–346 (2024)

Chapter 16
High-Power Narrow Linewidth Linearly Polarized Fiber Laser

Shuai Ren, Yisha Chen, Wei Liu, Pengfei Ma, and Pu Zhou

Abstract High-power narrow linewidth (HPNL) linearly polarized all-fiberized lasers have been widely used in industrial, medical, scientific research, and national security fields. Enhancing the brightness of such fiber lasers has always been a focus of researchers. Firstly, the physical factors that limit the power increase and beam quality maintenance of HPNL fiber amplifiers are analyzed, and corresponding mitigation measures are discussed. Then, the research progress of HPNL linearly polarized fiber lasers is reviewed, and some representative achievements are introduced in detail. Finally, the future development trends are discussed and outlooked.

Keywords High power · Ytterbium-doped fiber laser · Linearly polarized · Narrow linewidth · SBS · SRS · TMI

16.1 Introduction

Due to the advantages of compact structure, convenient thermal management and high electro-optic conversion efficiency, high-power narrow linewidth (HPNL) all-fiberized lasers with near-diffraction-limited beam quality have been widely used in industrial, medical, scientific research, and national security fields. The power growth of such lasers has always been one of the main goals pursued by researchers. With the invention of double-clad fibers and the enhancement of pumping

S. Ren
School of Information and Communications, National University of Defense Technology, Wuhan, Hubei, China

Y. Chen · W. Liu · P. Ma · P. Zhou (✉)
College of Advanced Interdisciplinary Studies, National University of Defense Technology, Changsha, Hunan, China

M. F. S. Ferreira, M. C. Paul (eds.), *Optical Fiber Lasers*, Springer Series in Optical Sciences 250, https://doi.org/10.1007/978-3-032-04110-4_16

capabilities [1, 2], fiber amplifiers using ytterbium-doped fibers (YDF) as gain medium have achieved single-mode fiber laser output with a power of 10 kW [3]. Within about 20 years, the output power of a single laser beam has increased by three orders of magnitude.

Meanwhile, based on multipath HPNL fiber lasers, beam combination has achieved significant results in recent years, with both coherent beam combining (CBC) and spectral beam combining (SBC) achieving tens of kilowatts of power output while maintaining excellent beam quality [4–6]. In addition, HPNL linearly polarized fiber lasers have been widely used in nonlinear frequency conversion due to their fine temporal coherence [7, 8]. For example, researchers have achieved kilowatt-level single-mode green laser output using narrow linewidth linearly polarized fiber lasers, which is currently the highest technical specification for single-mode green lasers [9, 10]. However, as the laser source for the aforementioned applications, limited by nonlinear effects (NLEs) such as stimulated Brillouin scattering (SBS), stimulated Raman scattering (SRS), and thermally induced mode instability (TMI) [11–13], the increase of output power without degradation of beam quality of HPNL fiber lasers has encountered a bottleneck period [14]. The limited brightness of single-beam HPNL fiber lasers has become one of the main factors affecting the power improvement of beam combination systems and nonlinear frequency conversion efficiency.

Researchers have conducted scientific research on HPNL fiber lasers in physical mechanisms, key technologies, system integration, and applications. Various techniques have been proposed to mitigate NLE and TMI in HPNL fiber amplifiers, which will be detailed in Sects. 16.2 and 16.3. Up to now, in the field of randomly polarized HPNL single-mode fiber lasers, output powers of 4 kW level (2018), 5 kW level (2020), 6 kW level (2021), and 7 kW level (2022) has been achieved successively [15–18]. However, the output power of polarization-maintained (PM) fiber amplifiers with narrow linewidth remains at 5 kW level [19, 20], which is much lower than that of randomly polarized ones. Research has shown that the SBS gain and SRS gain in PM fibers are usually higher than those in non-PM fibers [21, 22], and the TMI threshold in PM fiber lasers is significantly lower than that in non-PM ones [23]. Therefore, power scaling of HPNL linearly polarized fiber lasers faces more challenges.

This chapter focuses on traditional HPNL linearly polarized single-mode fiber lasers. Firstly, the classic limiting factors and mitigation methods for laser power enhancement in amplification structures are systematically reviewed. Then, the progress and representative achievements of PM narrow linewidth fiber lasers of conventional wavelength and special wavelengths are introduced in detail, respectively. Finally, combining the latest research results in theoretical analysis, device development, and experimental data, the future development trends are discussed and outlooked.

16.2 Classic Limitations and Solutions

At present, the most frequent way to obtain HPNL fiber lasers is through the main oscillator power amplifier (MOPA) configuration. The traditional NLEs, such as SBS and SRS in amplifiers, as well as the recently identified TMI effect, are important factors that limit the power increase of narrow linewidth fiber lasers while maintaining excellent beam quality [24]. Moreover, traditional methods for suppressing NLEs and TMI effects are mutually constrained, which further increases the difficulty of research. In 2016, Hans-Jürgen Otto et al. from the University Jena comprehensively considered the effects of TMI effect, pump brightness, SRS effect, and thermal lens effect, and studied the average power output that laser diode (LD) pumped fiber lasers can achieve at different core diameters and fiber lengths. They finally pointed out that disregarding possible practical constraints, the power limit of single-mode fiber lasers can theoretically reach 70 kW [25]. Currently, the maximum output power of broadband spectral near-single-mode fiber lasers remains at 10 kW [3], while the narrow linewidth fiber lasers are on the order of 7 kW [18]. To realize the goal of the power scaling of narrow linewidth fiber lasers, it is necessary to adopt technical means to comprehensively suppress the NLEs and TMI effect. This section will briefly illustrate the mechanisms and suppressing methods of SBS, SRS, and TMI.

16.2.1 SBS Effect and Suppressing Methods

The SBS effect arises from the interaction of light with acoustic waves propagating in the fibers. The following will briefly introduce its basic physical process. When Brillouin scattering occurs, the photon of the signal light is converted into a scattered photon and a phonon, and the scattered light has a certain frequency shift, which is called Brillouin Stokes light. The backward Brillouin Stokes light interferes with the signal light, generating acoustic waves through the electrostriction effect. As the intensity of the Brillouin Stokes light increases, the interference becomes stronger and the amplitude of the acoustic waves increases. At this point, the forward propagating acoustic waves resemble a Bragg grating and scatter more light backward [26]. The geometric structure of fibers determines that Brillouin Stokes light mainly propagates in the opposite direction to the signal light. When the signal light power reaches a certain level, the remarkable feature exhibited by the SBS effect is the effective energy conversion from signal light to Brillouin Stokes light. And it is characterized by a sharp power threshold above which energy starts being transferred from the signal to a frequency-downshifted beam propagating in the opposite direction to the signal. In high-power fiber laser systems, Brillouin Stokes light typically manifests as high peak power pulses, which can cause serious physical damage to fiber or fiber-based devices. Obviously, for fiber lasers, the SBS effect limits the average or peak power of the system.

Researchers have proposed various methods to alleviate the SBS effect, which can be summarized into two main aspects. One is to optimize the fiber parameters [27–29] to reduce the gain coefficient of the SBS effect. The most common methods include shortening the fiber length [30, 31], increasing the fiber core diameter [32, 33], and designing gain-tailoring fibers [34–39]. The other is to improve the parameters of the fiber amplifier system, mainly including broadening the seed laser linewidth [40–52], optimizing the pumping method [53–55], introducing gain competition [40, 56, 57], and applying stress or temperature gradients [35, 53, 58–61]. Among them, using phase modulation technology for linewidth broadening of single-frequency lasers is currently the preferred solution for narrow linewidth fiber amplifiers of several kilowatts [15, 17, 62–71] due to its advantages, such as spectral purity and stable timing [72–74].

It should be noted that research has shown that the SBS gain coefficient is related to the polarization state [21, 22, 27, 75, 76]. As early as 1979, Stolen from Bell Laboratories studied the relationship between SBS gain coefficient and polarization and experimentally verified it using a birefringent PM single-mode fiber. Both theoretical analysis and experimental results indicated that the SBS gain under linear polarization was twice that under random polarization [22]. Therefore, compared with non-PM narrow linewidth fiber lasers, it is more difficult to suppress the SBS effect and increase the power of linearly polarized narrow linewidth fiber lasers.

16.2.2 SRS Effect and Suppressing Methods

The formation of SRS in the fibers is similar to SBS. When Raman scattering occurs, the signal light is scattered to generate Raman Stokes scattering light with a certain frequency shift. The SRS effect also exhibits obvious power threshold features, that is, only when the input signal light power is increased to a certain level will significant SRS be excited [77, 78]. Unlike the scattering light excited by SBS, which mainly appears in the backward propagation direction, the scattering light excited by SRS propagates forward and backward along the fiber simultaneously. Therefore, when the SRS effect is triggered in the system, on the one hand, some of the energy of the signal light will be transferred to longer wavelength Raman scattering light, reducing the efficiency of the laser. On the other hand, the Raman scattering light propagated backward can cause physical damage to the fibers and components in the system, so the SRS effect also limits the power scaling of fiber lasers.

Regarding the mitigation of the SRS effect, currently reported research mainly includes the following two ways. Firstly, from the perspective of fiber design and amplifier structure optimization, certain technical means are used to reduce the

Raman Stokes optical gain, such as increasing the effective mode area of the fiber [79, 80], increasing the Raman Stokes optical loss [81, 82], reducing the overlap factor between signal light and Raman Stokes light [83–85], and optimizing the pump configuration of fiber lasers [86, 87]. The second is to effectively suppress the SRS effect by filtering out the Raman Stokes light in the seed laser [88, 89].

The SRS gain is also closely related to the laser polarization [22, 27, 90, 91]. Specifically, the SRS gain coefficient in birefringent single-mode fibers under linear polarization is twice that under random polarization [22]. In 2013, Haarlammert et al. from the University Jena investigated the dependence of laser polarization on NLEs in high-power amplifiers through experiments. The results showed that when the seed polarization state was changed from linear polarization to circular polarization, the power proportion of the Raman Stokes light was reduced by about 50% at most [91]. Overall, the SRS effect is more severe in linearly polarized lasers.

16.2.3 TMI Effect and Suppressing Methods

Compared to the classic SBS and SRS, the TMI effect is a newly discovered mechanism. In 2010, Christian et al. from the University Jena first observed the TMI effect in an experiment [92]. Since the TMI effect was reported, it has attracted widespread attention from researchers. The output laser pattern will change before and after the occurrence of TMI. When the output power is below the TMI threshold, the output beam exhibits a stable fundamental mode. When it exceeds the TMI threshold, rapid energy transfer occurs between the fundamental mode and higher-order modes, and the beam shape begins to fluctuate on the millisecond time scale, accompanied by stagnation of output power and degradation of beam quality [24, 93]. In the experiment, detecting the time-domain signal of the output beam is a simple and effective method to monitor the TMI effect. When the TMI effect occurs, the time-domain signal will demonstrate random fluctuations, and the corresponding frequency-domain signal will exhibit obvious noise-like frequency distribution in the kHz range. For different fiber lasers in practice, the TMI threshold power is usually between several hundred watts and several kilowatts. Since its discovery, TMI has become the main limiting factor for the power growth of narrow linewidth fiber lasers.

At present, the measures to increase the TMI threshold in high-power fiber lasers mainly include optimizing seed lasers and pumping methods, developing new types of fibers, and increasing loss of higher-order modes. Seed laser optimization technology includes increasing seed laser power and linewidth [13, 94–96]. The optimization of pumping methods mainly includes adopting a backward pumping scheme, a bidirectional pumping scheme with a certain proportion of forward and backward pumping power [54, 97], using a pump laser centered at 915 nm, and a tandem pumping scheme [54, 98–100]. The development of new optical fibers includes the design of tapered fibers [101], confined-doped fiber [100, 102], acoustic-field tailored fibers [35], photonic bandgap fibers [103], and multi-core fibers [104]. Among them, coiling the gain fiber to increase the higher-order modes loss is a simple and effective way to significantly improve the TMI threshold [69, 105–107], which is widely used in practical high-power fiber laser systems.

Research has found that the TMI threshold is also related to the polarization state of the signal laser. In 2014, Khush Brar et al. from Lockheed Martin Corporation constructed a kilowatt level narrow linewidth PM/non-PM fiber laser based on a forward pumping structure, and studied the effect of polarization on the TMI threshold. The results indicated that when constructing fiber amplifiers using non-PM YDF and PM YDF with core/cladding diameters of 25/400 μm, respectively, the TMI threshold of the former was approximately twice that of the latter [23]. Subsequent theoretical and experimental studies have also found that rotating the polarization direction of the signal laser relative to the main polarization axis of the PM gain fiber by a certain angle can reduce mode coupling and achieve an increase in TMI threshold [108–110]. However, this approach essentially changes the polarization state of the signal laser and is not suitable for systems that require strict linear polarization laser output. Therefore, improving the TMI threshold of fiber amplifiers while maintaining a high polarization extinction ratio (PER) still faces significant technical challenges.

16.2.4 Discussion on Limiting Factors

As mentioned above, for HPNL fiber lasers, their power growth is influenced by both NLEs and TMI effects, and researchers have proposed various targeted suppression methods. However, there is a certain contradiction between suppressing NLEs and TMI effects in practical applications. For example, increasing the core size of the gain fiber can improve the SBS threshold of the amplifier but can enhance the TMI effect at the same time. Increasing the seed laser injection power can effectively improve the TMI threshold, but it is adverse to suppressing the SBS effect. Moreover, both NLEs and TMI effects are related to the polarization state of the signal laser. Therefore, the contradiction between alleviating NLEs and TMI, as well as this polarization correlation, poses significant challenges for improving the power of narrow linewidth linearly polarized fiber lasers.

16.3 Recent Progress

As previously mentioned, as an important high-brightness fiber laser source, linearly polarized narrow linewidth lasers have higher combination (conversion) efficiency for applications such as beam combination and nonlinear frequency conversion. There are currently two main solutions to obtain linearly polarized laser output. One is to use PM components to build an all-PM fiber amplifier. The second is to implement active polarization control in a non-PM fiber amplifier. The following will review the research progress of the two schemes respectively.

16.3.1 HPNL Fiber Laser Based on All-Polarization-Maintained Design

Usually, the methods of generating narrow linewidth seeds for fiber amplifiers mainly includes narrow linewidth laser diodes, few-longitudinal-mode fiber oscillators, filtered superfluorescent sources (SFS), and phase-modulated single-frequency lasers [111].

Table 16.1 shows the recent progress of high-power linearly polarized fiber lasers based on narrow linewidth filtered PM SFS and PM fiber oscillators.

The use of narrow linewidth fiber oscillators as seed lasers for power amplification has the advantages of low cost and compact system design. In 2017, National University of Defense Technology (NUDT) used a compact one-stage amplification structure to amplify the power of a narrowband PM fiber oscillator seed, achieving a linearly polarized laser output with a polarization extinction ratio (PER) of 14 dB, a 3 dB linewidth of 0.3 nm, and an output power of 1.018 kW. The beam quality was $M^2_x = 1.18$ and $M^2_y = 1.24$ [43]. In 2020, the China Academy of Engineering Physics (CAEP) used a PM few-longitudinal-mode fiber oscillator to alleviate spectral broadening and built a HPNL fiber amplifier. The amplifier adopts a bidirectional pumped configuration, with a PM YDF of length 9 m and core/cladding diameter of 20/400 μm coiled in the water-cooled plate with a diameter of 10 cm to increase loss of higher-order modes. The pump power is provided by stable wavelength LDs centered at 976 nm. The maximum output power of the PM amplifier is 3.08 kW, and the 3 dB spectral linewidth is 0.2 nm. The PER during power amplification is greater than 11.6 dB, and the beam quality M^2 is less than 1.45. The SRS signal-to-noise ratio at the highest power is about 20 dB, and the power growth is limited by the SRS effect. This result is currently the highest laser power reported for a narrow linewidth linearly polarized single-mode fiber amplifier using a PM fiber oscillator as the seed laser [114].

Although notable progress has been made in achieving high-power linearly polarized laser output in recent years, due to the temporal characteristics of narrow linewidth oscillator seeds, there is usually spectral broadening as the output power increases. The narrow linewidth seeds obtained by phase modulation technology to broaden the linewidth of single-frequency lasers have characteristics such as temporal stability, high threshold for NLEs, and good spectral characteristics during power amplification. This is another important solution for obtaining HPNL fiber lasers. At present, sine-wave signal, pseudo-random bit sequence (PRBS), and

Table 16.1 Results of HPNL PM fiber lasers based on different types of seed lasers

Seed type	Year	Institution	Wavelength/nm	Power/kW	Linewidth/nm	M^2	PER/dB	Ref
filtered PM SFS	2015	NUDT	1064	0.8	0.2	1.28	12.2	[112]
	2017	NUDT	1080	1.01	0.21	<1.15	17	[113]
PM fiber oscillator	2017	NUDT	1064	1.018	0.3	<1.24	14	[43]
	2020	CAEP	1064	3.08	0.2	<1.45	11.6	[114]

white noise signal (WNS) have been regarded as typical phase modulation signals. Table 16.2 shows typical research results of narrow linewidth linearly polarized fiber lasers based on phase-modulated single-frequency lasers in recent years.

In 2016, researchers from Massachusetts Institute of Technology (MIT) used self-developed gold-coated specialty gain fibers to build a fiber amplifier. The gold-coated cladding can improve the heat loads of the fiber, thereby increasing the TMI threshold. Phase modulation technology is used to widen the linewidth of the seed laser, and the modulation signal format is PRBS. The amplifier adopts a bidirectional pumped structure and finally achieves a linearly polarized laser output with a power of 3.1 kW, a PER of 10 dB, a spectral linewidth of 12 GHz, and a beam quality $M^2 < 1.15$. However, the amplifier adopts a spatial coupling structure, which is not conducive to its application in beam combination systems.

In 2016, researchers from NUDT adopted a three-stage cascaded sine phase modulation system to broaden the linewidth of the seed laser and mitigate the SBS effect. The main amplifier used a PM YDF with a length of 8.5 m and a core/cladding diameter of 20/400 μm. The gain fiber was tightly coiled with a radius of 5.5 cm to enhance the TMI threshold. They ultimately achieved a linearly polarized laser output with a power of 1.89 kW, a linewidth of 45 GHz, and a PER of 15.5 dB [63]. In 2017, the influence of different bending methods of gain fibers in PM fiber amplifiers on the TMI threshold was compared through experiments. When a PM YDF with a length of 9 m and a core/cladding diameter of 20/400 μm was coiled on an aluminum cylindrical water-cooled plate with a diameter of 9 cm, the linearly polarized laser output power was increased to 2.43 kW, and the spectral linewidth and PER were 0.255 nm and 18.3 dB, respectively. There was no TMI effect in the system, and the power growth was limited by the available pump power [69]. In

Table 16.2 Results of HPNL linearly polarized fiber lasers based on phase modulation techniques

Modulation signal	Year	Institution	Wavelength/nm	Power/kW	Linewidth	M^2	PER/dB	Ref
Sine	2016	NUDT	1064	1.89	45 GHz	<1.3	15.5	[63]
	2019	NUDT	1064	1.082	7.6 GHz	1.14	14	[57]
PRBS	2016	MIT	1066	3.1	12 GHz	<1.15	10	[66]
	2019	GIST	1064	0.818	6.6 GHz	-	13	[115]
WNS	2017	NUDT	1064	2.43	0.255 nm	-	18.3	[69]
	2020	CAEP	1064	2.62	32 GHz	<1.3	14	[70]
	2022	NUDT	1064	3.96	0.62	<1.41	13.9	[116]
	2022	NUDT	1064	4.515	0.33	<1.55	10.3	[117]
	2022	NUDT	1064	5.023	0.38	<1.43	11.8	[19]
MPCS	2021	NUDT	1064	1.023	4.6 GHz	<1.29	13.3	[40]
unavailable phase modulation	2018	IPG	1064	2.0	30 GHz	1.07	>20	[118]
	2021	CAEP	1064	3.25	20 GHz	<1.22	15	[71]
	2023	CAEP	1064	5.043	0.2 nm	<1.3	16.5	[20]

2022, an integral alleviation of SBS and TMI was achieved by combining WNS phase modulation with bending loss optimization. Narrow linewidth linearly polarized near-single-mode fiber lasers achieved high-power outputs of 4 kW and 4.5 kW, respectively. Figure 16.1 shows the schematic diagram of the 4-kW power level near-single-mode narrow linewidth linearly polarized fiber amplifier of NUDT.

Power enhancement and linewidth reduction are constantly balanced and pursued goals in the development of HPNL fiber lasers. To obtain higher power and narrower linewidth linearly polarized laser output, researchers have also proposed new phase modulation techniques and applied them to high-power fiber lasers. In 2018, IPG Corporation widened the linewidth of a single frequency laser to 30 GHz and amplified its power, achieving a near-diffraction-limited linearly polarized laser output with a power of 2.0 kW and a PER greater than 20 dB [118]. In 2021, researchers from NUDT applied their proposed multi-phase coded signal (MPCS) phase modulation [119] to high-power linearly polarized fiber lasers and combined it with gain competition to improve the SBS threshold. The amplifier output power was 1.023 kW, with a PER of 13.3 dB, a spectral linewidth of only 4.6 GHz, and beam quality of $M^2_x = 1.29$ and $M^2_y = 1.28$ [40].

According to Table 16.2, WNS phase modulation plays an important role in obtaining high-power narrow-linewidth fiber lasers. However, self-pulsing will be a crucial limitation for the power scaling of this type of fiber amplifier with WNS phase modulation [120]. In 2021, CAEP optimized the phase modulation signal and generated a near top-hat-shaped spectrum, which effectively suppresses the self-pulsing effect. The results show that under the same linewidth, the self-pulsing threshold of the WNS phase modulation is doubled. Based on the optimized signal and high-power PM backward-pumped fiber amplifier, a narrow linewidth linearly polarized laser output of 3.25 kW was obtained, with a root mean square (RMS) linewidth of~20 GHz, a PER of approximately 15 dB, and M^2 less than 1.22 at the

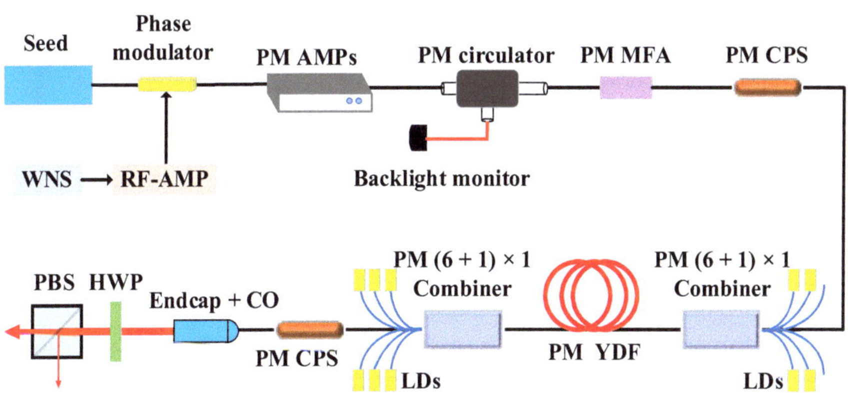

Fig. 16.1 Schematic of the 4 kW PM fiber laser of NUDT based on WNS phase modulation

maximum output power [71]. In 2023, researchers from CAEP achieved a breakthrough in output power of 5.043 kW by further optimizing phase modulation and TMI effect suppression techniques. At the highest power, the RMS linewidth of the signal laser was 0.2 nm, and the PER was 16.5 dB. This result is currently the highest output power of near-single-mode narrow linewidth linearly polarized fiber lasers [20].

The performance of PM YDF as the gain medium for fiber lasers has a crucial impact on the properties of narrow linewidth linearly polarized fiber lasers. In recent years, remarkable progress has been made in the development of homemade large-mode-area PM YDFs, which has also promoted the technological advancement of linearly polarized narrow linewidth lasers. In 2023, researchers from Huazhong University of Science and Technology (HUST) prepared three types of PM YDFs with numerical apertures (NA) of 0.068, 0.064, and 0.060, respectively. The core/inner cladding diameters were all 20/400 μm. The experimental results indicate that reducing the NA can improve the TMI threshold of PM YDF. Finally, the research group achieved a narrow linewidth linearly polarized laser output with a maximum power of 3.2 kW, a 3 dB spectral linewidth of 0.22 nm, and a PER of 16.6 dB using a homemade PM YDF with an NA of 0.060 [121].

In 2023, researchers from NUDT built an amplifier based on a homemade large-mode-area PM YDF with a core/cladding diameter of ~20/400 μm. A time-domain stable single-frequency WNS phase modulation seed was used to mitigate NLEs and high-order mode relative loss was increased through bending loss to enhance TMI threshold. Then a near-single-mode linearly polarized narrow linewidth fiber amplifier with a maximum power of 5.023 kW was successfully achieved. The time-domain and frequency-domain signals of the amplifier operating at 5.023 kW indicate that no TMI effect has occurred. The experimental results are shown in Fig. 16.2, which is currently the highest output power achieved based on homemade PM YDF [19].

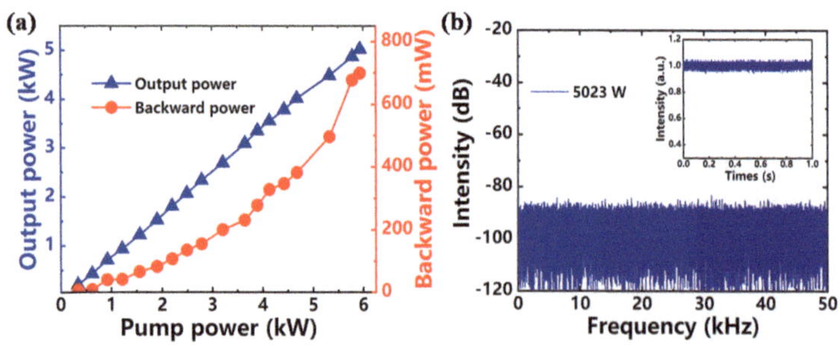

Fig. 16.2 Experimental results of a high-power narrow linewidth (HPNL) fiber laser based on homemade polarization-maintained fiber. (**a**) Curves of output power and backward power; (**b**) time domain signal (inset) and frequency domain signal of signal laser at 5023 W

16.3.2 HPNL Fiber Laser Based on Active Polarization Control

In addition to achieving linearly polarized laser output through HPNL fiber amplifiers based on all-PM fiber components, employing polarization control in non-PM fiber lasers has been developed as an alternative solution to obtain high-power linearly polarized fiber lasers. The basic principle of active polarization control is to inject a linearly polarized seed into a non-PM fiber amplifier and implement active polarization control through a feedback system, thereby achieving a linearly polarized laser output. Table 16.3 presents representative research results on the high-power polarization control technique.

In 2010, Goodno et al. from Northrop Grumman (NG) reported the application of polarization control technique for high-power fiber amplifiers in CBC systems. A polarization controller was used to control the polarization of a non-PM fiber amplifier with a power of 1.43 kW and a linewidth of 25 GHz in the system. However, the relevant technical details and polarization control results were not presented in the report [122]. In 2017, researchers from CAEP built a kilowatt-level fiber laser system using a PM seed laser and a non-PM fiber amplifier. The linewidth of the single-frequency PM seed laser was broadened to 6.5 GHz through WNS modulation. The seed laser was amplified by a non-PM pre-amplifier and a main amplifier before being output. Then, active polarization control was implemented based on the stochastic parallel gradient descent (SPGD) algorithm, and a linearly polarized laser output was obtained with a power of 1 kW and a linewidth of 6.5 GHz. The PER was measured to be 14.5 dB at the highest power, and the beam quality M2 was approximately 1.1 [120]. In 2018, IPG corporation demonstrated a 2.5 kW linear polarization fiber laser based on their non-PM fiber amplifier by controlling the polarization state of the input signal, and the PER reached 16–17 dB. Further power increase was limited by the TMI effect, and details of polarization control were not given [118].

In 2017, researchers from NUDT built an HPNL fiber laser system based on active polarization control. The polarization control algorithm, implementation method, and test results were introduced in detail for the first time. The

Table 16.3 Results of high-power narrow linewidth fiber lasers based on active polarization control

Year	Institution	Wavelength/nm	Power/kW	Linewidth	M^2	PER/dB	Ref
2010	NG	1064	1.43	25 GHz	–	–	[122]
2017	CAEP	1064	1	6.5 GHz	1.1	14.5	[120]
2017	NUDT	1064	1.43	0.17 nm	–	11.1	[123]
2018	IPG	1064	2.5	30 GHz	1.08	16–17	[118]
2023	NUDT	1071.5	3.378	0.32 nm	<1.50	12.2	[124]

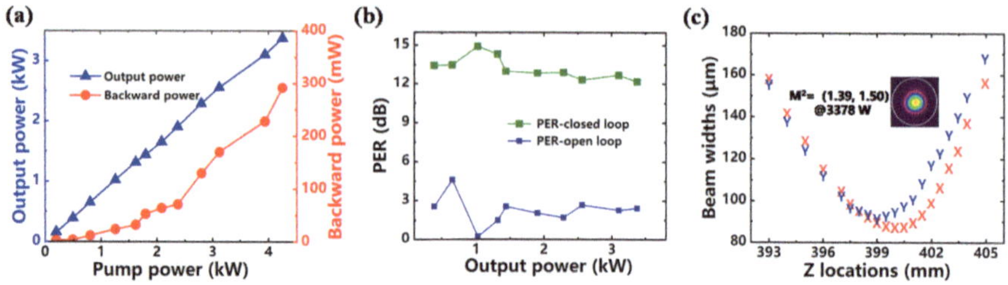

Fig. 16.3 Experimental results of a 3.378 kW fiber laser with RMS-Prop algorithm associated active polarization control reported by NUDT. (**a**) Curves of output power and backward power; (**b**) PER; (**c**) beam quality at maximum power

experimental structure consists of a narrow linewidth seed laser generated by WNS modulation, a polarization control system, and a three-stage non-PM fiber amplifier. The polarization control was implemented using the SPGD algorithm. Finally, a linearly polarized laser output with a maximum power of 1.43 kW, a PER of 11.1 dB, and a linewidth of 0.17 nm was obtained. During the power amplification, the beam quality remained excellent, but the power growth was limited by the available pump power.

In 2023, researchers from NUDT established a theoretical model for active polarization control based on the root mean square propagation (RMS-Prop) algorithm, simulated and analyzed the influence of main parameters in the algorithm on the polarization control effect, designed and implemented the algorithm control circuit independently, and built an HPNL fiber laser based on active polarization control. The experimental structure mainly consists of linearly polarized narrow linewidth seeds, non-PM cascaded fiber amplification links, and active polarization control systems [124]. The experimental results are shown in Fig. 16.3. After implementing polarization control in a high-power non-PM fiber amplifier using the RMS Prop algorithm, a linearly polarized laser output with a maximum power of 3.378 kW, a PER of 12.2 dB, and a 3 dB spectral linewidth of 0.32 nm was achieved. The beam quality at the highest power was $M^2_x = 1.39$ and $M^2_y = 1.50$. This is the first time that over 3 kW linearly polarized fiber laser output has been realized based on active polarization control in non-PM fiber amplifiers to the best of our knowledge.

16.3.3 HPNL Linearly Polarized Fiber Laser of Special Wavelength

The center wavelengths of the high-power fiber lasers introduced in Sects. 16.3.1 and 16.3.2 are mainly in the range of 1060 ~ 1080 nm, which is located in the area where the net gain cross-section of YDF is relatively large, which is conducive to power amplification. However, for practical applications, extending the wavelength of fiber lasers to other special wavelengths is of great significance. For example, high-power fiber lasers operating at 1018 nm wavelength have high brightness

Table 16.4 Results of high-power PM narrow linewidth fiber lasers with special wavelength

Year	Institution	Wavelength/nm	Power/kW	Linewidth	M^2	PER/dB	Ref
2019	TJU	1018	0.104	0.073 nm	<1.64	17.9	[126]
2019	Jena	1018	0.616	0.12 nm	<1.09	-	[127]
2018	TJU	1030	0.925	60 GHz	<1.3	15.2	[128]
2018	CAEP	1030	>1 kW	0.14 nm	<1.12	12	[129]

characteristics and are very suitable as pump sources for the tandem pump scheme, and a wavelength extension of 1030 nm means an increase in the number of single laser beams available for the SBC, which helps to expand the number of laser paths and increase the total power of the SBC system. Despite facing serious technical challenges such as laser reabsorption, intense gain competition with amplified spontaneous emission (ASE) and parasitic lasers in expanding output lasers to short wavelengths, significant progress has also been made in high-power special wavelength narrow linewidth linearly polarized fiber lasers due to application demand traction, fiber process improvement, and key technological breakthroughs [125]. Representative research results in recent years are shown in Table 16.4.

In 2019, researchers from Jena designed a 1018 nm high-power MOPA system with a two-stage amplifier based on spatial structure. The first stage pre-amplified the power of the phase-modulated seed laser and added filtering devices to suppress ASE. The main amplifier used an in-house designed and fabricated low NA ring-up-doped fiber with a core/cladding diameter of 32/260 μm. Then a near-single-mode 1018 nm linearly polarized laser output with a power of 616 W and a linewidth of 0.12 nm was achieved, with an ASE suppression ratio of up to 55 dB [127]. In 2018, researchers from Tianjin University (TJU) realized a 925 W linearly polarized laser with a center wavelength of 1030.1 nm, a PER of 15.2 dB, a beam quality factor $M^2 < 1.3$, and a linewidth of approximately 60 GHz using an all-fiber MOPA structure. The fiber laser consists of a narrow linewidth linearly polarized fiber oscillator and a two-stage PM fiber amplifier. Owing to the setting of the appropriate parameters for the laser, no indication of SBS was observed. The gain fiber of the main amplifier was bent and coiled to suppress higher-order modes, and TMI effect was successfully avoided [128]. In the same year, researchers from CAEP optimized the parameters of a fiber laser system based on MOPA structure through theoretical model analysis of the power balance equation. Under both forward pumping and backward pumping schemes, laser output of >1 kW was achieved. This is the first realization of a kilowatt-level narrow linewidth linearly polarized near-diffraction-limit fiber laser with a center wavelength of 1030 nm [129].

16.4 Discussion

HPNL linearly polarized fiber laser has developed rapidly in recent years, but the entire process is not spontaneous or isolated and is accompanied by the demand for beam combination high-energy laser systems, frequency conversion laser countermeasure systems, precision measurement, and related supporting technologies.

However, further power growth and narrower linewidth still face multiple challenges. As briefly analyzed in Sect. 16.2 of this chapter, the main limiting factors and mitigation methods for improving the power of narrow linewidth fiber lasers while maintaining good beam quality are discussed. The emerging functional active and passive optical fibers in recent years have exhibited tremendous advantages in balancing or reducing NLEs and the TMI effect that hinder high-power transmission. Then the concept of "functionalization" is proposed by researchers from NUDT based on the research progress of laser fibers and fiber-based components. The "functionalization" mentioned here is derived from the ability to functionalize the high-power fiber laser operations, usually referring to the selective control for its spatial mode, spectrum, or linewidth regime [130]. Thus, developing laser fibers with special functions and passive fiber-based components points out a new way to further increase the power of narrow linewidth fiber lasers [101, 131]. In addition to increasing power, achieving flexible control of the frequency domain, spatial domain, and temporal domain characteristics of HPNL fiber lasers to meet more application needs is also an important direction for technological development [132, 133].

Notably, as one of the main schemes for achieving HPNL fiber lasers, the single-frequency laser phase modulation (see Table 16.2) structure adopts the important optoelectronic device of the phase modulator. Its electro-optic modulation characteristics enable the laser system to have programming capabilities (pulse-operated narrow linewidth fiber laser systems usually have intensity modulators, which also have programming functions). Researchers can flexibly introduce relevant research results in the field of electronics into the laser system. The commonly used MPCS algorithm in the field of radar signal processing was applied to phase modulation, and the linewidth was narrowed under the same power conditions [40]. The programming capabilities brought by two types of modulators will introduce new ideas and perspectives for the development of high-power narrow linewidth fiber lasers [134, 135]. Furthermore, research and understanding of basic physical processes, such as how to define and describe narrow linewidth, and how linewidth, arises and evolves, will contribute to the development of HPNL fiber amplifiers [14].

16.5 Conclusion and Prospect

The power scaling of narrow linewidth linearly polarized fiber lasers has achieved encouraging results with the continuous efforts of researchers in recent years. A near-single-mode narrow linewidth laser output of 5 kW level was realized based on an all-PM fiber structure [19, 20]. Using the active polarization control technique, a linearly polarized laser output with a power of 3.378 kW was obtained in a non-PM fiber amplifier [124]. For practical application systems such as the CBC system and nonlinear frequency conversion, it is hoped that a single laser beam can maintain high brightness output while narrowing the spectral linewidth as much as possible. However, the narrower the linewidth, the lower the SBS threshold of the amplifier, and the TMI threshold will also show substantial degradation. In order to further obtain higher power and narrower linewidth near-single-mode linearly polarized narrow linewidth fiber lasers, research needs to be conducted in the development of basic physical concepts, new phase modulation technologies, functional fiber and fiber-based device design, and amplifier system structure optimization in the future.

References

1. Po, H., Snitzer, E., Tumminelli, R., Zenteno, L., Hakimi, F., Cho, N.M., Haw, T.: Double clad high brightness Nd fiber laser pumped by GaAlAs phased array. In: Optical Fiber Communication Conference, 1989 OSA Technical Digest Series. Optica Publishing Group (1989) PD7
2. Pu, Z., Liangjin, H., Jinyong, L., Hu, X., Jiangming, X., Tianfu, Y.: High-power double-cladding fiber lasers: a 30-year overview. Sci. Sin. Technologica. **50**, 123–135 (2020)
3. O'Connor, M., Gapontsev, V., Fomin, V., Abramov, M., Ferin, A.: Power scaling of SM fiber lasers toward 10kW. In: Conference on Lasers and Electro-Optics/International Quantum Electronics Conference, OSA Technical Digest (CD). Optica Publishing Group (2009) CThA3
4. Man, J., Pengfei, M., Rongtao, S., Can, L., Jian, W., Yanxing, M., Pu, Z.: Research progress and prospect of spectral beam combining (invited). Infrared Laser Eng. **49**, 172–189 (2020)
5. Pu, Z., Hongxiang, C., Rongtao, S., Xiaolin, W., Yanxing, M.: Research history and prospects of coherent beam combining of fiber lasers: from perspective of citations (invited). Chin J Lasers. **51**, 0121002 (2024)
6. Müller, M., Aleshire, C., Klenke, A., Haddad, E., Légaré, F., Tünnermann, A., Limpert, J.: 10.4 kW coherently combined ultrafast fiber laser. Opt. Lett. **45**, 3083–3086 (2020)
7. Wenchang, L., Pengfei, M., Hu, X., Wei, L., Can, L., Man, J., Jiangming, X., Rongtao, S., Jinyong, L., Yanxing, M., Pu, Z.: High-power narrow-linewidth fiber laser technology. High Power Laser Part Beams. **32**, 121001 (2020)
8. Tian, M., Fuquan, L., Honghuan, L.: Recent progress of high power green laser based on frequency doubling technology for fiber laser. High Power Laser Part Beams. **35**, 44–51 (2023)
9. Zhangpeng, Q., Nan, W., Zihao, C., Xiaoting, L., Zebiao, L., Jianqiang, L., Youhui, L., Qitao, L., Peiguang, Y.: Research progress in high-power continuous green lasers. Chin. Sci. Bull. **68**, 4618–4629 (2023)
10. Ahmadi, P., Creeden, D., Aschaffenburg, D., Mokan, V., Underwood, M., Caprara, A., Shu, Q.-Z., Spinelli, L., Minelly, J., Nikolov, I.: Generating kW laser light at 532 nm via second harmonic generation of a high power Yb-doped fiber amplifier (2020)
11. Lu, H.B., Zhou, P., Wang, X.L., Jiang, Z.F.: Theoretical and numerical study of the threshold of stimulated Brillouin scattering in multimode fibers. J. Lightwave Technol. **33**, 4464–4470 (2015)
12. Smith, A.V., Smith, J.J.: Mode instability in high power fiber amplifiers. Opt. Express. **19**, 10180–10192 (2011)
13. Ren, S., Lai, W.C., Wang, G.J., Li, W., Song, J.X., Chen, Y.S., Ma, P.F., Liu, W., Zhou, P.: Experimental study on the impact of signal bandwidth on the transverse mode instability threshold of fiber amplifiers. Opt. Express. **30**, 7845–7853 (2022)
14. Pu, Z., Pengfei, M., Shuai, R., Yisha, C., Wei, L., Tianfu, Y., Zhiyong, P., Jinbao, C.: High-power narrow linewidth fiber laser: progress and prospect. Inf. Countermeasure Tech. **2**, 16–36 (2023)
15. Huang, Z.M., Shu, Q., Tao, R.M., Chu, Q.H., Luo, Y., Yan, D.L., Feng, X., Liu, Y., Wu, W.J., Zhang, H.Y., Lin, H.H., Wang, J.J., Jing, F.: > 5kW record high power narrow linewidth laser from traditional step-index monolithic fiber amplifier. IEEE Photon. Technol. Lett. **33**, 1181–1184 (2021)

16. Zhou, P., Ma, P., Liu, W., Xiao, H., Ren, S., Song, J., Xu, J., Chen, Y., Lai, W.: High power, narrow linewidth all-fiber amplifiers. In: SPIE LASE, p. 11981, SPIE (2022)
17. Wang, G., Song, J., Chen, Y., Ren, S., Ma, P., Liu, W., Yao, T., Zhou, P.: Six kilowatt record all-fiberized and narrow-linewidth fiber amplifier with near-diffraction-limited beam quality. High Power Laser Sci. Eng. **10**, e22 (2022)
18. Pengfei, M., Tianfu, Y., Yisha, C., Guangjian, W., Shuai, R., Wei, L., Wei, L., Zhiyong, P., Liangjin, H., Huan, Y., Zilun, C., Pu, Z., Jinbao, C.: New progress of high-power narrow-linewidth fiber lasers. In: Proc. SPIE (2022) 123100E
19. Shuai, R., Pengfei, M., Yisha, C., Wei, L., Guangjian, W., Wei, L., Liangjin, H., Zhiyong, P., Tianfu, Y., Pu, Z.: 5 kW-level narrow linewidth fiber laser output realized by homemade polarization-maintained fiber. Infrared Laser Eng. **52**, 443–444 (2023)
20. Wang, Y., Peng, W., Liu, H., Yang, X., Yu, H., Yao Wang, J., Wang, Y., Feng, Y., Sun, Y., Ma, Q.G., Tang, C.: Linearly polarized fiber amplifier with narrow linewidth of 5 kW exhibiting a record output power and near-diffraction-limited beam quality. Opt. Lett. **48**, 2909–2912 (2023)
21. van Deventer, M.O., Boot, A.J.: Polarization properties of stimulated Brillouin scattering in single-mode fibers. J. Lightwave Technol. **12**, 585–590 (1994)
22. Stolen, R.: Polarization effects in fiber Raman and Brillouin lasers. IEEE J. Quantum Electron. **15**, 1157–1160 (1979)
23. Brar, K., Savage-Leuchs, M., Henrie, J., Courtney, S., Dilley, C., Afzal, R., Honea, E.: Threshold power and fiber degradation induced modal instabilities in high power fiber amplifiers based on large mode area fibers. In: Fiber Lasers Xi: Technology, Systems, and Applications (2014) 89611R
24. Jauregui, C., Limpert, J., Tünnermann, A.: High-power fibre lasers. Nat. Photonics. **7**, 861–867 (2013)
25. Otto, H.-J., Jauregui, C., Limpert, J., Tünnermann, A.: Average power limit of fiber-laser systems with nearly diffraction-limited beam quality. In: SPIE LASE, SPIE (2016) 97280E
26. Kobyakov, A., Sauer, M., Chowdhury, D.: Stimulated Brillouin scattering in optical fibers. Adv. Opt. Photon. **2**, 1–59 (2010)
27. Li, M.-J., Chen, X., Wang, J., Liu, A., Gray, S., Walton, D., Ruffin, A., Demeritt, J., Zenteno, L.: Fiber designs for high power lasers. Proc. SPIE. (2007) 64690H
28. Gray, S., Walton, D., Wang, J., Liu, A., Li, M.-J., Chen, X., Ruffin, A.B., DeMeritt, J., Zenteno, L.: Suppression of stimulated Brillouin scattering in high power, narrow linewidth fiber amplifiers. In: Nonlinear Optics: Materials, Fundamentals and Applications, OSA Technical Digest (CD), p. WB2. Optica Publishing Group (2007)
29. Ruffin, A.B., Li, M.-J., Chen, X., Kobyakov, A., Annunziata, F.: Brillouin gain analysis for fibers with different refractive indices. Opt. Lett. **30**, 3123–3125 (2005)
30. Dixneuf, C., Guiraud, G., Bardin, Y.-V., Rosa, Q., Goeppner, M., Hilico, A., Pierre, C., Boullet, J., Traynor, N., Santarelli, G.: Ultra-low intensity noise, all fiber 365 W linearly polarized single frequency laser at 1064 nm. Opt. Express. **28**, 10960–10969 (2020)
31. Jeong, Y., Nilsson, J., Sahu, J.K., Soh, D.B.S., Alegria, C., Dupriez, P., Codemard, C.A., Payne, D.N., Horley, R., Hickey, L.M.B., Wanzcyk, L., Chryssou, C.E., Alvarez-Chavez, J.A., Turner, P.W.: Single-frequency, single-mode, plane-polarized ytterbium-doped fiber master oscillator power amplifier source with 264 W of output power. Opt. Lett. **30**, 459–461 (2005)
32. Wang, X.L., Zhou, P., Xiao, H., Ma, Y.X., Xu, X.J., Liu, Z.J.: 310 W single-frequency all-fiber laser in master oscillator power amplification configuration. Laser Phys. Lett. **9**, 591–595 (2012)
33. Ma, P., Zhou, P., Ma, Y., Su, R., Xu, X., Liu, Z.: Single-frequency 332 W, linearly polarized Yb-doped all-fiber amplifier with near diffraction-limited beam quality. Appl. Opt. **52**, 4854–4857 (2013)
34. Gray, S., Liu, A., Walton, D.T., Wang, J., Li, M.-J., Chen, X., Ruffin, A.B., DeMeritt, J.A., Zenteno, L.A.: 502 watt, single transverse mode, narrow linewidth, bidirectionally pumped Yb-doped fiber amplifier. Opt. Express. **15**, 17044–17050 (2007)

35. Robin, C., Dajani, I., Pulford, B.: Modal instability-suppressing, single-frequency photonic crystal fiber amplifier with 811 W output power. Opt. Lett. **39**, 666–669 (2014)
36. Kobyakov, A., Kumar, S., Chowdhury, D.Q., Ruffin, A.B., Sauer, M., Bickham, S.R., Mishra, R.: Design concept for optical fibers with enhanced SBS threshold. Opt. Express. **13**, 5338–5346 (2005)
37. Robin, C., Dajani, I., Pulford, B., Vergien, C.: Single-frequency Yb-doped photonic crystal fiber amplifier with 800W output power. In: SPIE LASE, p. 896103, SPIE (2014)
38. Li, W., Yan, Z., Ren, S., Deng, Y., Chen, Y., Ma, P., Liu, W., Huang, L., Pan, Z., Zhou, P., Si, L.: Confined-doped active fiber enabled all-fiber high-power single-frequency laser. Opt. Lett. **47**, 5024–5027 (2022)
39. Li, W., Ma, P., Chen, Y., Song, J., Lai, W., Liu, W., Zhou, P., Si, L.: 694 W sub-GHz polarization-maintained tapered fiber amplifier based on spectral and pump wavelength optimization. Opt. Express. **30**, 26875–26885 (2022)
40. Lai, W.C., Ma, P.F., Song, J.X., Ren, S., Liu, W., Zhou, P.: Kilowatt-level, narrow linewidth, polarization-maintained all-fiber amplifiers based on multi-phase coded signal modulation and laser gain competition. Results Phys. **31**, 105050 (2021)
41. White, J.O., Harfouche, M., Edgecumbe, J., Satyan, N., Rakuljic, G., Jayaraman, V., Burgner, C., Yariv, A.: 1.6 kW Yb fiber amplifier using chirped seed amplification for stimulated Brillouin scattering suppression. Appl. Opt. **56**, B116–B122 (2017)
42. Lee, J., Lee, K.H., Jeong, H., Park, M., Seung, J.H., Lee, J.H.: 2.05 kW all-fiber high-beam-quality fiber amplifier with stimulated Brillouin scattering suppression incorporating a narrow-linewidth fiber-Bragg-grating-stabilized laser diode seed source. Appl. Opt. **58**, 6251–6256 (2019)
43. Jiang, M., Ma, P.F., Huang, L., Xu, J.M., Zhou, P., Gu, X.J.: kW-level, narrow-linewidth linearly polarized fiber laser with excellent beam quality through compact one-stage amplification scheme. High Power Laser Sci. Eng. **5** (2017)
44. Huang, Y.S., Yan, P., Wang, Z.H., Tian, J.D., Li, D., Xiao, Q.R., Gong, M.L.: 2.19 kW narrow linewidth FBG-based MOPA configuration fiber laser. Opt. Express. **27**, 3136–3145 (2019)
45. Huang, Z.H., Liang, X.B., Li, C.Y., Lin, H.H., Li, Q., Wang, J.J., Jing, F.: Spectral broadening in high-power Yb-doped fiber lasers employing narrow-linewidth multilongitudinal-mode oscillators. Appl. Opt. **55**, 297–302 (2016)
46. Xu, J.M., Liu, W., Leng, J.Y., Xiao, H., Guo, S.F., Zhou, P., Chen, J.B.: Power scaling of narrowband high-power all-fiber superfluorescent fiber source to 1.87 kW. Opt. Lett. **40**, 2973–2976 (2015)
47. Qi, Y., Yang, Y., Shen, H., He, B., Zhou, J.: 2.7 kW CW narrow linewidth Yb-doped all-fiber amplifiers for beam combining application. In: Laser Congress 2017 (ASSL, LAC), OSA Technical Digest (Online). Optica Publishing Group (2017) ATu3A.1
48. Xu, J.M., Ye, J., Xiao, H., Leng, J.Y., Liu, W., Zhou, P.: In-band pumping avenue based high power superfluorescent fiber source with record power andnear-diffraction-limited beam quality. High Power Laser Science and Engineering. **6**, e46 (2018)
49. Prakash, R., Vikram, B.S., Supradeepa, V.R.: Enhancing the efficacy of noise modulation for SBS suppression in high power, narrow linewidth fiber lasers by the incorporation of sinusoidal modulation. IEEE Photon. J. **13**, 1500106 (2021)
50. Goodno, G.D., Rothenberg, J.E.: Suppression of stimulated Brillouin scattering in high power fibers using nonlinear phase demodulation. Opt. Express. **27**, 13129–13141 (2019)
51. Shen, H., Lou, Q., Quan, Z., Li, X., Yang, Y., Chen, X., Li, Q., Bai, G., Qi, Y., He, B., Zhou, J.: Narrow-linewidth all-fiber amplifier with up to 3.01 kW output power based on commercial 20/400 mum active fiber and counterpumped configuration. Appl. Opt. **58**, 3053–3058 (2019)
52. Harish, A.V., Nilsson, J.: Optimization of phase modulation formats for suppression of stimulated Brillouin scattering in optical fibers. IEEE J. Sel. Top. Quantum Electron. **24**, 5100110 (2018)
53. Liu, A., Chen, X., Li, M.-J., Wang, J., Walton, D.T., Zenteno, L.A.: Comprehensive modeling of single frequency Fiber amplifiers for mitigating stimulated Brillouin scattering. J. Lightwave Technol. **27**, 2189–2198 (2009)

54. Tao, R.M., Ma, P.F., Wang, X.L., Zhou, P., Liu, Z.J.: Theoretical study of pump power distribution on modal instabilities in high power fiber amplifiers. Laser Phys. Lett. **14**, 025002 (2017)
55. Theeg, T., Sayinc, H., Neumann, J., Kracht, D.: All-Fiber counter-propagation pumped single frequency amplifier stage with 300-W output power. IEEE Photon. Technol. Lett. **24**, 1864–1867 (2012)
56. Zeringue, C., Vergien, C., Dajani, I.: Pump-limited, 203 W, single-frequency monolithic fiber amplifier based on laser gain competition. Opt. Lett. **36**, 618–620 (2011)
57. Meng, D., Ma, P., Wang, X., Ma, Y., Su, R., Zhou, P., Yang, L.: Kilowatt-level, high brightness, narrow-linewidth polarization-maintained fiber amplifiers based on laser gain competition. Jpn. J. Appl. Phys. **58**, 012007 (2019)
58. Zhang, L., Cui, S., Liu, C., Zhou, J., Feng, Y.: 170 W, single-frequency, single-mode, linearly-polarized, Yb-doped all-fiber amplifier. Opt. Express. **21**, 5456–5462 (2013)
59. Huang, L., Wu, H., Li, R., Li, L., Ma, P., Wang, X., Leng, J., Zhou, P.: 414 W near-diffraction-limited all-fiberized single-frequency polarization-maintained fiber amplifier. Opt. Lett. **42**, 1–4 (2017)
60. Hildebrandt, M., Buesche, S., Wessels, P., Frede, M., Kracht, D.: Brillouin scattering spectra in high-power single-frequency ytterbium doped fiber amplifiers. Opt. Express. **16**, 15970–15979 (2008)
61. Nikles, M., Thevenaz, L., Robert, P.A.: Brillouin gain spectrum characterization in single-mode optical fibers. J. Lightwave Technol. **15**, 1842–1851 (1997)
62. Engin, D., Lu, W., Akbulut, M., McIntosh, B., Verdun, H., Gupta, S.: 1 kW cw Yb-fiber-amplifier with < 0.5GHz linewidth and near-diffraction limited beam-quality, for coherent combining application. In: Fiber Lasers Viii: Technology, Systems, and Applications, p. 791407 (2011)
63. Ma, P.F., Tao, R.M., Su, R.T., Wang, X.L., Zhou, P., Liu, Z.J.: 1.89 kW all-fiberized and polarization-maintained amplifiers with narrow linewidth and near-diffraction-limited beam quality. Opt. Express. **24**, 4187–4195 (2016)
64. Flores, A., Robin, C., Lanari, A., Dajani, I.: Pseudo-random binary sequence phase modulation for narrow linewidth, kilowatt, monolithic fiber amplifiers. Opt. Express. **22**, 17735–17744 (2014)
65. Liu, M.Z., Yang, Y.F., Shen, H., Zhang, J.P., Zou, X.X., Wang, H.B., Yuan, L.C., You, Y., Bai, G., He, B., Zhou, J.: 1.27 kW, 2.2 GHz pseudo-random binary sequence phase modulated fiber amplifier with Brillouin gain-spectrum overlap. Sci. Rep. **10**, 629 (2020)
66. Yu, C.X., Shatrovoy, O., Fan, T.Y., Taunay, T.F.: Diode-pumped narrow linewidth multi-kilowatt metalized Yb fiber amplifier. Opt. Lett. **41**, 5202–5205 (2016)
67. Lin, H., Tao, R., Li, C., Wang, B., Guo, C., Shu, Q., Zhao, P., Xu, L., Wang, J., Jing, F., Chu, Q.: 3.7 kW monolithic narrow linewidth single mode fiber laser through simultaneously suppressing nonlinear effects and mode instability. Opt. Express. **27**, 9716–9724 (2019)
68. Ma, P.F., Xiao, H., Liu, W., Zhang, H.W., Wang, X.L., Leng, J.Y., Zhou, P.: All-fiberized and narrow-linewidth 5 kW power-level fiber amplifier based on a bidirectional pumping configuration. High Power Laser Sci. Eng. **9**, e45 (2021)
69. Su, R.T., Tao, R.M., Wang, X.L., Zhang, H.W., Ma, P.F., Zhou, P., Xu, X.J.: 2.43 kW narrow linewidth linearly polarized all-fiber amplifier based on mode instability suppression. Laser Phys. Lett. **14**, 085102 (2017)
70. Wang, Y.S., Feng, Y.J., Ma, Y., Chang, Z., Peng, W.J., Sun, Y.H., Gao, Q.S., Zhu, R.H., Tang, C.: 2.5 kW narrow linewidth linearly polarized all-Fiber MOPA with cascaded phase-modulation to suppress SBS induced self-pulsing. IEEE Photon. J. **12**, 1502815 (2020)
71. Wang, Y.S., Sun, Y.H., Peng, W.J., Feng, Y.J., Wang, J., Ma, Y., Gao, Q.S., Zhu, R.H., Tang, C.: 3.25 kW all-fiberized and polarization-maintained Yb-doped amplifier with a 20 GHz linewidth and near-diffraction-limited beam quality. Appl. Opt. **60**, 6331–6336 (2021)
72. Liu, W., Ma, P.F., Lv, H.B., Xu, J.M., Zhou, P., Jiang, Z.F.: General analysis of SRS-limited high-power fiber lasers and design strategy. Opt. Express. **24**, 26715–26721 (2016)

73. Liu, W., Ma, P.F., Lv, H.B., Xu, J.M., Zhou, P., Jiang, Z.F.: Investigation of stimulated Raman scattering effect in high-power fiber amplifiers seeded by narrow-band filtered superfluorescent source. Opt. Express. **24**, 8708–8717 (2016)
74. Ma, P., Miao, Y., Liu, W., Meng, D., Zhou, P.: Kilowatt-level ytterbium-Raman fiber amplifier with a narrow-linewidth and near-diffraction-limited beam quality. Opt. Lett. **45**, 1974–1977 (2020)
75. Shimizu, T., Nakajima, K., Shiraki, K., Ieda, K., Sankawa, I.: Evaluation methods and requirements for the stimulated Brillouin scattering threshold in a single-mode fiber. Opt. Fiber Technol. **14**, 10–15 (2008)
76. Aoki, Y., Tajima, K., Mito, I.: Input power limits of single-mode optical fibers due to stimulated Brillouin scattering in optical communication systems. J. Lightwave Technol. **6**, 710–719 (1988)
77. Haibin, L.: The Study on Dynamics of Stimulated Btillouin Scattering in fiber lasers. Doctor. National University of Defense Technology, Changsha (2015)
78. Wei, L.: Investigation on dynamics of stimulated Raman scattering in high-power fiber lasers. Doctor. National University of Defense Technology, Changsha (2018)
79. Shcherbakov, E.A., Fomin, V.V., Abramov, A.A., Ferin, A.A., Mochalov, D.V., Gapontsev, V.P.: Industrial grade 100 kW power CW fiber laser. In: Advanced Solid-State Lasers Congress, OSA Technical Digest (Online). Optica Publishing Group (2013) ATh4A.2
80. Shima, K., Ikoma, S., Uchiyama, K., Takubo, Y., Kashiwagi, M., Tanaka, D.: 5-kW single stage all-fiber Yb-doped single-mode fiber laser for materials processing. In: SPIE LASE, SPIE (2018) 105120C
81. Ma, X., Hu, I.N., Galvanauskas, A.: Propagation-length independent SRS threshold in chirally-coupled-core fibers. Opt. Express. **19**, 22575–22581 (2011)
82. Heck, M., Bock, V., Krämer, R., Richter, D., Goebel, T., Matzdorf, C., Liem, A., Schreiber, T., Tünnermann, A., Nolte, S.: Mitigation of stimulated Raman scattering in high power fiber lasers using transmission gratings. In: SPIE LASE, SPIE (2018) 105121I
83. Zenteno, L.A., Wang, J., Walton, D.T., Ruffin, B.A., Li, M.J., Gray, S., Crowley, A., Chen, X.: Suppression of Raman gain in single-transverse-mode dual-hole-assisted fiber. Opt. Express. **13**, 8921–8926 (2005)
84. Sahu, J.K., Dupriez, P., Kim, J., Codemard, C., Nilsson, J., Payne, D.N.: Suppression of stimulated Raman scattering in a high-peak-power pulsed 1060 nm fiber MOPA source with purely single-mode output using W-type fiber. In: Optical Fiber Communication Conference and Exposition and the National Fiber Optic Engineers Conference, Technical Digest (CD). Optica Publishing Group (2006) OWD5
85. Fini, J.M., Mermelstein, M.D., Yan, M.F., Bise, R.T., Yablon, A.D., Wisk, P.W., Andrejco, M.J.: Distributed suppression of stimulated Raman scattering in an Yb-doped filter-fiber amplifier. Opt. Lett. **31**, 2550–2552 (2006)
86. Wang, Y.: Stimulated Raman scattering in high-power double-clad fiber lasers and power amplifiers. Opt. Eng. **44**, 114202 (2005)
87. Ikoma, S., Nguyen, H.K., Kashiwagi, M., Uchiyama, K., Shima, K., Tanaka, D.: 3 kW single stage all-fiber Yb-doped single-mode fiber laser for highly reflective and highly thermal conductive materials processing. In: SPIE LASE, SPIE (2017)
88. Nodop, D., Jauregui, C., Jansen, F., Limpert, J., Tünnermann, A.: Suppression of stimulated Raman scattering employing long period gratings in double-clad fiber amplifiers. Opt. Lett. **35**, 2982–2984 (2010)
89. Wang, M., Zhang, Y., Wang, Z., Sun, J., Cao, J., Leng, J., Gu, X., Xu, X.: Fabrication of chirped and tilted fiber Bragg gratings and suppression of stimulated Raman scattering in fiber amplifiers. Opt. Express. **25**, 1529–1534 (2017)
90. Golovchenko, E.A., Mamyshev, P.V., Pilipetskii, A.N., Dianov, E.M.: Mutual influence of the parametric effects and stimulated Raman scattering in optical fibers. IEEE J. Quantum Electron. **26**, 1815–1820 (1990)

91. Haarlammert, N., Rekas, M., de Vries, O., Kliner, A., Liem, A., Schreiber, T., Eberhardt, R., Tünnermann, A.: Polarization dependent nonlinear limitations in continuous-wave high power fiber amplifiers. In: Proc. of SPIE, SPIE (2013) 86012W
92. Wirth, C., Schreiber, T., Rekas, M., Tsybin, I., Peschel, T., Eberhardt, R., Tünnermann, A.: High-power linear-polarized narrow linewidth photonic crystal fiber amplifier. In: SPIE LASE, SPIE (2010) 75801H
93. Eidam, T., Wirth, C., Jauregui, C., Stutzki, F., Jansen, F., Otto, H.-J., Schmidt, O., Schreiber, T., Limpert, J., Tuennermann, A.: Experimental observations of the threshold-like onset of mode instabilities in high power fiber amplifiers. Opt. Express. **19**, 13218–13224 (2011)
94. Stihler, C., Jauregui, C., Otto, H.-J., Limpert, J., Tünnermann, A.: Controlling mode instabilities at 628 W average output power in an Yb-doped rod-type fiber amplifier by active modulation of the pump power. In: SPIE LASE, SPIE (2017) 100830P
95. Tao, R., Wang, X., Zhou, P., Liu, Z.: Seed power dependence of mode instabilities in high-power fiber amplifiers. J. Opt. **19**, 065202 (2017)
96. Smith, J.J., Smith, A.V.: Influence of signal bandwidth on mode instability thresholds of fiber amplifiers. In: SPIE *(2015)* 93440L
97. Sanjabi Eznaveh, Z., López-Galmiche, G., Antonio-López, E., Amezcua Correa, R.: Bi-directional pump configuration for increasing thermal modal instabilities threshold in high power fiber amplifiers. In: SPIE LASE, SPIE (2015) 93442G
98. Naderi, S., Dajani, I., Grosek, J., Madden, T., Dinh, T.-N.: Theoretical analysis of effect of pump and signal wavelengths on modal instabilities in Yb-doped fiber amplifiers. In: SPIE LASE, SPIE (2014) 89641W
99. Jauregui, C., Otto, H.-J., Breitkopf, S., Limpert, J., Tünnermann, A.: Optimizing high-power Yb-doped fiber amplifier systems in the presence of transverse mode instabilities. Opt. Express. **24**, 7879–7892 (2016)
100. Wu, H., Li, R., Xiao, H., Huang, L., Yang, H., Pan, Z., Leng, J., Zhou, P.: High-power tandem-pumped fiber amplifier with beam quality maintenance enabled by the confined-doped fiber. Opt. Express. **29**, 31337–31347 (2021)
101. Song, J., Ren, S., Wang, G., Yang, H., Chen, Y., Ma, P., Liu, W., Huang, L., Pan, Z., Zhou, P.: High power narrow-linewidth Fiber laser based on double-tapered Fiber. J. Lightwave Technol. **40**, 5668–5672 (2022)
102. W. Li, S. Ren, Y. Deng, Y. Chen, Y. Lu, W. Liu, P. Ma, Z. Pan, Z. Chen, Lei Si, and P. Zhou, "Investigation of the confined-doped fiber on single-mode operating and power scaling in all-fiber single-frequency amplifiers," Frontiers in Physics **10**(2022)
103. Gu, G., Kong, F., Hawkins, T., Parsons, J., Jones, M., Dunn, C., Kalichevsky-Dong, M.T., Saitoh, K., Dong, L.: Ytterbium-doped large-mode-area all-solid photonic bandgap fiber lasers. Opt. Express. **22**, 13962–13968 (2014)
104. Klenke, A., Müller, M., Stark, H., Stutzki, F., Hupel, C., Schreiber, T., Tünnermann, A., Limpert, J.: Coherently combined 16-channel multicore fiber laser system. Opt. Lett. **43**, 1519–1522 (2018)
105. Tao, R.M., Su, R.T., Ma, P.F., Wang, X.L., Zhou, P.: Suppressing mode instabilities by optimizing the fiber coiling methods. Laser Phys. Lett. **14** (2017)
106. Lei, M., Qi, Y., Liu, C., Yang, Y., Zheng, Y., Zhou, J.: Mode controlling study on narrow-linewidth and high power all-fiber amplifier. In: Third International Symposium on Laser Interaction with Matter, SPIE (2015) 95431L
107. Tao, R., Ma, P., Wang, X., Zhou, P., Liu, Z.: 1.3 kW monolithic linearly polarized single-mode master oscillator power amplifier and strategies for mitigating mode instabilitics. Photon. Res. **3**, 86–93 (2015)
108. Lai, W., Ma, P., Liu, W., Huang, L., Li, C., Ma, Y., Zhou, P.: 550 W single frequency fiber amplifiers emitting at 1030 nm based on a tapered Yb-doped fiber. Opt. Express. **28**, 20908–20919 (2020)
109. Jauregui, C., Stihler, C., Kholaif, S., Tu, Y., Limpert, J.: Mitigation of transverse mode instability in polarization maintaining, high-power fiber amplifiers. In: SPIE LASE, SPIE (2021) 116650V

110. Jauregui, C., Kholaif, S., Tu, Y., Limpert, J.: Static modal energy transfer in high power, polarization maintaining fiber laser systems. In: SPIE LASE, p. 1198110, SPIE (2022)
111. M. Pengfei, "Study on coherent polarization beam combining system of high power fiber lasers. Doctor. National University of Defense Technology, 2016)
112. Ma, P., Tao, R., Wang, X., Zhou, P.: High-power narrow-band and polarization-maintained all fiber superfluorescent source. Photon. Technol. Lett. IEEE. **27**, 879–882 (2015)
113. Xu, J., Huang, L., Jiang, M., Ye, J., Ma, P., Leng, J., Wu, J., Zhang, H., Zhou, P.: Near-diffraction-limited linearly polarized narrow-linewidth random fiber laser with record kilowatt output. Photon. Res. **5**, 350–354 (2017)
114. Wang, Y., Ke, W., Peng, W., Chang, Z., Feng, Y., Sun, Y., Gao, Q., Ma, Y., Zhu, R., Tang, C.: 3 kW, 0.2 nm narrow linewidth linearly polarized all-fiber laser based on a compact MOPA structure. Laser Phys. Lett. **17**, 075101 (2020)
115. Jun, C., Jung, M., Shin, W., Yu, B.-A., Yoon, Y.S., Park, Y., Choi, K.: 818 W Yb-doped amplifier with <7 GHz linewidth based on pseudo-random phase modulation in polarization-maintained all-fiber configuration. Laser Phys. Lett. **16**, 015102 (2019)
116. Ren, S., Ma, P., Li, W., Wang, G., Chen, Y., Song, J., Liu, W., Zhou, P.: 3.96 kW all-fiberized linearly polarized and narrow linewidth Fiber laser with near-diffraction-limited beam quality. Nano. **12**, 2541 (2022)
117. Shuai, R., Yisha, C., Pengfei, M., Wei, L., Guangjian, W., Wei, L., Pu, Z.: 4.5 kW, 0.33 nm near-single-mode narrow linewidth polarization-maintained fiber laser. High Power Laser Part Beams. **34**, 065002 (2022)
118. Platonov, N., Gapontsev, V., Yagodkin, R., Cruz, J., Yusim, A.: Up to 2.5-kW on non-PM fiber and 2.0-kW linear polarized on PM fiber narrow linewidth CW diffraction-limited fiber amplifiers in all-fiber format. In: Fiber Lasers Xv: Technology & Systems (2018) 105120E
119. Lai, W., Ma, P., Liu, W., Li, C., Su, R., Ma, Y., Wu, J., Jiang, M., Zhou, P.: Seeding high brightness Fiber amplifiers with multi-phase coded signal modulation for SBS effect management. IEEE Access. **8**, 127682–127689 (2020)
120. Wang, Y., Feng, Y., Wang, X., Yan, H., Peng, J., Peng, W., Sun, Y., Ma, Y., Tang, C.: 6.5 GHz linearly polarized kilowatt fiber amplifier based on active polarization control. Appl. Opt. **56**, 2760–2765 (2017)
121. Liao, S., Luo, T., Xiao, R., Cheng, J., Shu, C., Zhang, Z., Zhou, Y., Xing, Y., Li, H., Peng, J., Dai, N., Li, J.: 3.2 kW, 0.22 nm narrow-linewidth MOPA configuration fiber laser with a homemade polarization-maintaining Yb-doped fiber. Front. Phys. **11** (2023)
122. Goodno, G.D., McNaught, S.J., Rothenberg, J.E., McComb, T.S., Thielen, P.A., Wickham, M.G., Weber, M.E.: Active phase and polarization locking of a 1.4 kW fiber amplifier. Opt. Lett. **35**, 1542–1544 (2010)
123. Su, R., Liu, Y., Yang, B., Ma, P., Wang, X., Zhou, P., Xu, X.: Active polarization control of a 1.43 kW narrow linewidth fiber amplifier based on SPGD algorithm. J. Opt. **19**, 045802 (2017)
124. Ren, S., Chang, H., Ma, P., Chen, Y., Wang, G., Li, W., Liu, W., Yao, T., Zhou, P.: 3.38 kW all-fiberized narrow linewidth fiber laser based on active polarization control using RMS-prop algorithm. Opt. Laser Technol. **166**, 109634 (2023)
125. Yan, P., Wang, X., Wang, Z., Huang, Y., Li, D., Xiao, Q., Gong, M.: A 1150-W 1018-nm Fiber laser bidirectional pumped by wavelength-stabilized laser diodes. IEEE J. Sel. Top. Quantum Electron. **24**, 1–6 (2018)
126. Xie, Z., Fang, Q., Xu, Y., Cui, X., Yao, J.: Hundred-watts-level monolithic narrow linewidth linearly-polarized fiber laser at 1018nm. Opt. Eng. **58**, 106106 (2019)
127. Palma-Vega, G., Walbaum, T., Heinzig, M., Kuhn, S., Hupel, C., Hein, S., Feldkamp, G., Sattler, B., Nold, J., Haarlammert, N., Schreiber, T., Eberhardt, R., Tünnermann, A.: Ring-up-doped fiber for the generation of more than 600 W single-mode narrow-band output at 1018 nm. Opt. Lett. **44**, 2502–2505 (2019)
128. Xu, Y., Fang, Q., Cui, X., Hou, B., Fu, S., Xie, Z., Shi, W.: 1 kW monolithic narrow linewidth linear-polarized fiber laser at 1030 nm. SPIE, Proc. SPIE. **105122S** (2018)

129. Chu, Q., Zhao, P., Lin, H., Liu, Y., Li, C., Wang, B., Guo, C., Tang, X., Tang, C., Jing, F.: kW-level 1030 nm polarization-maintained fiber laser with narrow linewidth and near-diffraction-limited beam quality. Appl. Opt. **57**, 2992–2996 (2018)
130. Chen, X., Yao, T., Huang, L., An, Y., Wu, H., Pan, Z., Zhou, P.: Functional fibers and functional Fiber-based components for high-power lasers. Adv. Fiber Mater. **5**, 59–106 (2023)
131. Wu, H., Song, J., Ma, P., Liu, W., Ren, S., Wang, G., Li, R., Xiao, H., Huang, L., Leng, J., Pan, Z., Zhou, P.: Bidirectional tandem-pumped high-brightness 6 kW level narrow-linewidth confined-doped fiber amplifier exploiting the side-coupled technique. Opt. Express. **30**, 21338–21348 (2022)
132. Liu, Y., Su, R., Ma, P., Wang, X., Zhang, H., Zhou, P., Si, L.: >1 kW all-fiberized narrow-linewidth polarization-maintained fiber amplifiers with wavelength spanning from 1065 to 1090 nm. Appl. Opt. **56**, 4213–4218 (2017)
133. Song, J., Xu, H., Wu, H., Huang, L., Xu, J., Zhang, H., Zhou, P.: High power narrow linewidth LP11 mode fiber laser using mode-selective FBGs. Laser Phys. Lett. **15**, 115101 (2018)
134. Jiang, M., Wu, H., An, Y., Hou, T., Chang, Q., Huang, L., Li, J., Su, R., Zhou, P.: Fiber laser development enabled by machine learning: review and prospect. PhotoniX. **3**, 16 (2022)
135. Ran, Y., Su, R., Ma, P., Wang, X., Lv, H., Zhou, P., Si, L.: 293 W, GHz narrow-linewidth, polarization maintaining nanosecond fiber amplifier with SBS suppression employing simultaneous phase and intensity modulation(article). Opt. Express. **23**, 25896–25905 (2015)

Chapter 17
Multielements-Doped Silica Glass-Based Specialty Optical Fibers for High-Power Laser and Optical Amplifiers

Mukul Chandra Paul, Shyamal Das, Anirban Dhar, S. K. Bhadra, Ramadas Pillai, Pius Varughese, and Alexander V. Kir'yanov

Abstract This study presents a new class of Er/Yb specialty optical fibers designed with a focus on high-power laser and optical amplifier applications. Utilizing an octagonal cladding design and a composition of multielement-doped silica glass (Al, Ge, P, Ce, B, F), we achieved remarkable lasing efficiency of over 41.0%, alongside stable output power and commendable radiation resistance. The optimization involved adjusting key parameters, including glass composition and the deposition temperature of the porous fluorinated phospho-silica core layer, through a modified chemical vapor deposition (MCVD) method combined with a solution doping technique to ensure uniform integration of the dopants, effectively mitigating clustering issues. The optimized doping ratios (7.5–11.1) of Yb_2O_3, and Er_2O_3 along with fluorine, GeO_2, CeO_2, Al_2O_3, B_2O_3, and P_2O_5 significantly enhanced fiber performance. We experimentally explored different cladding geometries (double-D, pentagonal, hexagonal, and octagonal) and evaluated their lasing capabilities. An octagonal cladding-pumped fiber demonstrated an output power of 21.0 W at 1558 nm, achieved with 52 W of absorbed pump power at 940 nm. This specific fiber structure contained 22.25 wt.% P_2O_5, 2.43 wt.% GeO_2, 0.5 wt.% Al_2O_3, 0.161 wt.% CeO_2, 2.41 wt.% B_2O_3, 0.25 wt.% F, 0.515 wt.% Er_2O_3, and 5.25 wt.% Yb_2O_3.

Performance characterization revealed a gain of approximately 28.0 dB with little variation (less than 0.45 dB flatness) over a wavelength range of 1540 to 1570 nm for a 10 dBm input signal. Additionally, under prolonged operation exceed-

M. C. Paul (✉) · S. Das · A. Dhar
CSIR-Central Glass and Ceramic Research Institute, Kolkata, India
e-mail: mcpal@cgcri.res.in

S. K. Bhadra
Raman Center for Atomic, Molecular and Optical Sciences, IACS, Kolkata, India

R. Pillai · P. Varughese
Vinvish Technologies Pvt. Ltd., Trivandrum, Kerala, India

A. V. Kir'yanov
Centro de Investigaciones en Óptica, León, Guanajuato, Mexico

M. F. S. Ferreira, M. C. Paul (eds.), *Optical Fiber Lasers*, Springer Series in Optical Sciences 250, https://doi.org/10.1007/978-3-032-04110-4_17

ing 150 h, the fiber maintained an average output power of around 15.0 W, with parasitic lasing below 1.0 μm and less than 10% degradation in output after exposure to 6.0 Krad of cumulative ^{60}Co gamma radiation. These results underscore the potential of this multielement-doped silica glass fiber in applications requiring robust performance and stability in challenging environments, such as free-space communication systems.

Keywords Rare-earth-doped materials · Fiber design and fabrication · Fluorescence life-time · ^{60}Co-ionizing radiation · High-power fiber laser and optical amplifiers

17.1 Introduction

Cladding-pumped Er/Yb-doped specialty fibers can serve as a suitable candidate for making of compact and efficient sources for high-power continuous-wave (CW) as well as high-energy short-pulse generation due to their robustness and high beam quality [1–4]. However, the development of an efficient Er-Yb co-doped fiber is associated with various technical bottlenecks, namely photodarkening at high pump power, degradation of output power under radiative environment, and origination of Yb^{3+} self-lasing around 1.0 μm which suppresses the output power at 1.55–1.56 μm, etc.

A lot of research work had been done regarding the development of Er-Yb codoped fibers with variation of the doping host composition as well as the fiber waveguide design and the fabrication method. G. G. Vienne et al. [5] developed ytterbium-erbium co-doped phosphor-aluminosilicate optical fibers containing 0.6–0.8 mol% Yb_2O_3, 11–13 mol% P_2O_5 with Yb/Er ratio 20 for efficient energy transfer from Yb to Er. They have investigated the effect of the Al to P ratio, which revealed that increasing Al concentration has no significant effect towards Yb/Er energy transfer with lasers emitting around 1.5 μm. N. Park et al. [6] made a high-power ErYb-doped fiber amplifier with multichannel gain flatness within 0.2 dB over 14 nm using ErYb co-doped fiber, which contains Er to produce 28 dB/m absorption at 1535 nm and 14 times more Yb ions to produce 800 dB/m at 980 nm and about 3 dB/m absorption at 1064 nm, containing 8 mol% P and small amounts of Al and Ge. This fiber exhibited a low noise figure of 5.2 dB at −1 dBm signal input and a high output power up to 24.6 dBm. In 1998, G. G. Vienne et al. [7] made Yb^{3+}:Er^{3+} phosphosilicate fibers for lasers by modified chemical vapor deposition (MCVD) coupled with solution doping (SD) technique through back-pass deposition method, where the core of the fiber contains 12 mol% P2Os with Yb^{3+}/Er^{3+} ratio of 3100, showing 2400 dB/m absorption loss at 975 nm to achieve uniformity of rare earth distribution in the final fiber. X. Orignac et al. [8] made SiO_2-TiO_2-Er_2O_3-Yb_2O_3-Al_2O_3 doped sol–gel layer-based waveguides for optical amplification at 1.5 μm which exhibited minimum propagation loss around 1.5 dB/cm for 5.7 cm length of fiber under total pump power of 100 mW. In 1999, L. Goldberg and

J. Koplow [9] made a high-power sidepumped Er/Yb fiber amplifier using Er/Yb fiber of 6.6 μm core diameter with 130 μm hexagonal inner cladding with a numerical aperture (NA) of 0.45 exhibited an attenuation co-efficient of 3.3 dB/m at 975 nm and produced 1.5 W and 1.1 W for the counter-propagating and co-propagating directions. L. Dong et al. [10] made Er/Yb codoped optical fiber comprising of cladding glass layer surrounding a glass core with an annular layer suitable for photosensitivity and a phosphosilicate core layer of size 5 μm composed of Er^{3+} (500 ppm) and Yb^{3+} (2000 ppm) surrounded by Ge/B doped photosensitivity regions and an index difference of 0.013 between the core and cladding region. S. Xi-Yi Huang and N. Sarma, [11] made highpower cladding-pumped broadband fiber source and amplifier based on erbium-ytterbium-aluminum (Er-Yb-Al) or erbium-ytterbium-phosphorous (Er-Yb-P) fibers with Er concentration ranging from 700 to 900 ppm, Yb 16,000–23,000 ppm and Al or P as much as possible such as greater than 6 mol% and 12 mol%, respectively, operating around 1550 nm. D. Bayart et al. [12] developed double-clad optical fiber and fiber amplifiers where the optical fiber core is doped with erbium ion (500–2500 ppm wt) and the Ytterbium ion concentration is 1 to 100 times of Er concentration, while Ce co-doping with concentration between 1 and 30 times of Er concentration.

J. K. Sahu et al. [13] had shown a 103 W erbium-ytterbium co-doped large-core fiber laser using Er-Yb co-doped optical fiber with a 24 μm core having D-shaped inner cladding with NA 0.20 fabricated by the MCVD process coupled with the solution doping technique. Such fiber exhibited small signal absorption of 2.5 dB/m containing 2 wt% Yb ion, which produced 103 W power at 1565 nm with slope efficiency of 30% with respect to the launched pump power and 34% with respect to the absorbed pump power, and for low power level efficiency it reached up to 40%. In 2003, N. Peyghambarian and S. Jiang [14] made erbium and ytterbium co-doped phosphate glass optical fiber amplifiers using short active fiber length where a phosphate glass fiber comprising P_2O_5 (30–80%), Yb_2O_3 (0–12%), Er_2O_3 (2.5–12%) R_2O (0–5%), L_2O_3 (5–30%), MO (5–30%) where R_2O is selected from alkali metal oxides like Li_2O, K_2O, Na_2O, Rb_2O, MO is selected from alkaline earth oxide consisting of BaO, BeO, MgO, SrO, CaO, ZnO, PbO, and La_2O_3 is selected from AhO_3, B_2O_3, Y_2O_3, and La_2O_3 prepared using a conventional glass melting technique process, which produces a gain per unit length up to 3 dB/cm at 1.54 μm when pumped with 980 nm light-emitting laser diode using few cm fiber lengths. In 2004, Z. G. Lu et al. [15] demonstrated erbium/ytterbium co-doped double-cladding fiber amplifier with 36.4 dBm output power wherein erbium ytterbium co-doped double-clad optical fiber amplifier.

In 2005, K. Tankala, [16] developed a method of fabricating preforms doped with rare earth metal for optical fibers wherein a porous silica soot is soaked with rare earth elements such as (Yb, Nd, or Er) along with other co-dopants, namely Al, P, Ge, or B. P. Nandi et al. [17] demonstrated super fluorescence from Yb- and Yb-Er-doped phospho-tellurite glass-based air-clad multimode fibers containing 25 mol% P_2O_5 and 0.25 mol% of Er_2O_3 and Yb_2O_3 exhibited ASE in the 1000–1100 and 1450–1650 nm range simultaneously.

In 2010, M. A Melkumov et al. [18] had shown the effects of Yb^{3+} and Er^{3+} concentrations and doping procedure on excitation transfer efficiency in ErYb-doped phosphosilicate fibers fabricated using MCVD coupled with the chelate delivery technique, which comprises Er concentrations from 0.07 to 0.4 wt% with Yb concentrations from 0.4 to 8 wt%, yielding around 38% lasing efficiency with respect to absorbed pump power and 33% with respect to launched pump power. W. Jing et al. [19] developed Er/Yb co-doped optical fiber using a heating-modified MCVD method, having the refractive index difference between core and cladding at about 0.75% and 588.8 dB/m Yb absorption at 976 nm.

M. Vivona et al. [20] made Ce-doped phosphosilicate Yb/Er-doped glass-based optical fiber containing 8 at% P, 0.008 at% Er, 0.165 at% Yb, and 0.009 at% Ce resulted in infrared emission around 1535 nm. A high-power, high-repetition-rate single-mode Er-Ybdoped fiber laser system was reported by I. Pavlov et al. [21] based on Er-Yb-co-doped phosphosilicate fiber amplifier it generated 10 W average output power with a pulse duration of 0.45 sec with M^2 value of 1.06. Later on, W. Chen et al. [22] made a high-power Er, Yb-doped superfluorescent fiber source of 16 W output power near 1.55 μm using Er/Yb co-doped fiber of core diameter 25 μm with NA 0.22 surrounded by pure silica cladding having a D-shaped coated with low RI polymer coating to achieve an effective NA of 0.4 with Er3+ concentration of 0.51 x 10^{20} cc and Yb^{3+}/ Er^{3+} doping ratio of 17 exhibited lasing efficiency of 30.7% and bandwidth (FWHM) of 17 nm.

In 2013, F. Sanchez et al. [23] presented a 10 W Er/Yb fiber amplifier, which resulted in 8 W power with 20% lasing efficiency. Liang Dong et al. [24] developed highly rare-earth-doped optical fibers for fiber lasers and amplifiers based on phosphosilicate glass containing about 50 mol% phosphorous, aluminum concentration variation within 0.5 to 1.5 mol%, boron in the range of 0 to 25 mol%, Er and Yb ion concentration within 0.5 to 1 mol% which resulted in gain within 100–500 dB/m and exhibited saturated photodarkening loss of less than 10 dB at emission wavelength.

L. Wang et al. [25] made Yb/Er co-doped phosphate glass-based all-solid single-mode photonic crystal fiber containing 6 wt% Yb^{3+} ions and 0.5 wt% Er^{3+} ions photonic crystal fiber shows 1.6 W output power with a lasing threshold of 150 mW and slope efficiency of 21%. B. Cadier et al. [26] reported radiation-resistant rare-earth-doped optical fiber and a method of radiation hardening rare-earth-doped optical fiber by MCVD process coupled with the solution doping technique with a core comprising up to 8 at% phosphorous, about 100–1000 ppm Er, 500–10,000 ppm Yb, with Yb to Er ratio between 12 and 25; Ce concentration between 500 and 10,000 ppm and Yb/Ce ratio between 0.1 and 10, surrounded by a fluorinated silica cladding with numerical aperture (NA) higher than 0.4 and tested in a ^{60}Co gamma radiation environment with a dose rate of the order of I krad/h showing degradation <20% of initial power.

E. Fujita et al. [27] had shown high-power narrow-linewidth linearly polarized 1610 nm Er:Yb all-fiber MOPA using Er-doped aluminosilicate fiber as the first amplifier attached with the second amplifier using Er/Yb co-doped phosphosilicate double-clad fiber comprising a core diameter of 6 μm, NA of 0.2 which was finally amplified by Er/Yb co-doped phosphosilicate fiber of core diameter 10 μm having NA of 0.2 to achieve emission from 1610 to 1627 nm. Y. Chu et al. [28] made Ce^{3+}/Yb^{3+}/Er^{3+} triply doped bismuth borosilicate glass for broadband near-infrared fiber amplifiers prepared by the conventional melt-quenching method to enhance emission around the 1.55-μ region.

In this book chapter, we have demonstrated the fabrication of multielements doped (Al, Ge, P, Ce, B, and F) Er/Yb co-doped phospho-silica glass-based optical preform employing the modified chemical vapour deposition (MCVD) process coupled with solution doping (SD) technique, varying the doping levels of such multielements followed by different shaping of the mother preform for the drawing of cladding pump fiber with on-line low RI resin coating. We have studied the lasing performances of double-D, pentagonal, hexagonal, and octagonal cladding-shaped low-RI coated fibers. The lasing performance of the pedestal design-based Er/Yb doped multielements glass-based octagonal-shaped cladding-pump fiber along with air-clad non-PM and Panda-type PM Er/Yb-doped multielement glass-based normal RI-coated fibers were also evaluated. We have achieved a maximum 21.0 W output power at 1558 nm under 52 W absorbed pump power of 940 nm using octagonal-shaped cladding pumped low RI coated 125 micron fiber containing 22.25 wt.% (P_2O_5), 2.43 wt.% (GeO_2), 0.5 wt.% (Al_2O_3), 0.161 wt.% (Ce_2O_3), 2.41 wt.% (B_2O_3), 0.25 wt.% (F), 0.515 wt.% (Er_2O_3), and 5.25 wt.% (Yb_2O_3). Such kind of multielement doped silica glass shows a stable average output power around 16.0 W for long-term operation more than 150 h without exhibiting parasitic lasing around 1.0 μm and strong radiation resistance behavior with degradation of output power below 10% under a 6.0 Krad cumulative radiation dose of ^{60}Co gamma-radiation environment, indicating its potential application in free-space communication. In terms of fiber performance, this novel glass composition provides a single complete solution showing high lasing efficiency (~ 41.0%) having an output power of 16.0 W without parasitic lasing around 1.0 μm, long-term stability of output power for more than 5 days with less than 1.5% degradation at room temperature. The fiber shows a gain of around 28.0 dB with a flatness of less than 0.45 dB in the wavelength range of 1540 to 1570 nm for a 10 dBm input signal power. Such kind of fiber exhibits the optical gain variation of 0.1 dB/Krad and radiation hardness behavior (< 10.0% of the initial power) under a 6.0 Krad cumulative dose of ^{60}Co gamma-radiation.

17.2 Multielements Doped Silica Glass-Based Er/Yb Co-Doped Optical Fibers

Multielements with Er/Yb codoped silica glass-based optical fiber preforms are made through deposition of a porous core layer of suitable vapor phase composition containing $SiCl_4$, $POCl_3$, SF_6, O_2, and He employing the MCVD process under the backpass deposition technique. The composition of the doping host of fibers is optimized based on their lasing and radiation hardening performance of fibers drawn from a series of preforms made through variation of the solution strength of doping precursors as well as the flow rate of $SiCl_4$, $GeCl_4$, and $POCl_3$. The optimization of the composition of the porous layer as well as the deposition temperature is accomplished through the formation of uniform porosity, ensuring good adhesion with the glass surface in order to avoid delimitation of the porous soot layer during the solution doping stage. Accordingly, such a porous core layer deposited at 1450 ± 25 °C followed by pre-sintering around 1200 ± 10 °C to make uniform porosity for the incorporation of other elements through the solution doping process [29, 30] shown in Fig. 17.1a. The deposited fluorinated phospho-silica soot layer was soaked for 1 h into an alcoholic solution comprised of a suitable strength of $AlCl_3.6H_2O$, $ErCl_3.6H_2O$, $YbCl_3.6H_2O$, $CeCl_3.7H_2O$, and H_3BO_3. The fabrication techniques involved optimization of various process parameters such as the individual concentration of dopants in soaking solution, P-doping levels as well as Yb:Er ratio (7.5–11.0) to improve the amplification performances with enhanced efficiency. The Er/Yb codoped multielement silica glass tube was converted into a solid optical preform of 10.0 ± 0.1 mm diameter under optimized collapsing conditions, as shown in Fig. 17.1b. The diameter of such an initial preform increased to 18.5 ± 0.1 mm through the overcladding process using a thick silica tube of suitable dimension. Finally, octagonal shaping of the overcladded preform is done through grinding and polishing process to draw octagonal cladding-shaped Er/Yb-doped fiber with a low RI resin (PC 375 L AP) coating. The novelty of fabricated fiber is the use of the multielement glass of suitable composition, all of which have their individual roles to improve the final fiber performance. F-doping helps to achieve radiation hardness by the elimination of Si- and P-related defect centers, Ce-doping helps to prevent the formation of Yb-related defect centers, which become precursor for the origination of photodarkening; and Al-doping levels require optimization to minimize the formation of Al-O hole defect centers without sacrificing Er and Yb ion solubility in the core glass matrix. On the other hand, Ge-doping is performed during sintering of soaked porous soot layer, which helps to control numerical aperture. B- along with P-doping enhances the energy transfer efficiency from Yb to Er by reducing the fluorescence lifetime of $^4I_{11/2}$ level of Er while increasing the fluorescence lifetime of $^4I_{13/2}$ level of Er to suppress 1.0 μm Yb self-lasing intrinsically. Here, we have selected four different normal design based octagonal-shaped cladding pump Er/Yb codoped fibers (MEY-1, MEY-2, MEY-3, and MEY-4) with varying doping levels of the multielements. The parameters of four such important fibers are given in Table 17.1. The field emission gun scanning electron microscope (FEGSEM) is utilized to investigate the surface morphology of multielement-doped Er/Yb codoped preform samples for optimization of the P_2O_5-doping level.

One of the fibers (MEY-4) reveals the formation of the phase-separated region above 25 wt% doping of P_2O_5 in the preform core (Fig. 17.1c). This phase-separated region serves as a scattering center of light, which induced high base loss of the fiber around 50.0 dB/Km at 1300 nm with reduced lasing slope efficiency. On the other hand, fiber preform (MEY-1) having P_2O_5 content less than 22.0 wt% does not form such kind of phase-separated glass shown in Fig. 17.1d having low background loss of 10.0–12.0 dB/km at 1300 nm as described later on in Fig. 17.4a.

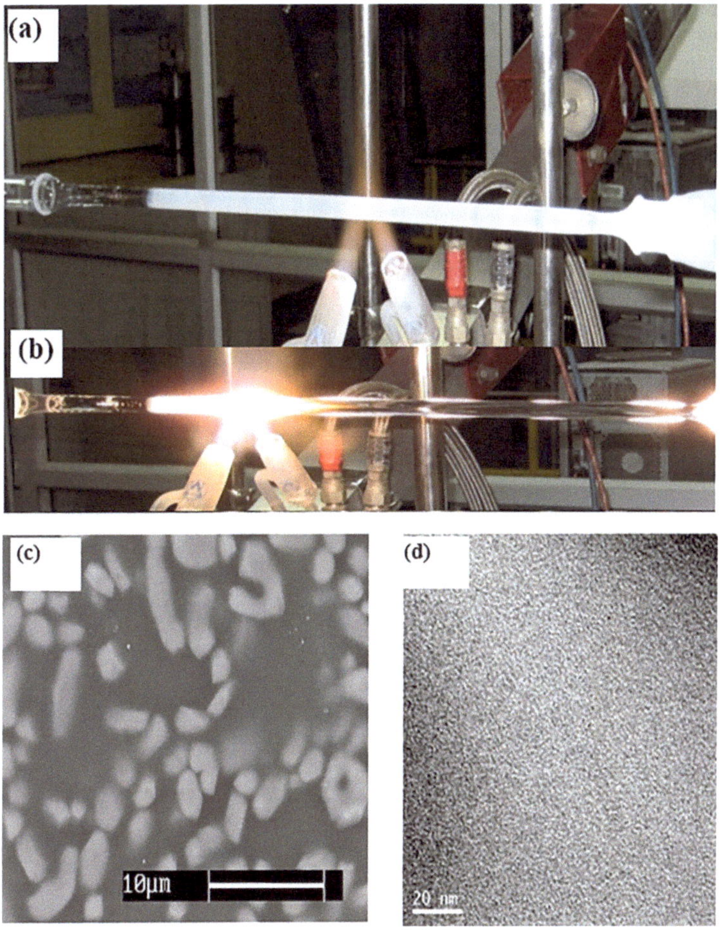

Fig. 17.1 (**a**) Backpass deposition of porous SiO_2-P_2O_5-F layer by MCVD process; (**b**) Collapsing of deposited silica tube into preform, (**c**) FEGSEM image of 25.0 wt% (MEY-4) and (**d**) 22.0 wt% (MEY-1) P_2O_5 with multielement codoped Er/Yb optical fiber preforms

Table 17.1 Fiber parameters

Fiber Number	NA	Core diameter (micron)	Yb/Er ratio	Cladding absorption at 915 nm (dB/m)	Core absorption at 1530 nm (dB/m)
MEY-1	0.22	9.70	10.15	2.60	50.5
MEY-2	0.19	13.0	8.5	3.15	51.0
MEY-3	0.24	10.5	11.5	3.5	51.6
MEY-4	0.25	10.75	12.5	3.75	35.7

The core region of preform MEY-1 containing 22.0 wt% P_2O_5 becomes optically transparent (Fig. 17.2a). Preform MEY-4 containing 25.0 wt% P_2O_5 shows a milky white region shown in Fig. 17.2b, signifies the formation of phase-separated regions which gives rise to high background loss of the fiber.

Fig. 17.2 Microscopic pictures of multielement co-doped Er/Yb optical fiber preforms (**a**) MEY-1 and (**b**) MEY-4

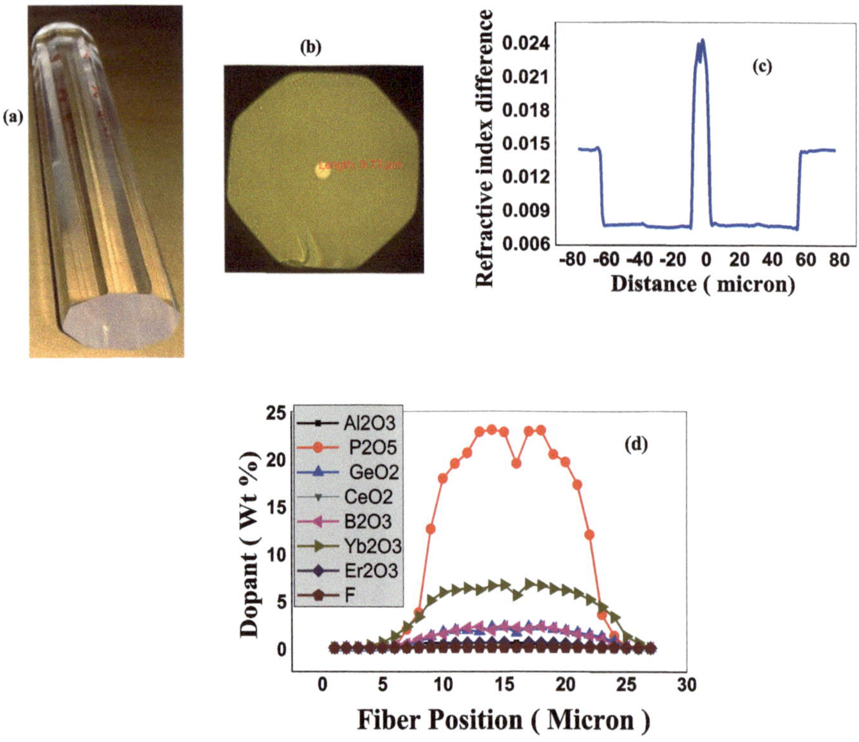

Fig. 17.3 Octagonal-shaped optical preform (**a**); cross-sectional view (**b**); refractive index profile (**c**); and EPMA curve of Er/Yb codoped fiber (MEY-1) (**d**)

The octagonal-shaped preform and cross-sectional view of fiber (MEY-1) having a core diameter of 9.7 μm drawn from such preform are presented in Fig. 17.3a, b, respectively. Fig. 17.3c shows the refractive index profile of a fiber having NA around 0.22 measured by a fiber analyzer. The nature of the profile is found to be very sharp from the core-cladding boundary to the center of the core region. The uniform distribution of different doping elements is evident from Fig. 17.3d which obtained from electron probe microanalyses (EPMA) of the fiber MEY-1 maintaining the spatial resolution of 1 μm. The different dopant concentrations evaluated

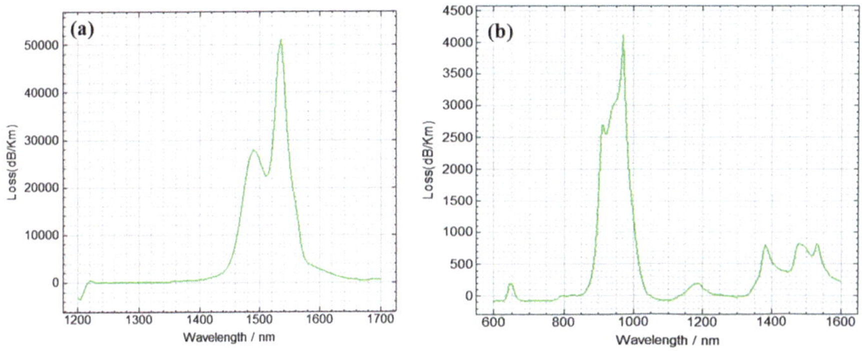

Fig. 17.4 Core absorption (**a**) and cladding absorption (**b**) loss curves of Octagonal shaped Er/Yb codoped fiber (MEY-1)

from EPMA indicate the presence of 22.25 wt.% (P_2O_5), 2.43 wt.% (GeO_2), 0.5 wt.% (Al_2O_3), 0.161 wt.% (Ce_2O_3), 2.41 wt.% (B_2O_3), 0.25 wt.% (F) 0.515 wt.% (Er_2O_3), and 5.25 wt.% (Yb_2O_3) (Fig. 17.4).

17.3 Pedestal Design-Based Multielements-Doped Silica Glass-Based Er/Yb Co-Doped Optical Fibers

Traverse mode instability (TMI) plays an important role in the stability of the output power of high-power fiber lasers and optical amplifiers. Such TMI was observed for single-mode and few-mode fibers but not for multimode fibers as the modal interference prevents the formation of necessary thermal grating within the core region. Therefore, the TMI raised due to the prominent overlapping of lower-order modes with the active rare-earths-elements-doped region. In this design, the index-raised pedestal results in distinct changes in the modal content beyond the core modes compared to the non-pedestal fibers. The reported data shows that the selection of appropriate pedestal diameter no longer influences the core modes by a change in pedestal size where the pedestal behaves like a cladding, considering the pedestal-to-core diameter ratios ranging from 1.5:1 to 4:1.

Here, we have made a pedestal design-based multielement doped silica glass-based Er/Yb co-doped optical preform through the MCVD process in conjunction with the SD technique. The inner ring core of the preforms surrounding the central core was formed by sintered alumina-silica silica layers which was followed by solution-soaked sintered SiO_2–GeO_2–Al_2O_3–P_2O_5–CeO_2–B_2O_3–Er_2O_3–Yb_2O_3–F core layer deposition into 20/17 mm diameter suprasil F-300 tubes. The design parameters of the fabricated pedestal design based on Er/Yb codoped fiber are given in the following table (Table 17.2).

Table 17.2 Fiber parameters

Fiber Number	NA	Thickness of aluminum doped ring core (micron)	Er/Yb doped core diameter (micron)	Yb/Er ratio	Cladding absorption at 915 nm (dB/m)	Core absorption at 1530 nm (dB/m)
PDEY-1	0.19	12.5 (Average)	13.25 (Average)	9.5	2.80	49

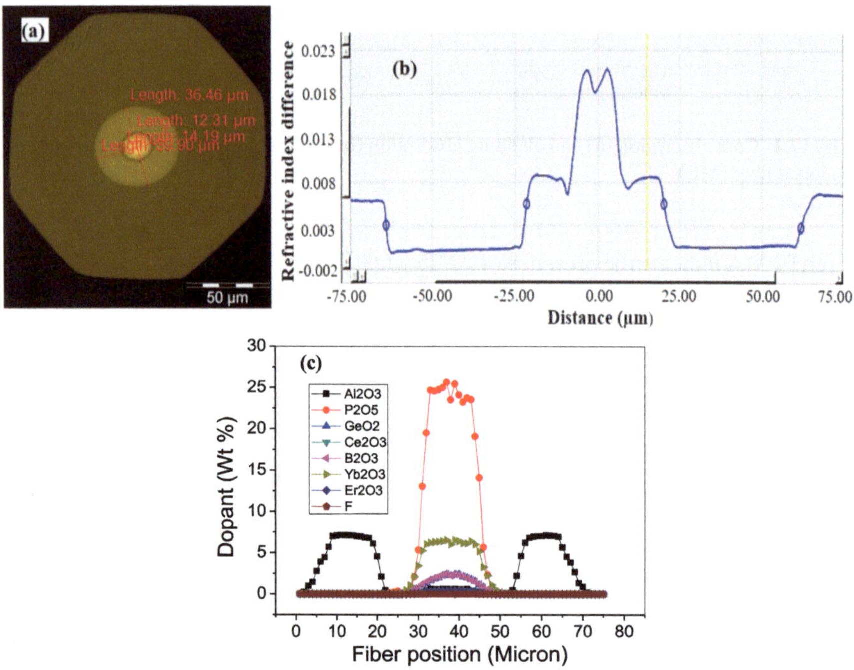

Fig. 17.5 Cross-sectional view (**a**), refractive index profile (**b**), and EPMA curve (**c**) of octagonal shaped pedestal design based multilelments doped Er/Yb optical fiber (PDEY-1)

The cross-sectional view along with the refractive index profile of this kind of fiber are shown in the following Fig. 17.5a, b, respectively. The doping levels of Al_2O_3 along the diameter of the alumina-silicate glass-based pedestal region, as well as the various multielements along the diameter of the central Er/Yb-doped region, are shown in Fig. 17.5c.

The undoped pedestal region of the fiber contains a maximum of 7.1 wt% Al_2O_3 only. On the other hand, the central doped core region consists of 0.6 wt% Al_2O_3, 24.5 wt% P_2O_5, 1.0 wt% GeO_2, 0.096 wt% CeO_2, 2.32 wt% B_2O_3, 0.029 wt% F, 2.97 wt% Yb_2O_3, and 0.55 wt% Er_2O_3. The measured core and cladding absorption loss curves of PDEY-1 fiber are given in Fig. 17.6a, b, respectively.

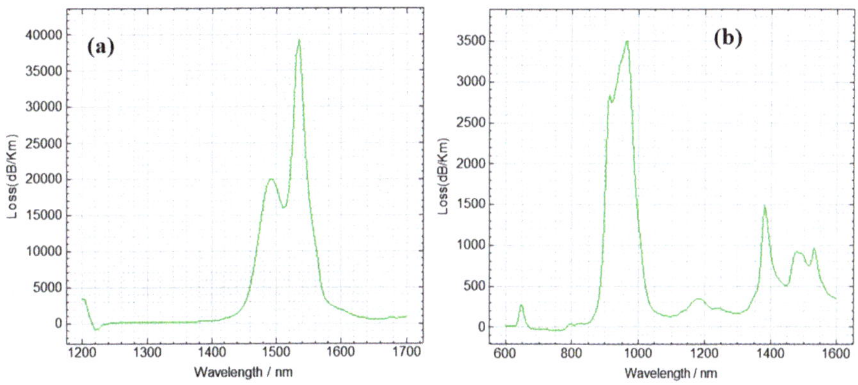

Fig. 17.6 Core absorption (**a**) and cladding absorption (**b**) loss curves of Octagonal shaped pedestal design based multielements doped Er/Yb optical fiber

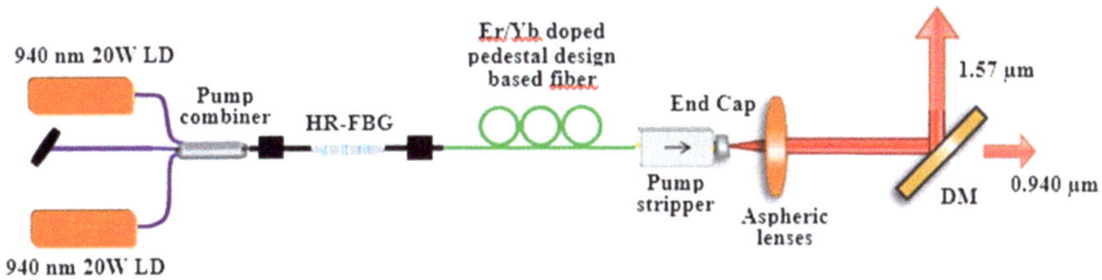

Fig. 17.7 The schematic diagram of experimental set-up for lasing characterization of Octagonal shaped pedestal design based multielements doped Er/Yb optical fiber

The CW lasing properties of such a type of pedestal design based on Er/Yb codoped fiber was characterized in a cladding pumping configuration as shown in Fig. 17.7.

CW lasing at ~1.57 µm has been tested, choosing a laser resonator formed by a high reflection fiber Bragg grating (HR FBG) ~99.9% at 1557 nm written in a 20/125 µm passive fiber with a 0.20 core NA to minimize splice loss with the fibers under test and cleaved normally at the active fiber ends, which provided 4% Fresnel reflection. Here, the fiber-pigtailed pump diode of 20.0 W output power at 940 nm was used in this case. The pump at 940 nm was coupled into the investigated fiber through the high-reflection fiber Bragg grating using a pump combiner unit (PC). The unabsorbed pump was filtered out at the output of the fiber end using a pump stripper made directly on an Er/Yb fiber under test a few cm away from its end. Additionally, a dichroic 1550/940 nm mirror (DM) was used to filter out the unabsorbed pump unfiltered by the pump stripper. The optimal fiber length of Er/Yb-doped fiber used was around 9.5 m which provided the maximum output power of 7.5 W with a slope efficiency of 25% as shown in Fig. 17.8a. The corresponding output lasing spectra was shown in Fig. 17.8b.

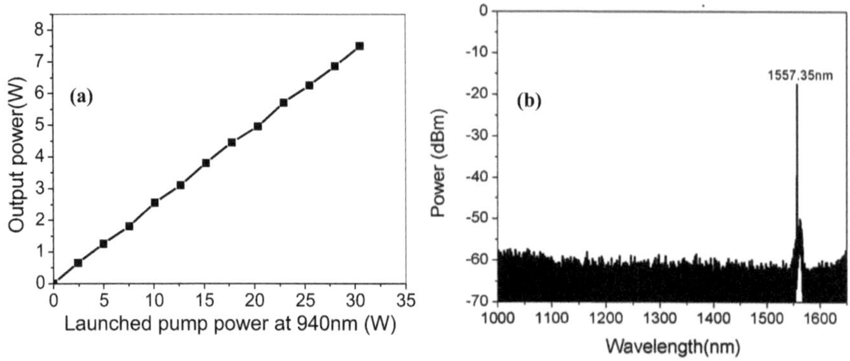

Fig. 17.8 Output power of EYPD-1 fiber versus the launched pump power using optimum 9.5 m fiber length (**a**); The corresponding lasing output spectrum (**b**)

17.4 Study of the Lasing Characterization of Various Asymmetric Structure of the Silica Cladding Glass

Here, we have studied the effect of cladding structure on the lasing properties of double-clad Er/Yb fiber having the same material composition of multielement doped silica glass. Here, we have taken double-D, pentagonal, hexagonal, and octagonal cladding-shaped low RI-coated fiber of 125-micron diameter. In all cases, the core NA becomes 0.20 ± 0.01, the core diameter varies slightly within 9.7–12.7 micron. All the fibers of different cladding-shaped structures contain almost the same doping levels as the initial circular preform becomes the same. The core glass of such fibers contains the same amount of various multielements such as Al_2O_3: 0.5 wt%, P_2O_5: 23.8 wt%; GeO_2: 2.4 wt%; CeO_2: 0.15 wt%; B_2O_3: 1.986 wt%; Yb_2O_3: 5.1 wt%; Er_2O_3:0.50 wt%; F: 0.025 wt% as estimated from EPMA. The microscopic pictures of different asymmetric cladding structures of Er/Yb-doped fibers are shown in Fig. 17.9.

The lasing property of various asymmetric cladding structures based low-RI-coated cladding pump fibers shown in Fig. 17.10 was measured using the lasing experimental setup as described in Fig. 17.7. The optimized length for the study of the lasing increased from 8.5 m for the octagonal-shaped structure to 14 m for the double-D-shaped structure. The lasing efficiency gradually decreases from octagonal cladding fiber to double-D-shaped cladding fiber. The maximum lasing efficiency of 41.8% and 40.1% observed for octagonal and hexagonal cladding-shaped fibers, respectively. On the other hand, pentagonal and double-D cladding-shaped fibers show 32.85% and 20.49% lasing efficiency, respectively. The different optimized lengths with maximum values of lasing efficiency of various asymmetric cladding structures arise due to different pump coupling efficiencies with the doped core region, though all four fibers contain the same doping levels of Er and Yb along with other elements.

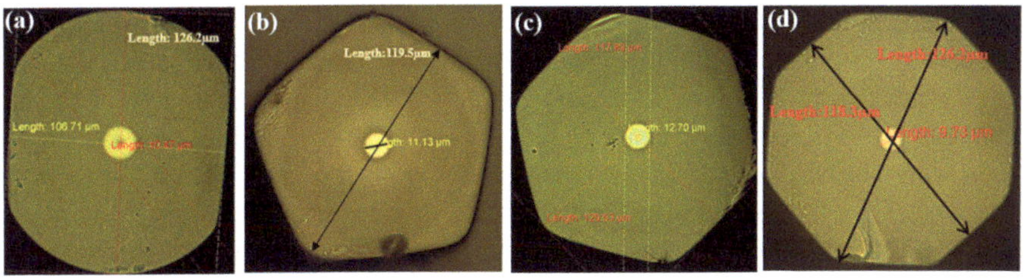

Fig. 17.9 Cross-sectional view various asymmetric cladding shaped Er/Yb doped fibers: (**a**) Double-D (**b**) Pentagonal (**c**) Hexagonal and (**d**) Octagonal

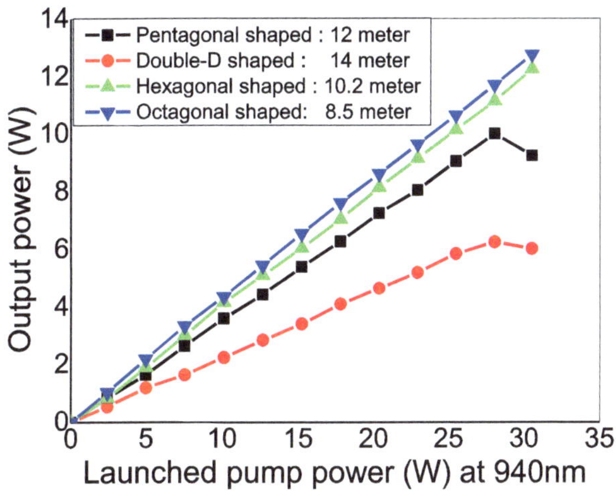

Fig. 17.10 Output power of various asymmetric cladding shaped double clad fiber versus the launched pump power under various optimised length

17.5 Fabrication of Air-Cladded Structure-Based Er/Yb Doped PM and Non-PM Fiber for High-Power Fiber Laser

The cladding-pumped air-clad silica fibers are very efficient at handling high power without damage [5]. The proper optimization of the structure of cladding pump air-clad rare-earths-doped fibers with respect to large core with low NA, shorter fiber length, high NA of pump cladding, and low residual pump power is required for efficient coupling of pump light along with minimization of nonlinear effects as well as high conversion of pump light and high heat handling capacity because they can be easily designed to have a large core size with low NA, short fiber length, high NA of the pump cladding, low residual pump power, and use of materials that can withstand high temperatures. Here, we are reporting the fabrication of circular air-clad-based Er/Yb-doped non-PM and PM fibers. Initially we have made the Er/Yb-doped multielement silica glass-based optical preform of 10.5 mm diameter by the MCVD process in combination with the solution doping technique as described earlier. The one circular array of 3.0 mm OD and 2.0 mm ID capillaries around the 10.5-mm diameter preform rod was inserted properly into the 20/17-mm silica tube in a proper tight way. The diameter of the jacketing tube will be chosen in such a way that the entire stack fits tightly inside it. The ends of capillaries outside the furnace are sealed using H_2/O_2 hand burner before cane drawing in order to sufficiently pressurize the air holes for attaining the air-clad structure. The inner side of the jacketing tube is evacuated in order to prevent the existence of any interstitial spaces in the final fiber structure. The fiber was drawn at a temperature of 1980–2000 °C with ACR preform feed rate of 6–8 mm/min and drawing speed of 30–35 m/min. The suction pressure was maintained at 275–300 mm of Hg in order to prevent interstitial spaces. The 125 ± 2 μm diameter fiber was drawn, having elongated and oval-shaped air holes. Dimensions of the air holes were 28.50 ± 0.05 and 17.25 ± 0.05 μm. The bridge width between two adjacent air holes was 450 ± 15 nm. For making an air hole-based Panda-type Er/Yb-doped ACR-based cladding pump Er/Yb fiber, we first made Er/Yb-doped multielement silica glass-based preform of a diameter around 10.5 mm by MCVD and solution doping techniques. The diameter of such a preform is reduced to 3.0 mm diameter by the chemical etching process using buffer HF solution containing NH_4F to maintain uniform etching along the whole length of the preform. The Er/Yb-doped core region is around 1.2 mm diameter within the chemically etched 3.0 mm preform rod. Once the Er/Yb-doped rod of 3.0 mm diameter is obtained, two thin-walled capillaries of 3.0 mm OD and 2.0 mm ID are placed on opposite sides of Er/Yb-doped rod and rest of the potion of first circle is filled by 3.0 mm diameter silica rods. Then suitable various diameter-based silica rods are staked further in such a way that doped central preform rod along with two silica capillaries and the outer stack of silica capillaries of OD of 3.0 mm and ID of 2.0 mm is put inside a 20/17 mm silica tube to form the final ACF preform. The diameter of the jacketing tube is chosen in such a way that the entire stack fits tightly inside it. The preform is then mounted on the furnace of a conventional fiber drawing tower with a special belt-tractor arrangement attached to it for drawing canes of uniform dimensions. Canes having a diameter of 5 mm are drawn from the preform, again inserted into a thick silica tube, sealed at the exposed end, mounted on the furnace of the drawing tower, and finally drawn into fibers of the required diameter and length (Tables 17.3 and 17.4).

Table 17.3 Fiber parameters

Fiber Number	NA	Hole diameter (μm)	Pitch (μm)	Er/Yb doped core diameter (μm)	Yb/Er ratio	Cladding absorption at 915 nm (dB/m)	Core absorption at 1530 nm (dB/m)
ACMEY	0.19	21.2	0.12	10.0	10.4	2.80	49

Table 17.4 Fiber parameters

Fiber Number	NA	Diameter of two air holes around the central core (μm)	Pitch (μm)	Er/Yb doped elliptical core diameter (μm)	Yb/Er ratio	Air hole length of the ACR (μm)	Separation distance between two air holes (μm)	Diameter of inner cladding (μm)
ACMEY-PM	0.20	12.20	0.95	13.25 8.15	10.45	28.57 14.95	15.25	52.94

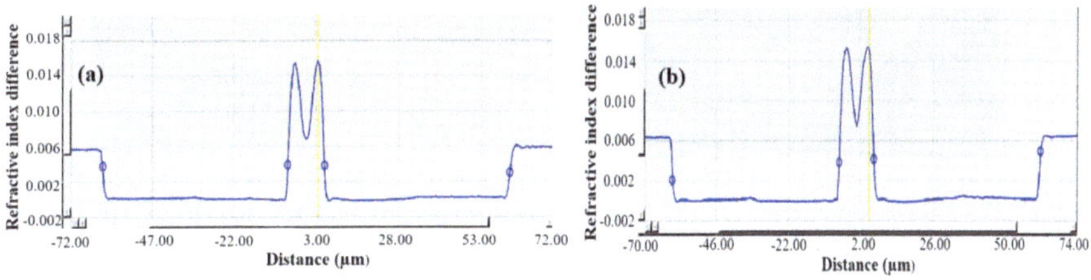

Fig. 17.11 Refractive index profile of initial Er/Yb doped multielement silica glass preforms of ACR fibers: (**a**) ACMEY and (**b**) ACMEY-PM

The refractive index profiles of initial Er/Yb-doped multielement silica glass based fibers drawn from their preforms, which are used for making ACMEY and ACMEY-PM fibers, are shown in Fig. 17.11 and their cross-sectional microscopic pictures in Fig. 17.12.

Both Er/Yb-doped ACR fibers contains the same doping levels of Er and Yb, having the core absorption of 49.0 dB/m at 1530 nm. The lasing properties of such ACR fiber were measured under pumping at a 940 nm wavelength. Here, a collimating lens is used to align the pumping beam, and a coupling lens with NA = 0.3 serves as the laser input couple. The cavity comprises a butt-coupled dichroic mirror with high reflectivity (>99%) at 1557 nm and a cleaved fiber end with ~5% Fresnel reflectivity. When pumped at 940 nm wavelength, lasing was obtained at 1557.3 nm from 7.7 m length of ACMEY fiber with a maximum average output power of 1.5 W under launched pump power of 9.0 W and a slope efficiency of 16.6% shown in Fig. 17.13. Further increasing the launched a pump power output decreases gradually due to sufficient leakage of pump power from the air-cladded structure. For ACMEY-PM fiber, the maximum output power obtained 2.1 W under 10.0 W launched pump power using 6.5-m fiber length and a slope efficiency of 21.0% shown in Fig. 17.13. Further increasing pump power, the fiber end burns due to sufficient leakage of pump power from the air-cladding region. The output spectra of both ACR fibers was shown in Fig. 17.14. The degree of polarization of such air-cladded Er/Yb-doped PM fiber will be measured in the next phase of work.

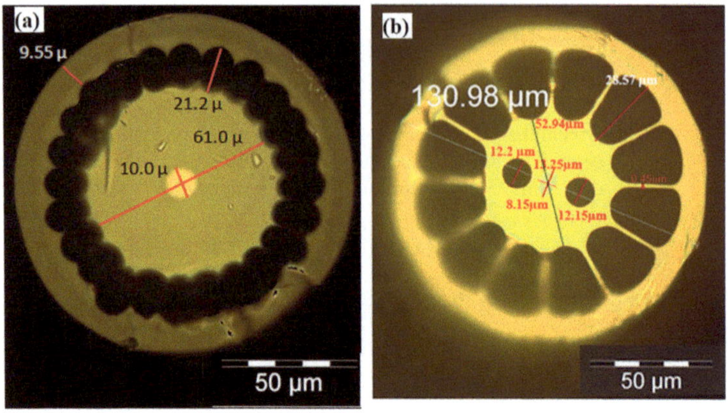

Fig. 17.12 Microscopic cross-sectional view of ACR fibers: (**a**) ACMEY and (**b**) ACMEY-PM

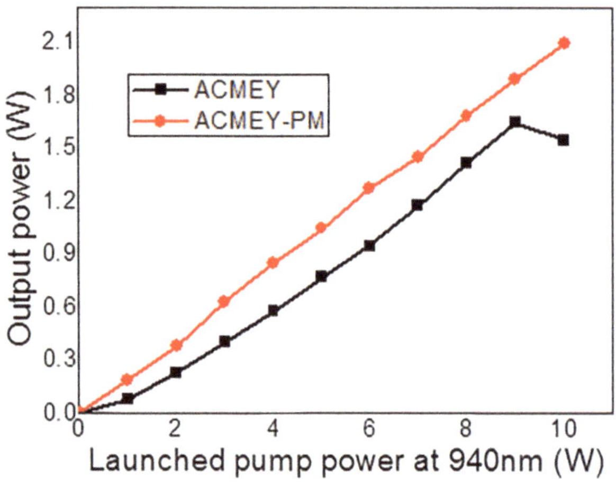

Fig. 17.13 Output power verses the launched pump power curves of ACMEY and ACMEY-PM fibers

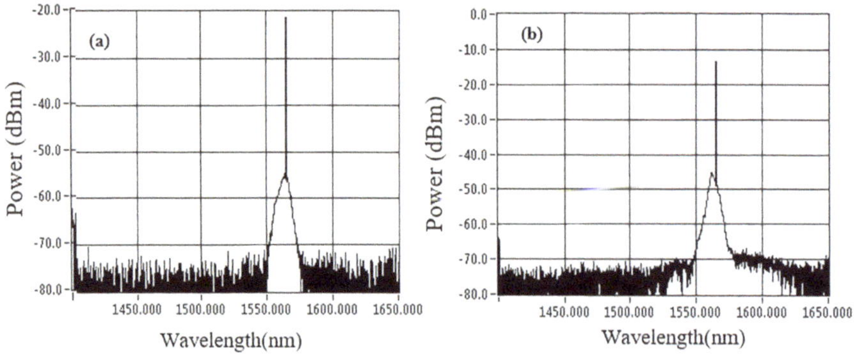

Fig. 17.14 Lasing output spectra of ACR fibers: (**a**) ACMEY and (**b**) ACMEY-PM

17.6 Experimental Set-Up for Study of Amplification Properties

The experimental setup for the study of amplification properties of Er/Yb codoped fibers is shown in Fig. 17.15. The pumping source is used at a 940 nm wavelength. Optical output is measured at the 99% port of the 1 × 2 high-power-rated tap coupler. The seed is an incoherent amplified stimulated emission (ASE) source at 1558 nm with a 0.5 nm spectral width. The backward pumping scheme has been adopted, which is favorable for optimizing the pump power to signal power conversion efficiency. Initially a signal with 600 mW power is used as the input optical signal to probe the high-power erbium-doped fiber amplifier (EDFA) module. At the third-stage amplification output, the C-band signal power of 3.0 W is achieved. With 0.6 W optical power as the input to the final, we have achieved 16.0 W output power at the fourth stage amplification using MEY-1 fiber and 3.2 W, as the input signal power, we have achieved maximum of 21.0 W output power as described later.

^{60}Co gamma radiation source is used to characterize the radiation sensitivity of optical amplification performance at low dose rates (~0.35 rad/s) and cumulative doses of up to 6.0 krad. Online optical gain and output power measurements have been conducted at room temperature, keeping the active fiber exposed to the Gamma-rays. All other parts, such as pump laser diode and detectors, remain in a radiation-free instrumentation zone for this set of experiments. The core absorption as well as cladding absorption losses of Er/Yb co-doped multi-element phospho-silica glass-based fiber is determined by applying a standard cut-back procedure using a white light source. The results shown in Fig. 17.16a reveal the clad absorption of 2.6 dB/m at 915 nm and core absorption of 50.5 dB/m at 1530 nm.

The fluorescence spectra of Er/Yb co-doped multi-element silica glass-based optical fiber is measured using in-core pumping of the fiber by a diode laser with single-mode fiber output (500 mW; 940 nm) in the lateral detecting arrangement shown in Fig. 17.16b. The fiber (MEY-1) shows broad fluorescence extending up to 1650 nm. The fluorescence decay time of $^4I_{11/2}$ levels of Er ions of fabricated Er/Yb codoped fibers MEY-1, MEY-2, and MEY-3 having P_2O_5 content of 22.25 wt%, 17.0 wt%, and 14.0 wt% found to be 0.52, 0.63, and 0.66 ms, respectively shown in Fig. 17.16c. We have also measured the fluorescence lifetime of $^4I_{13/2}$ levels of Er ions and $^2F_{5/2}$ level of Yb ions in different Er-Yb-doped optical fibers given in Table 17.5, The fluorescence lifetime of $^2F_{5/2}$ level of Yb ions almost remains

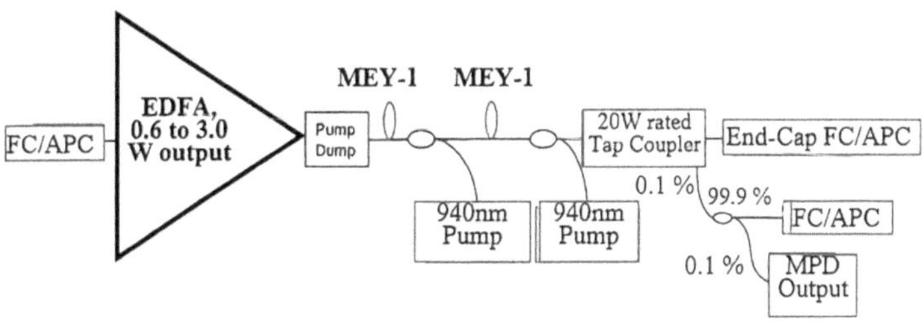

Fig. 17.15 Experimental set-up for studying amplification properties of Er/Yb co-doped fibers

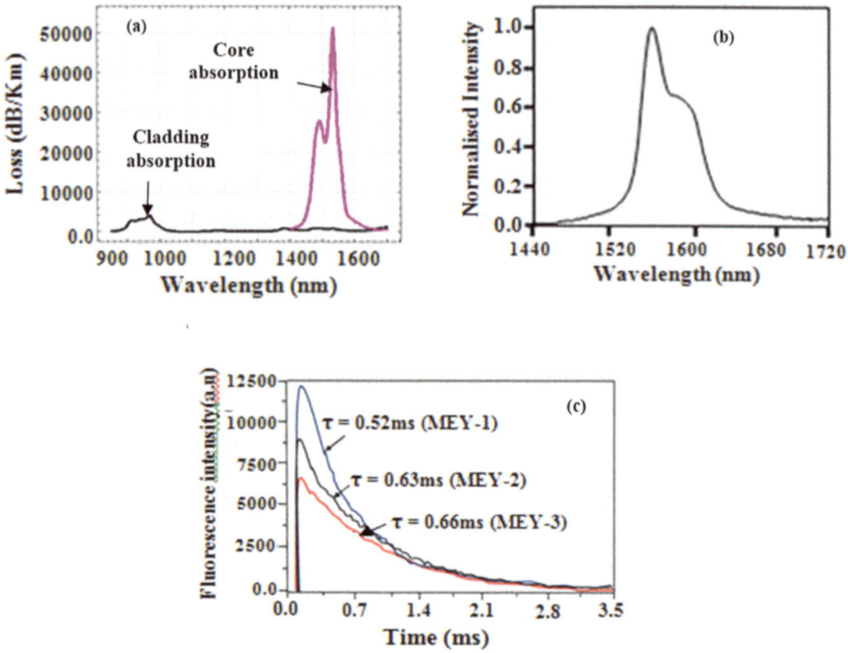

Fig. 17.16 Core & cladding absorption (**a**) along with fluorescence spectra (**b**) of Er/Yb codoped fiber (MEY-1) and fluorescence life time curves of $^4I_{11/2}$ level of Er ions of different Er/Yb codoped fibers

Table 17.5 Fluorescence lifetime data

Fiber Number	Fluorescence life-time of $^4I_{11/2}$ level	Fluorescence life-time of $^4I_{13/2}$ level	Fluorescence life-time of $^2F_{5/2}$ level of Yb^{3+}
MEY-3	0.66 ms	7.849 ms	1.079 ms
MEY-2	0.63 ms	8.2151 ms	1.089 ms
MEY-1	0.52 ms	9.49 ms	1.07 ms

constant with increasing the fluorescence lifetime of $^4I_{13/2}$ levels of Er ions from fiber MEY-3 to MEY-1.

The effect of variation of pump power at 940 nm pump wavelength on the fluorescence lifetime of $^4I_{13/2}$ levels of Er ions and $^2F_{5/2}$ level of Yb ions of two fibers, MEY-1 and MEY-3, are shown in Fig. 17.17a, and 17.17b respectively.

The fluorescence lifetime of $^4I_{13/2}$ levels of Er ions of MEY-1 fiber increases gradually with increasing pump power and reaches a maximum value of 9.49 ms, whereas for MEY-3 fiber, the fluorescence lifetime of $^4I_{13/2}$ levels of Er ions decreases with increasing pump power. On the other hand, the fluorescence lifetime of $^2F_{5/2}$ level of Yb ions of two fibers, MEY-1 and MEY-3, remains the same. This phenomenon indicates that the clustering effect of erbium ions is very more prominent in the case of MEY-3 fiber compared to MEY-1, as the P_2O_5 content of MEY-1 is much higher than MEY-3 fiber.

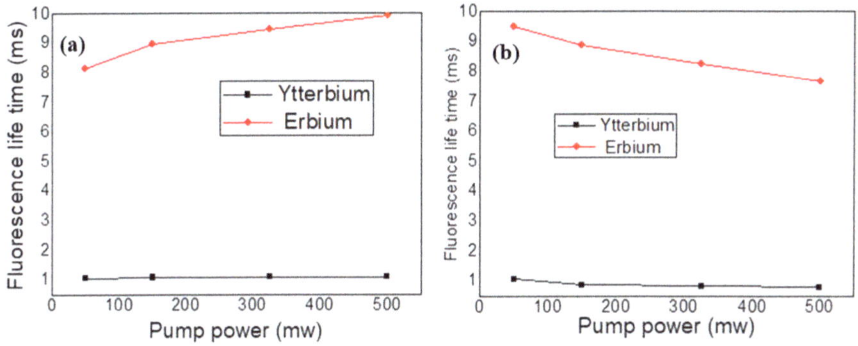

Fig. 17.17 The fluorescence lifetime curves of $^4I_{13/2}$ levels of Er ions and $^2F_{5/2}$ level of Yb ions of two fibers (**a**) MEY-1 and (**b**) MEY-3 with variation of pump power at 940 nm wavelength

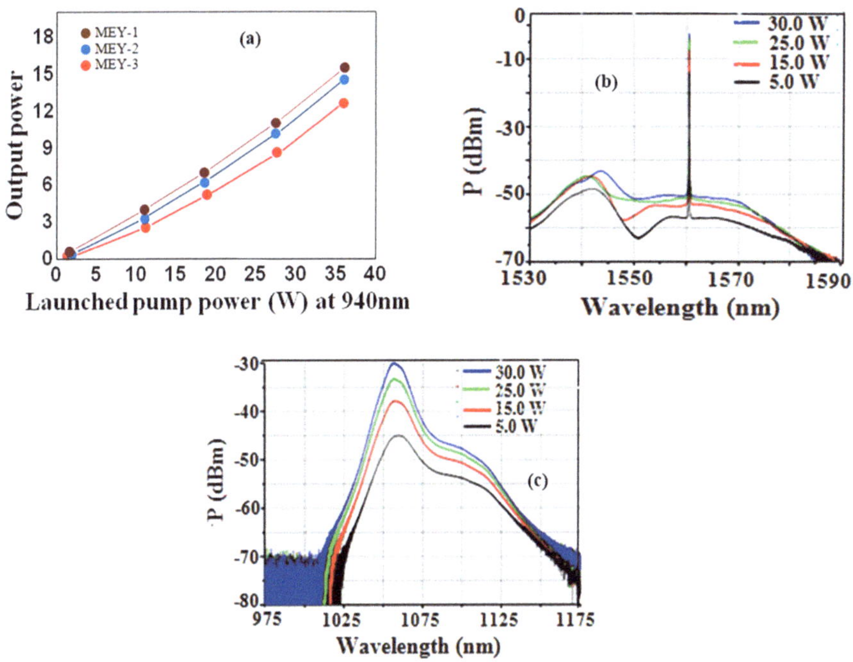

Fig. 17.18 Output power vs launched pump power curve (**a**) with input signal power of 0.6 W of three fibers; The lasing output spectra (**b**) and 1.0-micron Yb^{3+} ASE spectra of MEY-1 (**c**) fiber under various pump power

The fiber (MEY-1) having the optimized P_2O_5 doping levels of 22.25 wt% along with other multielements shows the lowest fluorescence lifetime of $^4I_{11/2}$ levels of Er ions, which indicates the most efficient energy transfer from Yb to Er ions as confirmed from the lasing output power curve shown in Fig. 17.18a. The calculated power conversion efficiency from measured output power against the combined pump power curve given in Fig. 17.18a found to be almost 41.0% for MEY-1, 40.0%

for MEY-2, and 33% for MEY-3. The output lasing spectra as well as 1.0-micron Yb^{3+} ASE spectra under various pump power shown in Fig. 17.18b, c, respectively. All three fibers do not show any self-lasing at 1.0 micron.

With 3.2 W as the input signal power, we have achieved a maximum 21.0 W output power using the optimized 8.5 m total length of MEY-1 fiber, which is shown in Fig. 17.19a. In the inset showing the value of output power. The amplified signal output spectra is shown in Fig. 17.19b. The MEY-1 fiber does not show any self-lasing at 1.0-micron up to 21.0 W amplification output power. The 1.0-micron ASE spectra of MEY-1 fiber at 52.0 W launched pump power is shown in Fig. 17.19c.

With the increase of input power (0.5–3.3 W) to the booster section, the output power goes up from 19.2 to 21.0 W, and the gain of the booster section is within the range of 38.4 down to 6.3 dB, as shown in Fig. 17.20.

The optical gain and noise figure curves of MEY-1 fiber under various signal powers (−10, 0, and 10 dBm) with 10.0 W pump power are shown in Fig. 17.21a and 17.21b, respectively. The fiber shows a gain of more than 28.0 dB and a gain flatness of less than 0.45 dB obtained in the wavelength range of 1545–1565 nm under 10 dBm input signal power. The noise fig. (NF) is about 4.5–5.5 dB within 1545–1565 nm.

^{60}Co gamma radiation source is used to characterize the radiation sensitivity of optical amplification performance at a low dose rate of 0.35 rad/s and a cumulative dose up to 6.0 Krad. The variation of output power under cumulative radiation dose

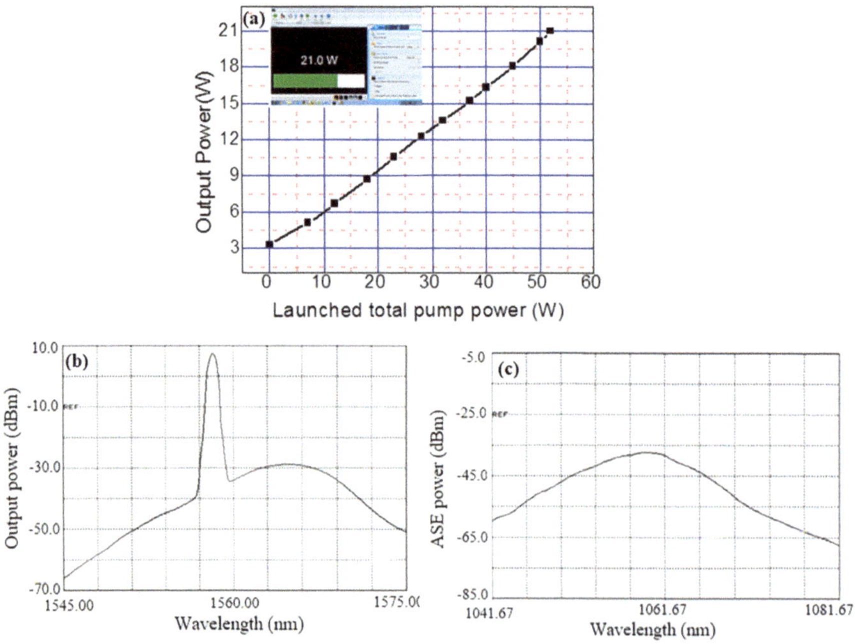

Fig. 17.19 The measured output power vs launched pump power curve with input signal power of 3.2 W at 1558 nm (**a**), Amplified signal output spectra (**b**) and 1.0 micron ASE spectra of MEY-1 fiber (**c**) at launched pump power of 52.0 W

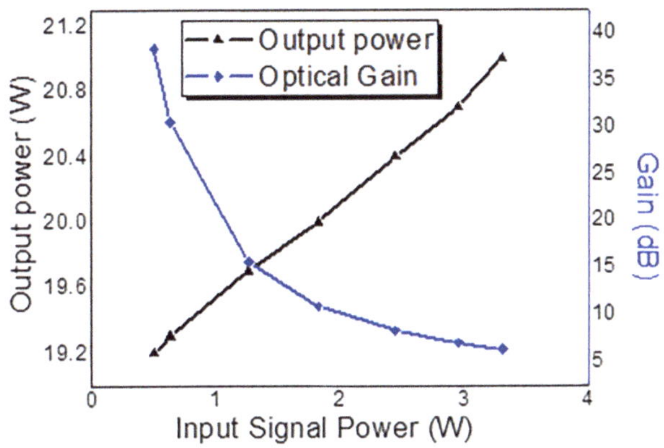

Fig. 17.20 With fixed absorb pump power, the booster output power vs input signal power and with fixed absorbed pump power, the booster gain vs input signal power of MEY-1 fiber

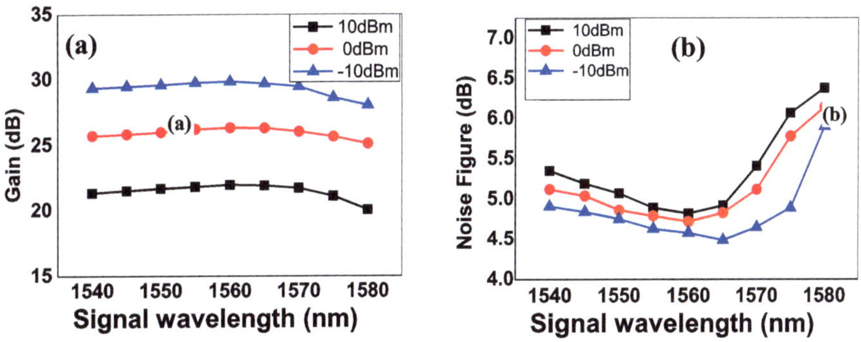

Fig. 17.21 Optical gain (a) and noise figure (b) curves of MEY-1fiber under different signal power

along with the microscopic view of the radiated fiber cross-section is shown in Fig. 17.22a. Fiber MEY-3 shows strong radiation degradation and revealed a decrease of 25% output power after 5.75 krad, whereas the fibers MEY-1 and MEY-2 show 9.5% and 16.0% decrease in output power. The result clearly reveals that such kind of multielement-doped silica glass-based Er/Yb codoped fiber (MEY-1) is able to suppress the degradation of lasing output power (<10.0% of the initial power). The brownish-red coloration of the fiber cross-section increases from fiber MEY-1 to MEY-3 as shown in the inset of Fig. 17.22a. The gain variation curves of such three fibers under cumulative radiation dose 6.0 Krad are shown in Fig. 17.22b. The gain degradation value per Krad is given in Table 17.6.

Fiber MEY-3 shows much higher gain variation of 0.285 dB/Krad under ^{60}Co-gamma irradiation due to the formation of P-related defect centers with the generation of Yb^{2+} ions. MEY-1 and MEY-2 show much less gain variation under ^{60}Co-gamma irradiation due to the presence of optimized doping levels of Ce (0.161 wt%) along with F (0.25 wt%) which prevents the formation of P-related defect centers. The induced loss curves of all three fibers are shown below, where it

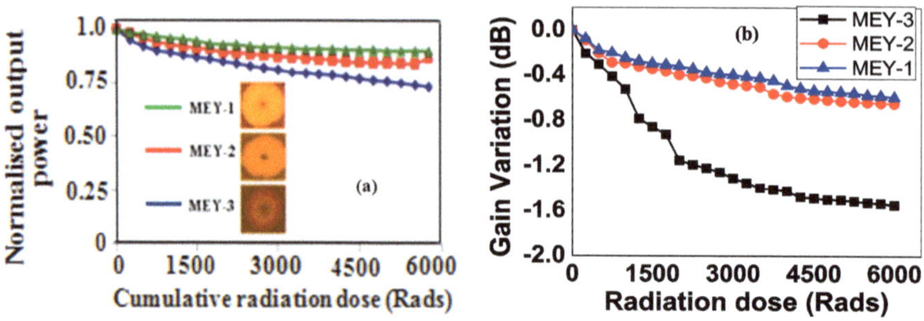

Fig. 17.22 Variation of output power vs cumulative radiation dose and cross-sectional view of MEY-1, MEY-2 and MEY-3 fibers after cumulative radiation dose of 5.75 Krad (**a**); Gain variation curves (**b**)

Table 17.6 Gain variation data

Fiber	Gain variation under ^{60}Co-gamma irradiation (dB/Krad)
MEY-1	0.1005
MEY-2	0.128
MEY-3	0.285

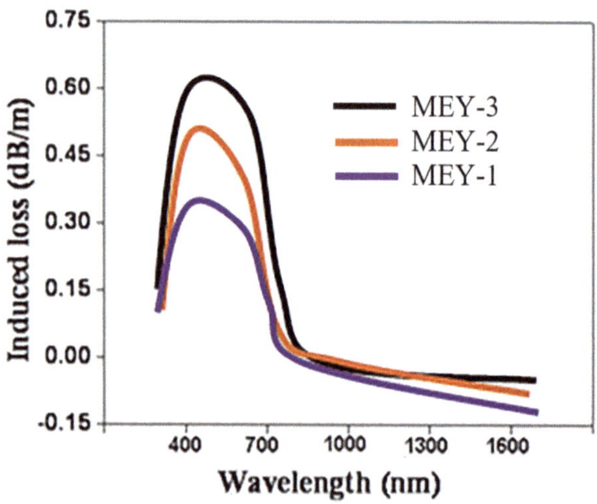

Fig. 17.23 Induced loss curve of fibers (MEY-1, MEY-2 and MEY-3) at Visible and NIR region

clearly indicates the strong induced absorption involved in visible region, showing broad radiation-induced absorptions in the 400–560 nm wavelength ranges. Such radiation-induced absorption in P_2O_5-doped SiO_2 glass arises due to phosphorous-oxygen hole defect center (POHC) [31]. Such POHC is a defect of the PO_4^{2-} type. It is a hole-type defect trapped on two nonbridging oxygens bonded to the same phosphorous. The induced absorption at 506 nm is associated with phosphorous-related hole center (Fig. 17.23).

Such induced absorption decreased from MEY-3 to MEY-1 fiber, showing the formation of such P-related defect centers minimum in the case of MEY-1 fiber. The improvement of radiation hardening behavior of Er/Yb codoped phospho-silica glass-based fiber was done through the incorporation of Ce along with F in the glass matrix, where Ce(IV) will be converted to Ce(III) through the trapping of released electrons from the glass matrix during ionizing irradiation and prevent the formation of P-related defect centers in the case of MEY-1, which contains more Ce along with F compared to MEY-2 and MEY-3 fibers.

17.7 Effect of Irradiation on the Coating Resin along with Core Glass Matrix

We have studied the effect of ^{60}Co ionization radiation on the coating resin along with the core glass matrix shown in Figs. 17.24 and 17.25, respectively. In this case, we have taken the microscopic picture of the UV-cured fiber coating as well as the core region of Er/Yb codoped fiber before and after irradiation. The following pictures are shown for fiber MEY-3. No major damage on the coating was observed under 50 Krad radiation, but the doping host of MEY-3 is damaged slightly. In the following figures, the color of the doping host arises due to the formation of a defect related to P-OHC which strongly gives rise to an absorption in the visible region.

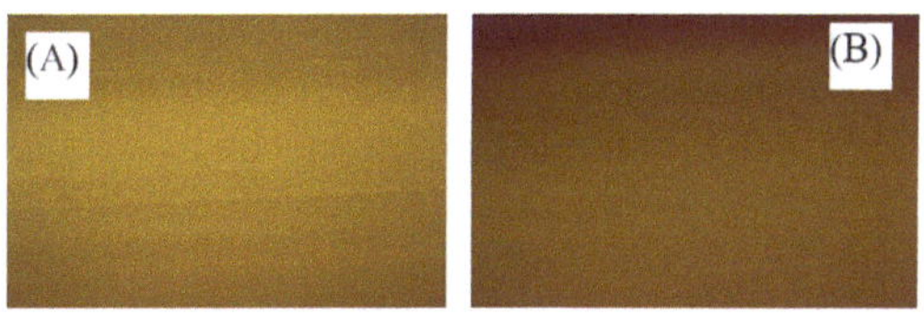

Fig. 17.24 Microscopic view of resin coating of MEY-3 fiber (**a**) before and (**b**) after gamma irradiation

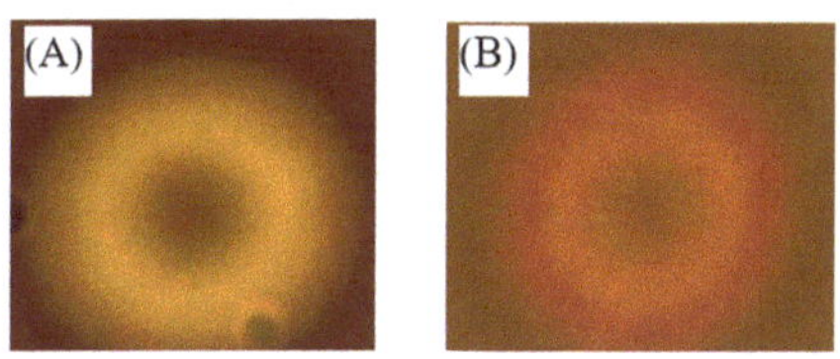

Fig. 17.25 Microscopic view of doping region of MEY-3 fiber (**a**) before and (**b**) after gamma irradiation

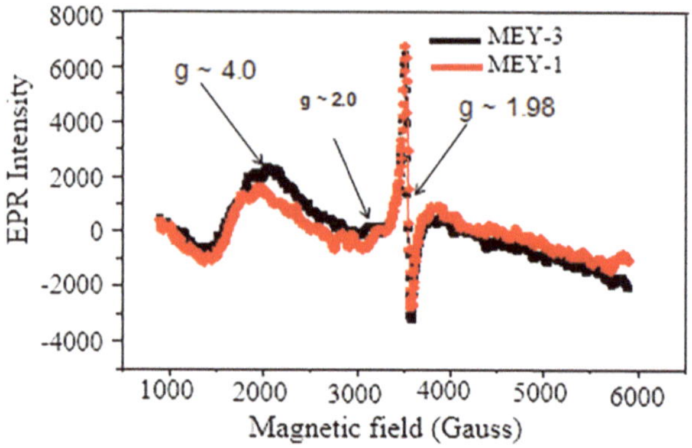

Fig. 17.26 ESR spectra of phospho-silica glass based optical fibers irradiated at dose rate of 100 cGy/h taken after 4 days

17.8 ESR Spectra of Er/Yb Doped Multielements Phospho-Silica Glass-Based Optical Fiber

We have measured the ESR spectra of phospho-silica glass-based optical fibers after ^{60}Co-gamma irradiation to detect the type of any color center developed, which becomes responsible for the high induced loss. The g values corresponding to 4.0 and 1.98 arise due to the formation of E$'$ defect center, which gives rise to absorption in the UV region. On the other hand, g values near 2.0 arise due to the formation of a defect center related to POHC, which gives rise to strong induced absorption at visible region (Fig. 17.26).

The presence of Ce (0.161 wt%) prevents the formation of P-related defect centers in MEY-1 showing no brownish-red coloration shown in the inset of Fig. 17.22a after radiation compared to fiber MEY-3 with low doping of Ce around (0.05 wt%). Such fiber shows strong brownish-red coloration after 5.75 krad cumulative radiation due to the formation of P-related defects in the center, showing strong degradation of output power. In our approach to fabricate the active Er/Yb fibers co-doped with multielements, which can be an effective solution to the task of making fibers stable against ionizing irradiations [32–35].

17.9 Testing of Output Power Stability of High-Power EYDFA

17.9.1 Output Power Stability Testing of 10.0 W High Power EYDFA

Finally, we have tested the long-term stability of 10.0 W and 15.0 W high-power EYDFAs using Er/Yb codoped multi-element silica glass-based optical fiber (MEY-1). The overall trend shows a negative slope for long-term stability testing of

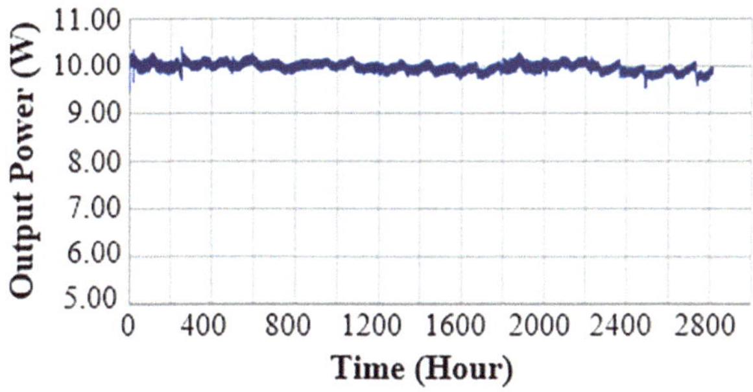

Fig. 17.27 The recorded output power vs time, over ~4 months' operation of EYDFA

10.0 W output power EYDFA running continuously for almost 4 months with a linear fit of $P_{out} = 0.000062 \times \text{Hour} + 10.073$ W.

Using this linear fit interpolation, at the end of the test, the average output power is 9.899 W. The output power reduction is less than 2%. The case temperature of the unit was not tightly maintained throughout the 4-month operation, and that contributes to the output power variation pattern. During the 4-month burn-in test, the 10 W output power doesn't show a significant sign of degradation (Fig. 17.27).

17.9.2 Output Power Stability Testing of 15.0 W High-Power EYDFA

The testing of output stability of 15.0 W high-power EYDFA was done continuously running for almost 154 h. During the first 5 h, the pump laser driving current is set at a lower value to provide an exact 15 W output power level. After the first 5 h, the pump laser driving current is increased by 3.7% so that the output power level is kept above the 15 W level. Starting from 5+ h, the average output power is 15.188 W, with minimum / maximum at 15.063 and 15.308, respectively. The peak-to-peak power variation is 0.245 W/15.188 W = 1.61%, or +/− 0.81%. During the test, the low / high end of the case temperature reaches 21 °C/32 °C, respectively. When the case temp is at the low end, such as 21 °C, the optical output power is on the high end (such as 15.3 W). When the case temp is at the high end, such as 32 °C, the optical output power is on the low end (15.06 W). During the 153.7 h burn-in test, the 15 W output power doesn't have sign of degradation yet. There are two reasons considered contributing to the excellent performance. One is that each section of the booster is with pump power of about 20 W. This helps totally eliminate the possibility of 1.0 μm lasing. The other reason is the case temp is kept relatively low, such as below 32 °C all the time during the 153.7 h burn-in test (Fig. 17.28).

The prototype module of a high-power Er/Yb-doped double-clad fiber-based optical amplifier using Er/Yb-codoped multielement silica glass-based optical fiber (MEY-1) developed jointly with M/s. Vinvish Technologies Pvt. Ltd., Trivandrum, Kerala.

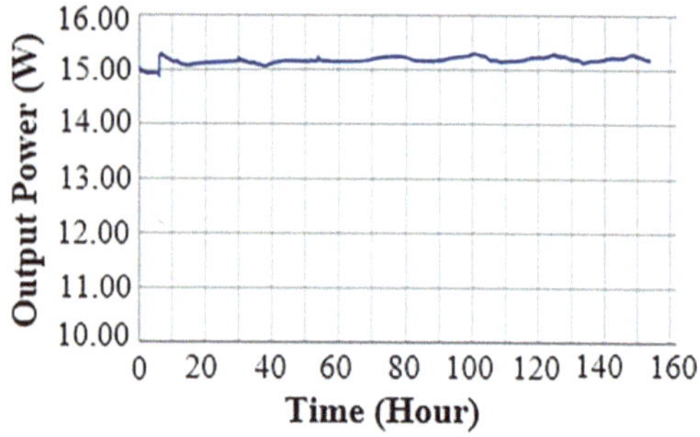

Fig. 17.28 The recorded output power vs. time, over 153.7 h operation of EYDFA

Table 17.7 Features of the high-power amplifier module

Parameters	Values
Operating wavelength	1540–1565 nm
Input signal power	0 to 10 dBm
Numbers of output ports	1,2,4,8, 16
Output power	+40 to +43 dBm
Noise figure	~6.0 dB
Amplified output	Fiber pigtailed connector/collimator
Optical gain medium	Er/Yb codoped double clad fiber
Dimension	380 mm × 254 mm × 115 mm

The features of such kind of high-power optical amplifier module are given in the following Table 17.7, which will be useful for free space communications, CATV Network, Fiber to the Home (FTTH), and LIDAR.

17.10 Conclusion

We have reported the fabrication of novel Er/Yb co-doped clad-pumped fibers with multi-elemental core-glass co-doped with Al, Ge, P, Ce, B, and F, suitable for high-power optical amplifier applications, showing <10% degradation of lasing output power under a cumulative radiation dose of ^{60}Co gamma radiation up to 6.0 Krad at low dose rates 0.35 rad/s. This dose rate is equivalent to the level of free space radiation. The doping levels of selected multielements such as F, GeO_2, CeO_2, Al_2O_3, B_2O_3, P_2O_5 along with Er_2O_3 and Yb_2O_3 as well as their ratio (7.5 to 11.1) optimized to improve the performance of fibers suitable for high-power fiber lasers and optical amplifiers. The octagonal cladding-shaped Er/Yb-doped multielement glass-based fibers show the best performance as high-power laser and optical amplifier applications compared to the double-D, pentagonal, and hexagonal cladding-shaped Er/Yb doped multielement glass-based fibers. The lasing performance of pedestal design based Er/Yb doped multielement glass-based octagonal-shaped cladding-pump fiber along with air-clad non-PM and Panda-type PM Er/Yb doped multielement glass-based normal RI-coated fibers was also evaluated. We have achieved maximum 21.0 W output power at 1558 nm under 52 W absorbed pump power of 940 nm using octagonal shaped cladding pumped low RI-coated 125-micron fiber containing 22.25 wt.% (P_2O_5), 2.43 wt.% (GeO_2), 0.5 wt.% (Al_2O_3), 0.161 wt.% (Ce_2O_3), 2.41 wt.% (B_2O_3), 0.25 wt% (F) 0.515 wt% (Er_2O_3) and 5.25 wt.% (Yb_2O_3). The fiber shows a gain of around 28.0 dB with a flatness of less than 0.45 dB in the wavelength range of 1540 to 1570 nm for a 10-dBm input signal power. Such kind of multi-elements doped silica glass shows a stable average output power around 15.0 W for the long-term operation more than 150 h without exhibiting parasitic lasing around 1.0 μm and strong radiation resistance behavior with degradation of output power below 10% under 6.0 Krad cumulative radiation dose of ^{60}Co gamma-radiation environment, indicating its potential application in free-space communication. The objective of our work is to provide information on valuable process parameters to fabricate a new kind of Er/Yb active fiber co-doped with a row of precursors capable of high-power laser operation at 1.5 microns (>15 W) and free from any spurious lasing at around 1 micron. At the same time, the fibers are found to be highly stable against ionizing Gamma radiation and free-space radiation environments.

Acknowledgments Author acknowledge with thanks the financial assistance from CSIR, Govt. of India under New Millennium Indian Technology Leadership Initiative (NMITLI) Scheme. And also thanks to Director and other staff members of FOPD. Authors from Vinvish Technologies Pvt. Ltd., Kerala thanks to CSIR for partial financial support as a soft loan.

References

1. Jeong, Y., Yoo, S., Codemard, C.A., Nilsson, J., Sahu, J.K., Payne, D.N., Horley, R., Turner, P.W., Hickey, L., Harker, A.: Erbium:ytterbium codoped large-core fiber laser with 297-W continuous-wave output power. IEEE J. Sel. Top. Quantum. **13**, 573–579 (2007)
2. Jeong, Y., Sahu, J.K., Soh, D.B.S., Codemard, C.A., Nilsson, J.: High-power tunable single-frequency single-mode erbium:ytterbium codoped large-core fiber master-oscillator power amplifier source. Opt. Lett. **30**, 2997–2999 (2005)
3. De Varona, O., Fittkau, W., Booker, P., Theeg, T., Steinke, M., Kracht, D., Neumann, J., Wessels, P.: Single-frequency fiber amplifier at 1.5m with 100W in the linearly-polarized TEM00 mode for next-generation gravitational wave detectors. Opt. Express. **25**, 24880–24892 (2017)
4. Shen, D.Y., Sahu, J.K., Clarkson, W.A.: Highly efficient Er,Yb-doped fiber laser with 188W free-running and > 100W tunable output power. Opt. Express. **13**, 4916–4921 (2005)
5. Vienne, G.G., Brocklesby, W.S., Brown, R.S., Chen, Z.J., Minelly, J.D., Roman, J.E., Payne, D.N.: Role of Aluminium in Ytterbium-Erbium co-doped phosphoaluminosilicate optical fibers. Optic. Fiber Technol. **2**, 387–393 (1996)
6. Park, N., Wysocki, P., Pedrazzani, R., Grubb, S., DiGiovanni, D., Walker, K.: High-power Er-Yb-doped fiber amplifier with multichannel gain flatness within 0.2 dB over 14 nm. IEEE Photon. Technol. Lett. **8**(9), 1148–1150 (1996)
7. Vienne, G.G., Caplen, J.E., Dong, L., Minley, J.D., Nilsson, J., Payne, D.N.: Fabrication and characterization of Yb^{3+}: Er^{3+} phosphosilicate fibers for lasers. J. Lightw. Technol. **16**(11), 1990–2001 (1998)

8. Orignac, X., Barbier, D., Du, X.M., Almeida, R.M., McCarthy, O., Yeatman, E.: Sol-gel silica/titania-on-silicon Er/Yb-doped waveguides for optical amplification at 1.5 μm. Opt. Mater. **12**, 1–18 (1999)
9. Goldberg, L., Koplow, J., [WA7-1/19-WA7-3/23] High power side-pumped Er/Yb fiber amplifier
10. Dong, L., Reekie, L., Payne, D. N.: US 5,881,197 B1 (Mar 1999) Optical fiber and optical fiber device
11. Xi-Yi Huang, S., Sarma, N.: US 6,263,003 B1 (July 2001) High-power cladding-pumped broadband fiber source and amplifier
12. Bayart, D., Leplingard, F., Gasca, L.: Double-Clad Optical Fiber and Fiber Amplifier. US Patent 2002/0126974 A1 (2002)
13. Sahu, J.K., Jeong, Y., Richardson, D.J., Nilsson, J.: A 103 W erbium-ytterbium co-doped large-core fiber laser. Optics Communications. **2227**, 159–163 (2003)
14. Peyghambarian, N., Jiang, S.: US 6,611,372 B1. Erbium and ytterbium co-doped phosphate glass optical fiber amplifiers using short active fiber length (Aug. 2003)
15. Lu, Z.G., Lavigne, A., Lin, P., Grover, C.P.: A erbium/ytterbium co-doped double-cladding fiber amplifier with 36.4 dBm output power. Photonics North (2004)
16. Tankala, K.: EP 1,043,281 B1. Method of fabricating preforms doped with rare earth metal for optical fibers (Oct. 2005)
17. Nandi, P., Jose, G.: Superfluorescence from Yb- and Yb-Er-doped phosphor tellurite glass fibres. Opt. Fiber Technol. **14**, 275–280 (2008)
18. Melkumov, M.A., Laptev, A.Y., Yashkov, M.V., Vechkanov, N.N., Guryanov, A.N., Bufetov, I.A.: Effects of Yb^{3+} and Er^{3+} concentrations and doping procedure on excitation transfer Efficiency in Er–Yb doped phosphosilicate fibers. Inorganic Mater. **46**(3), 299–303 (2010)
19. Jing, W., Chen Fang, Z., SiWen, Z., Zhen, L., ZeXin, K., Jiang, S., Chun Can, W., Shui Sheng, J.: Er^{3+}/Yb^{3+} co-doped optical fibers: Fabrication and Characteristics. ICEOE 2011
20. Vivona, M., Girard, S., Robin, T., Cadier, B., Vaccaro, L., Cannas, M., Boukenter, A., Ouerdane, Y.: Influence of Ce^{3+} codoping on the photoluminescence excitation channels of phosphosilicate Yb/Er-doped glasses. IEEE Photon. Technol. Lett. **24**(6), 509–511 (2012)
21. Pavlov, I., Ilbey, E., Dulgergil, E., Bayri, A., Omer Ilday, F.: High-power high repetition-rate single-mode Er-Yb-doped fiber laser system. Optics Express. **20**(9), 9471–9475 (2012)
22. Chen, W., Shen, D., Zhao, T., Yang, X.: High power Er, Yb-doped superfluroscent fiber source with over 16 W output near 1.55 μm. Optics Express. **20**(13), 14542–14546 (2012)
23. Sanchez, F., Amrani, F., Niang, A., Salhi, M., Komarov, A.: Characterization of high-power erbium-doped fiber laser. Int. J. Microwav. Appl. **2**(2), 89–92 (2013)
24. Dong, L., Fermann, M.E., McKay, H., Fu, L., Suzuki, S.: US Patent 8,902,493 B2. Highly rare-earth-doped optical fibers for fiber lasers and amplifiers (2014)
25. L. Wang, D. He, S. Feng, C. Yu, L. Hu, J. Qiu and D. Chen. Yb/Er co-doped phosphate all-solid single-mode photonic crystal fiber. Scientific Reports 20th August, 2014
26. Cadier, B., Laurent, A., Robin, T., Girard, S., Marcandella, C.: Radiation resistant rare-earth-doped optical fiber and method of radiation-hardening a rare-earth-doped optical fiber. US Patent 9, 025,925 B2 (2015)
27. Fujita, E., Mashiko, Y., Asaya, S., Musha, M., Tokurakawa, M.: High power narrow-linewidth linearly-polarized 1610 nm Er:Yb all-fiber MOPA. Optics Express. **22**(24), 26255–26260 (2016)
28. Y. Chu, J. Ren, J. Zhang, G. Peng, J. Yang, P. Wang and L. Yuan. $Ce^{3+}/Yb^{3+}/Er^{3+}$ triply doped bismuth borosilicate glass: a potential fiber material for broadband near-infrared fiber amplifiers. Scientific Reports 20th September, 2016
29. Poole, S.B., Payne, D.N., Fermann, M.E.: Fabrication of low-loss optical fibres containing rare-earth ions. Electron. Lett. **21**, 737–738 (1985)
30. Paul, M.C., Das, S., Dhar, A., Pal, M., Bhadra, S.K.: Process of fabrication of erbium and ytterbium-co-doped multi-elements silica glass based cladding-pumped fiber. US Patent, US 2019/0375672 A1 (2019)

31. Griscom, D.L., Friebele, E.J., Long, K.J.: Fundamental defect centers in glass: electron spin resonance and optical absorption studies od irradiated phosphorus-doped silica glass and optical fibers. J. Appl. Phys. **54**, 3743 (1983)
32. Di Francesca, D., Li Vecchi, G., Girard, S., Morana, A., Reghioua, I., Alessi, A., Hoehr, C., Robin, T., Cadi, Y.: Qualification and calibration of single-mode phosphosilicate optical fiber for dosimetry at CERN. IEEE J. Lightw. Techn. **37**, 4643–4649 (2019)
33. Di Francesca, D., Girard, S., Agnello, S., Alessi, A., Marchandella, C., Pailett, P., Ouerdane, Y., Kadi, Y., Brugger, M., Boukenter, A.: Combined temperature radiation effects and influence of drawing conditions on phosphorous-doped optical fibers. Phys. Stat. Sol. A. **216**, art. 1800553 (2019)
34. Girard, S., Ouerdane, Y., Marchandella, C., Boukenter, A., Quenard, S., Authier, N.: Feasibility of radiation dosimetry with phosphorousdoped optical fibers in the ultraviolet and visible domain. J. Non-Cryst. Solid. **357**, 1871–1874 (2011)
35. Girard, S., Marchandella, C., Morana, A., Perisse, J., Di Francesca, D., Mace, J.-R., Boukenter, A., Leon, M., Gailardin, M., Richard, N., Raine, M., Agnello, S., Cannas, M., Ouerdane, Y.: Combined high dose and temperature radiation effects on multimode silica-based optical fibers. IEEE Trans. Nucl. Sci. **60**, 4305–4313 (2013)

Chapter 18
The Journey, Current Status, and Challenges of High-Power Fiber Laser Spectral Beam Combining Technology

Hanbin Wang, Yinglin Song, Shuai Li, and Zijian Liang

Abstract Spectral beam combining (SBC) is a technique that enables the stable combination of high-power lasers while maintaining high beam quality. By using optical elements to combine beams of different wavelengths into a common-aperture output, SBC represents one of the most promising approaches for achieving high-brightness power scaling in high-power fiber lasers. In recent years, SBC technology has experienced rapid development, with significant advancements in the combined output power. This chapter provides a detailed review of the evolution of SBC technology, comparing different SBC schemes and system architectures. It also analyzes the limitations of various approaches and discusses the challenges in further increasing power levels. This chapter concludes by offering insights into the future development of SBC technology.

Keywords Spectral beam combining · Fiber laser · High power · Beam quality · Multilayer dielectric grating · Thermal deformation · Stimulated Brillouin scattering

---

H. Wang (✉)
Weihai Institute of Beijing Jiaotong University, Weihai, China

School of Physical Science and Engineering, Beijing Jiaotong University, Beijing, China
e-mail: hbwang1@bjtu.edu.cn

Y. Song
School of Physics, Harbin Institute of Technology, Harbin, China
e-mail: ylsong@hit.edu.cn

S. Li
Center for Advanced Laser Technology, Hebei University of Technology, Tianjin, China

Hebei Key Laboratory Advanced Laser Technology and Equipment, Tianjin, China
e-mail: lishuai@hebut.edu.cn

Z. Liang
Changchun Institute of Optics, Fine Mechanics and Physics, Chinese Academy of Sciences, Changchun, China
e-mail: liangzijian@ciomp.ac.cn

M. F. S. Ferreira, M. C. Paul (eds.), *Optical Fiber Lasers*, Springer Series in Optical Sciences 250, https://doi.org/10.1007/978-3-032-04110-4_18

18.1 Overview of the Development of Spectral Beam Combining Technology

Lasers possess several key advantages, including high coherence and excellent monochromaticity. Since their invention in 1960, there have seen rapid development and are now widely applied in fields such as energy exploration, communications, medical surgery, national defense, and industry [1–3]. However, in the past decade, the power of single-fiber lasers has been hindered by factors such as fiber nonlinearities [4–8]. To address this limitation, methods for combining multiple laser beams to enhance output power have emerged.

High-power laser combining techniques are generally classified into coherent and spectral beam combining (SBC) based on their principles [9–19]. Coherent beam combining requires phase-locking of the beams, necessitating precise phase control for each beam, which is then combined through spatial coherence or optical components. While this produces a nearly monochromatic beam, far-field side lobes are often an issue in practical systems. As the size of the beam array increases, the demands on phase control precision also grow. In contrast, incoherent combining allows beams of different wavelengths to be combined into a common-aperture output. This approach has a simpler optical setup and does not require phase control, but the combined beam is multiwavelength, limiting its applicability in certain scenarios.

Spectral beam combining (SBC) combines beams of different wavelengths by directing them into a dispersive element at specific angles. Based on the principle of optical path reversibility, this method achieves a common-aperture output. The dispersive elements commonly used in SBC schemes include dichromatic mirrors, edge filters, multilayer dielectric gratings (MDGs), and volume Bragg gratings (VBGs). Prisms have also been employed, but due to their limited dispersive ability, they restrict the number of beams that can be combined, making large-scale array combining difficult. This approach is widely discussed in patent literature [20–22]. There are many types of dispersive elements and a variety of combining architectures.

SBC technology has evolved over the past 20 years, undergoing three major transformations: system architectures have shifted from external-cavity structures to beam-stitching setups; the materials used for combining gratings have progressed from metals to dielectrics; and the combining elements have advanced from VBG to MDG. Table 18.1 summarizes the specific developments and key parameters.

Different SBC schemes impose varying requirements on individual beams. The following sections will provide a detailed review of the mainstream SBC schemes and the combining elements they employ.

Table 18.1 Research progress in spectral beam combining technology

Time	Institution	Combining devices	Combining efficiency	Output power	Beam quality
2002	Massachusetts Institute of Technology (MIT)	Metal Grating (MG)	—	1 W	$M^2 = 1.55$
2003	Massachusetts Institute of Technology (MIT)	Metal Grating (MG)	71.4%	8.4 W	$M^2 = 1.14$
2007	Aculight Company	Multilayer Dielectric Grating (MDG)	93%	500 W	—
2009	University of Jena	Multilayer Dielectric Grating (MDG)	98%	2 kW	$M_x^2 = 2.0$, $M_y^2 = 1.8$
2011	University of Jena	Multilayer Dielectric Grating (MDG)	–	8.2 kW	$M^2 = 4.3$
2013	Aculight Company	Multilayer Dielectric Grating (MDG)	97%	3 kW	$M^2 = 1.35$
2015	China Academy of Engineering Physics	Volume Bragg Grating (VBG)	73.7%	856 W	$M_x^2 = 7.9$, $M_y^2 = 2.7$
2015	China Academy of Engineering Physics	Multilayer Dielectric Grating (MDG)	91.2%	5.07 kW	$M_x^2 = 2.9$, $M_y^2 = 1.5$
2016	China Academy of Engineering Physics	Multilayer Dielectric Grating (MDG)	92%	9.6 kW	$M^2 = 2.9$
2016	Shanghai Institute of Optics and Fine Mechanics, Chinese Academy of Sciences	Multilayer Dielectric Grating (MDG)	94%	10.8 kW	–
2016	Aculight Company	Multilayer Dielectric Grating (MDG)	–	30 kW	$M_x^2 = 1.6$, $M_y^2 = 1.8$
2017	Nanjing University of Science and Technology	Edge Filter	98.9%	10 kW	$M_x^2 = 11$, $M_y^2 = 10$
2017	Lockheed Martin Company	Multilayer Dielectric Grating (MDG)	–	60 kW	–
2019	University of Jena	Multilayer Dielectric Grating (MDG)	–	20 kW	–
2019	China Aerospace Science and Technology Corporation (CASC)	Multilayer Dielectric Grating (MDG)	90%	10.6 kW	$M^2<2$
2022	Beijing Institute of Aerospace Control Devices	Dichromatic Mirrors	95%	11.34 kW	$M_x^2 = 2.9$, $M_y^2 = 3.2$
2023	China Electronics Technology Group Corporation	Dichromatic Mirrors	92%	9.6 kW	$M^2 = 1.7$
2024	Shanghai Institute of Optics and Fine Mechanics, Chinesc Academy of Sciences	Multilayer Dielectric Grating (MDG)	–	150 kW	–

18.1.1 *Spectral Beam Combining System Based on Dichromatic Mirrors*

The basic principle of this scheme, illustrated in Fig. 18.1, involves combining beams by leveraging the differing transmission and reflection properties of the coating layers for various wavelengths. Dichromatic mirrors selectively transmit light within a specific wavelength range at transmission rates exceeding 90%, while reflecting wavelengths outside this range with high reflectivity. By cascading multiple dichromatic mirrors, each designed for a specific wavelength range, beams of different wavelengths can be effectively combined.

Interference filters, a subtype of dichromatic mirrors, have been utilized in SBC. In 2009, Schmidt et al. demonstrated an SBC scheme (Fig. 18.2a) that achieved an average combined power of 208 W [23]. Figure 18.2b shows the transmission and reflection spectra for different wavelengths of light incident on the interference filters used in this system. In 2019, Ludewigt et al. implemented a dichroic mirror-based combining scheme that achieved a total power output of 5.9 kW with a beam quality factor (M^2) of 1.9 [24].

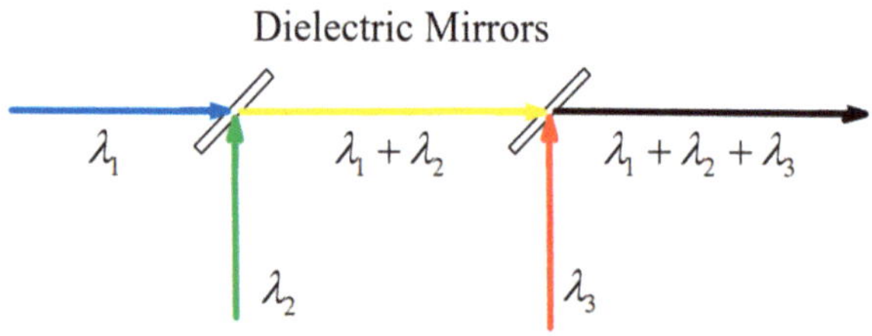

Fig. 18.1 Schematic diagram of the spectral beam combining scheme based on dichroic mirrors

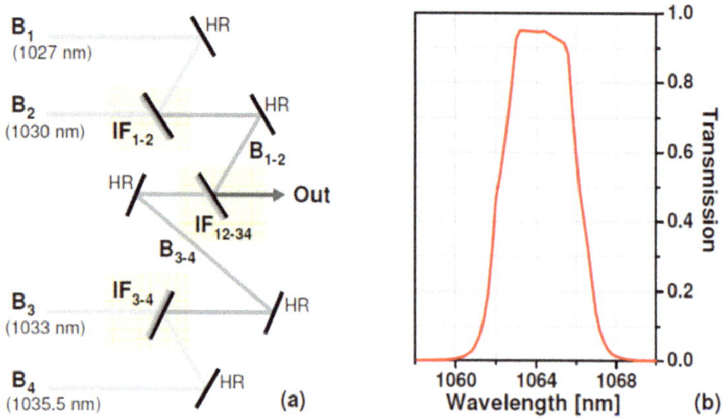

Fig. 18.2 (**a**) Schematic of the spectral combining scheme using interference filters. (**b**) Transmission rate in the 1058–1070 nm range

The principle of the dichromatic mirrors scheme suggests that there are no stringent requirements on the linewidth of the sub-beams involved in the combination process. However, as the number of combined beams increases, the design and fabrication demand for the coating layers become increasingly challenging. To accommodate more beams, the wavelength spacing between the combined lasers must be reduced, and the transmission spectrum of the dichromatic mirrors must shift from very high transmission rate to nearly zero transmission rate within just a few nanometers. This imposes extremely demanding requirements on the coating technology.

18.1.2 Spectral Beam Combining System Based on Edge Filters

Compared to interference filters, edge filters provide a broader wavelength range for beam combining. In 2017, Nanjing University of Science and Technology successfully achieved stable beam combining with a total power output of 10 kW. This scheme used edge filters as the combining elements, with the experimental setup shown in Fig. 18.3 [25]. Two laser beams, each with a power of 5 kW, were combined using an edge filter, as illustrated in Fig. 18.4a. Figure 18.4b presents the reflection spectrum of the edge filter. The two beams at wavelengths of 1070 and 1090 nm corresponded to the high-reflectivity and high-transmission ranges of the filter, respectively, resulting in a combining efficiency of 98.9% [26].

The same team in Nanjing University of Science and Technology also measured the beam quality of the combined output and observed that heat deposition within the edge filter caused internal refractive index changes, which negatively impacted beam quality. To address this, they proposed a compensation method [27]. Using the experimental setup shown in Fig. 18.5, they introduced a transmission element with a negative thermo-optic coefficient to offset the refractive index changes caused by

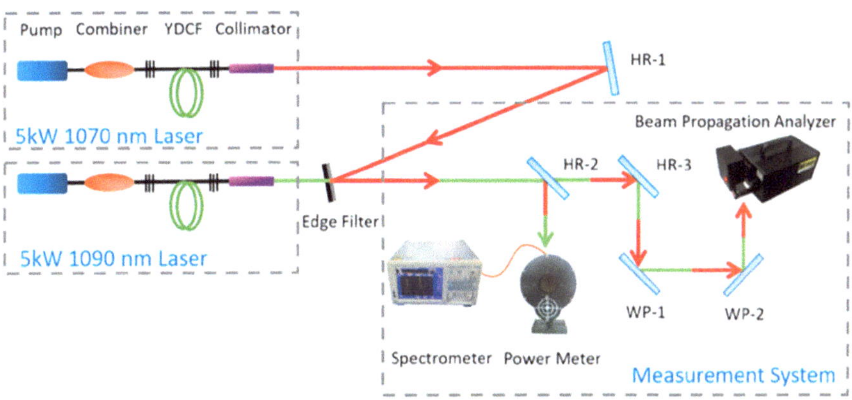

Fig. 18.3 Schematic of the spectral beam combining device using edge filters

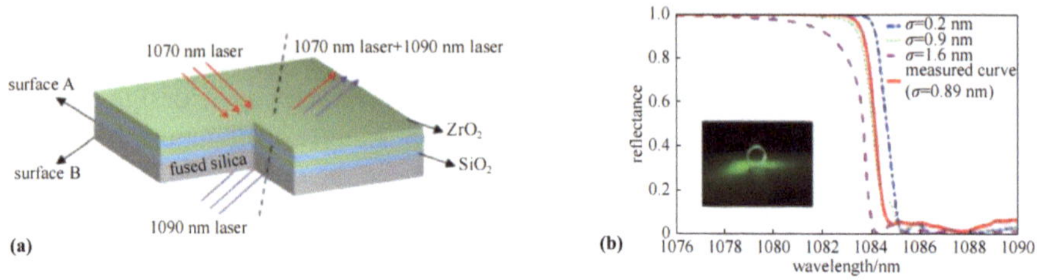

Fig. 18.4 (**a**) Structural diagram of the edge filter. (**b**) Reflection curve of the edge filter

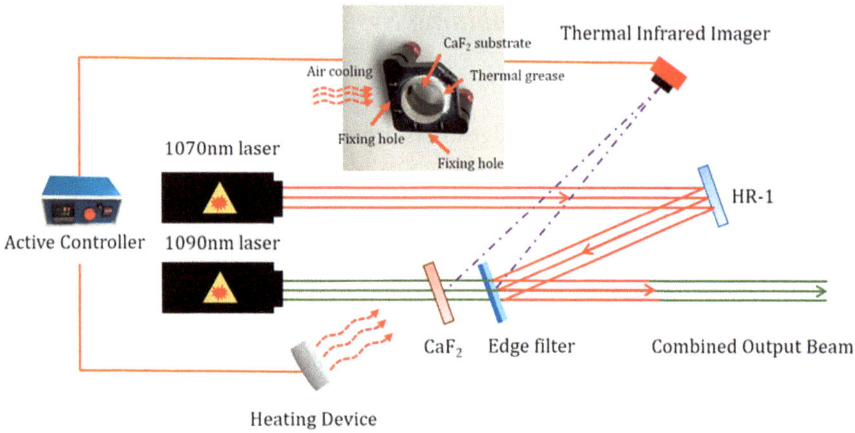

Fig. 18.5 Experimental setup of the thermal lens compensation scheme

thermal deposition. The experimental results demonstrated a significant improvement in beam quality, reducing the M^2 factor from 16 to 10 [25–27].

However, during the beam combining process, the high-power density of the lasers causes prolonged irradiation of the combining elements, such as dichromatic mirrors and edge filters, leading to inevitable temperature increases and thermal deformation. Experimental data indicate that thermal effects degrade the beam quality of the combined beam. While compensating elements can mitigate refractive index changes to some extent, the overall compensation effect remains suboptimal.

18.1.3 Spectral Beam Combining System Based on Volume Bragg Grating

Volume Bragg gratings (VBGs), made from photo-thermo-refractive (PTR) glass, offer high spectral resolution ability. When a laser beam with a wavelength matching the Bragg wavelength (λ_{Bragg}) irradiates the grating, the diffraction efficiency can reach nearly 100%. However, for beams with wavelengths at $\lambda_{\mathrm{Bragg}} \pm \Delta\lambda$, there is no diffraction, and the light passes through the grating unobstructed. Figure 18.6a shows the diffraction spectrum of a VBGs, and Fig. 18.6b illustrates the principle of spectral beam combining using VBGs [28]. By cascading multiple VBGs, beams of different wavelengths can be combined into a beam.

In 2003, Ciapurin et al. used VBGs as combining elements and achieved promising results [29, 30]. In 2008, using the setup shown in Fig. 18.7, they successfully combined five laser beams with an efficiency of 93%, reaching a stable output power of 773 W and a beam quality factor of M^2 = 1.11. In 2015, Liang et al. employed a similar setup to combine two laser beams, achieving a combining efficiency of approximately 73.7% and a combined power of 856 W. However, the beam quality was suboptimal, with M_x^2 = 7.9 and M_y^2 = 2.7 [31]. The team later optimized this setup, increasing the combined power to 2.5 kW [32]. They also explored using reflective VBGs, as shown in Fig. 18.8, achieving a combined power of 720 W, a combining efficiency of 94.7%, and a beam quality factor of M^2 = 1.54 [33].

Despite the promise of VBG, the use of PTR glass as the material for these elements introduces challenges when increasing the power and number of beams. As the irradiated power increases, the thermal effects in the VBGs intensify, leading to

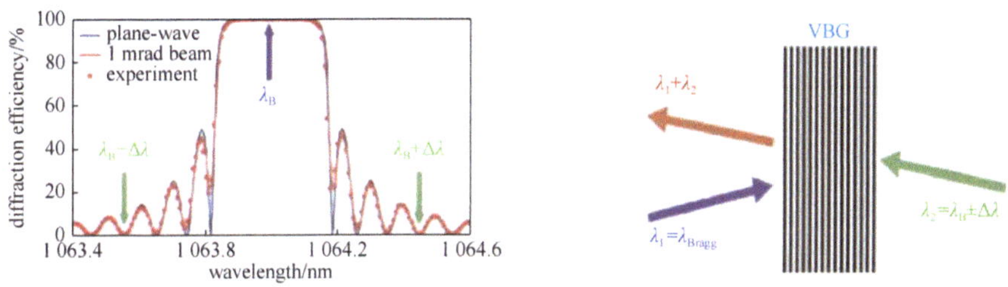

Fig. 18.6 (**a**) Diffraction spectrum of the volume Bragg gratings (**b**) Schematic of the combining principle

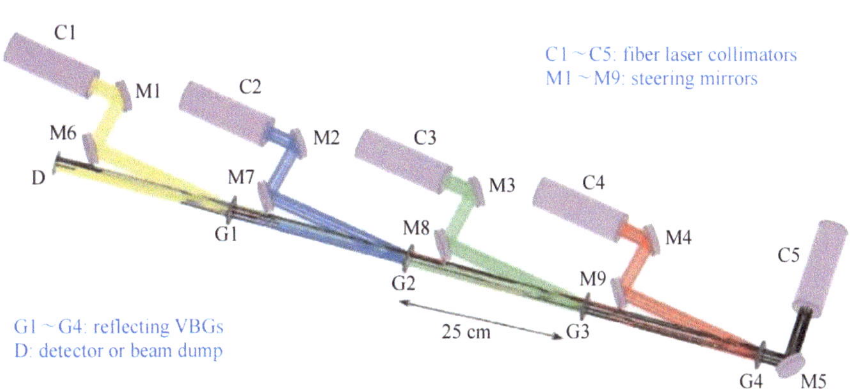

Fig. 18.7 Schematic of the VBG-Based SBC Scheme

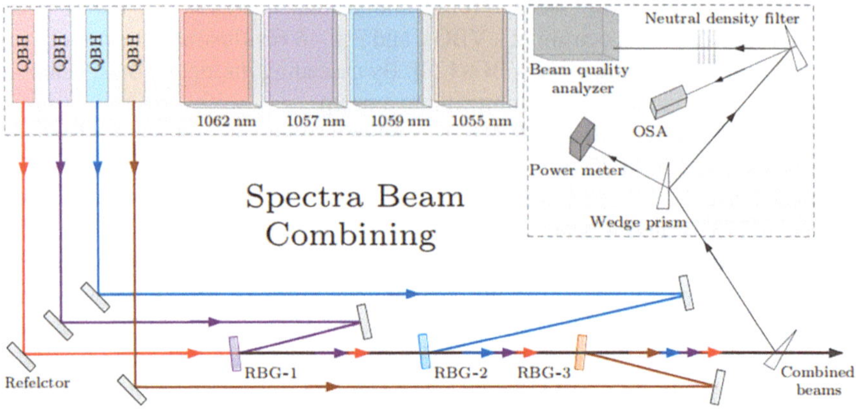

Fig. 18.8 Experimental setup for spectral beam combining based on VBG

changes in the internal refractive index of the PTR glass. This, in turn, affects the transmission and diffraction efficiency at specific wavelengths. To ensure high combining efficiency and maintain beam quality in practical applications, issues such as improving combining efficiency and addressing the thermal tuning of sub-beam wavelengths must be addressed [34–36]. As a result, achieving large-scale, high-power beam combining with VBG remains a challenge.

18.1.4 Spectral Beam Combining System Based on Diffraction Grating

The spectral beam combining system using diffraction gratings operates by diffracting multiple incident beams to achieve a common output angle and aperture. Depending on the type of grating, either transmissive or reflective gratings can be used for beam combining, as illustrated in Fig. 18.9. Both approaches are based on the same fundamental principle.

Currently, most combining schemes employ reflective gratings, governed by the grating equation shown in Eq. (18.1):

$$d(\sin\theta + \sin\phi) = m\lambda \tag{18.1}$$

where d is the grating spacing, θ is the angle of incidence, ϕ is the diffraction angle, λ is the wavelength of the incident light, and m is the diffraction order. When the incident and diffraction angles are equal, the grating equation simplifies to the form shown in Eq. (18.2):

$$2d\sin\theta = m\lambda \tag{18.2}$$

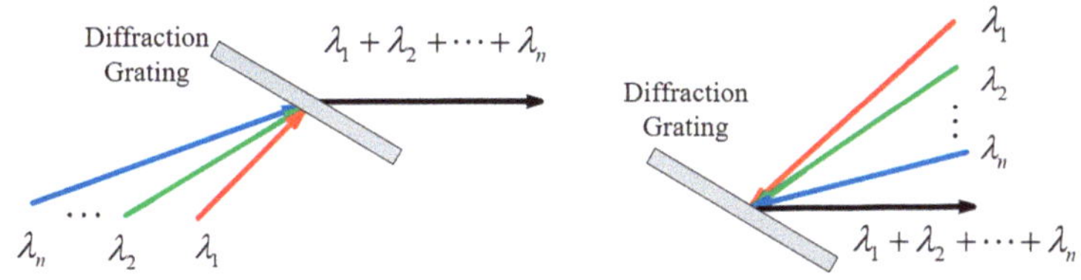

Fig. 18.9 (**a**) Schematic of spectral beam combining using transmissive gratings. (**b**) Schematic of spectral beam combining using reflective gratings

In this case, the diffracted light at a specific wavelength λ returns along the path of the incident beam, corresponding to the Littrow angle. Based on this principle, when light of different wavelengths is incident on a reflective diffraction grating at angles close to the Littrow angle, the beams can be combined to achieve a common output angle and aperture.

18.2 Evolution of Spectral Beam Combining Architectures

The system architecture of spectral beam combining technology has evolved through several phases. In efforts to increase combining power while maintaining system stability, researchers have explored various approaches. These can be broadly classified into three categories: external-cavity spectral beam combining, spectral combining based on beam stitching, and compact spectral beam combining using dense beam multiplexing. The following sections provide an overview of each approach.

18.2.1 External-Cavity Spectral Beam Combining Scheme

In 1999, Cook et al. explored an external-cavity spectral beam combining scheme [37], with the structure illustrated in Fig. 18.10. In this setup, laser light emitted from the fiber end is first collimated by a lens and then diffracted by a grating. The diffracted beam is reflected by a mirror and coupled back into the fiber along the original path, forming an external-cavity fiber laser. The team combined two fiber lasers at wavelengths of 1087 and 1120 nm, demonstrating that this architecture allows for wavelength tuning over a range of 50 nm, thereby validating its feasibility for spectral beam combining.

In 2002, Naval Research Laboratory of America conducted extensive theoretical research on external-cavity spectral combining and investigated the impact of different lens types on the performance of spectral beam combining [38]. In 2003, Massachusetts Institute of Technology (MIT) used this same architecture for beam combining, achieving a combined beam power of 6 W and a beam quality factor of $M^2 \approx 1.14$ [39]. In 2018, Zhang et al. performed simulations and

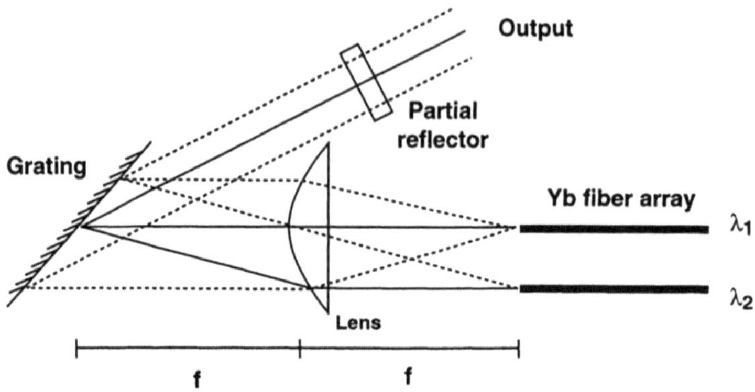

Fig. 18.10 Architecture of the external-cavity spectral beam combining system

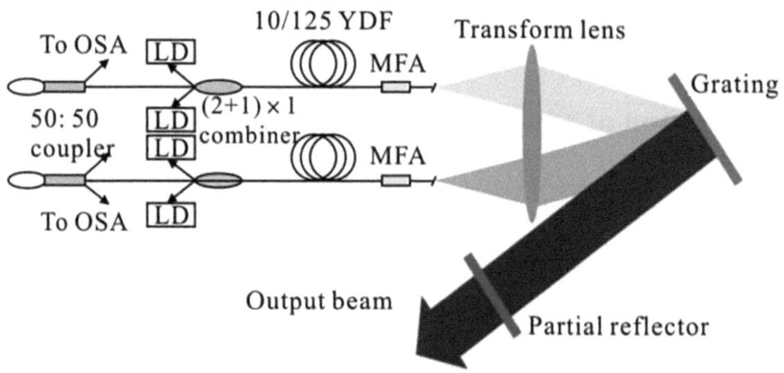

Fig. 18.11 Experimental setup for the external-cavity spectral combining system

experimental validations on this external-cavity combining system, as illustrated in Fig. 18.11. Their results indicated that under equivalent conditions, plano-convex aspheric lenses offered better scalability for combining multiple beams. Additionally, the spectral density capacity of the system increased with the focal length of the lens [40].

This external-cavity spectral combining architecture was widely used in early research on spectral beam combining. These studies laid a solid foundation for understanding the fundamental principles of spectral beam combining technology.

18.2.2 Spectral Beam Combining Scheme Based on Beam Stitching

As Master Oscillator Power Amplifier (MOPA) structured fiber lasers continue to increase higher power while maintaining high beam quality and narrow linewidth, spectral beam combining via beam stitching has developed rapidly. In this approach, each collimated sub-beam is amplified using a MOPA structure, and a reflector directs the sub-beams toward the grating at the appropriate angles, achieving common-aperture output. This method allows for significant power increasing.

In 2002, MIT demonstrated beam combining with an output power of 1 W, using a gold-coated grating as the combining element. This setup combined four laser beams, resulting in a beam quality factor of $M^2 = 1.55$. In 2003, the same team improved the system, achieving 8.4 W of combined power from five laser beams, with a combining efficiency of 71.4% and a beam quality factor of $M^2 = 1.14$ [41]. In 2007, Aculight company used a multilayer dielectric grating with 1740 lines/mm, achieving approximately 500 W of combined output power and a combining efficiency of 93% [42]. In 2009, Wirth et al. used a beam stitching spectral beam combining scheme to achieve 2 kW of combined power from four beams, with the experimental setup shown in Fig. 18.12. The diffraction efficiency of the grating exceeded 98%, and the output beam quality factors were $M_x^2 = 2.0$ and $M_y^2 = 1.8$ [43].

In 2011, the same team increased the output power to 8.2 kW, with a beam quality factor of $M^2 = 4.3$. This represented the highest reported spectral beam combining power at the time, with a notable improvement in beam quality [44]. Early research on spectral beam combining in China focused on simulations [45–47], but in recent years, diffraction grating-based spectral beam combining schemes have matured. Three representative schemes have emerged: the reflective polarization-maintaining grating-based spectral beam combining scheme [48, 49], the reflective nonpolarization-maintaining grating-based spectral beam combining scheme [50], and the transmissive diffraction grating-based spectral beam combining scheme [51].

In 2015, Yan et al. combined five beams using four polarization-maintaining gratings, achieving a total output power of 5.07 kW and a combining efficiency of 91.2% [48]. The setup, shown in Fig. 18.13, employed a dual-grating structure. In 2016, the team increased the number of beams to 10, raising the power to 9.6 kW, with a combining efficiency of 92% and a beam quality factor of $M^2 = 2.9$ [49]. However, the system included many optical elements, and thermal deformation from these elements caused wavefront distortion, affecting beam quality. Additionally, the sub-beams were unpolarized, but the combined beam was

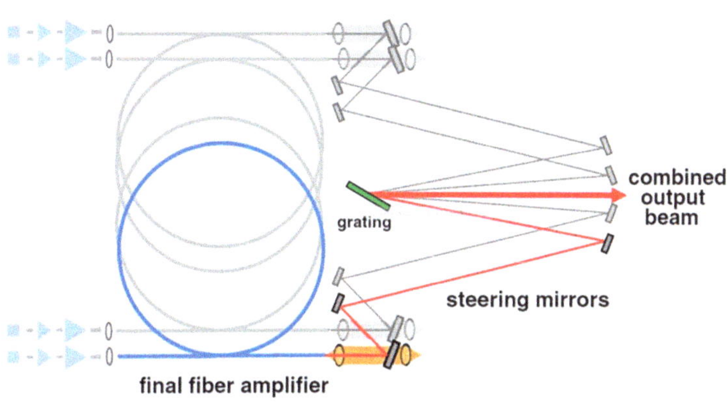

Fig. 18.12 Schematic of the fiber laser spectral beam combining architecture

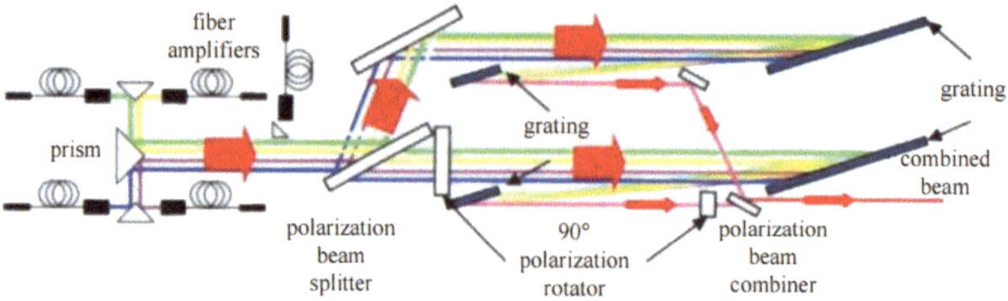

Fig. 18.13 Schematic of the fiber laser spectral combining scheme from the China Academy of Engineering Physics

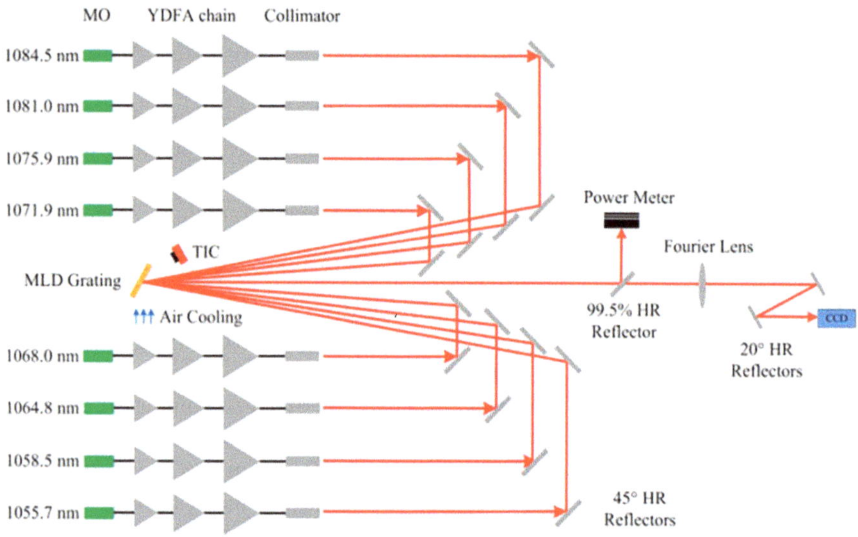

Fig. 18.14 Schematic of the fiber laser spectral beam combining scheme from the Shanghai Institute of Optics and Fine Mechanics, Chinese Academy of Sciences

polarized, which led to errors during polarization process and further degraded the beam quality.

In 2016, Zheng et al. achieved a combined power of 10.8 kW, the highest reported spectral beam combining power at the time. The structure, shown in Fig. 18.14, used a homemade developed reflective multilayer dielectric grating with a line density of 960 lines/mm [50].

Because transmissive optical components offer a larger light-transmitting volume, they absorb more laser energy over the same irradiation time, resulting in higher temperatures and greater thermal deformation, which negatively affects beam quality. Consequently, transmissive components are less frequently used in high-power spectral combining systems. However, with recent advances in transmissive grating technology, their usage has increased. In 2019, Zheng et al. used a

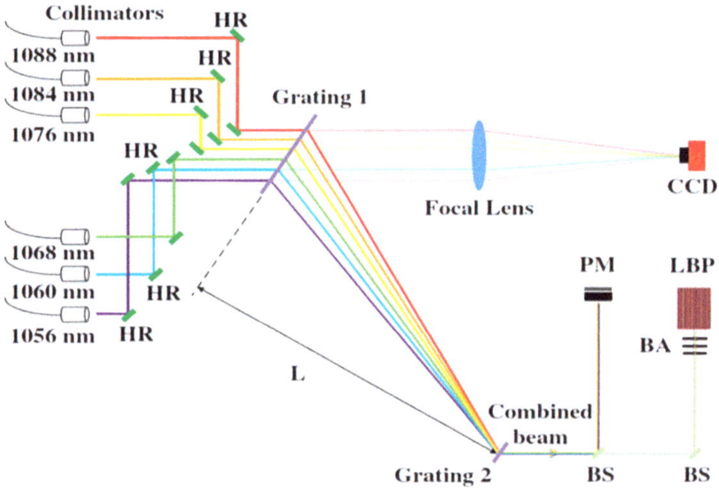

Fig. 18.15 Schematic of the fiber laser spectral beam combining system from China Aerospace Science and Technology Corporation

dual-grating structure for spectral beam combining, as shown in Fig. 18.15 [51]. This scheme employed two gratings for dispersion compensation, combining six beams to achieve a total output power of 10.6 kW, with a beam quality $M^2 < 2$.

18.2.3 Compact Spectral Beam Combining Scheme Based on Dense Beam Multiplexing

In recent years, while spectral beam combining using beam stitching has continued to evolve, enabling laser outputs exceeding 10 kW, increasing the power of individual beams further remains challenging, and the number of beams used in these systems is typically limited. To achieve higher system output power, it has become crucial to optimize the spectral beam combining scheme, increase the number of combined beams, and enhance the compactness of the system.

In 2013, Aculight company combined 12 narrow-linewidth fiber lasers using an multilayer dielectric grating with a line density of 1740 l/mm, achieving a total output power of 3 kW and a beam quality factor of $M^2 = 1.35$ [52]. In 2016, the team further expanded the number of beams to 96, increasing the combined output power to 30 kW. The structural diagram of the system is presented in Fig. 18.16a [53]. The resulting beam quality was $M_x^2 = 1.6$ and $M_y^2 = 1.8$, with the spectral distribution of the combined beams shown in Fig. 18.16b.

An analysis of the combining bandwidth and individual beam power suggests that increasing sub-beam power and adopting a more compact combining scheme (integrating lasers in the 1073 nm to 1090 nm wavelength range) could further

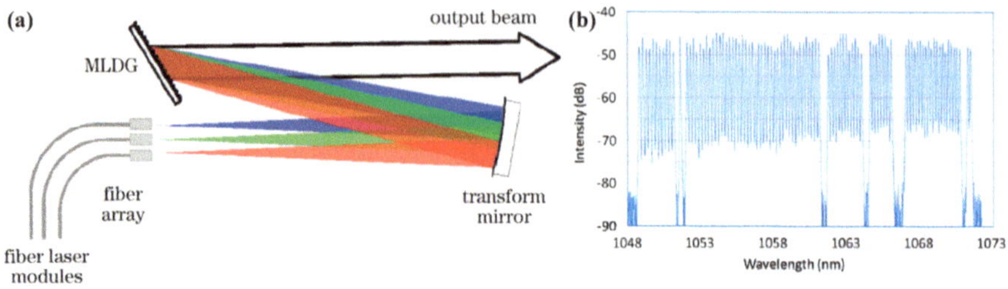

Fig. 18.16 (**a**) Schematic of the fiber laser spectral beam combining system. (**b**) Spectrum of the combined output beam

enhance output power. In March 2017, Lockheed Martin company reported developing a combining system with an output power of 60 kW [54]. Currently, the highest output power of a diffraction grating-based spectral beam combining system for fiber lasers reported in publicly available literature is 150 kW [55], as mentioned in a paper published by the Shanghai Institute of Optics and Fine Mechanics (SIOM), Chinese Academy of Sciences, in May 2024.

18.3 Challenges in the Development of Spectral Beam Combining Technology

The steady increase in output power for spectral beam combining technology primarily depends on the rising power of individual fiber lasers and advancements in the fabrication of combining elements. However, due to the spectral bandwidth constraints mentioned earlier, fiber lasers used in these systems must maintain a narrow linewidth, making further power scaling challenging. This is largely due to the impact of nonlinear effects and mode instability. Additionally, as the combined power continues to grow, even though multilayer dielectric gratings exhibit low levels of laser energy absorption, the increasing power density and longer irradiation times have caused the thermal deformation of the gratings to reach significant levels. In the following section, we will provide a detailed discussion of the key obstacles to the further development of spectral beam combining technology.

18.3.1 Thermal Effects of Combining Gratings

As the power of spectral beam combining systems continues to increase, the temperature rise in diffraction gratings due to laser energy absorption becomes more pronounced, leading to significant thermal deformation. This deformation degrades beam quality, limits further power scaling, and, in severe cases, can damage the gratings. This section reviews the current theoretical and experimental research on grating thermal deformation and analyzes experimental measurements of these effects.

In 2004, Anping Liu et al. observed thermal deformation in gold-coated gratings under 30 W laser irradiation using a Michelson interferometer [56]. They identified the photoresist layer between the gold coating and the substrate as highly sensitive to temperature, causing significant thermal deformation. This deformation led to wavefront distortion, with interference fringes becoming distorted, as shown in Fig. 18.17. In 2006, Thomas H. Loftus, using the optical system in Fig. 18.18a, observed interference fringe distortion in an all-dielectric grating at a laser power density of 1.5 kW/cm^2, as shown in Fig. 18.18b [57]. They reported a peak thermal deformation of 126.6 nm at this power density. However, because the power density handled by the grating at that time was relatively low, the fringe distortion was minimal, and the issue of thermal deformation did not receive sufficient attention.

In 2007, Jens Limpert et al. used finite element analysis software to simulate the temperature distribution and thermal deformation of both transmissive and reflective gratings under high-power laser irradiation [58]. The study showed that the absorption coefficient of the coating significantly affects the grating's temperature rise. As shown in Fig. 18.19, transmissive gratings experienced higher temperature rise and greater thermal deformation than reflective gratings under steady-state conditions. The study also noted that thermal deformation causes issues such as thermal lensing, changes in grating line spacing, and surface profile alterations, all of which impact the near-field phase distribution and beam propagation. The research recommended using reflective gratings in spectral beam combining systems to mitigate thermal deformation.

In 2016, Zhang et al. developed a thermal distortion model for reflective multilayer dielectric gratings under laser irradiation and conducted numerical simulations on thermal deformation in gratings [59]. The study simulated the phase distortion of beams under varying laser power densities and examined the resulting

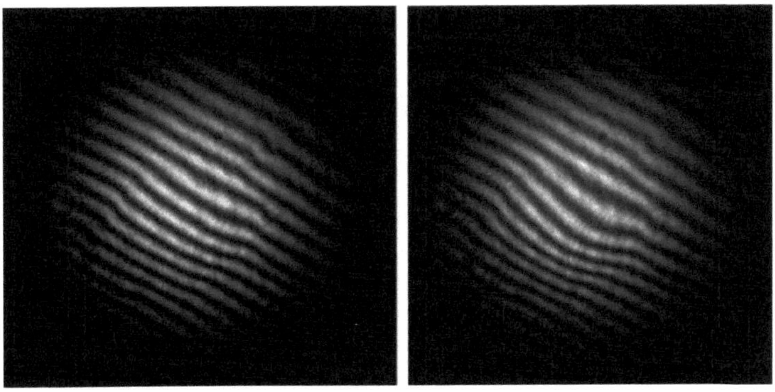

Fig. 18.17 Interference measurement results. Left: Interference fringe without laser irradiation; Right: Interference fringe under 30 W laser irradiation

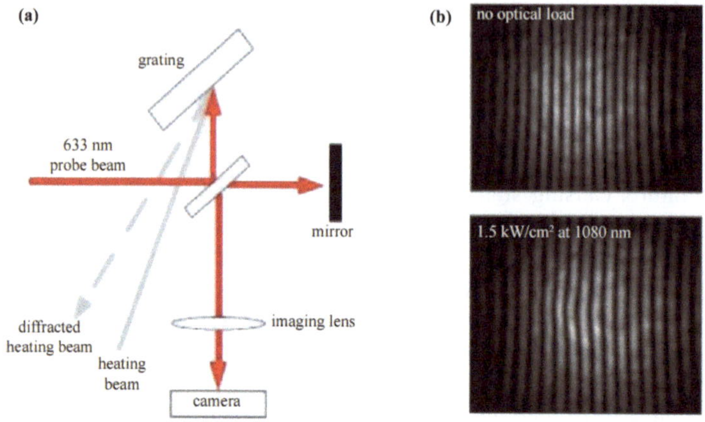

Fig. 18.18 (**a**) Test setup for measuring deformation of the multilayer dielectric gratings (**b**) Comparison of grating interference fringes before and after laser irradiation

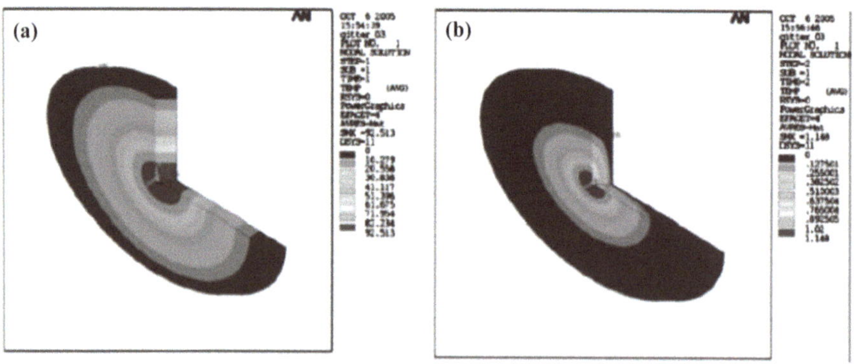

Fig. 18.19 Temperature distribution of gratings under 100 kW laser irradiation: (**a**) Transmissive grating; (**b**) reflective grating

far-field beam quality and intensity distribution, as shown in Fig. 18.20. The results indicated that as the laser power increased, sidelobe energy in the intensity distribution increase and beam quality degrade.

In 2017, Li et al. used finite element analysis software to simulate the thermal effects of multilayer dielectric gratings. Their simulations showed that when the surface temperature reached 192 °C, the surface deformation was 139.35 nm. The experimental gratings and the simulation results are shown in Fig. 18.21 [60]. The study found that as power density increased, wave front distortion became more pronounced. They also conducted validation experiments, shown in Fig. 18.22, by separating the heating and detection beams to isolate the degradation in beam quality caused by thermal deformation. Both the simulation and experimental results confirmed that as laser power increased, the temperature and thermal deformation

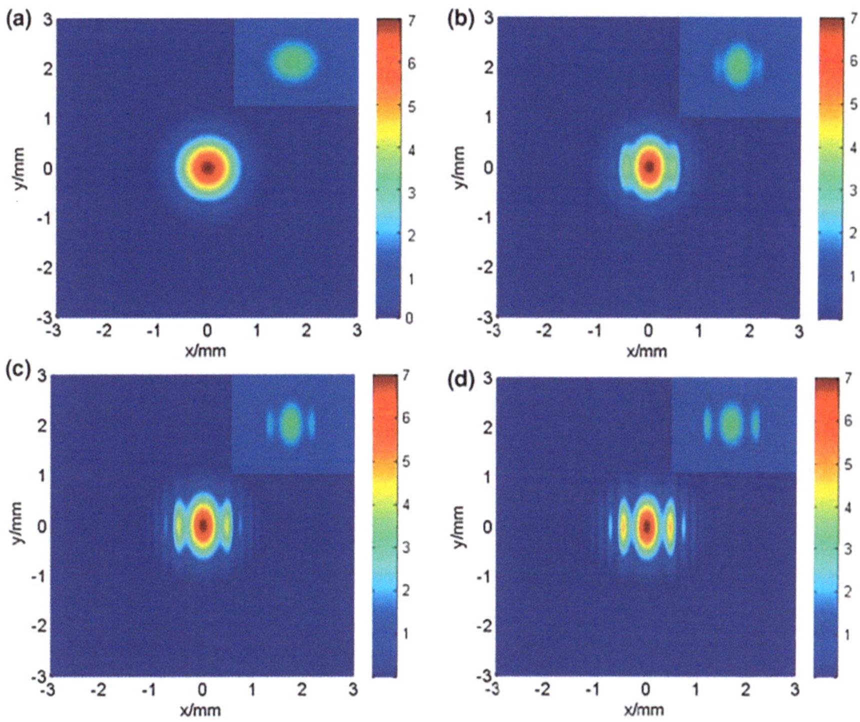

Fig. 18.20 Intensity distribution of combined and single beams at different irradiation power densities: (a) No deformation, (b) 1 kW/cm^2, (c) 2 kW/cm^2, and (d) 3 kW/cm^2

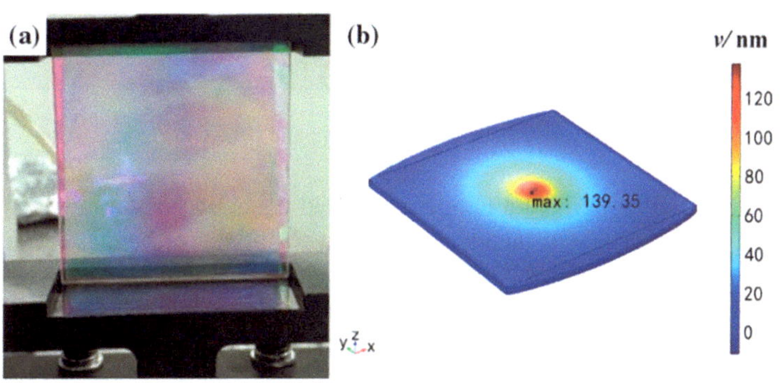

Fig. 18.21 (a) MDG used in the experiment. (b) Surface deformation of the grating after 1000 s of laser irradiation

of the grating also increased, leading to reduced beam quality. The central beam intensity decreased, and sidelobe intensity increased, as shown in Fig. 18.23. The study concluded that the primary cause of thermal deformation was the thermal expansion of the grating substrate.

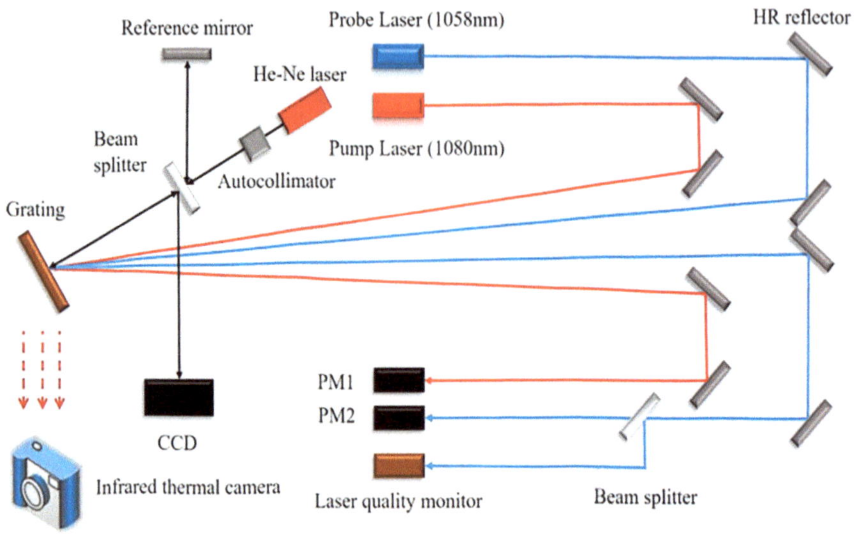

Fig. 18.22 Experimental setup for grating thermal deformation

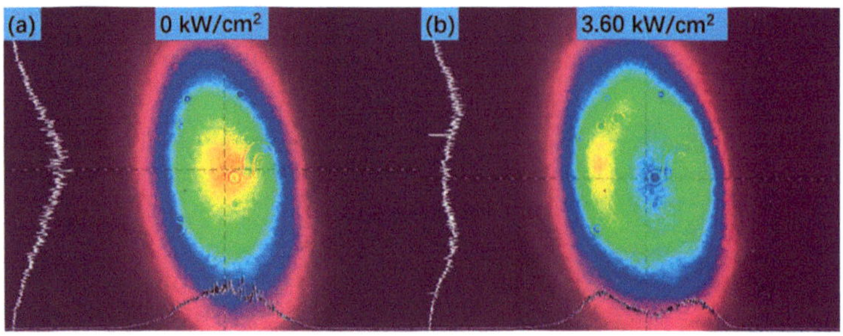

Fig. 18.23 Far-field intensity distribution: (**a**) 0 kW/cm^2 and (**b**) 3.6 kW/cm^2

In 2018, Xu et al. derived a temperature field equation for gratings under laser irradiation and investigated factors influencing temperature changes in multilayer dielectric gratings, including laser power, beam radius, and substrate size. They found that using substrates with higher thermal conductivity could effectively reduce grating temperature [61]. In 2019, based on their model, the group proposed increasing substrate thickness to reduce grating temperature and thermal distortion, which improved beam quality. The simulation results were experimentally verified, and the comparison between simulation and experimental data is shown in Fig. 18.24 [62].

In 2017, Zhou et al. conducted experiments using the setup shown in Fig. 18.25 to study the temperature, deformation, and beam quality of multilayer dielectric gratings under high-power laser irradiation [63]. The results, shown in Fig. 18.26, indicated that as irradiation power increased, both grating temperature and thermal

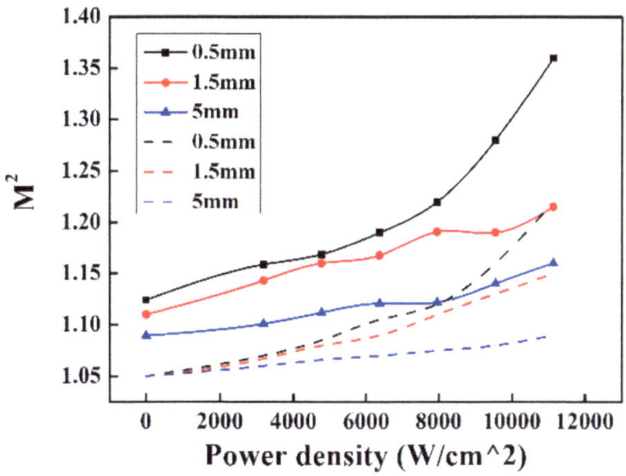

Fig. 18.24 Change in beam quality with different laser power

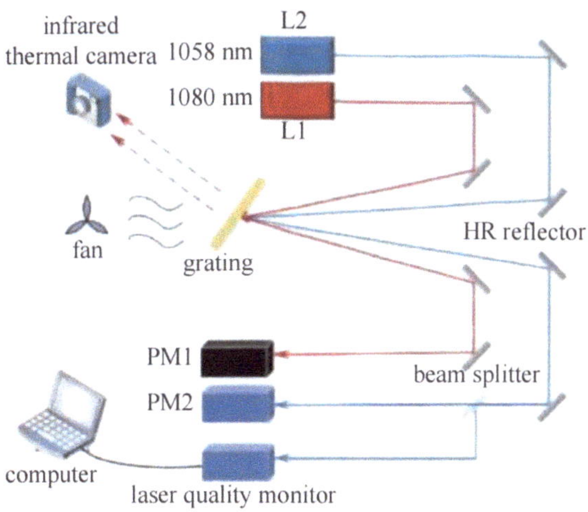

Fig. 18.25 Experimental setup for monitoring grating diffraction characteristics

deformation increased. When the laser power density reached 3.6 kW/cm^2, the beam quality factor M^2 deteriorated to 2.56. However, the thermal deformation had little effect on the grating's diffraction efficiency, with fluctuations of less than 1.8% during the experiment.

In 2020, Wang et al. further examined the absorption of laser energy by the substrate. Combining Lambert-Beer's law with the heat conduction equation, they developed a new laser absorption model [64]. The model divided the grating into a laser irradiation zone and a heat transfer zone, demonstrating that a dynamic equilibrium exists between the energy absorbed in the irradiation zone and the energy transferred in the heat transfer zone when the maximum temperature remains

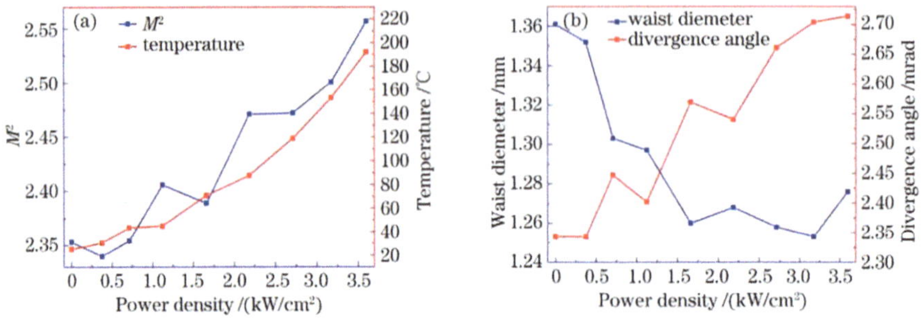

Fig. 18.26 (**a**) Changes in grating surface temperature and beam quality with power density. (**b**) Changes in beam waist diameter and far-field divergence with power density

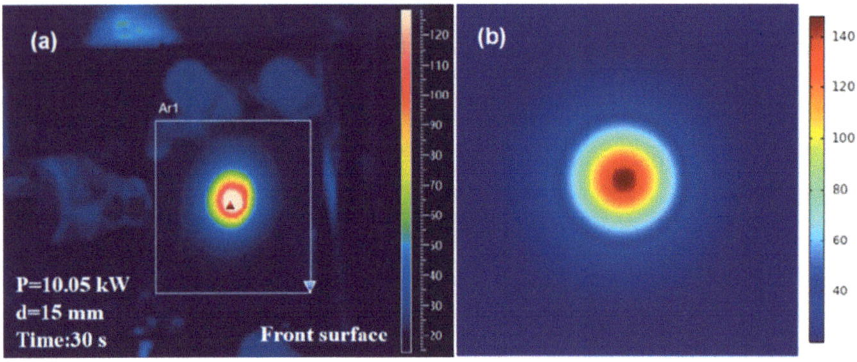

Fig. 18.27 Surface temperature distribution of gratings after 30 s of 10.05 kW laser irradiation: (**a**) Experimental data and (**b**) simulation data

constant. This model was experimentally verified, with the results shown in Figs. 18.27 and 18.28.

In 2020, Wang et al. constructed a thermal deformation model by combining the thermal elasticity equation and stress-strain, and subsequently, they studied the evolution of internal stress and thermal deformation within the grating, as shown in Fig. 18.29 [65].

Based on research, both domestically and internationally, it is clear that thermal deformation in gratings was first identified in 2004. However, as the total power of spectral beam combining systems was relatively low at the time and other factors such as linewidth and sub-beam misalignment also affected combining efficiency, the thermal deformation issue was not severe and did not receive much attention. In recent years, with the rapid development of spectral beam combining, where systems feature more combined beams and higher power, grating thermal deformation has become a significant issue. It is now one of the main factors limiting further power scaling. Simulations and experiments have consistently shown that thermal deformation alters the grating surface profile, impacting beam propagation and

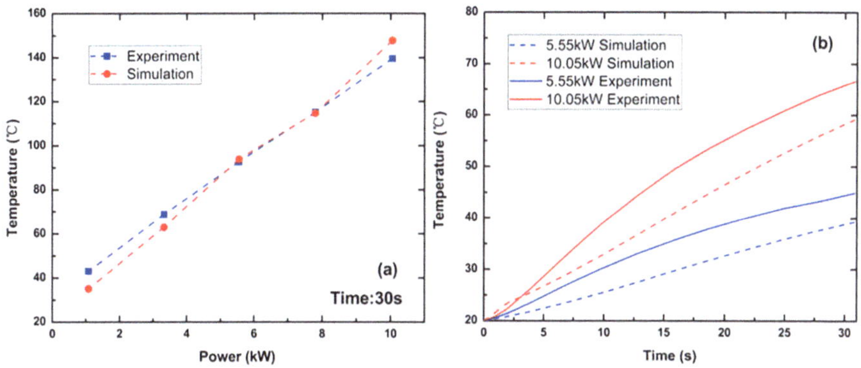

Fig. 18.28 Comparison of experimental and simulated maximum temperatures of gratings under different laser powers: (**a**) Surface and (**b**) back surface

beam quality. To achieve good beam quality output, effective measures must be taken to reduce grating thermal deformation. Without such measures, thermal accumulation could lead to grating damage.

18.3.2 Challenges in Improving Combined Power

The increase in output power of the spectral beam combining system is influenced by multiple factors. In addition to the grating thermal effects discussed in previous sections, the power and beam quality of sub-beam array lasers also directly impact combining performance. Laser power increase is mainly constrained by nonlinear effects and transverse mode instability (TMI), both of which degrade beam quality. Furthermore, due to limitations in grating coating design, the wavelengths of lasers involved in spectral synthesis must be within a certain range. To accommodate as many beams as possible, the linewidth of these lasers must be kept as narrow as possible.

For narrow-linewidth fiber lasers, output power is limited by thermal and nonlinear effects intrinsic to the fiber, resulting in spectral broadening and a reduction in the signal-to-noise ratio. Large-mode-area fibers can effectively suppress nonlinear effects within the fiber; however, they support multiple transverse modes, which can lead to mode deterioration in the fiber laser, reducing beam quality and even causing TMI [66–68], further limiting power scaling. However, when using optical fibers with small core diameters, due to the small core size of the fiber, strong nonlinear effects can occur during laser transmission.

Under high-power conditions, the primary nonlinear effects in fibers include stimulated Brillouin scattering (SBS), stimulated Raman scattering (SRS), and four-wave mixing (FWM) [69]. In narrow-linewidth, high-power lasers, SBS is the main factor that limits power scaling. Research on the SBS mechanism began early, with initial studies thoroughly describing the SBS process and deriving its threshold formula, which is influenced by factors such as fiber effective length, mode field area, seed source linewidth, and Brillouin gain bandwidth [70].

Various methods have been proposed to suppress SBS, broadly categorized into three strategies: (1) modifying the gain fiber structure and dopant concentration, (2) optimizing the pump scheme, and (3) applying phase modulation to the seed light. Specific techniques include seed-light phase modulation for spectrum broadening [71–73], backward pumping schemes [74], laser gain competition [75], and control of fiber temperature and stress gradients [76–78]. These techniques have effectively suppressed SBS and enabled increased power in narrow-linewidth fiber lasers.

In 2012, Theeg et al. constructed a single-frequency fiber amplifier using a backward-pumping configuration based on spatial structuring. By employing a gain fiber with a 25 μm core diameter and applying a temperature gradient, they successfully increased the SBS threshold, achieving a single-frequency fiber laser output with 301 W of power and an optical-to-optical conversion efficiency of 68.4% [79]. The experimental setup is illustrated in Fig. 18.30. In 2016, the Shanghai Institute of Optics and Fine Mechanics (SIOM) utilized noise modulation to control spectral linewidth, resulting in a laser output of 50 GHz and 2.5 kW, with beam quality values of $M_x^2 = 1.191$ and $M_y^2 = 1.186$ [80]. In 2017, SIOM further achieved a triple single-frequency, 302-W single-mode laser output using sinusoidal modulation and fiber stress gradient control, with a polarization extinction ratio of 18 dB and beam quality $M^2 = 1.04$, as shown in Fig. 18.31 [81].

In 2019, Shen et al. applied white noise and sinusoidal cascade modulation with a backward pumping configuration, using a commercial 20/400 μm fiber, and achieved a laser output of 48 GHz and 3.01 kW, with improved beam quality ($M^2 = 1.17$) [82]. The high-power, narrow-linewidth fiber laser setup and test results are depicted in Fig. 18.32.

In 2020, Lai et al. amplified a single-frequency fiber laser at a wavelength of 1030 nm using a tapered fiber. Exploiting the high SBS threshold of the tapered fiber structure, they tested the output power in both polarization-maintaining and nonpolarization-maintaining amplification systems. By adjusting the polarization state of the input light, they achieved a single-frequency fiber laser output of 550 W, with a beam quality factor of $M^2 = 1.47$ [83]. The experimental setup is shown in Fig. 18.33.

In 2022, Shen et al. employed multistage phase modulation to broaden and optimize the laser spectrum [84]. Using a low-numerical-aperture 25/400-μm fiber and an optimized seed spectrum control scheme, they achieved a single-mode laser output with a center wavelength of 1050.5 nm and a power of 4.23 kW, as demonstrated in Fig. 18.34.

In 2022, Zhou et al. combined white-noise phase modulation with bending loss optimization to suppress both SBS and TMI, achieving a 4-kW laser output [85]. The experimental setup is depicted in Fig. 18.35. In the same year, Shi et al. optimized the cascaded gain fiber structure and hybrid wavelength pumping scheme, as shown in Fig. 18.36. By using gain fibers with different doping concentrations, they effectively mitigated thermal effects. And in combination with hybrid wavelength

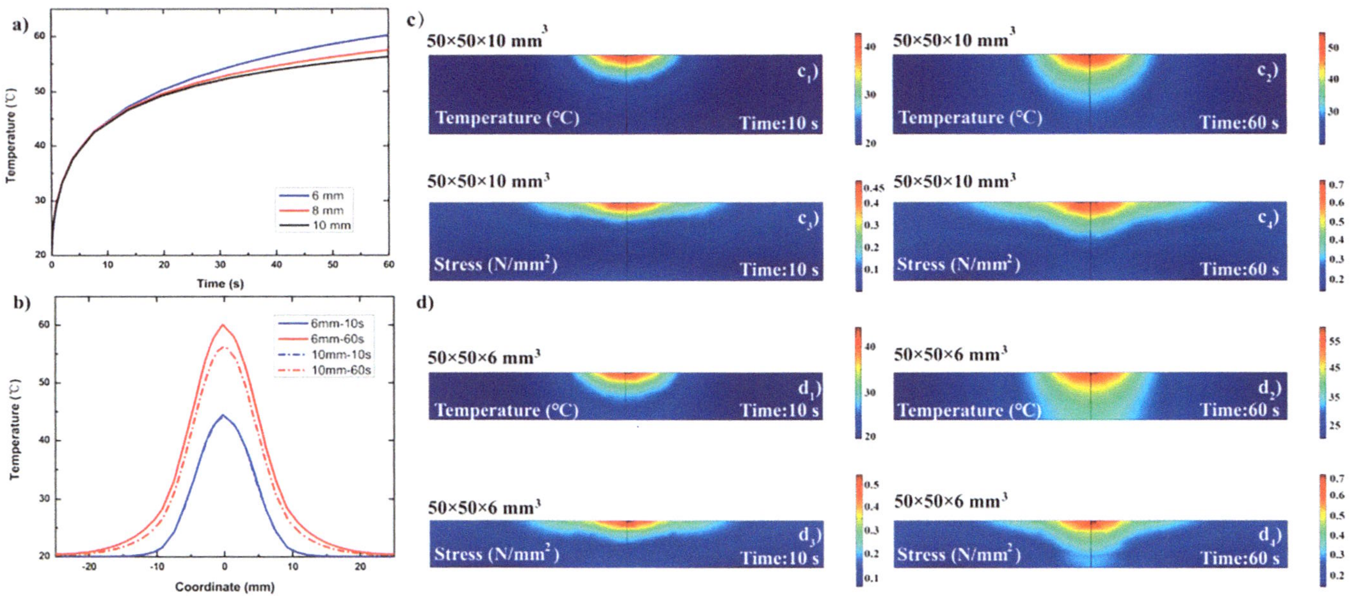

Fig. 18.29 (**a**) Maximum temperature of the grating versus irradiation time for varied substrate thicknesses. (**b**) Maximum front surface temperature distribution of the grating along the *x*-axis for different substrate thicknesses and irradiation times. Stress and temperature distributions for different substrate thicknesses at varying irradiation times: (**c**) 10 mm substrate and (**d**) 6 mm substrate

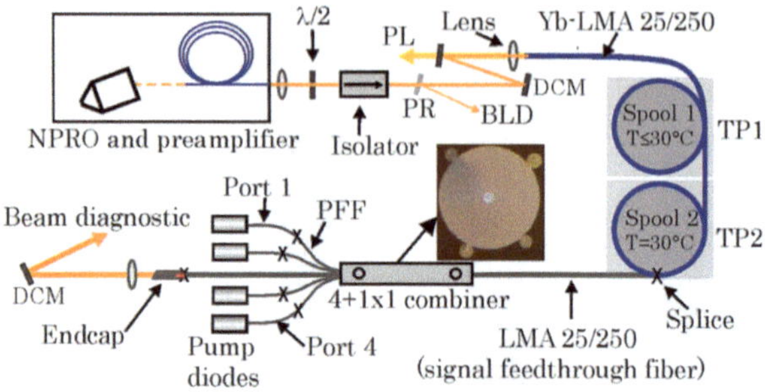

Fig. 18.30 Schematic of the experimental setup for a high-power fiber laser using a backward pumping configuration

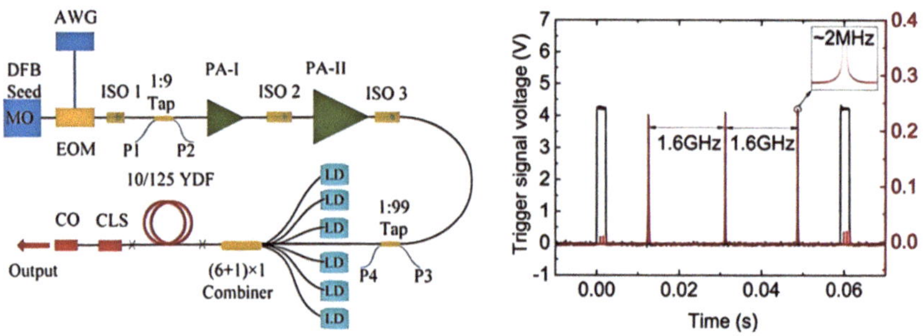

Fig. 18.31 High-power triple single-frequency laser achieved through sinusoidal modulation and fiber stress gradient control

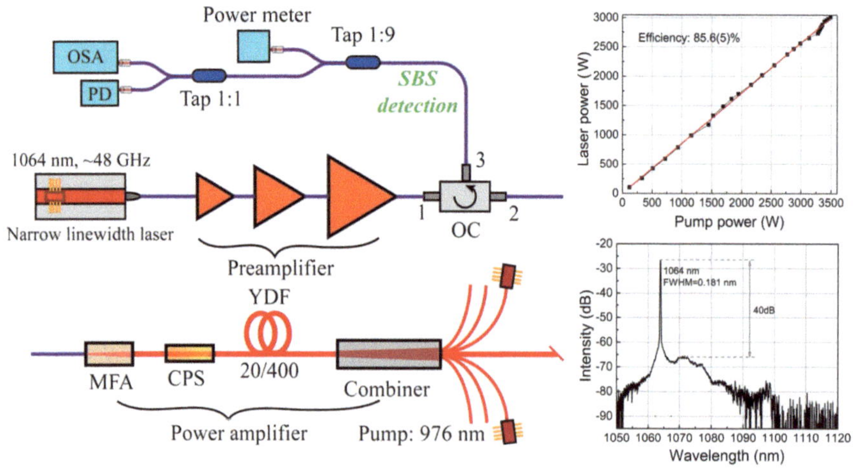

Fig. 18.32 High-power narrow-linewidth fiber laser based on spectral modulation

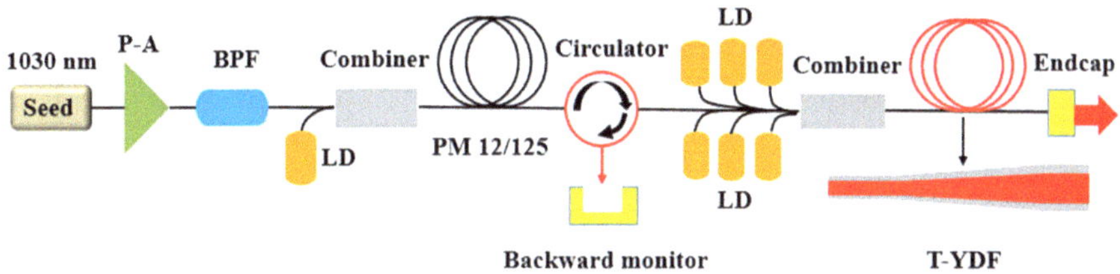

Fig. 18.33 Schematic of the fiber laser experimental setup

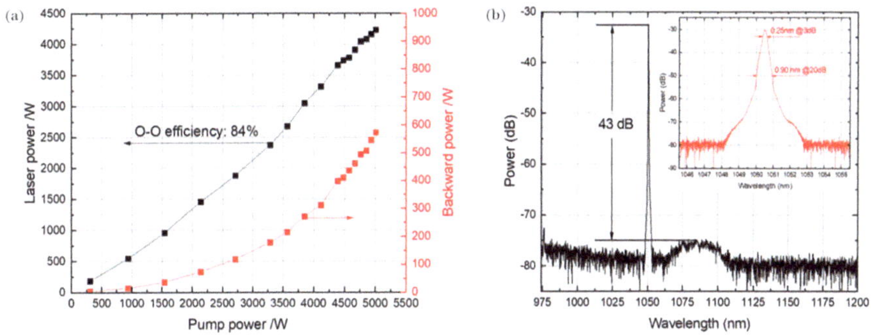

Fig. 18.34 High-power narrow-linewidth fiber laser based on spectral modulation

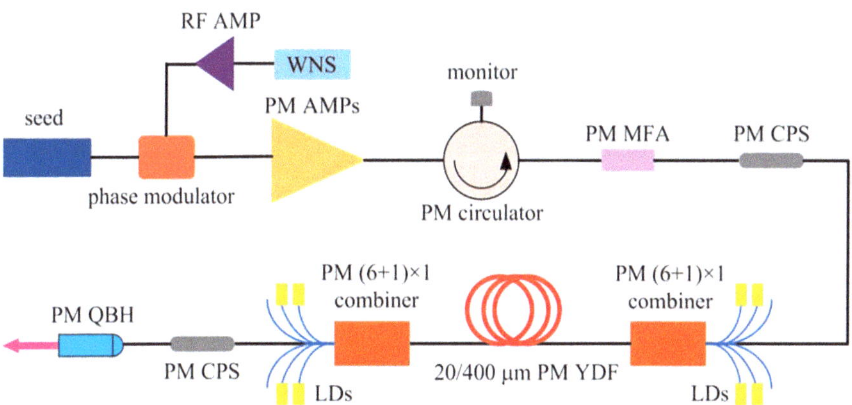

Fig. 18.35 Schematic of the fiber laser structure using WNS phase modulation

pumping, they suppressed the TMI effect, achieving a laser output with a center wavelength of 1064 nm, 703 W of power, a 2 kHz laser linewidth, and beam quality $M^2 = 1.4$ [86].

In 2023, Cao et al. combined sinusoidal signals with pseudorandom binary sequence (PRBS) modulation, as shown in Fig. 18.37 [87]. They achieved a laser output at 1067.5 nm with an RMS linewidth of 46 GHz and a power of 4.93 kW, with a beam quality $M^2 < 1.2$.

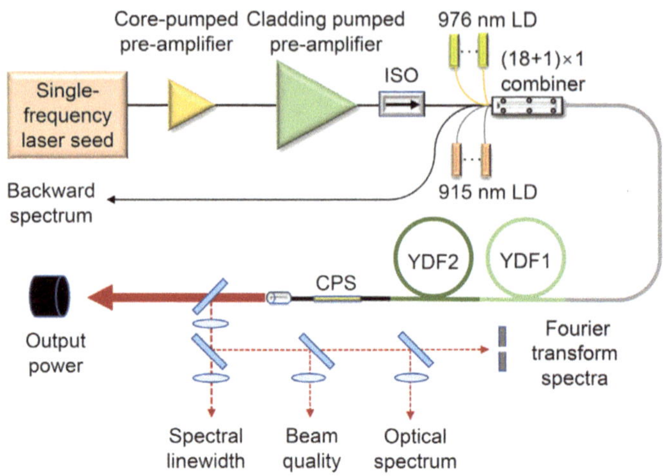

Fig. 18.36 Experimental setup for the single-frequency fiber amplifier

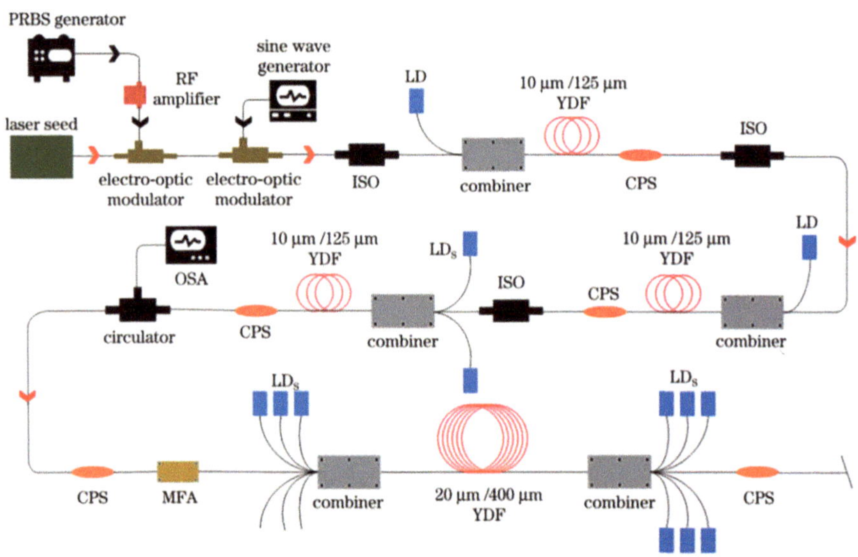

Fig. 18.37 Schematic of the narrow-linewidth fiber laser system

Despite advances in increasing the power of narrow-linewidth fiber lasers, nonlinear effects remain a significant challenge. To further increase fiber laser power, three areas of research require breakthroughs: first, the development of new fiber materials and advanced doping techniques to raise the SBS threshold; second, the creation of novel fiber structures, such as photonic crystal fibers, to mitigate TMI and suppress higher-order modes; and third, the advancement of modulation techniques based on existing sinusoidal, WNS, and PRBS modulation technologies to further increase the SBS threshold and boost the output power of single-frequency fiber lasers.

18.4 Conclusion

As high-power fiber laser technology continues to advance and its applications expand, spectral beam combining has become a key technology for achieving high-brightness, high-power laser output. In recent years, it has developed rapidly, reaching laser power levels beyond 10 kW. This chapter provides a comprehensive review of the development of spectral beam combining technology and analyzes various combining schemes and system architectures. Additionally, it discusses the current challenges in further enhancing the power of spectral beam combining systems.

This chapter specifically addresses the issues of grating thermal deformation and nonlinear effects, which are significant bottlenecks in the current progress of spectral beam combining technology and important directions for future research and innovation. In recent years, an increasing number of researchers have turned their focus to these challenges. It is expected that, in the near future, with continued efforts from the scientific community, spectral beam combining technology will overcome these technical limitations. This progress will lay a solid foundation for achieving higher performance and broader applications of high-power laser systems, playing a crucial role in defense, industry, and scientific research.

References

1. Shi, W., et al.: Fiber lasers and their applications. Appl. Opt. **53**(28), 6554–6568 (2014)
2. Zhou, P., et al.: Review on recent progress on Yb-doped fiber laser in a variety of oscillation spectral ranges. Laser Phys. **22**(5), 823–831 (2012)
3. Pask, H.M., et al.: Ytterbium-doped silica fiber lasers: versatile sources for the 1-1.2 μm region. IEEE J. Sel. Top. Quantum Electron. **1**(1), 2–13 (1995)
4. Dawson, J.W., et al.: Analysis of the scalability of diffraction-limited fiber lasers and amplifiers to high average power. Opt. Express. **16**(17), 13240–13266 (2008)
5. Zhu, J., et al.: Power scaling analysis of tandem-pumped Yb-doped fiber lasers and amplifiers. Opt. Express. **19**(19), 18645–18654 (2011)
6. Dajani, I., et al.: Stimulated Brillouin scattering suppression through laser gain competition: scalability to high power. Opt. Lett. **35**(18), 3114–3116 (2010)
7. White, J.O., et al.: Suppression of stimulated Brillouin scattering in optical fibers using a linearly chirped diode laser. Opt. Express. **20**(12), 15872–15881 (2012)
8. Jeong, Y., et al.: Ytterbium doped large-core fiber laser with 1:36 kW continuous-wave output power. Opt. Express. **12**, 6088–6092 (2004)
9. Fridman, M., et al.: Phase locking of two coupled lasers with many longitudinal modes. Opt. Lett. **35**(4), 526–528 (2010)

10. Bourdon, P., et al.: Coherent combining of fiber lasers[C]//XXI International Symposium on High Power Laser Systems and Applications 2016. Int. Soc. Optics Photon. **10254**, 1025402 (2017)
11. Wang, X., et al.: Current situation, tendency and challenge of coherent combining of high power fiber lasers. Chin. J. Lasers. **44**(2), 0201001 (2017)
12. Wang, B., Sanchez, A.: All-fiber passive coherent beam combining of fiber lasers and challenges [C]//Fiber Laser Applications. Optic. Soc. Am., FTh3A. 2 (2012)
13. He, B., et al.: Experimental demonstration of phase locking of a two-dimensional fiber laser array using a self-imaging resonator. Appl. Phys. Lett. **92**(25), 43–45 (2008)
14. Zhou, P.: Beam Quality and Power Scalability of Fiber Laser Array in a S-F Cavity[C]// International Conference on Optical Instruments & Technology: Optoelectronic Devices & Integration. International Society for Optics and Photonics (2009)
15. Loftus, T.H., et al.: Four-Channel, High Power, Passively Phase Locked Fiber Array[C]/ Advanced Solid-State Photonics. Optical Society of America (2008)
16. Xue, Y.H., et al.: High power passive phase locking of four Yb-doped fiber amplifiers by an all-optical feedback loop. Chin. Phys. Lett. **28**(5), 054212 (2011)
17. Ma, P.F., et al.: Passive coherent polarization beam combination of a four-fiber amplifier Array. IEEE Photon. J. **5**(6), 7101307–7101307 (2013)
18. Chang, W., et al.: Array size scalability of passively coherently phased fiber laser arrays. Opt. Express. **18**(9), 9634–9642 (2010)
19. Daneu, V., et al.: Spectral beam combining of a broad-stripe diode laser array in an external cavity. Opt. Lett. **25**(6), 405–407 (2000)
20. Von Elm et al. Beam-combiner for fiber-delivered laser-beams of different wavelengths[P]. U.S. Patent No. 8, 599, 487, 2013-12-3.
21. Gold et al. Beam combining/splitter cube prism for color polarization [P]. U.S. Patent No.5, 067, 799, 1991-11-26
22. R. D. Pickering. Beam combining prism [P]. U.S. Patent No. 2, 983, 183, 1961-5-9.
23. Schmidt, O., et al.: Spectral beam combination of fiber amplified ns-pulses by means of interference filters. Opt. Express. **17**(25), 22974–22982 (2009)
24. Ludewigt, K., et al.: High-power laser development for laser weapons [C]//High Power Lasers: Technology and Systems, Platforms, Effects III. Int. Soc. Optics Photon. **11162**, 1116207 (2019)
25. Chen, F., et al.: Beam quality analysis and optimization for 10 kW-level spectral beam combination system. Opt. Commun. **444**, 45–55 (2019)
26. Chen, F., et al.: 10 kW-level spectral beam combination of two high power broad-linewidth fiber lasers by means of edge filters. Opt. Express. **25**(26), 32783–32791 (2017)
27. Ma, J., et al.: Modeling and analysis of the influence of an edge filter on the combining efficiency and beam quality of a 10-kW-class spectral beam-combining system. Appl. Sci. **9**(10), 2152 (2019)
28. Divliansky, I.: Volume Bragg gratings: fundamentals and applications in laser beam combining and beam phase transformations. Hologr. Mater. Opt. Syst., 51 (2017)
29. Ciapurin, I.V., et al.: Incoherent combining of 100-W Yb-fiber laser beams by PTR Bragg grating[C]//advances in fiber lasers. SPIE. **4974**, 209–219 (2003)
30. Sevian, A., et al.: Efficient power scaling of laser radiation by spectral beam combining. Opt. Lett. **33**(4), 384–386 (2008)
31. Liang, X., et al.: High average power spectral beam combining employing volume Bragg gratings. High Power Laser Part. Beams. **27**(7), 071012 (2015)
32. Zhou, T., et al.: 2.5 kW average power, two-channel spectral-beam-combined output based on transmitting volume Bragg grating. Acta Phys. Sin. **66**(8), 84204–084204 (2017)
33. Zhou, T., et al.: Spectral beam combining of fiber lasers by using reflecting volume Bragg gratings. Chin. Phys. Lett. **33**(12), 124205–124205 (2016)
34. Drachenberg, D.R., et al.: Thermal tuning of volume Bragg gratings for spectral beam combining of high-power fiber lasers. Appl. Opt. **53**(6), 1242–1246 (2014)

35. Ott, D., et al.: Scaling the spectral beam combining channels in a multiplexed volume Bragg grating. Opt. Express. **21**(24), 29620 (2013)
36. Ingersoll, G.B., Leger, J.R.: Channel density and efficiency optimization of spectral beam combining systems based on volume Bragg gratings in sequential and multiplexed arrangements. Appl. Opt. **54**(20), 6244–6253 (2015)
37. Cook, C.C., Fan, T.Y.: Spectral beam combining of Yb-doped fiber lasers in an external cavity [J]. Opt. Photon. News. **10**(10), 411–411 (1999)
38. Bochove, E.J.: Theory of spectral beam combining of fiber lasers. IEEE J. Quantum Electron. **38**(5), 432–445 (2002)
39. Augst, S.J., et al.: Wavelength beam combining of ytterbium fiber lasers. Opt. Lett. **28**(5), 331–333 (2003)
40. Zhang, J., et al.: Spectral beam combining system of fiber laser by external-cavity fiber oscillator. Infrared Laser Eng. **47**(1), 103008–0103008 (2018)
41. Augst S. J., et al. Wavelength beam combining of ytterbium fiber lasers in a MOPA configuration[C]//Conference on Lasers and Electro Optics, 2002: 594–595
42. Loftus, T.H., et al.: 522 W average power, spectrally beam-combined fiber laser with near-diffraction-limited beam quality. Opt. Lett. **32**(4), 349–351 (2007)
43. Wirth, C., et al.: 2 kW incoherent beam combining of four narrow-linewidth photonic crystal fiber amplifiers. Opt. Express. **17**(3), 1178–1183 (2009)
44. Christian, W., et al.: High average power spectral beam combining of four fiber amplifiers to 8.2 kW. Opt. Lett. **36**(16), 3118–3120 (2011)
45. Zhang, Y., et al.: Analysis of the property of the beam after spectral beam combining. Acta Phys. Sin. **56**(8), 4590–4595 (2007)
46. Kong, W., et al.: Diffraction property of multi-layer dielectric gratings studied by rigorous coupled-wave analysis. Acta Phys. Sin. **57**(8), 4904–4910 (2008)
47. Zhang, Y., Zhang, B.: Analysis of beam quality for the laser beams after spectral beam combining. Optics. **121**(13), 1236–1242 (2010)
48. Hong, Y., et al.: Scalable hybrid beam combining of kilowatt fiber amplifiers into a 5-kW beam. Opt. Commun. **397**, 95–99 (2017)
49. Ma, Y., et al.: 9.6 kW common aperture spectral beam combination system based on Multi-Channel narrow-linewidth fiber lasers. Chin. J. Lasers. **43**(9), 0901009 (2016)
50. Zheng, Y., et al.: 10.8 kW spectral beam combination of eight all-fiber superfluorescent sources and their dispersion compensation. Opt. Express. **24(11):12063** (2016)
51. Zheng, Y., et al.: High-power, high-beam-quality spectral beam combination of six narrow-linewidth fiber amplifiers with two transmission diffraction gratings. Appl. Opt. **58**(30), 8339–8343 (2019)
52. Eric, H., et al.: Spectrally beam combined fiber lasers for high power, efficiency, and brightness [C]//fiber lasers X: technology, systems, and applications. Int. Soc. Optics Photon. **8601**, 860115 (2013)
53. Eric, H., et al.: Advances in fiber laser spectral beam combining for power scaling [C]// Components and packaging for laser systems II. Int. Soc. Optics Photon. **9730**, 97300Y (2016)
54. Martin, L.: Lockheed Martin to deliver world recordsetting 60 kw laser to U.S. Army. https://phys.org/news/2017-03-lockheed-martin-world-record-setting-60kw.html
55. Zhou, J., et al.: High-power fiber laser technology [J]. Chin. J. Lasers. **51**(11), 1101021 (2024)
56. Liu, A., et al.: Spectral beam combining of high-power fiber lasers [C]. Proc. SPIE. **5335**, 81–88 (2004)
57. Loftus, T.H., et al.: 258 W of spectrally beam combined power with near-diffraction limited beam quality [J]. Proc. SPIE. **6102**, 6102S (2006)
58. Limpert, J., et al.: The rising power of fiber lasers and amplifiers. IEEE J. Sel. Top. Quantum Electron. **13**(3), 537–545 (2007)
59. Yang, L., et al.: Influence of thermal deformation of a multilayer dielectric grating on a spectrally combined beam. Appl. Opt. **55**(32), 9091–9100 (2016)

60. Li, L., et al.: Beam modulation due to thermal deformation of grating in a spectral beam combining system. Appl. Opt. **56**(19), 5511–5519 (2017)
61. Xu, J., et al.: Study of the key factors affecting temperature of spectral-beam-combination grating. Opt.Express. **26**(17), 21675–21684 (2018)
62. Xu, J., et al.: Dependence of temperature and far-field beam quality on substrate thickness of a spectral beam combining grating with 13.4 kW/cm^2 laser irradiation. Appl. Opt. **57**(18), D165 (2018)
63. Gong, W., et al.: Research on characteristic of multilayer dielectric diffraction grating under high power laser irradiation. Chin. J. Lasers. **44**(05), 0504003 (2017)
64. Wang, H., et al.: Thermal analysis of multilayer dielectric grating with high power laser irradiation. AIP Adv. **10**, 055207 (2020)
65. Wang, H., et al.: Simulation and experimental study of laser-induced thermal deformation of spectral beam combination grating. Opt. Express. **28**(22), 33334–33345 (2020)
66. Jauregui, C., et al.: Physical origin of mode instabilities in high-power fiber laser systems. Opt. Express. **20**(12), 12912–12925 (2012)
67. Jauregui, C., et al.: Transverse mode instability. Adv. Opt. Photon. **12**(2), 429–484 (2020)
68. Otto, H.J., et al.: Temporal dynamics of mode instabilities in high-power fiber lasers and amplifiers. Opt. Express. **20**(14), 15710–15722 (2012)
69. Boyd, R.W.: Nonlinear Optics [M], 4th edn. Academic Press, London (2020)
70. Aoki, Y., et al.: Input power limits of single-mode optical fibers due to stimulated Brillouin scattering in optical communication systems. J. Lightwave Technol. **6**(5), 710–719 (1988)
71. Zeringue, C., et al.: A theoretical study of transient stimulated Brillouin scattering in optical fibers seeded with phase-modulated light. Opt. Express. **20**(19), 21196–21213 (2012)
72. Supradeepa, V.R.: Stimulated Brillouin scattering thresholds in optical fibers for lasers linewidth broadened with noise. Opt. Express. **21**(4), 4677–4687 (2013)
73. Anderson, B., et al.: Comparison of phase modulation schemes for coherently combined fiber amplifiers. Opt. Express. **23**(21), 27046–27060 (2015)
74. Belgabad, A.A., et al.: Comparison between threshold and sensitivity of stimulated Brillouin scattering at different pumping configurations in high-power double-clad fiber lasers. Optik. **126**(24), 4924–4928 (2015)
75. Naderi, N.A., et al.: Beam combinable, kilowatt, all-fiber amplifier based on phase-modulated laser gain competition. Opt. Lett. **41**(17), 3964–3967 (2016)
76. Boggio, J.C., et al.: Experimental and numerical investigation of the SBS-threshold increase in an optical fiber by applying strain distributions. J. Lightwave Technol. **23**(11), 3808–3814 (2005)
77. Kovalev, V.I., Harrison, R.G.: Suppression of stimulated Brillouin scattering in high-power single-frequency fiber amplifiers. Opt. Lett. **31**(2), 161–163 (2006)
78. Liu, A.P.: Suppressing stimulated Brillouin scattering in fiber amplifiers using nonuniform fiber and temperature gradient. Opt. Express. **15**(3), 977–984 (2007)
79. Theeg, T., et al.: All-fiber counter-propagation pumped single frequency amplifier stage with 300-W output power. IEEE Photon. Technol. Lett. **24**(20), 1864–1867 (2012)
80. Yang, Y., et al.: Realization of 2.5 kW near-diffraction-limited output from an all-fiber high-efficiency, narrow-linewidth fiber laser[J]. Chin. J. Lasers. **43**(4), 0419004 (2016)
81. Zhao, X., et al.: 302 W triple-frequency, singlemode, linearly polarized Yb-doped all-fiber amplifier. High Power Laser Sci. Eng. **5**, e31 (2017)
82. Shen, H., et al.: Narrow-linewidth all-fiber amplifier with up to 3.01 kW output power based on commercial 20/400 μm active fiber and counterpumped configuration. Appl. Opt. **58**(12), 3053–3058 (2019)
83. Lai, W., et al.: 550 W single frequency fiber amplifiers emitting at 1030 nm based on a tapered Yb-doped fiber. Opt. Express. **28**(14), 20908–20919 (2020)
84. Shen, H., et al.: Realization of 0.25 nm, 4.23 kW all-fiber single-mode laser based on self-developed 25 μm/400 μm active fiber[J]. Chin. J. Lasers. **49**(11), 1116001 (2022)

85. Ren, S., et al.: 3.96 kW all-fiberized linearly polarized and narrow linewidth fiber laser with near diffraction limited beam quality. Nano. **12**(15), 2541 (2022)
86. Shi, C., et al.: 700 W single-frequency all-fiber amplifier at 1064 nm with kHz-level spectral linewidth. Frontiers. Physics. **10**, 758 (2022)
87. Cao, J., et al.: 4.9-kW narrow-linewidth single-fiber laser based on cascaded phase modulation [J]. Chin. J. Lasers. **50**(10), 1001005 (2023)

Chapter 19
High-Power Fiber Lasers in the AI Era

Jun Li, Yi An, Min Jiang, Junrui Liang, Liangjin Huang, and Pu Zhou

Abstract The chapter reviews the significant role that artificial intelligence (AI) has played in advancing high-power fiber lasers (HPFL). Instead of providing a comprehensive summary, we emphasize the contributions of AI from three key perspectives: optical fiber design, laser properties measurement, and coherent beam combining, to provide an overview of the development trajectory of HPFL in the AI era.

Keywords High-power fiber laser · Artificial intelligence · Optical fiber design · Laser properties measurement · Coherent beam combining

19.1 Introduction

During the past decade, the rapid evolution of deep learning (DL) technology and the explosive emergence of large-scale neural network (NN) models have marked an unprecedented era in artificial intelligence (AI). In the domain of high-power fiber lasers (HPFL), there is a continual inquiry into the future advancements associated with the AI era, along with devoted efforts to study AI's role in driving innovation. This chapter reviews the contributions of AI in the field of HPFL, focusing on the dimensions of design, measurement, and control. Due to space limitations, this chapter does not aim for an exhaustive survey of the body of research. Instead, we highlight representative studies and novel insights to illustrate the transformative impact of AI. More specifically, toward HPFL, we concentrate on methodically

J. Li · J. Liang · L. Huang · P. Zhou (✉)
College of Interdisciplinary Studies, National University of Defense Technology, Changsha, Hunan, China

Y. An · M. Jiang
Test Center, National University of Defense Technology, Xi'an, Shaanxi, China

M. F. S. Ferreira, M. C. Paul (eds.), *Optical Fiber Lasers*, Springer Series in Optical Sciences 250, https://doi.org/10.1007/978-3-032-04110-4_19

categorizing and summarizing AI-related works from three primary perspectives: optical fiber design, laser properties measurement, and coherent beam combining (CBC), providing a snapshot of the development trajectory of high-power fiber lasers in the age of artificial intelligence.

19.2 Optical Fiber Design

Currently, the primary factors limiting the power scaling of fiber lasers are nonlinear effects [1] and transverse mode instability (TMI) [2]. To mitigate these issues, researchers have explored various external solutions. For example, nonlinear effects can be reduced by shortening the fiber length [3], or optimizing the pump direction [4]. On the other hand, TMI can be suppressed by altering the pump wavelength [5], optimizing the bending radius [6], or minimizing laser noise [7]. However, it is difficult to address both issues simultaneously.

The design of the fiber itself can provide an internal solution to the challenges faced by high-power fiber lasers [8].It is known that nonlinear effects are influenced by the mode area of the fiber. Meanwhile, the TMI threshold is related to the content of higher-order modes (HOMs) within the fiber. If the fiber is optimally designed with a large mode area and the desired HOM content, both issues can be simultaneously addressed. Then, the higher power of fiber lasers can be expected.

19.2.1 Traditional Fiber Design

Calculating the optical properties under different structural parameters of the fiber is the common approach to fiber design. Traditional calculation approaches include the finite difference method (FDM) [9], the finite element method (FEM) [10], and the transmission matrix method [11].

Both the FDM and the FEM require dividing the fiber cross-section into discrete grids for analysis. As the cross-sections of almost all kinds of fibers can be meshed, these methods have been successfully applied in the design of various types of fibers, including photonic crystal fibers(PCFs) [12–14] and multicore fibers [15–17].

Unlike FDM and FEM, the transmission matrix method does not require splitting grids. Instead, it relies on the boundary conditions of the electric and magnetic fields at the interfaces between adjacent refractive index layers of the fiber. These boundary conditions are then converted into a transmission matrix for calculation. At present, the transmission matrix method is mainly applied to the evaluation and design of the multilayer fibers [18, 19].

To obtain the optical properties of fibers, the above traditional approaches commonly employ finite discrete grids or boundary conditions to solve Maxwell's equations. Typically, researchers evaluate the properties of the fiber by testing various structural parameters to identify optimal parameters. As a result, Maxwell's equations need to be solved iteratively, which demands substantial computational resources, especially when dealing with fiber structural parameters spanning wide ranges with small increments. Undoubtedly, an AI-enabled approach that offers a faster and more cost-effective fiber design would be a valuable complement to traditional methods.

19.2.2 AI-Enabled Fiber Design

There are two AI-enabled ideas for fiber design. One is the forward design, while the other is the inverse design.

For forward design, AI approaches such as artificial neural networks (ANNs) [20] can be adopted to learn the relationship between the fiber input structural parameters and the output optical properties of the training samples. After training, AI approaches can quickly evaluate optical properties for different input fiber parameters, without the need to solve Maxwell's equations repeatedly. Then, the optimal designing parameters can be selected by observing and summarizing the corresponding optical properties under different structural parameters. Recent progress in AI-enabled forward fiber design is listed in Table 19.1.

For the inverse design, pretrained NN models are usually combined with some optimizing algorithms such as particle swarm optimization (PSO) [30] and genetic algorithm (GA) [31]. These optimization algorithms usually adopt an iterative process within the specified structural parameter space to find the parameters

Table 19.1 Recent Progress of the AI-enabled forward fiber design

Fiber type	Year	AI method	Output optical properties
PCF	2008 [21]	ANN with 1 hidden layer	Effective index, etc.
PCF	2013 [22]	ANN with 2 hidden layers	Chromatic dispersion
PCF	2019 [23]	ANN with 3 hidden layers	Effective mode area, etc.
PCF	2022 [24]	ANN with 3 hidden layers	Birefringence, etc.
PCF	2021 [25]	Generative adversarial network	Confinement loss
PCF	2022 [26]	k-nearest neighbor regression	Bending loss
PCF	2023 [27]	Support vector machine	Confinement loss, etc.
Suspended core fibers	2022 [28]	ANN with 3 hidden layers	Nonlinear coefficient, etc.
Orbital angular momentum fibers	2024 [29]	ANN with 1 hidden layer	Overlap factor, etc.

Table 19.2 Recent progress of the AI-enabled inverse fiber design

Fiber type	Year	AI method	Target properties
Orbital angular momentum fibers	2022 [32]	ANN and PSO	Index difference
PCF	2023 [33]	ANN and GA	Confinement loss
PCF	2023 [34]	ANN and Grey wolf optimizer	Confinement loss
PCF	2023 [35]	ANN and differential evolution algorithm	Dispersion
PCF	2024 [36]	ANN and simulated annealing algorithm	Confinement loss
Few-mode fiber	2020 [37]	ANN with 3 hidden layers	Index difference
Few-mode fiber	2024 [38]	ANN with 5 hidden layers	Index difference

corresponding to the optimal target properties. The iterative process can replace the manual selection of the forward design approach in a more efficient and accurate way. Recent progress in AI-enabled inverse fiber design is listed in Table 19.2.

The following will introduce in detail the recent work of these two design types.

19.2.2.1 AI-Enabled Forward Design of Fiber

Figure 19.1 presents an example diagram for AI-enabled forward design of fiber using ANN. This ANN consists of an input layer, three hidden layers, and an output layer. The nodes I1-Im of the input layer represent the fiber structural parameters, and the nodes O1-On of the output layer denote the optical properties of the fiber.

As a typical kind of fiber, PCF has attracted much attention in the field of forward design. As early as 2008, M. F. O. Hameed et al. from the University of Glamorgan used an ANN to predict the optical properties of a PCF [21]. In this PCF, the air holes are arranged in a triangular lattice of a hole pitch Λ with the same diameter d. The input parameters of the ANN include the wavelength and the ratio of Λ/d. The output optical properties include effective index, dispersion, and confinement losses. Although this ANN only has one hidden layer, its predicting results agree well with the FDM results. With the help of the proposed ANN model, this group further performed a calculation and found the desired PCF design with zero and ultraflat dispersion [39].

In 2013, V. F. Rodriguez-Esquerre et al. from Federal University of Bahia adopted an ANN to calculate the chromatic dispersion for PCFs [22]. This ANN has two hidden layers, which exhibit quick convergence and high efficiency during the training process. In 2019, Sunny Chugh et al. from the University of London used an ANN containing three hidden layers to perform a more comprehensive forward

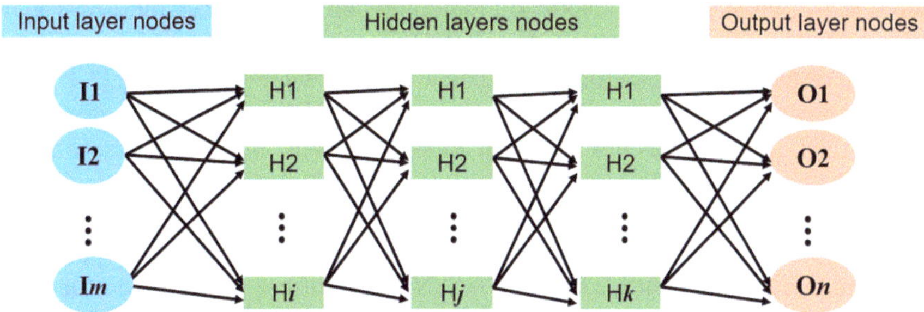

Fig. 19.1 An example diagram for AI-enabled forward design of fiber using ANN

prediction for PCFs [23]. The ANN takes into account PCF structural parameters, such as hole diameter, pitch, core refractive index, wavelength, and the number of rings, as input. Optical properties, including the effective index, the effective mode area, dispersion, and confinement loss, act as labels and are calculated using Lumerical Mode Solutions. Quantitative simulation analyses demonstrated the effectiveness of this method in accurately and fast predicting optical properties under different PCF structural parameters. The predicted optical properties of the trained ANN model exhibit a small mean square error (MSE) compared to their labels. In terms of computation efficiency, Lumerical Mode Solutions requires a few minutes to compute each structural parameter, whereas the ANN model completes calculations in just a few milliseconds.

Based on the above studies, Md. Asaduzzaman Jabin et al. from the University of Georgia used a similar ANN architecture to predict more optical properties of the PCF [24]. This work expands the types of optical properties of PCF from 4 to 14. Among them, 12 properties can be accurately predicted. These properties include effective refractive index, birefringence, effective area, chromatic dispersion, numerical aperture, power fraction, nonlinear coefficient, relative sensitivity, V parameter, confinement loss, effective material loss, and scattering loss. However, the other two properties including bending loss and transmittance cannot be accurately predicted. The authors argue that the reason why ANN cannot accurately predict these two parameters is that the last two properties are mostly nonlinear across different wavelengths. The nonlinear variation might produce out-of-range samples, and thus prevents the ANN from converging.

Aimen Zelaci et al. from the European University of Lefke proposed an approach to handle the out-of-range samples [25]. They combined ANN with Generative Adversarial Network (GAN) to predict the confinement loss of the PCF. In their work, the collected samples are fed into the GAN, then the GAN filters out-of-range samples and generates new samples. Both the collected samples and new samples can be used to train ANN. This method can obtain accurate predictions even when the collected samples are limited.

Besides ANN, some other algorithms such as k-Nearest Neighbor Regression (kNNR) and Support Vector Machine (SVM) have also been adopted in the forward design of PCFs. For example, Cem Kalyoncu et al. from European University of

Lefke successfully adopted kNNR to predict the confinement loss for a bent PCF [26], while Hongwei Li et al. from Yanshan University utilized SVM to predict the effective refractive index, chromatic dispersion, and confinement loss of PCF [27].

The above work focuses on the forward design of PCF, and the key designing tool is ANN. In fact, ANN has also been adopted for other kinds of fibers. In 2022, Shuyu Yuan et al. from Hainan University used an ANN to predict the optical properties of suspended-core fibers [28]. In 2023, Yi An et al. from National University of Defense Technology adopted a shallow ANN to predict the mode properties of active multilayer fibers [40]. This shallow ANN was further applied to evaluate the optical properties of orbital angular momentum (OAM) fibers in 2024 [29].

19.2.2.2 AI-Enabled Inverse Design of Fiber

For the inverse design of fiber, optimizing algorithms such as the mentioned PSO and GA usually play important roles for searching optimal parameters. Although these optimizing methods can also work with traditional methods (FDM, FEM, etc.) [31, 41–44], a pretrained ANN model will efficiently save the optimizing time.

In 2022, Haibo Qin et al. from Tianjin University of Technology used ANN and PSO to inversely design the OAM fiber [32]. In their work, a prediction ANN model was first established and trained. Then, the PSO algorithm was applied to conduct a global search for high effective index difference and low chromatic dispersion.

In 2023, Dan Yang et al. from Northeastern University performed an inverse design for PCF based on ANN and GA [33]. The typical GA inverse design process is plotted in Fig. 19.2. In the process, the gene parameters are fed into the simulation software to evaluate the fiber performance. Then, the inverse iteration including the selection, crossover, and mutation of the genes can continuously update the fiber structure, until the targets are met [31]. When GA is combined with ANN for the inverse design of PCF [33], the initial random gene parameters are input into the trained ANN model to predict the crosstalk, bandwidth, and resonance wavelength of a PCF in a matter of milliseconds. After that, the fitness value of each chromosome is iteratively evaluated based on the gene's optical properties using the defined

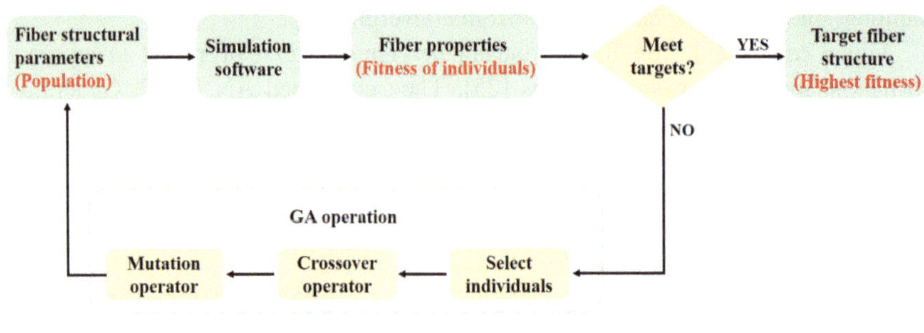

Fig. 19.2 Flow chart of the typical GA inverse design process

fitness function. Then the largest confinement loss and bandwidth of the y polarization core mode can be found to ensure the polarization filtering effects.

Besides the PSO and GA, there are some other optimizing algorithms that can be combined with ANNs for inverse fiber design. These algorithms include the Grey Wolf Optimizer [34], differential evolution [35], and simulated annealing [36].

The above inverse approaches can be attributed to hybrid methods, in which the individual ANN cannot complete the fiber design process. It should be noted that the sole ANN can also perform inverse design in some special cases. For example, Zhiqin He et al. from Shanghai Jiao Tong University utilized an ANN to perform inverse design for weak-coupling few-mode fiber (FMF) in 2020 [37]. The inputs of the ANN are the effective index values of modes while the outputs of the ANN are the fiber structural parameters containing core radius and waveguide refractive index differences. In this way, when the desired mode effective index values are input into the trained ANN, the corresponding structural parameters can be quickly obtained.

In 2024, Yinghao Guo et al. from Beijing Jiaotong University employed similar ANN architectures for the inverse design of vector modes-based weak-coupling FMF [38]. They also utilized this type of ANN for the inverse design of flattop beam fibers [45]. As no other optimization method is involved in these inverse design studies, this sole-ANN approach is free from iterative optimizations.

19.2.3 Discussion

From the above analysis, we can find that ANNs have been powerful tools to learn the input and output relations of various fiber models. As a result, pretrained ANNs can accurately and rapidly predict the fiber properties for giving structural parameters. When these pretrained ANNs are combined with target-optimizing algorithms, the inverse design of fiber can be conducted. Looking forward, the design of high-power laser fibers is expected to leverage more powerful neural networks(Physics-informed NNs [46, 47] might be a new choice) for forward prediction and more efficient target-optimizing architectures for inverse design.

19.3 Laser Properties Measurement

The swift advancement of high-power laser technology has made the measurement of typical laser properties become a significant scientific concern in contemporary laser research. The measurement of laser properties using AI technology is based on interdisciplinary research, creatively integrating artificial intelligence with the evaluation of the laser spectrum, mode, beam quality, and other parameters. As illustrated in Fig. 19.3, a typical unified measurement framework for common laser properties is established, based on NNs with observed intensity images as input.

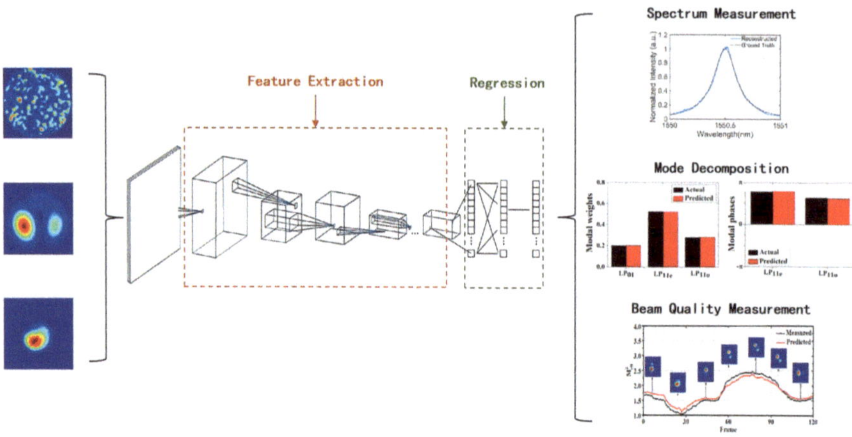

Fig. 19.3 A unified measurement framework for laser spectrum, mode, and beam quality

Compared with traditional measurement methods, this significantly improves real-time performance, anti-interference ability, and usability. In this section, we mainly focus on reviewing the advancement of AI in laser spectrum measurement and mode decomposition. Other measurements, such as beam quality [48], can be addressed using similar methods.

19.3.1 Spectrum Measurement

Optical spectrum is a crucial parameter of high-power lasers, holding important research value in spectral beam combining systems [49], optoelectronic countermeasures [50], power amplification [51], and other fields [52, 53]. Optical spectrometers serve as key tools for measuring spectra. However, traditional setups are often bulky and expensive, requiring high-precision manual alignment. In addition, there is a conflict between resolution and compactness, and increasing the number of spectral points results in slow measurement speeds. These issues limit their application in specific circumstances.

In recent years, with the improvement in computing power and algorithms, reconstructive spectrometers (RSs) have garnered widespread attention [54]. This technology leverages powerful computing to share the workload of dispersion elements, approximating or reconstructing incident light spectra from precalibrated data encoded within a group of observed values [55]. Specifically, the term "reconstruction" refers to solving a system of linear equations. Currently, most RSs are based on two strategies: spectral-to-spatial mapping and spectral response engineering [55].

19.3.1.1 RSs Based on Spectral-to-Spatial Mapping

RSs based on spectral-to-spatial mapping reconstruct spectra by injecting light into disordered media and obtaining the output speckle pattern. Based on existing reports, this approach offers several advantages over the second strategy, including high resolution, fast speed, relatively low cost, and high-power tolerance. It is particularly well-suited for spectral characterization with high-power lasers, which is why this approach will be highlighted here.

General algorithms used in speckle RSs include methods based on transmission matrix [56, 57], principal component analysis (PCA) [58, 59], and AI [60, 61], among which AI-based methods have become a promising direction. A large body of research has realized single-wavelength classification with a primary focus on achieving high spectral resolution. In 2020, Gupta et al. [60] demonstrated that convolutional neural networks (CNNs) are optimal for analyzing wavelength variations from a speckle image. This approach excelled in identifying patterns and mitigating the challenges posed by low signal-to-noise ratios in intricate datasets. Attometer-level wavelength resolution across an operational range from 488 nm to 976 nm was presented. In 2023, Collins et al.[62] realized a resolution of 1 pm on a 0.1 m-long multimode fiber (MMF) using a CNN, whereas previously, this level of resolution required a 100-m-long MMF [57]. This shows that CNNs can greatly enhance the robustness of speckle-based wavemeters. Additionally, AI can lower the cost by eliminating the need for millions of pixels found in conventional cameras. In 2024, Wang et al. [63] introduced a new demodulation algorithm called the shallow residual network, which distinguished wavelengths from compressed speckles captured by a quadrant detector. A wavelength precision of 4 fm ($\sim$ 0.5 MHz) with an updating speed of $\sim$ 1 kHz was achieved.

In these wavemeters, the measurable wavelengths are usually limited, and the identification of wavelengths is treated as a classification task in machine learning. However, in practical applications, the wavelengths under test may be massive, and their values continuously change, so treating the wavelength identification as a regression task is more appropriate. In a recent study, Gao et al. [64] utilized the wavelength-dependent pulse sequences as their datasets, which were analyzed using a multilayer perceptron (MLP) to estimate the absolute wavelength. A measurement rate of 100 MHz with an accuracy of 2.7 pm in a 1 nm operation bandwidth was achieved. To summarize, recent progress in the AI-enabled single-wavelength measurement is shown in Table 19.3.

Table 19.3 Recent progress in the AI-enabled single-wavelength measurement

Year	Model	Task	Special Feature
2020 [60]	CNN	Classification	Noise rejection, precision of 2 am
2023 [62]	CNN, VGG16, and ResNet50	Classification	High resolution with short MMF
2024 [63]	Shallow residual network	Classification	Camera-free, high precision
2024 [64]	MLP	Regression	Spectra-space-time mapping

Combining deep learning with speckle patterns not only measures single wavelengths but also enables the detection of spectra with multiple wavelengths. In 2022, Feng et al. [65] utilized Residual Neural Networks (ResNet) to achieve robust spectral sequence regression with a resolution of 0.1 pm and a working range of 1 pm. A strategy of training multiple ResNets in parallel was also proposed to expand the operating bandwidth. Along with resolution and bandwidth requirements, lightweight and low-latency networks are gaining attention for developing spectral measurement technologies suitable for mobile devices. In 2023, Ma et al. [66] used a fiber taper tip to obtain leaky modes and employed the MobileViT model for spectral reconstruction, achieving a range of 450-1100 nm, with a maximum resolution of 1.5 pm. In addition to using CNNs, Wang et al. [67] found that an MLP has higher efficiency and accuracy. The above-mentioned data-driven AI methods usually face challenges in generalization, meaning that test results highly depend on the training data. Therefore, when researchers measure different types of spectra, they need to use distinct neural networks—one for broadband and another for narrowband [68]. Moreover, data-driven AI methods often require large-scale training datasets and lengthy training times when computing spectral outputs on a large number of channels. Addressing this issue, Fan et al. [61] proposed the PDNet, a spectral reconstruction network combining physical modeling with data. Its loss function, based on the common mean square between the real spectrum and the reconstructed spectrum, adds the mean square error term of the observed speckle and the reconstructed speckle calculated according to the transmission matrix. This approach enabled reconstruction of up to 2,300 spectral channels using small-scale datasets. Recent progress in the AI-enabled multiwavelength measurement is summarized in Table 19.4.

Another disruptive technology brought by AI is the sensing of multidimensional light field parameters. In the past, in spectral measurement scenarios that relied on the transmission matrix, it was necessary to convert the detected light into a fundamental spatial mode. Moreover, this light had to be linearly polarized and normally incident to ensure that the speckles produced after entering a disordered scattering medium at the same wavelength were consistent. However, AI allows for the relaxation of the requirement for consistency in this incident state. It can learn the abstract features from a single speckle image and their association with various light field parameters, thus achieving a single-shot perception of high-dimensional light field parameters (such as polarization, wavelength, phase, and intensity). In recent years,

Table 19.4 Recent progress in the AI-enabled multiwavelength measurement

Year	Model	Dataset type	Spectrum lengthy	Special feature
2022 [65]	ResNet50	Experimental	10	Robust to environment
2023 [66]	MobileVit	Synthetic	~ 100	Lightweight, low-latency
2023 [61]	U-Net	Synthetic	2300	Physics-informed
2024 [67]	MLP	Experimental	1121	Higher accuracy and efficiency

some works have emerged in this area. In 2023, Xiong et al. [71] utilized a CNN to establish the mapping relationship among MMF output speckles, incident light wavelength, and Stokes parameters. They achieved a wavelength resolution of 0.5 nm and distinguished two polarization states on the Poincaré sphere with a vector difference of 0.365. In the same year, Zhu et al. [72] utilized a multitask learning deep NN to process single-shot light-field-encoded speckle images, achieving high-accuracy recognition of the full Stokes vector, multiple orbital angular momentum (OAM), wavelength, and power. The measurable state space of a four-dimensional light field was as high as $1680 = 4$ (multiple-OAM) $\times$ 2 (OAM power spectra) $\times$ 15 (multiple-wavelength) $\times$ 14 (polarizations). Recently, Fan et al. [73] also showed that using spatial and frequency dispersion in simple thin-film interfaces can project and customize polarization and spectrum responses in the wavevector domain. By doing so, high-dimensional light information can be encoded into a single-shot image and decoded using a ResNet. It should be pointed out that deep learning has opened up new avenues for miniaturized and high-dimensional photodetection.

19.3.1.2 RSs Based on Spectral Response Engineering

In the previous section, we focused on using AI for spectral-to-spatial mapping RSs. Here, we will review AI-enabled spectral response RSs. Given that the main development trends for this type of RS are miniaturization and daily applications, which might differ significantly from high-power laser beam characterization scenarios, we will provide only a brief overview as an additional reference. The essence of this method lies in designing a set of randomly distributed spectral response curves. By detecting a series of light intensities as the test light passes through various spectral encoders, researchers can reconstruct the incoming components of the test spectrum [74]. Song et al. [75] proposed two deep learning-based optimization methods for spectral encoding filters, termed "spectral encoding and decoding (SED)-inv" and "parameter constrained spectral encoder and decoder (PCSED)," addressing the lack of quantifiable optimization in previous studies using spectral encoding filters. These methods leverage the equivalence between filters in spectral encoding and neural networks, constructing a neural network architecture for SED to quantitatively optimize the filters using the neural network training process. SED-inv introduces a smoothness constraint to SED and then employs a deep learning-based filter inverse design method to design the filters. On the other hand, PCSED directly integrates the filter design and quantitative optimization processes into the same training process, overcoming the convergence issues of inverse design methods. Simulation results demonstrate that PCSED achieves 2–10 times higher filter design accuracy compared to SED-inv, leading to over 30 times improvement in spectral reconstruction accuracy. Additionally, PCSED's design results exhibit higher tolerance to processing errors, improving spectral reconstruction accuracy by approximately 7.5 times.

19.3.2 Mode Decomposition

The growing utilization of multi-mode fibers in various applications such as optical sensing, optical telecommunication, and optical imaging has increased the need for beam mode measurement [76]. Mode decomposition is an important laser spatial property measurement technique [77]. The transverse distribution of an electric field in a fiber can be represented as a linear superposition of eigenmodes, $E(x, y) = \sum_k C_k \Psi_k(x, y)$, where $\Psi_k(x, y)$ denotes the electric field of the kth eigenmode of the fiber, and $C_k = \rho_k e^{i\vartheta_k}$ represents its modal coefficient, including the amplitude ρ_k (modal wights ρ_k^2) and phase ϑ_k of the kth eigenmode. The aim of mode decomposition is to determine the modal coefficients, which can be addressed by coherent and noncoherent methods, based on the necessity of coherent radiation. Typical coherent methods include the spatially and spectrally resolved imaging (S^2) method [78–80], the cross-correlated (C^2) imaging method [81–83], and correlation filter method [84]. Noncoherent methods include numerical methods and AI-enabled methods, both of which require intensity distribution observed at the fiber end facet (near-field image) or at the infinite distance from the fiber end facet (far-field image). Numerical methods rely on different numerical algorithms, such as optimization algorithms (Gerchberg-Saxton algorithm [85], stochastic parallel gradient descent algorithm [86, 87]), pseudoinverse matrix algorithm [88–90], and so on.

This part focuses on the AI-enabled mode decomposition. As Fig. 19.4 shows, NNs can be used to complete mode decomposition in different kinds of forms. On the one hand, NNs can help to find the relationship between the beam intensity image and mode coefficients. The most popular form is supervised learning with deep NNs, where the input are single near-field images, and the labels are either modal amplitudes and cosine values of the modal phases [91], or a combination of modal weights, modal phases, near-field, and far-field images [92]. The shallow NN can also be used for mode decomposition under supervised learning, which takes a low-dimensional representation of the near-field image as input [93]. Unsupervised learning with deep NNs is another form that can directly compute accurate modal weights and phases with a near-field image as input [94]. On the other hand, a training-free shallow NN (TFSNN-MD) can act as an optimizer to find modal amplitudes and phases with a physics model that represents the eigenmode superposition mechanism [95]. Recent progress in the AI-enabled mode decomposition is listed in Table 19.5.

19.3.2.1 Supervised Learning with Deep Neural Network

In 2019, Yi An et al. used a CNN modified from VGG-16 with supervised training to predict modal weights and phases for the first time [96]. The input of the CNN is a near-field image, and the label is the modal weights and cosine values of the phases (rather than phases themselves to ensure the network can converge under a one-to-one mapping). After training, this CNN can predict modal weights and cosine

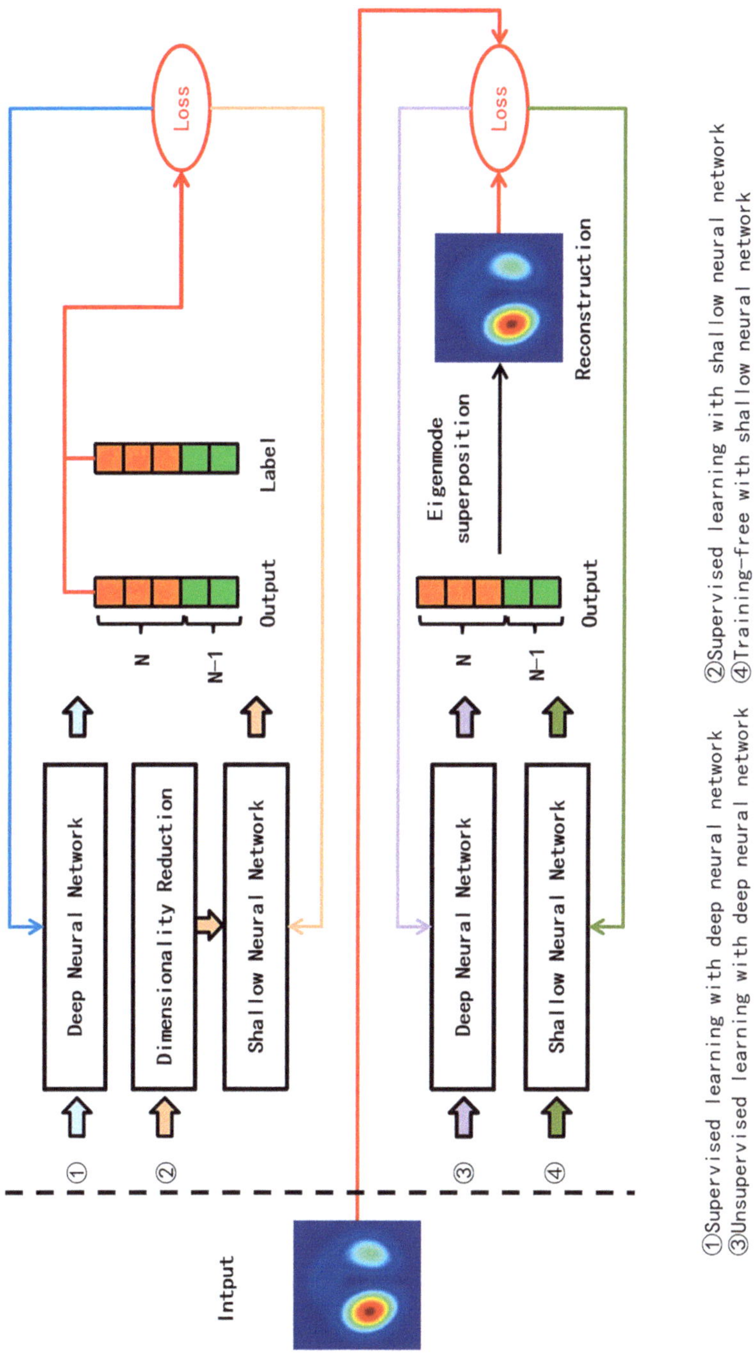

Fig. 19.4 AI-enabled mode decomposition

Table 19.5 Recent progress in the AI-enabled mode decomposition

Year	AI method	Maximum number of modes
2019 [96]	Supervised learning with VGG 16	10 (simulation), 6 (experiment)
2020 [97]	Supervised learning with VGG 16	6 (simulation), 6 (real-time experiment)
2020 [98]	Supervised learning with DenseNet 121	10 (simulation), 10 (experiment)
2021 [93]	Supervised learning with shallow neural network	10 (simulation), 3 (experiment)
2022 [94]	Unsupervised learning with VGG 16	10 (simulation), 6 (real-time experiment)
2023 [95]	Training-free with shallow neural network	3 (simulation), 3 (experiment)

values of phase, while the signs of modal phase can be determined by evaluating the difference between the corresponding reconstructed near-field image and the CNN input. The time taken for sign determination grows with the increase in the number of modes. Restricted by the capturing speed of the CCD (30 Hz), the real-time decomposition rate is experimentally limited to 29.9 Hz for 3-mode and 6-mode cases if it only needs modal weight determination [97]. When predicting both modal weights and phases, the real-time decomposing rate is 29.9 Hz for 3-mode and 24 Hz for 6-mode cases. Further, this work was extended to modal analysis for the Hermite–Gaussian beams emitting from solid-state lasers [99].

Later, in 2021, with more powerful computational units, Stefan Rothe et al. presented mode decomposition with another type of CNN, DenseNet, achieving ten modes experimentally [98]. Like in [96, 97, 99], they also used the cosine values of phase to make the label of CNN input.

In 2024, Baorui Yan et al. presented a novel joint classification-and-regression-task CNN (JCR-CNN) architecture, which can perform the identification of different classes of hybrid mode groups when different modes are excited, and predict corresponding mode coefficients [100]. Simulations and experiments were completed under 3-mode, 5-mode, and 6-mode cases.

In 2020, Xiaojie Fan et al. handled the phase ambiguity of modal coefficients using three labels: modal coefficients, near-field images, and far-field images [92]. A combined loss to train a CNN considers the prediction loss of modal coefficients and reconstruction loss of near-field and far-field images. Simulation results under 3-mode, 5-mode, 6-mode, 8-mode, and 10-mode scenarios show that the training accuracy can be improved at the cost of an increased number of labels in the datasets.

19.3.2.2 Supervised Learning with Shallow Neural Network

In 2021, Han Gao et al. used optimized datasets for network training to reduce the calculation complexity [93]. The PCA, a dimensionality reduction algorithm, was adopted to remove redundant information and noise in the near-field image. Two shallow NNs were trained to map the near-field image after dimensionality reduction to modal weights and the cosine values of the modal phase, respectively. In a

3-mode simulation, dataset optimization can help reduce the speed of complete modal demodulation from 40 ms per frame to 5 ms per frame. In the 3-mode experiment test, the averaged correlation between reconstructed images and target images of 300 samples is 0.9224.

19.3.2.3 Unsupervised Learning with Deep Neural Network

In 2022, Min Jiang et al. demonstrated a deep NN with unsupervised learning to predict modal weights and phases simultaneously for the first time [94]. A trained CNN modified from VGG-16 can directly compute the accurate modal coefficients of the eigenmodes propagating in the fibers, including the modal weights and phases with only one near-field image as input. The training of CNN requires a reconstruction loss of near-field image. Regardless of the number of modes, the decomposing rate is similar in simulations, around 571 Hz, benefiting real-time applications.

19.3.2.4 Training-Free with Shallow Neural Network

In 2023, Zichen Tian et al. proposed a mode decomposition method based on a training-free shallow neural network (TFSNN-MD) [95]. The TFSNN-MD is based on the integration of a physic model and a NN, eliminating the massive beam intensity image datasets and training process. It has obvious advantages with high accuracy, fast operation, and great versatility. In the case of using only a shallow NN with only two convolutional layers, the errors of modal weights and modal phases are lower than 0.50% and 0.60%, theoretically. Furthermore, it performs well in generalization regardless of the variation of the FMF's core radius and NA. In the experiments, the correlation between the real and reconstructed near-field image is higher than 95%.

19.3.3 Discussion

AI-enabled laser properties measurement has achieved impressive results, typically in spectrum measurement and mode decomposition. In comparison to conventional algorithms, AI-enabled spectrum measurement has two advantages [101]: first, the speed of spectrum reconstruction can be several orders of magnitude faster, and second, well-trained deep neural networks have better noise reduction capabilities because they can add more diverse noise samples at different levels during training. However, deep learning methods still encounter several challenges in spectrum measurement. For example, they have a strong reliance on the data distribution of the training dataset, which can result in bias toward those data and hinder generalization to new datasets. Furthermore, generating and labeling spectral datasets from natural sources is relatively difficult, particularly at high resolutions. Addressing these issues will be crucial for the future development of AI-based spectrum measurement. AI-enabled mode decomposition using beam intensity images has attracted significant attention due to its straightforward experimental setup, accurate results, and fast processing speed. There also remains potential for enhancement in the AI-enabled mode decomposition. For example, AI-enabled mode decomposition is currently very limited in terms of the number of modes in the fiber and is characterized by relatively long processing time compared with mode decomposition based on pseudo inverse matrix algorithm [88–90]. More powerful AI algorithms are required to achieve mode decomposition for a larger number of modes with less time.

19.4 Coherent Beam Combining

Coherent beam combining (CBC) has proven to be an effective method to break the inherent power limitations of single-fiber lasers [102, 103]. Based on the type of beam splitter/combiner that is employed, CBC can be divided into two categories: tiled aperture and filled aperture. Currently, artificial intelligence is mainly utilized to explore the first scenario, concentrating on the precise and efficient regulation of the properties of each beam in tiled aperture configuration. However, AI is also expected to be incorporated with the CBC of filled aperture in the future.

Due to the presence of stochastic phase noise, one of the primary challenges in CBC systems, particularly aggravated in high-power configurations, lies in fast and accurate detection of phases. The alignment of the phase of each individual unit within the laser array facilitates the coherent combination of sub-beams, thereby enabling high-power output while preserving beam quality [104–106]. As the demand for array size and output power increases, traditional optimization approaches, including stochastic parallel gradient descent (SPGD) [107–109], face significant challenges when addressing phase noises in large-scale and high-power CBC systems, as the phase locking speed fails to match the rate of phase variations. Chang et al. designed a phase locking module employing an interferometric technique for a CBC system consisting of 397 beams [110], achieving a control frequency of 2000 Hz. However, integrating a reference light beam and ensuring its exact alignment with the entire setup is crucial, which adds additional complexity to the system. Recent advancements have highlighted the significant potential of deep learning (DL) approaches employing deep neural networks for phase locking. This potential is largely due to their remarkable capabilities in rapid reasoning and straightforward setup [111]. Common methodologies for learning-based CBC include supervised learning and reinforcement learning. In supervised learning, neural networks are usually trained to learn a nonlinear map- ping between observed intensity patterns and their corresponding phase distributions using labeled datasets. After training, these networks are able to accurately predict the phases of each beam from diverse intensity images and thus can guide phase compensation. Reinforcement learning does not depend on labeled data; instead, the success of the learning process relies on the design of a suitable reward model. In reinforcement learning-based CBC, the neural networks directly generate an action to modify the phases, rather than predicting phase values as in supervised learning.

19.4.1 CBC Based on Supervised Deep Learning

In the last 5 years, many investigations of supervised learning methods have been presented for CBC systems [112–124]. In 2019, the supervised learning-based DL method was incorporated into CBC for the first time by Hou et al. [112]. A well-trained convolutional neural network (CNN) VGG-16 was utilized to predict the phase error of 7-element and 19-element CBC systems. The primary challenge that constrains network training is the phase ambiguity within the far-field intensity profile at the focal plane. More specifically, due to the fundamental attributes of the Fourier transform, various phase distributions such as global phase shifts and conjugate inversions can result in identical far-field intensity patterns at the focal plane. This feature poses a challenge for the inverse problem as it introduces one-to-many mappings, which indeed make the network confused during training and fail to get convergence. Therefore, Hou et al. annotated their dataset using phases relative to the central beam to eliminate the effect of global phase shifts. Furthermore, every far-field pattern is obtained at a non-focal plane location to ensure the training data is "clean." A standard experimental configuration for image-based CBC with supervised deep learning is illustrated in Fig. 19.5.

Rather than collecting the intensity profiles at a non-focal plane, Chang et al. adopted a diffuser to avoid the one-to-many inverse mapping at the focal plane [113]. Zuo et al. employed a spiral phase shape modulation to resolve the phase ambiguity, thereby guaranteeing a one-to-one rather than one-to-many mapping for

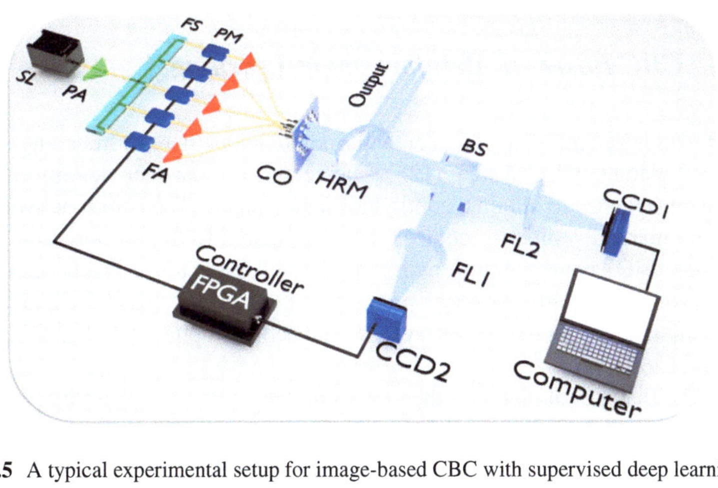

Fig. 19.5 A typical experimental setup for image-based CBC with supervised deep learning

network training [118]. Given that the standard MSE loss fails to accurately capture the periodic variations in phase errors and tends to be more sensitive to outliers, Jia et al. combined spiral phase shape modulation and MSE-NPCD loss to achieve enhanced training stability [116]. Wang et al. demonstrated that a diffractive optical element is capable of generating interference patterns that facilitate phase identification. However, the periodic characteristics of the interference fringes posed challenges for the neural network in processing the entire spectrum of phase values. Despite this, the authors confirmed that the method consistently achieved convergence, and minor phase discrepancies could be rectified in a single iteration [115]. In 2022, Milles et al. proposed a theoretical framework for achieving one-step custom beam shaping in a CBC system with CNNs trained from pairs of phase images and focal intensity images, addressing the challenge of the nonuniqueness between a focal intensity profile and the associated phase profiles [117]. Subsequently, they presented an experimental validation of these theoretical results using a spatial light modulator [123] and also demonstrated that a neural network, which was trained using intensity patterns of simulated fibers with uniform amplitude profiles, was able to successfully identify phase information for intensity patterns arising from fibers with varying amplitude profiles [124].

In a supervised learning framework, a significant hurdle in training networks is to gather a labeled dataset of adequate scale. Collecting labeled training samples from actual experiments is both difficult and expensive. Although training with simulated data is significantly more cost-effective and straightforward, there is a transfer issue when implementing the trained network in a real-world scenario. The optical hardware parameters such as beam center spacings, focal lengths, and the placement of devices might not exactly match the simulated configuration, consequently leading to unexpected performance degradation of the trained networks in real experiments. Thus, addressing the domain transfer issue is important for supervised learning methodologies, which requires a careful alignment of real and simulated profiles.

19.4.2 CBC Based on Reinforcement Learning

In 2019, Tünnermann et al. pioneered the application of reinforcement learning (RL) to a 2-element CBC system [125]. They employed the deep deterministic policy gradient (DDPG) to obtain a continuous action space. The neural network demonstrated a markedly superior control effect compared to the proportional integral differential (PID) controller in environments with high noise levels, and it also exhibited some capability for phase prediction. In 2021, they used the DDPG algorithm to simulate a 7-element CBC system, and achieved phase locking after 6 to 12 iterations [126].

In 2021, Zhang et al. adopted the Q learning algorithm for phase compensation in CBC [127]. For continuous coherent synthesis using Q-learning, they restructured the Q-table by using small phase adjustments as the action space, optimizing state space to minimize its dimensions, and dividing the action space for multichannel synthesis.

In 2021, Shpakovych et al. introduced a quasi-reinforcement learning method by utilizing a neural network for agent strategies [128]. The neural network was trained with diffused images, using a reward system similar to supervised learning loss functions, to map intensity distribution to the array phases over iterations, achieving a phase residual of λ/30. In 2022, they introduced a target adaptive neural network capable of facilitating on-demand adjustments to the target phase during operation. When applied to a coherent synthesis system comprising seven fiber amplifiers with a power output less than 7 W, the system achieved a control residual of λ/20 and a bandwidth greater than 1 kHz [129].

Generally speaking, for reinforcement learning in CBC systems, the main objective is to predict an appropriate action for phase adjustment based on the present observations. In contrast to supervised learning, reinforcement learning exhibits greater flexibility for training in real-world environments as it does not require labeled data. However, reinforcement learning often requires costly interaction with the environment and is typically more challenging to train the model and reach convergence.

19.4.3 Discussion

Although there have been a significant number of studies on AI-enabled CBC, the combined scale remains limited to a few dozen elements. The question of the upper limit of channels that a laser array with AI support can effectively handle is still open for exploration. To advance the development of AI-enabled CBC systems, it is important to conduct in-depth investigations into learning strategies, such as unsupervised learning, transfer learning, and online learning. Moreover, the simultaneous control of multiple parameters within CBC systems represents an excellent avenue for exploring the potential of artificial intelligence in the future.

19.5 Conclusion and Prospect

In the domain of high-power fiber lasers, artificial intelligence has demonstrated superior attributes compared to traditional methods, such as faster response speed, increased robustness against interference, and a more simplified experimental setup. In the future, we believe the generative networks which are widely employed in popular large-scale models such as ChatGPT and Sora would further drive innovations in HPFL. Furthermore, the rapid advancement of ChatGPT technology significantly improves the effectiveness of human interactions with HPFL systems through auditory and textual modalities, thereby enabling them to function as smart agents that can interpret human instructions and autonomously generate appropriate actions to execute.

References

1. Zervas, M.N., Codemard, C.A.: High power fiber lasers: a review. IEEE J. Sel. Top. Quant. Electron. **20**(5), 219–241 (2014). https://doi.org/10.1109/JSTQE.2014.2321279
2. Jauregui, C., Limpert, J., Tünnermann, A.: High-power fibre lasers. Nat. Photon. **7**(11), 861–867 (2013). https://doi.org/10.1038/nphoton.2013.273
3. Huang, Z.-M., Shu, Q., Tao, R.-M., Chu, Q.-H., Luo, Y., Yan, D.-L., Feng, X., Liu, Y., Wu, W.-J., Zhang, H.-Y., et al.: 5kw record high power narrow linewidth laser from traditional step-index monolithic fiber amplifier. IEEE Photon. Technol. Lett. **33**(21), 1181–1184 (2021). https://doi.org/10.1109/LPT.2021.3112270
4. Li, R., Wu, H., Xiao, H., Leng, J., Huang, L., Zhou, P.: More than 6 kw near single-mode fiber amplifier based on a bidirectional tandem pumping scheme. Appl. Optics. **61**(23), 6804–6810 (2022). https://doi.org/10.1364/AO.465076
5. Jauregui, C., Otto, H.-J., Breitkopf, S., Limpert, J., Tünnermann, A.: Optimizing high-power yb-doped fiber amplifier systems in the presence of transverse mode instabilities. Optics Express. **24**(8), 7879–7892 (2016). https://doi.org/10.1364/OE.24.007879
6. Tao, R., Ma, P., Wang, X., Zhou, P., Liu, Z.: 1.3 kw monolithic linearly polarized single-mode master oscillator power amplifier and strategies for mitigating mode instabilities. Photon. Res. **3**(3), 86–93 (2015). https://doi.org/10.1364/PRJ.3.000086
7. Chu, Q., Guo, C., Yang, X., Li, F., Yan, D., Zhang, H., Tao, R.: Increasing mode instability threshold by mitigating driver noise in high power fiber amplifiers. IEEE Photon. Technol. Lett. **33**(24), 1515–1518 (2021). https://doi.org/10.1109/LPT.2021.3124641
8. Chen, X., Yao, T., Huang, L., An, Y., Wu, H., Pan, Z., Zhou, P.: Functional fibers and functional fiber-based components for high-power lasers. Advanced Fiber Materials. **5**(1), 59–106 (2023). https://doi.org/10.1007/s42765-022-00219-7
9. Fallahkhair, A.B., Li, K.S., Murphy, T.E.: Vector finite difference modesolver for anisotropic dielectric waveguides. J. Lightw. Ttechnol. **26**(11), 1423–1431 (2008). https://doi.org/10.1109/JLT.2008.923643
10. Cucinotta, A., Selleri, S., Vincetti, L., Zoboli, M.: Holey fiber analysis through the finite-element method. IEEE Photon. Technol. Lett. **14**(11), 1530–1532 (2002). https://doi.org/10.1109/LPT.2002.803375
11. Garcia, S., Gasulla, I.: Universal characteristic equation for multi-layer optical fibers. IEEE J. Sel. Top. Quant.Electron. **26**(4), 1–11 (2020). https://doi.org/10.1109/JSTQE.2020.2996375
12. Yu, C.-P., Chang, H.-C.: Applications of the finite difference mode solution method to photonic crystal structures. Optic. Quant. Electron. **36**, 145–163 (2004). https://doi.org/10.1023/B:OQEL.0000015636.20125.7e
13. Paul, B.K., Ahmed, K.: Analysis of terahertz waveguide properties of q-pcf based on fem scheme. Opt. Mater. **100**, 109634 (2020). https://doi.org/10.1016/j.optmat.2019.109634
14. Mamtaz, R., Ahmed, K., Paul, B.K., Mollah, M.A., Aktar, M.N., Uddin, M.S., Dhasarathan, V.: Design and fem analysis of pentagonal photonic crystal fiber for highly non-linear applications. Optic. Quant. Electron. **52**, 1–13 (2020). https://doi.org/10.1007/s11082-020-02562-8
15. Van Newkirk, A., Antonio-Lopez, E., Salceda-Delgado, G., Amezcua-Correa, R., Schülzgen, A.: Optimization of multicore fiber for high-temperature sensing. Opt. Lett. **39**(16), 4812–4815 (2014). https://doi.org/10.1364/OL.39.004812
16. Ren, W., Tan, Z., Ren, G.: Analytical formulation of supermodes in multicore fibers with hexagonally distributed cores. IEEE Photon. J. **7**(1), 1–11 (2015). https://doi.org/10.1109/JPHOT.2015.2395993
17. Xie, Y., Pei, L., Zheng, J., Zhao, Q., Ning, T., Li, J.: Impact analysis of a dense hole-assisted structure on crosstalk and bending loss in homogeneous few-mode multi-core fibers. Opt. Express. **28**(16), 23806–23819 (2020). https://doi.org/10.1364/OE.400461
18. Brunet, C., B'elanger, P.-A., Rusch, L.A.: Exact expressions for vector mode cutoff in three-layer step-index fibers. J. Lightw. Technol. **34**(13), 3094–3102 (2016). https://doi.org/10.1109/JLT.2016.2561975

19. Luo, H., Sun, C., Geng, T., Yuan, L.: A dark hollow beam generator based on special optical fiber with long period fiber grating. Optics Laser Technol. **134**, 106598 (2021). https://doi.org/10.1016/j.optlastec.2020.106598
20. Jiang, J., Chen, M., Fan, J.A.: Deep neural networks for the evaluation and design of photonic devices. Nat. Rev. Mater. **6**, 679–700 (2021). https://doi.org/10.1038/s41578-020-00260-1
21. Hameed, M., Obayya, S., Al-Begain, K., Nasr, A., Maaty, M.: Accurate radial basis function based neural network approach for analysis of photonic crystal fibers. Optical and Quantum Electronics **40**, 891–905 (2008). https://doi.org/10.1007/s11082-009-9290-5
22. Rodr'ıguez-Esquerre, V.F., Is'ıdio-Lima, J., Dourado-Sisnando, A., Simoes Silva, F.: Artificial neural networks for the chromatic dispersion prediction of photonic crystal fibers. Microw. Optical Technol. Lett. **55**(9), 2179–2181 (2013). https://doi.org/10.1002/mop.27753
23. Chugh, S., Gulistan, A., Ghosh, S., Rahman, B.: Machine learning approach for computing optical properties of a photonic crystal fiber. Opt. Express. **27**(25), 36414–36425 (2019). https://doi.org/10.1364/OE.27.036414
24. Jabin, M.A., Fok, M.P.: Prediction of 12 photonic crystal fiber optical properties using mlp in deep learning. IEEE Photon. Technol. Lett. **34**(7), 391–394 (2022). https://doi.org/10.1109/LPT.2022.3157266
25. Zelaci, A., Yasli, A., Kalyoncu, C., Ademgil, H.: Generative adversarial neural networks model of photonic crystal fiber based surface plasmon resonance sensor. J. Lightw. Technol. **39**(5), 1515–1522 (2020). https://doi.org/10.1109/JLT.2020.3035580
26. Kalyoncu, C., Yasli, A., Ademgil, H.: Machine learning methods for estimating bent photonic crystal fiber based spr sensor properties. Heliyon. **8**(11) (2022). https://doi.org/10.1016/j.heliyon.2022.e11582
27. Li, H., Chen, H., Li, Y., Chen, Q., Fan, X., Li, S., Ma, M.: Prediction of the optical properties in photonic crystal fiber using support vector machine based on radial basis functions. Optik. **275**, 170603 (2023). https://doi.org/10.1016/j.ijleo.2023.170603
28. Yuan, S., Chen, S., Yang, J., Yang, Q., Ren, S., Wang, G., Yu, B.: Efficient calculation of optical properties of suspended-core fiber via a machine learning algorithm. Appl. Optics. **61**(19), 5714–5721 (2022). https://doi.org/10.1364/AO.458048
29. An, Y., Jiang, M., Li, J., Su, R., Huang, L., Leng, J., Pan, Z., Jiang, Z., Zhou, P.: Shallow neural network boosts the evaluation of oam fibers. J. Lightw. Technol. **42**(7), 2499–2505 (2024). https://doi.org/10.1109/JLT.2023.3334564
30. Wang, D., Tan, D., Liu, L.: Particle swarm optimization algorithm: an overview. Soft Comput. **22**, 387–408 (2018). https://doi.org/10.1007/s00500-016-2474-6
31. Zhang, F., Liu, Z., Du, H., Shao, Y., Shen, L., Yang, L., Yan, C., Zhao, Z., Tang, M.: Genetic algorithm assisted bridge fiber design and fabrication for few- mode multi-core fiber fan-in/fan-out device. Opt. Express. **30**(11), 19042–19054 (2022). https://doi.org/10.1364/OE.457374
32. Qin, H., Huang, W., Song, B., Chen, S.: Hybrid method for inverse design of orbital angular momentum transmission fiber based on neural network and optimization algorithms. J. Lightw. Technol. **40**(17), 5974–5985 (2022). https://doi.org/10.1109/JLT.2022.3185059
33. Yang, D., Qin, H., Li, Y., Tang, C., Xu, B., Cheng, T.: A hybrid method for photonic crystal fiber polarization filter based on artificial neural network and genetic algorithms. Optical Fiber Technol. **80**, 103426 (2023). https://doi.org/10.1016/j.yofte.2023.103426
34. Yang, D., Huang, J., Xu, B., Lv, G., Li, Y., Cheng, T.: Design of high-performance photonic crystal fiber polarization filter by Grey Wolf Optimizer with convolutional neural network. Optik. **283**, 170925 (2023). https://doi.org/10.1016/j.ijleo.2023.170925
35. Ma, F., Ma, Y., Li, P., Shi, W.: Inverse design of broadband dispersion compensation fiber based on deep learning and differential evolution algorithm. IEEE Photon. J. **15**(3),7201307 (2023). https://doi.org/10.1109/JPHOT.2023.3277129
36. Li, H., Chen, H., Li, Y., Li, S., Ma, M.: Reverse-designed photonic crystal fiber-based polarization filter with optimal performance. Optics Laser Technol. **168**, 109909 (2024). https://doi.org/10.1016/j.yofte.2024.103718

37. He, Z., Du, J., Chen, X., Shen, W., Huang, Y., Wang, C., Xu, K., He, Z.: Machine learning aided inverse design for few-mode fiber weak-coupling optimization. Opt. Express. **28**(15), 21668–21681 (2020). https://doi.org/10.1364/OE.398157
38. Guo, Y., Cheng, Y., Gao, J., Ren, W., Ren, G.: Machine learning aided inverse design for vector modes-based weak-coupling few-mode fiber. Optical Fiber Technol. **82**, 103627 (2024). https://doi.org/10.1016/j.yofte.2023.103627
39. El-Mosalmy, D.D., Hameed, M., Areed, N.F., Obayya, S.: Novel neural network based optimization approach for photonic devices. Optic. Quant. Electron. **46**, 439–453 (2014). https://doi.org/10.1007/s11082-013-9869-8
40. An, Y., Jiang, M., Chen, X., Li, J., Su, R., Huang, L., Pan, Z., Leng, J., Zongfu, J., Zhou, P.: Machine learning predicting mode properties for multi-layer active fiber. Chin. J. Lasers **50**(11), 11010131101,013 (2023). https://doi.org/10.3788/CJL230476
41. Chang, J.H., Corsi, A., Rusch, L.A., LaRochelle, S.: Design analysis of oam fibers using particle swarm optimization algorithm. J. Lightw. Ttechnol. **38**(4), 846–856 (2019). https://doi.org/10.1109/JLT.2019.2945870
42. Mirjalili, S.M., Taleb, H., Kabir, M., Bianucci, P.: Design optimization of orbital angular momentum fibers using the gray wolf optimizer. Appl. Optics. **59**(20), 6181–6190 (2020). https://doi.org/10.1364/AO.391731
43. Meng, F., Zhao, X., Ding, J., Niu, Y., Zhang, X., Yang, L., Wang, X., Lou, S., Sheng, X., Tao, G., et al.: Discovering extremely low confinement-loss anti- resonant fibers via swarm intelligence. Opt. Express. **29**(22), 35544–35555 (2021). https://doi.org/10.1364/OE.440949
44. Yang, D., Wei, Z., Qin, H., Xu, B., Cheng, T.: Optimization of structural parameters of pcf polarization filter by a genetic algorithm. IEEE Photon. J. **14**(6), 1–9 (2022). https://doi.org/10.1109/JPHOT.2022.3221095
45. Guo, Y., Cheng, Y., Jiang, Y., Cao, M., Tang, M., Ren, W., Ren, G.: Machine learning aided inverse design for flattop beam fiber. Optics Commun. **524**, 128814 (2022). https://doi.org/10.1016/j.optcom.2022.128814
46. Raissi, M., Perdikaris, P., Karniadakis, G.E.: Physics-informed neural networks: A deep learning framework for solving forward and inverse problems involving nonlinear partial differential equations. J. Comput. Phys. **378**, 686–707 (2019). https://doi.org/10.1016/j.jcp.2018.10.045
47. Wang, D., Jiang, X., Song, Y., Luo, X., Dong, J., Zhang, M.: Physics-informed machine learning for optical fiber communications: opportunities and challenges. In: 2023 Opto-Electronics and Communications Conference (OECC), pp. 1–6. IEEE (2023). https://doi.org/10.1109/OECC56963.2023.10209949
48. An, Y., Li, J., Huang, L., Leng, J., Yang, L., Zhou, P.: Deep learning enabled superfast and accurate M 2 evaluation for fiber beams. Opt. Express. **27**(13), 18683 (2019). https://doi.org/10.1364/OE.27.018683. Accessed 2024-04-09
49. Zhang, J., Peng, H., Zhang, J., Wang, J., Han, J., Fu, X., Lei, Y., Qin, L., Ning, Y., Wang, L.: Polarization-separated feedback spectral beam combining source. Opt. Lett. **49**(2), 306 (2024). https://doi.org/10.1364/OL.511706
50. Titterton, D.H.: Development of infrared countermeasure technology and systems. In: Krier, A. (ed.) Mid-Infrared Semiconductor Optoelectronics, vol. 118, pp. 635–671. Springer, London (2006). https://doi.org/10.1007/1-84628-209-820
51. Liu, W., Kuang, W., Huang, L., Zhou, P.: Modeling of the spectral properties of CW Yb-doped fiber amplifier and experimental validation. Laser Phys. Lett. **12**(4), 045104 (2015). https://doi.org/10.1088/1612-2011/12/4/045104
52. Liang, J., Xu, J., Zhang, Y., Ye, J., Li, S., Ma, X., Ke, Y., Leng, J., Zhou, P.: Hundred-watt-level, linearly polarized multi- wavelength fiber oscillator with wavelength, interval, and intensity tunability. J. Lightw. Technol. **42**(2), 882–890 (2024). https://doi.org/10.1109/JLT.2023.3317427

53. Gaida, C., Gebhardt, M., Heuermann, T., Wang, Z., Jauregui, C., Limpert, J.: Transverse mode instability and thermal effects in thulium-doped fiber amplifiers under high thermal loads. Opt. Express. **29**(10), 14963 (2021). https://doi.org/10.1364/OE.421954
54. Gao, L., Qu, Y., Wang, L., Yu, Z.: Computational spectrometers enabled by nanophotonics and deep learning. Nanophotonics. **11**(11), 2507–2529 (2022). https://doi.org/10.1515/nanoph-2021-0636
55. Yang, Z., Albrow-Owen, T., Cai, W., Hasan, T.: Miniaturization of optical spectrometers. Science. **371**(6528), 0722 (2021). https://doi.org/10.1126/science.abe0722
56. Liang, J., Ye, J., Ke, Y., Zhang, Y., Ma, X., He, J., Li, J., Xu, J., Leng, J., Zhou, P.: Polarization transmission matrix enabled high-accuracy, large-bandwidth speckle-based reconstructive spectrometer. Appl. Phys. Lett. **124**(7), 071111 (2024). https://doi.org/10.1063/5.0197612. Accessed 2024-02-15
57. Redding, B., Alam, M., Seifert, M., Cao, H.: High-resolution and broadband all- fiber spectrometers. Optica. **1**(3), 175 (2014). https://doi.org/10.1364/OPTICA.1.000175. Accessed 2022-10-06
58. Metzger, N.K., Spesyvtsev, R., Bruce, G.D., Miller, B., Maker, G.T., Malcolm, G., Mazilu, M., Dholakia, K.: Harnessing speckle for a sub-femtometre resolved broadband wavemeter and laser stabilization. Nat. Commun. **8**(1), 15610 (2017). https://doi.org/10.1038/ncomms15610. arXiv:1706.02378 [physics]. Accessed 2023-06-28
59. Bruce, G.D., O'Donnell, L., Chen, M., Dholakia, K.: Overcoming the speckle correlation limit to achieve a fiber wavemeter with attometer resolution. Opt. Lett. **44**(6), 1367–1370 (2019). https://doi.org/10.1364/OL.44.001367. Publisher: Optica Publishing Group
60. Gupta, R.K., Bruce, G.D., Powis, S.J., Dholakia, K.: Deep learning enabled laser speckle wavemeter with a high dynamic range. Laser Photon. Rev. **14**(9), 2000120 (2020). https://doi.org/10.1002/lpor.202000120. arXiv:1910.10702 [physics]. Accessed 2023-06-28
61. Wan, Y., Fan, X., Xu, B., He, Z.: Reconstructive spectrum analyzer with high- resolution and large-bandwidth using physical-model and data-driven model combined neural network. Laser Photon. Rev. **17**(7), 2201018 (2023). https://doi.org/10.1002/lpor.202201018
62. Inalegwu, O.C., Ii, R.E.G., Huang, J.: A machine learning Specklegram Wavemeter (MaSWave) based on a short section of multimode fiber as the dispersive element. Sensors. **23**(10), 4574 (2023). https://doi.org/10.3390/s23104574. Accessed 2024-05-21
63. Wang, T., Tao, J., Wang, X., Liang, Q., Tian, H., Zhou, P., Qiu, Y., Mao, B., Meng, Y., Zhao, C., Kang, J., Wang, L., Li, Y.: Decoding wavelengths from compressed speckle patterns with deep learning. Opt. Lasers Eng. **180**, 108268 (2024). https://doi.org/10.1016/j.optlaseng.2024.108268. Accessed 2024-05-31
64. Gao, Z., Jiang, T., Zhang, M., Xiong, Y., Wu, H., Tang, M.: Breaking the speed limitation of wavemeter through spectra-space-time mapping. Light. Adv. Manuf. **4**(0), 1 (2024). https://doi.org/10.37188/lam.2024.013. Accessed 2024-06-03
65. Feng, F., Gan, J., Chen, P., Lin, W., Chen, G., Min, C., Yuan, X., Somekh, M.: AI-assisted spectrometer based on multi-mode optical fiber speckle patterns. Optics Commun. **522**, 128675 (2022). https://doi.org/10.1016/j.optcom.2022.128675
66. Cen, Q., Pian, S., Liu, X., Tang, Y., He, X., Ma, Y.: Microtaper leaky-mode spectrometer with picometer resolution. eLight. **3**(1), 9 (2023). https://doi.org/10.1186/s43593-023-00041-7
67. Wang, J., Zhang, F., Zhou, X.-h., Xiao, S., Niu, Q., Yang, T.: Miniaturized spectrometer based on MLP neural networks and a frosted glass encoder. Opt. Express. (2024)
68. Mondal, A., Debnath, K.: Design of resolution-tunable neural network-based integrated reconstructive spectrometer. IEEE Sens. J. **22**(3), 2630–2636 (2022). https://doi.org/10.1109/JSEN.2021.3139325
69. Feng, F., Gan, J., Chen, P., Lin, W., Chen, G., Min, C., Yuan, X., Somekh, M.: AI-assisted spectrometer based on multi-mode optical fiber speckle patterns. Optics Commun. **522**, 128675 (2022). https://doi.org/10.1016/j.optcom.2022.128675. Accessed 2023-06-28

70. Cen, Q., Pian, S., Liu, X., Tang, Y., He, X., Ma, Y.: Microtaper leaky-mode spectrometer with picometer resolution. eLight. **3**(1), 9 (2023). https://doi.org/10.1186/s43593-023-00041-7. Accessed 2023-09-11
71. Xiong, Y., Jiang, T., Gao, Z., Wu, H., Zhou, S., Tang, M.: Wavelength and polarization state synchronization measurement based on mmf scattering pattern. In: 2023 Asia Communications and Photonics Conference/2023 International Photonics and Optoelectronics Meetings (ACP/POEM), pp. 1–3 (2023). https://doi.org/10.1109/ACP/POEM59049.2023.10369635
72. Zhu, S.-k., Zheng, Z.-h., Meng, W., Chang, S.-s., Tan, Y., Chen, L.-J., Fang, X., Gu, M., Chen, J.-h. Harnessing disordered photonics via multi-task learning towards intelligent four-dimensional light field sensors. PhotoniX 4(1), 26 (2023) https://doi.org/10.1186/s43074-023-00102-7. Accessed 2024-06-04.
73. Fan, Y., Huang, W., Zhu, F., Liu, X., Jin, C., Guo, C., An, Y., Kivshar, Y., Qiu, C.-W., Li, W.: Dispersion-assisted high-dimensional photodetector. Nature (2024). https://doi.org/10.1038/s41586-024-07398-w. Accessed 2024-06-04.
74. A. Li, C. Wang, F. Bao, W. Fang, Y. Liang, R. Cheng, and S. Pan, "An integrated single-shot spectrometer with large bandwidth-resolution ratio and wide operation temperature range," PhotoniX **4**(1), 29 (2023).
75. Song, H., Ma, Y., Han, Y., Shen, W., Zhang, W., Li, Y., Liu, X., Peng, Y., Hao, X.: Deep-learned broadband encoding stochastic filters for computational spectroscopic instruments. Adv. Theory Simulat. **4**(3), 2000299 (2021). https://doi.org/10.1002/adts.202000299. eprint: https://onlinelibrary.wiley.com/doi/pdf/10.1002/adts.202000299
76. An, Y., Li, J., Huang, L., Li, L., Leng, J., Yang, L., Zhou, P.: Numerical mode decomposition for multimode fiber: From multi-variable optimization to deep learning. Optical Fiber Technol. **52**, 101960 (2019). https://doi.org/10.1016/j.yofte.2019.101960
77. Brüning, R., Gelszinnis, P., Schulze, C., Flamm, D., Duparr'e, M.: Comparative analysis of numerical methods for the mode analysis of laser beams. Appl. Optics. **52**(32), 7769 (2013). https://doi.org/10.1364/ao.52.007769
78. Nicholson, J.W., Yablon, A.D., Ramachandran, S., Ghalmi, S.: Spatially and spectrally resolved imaging of modal content in large-mode-area fibers. Opt. Express. **16**(10), 7233 (2008). https://doi.org/10.1364/OE.16.007233
79. Gray, D., Petrovich, M., Sandoghchi, S.R., Wheeler, N., Baddela, N., Jasion, G., Bradley, T., Richardson, D., Poletti, F.: Real-time modal analysis via wavelength- Swept Spatial and Spectral (S^2) Imaging. IEEE Photon. Technol. Lett. **28**(9), 1034-1037 (2016). https://doi.org/10.1109/lpt.2016.2524592
80. Bock, V., Plötner, M., De Vries, O., Nold, J., Haarlammert, N., Schreiber, T., Eberhardt, R., Tünnermann, A.: Modal content measurements (S^2) of negative curvature hollow-core photonic crystal fibers. Opt. Express. **25**(4), 3006 (2017). https://doi.org/10.1364/oe.25.003006
81. Schimpf, D.N., Barankov, R.A., Ramachandran, S. Cross-correlated (c^2) imaging of fiber and waveguide modes. Opt. Express **19**(14), 13008 (2011). https://doi.org/10.1364/OE.19.013008
82. Demas, J., Ramachandran, S.: Sub-second mode measurement of fibers using C^2 imaging. Opt. Express. **22**(19), 23043 (2014). https://doi.org/10.1364/OE.22.023043
83. Carpenter, J., Eggleton, B.J., Schröder, J.: Polarization-resolved cross-correlated (C^2) imaging of a photonic bandgap fiber. Opt. Express. **24**(24), 27785 (2016). https://doi.org/10.1364/OE.24.027785
84. Xie, K., Liu, W., Zhou, Q., Huang, L., Jiang, Z., Xi, F., Xu, X.: Adaptive phase correction of dynamic multimode beam based on modal decomposition. Opt. Express. **27**(10), 13793 (2019). https://doi.org/10.1364/OE.27.013793
85. Shapira, O., Abouraddy, A.F., Joannopoulos, J.D., Fink, Y.: Complete modal decomposition for optical waveguides. Phys. Rev. Lett. **94**(14), 143902 (2005). https://doi.org/10.1103/physrevlett.94.143902

86. Huang, L., Guo, S., Leng, J., Lü, H., Zhou, P., Cheng, X.: Real-time mode decomposition for few-mode fiber based on numerical method. Opt. Express. **23**(4), 4620 (2015). https://doi.org/10.1364/oe.23.004620
87. Choi, K., Jun, C.: Sub-sampled modal decomposition in few-mode fibers. Opt. Express. **29**(20), 32670 (2021). https://doi.org/10.1364/oe.438533
88. Manuylovich, E.S., Dvoyrin, V.V., Turitsyn, S.K.: Fast mode decomposition in few-mode fibers. Nat. Commun. **11**(1), 5507 (2020). https://doi.org/10.1038/s41467-020-19323-6
89. Manuylovich, E., Donodin, A., Turitsyn, S.: Intensity-only-measurement mode decomposition in few-mode fibers. Opt. Express **29**(22), 36769 (2021). https://doi.org/10.1364/oe.437907
90. Deng, Y., Chang, Q., Chang, H., Liu, W., Ma, P., Zhou, P., Jiang, Z.: Analysis of an image noise sensitivity mechanism for matrix-operation-mode-decomposition and a strong anti-noise method. Opt. Express. **31**(8), 12299 (2023). https://doi.org/10.1364/OE.482552
91. An, Y., Huang, L., Li, J., Leng, J., Yang, L., Zhou, P.: Deep learning enabled real-time modal analysis for fiber beams. In: Tanida, J., Jiang, Y., Liu, D., Greivenkamp, J., Gong, H., Lu, J. (eds.) AOPC 2019: AI in Optics And Photonics, p. 2. SPIE, Beijing, China (2019). https://doi.org/10.1117/12.2541268
92. Fan, X., Ren, F., Xie, Y., Zhang, Y., Niu, J., Zhang, J., Wang, J.: Mitigating ambiguity by deep-learning-based modal decomposition method. Optics Commun. **471**, 125845 (2020). https://doi.org/10.1016/j.optcom.2020.125845
93. Gao, H., Chen, Z., Zhang, Y.-X., Zhang, W.-G., Hu, H.-F., Yan, T.-Y.: Rapid mode decomposition of few-mode fiber by artificial neural network. J. Lightw. Technol. **39**(19), 6294–6300 (2021). https://doi.org/10.1109/JLT.2021.3097501
94. Jiang, M., An, Y., Su, R., Huang, L., Li, J., Ma, P., Ma, Y., Zhou, P.: Deep mode decomposition: real-time mode decomposition of multimode fibers based on unsupervised learning. IEEE J. Sel. Top. Quant.Electron. **28**(4), 0900207 (2022). https://doi.org/10.1109/jstqe.2022.3195203
95. Tian, Z., Pei, L., Wang, J., Hu, K., Zheng, J., Li, J., Ning, T.: Physics-driven mode decomposition based on a training-free shallow neural network. Optical Fiber Technol. **76**, 103239 (2023). https://doi.org/10.1016/j.yofte.2023.103239
96. An, Y., Huang, L., Li, J., Leng, J., Yang, L., Zhou, P.: Learning to decompose the modes in few-mode fibers with deep convolutional neural network. Opt. Express. **27**(7), 10127 (2019). https://doi.org/10.1364/oe.27.010127
97. An, Y., Huang, L., Li, J., Leng, J., Yang, L., Zhou, P.: Deep learning-based real-time mode decomposition for multimode fibers. IEEE J. Sel. Top. Quant.Electron. **26**(4), 440806 (2020). https://doi.org/10.1109/jstqe.2020.2969511
98. Rothe, S., Zhang, Q., Koukourakis, N., Czarske, J.W.: Deep learning for computational mode decomposition in optical fibers. Applied Sci. **10**(4), 1367 (2020). https://doi.org/10.3390/app10041367
99. An, Y., Hou, T., Li, J., Huang, L., Leng, J., Yang, L., Zhou, P.: Fast modal analysis for Hermite–Gaussian beams via deep learning. Appl. Optics **59**(7), 1954 (2020) https://doi.org/10.1364/AO.377189
100. Yan, B., Zhang, J., Mi, S., Fan, G., Wang, C., Wang, M., Yan, F.: Simulation and experimental validation of the JCR-CNN model for hybrid modes recognition and decomposition in few-mode fibers. Optics Laser Technol. **175**, 110780 (2024). https://doi.org/10.1016/j.optlastec.2024.110780
101. Zhang, W., Song, H., He, X., Huang, L., Zhang, X., Zheng, J., Shen, W., Hao, X., Liu, X.: Deeply learned broadband encoding stochastic hyperspectral imaging. Light Sci Appl. **10**(1), 108 (2021). https://doi.org/10.1038/s41377-021-00545-2
102. Leger, J.: Laser beam combining: theory and methods. In: Conference on Lasers and Electro-Optics, p. 1. Optica Publishing Group (2010)
103. Fathi, H., Närhi, M., Gumenyuk, R.: Towards ultimate high-power scaling: coherent beam combining of fiber lasers. Photonics. **8**, 566 (2021) MDPI

104. Fan, T.Y.: Laser beam combining for high-power, high-radiance sources. IEEE J. Sel. Top. Quant. Electron. **11**(3), 567–577 (2005)
105. He, B., Lou, Q., Zhou, J., Dong, J., Wei, Y., Xue, D., Qi, Y., Su, Z., Li, L., Zhang, F.: High power coherent beam combination from two fiber lasers. Opt. Express. **14**(7), 2721–2726 (2006)
106. Goodno, G.D., Asman, C.P., Anderegg, J., Brosnan, S., Cheung, E.C., Ham-mons, D., Injeyan, H., Komine, H., Long, W.H., McClellan, M., et al.: Brightness-scaling potential of actively phase-locked solid-state laser arrays. IEEE J. Sel. Top. Quant.Electron. **13**(3), 460–472 (2007)
107. Zhou, P., Liu, Z., Wang, X., Ma, Y., Ma, H., Xu, X., Guo, S.: Coherent beam combining of fiber amplifiers using stochastic parallel gradient descent algorithm and its application. IEEE J. Sel. Top. Quant. Electron. **15**(2), 248–256 (2009)
108. Vorontsov, M.A., Sivokon, V.P.: Stochastic parallel-gradient-descent technique for high-resolution wave-front phase-distortion correction. JOSA A. **15**(10), 2745–2758 (1998)
109. Chang, H., Chang, Q., Xi, J., Hou, T., Su, R., Ma, P., Wu, J., Li, C., Jiang, M., Ma, Y., et al.: First experimental demonstration of coherent beam combining of more than 100 beams. Photon. Res. **8**(12), 1943–1948 (2020)
110. Chang, Q., Hou, T., Long, J., Deng, Y., Chang, H., Ma, P., Su, R., Ma, Y., Zhou, P.: Experimental phase stabilization of a 397-channel laser beam array via image processing in dynamic noise environment. J. Lightw. Technol. **40**(8), 6542–6547 (2022)
111. Jiang, M., Wu, H., An, Y., Hou, T., Chang, Q., Huang, L., Li, J., Su, R., Zhou, P.: Fiber laser development enabled by machine learning: review and prospect. PhotoniX. **3**(1), 16 (2022)
112. Hou, T., An, Y., Chang, Q., Ma, P., Li, J., Zhi, D., Huang, L., Su, R., Wu, J., Ma, Y., Zhou, P.: Deep-learning-based phase control method for tiled aperture coherent beam combining systems. High Power Laser Sci. Eng. **7**(e59) (2019)
113. Chang, Q., An, Y., Hou, T., Su, R., Ma, P., Zhou, P.: Phase-locking system in fiber laser array through deep learning with diffusers. In: Asia Communications and Photonics Conference, pp. 4–96. Optica Publishing Group (2020)
114. Liu, R., Peng, C., Liang, X., Li, R.: Coherent beam combination far-field measuring method based on amplitude modulation and deep learning. Chin. Optics Lett. **18**(4), 041402 (2020)
115. Wang, D., Du, Q., Zhou, T., Li, D., Wilcox, R.: Stabilization of the 81-channel coherent beam combination using machine learning. Opt. Express. **29**(4), 5694–5709 (2021)
116. Jia, H., Zuo, J., Bao, Q., Geng, C., Li, X., Li, F., Li, Z., Jiang, J., Liu, J., Xia,Y., et al.: A phase-error prediction method for coherent beam combining via convolutional neural network. Optik 246, 902–916 (2021).
117. Mills, B., Grant-Jacob, J.A., Praeger, M., Eason, R.W., Nilsson, J., Zervas, M.N.: Single step phase optimisation for coherent beam combination using deep learning. Sci. Rep. **12**(1), 5188 (2022)
118. Zuo, J., Jia, H., Geng, C., Bao, Q., Zou, F., Li, Z., Jiang, J., Li, F., Li, B., Li, X.: Deep learning piston aberration control of fiber laser phased array by spiral phase modulation. J. Lightw. Technol. **40**(12), 3980–3991 (2022)
119. Du, Q., Wang, D., Zhou, T., Gilardi, A., Kiran, M., Mohammed, B., Li, D., Wilcox, R.: Experimental beam combining stabilization using machine learning trained while phases drift. Opt. Express. **30**(8), 12639–12653 (2022)
120. Wang, D., Leng, Y.: Simultaneous wavefront sensing of multiple beams using neural networks. Applied Phys. B. **128**(1), 8 (2022)
121. Wang, D., Du, Q., Zhou, T., Gilardi, A., Kiran, M., Mohammed, B., Li, D., Wilcox, R.: Machine learning pattern recognition algorithm with applications to coherent laser combination. IEEE J. Quant. Electron. **58**(6), 1–9 (2022)
122. Li, X., Peng, C., Liang, X.: Far-field phasing method based on deep learning for tiled-aperture coherent beam combination. Optics Commun. **527**, 128928 (2023)
123. Xie, Y., Chernikov, F., Ben, M., Liu, Y., Praeger, M., Grant-Jacob, J.A., Zer-vas, M.N.: Single-step phase identification and phase locking for coherent beam combination using deep learning. Sci. Rep. **14**, 7501 (2024)

124. Grant-Jacob, J.A., Mills, B., Zervas, M.N.: Phase identification despite amplitude variation in a coherent beam combination using deep learning. Opt. Continuum. **2**(4), 902–916 (2023)
125. Tünnermann, H., Shirakawa, A.: Deep reinforcement learning for coherent beam combining applications. Opt. Express. **27**(17), 24223–24230 (2019). https://doi.org/10.1364/OE.27.024223
126. Tünnermann, H., Shirakawa, A.: Deep reinforcement learning for tiled aperture beam combining in a simulated environment. J. Phys. Photon. **3**(1), 015004 (2021) https://doi.org/10.1088/2515-7647/abcd83
127. Zhang, X., Li, P., Zhu, Y., Li, C., Yao, C., Wang, L., Dong, X., Li, S.: Coherent beam combination based on q-learning algorithm. Optics Commun. **490**, 126930 (2021). https://doi.org/10.1016/j.optcom.2021.126930
128. Shpakovych, M., Maulion, G., Kermene, V., Boju, A., Armand, P., Desfarges- Berthelemot, A., Barth'elemy, A.: Experimental phase control of a 100 laser beam array with quasi-reinforcement learning of a neural network in an error reduction loop. Opt. Express. **29**(8), 12307–12318 (2021). https://doi.org/10.1364/OE.419232
129. Shpakovych, M., Maulion, G., Boju, A., Armand, P., Barth'el'emy, A., Desfarges-Berthelemot, A., Kermene, V.: On-demand phase control of a 7-fiber amplifiers array with neural network and quasi-reinforcement learning. Photonics. **9**(4), 243 (2022)

Chapter 20
Advancement in Plasmonic Fiber Random Lasers

Naeem Iqbal, Aziz Gohar, Junhua Tong, Xiaoyu Shi, and Tianrui Zhai

Abstract Random lasers are an advanced type of laser device that fundamentally differ from traditional lasers in both concept and operation. They show significant potential for applications in imaging, sensing, and photonic chips due to their unique light field properties, simple design, ease of fabrication, and low cost. Unlike conventional lasers, random lasers rely on disordered structures for optical feedback, resulting in distinctive features such as low spatial coherence and random emission wavelengths and directions. Directionality is crucial in many applications, and integrating random lasers with optical fibers enhances their output directionality while enabling miniaturization and integration. Additionally, metal nanoparticles serve as highly effective scatterers, offering a larger scattering cross-section compared to dielectric nanoparticles of similar size. Their localized surface plasmon resonance further enhances light confinement and significantly boosts optical gain, making them instrumental in the generation of random lasers. This chapter delves into the fundamental concepts, key characteristics, and control mechanisms of random lasers. It also examines design strategies, fabrication techniques, emission properties, and the applications of random lasers with fiber-embedded cavities, providing examples in imaging and sensing. Furthermore, it briefly discusses WGM-coupled random hybrid cavities and their applications in information security.

Keywords Plasmonic coupling · Fiber random lasers · Compound cavity · Speckle-free imaging · Sensing · Dual chaotic encryption

---

N. Iqbal (✉) · A. Gohar · J. Tong · X. Shi · T. Zhai
School of Physics and Optoelectronic Engineering, Beijing University of Technology, Beijing, China
e-mail: naeem_iqbal@bjut.edu.cn; trzhai@bjut.edu.cn

M. F. S. Ferreira, M. C. Paul (eds.), *Optical Fiber Lasers*, Springer Series in Optical Sciences 250, https://doi.org/10.1007/978-3-032-04110-4_20

20.1 Introduction

The dynamic growth of photonic information technology and the unceasing development of nanotechnology such as the size miniaturization, facile integration, and efficient laser light sources have been evolved gradually. Among the various light sources, random lasers are the new types of microcavity lasers that have gained considerable attention from the research community. In 1966, first time a nonresonant feedback-based stochastic laser concept was proposed [1]. Two years later, a theoretical investigation about a mirrorless cavity for laser light source was developed by providing the feedback mechanism through scatters in combination with gain media responsible for amplification [2]. Generally, scattering introduced the losses which could restrict the laser action by increasing its threshold, thereby in the perspective of laser performance enhancement, it is vital to reduce the scattering losses. In principle, for stochastic lasers, the scattering facilitates with feedback directly which in turn enhances its strength, as a result, the light gain length and gain volume both increase which are favorable in developing the stochastic lasers. The theoretical idea of laser was verified experimentally in a strong scattering embedded gain medium and was known as random laser (RL) [3, 4]. The random coherent lasing was performed through zinc oxide nanoparticles which laid the foundation for new research direction [5].

During the last two decades, plenty of research has been made on the development of RLs owing to their simple resonant structure and relaxed manufacturing procedure as compared to traditional resonators. RLs have drawn considerable attention for several applications such as speckle-free imaging [6–9], information security [10–12], sensing [13–15], and biomedicine [16–18]. In this context, the adopted relevant manufacturing mechanisms included but not limited to liquid/solid phase [19–21], rigid/flexible [22, 23], and optical/electrical pumping [24–27]. Furthermore, the majority of RL generation systems involve the single-photon/multi-photons excitation under the optical pumping [28, 29]. The gain materials encompass semiconductors [30], dyes [31, 32], quantum dots [33–35], polymers [36, 37], and perovskite [38–40], while scatters include the dielectric/metallic nanoparticles [41–43], biological tissues [44, 45], and liquid crystals [46–48].

However, the efficient controlled RLs are demanding for application prospects; in this context, plenty of research have devoted in enhancing its performance by exploring several different methods such as the adjustment of pump light's shape and position [49–51]. Further, wavelength tunability, optimum and control linewidth, and threshold can be realized by scatters-gain modification in the lasing system, where respective choice of materials nature and their precise concentration ratios are vital [51, 52]. Apart from this, the external stimuli such as temperature, mechanical force, humidity, and electromagnetic field are also significantly influenced the system characteristics [53–55]. The directionality of the random laser is controlled by employing the directional elements including the plane waveguides, spherical cavities, optical fibers, etc. [56–58]. The polarized RLs are obtained by incorporating the anisotropic material and polarization components [59, 60]. Last but not least the beam shaping and mode conversion are achieved by utilizing the grating and hybrid structures [61, 62].

This chapter will focus solely on the use of fiber waveguides to control the emission directionality and enhance the performance of random lasers. We begin by outlining the fundamental principles of random lasers, followed by a review of the research advancements and applications of fiber-integrated random lasers. Additionally, we touch upon design strategies, fabrication methods, emission characteristics, and application examples of random lasers in areas such as speckle-free imaging and sensing.

20.2 RLs Working Principle

The working arrangement of conventional lasers is different from the RLs, where a resonator composed of different mirror arrangements is responsible for the feedback of paraxial rays propagation, which can amplify over each back-and-forth passage through gain medium, and this amplification after exceeding the resonator losses yields lasing as shown in Fig. 20.1a [63]. However, in the RLs, resonator is replaced with scattering system, where light under long random walk due to presence of scatters continuously intermingles with gain medium, thereby enforces the stimulated radiation amplification of the light as shown in the Fig. 20.1b [64]. Here, nonresonant feedback (green color) and the resonant feedback (i.e., random walk forming a loop represented by blue color) provide the noncoherent (Fig. 20.1c) and coherent (Fig. 20.1d) random lasing, respectively. The coherent random lasing is composed of several spikes (modes) as a result of interference effect contrary to noncoherent lasing. Moreover, in this kind of lasing, the mode selection, feedback functions, and lasing direction are not as strong as in conventional lasers attributed to its stochastic nature [65].

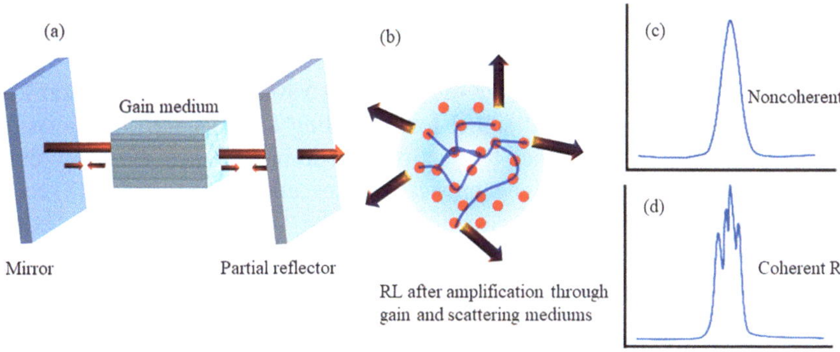

Fig. 20.1 (**a**) Schematic diagram of conventional lasers. (**b**) The RLs functioning principle, green color line and arrow indicate the incoherent feedback process while blue color line and arrow represent coherent feedback. Noncoherent (**c**) and coherent (**d**) RLs emission [65]

As mentioned earlier, scatters and gain media are two important elements for random lasing mechanism, where the doped nanoparticles induce the scattering phenomenon. The nanoparticles size and wavelength are important for description of elastic scattering theories which can be classified as Rayleigh scattering [66] and Mie Scattering [67]; in the former case, the size of nanoparticles is much smaller than incident wavelength, while in latter case, sizes of both scatters and wavelength are similar. In combining the scatters with active medium derived from the gain, light amplification occurs as a result of multiple scattering in the system without affecting the involved system and leading to a lasing action. The scattering mean free path is illustrated as the light average distance traveled between two successive scatters and can be determined by using the expression [68, 69]; $l_s = 1/\rho\sigma_s$, where the symbols ρ and σ_s represent particles density and scattering cross-section of scatters, respectively. Further, the propagation mean free path denotes the freely propagated average distance of light and can be determined using the relation: $l_t = l_s/(1 - \cos\theta)$; for Rayleigh scattering, $\cos\theta = 0$, which means $l_t = l_s$. Moreover, the stimulated emission amplification mechanism in a random system is defined by gain length l_g, i.e., the length during which photon emission intensity amplified to e times, finally random lasing occurs when gain overcomes the losses by meeting the condition $l_s l_g$.

RLs, due to the presence of scatters, typically exhibit nondirectional emission. However, directional emission can be achieved by employing various techniques. One such method involves introducing a distributed Bragg reflector (DBR) structure above and below the random lasing system (which consists of gain material and scatters), resulting in directional emission with a divergence angle of approximately 3° to 5° [57]. Additionally, the combination of polymer nanowires dispersed on an aluminum thin film, which exhibits random scattering features, enables random lasing along the long axis of the nanowires. In this context, optical fiber-based random laser approaches have remained effective, and a detailed discussion of these kinds of developed lasers can be found in the subsequent sections.

20.3 Fiber Random Laser

Optical fiber, as a quintessential one-dimensional waveguide, offers several key advantages: (1) minimal optical signal loss, enabling long-distance transmission with high capacity; (2) resistance to electromagnetic interference, ensuring high-quality transmission with low signal crosstalk and excellent data security; (3) compact and lightweight, making it easy to install and transport, while also being cost-effective due to its potential for mass production and having a long operational lifespan; (4) a well-established integration technology with other devices, allowing for straightforward compatibility and connection. Attaching random laser structures to the inner or outer walls, and end faces of optical fibers not only enables the directional emission of RLs but also fulfills the needs for miniaturization, integration, and data transmission. This approach accelerates the progress of developing high-brightness, compact, and portable laser devices.

The described system is a packed fiber-based random laser (RFL), consisting of a hollow-core optical fiber filled with a colloidal solution of rhodamine 6G (the gain medium) and TiO_2 nanoparticles (the scattering agents). This setup enables directional emission of the random laser through side pumping. Such RLs, which feature a disordered structure within the fiber, are often referred to in the literature as random fiber lasers (RFLs). They involve the use of commercial hollow or polymer fibers filled with a gain-scattering medium [70]. A coherent random laser with weak scattering and directional emission was realized by filling a hollow fiber with a gain-scattering medium, as shown in Fig. 20.2a. However, this approach necessitates the use of a specialized commercial hollow fiber, which adds to the overall cost of the device [71].

In another approach, a three-capillary coupled channel made of dyes of three different emission wavelengths when each mixed with solution of PS microspheres separately for development of scattering system resulted in generation of red, green, and blue RLs. By controlling the flow of the solution within the capillaries, the wavelength of the random laser could be switched dynamically without altering the pump position. In contrast to hollow fiber-based RLs, this capillary-based system is a fluid-type device, which typically leads to lower laser stability [72]. Other than dyes, polymer-based fiber random laser were also utilized as packed fiber random laser. In this context, a PM597 dye incorporated into disordered polymer fibers to create a random laser using these fibers. They employed end-pumping to get directional random lasing [73]. Following this, a series of pioneering studies on the design, optimization, and practical use of polymer fiber-based RLs were published in literature [74–76]. In one such key study, Morello et al. [77] developed a random

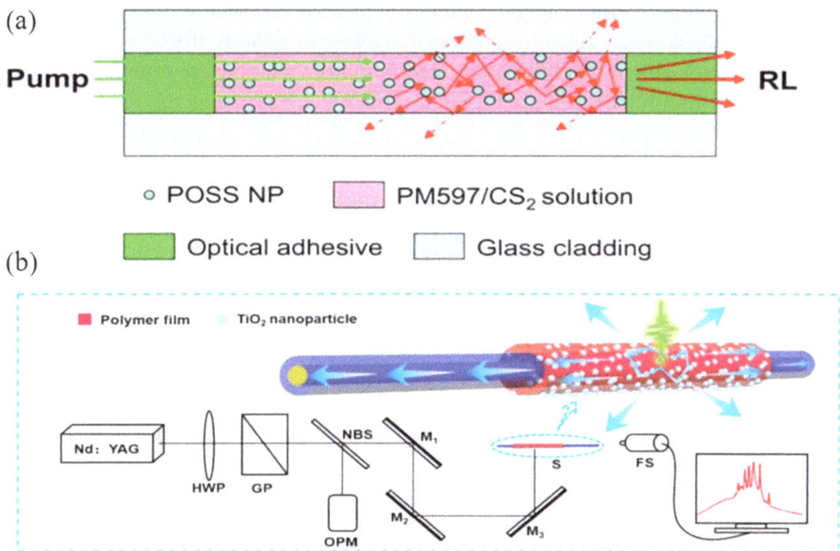

Fig. 20.2 The working principle demonstration of fiber random laser [71]. (**a**) The schematic diagram for random fiber laser emission under the side pumping scheme [82]

fiber laser by combining the disordered polystyrenes with a laser dye under electrospinning method. By carefully tuning the dye concentration, they were able to control both the wavelength and threshold behavior of the laser. Several review articles were published focusing on the fiber-based random laser emission [78–81].

In another method, a stochastic structure is incorporated along the sidewall of the optical fiber, facilitating easier fabrication and making it suitable for speckle-free imaging applications. The structure of this fiber, illustrated in Fig. 20.2b, consists of DCJT dye-doped TiO_2 nanoparticles, created by simply immersing a commercially available optical waveguide into a solution of the random system. Additionally, three-color random laser emission with dynamic wavelength switching was achieved along the side of the same fiber by using a disordered system containing three dyes—DCJTB, PM567, and C153—each doped at an optimal concentration with TiO_2 nanoparticles. By mechanically adjusting the pump position, red, green, and blue light emissions were obtained. The light emitted from the fiber end exhibited a divergence angle, and a ring-shaped beam was observed in momentum space [82]. A polymer-based microribbon capable of three-color (red-green-blue) random lasing was recently fabricated. The custom-designed PMMA polymer fiber embedded with RhB, Uranin, and S420 as gain media, and mixed with TiO2 nanoparticles as scatterers, demonstrated uniform winding and efficient energy transfer between the dyes. This resulted in multicolor directional emission under single-pump excitation, successfully producing a speckle-free white image [83]. Notably, the overlap between absorption and emission spectra is essential for energy transfer between different dyes. Relevant information can be found in Table 20.1. However, in this type of random fiber laser, light emission through the fiber face can be directional only under certain conditions, potentially leading to reduced fiber efficiency.

Recently, photo-isomerization activated fiber random laser was proposed [85]. This study introduced a reversible, switchable-wavelength fiber random laser, developed using a blend of highly polarized intramolecular charge transfer dye molecules and optimally concentrated titanium dioxide which act as the gain and scattering mediums, respectively, within a polyvinyl alcohol (PVA) matrix. The mixture is applied to a fiber using a dip-coating technique, followed by encapsulation with polydimethylsiloxane (PDMS) to provide thermal stability due to its high thermal expansion coefficient (Fig. 20.3a). The system enables random laser emission with dynamically switchable wavelengths, driven by the excited-state intramolecular proton transfer mechanism during photo-isomerization.

Table 20.1 Absorption and fluorescence characteristics of diverse laser dyes [84]

Dye	Absorption band (nm)	Fluorescence band (nm)
C440	320–400	401–548
C153	324–480	483–648
C460	308–417	418–547
C6	348–496	498–652
R6G	436–553	550–695
Oxazine	481–512	622–765

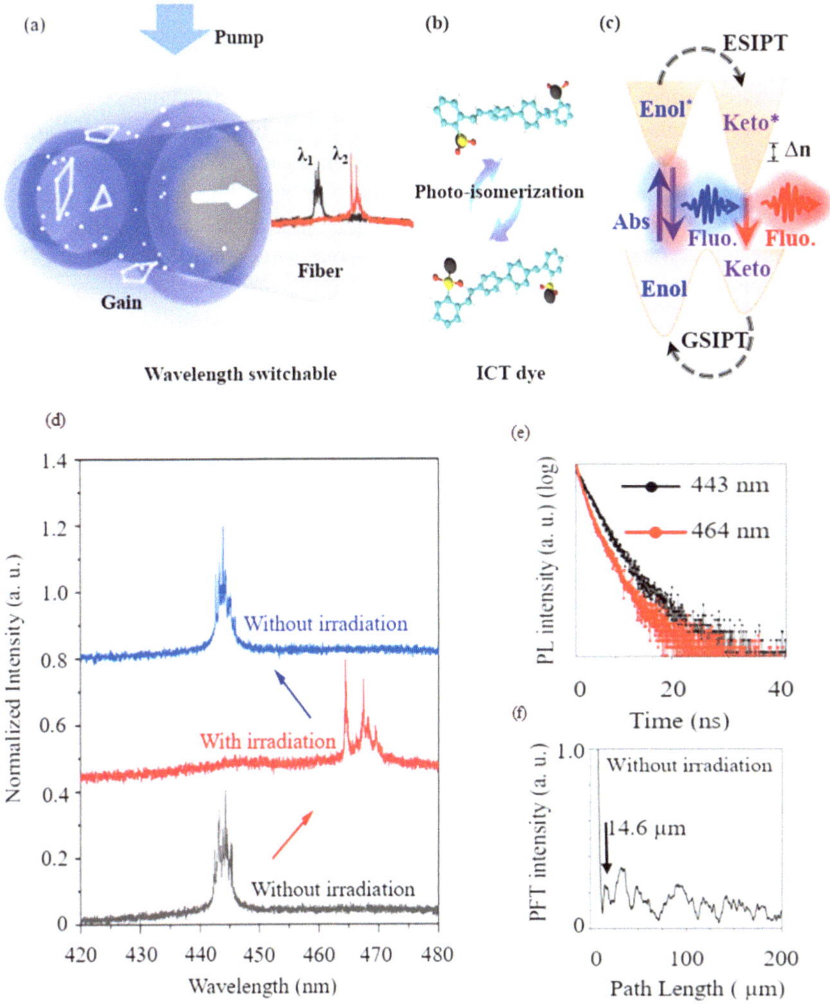

Fig. 20.3 (**a**) Schematic diagram of ICT-based fiber random laser for reversible wavelength switching. (**b**) Photo-isomerization effect. (**c**) Energy diagram representation for ESIPT process and wavelength switch. (**d**) Wavelength shift in emission spectra under photo-isomerization. (**e**) PL decay. (**f**) Power Fourier transform analysis [85]

Figure 20.3b illustrates the effect of photo-isomerization on ICT dye molecular structure. Upon light absorption, the dye undergoes a molecular rearrangement, forming a new isomeric state and altering its optical properties. As shown in Fig. 20.3c, the dye follows a four-level photophysical process, starting with the Enol state as the ground state. Upon excitation, the Enol state transitions to an excited Enol* state, which rapidly converts to an excited Keto* state via photo-isomerization through an excited-state intramolecular proton transfer (ESIPT) process. The Keto* state then radiatively relaxes to the ground Keto state, which subsequently reverts to the original Enol state through ground-state intramolecular proton transfer (GSIPT).

Next, the effect of photo-isomerization on the emission properties of an optimized fiber random laser is examined (as shown in Fig. 20.3d). The device, pumped with a 343-ns pulse laser at a fluence of 41 μJ/cm^2, exhibits multimode random lasing peaks due to the disordered nature of the fiber scatters. Without irradiation, the laser emits multimode lasing around 443 nm, corresponding to the excited state (E-state) of ICT dye molecules [86]. Upon irradiation, photo-isomerization causes a shift to the Z-state, resulting in red-shifted lasing peaks centered at 464 nm. This reversible tuning of emission, enabled by the fiber's structure, returns to 443 nm when irradiation ceases, demonstrating the reversibility of the process. Further, the potential of lasing systems to maintain high reliability and durability under cyclic operating conditions was investigated, making them suitable for applications requiring controlled intensity modulation, such as optical data storage, dynamic light filters, and sensors [87–90]. Moreover, the PL decay time as predicted in Fig. 20.3e is reduced under the situation of involvement of photo-isomerization, supported another path for the energy release predominantly via enhancement in radiation transitions or a reduction in nonradiative losses [91, 92].

Power Fourier transform (PFT) analysis can be used to explore structural changes in the lasing cavity and track light path alterations in the random laser system. PFT effectively analyzes RLs by assessing light scattering strength [93]. Here, only the lasing spectra from a sample excited at 41 μJ/cm^2 without irradiation was examined (Fig. 20.3f). The PFT spectra's peaks correspond to Fourier components, allowing calculation of the cavity length using $Lc = \pi\, p_m / m\, n$, where $n = 1.50$, $m = 1$, and the optical path length (p_m) is approximately 14.6 μm. This yields a cavity length of 30.6 μm, significantly smaller than the pump spot size, indicating strong light scattering, and supporting localized random lasing modes in the system. It could be effective for applications of the information security systems.

20.4 Plasmonic Fiber Random Laser

As discussed earlier, the optimization of both the gain media and scatters concentration is crucial for enhancing the performance of RLs in random systems [94]. However, attaining a large gain through strong scattering strength is challenging when using dielectric media as scatterers. The use of metal nanoparticles as scatterers offer a solution to this issue, attributed to larger scattering cross-section compared to dielectric nanoparticles of the same size. Moreover, metallic nanoparticles exhibit localized surface plasmon resonance effects, which confine light near the particles and significantly enhance optical gain. In essence, metal nanoparticles facilitates the generation of RLs by boosting optical gain.

In this context, three adjacent capillaries were arranged to function as microfluidic channels, filled with colloidal solutions of different laser dyes, each dye combined with gold-silver alloy nanowires. By leveraging an energy cascade transfer process between the multiple dyes, a three-color random laser (red, green, and blue) under a single pump excitation was generated. Additionally, a direction-dependent white light random laser through the interaction of the three coupled microfluidic channels was achieved [84, 95]. It is worth mentioning that such fiber fabrication techniques required the electrospinning method and other equipment. The fiber-based plasmonic RLs supported the directional random lasing along with significantly lower threshold as compared to planar-structured RLs.

20.4.1 Plasmonic Fiber Random Laser at the End Face

The integration of a disordered system at the fiber's end face also generates random lasing, referred to as a fiber facet plasmonic random laser (FFPRL). This type of device offers excellent waveguide transmission properties and features a more compact, cost-effective design with fewer constraints on fiber configuration. These advantages contribute to the development of miniaturized and informatized random laser systems. Such lasers hold great potential for applications in integrated light sources, medical technology, and communications. One example of a disordered fiber random laser involved introducing a stochastic system at the fiber's cross-sectional face (as shown in Fig. 20.4a), resulting in random lasing with a small angular deviation [96]. The electric field distribution shown in Fig. 20.4b reveals areas of strong field intensity, driven by the coupling of localized surface plasmons, which play a key role in enhancing device performance as mentioned earlier. Moreover, several studies utilizing this strategy have been reported [97, 98], including a stochastic system at the fiber end face composed of gold nanostars (scatterers) combined with R6G dye (gain), resulting in a random laser with a low threshold and narrow full-width at half maximum [99].

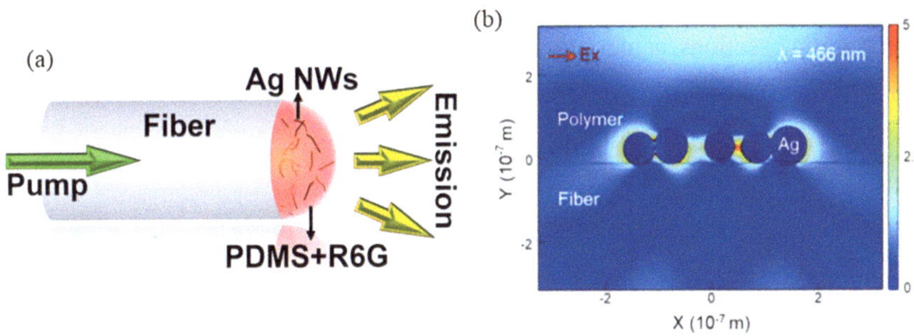

Fig. 20.4 (**a**) Schematic representation of end-facet plasmonic fiber random laser [96]. (**b**) Electric field distribution indicating the strong location of field as a result of localized surface plasmons coupling [98]

Design and Execution
The integration of a random system at the end of an optical fiber can be achieved through various methods, such as the film method [100–102], laser-induced transfer [103, 104], and dip-coating method [98, 105], with an emphasis on the device's solid phase. The following sections will discuss the film and dip-coating methods, covering device design and performance analysis.

Figure 20.5 illustrates the step-by-step process for fabricating a free-standing film designed for use in a random system. The process begins with attaching a piece of PE film to a glass substrate, followed by oxygen plasma treatment. PEDOT:PSS solution is spin-coated onto the surface of the PE film and annealed. This spin-coating and annealing cycle is repeated ten times to enhance the hydrophilicity of the PEDOT:PSS film and strengthen its mechanical properties through multilayer deposition.

At the same time, a glass substrate is coated with a PVA layer, serving as a water-soluble layer. A xylene solution containing gain polymer doped with scatterers is spin-coated onto the PVA layer, forming a free-standing gain layer once the PVA dissolves in water. The resulting free-standing random system layer can be combined with a free-standing active membrane made of PEDOT:PSS composite. After peeling off the layers, the structure exhibits characteristics suitable for humidity sensing [13]. Additionally, this thin film can function independently as a self-supporting random exciter and can be transferred to the fiber end face to achieve an end-face plasmonic fiber random laser.

Alternatively, the dipping method can be employed by immersing the fiber tip into a mixture solution containing the random system, followed by annealing. In both approaches, the selection of gain materials and scatterers can be tailored to meet the specific requirements of the microcavity.

20.4.2 *Emission Wavelength Management*

Now, we will focus on the emission wavelengths of single or multicolor RLs in the upcoming subsections, achieved by integrating the corresponding stochastic system at the tip of the fiber.

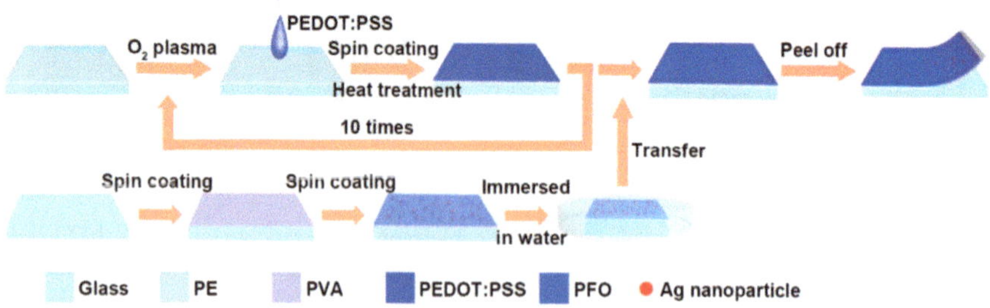

Fig. 20.5 Schematic demonstration for fabrication of random system-based free-standing film [13]

20.4.3 Red Emission Wavelength

A fiber end-facet random laser with a wavelength in the red (visible) emission band is fabricated by integrating an active gain organic dye with metallic NPs as scatterers and PDMS as the waveguide layer at the fiber tip, utilizing cross-linking polymerization and an annealing process. Details on the fabrication and operating principle of such laser can be found in reference [96].

When the fabricated device was pumped with nanosecond pulse laser light source along the fiber axis, opposite to the face connected with the random system, the measured intensity spectra and corresponding full width at half maximum (FWHM) are shown in Fig. 20.6 (a) and (b). The emitted random laser is centered around 571 nm with a FWHM of approximately 1 nm, which falls within the emission band of the dye. The measured threshold value was 4.2 nJ/cm^2, attributed to the

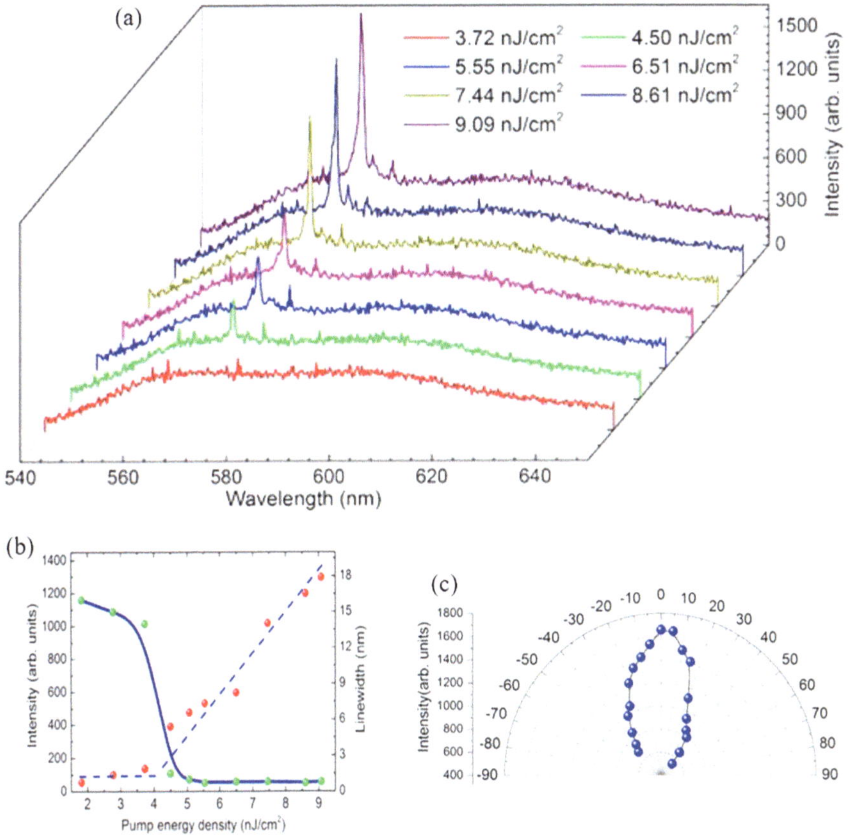

Fig. 20.6 (a) Intensity spectra versus wavelength for different input pump powers. (b) Output intensity (left axis) and linewidth (right axis) as a function of pump fluence. (c) Polar plot illustrating the small deviation of the random laser emission from the fiber's axial direction, with pumping applied along the axial direction [96]

interaction between the plasmonic resonance of silver nanoparticles and the photoluminescence spectrum of the dye.

Next, we discuss the effect of waveguide thickness, facilitated by the plasmonic gain channel, on the intensity of random lasing. The PDMS dipping process forms a hemispherical layer at the fiber tip, which influences the random laser intensity. Fig 20.6c presents the intensity measurement as a function of deviation from the fiber's axial direction, showing a symmetrical trend. Additionally, the intensity deviation was found to be less than 20°, with most of the intensity concentrated around the fiber's axial direction. This is significantly better than planar-structured RLs, where the deviation is much larger. This design offers improved control over the emission direction of the RLs.

20.4.4 RGB Characteristics

FFPRL for multi-wavelength emission can typically be achieved by integrating with optical sources through either the dipping or film method. Multi-wavelength random emissions are primarily driven by energy transfer processes, making the selection of gain materials and the stacking of the random system critical for optimal performance.

Multi-wavelength random lasing can be achieved using either single pump fluence or multiple pump sources. In this section, we will focus on the single pump source. Single pump fluence can be categorized into two types: direct pumping [97, 106] and cascaded pumping [107]. In direct pumping, energy transfer is not essential if the absorption and emission peaks of different gain materials do not overlap. However, if these peaks do overlap, the stacking order of the gain materials becomes crucial. In such cases, the material with the shorter wavelength emission should be positioned farther from the fiber end face. For cascaded energy transfer under single-pump excitation, overlapping absorption and emission spectra of different gain materials is necessary, while their absorption peaks should not coincide. Efficient energy transfer can be achieved by integrating donor and acceptor molecules in a cascading configuration, optimizing the overall performance of the system.

In this study, to achieve a red-green-blue FFPRL, three dyes were selected: red (MDMO-PPV), green (F8BT), and blue (PFO); these polymer dyes served as the gain materials. While for resonant feedback, silver nanoparticles were chosen as scatters (SEM image is given in Fig. 20.7a). A homogeneous mixture was prepared using an ultrasonic method, with the optimal concentration and proportion of each component. Once the red, green, and blue random system mixtures were prepared, the fiber was sequentially dipped into each solution, starting with the blue and finishing with the red, as previously described. To prevent color dissolution, a PVA isolation layer was applied. The final structure obtained is shown in Fig. 20.7b.

The absorption spectra of the three polymer films span the 400-nm wavelength range, ensuring that a single pump source is sufficient for laser emission. Additionally, the extinction spectrum of the red polymer overlaps with those of the

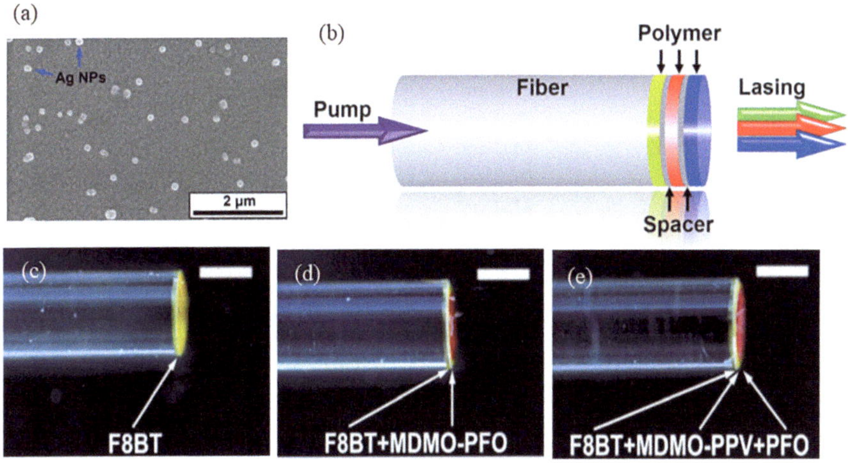

Fig. 20.7 RGB plasmonic fiber random laser: (**a**) SEM image showing the Ag nanoparticles acting as scatterers. (**b**) Schematic illustrating the multi-wavelength emission of the fiber random laser at its end face. Microscopic images of the plasmonic random laser at the fiber's end face: (**c**) displays green emission; (**d**) shows both green and red emissions; and (**e**) depicts the full RGB plasmonic random laser emissions at the fiber tip, with scale bar of 300 μm [97]

green and blue polymers, promoting efficient energy transfer between them. For optimal output from the three-color random laser, careful design of the polymer film stacking sequence is essential. Specifically, to achieve simultaneous emission of all three colors, it is crucial to prevent the laser energy from the green and blue polymers from being fully absorbed by the red and green polymers. Therefore, the stacking order in the experimental setup is green, red, and blue. Furthermore, energy transfer occurs in a cascade manner due to the shift effect, where backscattered random laser light from the blue and yellow polymers is absorbed by the green and red polymers, respectively. This process enhances the overall output efficiency of the three-color plasmonic random laser in the stacked structure. In this context, microscopic images of the random laser systems at the fiber tips are shown in Figs. 20.7c–e. Figure 20.7c displays random emission in green, Fig. 20.7d shows random laser emissions in both green and red, and Fig. 20.7e depicts the RGB emissions. The corresponding polymers used are also indicated.

Next, spectroscopic characterization of the RGB random laser was conducted using an optical spectrometer, with the emission spectra measured at the fiber tip (Fig. 20.8a). Three distinct lasing peaks were observed at 466 nm (blue), 572 nm (green), and 638 nm (red), indicating that PVA spacers prevent optical coupling between the waveguides. The five-layer structure can be treated as a combination of the individual RGB RLs. Notably, the FWHM for all peaks is remained under 10 nm, with threshold values at 10 mJ/cm^2 for blue, 60 mJ/cm^2 for green, and 50 mJ/cm^2 for red.

The emission spectra evolve with increasing pump power density, enabling dynamic tuning of the mixed colors. The chromaticity of each emission spectrum,

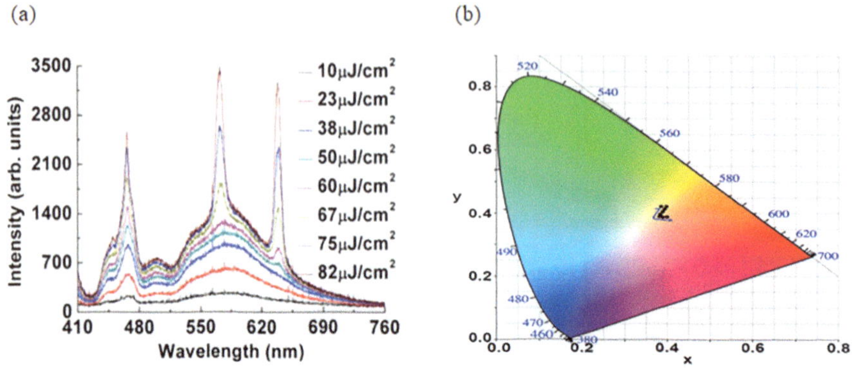

Fig. 20.8 (**a**) Emission intensities of RGB FFPRL at several pump influences. (**b**) Chromaticity of the respective random laser [97]

calculated using the emission wavelengths and FWHM values, is plotted on the CIE 1931 color diagram (Fig. 20.8b). The positions of the mixed colors correlate with the emission spectra, and the blue arrow in the diagram shows the evolution of the CIE coordinates with pump power, illustrating the extended chromaticity range of the RGB random laser.

20.4.5 FFPRL for Speckle-Free Imaging

Speckle-fre high-quality imaging can be obtained by employing a RL as the laser source positioned at the edge of the fiber [56]. As previously noted, the random laser's directional properties enhance the efficiency of laser energy utilization, making it suitable as the illumination source for the imaging system. Additionally, the random laser's low coherence plays a crucial role in enabling high-quality, speckle-free imaging.

The feedback mechanism inherent in the random laser results in a nonhomogeneous spatial pattern and a distinct, structured wavefront. According to Young's double-slit interferometry experiment, the spatial coherence of this **FFPRL** light source is $\gamma = 0.6279$, which is lower than that of a traditional 532 nm CW laser, where $\gamma = 0.9344$. Using such a low-coherence laser as the illumination source for the imaging system helps to suppress speckle creation and improves image eminence. The optical path for speckle-free imaging is illustrated in Fig. 20.9a. Here, the random laser light is sequentially filtered and focused through the filter and lens, then projected onto the ground glass. From the speckle contrast formula $C = \sigma/I$, where σ indicates the standard deviation of intensity and I represents the average intensity of the image, the speckle contrast for the **FFPRL** is calculated to be $C = 0.05$. In contrast, when it is replaced with a commercial 532-nm CW laser, the recorded image exhibits a significant number of speckles, and the speckle contrast increases to $C = 0.73$.

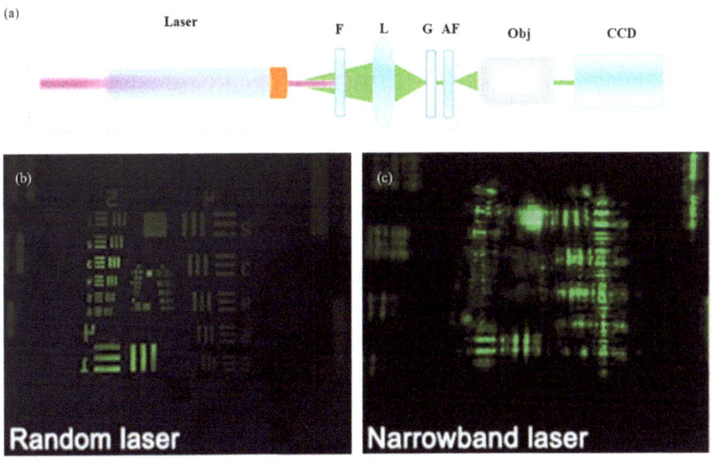

Fig. 20.9 The use of FFPRL for speckle-free imaging application. (**a**) Experimental setup. AF resolution test chart image using (**b**) FFPRL and (**c**) a conventional 532 nm continuous laser light source [58]

To further evaluate the quality of speckle-free imaging using a random laser as light source especially **FFPRL**, AF resolution test chart was placed between the ground glass and the object lens, as shown in Fig. 20.9a. In case of **FFPRL** as the light source, no visible speckles are observed, as shown in Fig. 20.9b. However, illumination with a 532-nm CW laser leads to the realization of numerous speckles, significantly degrading the image quality, as shown in Fig. 20.9c. Similar imaging results were obtained when a spherical nonresonant scattering cavity, coupled with a random laser at the fiber's end, was used as the light source [56]. Therefore, **FFPRL** can serve as compact light sources for speckle-free imaging.

20.5 Fiber-Integrated Microbelts

Miniaturized lasers capable of on-demand color emission hold significant promise for advancing integrated imaging sources [108, 109]. However, achieving a broad tuning range and precise, reversible control of random lasing colors with a defined emission direction remains a challenge. This section demonstrates random lasing with customizable color emission and directional control using fiber-integrated microbelts enabled by programmable pumping. Individual RL microbelts, doped with blue (B), green (G), and red (R) emissive dyes along with titanium dioxide nanoparticles, are assembled on an optical fiber. This setup produces high-performance B/G/R lasing and facilitates the directional emission of multicolor RL. The details on design and fabrication process can be found in reference [83].

By leveraging the high emission performance and flexible structure of individual blue, green, and red emissive RL units, random lasing with arbitrary colors across the visible spectrum can be achieved by adjusting the arrangement of these R/G/B microbelt units. The fiber-integrated arrays of R/G/B RL microbelts are fabricated in various configurations, such as BGRBGR, BGRGR, and others. Programmable color emission is realized by tailoring the structural arrangement of the RL units on the fiber. For example, white lasing emission is achieved when the BGRBGR array is simultaneously excited. This demonstrates the capability to produce and fine-tune a wide range of laser emission colors through a flexible structural design [83].

The tunable color arrangement offers a novel approach to achieving on-demand random lasing by selectively pumping B/G/R color units to define the final emission color. This enables dynamic color control without altering the laser cavity, a crucial aspect for integrated imaging source design. Different pumping strategies, shaped by masks into various beam arrays (e.g., RRR, GRR, BGR, etc.), produced multi-color lasing from waveguide-integrated RL arrays as shown in Fig. 20.10a. Fluorescent microbelt images and lasing spectra demonstrated the emission of various colors by selectively exciting RL units above their thresholds (Fig. 20.10b). The emission spectra revealed color-specific lasing, such as red (~620 nm) for RRR unit arrays, green for GGG units, and blue for BBB units. Dual-color lasing was achieved

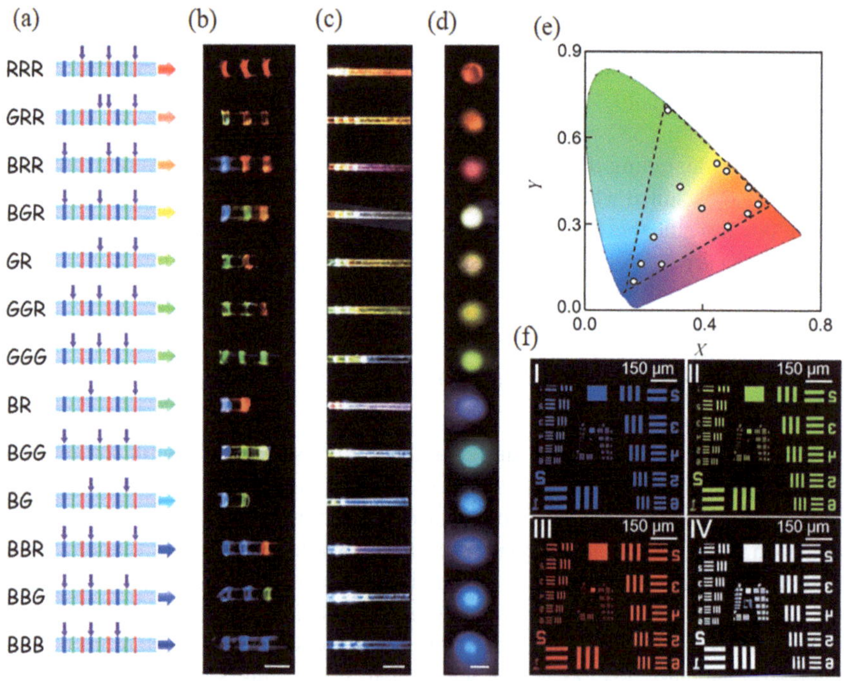

Fig. 20.10 (**a**) Schematic diagram for the active pumping scheme for color tunable fiber random laser. (**b**) The fluorescent images corresponding to scheme adopted in (**a**). (**c**) Corresponding far-field photographs. (**d**) Emitted spot images corresponding to part (**c**). (**e**) Chromaticity chart of respective lasing peaks. (**f**) AF resolution test chart shined by fiber random laser under the situations (I) BBB unit arrays, (II) GGG unit arrays, (III) RRR unit arrays, and (IV) BGR unit arrays [83]

by simultaneously exciting two color units, while mixed-color lasing was generated by combining primary colors in programmable configurations. Photographs (Fig. 20.10c) and spectra (Fig. 20.10d) illustrate the emitted colors, and the calculated chromaticity on a CIE1931 diagram (Fig. 20.10e) confirmed that any color within the triangle of the B/G/R primary colors can be achieved by mixing appropriate proportions. This precise and dynamic color control, comparable to "playing the piano" on RL array "keys," allows programmable manipulation of RL colors without structural modifications, providing a versatile platform for integrated optical applications [83].

The color-tunable FRLs serve as versatile light sources for stimulated emission, enabling rapid and flexible color changes. They also improve imaging quality by addressing issues of small spatial coherence. When used as speckle-free imaging sources, these FRLs with blue, green, red, and white emissions illuminated an AF resolution test chart through ground glass. Clear, speckle-free images were captured by a CCD behind an object lens, as shown in Fig. 20.10 (fI)–(fIV). This RGB lasing technology offers potential for high-resolution, long-distance full-color imaging, with promising applications in medical diagnostics and industrial inspection [83].

20.6 Applications

Fiber random lasers enable long-distance pumping and recognition, offering advantages of being compact, portable, and simple to integrate. They hold significant potential for applications in the sensing field, medicine, information security etc., some of them will be discussed here shortly.

20.6.1 FFPRL for Biosensors

In this context, a cost-effective, single-wavelength FFPRL is introduced and its application in biosensing is demonstrated [98]. The random scattering structure of silver nanoparticles was self-assembled on the fiber end-face through the thermal reduction of $AgNO_3$. Subsequently, the photoluminescent film of the gain medium (PFO) was transferred onto the scattering structure using the film method, as shown in Fig. 20.11a. The absorption spectrum of PFO (not shown here) spans a band range from 320 nm to 430 nm, overlapping with the extinction band of Ag NPs. This overlap suggests that Ag NPs can enhance the excitation of the random laser, which is evident from the significant increase in fluorescence intensity [31]. Further investigation of the photoluminescence lifetime, by comparing the decay dynamics of PFO with (red dots) and without (blue dots) Ag NPs, shows that the interaction with Ag NPs reduces the photoluminescence lifetime from 0.8 to 0.5 ns, as illustrated in Fig. 20.11b [41].

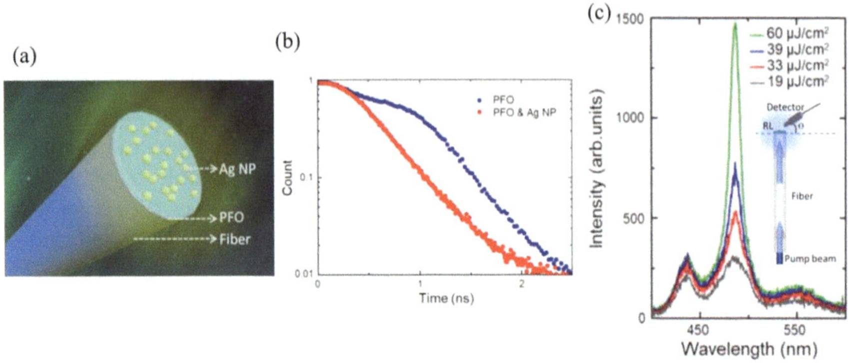

Fig. 20.11 (**a**) Design of the FFPRL. (**b**) Photoluminescence lifetime of the gain medium, with and without silver NPs, indicated by red and blue dots, respectively. (**c**) Emission spectra of the FFPRL at various pump power fluences. The inset shows the experimental setup used for measurement, with output detection at a 45° angle relative to the axial direction [98]

Next, under pumping conditions, when the pump energy density exceeds the threshold of 30.5 μJ/cm^2, a sharp increase in the spectral peak intensity is observed, accompanied by the appearance of numerous small spikes with a half-height full width of less than 0.5 nm, as seen in Fig. 20.11c. Additionally, it is noteworthy that the intensity measurements were conducted with the spectrometer positioned at a 45° angle relative to the fiber's axial direction.

Now, we discuss the refractive index sensing behavior of a FFPRL. The random laser operates through multiple scattering between silver nanoparticles (Ag NPs) on the fiber and the amplifying material, polyfluorene (PFO). The lasing mode satisfies the equation [110, 111]: $2nL = m\lambda$, where the symbols n, L, m, and λ represent the effective refractive index of the gain material, the resonant cavity length, the mode number, and the lasing wavelength respectively. The effective refractive index n is influenced by the surrounding medium and the PFO material. The study showed that exposing the RL to different solutions with varying refractive indices (e.g., methanol, deionized water, ethanol, etc.) results in a linear shift of the laser emission wavelength as shown in Fig. 20.12a. A refractive index sensitivity of $\Delta\lambda/\Delta n = 1.24$ nm/RIU was found. The laser's sensitivity to near-surface changes is also demonstrated by monitoring wavelength shifts caused by the absorption of a polymer monolayer on the RL surface [112].

The surface sensing capability of RLs enables biomolecule detection via surface adsorption [113, 114]. The sensing process is demonstrated in Fig. 20.12b. The process begins with establishing a baseline emission wavelength, followed by functionalizing the surface with protein A through noncovalent hydrophobic interactions. After rinsing with phosphate buffered saline (PBS), solutions of human IgG antibodies at 34 μM are introduced to the surface, where they specifically bind to protein A through immunological interactions (Fig. 20.12c). The sensing kinetics of this binding process are monitored spectrally at 5-min intervals.

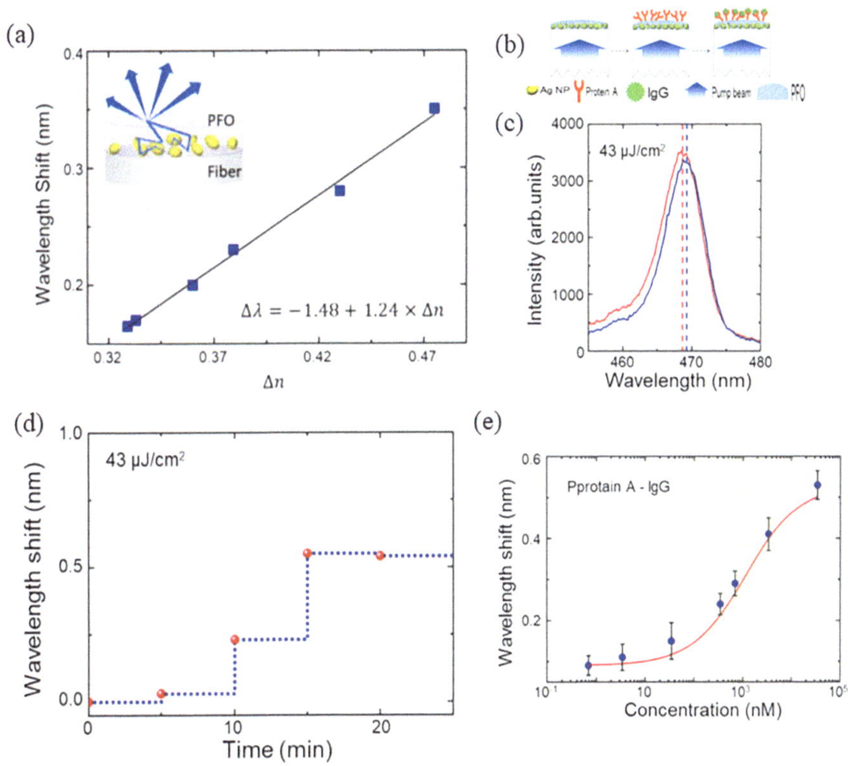

Fig. 20.12 (**a**) Refractive index-based sensing. (**b**) Schematic representation of the capture immunoassay used for IgG detection. (**c**) The intensity spectra of the FFPRL-based sensor before functionalization with protein A (red line) and after exposure to IgG (blue line). (**d**) Temporal behavior showing the wavelength shift. (**e**) The wavelength shift as a function of different concentrations of IgG solution [98]

The wavelength shift stabilizes after 15 min, indicating that the sensing system reaches dynamic equilibrium during the protein binding process. After 15 min of cultivation, the random laser exhibits a central wavelength shift of approximately 0.53 nm, demonstrating that FFPRL can detect biomolecules by changes in the effective refractive index (Fig. 20.12d). Further, six RL sensors decorated with protein A were exposed to human IgG solutions of varying concentrations, with the wavelength shift showing a concentration-dependent behavior as shown in Fig. 20.12e. For example, at 43 μJ/cm^2, the wavelength shifts were 0.41 nm (for 3.4 μM), 0.29 nm (for 0.86 μM), and 0.09 nm (for 0.68 nM). As the protein A binding sites become saturated at higher IgG concentrations, the detection limit of human IgG is 0.68 nM. This detection limit is comparable to traditional resonance cavity-based biosensors. The results also show that RL-based biosensors have a broad linear detection range between 84 nM and 12 μM [113, 115].

20.6.2 3D-Curved Microcavity-Based Gas Sensing

Drop-based microcavity lasers have gained attention as a promising tool for modern physics research and chemical detection due to their low-cost fabrication, high luminescence, and sensitivity to molecular sensing. However, achieving highly directional emission and high quality factors remains a significant challenge when relying on the traditional droplet self-assembly process of the planar-structured gain medium.

The successful fabrication of self-assembled 3D-curved WGM microcavities, created by dye-doped droplets on silica fibers through a self-assembly process, is demonstrated. These microcavities exhibit direction-adjustable single-WGM mode lasing with a high Q-factor and low lasing threshold under optical pumping, as well as trichromatic single-mode lasing with narrow linewidths. This system can be used to monitor chemical reactions via wavelength shifts, as detailed in reference [116].

Gas sensing plays a crucial role in agricultural analysis, environmental monitoring, and industrial processes. A single-mode 3D-curved fabricated microcavity was uncovered to an acetic acid (AcOH) gas atmosphere generated from a 30 wt% AcOH solution. The interaction between the R6G-doped single-mode microcavity laser and the AcOH gas resulted in a decrease in the effective refractive index (neff), affecting the lasing behavior, as shown in the schematic of Fig. 20.13a. Indeed, the change in neff caused a blue shift in the emission peak, indicating a successful gas response [116]. The mechanism involves the protonation of R6G and the formation of acetate, as illustrated in Fig. 20.13. This interaction causes a reduction in the refractive index from 1.45 to 1.43, which in turn shifts the resonant wavelength (Fig. 20.13c). Lasing peaks observed at 581.73, 579.74, 578.07, and 574.26 nm correspond to reaction times of 0, 5, 20, and 40 minutes, respectively, showing a clear blue shift with increased exposure to AcOH gas. This shift demonstrates a linear relationship between resonant wavelength and reaction time (Fig. 20.12d), providing a direct correlation between emission wavelength and reaction duration. The results also highlight the sensor's limit of detection (LoD), an important metric for its performance. The value of LoD can be depicted using the relation [117]: $oD = 3.3\sigma/S$, with the parameters S and σ representing the sensitivity and standard deviation, $\sigma = \sqrt{\frac{1}{N}\Sigma_1^N \left(x_i - \mu\right)^2}$, with μ indicating the average value. The calculated value of sensitivity is 186.5 pm/min, thereby LoD of the sensor is 5 min. This time is sufficient to distinguish the shift in the lasing signal of the proposed sensor.

20.6.3 Enantiomeric Excess Detection Using Optofluidic Microlaser

Enantiomers, which are isomers with opposite chirality, share similar physical and chemical properties due to their identical functional groups and compositions but differ significantly in their biological effects [118, 119]. While one enantiomer may

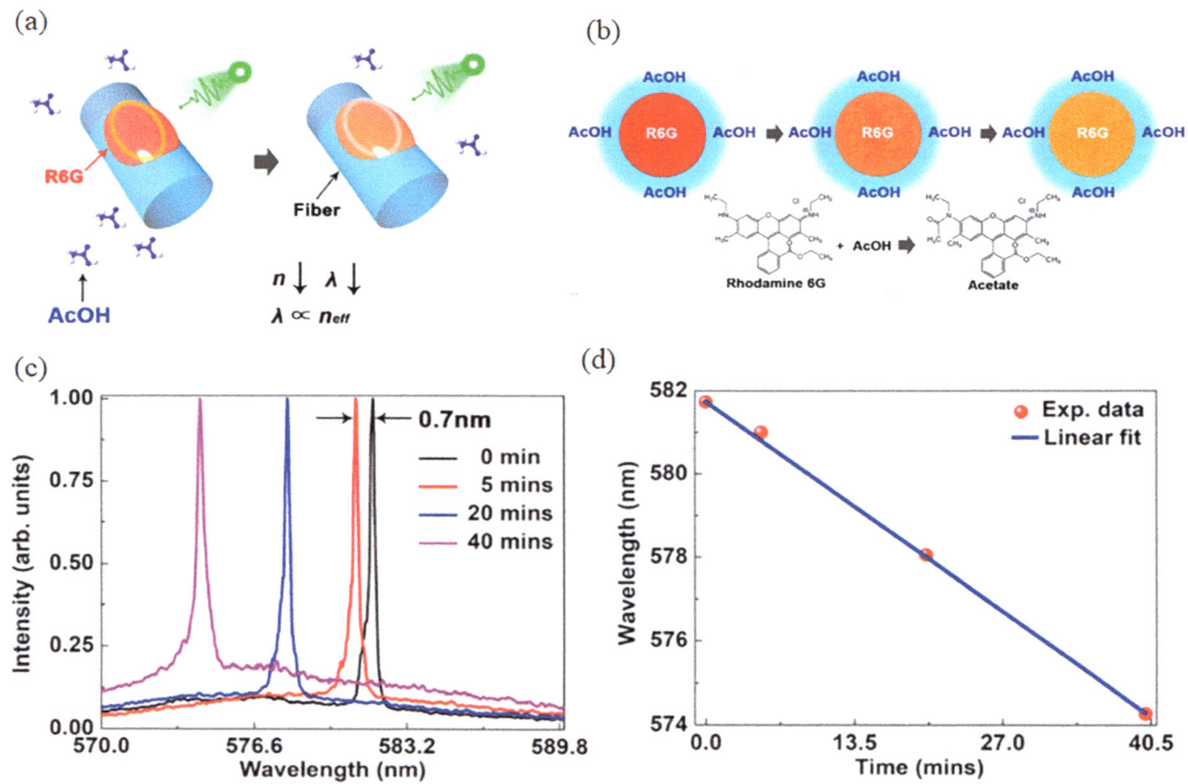

Fig. 20.13 3D curved microcavity single-mode laser for AcOH gas sensing. (**a**) Operational scheme. (**b**) Reaction process between R6G fiber laser and AcOH gas. (**c**) Dynamical evolution of emission spectra. (**d**) Wavelength shift during reaction time, red spheres are corresponding to the experimental results [116]

act as an effective therapeutic agent, the other can potentially cause severe adverse effects [120]. Therefore, determining enantiomeric excess (ee)—a measure of the purity of mixed enantiomers—is critical during every stage of chiral compound synthesis, particularly in the pharmaceutical industry [121, 122]. The ee of enantiomers is generally evaluated based on their chirality and concentration.

Optofluidic-based whispering-gallery mode (WGM) microlasers have garnered significant research attention owing to their high-quality factors, empowering strong light-matter interactions [123–126]. However, these microlasers assist chiral symmetry-based both clockwise and counterclockwise propagating wave modes [127]. This symmetry causes the chiroptical signals of enantiomers within the microfluidic system to cancel each other out, limiting their potential applications in detecting the chirality of biomolecules [128].

In this section, a WGM optofluidic-based microlaser operating near the exceptional point (EP) to enable simultaneous detection of enantiomer chirality and concentration. By providing asymmetric backscattering, the microlaser achieves unidirectional mode operation at the EP [129–132]. The polarization of the WGM lasing, represented by the relative intensity of transverse magnetic (TM) and transverse electric (TE) modes, reflects the chirality of enantiomers, offering a novel detection capability. Additionally, the enantiomer concentration is determined from the lasing wavelength, enabling one-step enantiomeric excess (ee) determination without the need for separate concentration measurements [133]. Importantly, this strategy was used to inspect the available key amino acids, demonstrating versatility of this approach.

The schematic of optofluidic microlasers operating near the exceptional point (EP) is shown in Fig. 20.14a. In this design, a hollow optical fiber (HOF) integrates a microfluidic channel-embedded optical microcavity and enables strong light-enantiomer interactions and analytes injection via capillary action. To prevent the chiroptical signals of enantiomers from being negated by counter-propagating modes (CW and CCW), the microlaser was driven near the EP using two silica nanotips as scatterers, ensuring lasing occurs only in the CW direction.

The primary polarization of the unidirectional lasing is determined by the pump source's polarization, adjustable via a linear polarizer and a quarter-wave plate. The lasing mode, confined within the WGM resonator, interacts with chiral molecules for extended durations, causing polarization rotation due to circular birefringence [134]. This birefringence induces a phase difference between left and right circularly polarized light (Fig. 20.14b), leading to a shift in the relative intensity of TM and TE modes in the lasing spectrum (Fig. 20.14c). By analyzing these changes, the chirality of enantiomers can be detected.

Next, to evaluate the optofluidic microlaser's capability for chiral recognition and enantiomeric excess (ee) determination, the microlaser was tuned to an exceptional point (EP) to enable unidirectional lasing in the CW direction. Glucose enantiomers (L- and D-glucose) were used as samples, and eight critical amino acids were tested to demonstrate the technique's generality. The lasing spectra were recorded by varying the pump source's polarization angle in 30° increments at a fixed pump fluence of 96 μJ/cm^2. The normalized TE and TM mode intensities for

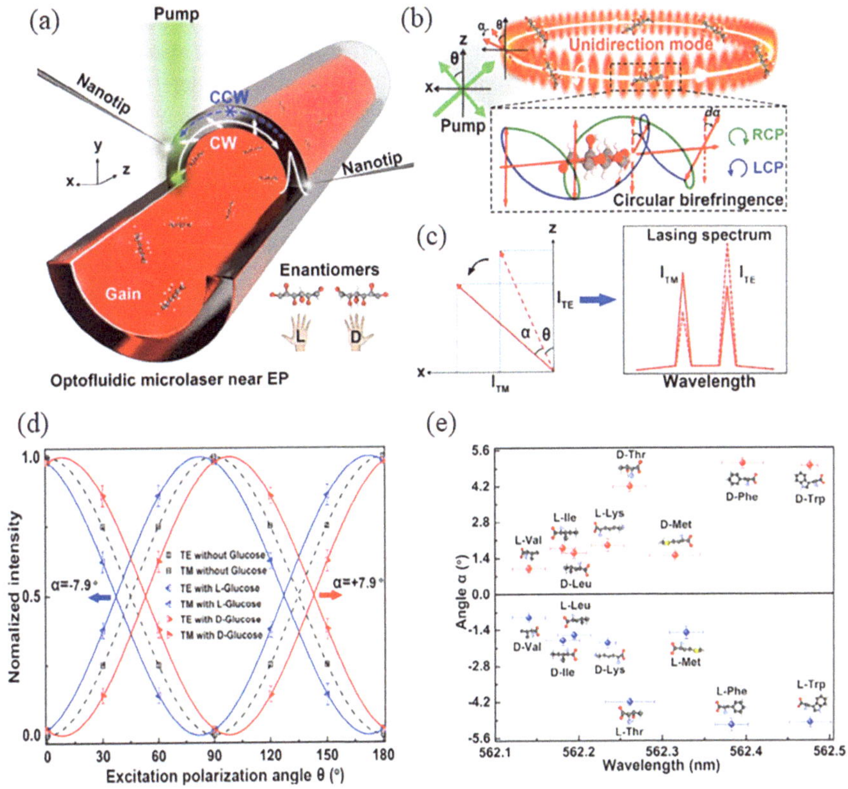

Fig. 20.14 (**a**) Unidirectional lasing action in the optofluidic-loaded WGM microlaser when driven to an EP by nanotips. (**b**) Optical rotation effect, θ and α represent the polarization angle of the pump and lasing rotation angle respectively. (**c**) The alternation in the intensities of TE and TM modes corresponding to laser polarization rotation. (**d**) The intensity of TM and TE modes of optofluidic microlaser under different situations such that with D-glucose, L-glucose, and without glucose versus polarization angle of excitation. (**e**) Chirality prediction of all crucial amino acids enantiomers, respective detected rotation angle, and emission wavelength [133]

D-glucose, L-glucose, and without glucose were plotted against the excitation polarization angle (Fig. 20.14d). Compared to the without glucose situation (black dashed lines), the intensity curves for L-glucose (blue lines) showed a polarization shift of −7.9°, while those for D-glucose (red lines) displayed a similar shift in the opposite direction. This demonstrates the microlaser's sensitivity to enantiomer chirality. Notably, enantiomeric excess detection is only possible if laser is driven to exceptional point.

Further, the applicability of this approach was demonstrated by applying it to the chiral detection of enantiomers of eight essential amino acids (EAAs) namely Val, Ile, Leu, Lys, Thr, Met, Phe, and Trp. Using the optofluidic microlaser near the exceptional point with 5 mg/mL solutions of each EAA, significant alterations in rotation angle and lasing wavelength were observed (Fig. 20.14e). Enantiomers of the same EAA species resulted in similar lasing wavelengths but opposite rotation angles, while different EAA species exhibited distinct lasing wavelengths. This indicates that the optofluidic microlaser near the EP enables ee determination for all eight EAAs and is broadly applicable to various chiral chemicals without requiring molecule-specific designs.

20.6.4 Microtubular-Structured RLs for Multifunctional Sensing

Microfluidic channels are widely used in microfluidic chip systems as highly precise sensors for biomedical and environmental nursing [135, 136]. They handle nanoliter-scale liquid volumes, enabling reduced feature sizes and analyte volumes while allowing multiple analysis units to be integrated onto a small chip [137]. These features make microfluidic sensors a valuable reference for designing random laser (RL) liquid sensors, highlighting the potential of RL sensors based on microfluidic channel structures.

Stimulated by microfluidic sensors, a new type of microtubule-structured RLs, functioning as microfluidic channels for multifunctional sensing, was introduced [138]. These RLs were created using a curly polymethylmethacrylate (PMMA) film implanted with gain media and scatterers (Fig. 20.15a). The whispering-gallery modes can couple with weak random scattering at the interface between the random laser (RL) layer and capillary wall, yielding a coherent laser with a low threshold and large quality factor. The sensing capability is explored by injecting analyte liquids into the RL microfluidic channels, showing that the laser's intensity decay and wavelength shift correspond to changes in the analyte's properties, driven by variations in gain and scattering (Fig. 20.15b). This innovative sensor offers plenty of features, such as excellent detection performance, simple process, cost-effective, and rapid response, making it suitable for applications in microfluidic chips and portable diagnostic instruments.

Sucrose, an essential substance for maintaining normal biological functions, serves as a model analyte in this study. A sucrose solution with varying concentrations was injected into the presented random laser (RL) tubule sensor. Thirteen RL sensors with identical parameters were exposed to sucrose solutions of different concentrations. Emission spectra were recorded, with an initial baseline noted before injection (Fig. 20.15c). As the sucrose concentration raised, the emission intensity reduced proportionally, while the central wavelength remained nearly constant (Fig. 20.15d). This behavior contrasts with traditional laser sensors, where emission spectra typically red-shift with increasing refractive index. Although the refractive index of the sucrose solution increased linearly with concentration, it did not cause a corresponding wavelength shift in the tubular RL sensor. Instead, the intensity variation was quantitatively characterized by calibrating against the baseline, showing a nonlinear reduction with increasing concentration. The sensor

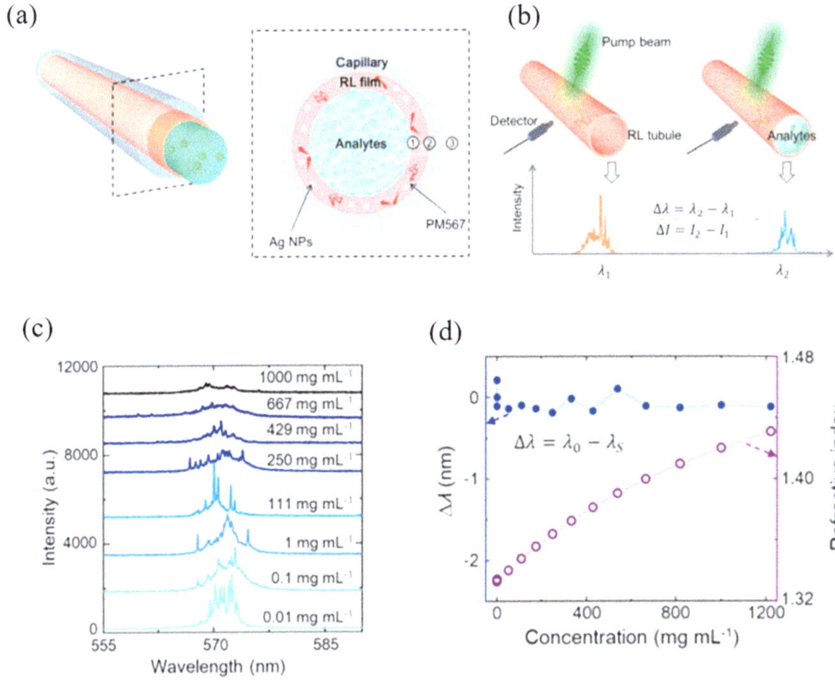

Fig. 20.15 (**a**) Fabrication of microtubular-based RL sensor, inset indicates the zoom cross-sectional view. (**b**) Sensing scheme. (**c**) Intensity spectra as a function of sucrose solution concentration. (**d**) Concentration variation effects on wavelength (right axis) and refractive index (left axis) [138]

demonstrated a detection limit of 0.001 mg/mL and a reliable linear sensing range from 0.01 mg/mL to 600 mg/mL, with consistent repeatability [138]. The results indicate that the tubular RL sensor is highly effective for detecting biochemical solutions that absorb visible light, offering excellent sensitivity, a broad detection range, and reliable performance.

20.7 Hybrid Microcavity Lasers

The control of coupling and synchronization in lasers has gained significant interest due to its applications in spectroscopy, sensing, secure communication, imaging, and optical networks. With the advent of complex microcavities, more intricate light-matter interactions arise, leading to unique lasing emission behaviors within these hybrid microcavities.

Optical microcavities, or optical resonators, can be categorized based on their operating principles or the attributes of light they manipulate. They include modulators such as all-optical, electro-optic, thermo-optic, etc., and are further classified into types like amplitude, phase, polarization, and wavelength modulators [139].

Various optical cavities, such as Fabry-Pérot (F-P), whispering-gallery mode (WGM), distributed Bragg reflector (DBR), distributed feedback (DFB), and scattering systems, each offers unique strengths in modulating laser output, including wavelength, amplitude, and polarization. However, conventional single optical resonators rarely achieve phase and mode modulation. While single-parameter modulation suffices for some applications, simultaneous modulation of multiple parameters is often desirable. To address this, hybrid microcavities, which integrate multiple resonator types, show significant promise (Fig. 20.16). By tailoring the dimensions and material properties of these hybrid systems, overlapping mode fields can be achieved, enabling enhanced resonant control and stronger light-matter interactions. This advancement allows comprehensive modulation of multiple optical parameters, making hybrid microcavities valuable for lasers and various optoelectronic devices [72].

20.7.1 For Information Security

To strengthen secure communication, optical encryption has become a vital approach, particularly through high-complexity methods. Recent studies demonstrate that the phase correlation of random lasing modes can be manipulated via mode interaction. Through the mode-locking mechanism, coherent lasing modes with multiple peaks can transition into incoherent modes with smooth spectra, effectively hiding diverse coherent modes within static incoherent spectra. This intrinsic complexity in mode-locked RLs enhances their security, offering significant potential for advanced multiplexing encryption [140].

20.7.2 WGM-Coupled Random Modes for Information Security

In this section, a novel spectral encryption strategy is proposed, utilizing selectively visualized spectral modes in RLs. By coupling RLs with smooth single-peak spectra to an external whispering-gallery mode cavity, hidden localized lasing modes with distinctive multi-peak spectra are unveiled (as shown in Fig. 20.17a). Further details on the design principles can be found in reference [141].

Encryption Method

Here, these localized modes are employed for spectral encryption to ensure secure information encoding. The encrypted modes, concealed within phase-correlated spectra, can be decoded only with a specialized external coupler. A strategy for

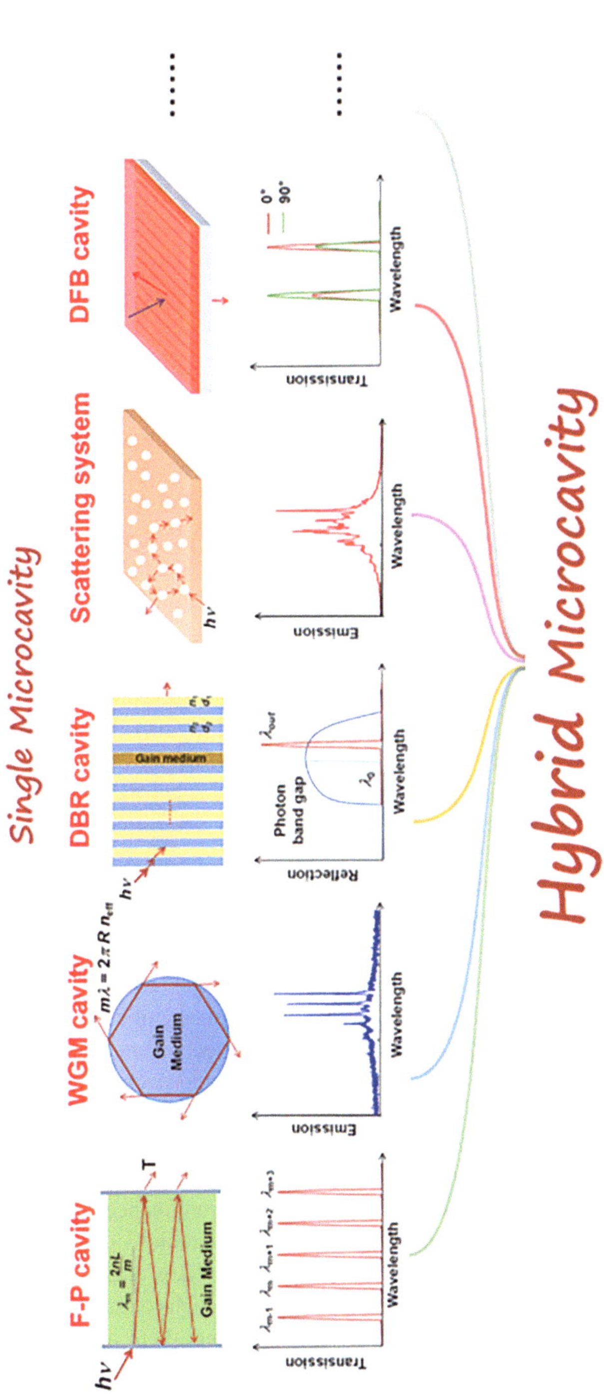

Fig. 20.16 The diagram shows various single microcavities, scattering systems, and their combinations used to generate hybrid microcavities for specific applications [72]

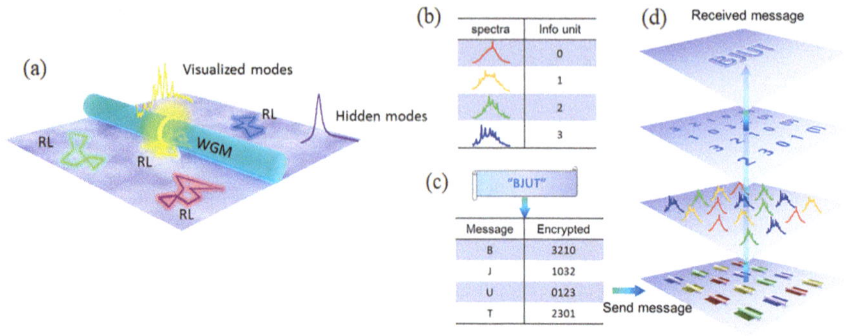

Fig. 20.17 (**a**) Working principle for selecting the hidden modes in phase correlated random laser. (**b**) The algorithm technique of converting the emission spectra to information units. (**c**) The decoding process for "BJUT" cryptograph. (**d**) The decryption procedure [141]

information encryption and hiding by leveraging the relationship between spectral modes and random scattering units in RLs is introduced. By encoding text messages into specific spectral signals, the system ensures secure data transmission. Random laser inks, composed of nanostructures and gain materials, serve as the building blocks for security codes. Four distinct random laser inks, each with unique spectral modes, are used to represent digital signs "0," "1," "2," and "3" by varying the doping concentrations of Ag NWs. The emission spectra corresponding to these modes are linked to digital signs for information recording (Fig. 20.17b). To enhance security, the spectral sequences are further encrypted using Morse code. The initial message is transformed into ciphertext using a predefined encryption algorithm (shown in Fig. 20.17c), and the encrypted spectral patterns are printed with RL inks to transmit the data [141].

Decryption involves measuring the emission spectra using a specialized external whispering-gallery mode (WGM) resonator. The spectra are translated back into digital data based on the encryption algorithm, retrieving the hidden message as given in Fig. 20.17d. This dual-layer security, relying on both the unique conditions of the external cavities and the encryption algorithm, significantly increases data confidentiality. RLs, with their unique feedback mechanisms and unpredictable emissions, add an extra layer of security, making this approach difficult to replicate. This encryption system offers advantages such as simplicity, low cost, and resistance to counterfeiting. It holds promise for applications in military communications, secure data transmission, and anticounterfeiting technologies [141].

The WGM-random hybrid cavities described above enhance random lasing but are typically implemented using externally coupled cavities or liquid gain media, which limit the compactness and scalability of dual chaos encryption technology. Therefore, developing a more integrated design approach for solid, internally coupled WGM-random microcavities and understanding their underlying physical mechanisms are of great importance. Such advancements could significantly advance integrated photonic applications, though they remain rarely reported.

20.7.3 Dual Chaotic Encryption System

This section introduces an internally coupled WGM-random hybrid cavity fabricated using an electrospinning instrument as shown in Fig. 20.8a [62]. High voltages are applied to the syringe needle (positive) and the metal roller (negative). The spinning solution (mixture of PMMA, EPN, Ag NPs, Rh6G, and dichloromethane), ejected from the syringe, forms fibers under the influence of the potential difference, which are collected on a rotating roller and transferred to a glass frame. Random lasing arises from the Ag NPs and Rh6G, while the fiber's smooth inner wall supports WGM laser output. Figure 20.18b shows the setup for pump light irradiation and a scanning electron microscopy (SEM) image of the fiber cross-section, highlighting the presence of Ag nanoparticles (Ag NPs). The pump light is a 532-nm emission from a frequency-doubled nanosecond laser (532 nm, 5–7 ns, 10 Hz). Notably, energy transfer from the WGM laser to Ag NPs enhances localized surface plasmon resonance (LSPR), which couples with random lasing. Along with total internal reflection (TIR) in the WGM cavity, dual enhancements facilitate efficient WGM and random lasing interactions.

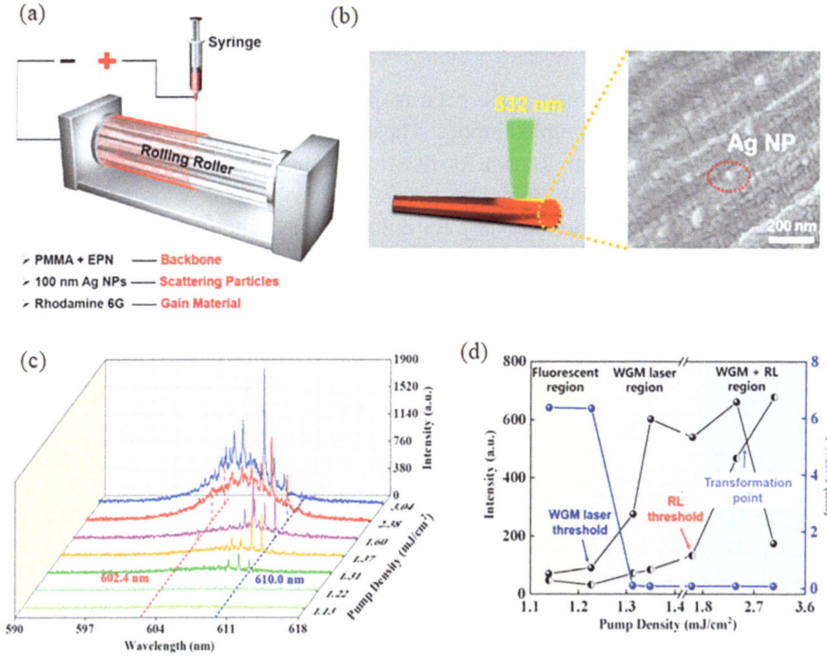

Fig. 20.18 (a) Electrospinning method. (b) Fabricated WGM-coupled random laser excited with pump power of 532 nm, inset figure indicates the presence of scatters. (c) Intensity spectra of WGM-coupled random laser under the variation of pump power density. (d) The pump power influence on the intensity and FWHM of hybrid cavity [62]

A WGM-random hybrid cavity was analyzed, with its lasing spectra shown in Fig. 20.18c. Initially, fluorescence appears in the range of 598.2–612.7 nm as pump density increases. At pump powers above 1.22 mJ/cm^2, WGM lasing emerges between 607.8 and 610.1 nm. At 1.31 mJ/cm^2, the WGM lasing exhibits three modes with a high-quality factor of 6100 at 610.0 nm. The system then transitions to a stage where WGM and random laser modes coexist in the 602.1–607.2 nm range. Here, random lasing mode numbers increase, with a Q factor of 1220 at 602.4 nm, while the WGM Q factor rises to 8590 at 610.0 nm. WGM lasing precedes random lasing due to the fiber's smooth inner wall, which provides high Q factors and tight optical confinement at low pump energy. However, Ag NP-induced scattering causes optical leakage, leading to random lasing at a higher threshold. As pump density exceeds 2.38 mJ/cm^2, WGM laser intensity decreases sharply, with some modes disappearing, while random lasing becomes dominant. The hybrid cavity demonstrates three distinct regions: fluorescence, WGM lasing, and a WGM + random lasing coexistence stage, where a transformation between the two modes occurs. Notably, WGM emission exhibits a larger redshift compared to random lasing, as predicted in Fig. 20.18d [62]. Additionally, noted that the efficient energy transfer from the WGM cavity to the Ag nanoparticles (Ag NPs). As the pump power increases, the disappearance of certain WGM modes further validates this energy transfer process. This effective energy transfer enhances the localized surface plasmon resonance (LSPR) effect of the Ag NPs, thereby promoting random lasing [142].

Two types of lasing modes have been realized from the hybrid cavity yielding WGM and random laser coupled modes. These modes are applied in a dual chaotic logistic and Hénon system, known for its strong robustness in altering the spatial and grayscale distributions of image pixels. This system is widely used as an advanced chaotic encryption strategy for color images. The mathematical expressions for the one-dimensional (1D) logistic system (Eq. (20.2)) and the two-dimensional (2D) Hénon system (Eqs. (20.3) and (20.4)) are provided below [62]:

$$x_{m+1} = u x_m \left(1 - x_m\right) \tag{20.1}$$

$$y_{n+1} = 1 + y_n - a x_n^2 \tag{20.2}$$

$$z_{n+1} = b x_n \tag{20.3}$$

In the 1D logistic system, the initial value is denoted as x_0, and the control parameter is u. The system exhibits chaotic behavior when $0 < x_0 < 1$ and $3.569 < u < 4$, with values constrained to $0 < x_n < 1$. In the 2D Hénon system, the initial values are y_0 and z_0, and the control parameters include a and b. The system becomes chaotic when $a = 1.4$ and $b = 0.3$, with values satisfying $y_n > 0$ and $z_n < 1$. In this approach, random lasing generates a true random value for x_0, while the WGM lasing in the hybrid cavity provides the initial value for u. For encrypting an $M \times N$ image, the process involves three steps: (1) WGM-random hybrid cavity laser coding, using initial values derived from the lasing modes. (2) Generation of a chaotic sequence, employing the dual chaotic system. (3) Scrambling for encryption by reorganizing

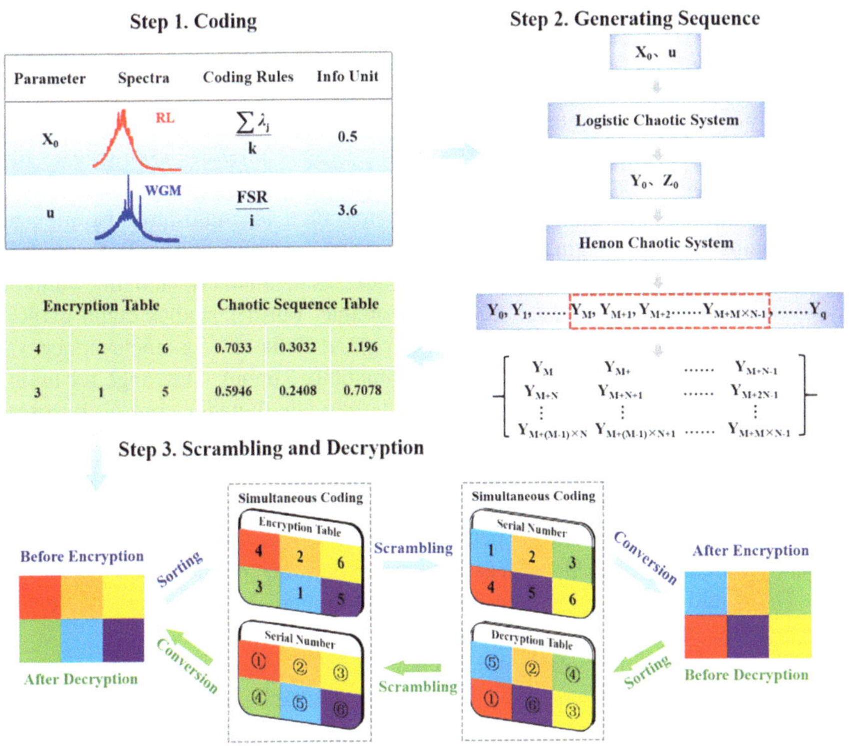

Fig. 20.19 Image encryption process by using dual chaotic system [62]

image data, as illustrated in Fig. 20.19 [62]. This method ensures robust and efficient image encryption.

A dual chaotic encryption system for color images is successfully developed using the WGM-coupled hybrid cavity. This approach leverages a combined multi-chaotic system, effectively mitigating the risk of frequency analysis and meticulous attacks. By making the cipher more complex and challenging to decode, it significantly enhances the security of the encryption system.

20.8 Summary and Prospect

RLs hold significant potential for applications in sensing, imaging, and photonic chips. Fiber-integrated RLs address challenges related to directionality and integration. This chapter presented the fundamental concepts and principles of RLs, explores advancements in single- and multi-wavelength RLs exhibiting plasmonic effects integrated with fiber, and highlights the role of energy transfer in multi-wavelength systems. It also presents examples of their use in imaging and sensing, offering insights for the development of high-performance plasmonic-resonance coupled optical fiber random laser sources. WGM-coupled random lasers were also explored in the context of information security applications.

Recent advancements in random laser research at the end face of optical fibers have garnered significant attention. Future research is expected to focus on several key areas such as (1) Improving Performance: Despite progress, core metrics like output power and lifetime of random laser devices remain suboptimal, especially when placed on the fiber end. Enhancements such as multi-laser parallelism, Q-switching, and packaging could improve device practicality. (2) Control of Output Characteristics: The coherence, polarization, intensity, and linewidth of fiber-end random lasers are not fully controllable. Further studies could explore techniques like mode locking and the use of optical elements (e.g., metasurfaces) to better manage microcavity control. (3) Chromaticity Tuning: The range for tuning multicolor random lasers is still limited for large-scale dynamic control. Research into novel tuning technologies, including new material systems for controlling various parameters (mechanical, thermal, optical, electrical), is needed. (4) Expanding Applications: The use of random lasers is still in its early stages, and their application area is narrow. Optimizing performance and systematizing the device could expand applications in fields such as storage, encryption, communication, and sensing. This could include improvements in multichromatic light generation, speckle-free imaging, and high-sensitivity sensing.

Hybrid microcavities hold significant potential for enhancing lasing modulation; however, several challenges persist. First, further research is needed to better understand the coupling mechanisms between homo- and heteromorphic microcavities, especially in strong coupling regimes, where both electromagnetic and material properties can be modified to enhance performance. Second, the need for compact, efficient, and broadband optical modulators calls for advances in design, materials, and fabrication to reduce optical loss and improve the quality factor (Q). Finally, more cost-effective and scalable fabrication methods, such as lithography and nanoimprinting, are crucial for the commercialization of hybrid microcavities. Despite these challenges, hybrid microcavity lasers show great promise, particularly in quantum information dispensation, single-photon sources, and sensing applications.

New research directions focus on enhancing sensing capacity through spatial-division and wavelength-division multiplexing, as well as exploring distributed optical fiber sensing [89, 90, 143]. The development of frequency tripling and cascaded frequency doubling could facilitate random lasing in the blue, violet, and ultraviolet spectral regions, further diversifying applications. Random lasing in visible and mid-infrared spectra presents opportunities for applications requiring low-coherence light sources [144]. Advances in fiber design, operational strategies, and application development will significantly enhance the performance and versatility of RFLs across various domains.

References

1. Ambartsumyan, R., et al.: A laser with nonresonant feedback. Sov. Phys. JETP. **24**, 481–485 (1967)
2. Letokhov, V.: Generation of light by a scattering medium with negative resonance absorption. Sov. Phys. JETP. **26**, 835–840 (1968)
3. Lawandy, N.M., et al.: Laser action in strongly scattering media. Nature. **368**, 436–438 (1994)
4. Wiersma, D.S., et al.: Random laser? Nature. **373**, 203–204 (1995)
5. Cao, H., et al.: Random laser action in semiconductor powder. Phys. Rev. Lett. **82**, 2278 (1999)
6. Redding, B., et al.: Speckle-free laser imaging using random laser illumination. Nat. Photonics. **6**, 355–359 (2012)
7. Lee, Y.-J., et al.: A curvature-tunable random laser. Nanoscale. **11**, 3534–3545 (2019)
8. Xie, Z., et al.: Multi-wavelength coherent random laser in bio-microfibers. Opt. Express. **28**, 5179–5188 (2020)
9. Mallick, S.P., Sung, Z.: Holographic image denoising using random laser illumination. Ann. Phys. **532**, 2000323 (2020)
10. Cui, L., et al.: Retrieval of contaminated information using random lasers. Appl. Phys. Lett. **106** (2015)
11. Ni, Y., et al.: Random lasing carbon dot fibers for multilevel anti-counterfeiting. Nanoscale. **13**, 16872–16878 (2021)
12. Sznitko, L., et al.: Bichromatic laser dye as a photonic random number generator. ACS Photon. **8**, 1630–1638 (2021)
13. Tong, J., et al.: Flexible plasmonic random laser for wearable humidity sensing. Sci. China Inf. Sci. **64**, 1–7 (2021)
14. Wang, Z., et al.: Long-distance fiber-optic point-sensing systems based on random fiber lasers. Opt. Express. **20**, 17695–17700 (2012)
15. Azmi, A.N., et al.: Review of open cavity random lasers as laser-based sensors. ACS Sensors. **7**, 914–928 (2022)
16. Xu, Z., et al.: Random lasing from label-free living cells for rapid cytometry of apoptosis. Nano Lett. **22**, 172–178 (2022)
17. Song, Q., et al.: Detection of nanoscale structural changes in bone using random lasers. Biomed. Opt. Express. **1**, 1401–1407 (2010)
18. Polson, R.C., Vardeny, Z.V.: Random lasing in human tissues. Appl. Phys. Lett. **85**, 1289–1291 (2004)
19. Chang, Q., et al.: Broadband plasmonic silver nanoflowers for high-performance random lasing covering visible region. Nanophotonics. **6**, 1151–1160 (2017)
20. Hou, Y., et al.: Full-color flexible laser displays based on random laser arrays. Sci. China Mater. **64**, 2805–2812 (2021)
21. Cao, M., et al.: Plasmonically enhanced colloidal quantum dot/graphene doped polymer random lasers. Materials. **15**, 2213 (2022)
22. Lau, S.P., et al.: Flexible ultraviolet random lasers based on nanoparticles. Small. **1**, 956–959 (2005)
23. Shi, X., et al.: Dissolvable and recyclable random lasers. ACS Nano. **11**, 7600–7607 (2017)
24. Li, X., et al.: Lotus-leaf-inspired flexible and tunable random laser. ACS Appl. Mater. Interfaces. **12**, 10050–10057 (2020)
25. Gaio, M., et al.: A nanophotonic laser on a graph. Nat. Commun. **10**, 226 (2019)
26. Caselli, N., et al.: Networks of mutually coupled random lasers. Optica. **8**, 193–201 (2021)
27. Shen, T.-L., et al.: Coherent Förster resonance energy transfer: a new paradigm for electrically driven quantum dot random lasers. Sci. Adv. **6**, eaba1705 (2020)

28. Gomes, A.S., et al.: Direct three-photon excitation of upconversion random laser emission in a weakly scattering organic colloidal system. Opt. Express. **22**, 14305–14310 (2014)
29. Weng, G., et al.: Picosecond random lasing based on three-photon absorption in organometallic halide CH3NH3PbBr3 perovskite thin films. ACS Photon. **5**, 2951–2959 (2018)
30. Nakamura, T., et al.: Gold-nanoparticle-assisted random lasing from powdered GaN. Opt. Express. **19**, 467–475 (2011)
31. Shi, X., et al.: Random lasing with a high quality factor over the whole visible range based on cascade energy transfer. Adv Opt Mater. **2**, 88–93 (2014)
32. Ye, L., et al.: Coherent random lasing from dye aggregates in polydimethylsiloxane thin films. ACS Appl. Mater. Interfaces. **9**, 27232–27238 (2017)
33. Chen, Y., et al.: Colloidal quantum dot random laser. Opt. Express. **19**, 2996–3003 (2011)
34. Cao, M., et al.: Random lasing in a colloidal quantum dot-doped disordered polymer. Opt. Express. **24**, 9325–9331 (2016)
35. Tong, J., et al.: Dual-color plasmonic random lasers for speckle-free imaging. Nanotechnology. **31**, 465204 (2020)
36. Xia, J., et al.: The transition from incoherent to coherent random laser in defect waveguide based on organic/inorganic hybrid laser dye. Nano. **7**, 1341–1350 (2018)
37. Zhai, T., et al.: Ultra-thin plasmonic random lasers. Opt. Express. **24**, 437–442 (2016)
38. Zhang, D.-G., et al.: Investigation of random lasing from all-inorganic halide perovskite quantum dots prepared under ambient conditions. Nanoscale. **13**, 3246–3251 (2021)
39. Gao, W., et al.: Robust and flexible random lasers using perovskite quantum dots coated nickel foam for speckle-free laser imaging. Small. **17**, 2103065 (2021)
40. Wang, L., et al.: Perovskite random lasers, process and prospects. Micromachines. **13**, 2040 (2022)
41. Zhai, T., et al.: Random laser based on waveguided plasmonic gain channels. Nano Lett. **11**, 4295–4298 (2011)
42. Wang, Y., et al.: Cascade-pumped random lasers with coherent emission formed by Ag–Au porous nanowires. Opt. Lett. **39**, 5–8 (2013)
43. Meng, X., et al.: Metal–dielectric core–shell nanoparticles: advanced plasmonic architectures towards multiple control of random lasers. Adv Opt Mater. **1**, 573–580 (2013)
44. dos Santos, M.V., et al.: Random laser action from flexible biocellulose-based device. J. Appl. Phys. **115** (2014)
45. Song, Q., et al.: Random lasing in bone tissue. Opt. Lett. **35**, 1425–1427 (2010)
46. Bian, Y., et al.: External electric field tailored spatial coherence of random lasing. Crystals. **12**, 1160 (2022)
47. Strangi, G., et al.: Random lasing and weak localization of light in dye-doped nematic liquid crystals. Opt. Express. **14**, 7737–7744 (2006)
48. Xie, R.-C., et al.: Manipulation of plasmonic random laser from dye-doped liquid crystals inside photonic crystal fiber by the electric field. Opt. Laser Technol. **151**, 108013 (2022)
49. Kumar, B., et al.: Localized modes revealed in random lasers. Optica. **8**, 1033–1039 (2021)
50. Saxena, D., et al.: Sensitivity and spectral control of network lasers. Nat. Commun. **13**, 6493 (2022)
51. Bachelard, N., et al.: Taming random lasers through active spatial control of the pump. Phys. Rev. Lett. **109**, 033903 (2012)
52. Li, X., et al.: A tunable blue random laser based on solid waveguide gain films with plasmonics and scatters. J. Alloys Compd. **790**, 558–562 (2019)
53. Wiersma, D.S., Cavalieri, S.: A temperature-tunable random laser. Nature. **414**, 708–709 (2001)
54. Caixeiro, S.C., et al.: Silk-based biocompatible random lasing. Adv Opt Mater. **4**, 998–1003 (2016)
55. Zhai, T., et al.: A plasmonic random laser tunable through stretching silver nanowires embedded in a flexible substrate. Nanoscale. **7**, 2235–2240 (2015)

56. Lee, K., et al.: Non-resonant power-efficient directional Nd: YAG ceramic laser using a scattering cavity. Nat. Commun. **12**, 8 (2021)
57. Liu, Y., et al.: Surface-emitting perovskite random lasers for speckle-free imaging. ACS Nano. **13**, 10653–10661 (2019)
58. Zhang, X., et al.: Perovskite random lasers on fiber facet. Nano. **9**, 935–941 (2020)
59. Lee, S., et al.: Random lasing with a high degree of circular dichroism by chiral plasmonic gold nanoparticles. Acs Photonics. **9**, 613–620 (2022)
60. Du, W., et al.: Thermal treatment effect on the random lasing polarization of polymer optical fiber. Opt. Laser Technol. **149**, 107855 (2022)
61. Ma, X., et al.: Vortex random fiber laser with controllable orbital angular momentum mode. Photon. Res. **9**, 266–271 (2021)
62. Zhai, T., et al.: Dual chaos encryption for color images enabled in a WGM–random hybrid microcavity. Nanoscale. **14**, 11252–11260 (2022)
63. Wiersma, D.: The smallest random laser. Nature. **406**, 133–135 (2000)
64. Cao, H.: Lasing in random media. Waves Random Media. **13**, R1 (2003)
65. Gayathri, R., et al.: Lasing from micro-and nano-scale photonic disordered structures for biomedical applications. Nano. **13**, 2466 (2023)
66. Young, A.T.: Rayleigh scattering. Appl. Opt. **20**, 533–535 (1981)
67. Du, H.: Mie-scattering calculation. Appl. Opt. **43**, 1951–1956 (2004)
68. Luan, F., et al.: Lasing in nanocomposite random media. Nano Today. **10**, 168–192 (2015)
69. Gohar, A., et al.: Tunable random laser based on hybrid plasmonic enhancement. Opt. Express. **31**, 36150–36160 (2023)
70. de Matos, C.J., et al.: Random fiber laser. Phys. Rev. Lett. **99**, 153903 (2007)
71. Hu, Z., et al.: Coherent random fiber laser based on nanoparticles scattering in the extremely weakly scattering regime. Phys. Rev. Lett. **109**, 253901 (2012)
72. Liang, N., et al.: Hybrid microcavity lasers: principle, design, and practical application. Laser Photonics Rev. **17**, 2300343 (2023)
73. Hu, Z., et al.: Disordered microstructure polymer optical fiber for stabilized coherent random fiber laser. Opt. Lett. **38**, 4644–4647 (2013)
74. Hu, Z., et al.: Polarized random laser emission from an oriented disorder polymer optical fiber. Opt. Lett. **41**, 2584–2587 (2016)
75. Hu, Z., et al.: Tunable random polymer fiber laser. Opt. Express. **25**, 18421–18430 (2017)
76. Chen, X., et al.: Polymer-fiber random lasers based on pumping radiation effect. Phys Scr. **94**, 115509 (2019)
77. Morello, G., et al.: Tuneable optical gain and broadband lasing driven in electrospun polymer fibers by high dye concentration. J. Mater. Chem. C. **10**, 2042–2048 (2022)
78. Churkin, D.V., et al.: Recent advances in fundamentals and applications of random fiber lasers. Adv. Opt. Photon. **7**, 516–569 (2015)
79. Gomes, A.S., et al.: Recent advances and applications of random lasers and random fiber lasers. Prog Quantum Electron. **78**, 100343 (2021)
80. Chen, H., et al.: Advances in random fiber lasers and their sensing application. Sensors. **20**, 6122 (2020)
81. Yang, X., et al.: A review of fiber optofluidic microlasers: structures, characteristics, and applications. Laser Photon. Rev. **16**(1) (2022)
82. Bian, Y., et al.: A ring-shaped random laser in momentum space. Nanoscale. **12**, 3166–3173 (2020)
83. Shi, X., et al.: Programmable complex pumping field induced color-on-demand random lasing in fiber-integrated microbelts for speckle free imaging. SCIENCE CHINA Inf. Sci. **66**, 222401 (2023)
84. Shi, X., et al.: Chromaticity-tunable white random lasing based on a microfluidic channel. Opt. Express. **28**, 13576–13585 (2020)
85. Aziz, G., et al.: Photo-isomerization enabled reversible wavelength switching in fiber random laser for color image encryption. Opt. Express. **32**, 30380–30392 (2024)

86. Pan, A., et al.: Continuous alloy-composition spatial grading and superbroad wavelength-tunable nanowire lasers on a single chip. Nano Lett. **9**, 784–788 (2009)
87. W. Yi, et al., "Dynamic Behavior of Molecular Switches in Crystal under Pressure and Its Reflection on Tactile Sensing," (2015)
88. Wan, T., et al.: In-sensor computing: materials, devices, and integration technologies. Adv. Mater. **35**, 2203830 (2023)
89. Han, B., et al.: Ultralong single-ended random fiber laser and sensor. Laser Photonics Rev. **17**, 2200797 (2023)
90. Lin, S., et al.: Wideband remote-sensing based on random fiber laser. J. Lightwave Technol. **40**, 3104–3110 (2022)
91. Rong, K., et al.: Configurable integration of on-chip quantum dot lasers and subwavelength plasmonic waveguides. Adv. Mater. **30**, 1706546 (2018)
92. Liu, S.-Y., et al.: Simultaneous excitation and emission enhancement of fluorescence assisted by double plasmon modes of gold nanorods. J. Phys. Chem. C. **117**, 10636–10642 (2013)
93. Yu, G., et al.: Continuous dynamic measurement of frequency scanning interferometry based on motion phase synchronization compensation and calibration. Opt. Express. **31**, 30974–30992 (2023)
94. Dice, G., et al.: Plasmonically enhanced diffusive and subdiffusive metal nanoparticle-dye random laser. Appl. Phys. Lett. **86** (2005)
95. Deheri, R., et al.: Cascaded Raman fiber lasers with ultrahigh spectral purity. Opt. Lett. **47**, 3499–3502 (2022)
96. Li, S., et al.: Plasmonic random laser on the fiber facet. Opt. Express. **23**, 23985–23991 (2015)
97. Zhai, T., et al.: A RGB random laser on an optical fiber facet. RSC Adv. **7**, 45852–45855 (2017)
98. Shi, X., et al.: Low-cost biosensors based on a plasmonic random laser on fiber facet. Opt. Express. **28**, 12233–12242 (2020)
99. Khatri, D.S., et al.: Plasmonic random laser on an optical fiber tip. arXiv preprint arXiv:2002.11797. (2020)
100. Zhai, T., et al.: Distributed feedback lasing in a metallic cavity. Appl. Phys. Lett. **111** (2017)
101. Chen, C., et al.: Tunable polymer lasers based on metal-dielectric hybrid cavity. Opt. Express. **26**, 32048–32054 (2018)
102. Xu, Z., et al.: Multifunctional sensing based on an ultrathin transferrable microring laser. ACS Appl. Mater. Interfaces. **13**, 19324–19331 (2021)
103. Zhai, T., et al.: Controllable plasmonic transfer through laser-induced nanostructuring. Appl. Phys. Express. **7**, 065201 (2014)
104. Zhai, T., et al.: Direct writing of flexible bimetallic nanoparticles for hybrid plasmon response. Appl. Phys. Lett. **105** (2014)
105. Su, D., et al.: WGM lasing in irregular cavities with arbitrary boundaries. Nanoscale. **13**, 18349–18355 (2021)
106. Zhai, T., et al.: Red-green-blue plasmonic random laser. Opt. Express. **25**, 2100–2106 (2017)
107. Li, S., et al.: Red-green-blue plasmonic random lasing from cascaded polymer slices. Laser Phys. Lett. **15**, 085803 (2018)
108. Jiang, L., et al.: Electrons dynamics control by shaping femtosecond laser pulses in micro/nanofabrication: modeling, method, measurement and application. Light Sci Appl. **7**, 17134–17134 (2018)
109. Fan, F., et al.: A monolithic white laser. Nat. Nanotechnol. **10**, 796–803 (2015)
110. Chen, Y., et al.: Enhancement of random lasing based on the composite consisting of nanospheres embedded in nanorods template. Opt. Express. **17**, 12706–12713 (2009)
111. Nathan, M.I., et al.: Oscillations in GaAs spontaneous emission in Fabry-Perot cavities. Phys. Rev. Lett. **11**, 152 (1963)
112. Watanabe, K., et al.: Label-free and spectral-analysis-free detection of neuropsychiatric disease biomarkers using an ion-sensitive GaInAsP nanolaser biosensor. Biosens. Bioelectron. **117**, 161–167 (2018)

113. Chiavaioli, F., et al.: Characterisation of a label-free biosensor based on long period grating. J. Biophotonics. **7**, 312–322 (2014)
114. Vafapour, Z.: Polarization-independent perfect optical metamaterial absorber as a glucose sensor in food industry applications. IEEE Trans. Nanobioscience. **18**, 622–627 (2019)
115. Pan, T., et al.: The application of functional oxide thin films in flexible sensor devices. Sci. China Inf. Sci. **48**, 635 (2018)
116. Zhang, S., et al.: Direction-adjustable single-mode lasing via self-assembly 3D-curved microcavities for gas sensing. ACS Appl. Mater. Interfaces. **13**, 45916–45923 (2021)
117. Zhang, Y., et al.: A cost-effective relative humidity sensor based on side coupling induction technology. Sensors. **17**, 944 (2017)
118. Ma, W., et al.: Chiral inorganic nanostructures. Chem. Rev. **117**, 8041–8093 (2017)
119. Long, J., et al.: Nickel/Brønsted acid dual-catalyzed regio-and enantioselective hydrophosphinylation of 1, 3-dienes: access to chiral allylic phosphine oxides. Chem. Sci. **13**, 1390–1397 (2022)
120. Ito, T., et al.: Identification of a primary target of thalidomide teratogenicity. Science. **327**, 1345–1350 (2010)
121. Cai, T., et al.: Cell-free chemoenzymatic starch synthesis from carbon dioxide. Science. **373**, 1523–1527 (2021)
122. P. L. Polavarapu, "Chiral Analysis: Advances in Spectroscopy, Chromatography and Emerging Methods," (2018)
123. He, L., et al.: Whispering gallery microcavity lasers. Laser Photonics Rev. **7**, 60–82 (2013)
124. Toropov, N., et al.: Review of biosensing with whispering-gallery mode lasers. Light Sci Appl. **10**, 42 (2021)
125. Yu, X.-C., et al.: Single-molecule optofluidic microsensor with interface whispering gallery modes. Proc. Natl. Acad. Sci. **119**, e2108678119 (2022)
126. Yang, X., et al.: Phase-transition microcavity laser. Nano Lett. **23**, 3048–3053 (2023)
127. Kippenberg, T., et al.: Modal coupling in traveling-wave resonators. Opt. Lett. **27**, 1669–1671 (2002)
128. Gelens, L., et al.: Exploring multistability in semiconductor ring lasers: theory and experiment. Phys. Rev. Lett. **102**, 193904 (2009)
129. Zhu, J., et al.: Controlled manipulation of mode splitting in an optical microcavity by two Rayleigh scatterers. Opt. Express. **18**, 23535–23543 (2010)
130. Kim, M., et al.: Partially directional microdisk laser with two Rayleigh scatterers. Opt. Lett. **39**, 2423–2426 (2014)
131. Miri, M.-A., Alu, A.: Exceptional points in optics and photonics. Science. **363**, eaar7709 (2019)
132. Longhi, S., Feng, L.: Photonics Res., 2017. **5**, B1
133. Xu, Z., et al.: Determination of enantiomeric excess by Optofluidic microlaser near exceptional point. Adv. Sci. **11**, 2308362 (2024)
134. Tang, Y., Cohen, A.E.: Optical chirality and its interaction with matter. Phys. Rev. Lett. **104**, 163901 (2010)
135. Liu, X., et al.: Nanozyme-based pump-free microfluidic chip for colorectal cancer diagnosis via circulating cancer stem cell detection. ACS Sensors. **9**, 5090–5098 (2024)
136. Stollmann, A., et al.: Molecular fingerprinting of biological nanoparticles with a label-free optofluidic platform. Nat. Commun. **15**, 4109 (2024)
137. Walter, V., et al.: Real-time signal processing via chemical reactions for a microfluidic molecular communication system. Nat. Commun. **14**, 7188 (2023)
138. Shi, X., et al.: Microtubular structured random lasers as microfluidic channels for multifunctional sensing. Opt. Lett. **50**, 455–458 (2025)
139. Sun, Z., et al.: Optical modulators with 2D layered materials. Nat. Photonics. **10**, 227–238 (2016)
140. Khurgin, J.B.: Exceptional points in polaritonic cavities and subthreshold Fabry–Perot lasers. Optica. **7**, 1015–1023 (2020)

141. Shi, X., et al.: Selectively visualizing the hidden modes in random lasers for secure communication. Laser Photonics Rev. **15**, 2100295 (2021)
142. Gökbulut, B., et al.: Enhanced light–matter interaction in a hybrid photonic–plasmonic cavity. Appli. Phys. A. **127**, 1–13 (2021)
143. DeMiguel-Soto, V., et al.: Ultra-long (290 km) remote interrogation sensor network based on a random distributed feedback fiber laser. Opt. Express. **26**, 27189–27200 (2018)
144. Wu, H., et al.: Temporal ghost imaging using wavelength conversion and two-color detection. Optica. **6**, 902–906 (2019)

MIX
Papier aus verantwortungsvollen Quellen
Paper from responsible sources
FSC® C105338

If you have any concerns about our products,
you can contact us on
ProductSafety@springernature.com

In case Publisher is established outside the EU,
the EU authorized representative is:
Springer Nature Customer Service Center GmbH
Europaplatz 3, 69115 Heidelberg, Germany

Printed by Libri Plureos GmbH
in Hamburg, Germany